Adaptive Control Design and Analysis

Adaptive and Learning Systems for Signal Processing, Communications, and Control

Editor: Simon Haykin

Beckerman / ADAPTIVE COOPERATIVE SYSTEMS

Chen and Gu / CONTROL-ORIENTED SYSTEM IDENTIFICATION: An $\mathcal{H}_\infty$ Approach

Cherkassky and Mulier / LEARNING FROM DATA: Concepts, Theory, and Methods

Diamantaras and Kung / PRINCIPAL COMPONENT NEURAL NETWORKS: Theory and Applications

Haykin / UNSUPERVISED ADAPTIVE FILTERING: Blind Source Separation

Haykin / UNSUPERVISED ADAPTIVE FILTERING: Blind Deconvolution

Haykin and Puthussarypady / CHAOTIC DYNAMICS OF SEA CLUTTER

Hrycej / NEUROCONTROL: Towards an Industrial Control Methodology

Hyvärinen, Karhunen, and Oja / INDEPENDENT COMPONENT ANALYSIS

Kristić, Kanellakopoulos, and Kokotović / NONLINEAR AND ADAPTIVE CONTROL DESIGN

Mann / INTELLIGENT IMAGE PROCESSING

Nikias and Shao / SIGNAL PROCESSING WITH ALPHA-STABLE DISTRIBUTIONS AND APPLICATIONS

Passino and Burgess / STABILITY ANALYSIS OF DISCRETE EVENT SYSTEMS

Sánchez-Peña and Sznaier / ROBUST SYSTEMS THEORY AND APPLICATIONS

Sandberg, Lo, Fancourt, Principe, Katagiri, and Haykin / NONLINEAR DYNAMICAL SYSTEMS: Feedforward Neural Network Perspectives

Spooner, Maggiore, Ordóñez, and Passino / STABLE ADAPTIVE CONTROL AND ESTIMATION FOR NONLINEAR SYSTEMS: Neural and Fuzzy Approximator Techniques

Tao / ADAPTIVE CONTROL DESIGN AND ANALYSIS

Tao and Kokotović / ADAPTIVE CONTROL OF SYSTEMS WITH ACTUATOR AND SENSOR NONLINEARITIES

Tsoukalas and Uhrig / FUZZY AND NEURAL APPROACHES IN ENGINEERING

Van Hulle / FAITHFUL REPRESENTATIONS AND TOPOGRAPHIC MAPS: From Distortion- to Information-Based Self-Organization

Vapnik / STATISTICAL LEARNING THEORY

Werbos / THE ROOTS OF BACKPROPAGATION: From Ordered Derivatives to Neural Networks and Political Forecasting

Yee and Haykin / REGULARIZED RADIAL BIAS FUNCTION NETWORKS: Theory and Applications

Adaptive Control Design and Analysis

GANG TAO
University of Virginia

WILEY-INTERSCIENCE
A JOHN WILEY & SONS, INC., PUBLICATION

This text is printed on acid-free paper. ♾

Published by John Wiley & Sons, Inc., Hoboken, New Jersey.
Published simultaneously in Canada.

Library of Congress Cataloging-in-Publication Data:

Tao, Gang.
Adaptive control design and analysis / Gang Tao.
p. cm — (Adaptive and learning systems for signal processing, communications, and control)
Includes bibliographical references and index.
ISBN 0-471-27452-6 (cloth)
1. Adaptive control systems. I. Title. II. Series.

TJ217.T34 2003
629.8'36—dc21 2003041152

Printed in the United States of America.

10 9 8 7 6 5 4 3 2 1

To my parents, my wife Lanlin,

and my sons Kai and Kwin.

Contents

Preface

Adaptive control is becoming popular in many fields of engineering and science as concepts of adaptive systems are becoming more attractive in developing advanced applications. Adaptive control theory is a mature branch of control theories, and there is a vast amount of literature on design and analysis of various adaptive control systems using rigorous methods based on different performance criteria. Adaptive control faces many important challenges, especially in nontraditional applications, such as real-time systems, which do not have precise classical models admissible to existing control designs, or a physiological system with an artificial heart, whose unknown parameters may change at a heart beat rate which is also a controlled variable. To meet the fast growth of adaptive control applications and theory development, a systematic and unified understanding of adaptive control theory is thus needed.

In an effort to introduce such an adaptive control theory, this book presents and analyzes some common and effective adaptive control design approaches, including model reference adaptive control, adaptive pole placement control, and adaptive backstepping control. The book addresses both continuous-time and discrete-time adaptive control designs and their analysis; deals with both single-input, single-output and multi-input, multi-output systems; and employs both state feedback and output feedback. Design and analysis of various adaptive control systems are presented in a systematic and unified framework. The book is a collection of lectures on system modeling and stability, adaptive control formulation and design, stability and robustness analysis, and adaptive system illustration and comparison, aimed at reflecting the state of the art in adaptive control as well as at presenting its fundamentals. It is a comprehensive book which can be used as either an academic textbook or technical reference for graduate students, researchers, engineers, and interested undergraduate students in the fields of engineering, computer science, applied mathematics and others, who have prerequisites in linear systems and feedback control at the undergraduate level.

In this self-contained book, basic concepts and fundamental principles of adaptive control design and analysis are covered in 10 chapters. As a graduate textbook, it is suitable for a one-semester course: lectures plus reading may cover most of the book without missing essential material. To help in understanding the topics, at the end of each chapter, there are problems related to that chapter's materials as well as technical discussions beyond the covered topics. A separate manual containing solutions to most of these problems is also available. At the end of most chapters, there are also some advanced topics for further study in adaptive control.

Chapter 1 compares different areas of control theory, introduces some basic concepts of adaptive control, and presents some simple adaptive control systems, including direct and indirect adaptive control systems in both continuous and discrete time, as well as an adaptive backstepping control design for a nonlinear system in continuous time.

Chapter 2 presents some fundamentals of dynamic system theory, including system models, system characterizations, signal measures, system stability theory (including Lyapunov stability and input–output operator stability), signal convergence lemmas, and operator norms. In particular, it gives a thorough study of the Lyapunov direct method for stability analysis, some time-varying feedback operator stability properties, several important inequalities for system analysis, some detailed input–output L^p stability results, various analytical L^p signal convergence results, some simplified analytical tools for discrete-time system stability, and multivariable operator norms. These results, whose proofs are given in detail and are easy to understand, clarify several important signal and system properties for adaptive control.

Chapter 3 addresses adaptive parameter estimation for a general linear model illustrated by a parametrized linear time-invariant system in either continuous or discrete time. Detailed design and analysis of a normalized gradient algorithm and a normalized least-squares algorithm in either continuous or discrete time are given, including structure, stability, robustness, and convergence of the algorithms. A collection of commonly used robust adaptive laws are presented which ensure robust stability of the adaptive schemes in the presence of modeling errors. An $L^{1+\alpha}$ $(\alpha \geq 1)$ theory is developed for adaptive parameter estimation for a linear model, revealing some important inherent robustness properties of adaptive parameter estimation algorithms.

Chapter 4 develops two types of state feedback adaptive control schemes: for state tracking and for output tracking (and its discrete-time version). For both continuous- and discrete-time systems, adaptive state feedback for output tracking control, based on a simple controller structure under standard model reference adaptive control assumptions, is used as an introduction to adaptive control of general linear systems. Adaptive disturbance rejection under different conditions is addressed in detail; in particular, adaptive output rejection of unmatched input disturbance is developed based on a derived property of linear systems. Another development is a derived parametrization of state feedback using a full- or reduced-order state observer, leading to the commonly used parametrized controller structures with output feedback.

Chapter 5 deals with continuous-time model reference adaptive control using output feedback for output tracking. The key components of model reference adaptive control theory—*a priori* plant knowledge, controller structure, plant–model matching, adaptive laws, stability, robustness, and robust adaptation—are addressed in a comprehensive formulation and, in particular, stability and robustness analysis is given in a simplified framework. The plant–model matching equation for a standard model reference controller structure is studied in a tutorial formula. Design and analysis of model reference adaptive control schemes are given for plants with relative degree 1 or larger, using a Lyapunov or gradient method based on a standard quadratic or nonquadratic cost function. For the relative degree 1 case, an $L^{1+\alpha}$ $(0 < \alpha < 1)$ adaptive control design is proposed for reducing output tracking errors. An $L^{1+\alpha}$ $(\alpha \geq 1)$ theory is developed for adaptive control with inherent robustness with respect to certain modeling errors. Robust adaptive control is formulated and solved in a compact framework. Assumptions on plant unmodeled dynamics are clarified, and robust adaptive laws are analyzed. Closed-loop signal boundedness and mean tracking error properties are proved. To develop adaptive control schemes without using the sign of the high frequency gain of the controlled plant, a modified controller parametrization leads to a framework of adaptive control using a Nussbaum gain for stable parameter adaptation and closed-loop stability and asymptotic output tracking.

Chapter 6 develops a model reference adaptive control theory for discrete-time linear time-invariant plants. A unique plant–model matching equation is derived, with unique controller parameters specified to ensure exact output tracking after a finite number of steps. A stable adaptive control scheme is designed and analyzed which ensures closed-loop signal boundedness and asymptotic output tracking. It is shown that the model reference adaptive control system is robust with respect to L^2 modeling errors and with modification is also robust with respect to $L^{1+\alpha}$ ($\alpha > 1$) modeling errors. Thus an $L^{1+\alpha}$ ($\alpha \geq 1$) robustness theory is developed for discrete-time adaptive control. Robust adaptive laws are derived for discrete-time adaptive control in the presence of bounded disturbances.

Chapter 7 presents two typical designs (and their analysis) of indirect adaptive control schemes: indirect model reference adaptive control and indirect adaptive pole placement control in both continuous and discrete time. Examples are used to illustrate the design procedures and analysis methods. For indirect model reference adaptive control in continuous or discrete time, a concise closed-loop error model is derived based on which the proof of signal boundedness and asymptotic output tracking is formed in a feedback and small-gain setting similar to that for the direct model reference adaptive control scheme of Chapters 5 and 6. For indirect adaptive pole placement control, a singularity problem is addressed, and closed-loop stability and output tracking are analyzed in a unified framework for both continuous and discrete time. As a comparison, a direct adaptive pole placement control scheme is presented and discussed for its potential to avoid the singularity problem.

Chapter 8 conducts a comparison study of several adaptive control schemes applied to a benchmark two-body system with joint flexibility and damping, including direct state feedback, direct output feedback, indirect output feedback, direct–indirect state feedback, and backstepping state feedback designs, with detailed design and analysis for the last two designs. With different complexity, they all ensure closed-loop signal boundedness and asymptotic output tracking. The design and analysis of the direct–indirect adaptive control scheme demonstrate some typical time-varying operations on signals in time-varying systems.

Chapter 9 first gives the design and analysis of adaptive state feedback state tracking control for multi-input systems. A multivariable state feedback adaptive control scheme is derived using LDU decomposition of a plant gain matrix. Multivariable adaptive control is applied to system identification. This chapter then develops a unified theory for robust model reference adaptive control of linear time-invariant multi-input, multi-output systems in both continuous and discrete time. Key issues such as *a priori* plant knowledge, plant and controller parametrizations, design of adaptive laws, stability, robustness, and performance are clarified and solved. In particular, an error model for a coupled tracking error equation is derived, a robust adaptive law for unmodeled dynamics is designed, a complete stability and robustness analysis for a general multivariable case is given, and a unified multivariable adaptive control theory is established in a form applicable in both continuous and discrete time. The chapter presents some recent results in reducing *a priori* plant knowledge for multivariable model reference adaptive control using LDU parametrizations of the high frequency gain matrix of the controlled plant. Model reference adaptive control designs for multivariable systems with input or output time delays are also derived. Different adaptive control schemes, including a variable structure design, a backstepping design, and a pole placement control design for multivariable systems, are presented. Finally, robust adaptive control theory is applied to adaptive control of robot manipulator systems in the presence of parameter variations and unmodeled dynamics.

Chapter 10 presents a general adaptive inverse approach for control of plants with uncertain nonsmooth actuator nonlinearities such as dead-zone, backlash, hysteresis, and other piecewise-linear characteristics which are common in control systems and often limit system performance. An adaptive inverse is employed for cancelling the effect of an actuator nonlinearity with unknown parameters, and a linear or nonlinear feedback control law is used for controlling a linear or smooth nonlinear dynamics following the actuator nonlinearity. This chapter gives an overview of various state feedback and output feedback control designs for linear, nonlinear, single-input and single-output, and multi-input and multi-output plants as well as open problems in this area of major theoretical and practical relevance. A key problem is to develop linearly parametrized error models suitable for developing adaptive laws to update the inverse and feedback controller parameters, which is solved for various considered cases. The chapter shows that control systems with commonly used linear or nonlinear feedback controllers such as a model reference, PID, pole placement, feedback linearization, or backstepping can be combined with an adaptive inverse to handle actuator nonlinearities.

The book is focused on adaptive control of deterministic systems with uncertain parameters, dynamics and disturbances. It can also be useful for understanding the adaptive control algorithms for stochastic systems (see references for "Stochastic Systems" in Section 1.4 for such algorithms). The material presented has been used and refined in a graduate course on adaptive control which I have taught for the past ten years at the University of Virginia to engineering, computer science, and applied mathematics students. Comments and modifications to the book can be found at

http://www.people.virginia.edu/~gt9s/wiley-book.

If used as a reference, this book can be followed in its chapter sequence for both continuous- and discrete-time adaptive control system design and analysis. The discrete-time contents are mainly in Sections 1.5.3 (adaptive control system examples), 2.7 and 2.8 (systems and signals), 3.6 (adaptive parameter estimation), 3.7.2 (robustness of parameter estimation), 3.8.2 (robust parameter estimation), 4.5 (state feedback adaptive control), Chapter 6 (model reference adaptive control), Sections 7.3 (indirect model reference adaptive control and adaptive pole placement control), 9.2 (multivariable model reference adaptive control), and 10.2–10.5 (adaptive actuator nonlinearity inverse control) (both in a unified continuous- and discrete-time framework). The rest of the book is for continuous-time adaptive control design and analysis.

If used as a textbook for students with knowledge of linear control systems, as a suggestion based on experience at the graduate level, the instruction may start with Sections 1.4 and 1.5 as an introduction to adaptive control (one or two lectures, 75 minutes each). Some basic knowledge of systems, signals, and stability may be taken from Sections 2.1–2.6 (system modeling, signal norms, Lyapunov stability, Gronwall-Bellman lemma, small-gain lemma, strictly positive realness and Lefschetz-Kalman-Yakubovich lemma, signal convergence lemmas including Lemmas 2.14, 2.15, and 2.16 (Barbălat lemma) for four or five lectures). Adaptive parameter estimation can be taught using Sections 3.1–3.6 in four or five lectures, including some reading assignments of robustness results from Sections 3.7 and 3.8. The design and analysis of adaptive control schemes with state feedback are presented in Sections 4.1–4.4 (three lectures), while the discrete-time results in Section 4.5 can be used as reading materials. Continuous-time model reference adaptive control in Chapter 5 can be covered in seven or eight lectures (Sections 5.1–5.5, with Section 5.6 as a reading assignment). Indirect adaptive control in Chapter 7 may need four lectures. One lecture plus reading is recommended for Chapter 8. Chapters 9 and 10 are for advanced study as either extended reading or project assignments. Further reading can be selected from the included extensive list of references on adaptive systems and control.

In this book, for a unified presentation of continuous- and discrete-time adaptive control designs in either the time or frequency domain, the notation $y(t) = G(D)[u](t)$ (or $y(D) = G(D)u(D)$) represents, as the case may be, the time-domain output at time t (or frequency-domain output) of a dynamic system characterized by a dynamic operator (or transfer function) $G(D)$ with input $u(\tau)$, $\tau \leq t$ (or $u(D)$), where the symbol D is used, in the continuous-time case, as the Laplace transform variable or the time differentiation operator $D[x](t) = \dot{x}(t)$, $t \in [0, \infty)$, or, in the discrete-time case, as the z-transform variable or the time advance operator $D[x](t) = x(t+1)$, $t \in \{0, 1, 2, 3, \ldots\}$, with $x(t) \stackrel{\triangle}{=} x(tT)$ for a sampling period $T > 0$.

Adaptive control as knowledge has no limit and as theory is rigorous. Adaptive control is a field of science. The universe is mysterious, diverse, and vigorous. The world is complicated, uncertain, and unstable. Adaptive control deals with complexity, uncertainty, and instability of dynamic systems. Taoist philosophy emphasizes simplicity, balance, and harmony of the universe. A goal of this book is to give a simplified, balanced, and harmonious presentation of the fundamentals of adaptive control theory, aimed at improving the understanding of adaptive control, which, like other control methodologies, brings more simplicity, balance, and harmony to the dynamic world.

This book has benefited from many people's help. First, I am especially grateful to Professors Petros Ioannou and Petar Kokotović. I was introduced to the field of adaptive control by Professor Ioannou, and his continuous support and vigorous instruction were most helpful to my study and research in adaptive control. Professor Kokotović has been a great mentor, and his persistent enthusiasm and continual encouragement have been most valuable to me in the writing of this book. Their robust adaptive control theory has been most influential to my research in adaptive control.

I would like to particularly acknowledge Professors Karl Åström, Graham Goodwin, Bob Narendra, and Shankar Sastry for their work on adaptive control, which inspired me in research and in writing this book.

I would like to thank Professors Brian Anderson, Anu Annaswamy, Er-Wei Bai, Bob Bitmead, Stephen Boyd, Marc Bodson, Carlos Canudas de Wit, Han-Fu Chen, Aniruddha Datta, Michael Demetriou, Manuel De la Sen, Gang Feng, Li-Chen Fu, Sam Shu-Zhi Ge, Lei Guo, Lui Hsu, Alberto Isidori, Zhong-Ping Jiang, Dr. Ioannis Kanellakopoulos, Professor Hassan Khalil, Dr. Bob Kosut,

Professors Gerhard Kreisselmeier, P. R. Kumar, Yoan Landau, Frank Lewis, Lennart Ljung, Wei Lin, Rogelio Lozano, David Mayne, Iven Mareels, Rick Middleton, Steve Morse, Romeo Ortega, Marios Polycapou, Laurent Praly, Drs. Darrel Recker, Doug Rhode, Professors Gary Rosen, Jack Rugh, Ali Saberi, Mark Spong, Yu Tang, T. J. Tarn, David Taylor, Chang-Yun Wen, John Ting-Yung Wen, and Erik Ydstie, whose knowledge of adaptive systems and controls helped my understanding of the field.

I especially thank Professors Murat Arcak, Ramon Costa, Dr. Suresh Joshi, Professor Miroslav Krstić, Dr. Jing Sun, and Professor Kostas Tsakalis for their knowledge and comments, which helped me in writing this book.

I am thankful to my graduate students Michael Baloh, Lori Brown, Jason Burkholder, Shu-Hao Chen, Tinya Coles, Warren Dennis, Emin Faruk Kececi, Yi Ling, Xiao-Li Ma, Raul Torres Muniz, Nilesh Pradhan, Gray Roberson, Min-Yan Shi, Xi-Dong Tang, Avinash Taware, Ming Tian, Timothy Waters, and Xue-Rui Zhang, and to computer scientists Chen-Yang Lu and Ying Lu, and engineer Yi Wu, for their earnest study, stimulating discussion, and interesting applications of adaptive control.

I would also like to express my thanks to my colleagues at the University of Virginia for their support, in particular, to Professors Milton Adams, Paul Allaire, Jim Aylor, Zong-Li Lin, Jack Stankovic, Steve Wilson, and Houston Wood, for their collaboration and help in my teaching and research.

Finally, I gratefully acknowledge that my study and research on adaptive control, which led to many of the results in this book, were supported by grants from the U.S. National Science Foundation and by a scholarship from the Chinese Academy of Sciences.

GANG TAO
Charlottesville, Virginia

Chapter 1

Introduction

Adaptive control is based on feedback of signals in a controlled system for control adaptation to effectively handle system uncertainties. Feedback is crucial for performance guarantees in control systems, including adaptive control systems, which can be designed in either continuous or discrete time. This chapter presents some basic concepts of feedback in control systems, system modeling in continuous time and discrete time, and feedback control design and some prototypes of adaptive control systems with illustrative examples.

1.1 Feedback in Control Systems

A system is a set of interconnected functional components organized for certain specific tasks in a physical world. Various types of systems are all around us. A control system is a system whose behavior can be influenced by some externally acting signals. A signal which describes a system's behavior is the output of the system, while an externally acting signal is a control input to the system. There are two types of control systems: *open loop* and *closed loop* (feedback). In an open-loop control system, the input signals are prespecified, assuming an ideal situation of system operation (e.g., without any uncertainties in the system), and no system output information is used in generating the control input signal. An open-loop system is unable to adapt to system changes and is not effective for sophisticated control tasks (in control theory, an open-loop system model usually serves as a system to be controlled). A closed-loop control system utilizes its output signals for feedback to generate a control input and is much more powerful than an open-loop control system. A closed-loop control system is capable of adapting to system changes and uncertainties and achieving high performance. Almost all control systems use certain feedback and thus operate in a closed loop.

Feedback is the key for automatic control that does not rely on human interference. To fulfill a control task, the controlled system variables are sensed and fed back, generating control signals which are applied to the system. Control signals are generated by actuators from control algorithms derived based on the actual and desired system dynamics. Control algorithms, actuators, and sensors are three key components of control systems.

Control is a physical concept as the results of control are usually seen in changes in some physical variables. For example, robot manipulators are controlled to reach desired positions to grasp chosen objects with desired forces. Control theory is based on firm mathematical foundations. A controlled system can be described mathematically by its dynamic equations, which makes its behavior analysis easy and its control design convenient. Control engineers derive system models, understand control methods, and design and analyze control algorithms mathematically and implement control designs physically. System performance analysis is a main part of control systems research.

Feedback control was used in early human history as humans learned to make tools and change their environment. An example is the float regulator mechanism used more than 2000 years ago to control the liquid flow rate in a water clock or the liquid level in a liquid tank. The key element used is a float valve between two liquid tanks. To regulate the liquid level of the lower tank, the float valve falls as the liquid level of the lower tank falls and more liquid from the upper tank flows into the lower tank. A constant liquid level can produce a constant flow rate when the lower tank liquid is used to maintain the clock's accuracy. In this case, the float valve is so designed that it can measure as well as control the liquid level, that is, it acts as a sensor as well as an actuator. Such a liquid level regulator is still popularly used today.

Another famous example is James Watt's centrifugal fly-ball governor, invented around 1788 for controlling the speed of a steam engine in an industrial process. The governor is so designed that when the engine speed is increased, the fly-ball moves away from its shaft axis so that the steam valve decreases the amount of steam driving the engine, which reduces the speed of the engine, and vice versa. With this mechanism, the engine speed is regulated at a constant one determined by the mechanical design.

A major development in feedback control was a feedback amplifier invented by H. S. Black in 1927 and analyzed by H. W. Bode and H. Nyquist later. Such devices are based on a negative-feedback principle and have desirable properties of stability of a closed-loop system and robustness with respect to system errors such as parameter variations and external noise. Feedback control was extensively used and significantly developed during World War II when it was needed to design and build automatic pilot systems for aircraft, gun positioning systems, and radar antenna control systems.

Nowadays feedback control systems are used almost everywhere in our daily life and production, such as aircraft flight control systems, automobile control systems, robot control systems, temperature control systems, process control systems, control systems for electronic devices, navigation control systems, real-time computer systems, and so on. Feedback control plays an important role in the development of science and technology.

Descriptions of classical control theory, control applications, and the history of feedback control can be found in [82], [104], [216].

1.2 System Modeling

Almost all physical systems operate in continuous time. However, many control systems are designed and implemented in discrete time to make use of computers for control implementation, and with sampling many physical systems are expressed in discrete time by difference equations or z-domain transfer functions, on which the design of a discrete-time controller is based.

1.2.1 Continuous-Time Systems

There are many physical laws which govern the motions of systems to be controlled. The most famous ones are Newton's laws for mechanical systems and Kirchhoff's laws for electrical systems. There are also other physical laws for electromechanical systems, thermodynamic systems, hydraulic systems, and so on. Based on these physical laws, one can first write a set of differential equations to describe a system and then derive an nth-order differential equation to describe the same system, where n is called the system order and is determined by the number of energy-storing elements in the system, such as a capacitor or inductor of an electric circuit.

A dynamic system can be described by a differential equation of the form

$$F(y^{(n)}(t), \ldots, y^{(1)}(t), y(t), u^{(n)}(t), \ldots, u^{(1)}(t), u(t), t) = 0, \ t \geq t_0, \tag{1.1}$$

where t is the time variable with initial time t_0 and, from a control system point of view, $y(t)$ is the system output and $u(t)$ is the system input and $y^{(i)}(t)$ and $u^{(i)}(t)$ denote the ith time derivatives $\frac{d^i y(t)}{dt^i}$ and $\frac{d^i u(t)}{dt^i}$ of $y(t)$ and $u(t)$, with a common notation $\dot{y}(t) = y^{(1)}(t)$, $\dot{u}(t) = u^{(1)}(t)$ and $\ddot{y}(t) = y^{(2)}(t)$, $\ddot{u}(t) = u^{(2)}(t)$. A specific form of the function F depends on a specific system under consideration. For a single-input, single-output system, both $y(t)$ and $u(t)$ are scalar signals, denoted as $y \in R$ and $u \in R$. For example, the differential equation for a pendulum with length l and mass m is $ml^2\ddot{\theta} + mgl\sin\theta = \tau$, where g is the acceleration of gravity, θ is the pendulum angle (output), and τ is the applied torque (input) [104]. When θ is small, $\sin\theta \approx \theta$ may be used to linearize the system equation, leading to $ml^2\ddot{\theta} + mgl\theta = \tau$.

For an nth-order system, there exist n state variables $x_i(t)$, $i = 1, 2, \ldots, n$, physical or artificial, to completely express the system behavior, such that system (1.1) can be expressed as

$$\dot{x} = f(x, u, t), \; y = h(x, u, t), \; t \geq t_0, \tag{1.2}$$

for some functions $f \in R^n$ and $h \in R$, where $x = [x_1, \ldots, x_n]^T \in R^n$ is the system state vector and $u(t)$ is the control input and $y(t)$ is the system output. For a linear time-invariant system, equation (1.2) has the form

$$\dot{x}(t) = Ax(t) + Bu(t), \; y(t) = Cx(t) + Du(t), \; t \geq 0, \tag{1.3}$$

for some constant matrices $A \in R^{n \times n}$, $B \in R^n$, and $C \in R^{1 \times n}$ and scalar $D \in R$. The systems models (1.1)–(1.3) also find their extensive applications in signal processing, communications, real-time computing, semiconductor manufacture, and biological and other systems.

For a linear time-invariant system, its differential equation is

$$\begin{aligned} &p_n y^{(n)}(t) + p_{n-1} y^{(n-1)}(t) + \cdots + p_1 \dot{y}(t) + p_0 y(t) \\ &= z_n u^{(n)}(t) + z_{n-1} u^{(n-1)}(t) + \cdots + z_1 \dot{u}(t) + z_0 u(t), \; t \geq 0, \end{aligned} \tag{1.4}$$

where p_i and z_j are some constant coefficients. Such a differential equation describes a wide class of control systems in real life. For example, an electric circuit consisting of a resistor R in series with an inductor L is described by $L\dot{y}(t) + Rx(t) = u(t)$, where $u(t)$ is the applied control voltage and $y(t)$ is the circuit current. A mechanical system with mass m, spring k, and damping b has the differential equation $m\ddot{y}(t) + b\dot{y}(t) + ky(t) = u(t)$, where $u(t)$ is the applied control force and $y(t)$ is the controlled mass displacement.

The quantitative description of system (1.4) is its solution

$$y(t) = y(t; u(\cdot), y^{(i)}(0), u^{(i)}(0), i = 0, 1, \ldots, n-1), \; t \geq 0, \tag{1.5}$$

where $y^{(i)}(0)$, $u^{(i)}(0)$, $i = 0, 1, \ldots, n-1$, are the initial conditions of the system, that is, the initial values of $y^{(i)}(t)$, $u^{(i)}(t)$. Such a solution $y(t)$ depends on the values of p_i and z_j, as well as on the control input $u(t)$ and initial conditions

$y^{(i)}(0)$, $u^{(i)}(0)$, $i = 0, 1, \ldots, n-1$. For example, for $n = 1$ and $z_1 = 0$, $z_0 = 1$, $p_1 = 1$, and $p_0 = a$, system (1.4) is $\dot{y}(t) + ay(t) = u(t)$, whose solution is

$$y(t) = e^{-at}y(0) + \int_0^t e^{-a(t-\tau)}u(\tau)\, d\tau,\ t \geq 0. \tag{1.6}$$

This simple example indicates that the system behavior, characterized by $y(t)$, is determined by the system structure and parameters and the control input $u(t)$ as well. The task of control is to generate an input signal $u(t)$ using feedback to modify the system structure and parameters to result in an output $y(t)$, which tracks a given desired reference output.

1.2.2 Discrete-Time Systems

Today, sophisticated control systems implement their control laws using digital computers which calculate desired control signals in digital form. Digital control systems are easy to build, flexible to change, less sensitive to noise and environmental variations, more compact and lightweight, more versatile, and less expensive. A digital controller has an analog-to-digital converter, which transforms analog signals from a controlled process to digital signals for a digital computer; a digital computer, which realizes a control algorithm; and a digital-to-analog converter, which transforms the digital signals generated by the digital computer to analog signals for controlling a process.

There are systems that operate in discrete time, such as a bank account balance model $y(k) = y(k-1) + ry(k-1) + u(k)$, where $u(k)$ is the deposit at $t = kT$, $y(k)$ is the account balance at $t = kT$ after $u(k)$ is made, and r is the interest per dollar per period T. However, most systems operate in continuous time. A discrete-time system model is crucial for digital control of a continuous-time system. A digital controller is designed based on a discrete-time model of a controlled process operating in continuous time with analog signals. In digital control, the input signal to the controlled process is kept constant over the sampling intervals of time, over which the control signal is computed. This is needed for control implementation and is useful for discrete-time system modeling as well. As an example, consider an electric circuit: a resistor of a ohms in series with an inductor $L = 1$ H. With a voltage source $u(t)$, the circuit current $x(t)$ is described by $\dot{x}(t) + ax(t) = u(t)$. If $u(t) = u(kT)$ for all $t \in [kT, (k+1)T)$ ($T > 0$ is called the sampling period), then $x((k+1)T)$ satisfies the difference equation

$$\begin{aligned} x((k+1)T) &= a_d x(kT) + b_d u(kT),\ k \in \{0, 1, 2, \ldots\}, \\ a_d &= e^{-aT}, b_d = (1 - e^{-aT})/a. \end{aligned} \tag{1.7}$$

This process is called *discretization* of a continuous-time system and can be performed for a general linear continuous-time system $\dot{x}(t) = Ax(t) + Bu(t), y(t) = Cx(t)+Du(t)$, where $A \in R^{n\times n}$, $B \in R^{n\times p}$ are constant matrices, $C \in R^{q\times n}$, $D \in R^{q\times p}$, and $u(t) = u(kT)$ for all $t \in [kT, (k+1)T)$ with $T > 0$, to obtain its discrete-time system representation:

$$\begin{aligned} x((k+1)T) &= A_d x(kT) + B_d u(kT), \\ y(kT) &= Cx(kT) + Du(kT),\ k \in \{0, 1, 2, \ldots\}, \end{aligned} \tag{1.8}$$

for $A_d \in R^{n\times n}$ and $B_d \in R^{n\times p}$ depending on A, T, and B, that is,

$$A_d = e^{AT},\ B_d = e^{AT}\int_0^T e^{-A\sigma}B\,d\sigma, \tag{1.9}$$

where $e^{At} = \mathcal{L}^{-1}[(sI - A)^{-1}]$ [209].

A discrete-time system can also be expressed by a difference equation

$$\begin{aligned} &y((k+n)T) + p_{n-1}y((k+n-1)T) + \cdots + p_1 y((k+1)T) + p_0 y(kT) \\ &= z_n u((k+n)T) + z_{n-1}u((k+n-1)T) + \cdots + z_0 u(kT), \end{aligned} \tag{1.10}$$

$k \in \{0, 1, 2, \ldots\}$, where p_i, z_j, $i, j = 1, \ldots, n-1$, and z_n are parameters.

In a discrete-time system expression, the time variable kT is from the above discretization of the continuous-time system: $t = kT$, $k = 0, 1, 2, \ldots$. Since the sampling period T is a fixed constant, we can simplify the expression of a discrete-time system by using $x(t)$, $t = 0, 1, 2, \ldots$, to represent $x(kT)$, $k = 0, 1, 2, \ldots$, whenever no confusion exists.

1.3 Feedback Control

For a linear time-invariant system described by the time-domain differential equation (1.3) (with $p_n = 1$), its frequency s-domain expression is

$$y(s) = G(s)u(s), \tag{1.11}$$

where $y(s)$ and $u(s)$ are the Laplace transforms of $y(t)$ and $u(t)$, respectively:

$$y(s) = \int_0^\infty y(t)e^{-st}\,dt,\ u(s) = \int_0^\infty u(t)e^{-st}\,dt \tag{1.12}$$

with s being the complex variable, and $G(s)$ is the system transfer function:

$$G(s) = \frac{z_n s^n + \cdots + z_1 s + z_0}{s^n + p_{n-1}s^{n-1} + \cdots + p_1 s + p_0}, \tag{1.13}$$

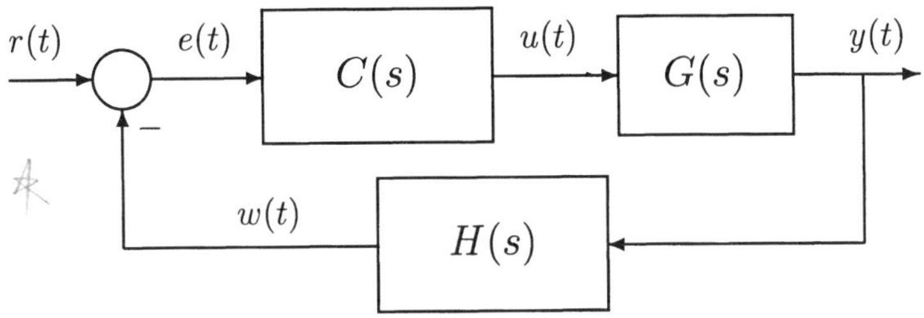

Figure 1.1: A typical structure of feedback control systems.

which, especially its poles (those complex numbers s_p such that $G(s_p) = \infty$), determines such system performance as stability and transient response.

The task of control is to generate a control input signal $u(t)$ (or $u(s)$ in the frequency s-domain) for system (1.4) (or (1.11)) so that the system output $y(t)$ has the desired behavior. Such a task is fulfilled by a feedback controller. A typical feedback control system block diagram is shown in Figure 1.1, where the input signal $u(t)$ is generated based on the error signal $e(t) = r(t) - w(t)$, where $r(t)$ is a reference signal and $w(t)$ is a feedback signal. A controller consists of a feedforward compensator $C(s)$ (which itself is a system whose input is $e(t)$ and output is $u(t)$ and is characterized by its own transfer function $C(s)$, that is, $u(s) = C(s)e(s)$) and a feedback compensator $H(s)$ (which generates the feedback signal $w(t)$ from the system output $y(t)$, i.e., $w(s) = H(s)y(s)$). Combining the three subsystems $G(s)$, $C(s)$, and $H(s)$ together, we have the closed-loop system in the frequency s-domain:

$$y(s) = G_c(s)r(s), \; G_c(s) = \frac{G(s)C(s)}{1 + H(s)G(s)C(s)}. \tag{1.14}$$

Now the closed-loop system performance is determined by the closed-loop transfer function $G_c(s)$ which can be modified by different choices of $C(s)$ and $H(s)$. Various design methods based on different control objectives and system conditions have been developed and verified in theory and practice.

PID Control

A popular feedback controller $C(s)$ is the proportional-integral-derivative (PID) controller whose s-domain representation is

$$C(s) = K_P + \frac{K_I}{s} + K_D s, \tag{1.15}$$

which in the time domain means

$$u(t) = K_P\, e(t) + K_I \int_0^t e(\tau)\, d\tau + K_D\, \dot{e}(t), \tag{1.16}$$

where K_P, K_I, K_D are constant proportional, integral, and derivative gains, respectively. Such a controller, simple and yet powerful for many practical systems, has been extensively studied for different types of systems and widely used in many industrial processes [65], [104], [364].

Pole Placement Control

The basic idea of pole placement control can be illustrated for the linear time-invariant system: $\dot{x}(t) = Ax(t) + Bu(t)$, $x(t) \in R^n$, $u(t) \in R^m$. The eigenvalues of the matrix $A \in R^{n\times n}$ determines the system stability and performance. A pole placement control design is to find a gain matrix $K \in R^{m\times n}$ such that the eigenvalues of $A - BK$ are placed at some desired values. Then the feedback control law is $u(t) = Kx(t) + r(t)$, where $r(t) \in R^m$ is a reference input which leads to a desired closed-loop system: $\dot{x}(t) = (A+BK)x(t)+Br(t)$. The necessary and sufficient condition for arbitrary pole placement is that (A, B) is controllable. The physical meaning of controllability is that for any given initial state x_0 and final state $x_f = 0$, a control $u(t)$ can be found to derive the system state $x(t)$ from $x(0) = x_0$ to $x(t_f) = x_f$ over the final interval $[0, t_f]$. Mathematically, controllability is equivalent to the condition: $\text{rank}[B|AB|A^2B|\cdots|A^{n-1}B] = n$. Study of pole placement control designs for a linear system has been extensively reported in the literature, including designs using observers [13], [335], which provide asymptotic estimates of $x(t)$ (when not available) from an output $y(t) = Cx(t)$. For nonlinear systems, the idea of pole placement needs further study.

Optimal Control

Optimal control theory was pioneered by R. Bellman (dynamic programming, 1957), L. S. Pontryagin (maximum principle, 1958), and R. E. Kalman (linear quadratic regulation, 1960). The basic idea is to find a control $u(t)$ for a system $\dot{x} = f(x, u, t)$ such that the cost

$$J = \phi(x(T), T) + \int_{t_0}^{T} L(x(t), u(t), t)\, dt \tag{1.17}$$

is minimized over the interval $[t_0, T]$ (T may be ∞) for some nonnegative functions ϕ and L. Typical applications include minimum-time and minimum-control-effort problems [24], [183], [216]. The linear quadratic case is with $f(x, u, t) = A(t)x(t) + B(t)u(t)$, $\phi(x(T), T) = x^T(T)Sx(T)$, and

$$L(x(t), u(t), t) = x^T(t)Q(t)x(t) + u^T(t)R(t)u(t) \tag{1.18}$$

for some matrices S, $Q(t)$, $t \geq t_0$, both positive semidefinite, and $R(t)$, positive definite, and to construct optimal control solutions, certain Riccati equations play important roles. A dual optimal estimation theory (the Kalman filter) was developed by Kalman in 1960 for estimating the state $x(t)$ of the system $\dot{x}(t) = Ax(t) + Bu(t) + w$, $y(t) = Cx(t) + v$, from the output $y(t)$, subject to certain system noises $w(t)$ and $v(t)$.

Recently, in the 1980s, a new optimal control problem, H_∞ control, was formulated and solved, with a vast amount of literature available. A basic problem may be illustrated as follows [244]: A system is described by $z(s) = P_{11}(s)w(s) + P_{12}(s)u(s)$, $y(s) = P_{21}(s)w(s) + P_{22}(s)u(s)$, where w is an external disturbance, z is a tracking error signal to be minimized, y is the controlled output, and u is the applied input. To achieve good tracking, we use a controller $u(s) = K(s)y(s)$, which results in $z(s) = (P_{11}(s) + P_{12}(s)K(s)(I - P_{22}(s)K(s))^{-1}P_{21}(s))w(s)$, to solve the problem of minimizing

$$J(K) = \|P_{11} + P_{12}K(I - P_{22}K)^{-1}P_{21}\|_\infty \tag{1.19}$$

over all possible stabilizing and realizable $K(s)$, where $\|G\|_\infty = \sup_\omega \bar{\sigma}(G(j\omega))$ with $\bar{\sigma}(G(j\omega))$ being the maximum singular value of $G(j\omega)$.

Robust Control

Robust control deals with systems with modeling errors. A linear time-invariant system with modeling errors may be expressed as

$$y(s) = (G_0(s)(1 + \Delta_m(s)) + \Delta_a(s))u(s) + d(s), \tag{1.20}$$

where $G_0(s)$ is a nominal dynamics, $\Delta_m(s)$ and $\Delta_a(s)$ are multiplicative and additive unmodeled dynamics due to system parameter and structure uncertainties, and $d(s)$ represents a disturbance due to environment uncertainties. A robust controller is usually a design for the worst-case system uncertainty, which ensures an attainable system performance for all uncertainties less severe than the worst case. A robust controller designed with fixed parameters works for a class of uncertain systems [244], [323], [459]. Robustness is a major issue for control system designs. Moreover, it is also helpful for system performance improvement if certain available qualitative knowledge about unmodeled dynamics is used for control system designs [188].

Nonlinear Control

A nonlinear controller makes use of the nonlinear dynamics information of a system to be controlled, which can be done in many ways. For example, for a nonlinear system $\dot{x} = f(x,t) + g(x,t)u$, $y = h(x,u,t)$, a feedback linearization method [157] uses a transformation $z = T(x)$ and a feedback law $u = \alpha(x) + \beta(x)v$ to linearize the system of a class as $\dot{z} = Az + Bv$ so that a desired linear feedback law can be designed for this resulting linear system. A backstepping method [206] is also a powerful design tool for some classes of nonlinear systems. A nonlinear design can also be applied to a linear system for improved system performance [206]. Control of systems with smooth and nonsmooth nonlinearities is an important research area.

An important and distinct class of nonlinear controllers is comprised of variable structure controllers which use control switching to reject the effects of system modeling errors and disturbance on system behavior to enhance robustness of system performance [424].

Adaptive Control

Adaptive control provides adaptation mechanisms that adjust a controller for a system with parametric, structural, and environmental uncertainties to achieve desired system performance. Payload variation or component aging causes parametric uncertainties, component failure leads to structural uncertainties, and external noises are typical environmental uncertainties. Such uncertainties often appear in airplane and automobile engines, electronic devices, and industrial processes. Adaptive control has experienced many successes in both theory and applications and is developing rapidly with the emergence of new challenging problems and their encouraging solutions. Typical adaptive control applications reported in the literature include temperature control, chemical reactor control, pulp dryer control, rolling mill control, automobile control, ship steering control, blood presure control, artificial heart control, robot control, and physiological control.

Unlike other controllers using PID, pole placement, optimal, robust, or nonlinear control methods, as described above, whose designs are based on certain knowledge of the system parameters, adaptive controllers do not need such knowledge; they are adapted to parameter uncertainties by using performance error information on-line.

Various theoretical issues and different design methods of adaptive control will be systematically addressed and presented in this book.

Discrete-Time Control

Similar to their continuous-time counterparts, for systems (1.8) and (1.10), there are many issues related to stability, performance, and control which have been extensively studied in the literature. Control methods such as PID, pole placement, optimal, robust, nonlinear, and adaptive can also be developed for discrete-time systems. Based on the z-transforms

$$y(z) = \sum_{k=0}^{\infty} y(kT)z^{-k}, \; u(z) = \sum_{k=0}^{\infty} u(kT)z^{-k}, \tag{1.21}$$

system (1.10) has the transfer function

$$G(z) = \frac{z_n z^n + \cdots + z_1 z + z_0}{z^n + p_{n-1} z^{n-1} + \cdots + p_1 z + p_0}, \tag{1.22}$$

such that the z-transforms $u(z)$ and $y(z)$ of the system input $u(kT)$ and output $y(kT)$ are related in the frequency z-domain by the expression

$$y(z) = G(z)u(z). \tag{1.23}$$

A feedback control system can be represented by the same block diagram in Figure 1.1, with the s-domain transfer functions $G(s)$, $C(s)$, and $H(s)$ replaced by their corresponding z-domain transfer functions $G(z)$, $C(z)$, and $H(z)$.

More references can be found for PID control [20], [21], [65], [364]; pole placement control [13], [44], [171], [335], [442]; optimal control [9], [24], [38], [183], [216]; robust control [43], [106], [107]; nonlinear control [107], [179], [251], [351], [426]; and discrete-time control [22], [105], [209], [308].

1.4 Adaptive Control System Prototypes

A typical adaptive control system consists of a system (process) to be controlled (which is called a *plant*; for adaptive control, the plant parameters are *unknown*), a controller with parameters, and an adaptive law to update the controller parameters to achieve some desired system performance.

A single-input, single-output linear continuous-time time-invariant system is described by a differential equation compactly expressed as

$$P(s)[y](t) = k_p Z(s)[u](t), \; t \geq 0, \tag{1.24}$$

where $u(t)$ and $y(t)$ are the system input and output, respectively, and $P(s)$ and $Z(s)$ are monic (i.e., with leading coefficient 1) polynomials of s:

$$P(s) = s^n + p_{n-1}s^{n-1} + \cdots + p_1 s + p_0, \tag{1.25}$$

$$Z(s) = s^m + z_{m-1}s^{m-1} + \cdots + z_1 s + z_0, \tag{1.26}$$

with constant coefficients p_i and z_j, and k_p is a constant gain. The symbol s is used to denote the time differentiation operator: $s[x](t) = \dot{x}(t)$. For example, with $P(s) = s^2 + p_1 s + p_0$ and $Z(s) = s + z_0$, system (1.24) is

$$\ddot{y}(t) + p_1\dot{y}(t) + p_0 y(t) = k_p(\dot{u}(t) + z_0 u(t)), \; t \geq 0. \tag{1.27}$$

In this expression, the polynomials $P(s)$ and $Z(s)$ are seen as operators on the signals $y(t)$ and $u(t)$ to generate the signals $P(s)[y](t)$ and $Z(s)[u](t)$, respectively. From its differential equation, one can obtain the transfer function of the system as $G(s) = P^{-1}(s)k_pZ(s)$, that is,

$$y(s) = G(s)u(s), \quad G(s) = k_p\frac{Z(s)}{P(s)}, \tag{1.28}$$

where the symbol s is used to denote the Laplace transform variable and $y(s)$ and $u(s)$ are the Laplace transforms of $y(t)$ and $u(t)$, respectively, when the effect of the system initial conditions is neglected.

In operator form, system (1.24) may be expressed as

$$y(t) = G(s)[u](t) \triangleq \mathcal{L}^{-1}[G(s)u(s)], \tag{1.29}$$

where $\mathcal{L}^{-1}[\cdot]$ is the inverse Laplace transform operator and $G(s)$ is considered as the operator which maps $u(t)$ to $y(t)$.

A linear time-invariant system can be described in a state-space form as

$$\dot{x}(t) = Ax(t) + Bu(t), \; y(t) = Cx(t), \; t \geq 0, \tag{1.30}$$

where $A \in R^{n\times n}$, $B \in R^n$ are constant parameter matrices, $x(t) \in R^n$ is the state vector, $u(t) \in R$ is the control input, and $y(t) \in R$ is the output.

An adaptive controller usually consists of an output or state feedback compensator and an input feedforward compensator (a compensator is a designer's parametrized dynamic system for generating a control signal). A set of nominal controller parameters can be calculated from some design equations based on the plant parameters, with which some desired system performance can be defined and achieved. In adaptive control, plant parameters such as p_1, p_0, k_p, and z_0 in (1.25)–(1.26) are *unknown* so that the nominal controller parameters are also unknown and their estimates have to be used for control. The main task of adaptive control is to develop an adaptive law to update those parameter estimates, based on system performance errors, so that the desired system performance can still be achieved asymptotically.

Output Feedback Design

A typical output feedback adaptive control system is shown in Figure 1.2, where the controller for the plant $P(s)[y](t) = k_pZ(s)[u](t)$ consists of the output feedback compensator $\theta_2^T\frac{a(s)}{\Lambda(s)} + \theta_{20}$ and the input feedforward compensator

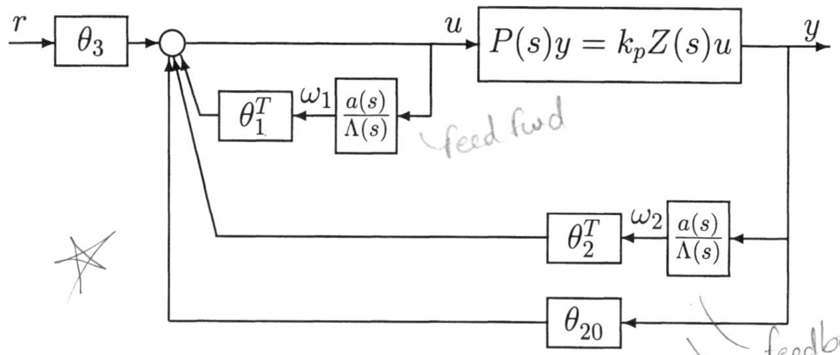

Figure 1.2: Output feedback control system.

$\theta_1^T \frac{a(s)}{\Lambda(s)}$, where θ_1 and θ_2 are parameter vectors, θ_{20} and θ_3 are parameters, and $\frac{a(s)}{\Lambda(s)}$ is a stable vector transfer function. The control input is

$$u(t) = \theta_1^T \omega_1(t) + \theta_2^T \omega_2(t) + \theta_{20} y(t) + \theta_3 r(t), \tag{1.31}$$

where $\omega_1(t) = \frac{a(s)}{\Lambda(s)}[u](t)$, $\omega_2(t) = \frac{a(s)}{\Lambda(s)}[y](t)$, and $r(t)$ is a reference input signal.

If the parameters of $G(s) = k_p \frac{Z(s)}{P(s)}$ were known, they could be used to calculate some ideal controller parameters θ_1^*, θ_2^*, θ_{20}^*, θ_3^* from some well-developed control design equations. The implementation of the controller, with $\theta_1 = \theta_1^*$, $\theta_2 = \theta_2^*$, $\theta_{20} = \theta_{20}^*$, $\theta_3 = \theta_3^*$, would lead to some desired system performance (behavior of $y(t)$) characterized by a reference output $y_m(t)$.

In the case of adaptive control when the parameters of $G(s)$ are unknown, a controller with θ_1^*, θ_2^*, θ_{20}^*, θ_3^* can no longer be available as these parameters depend on the parameters of $G(s)$ and thus are unknown. An adaptive control solution is to implement the controller with parameters $\theta_1(t)$, $\theta_2(t)$, $\theta_{20}(t)$, $\theta_3(t)$, which are the estimates of θ_1^*, θ_2^*, θ_{20}^*, θ_3^*. These estimates are obtained from some adaptive laws, that is, $\theta_1(t)$, $\theta_2(t)$, $\theta_{20}(t)$, $\theta_3(t)$ are updated on line as the control system is operating. The adaptation of the controller parameters is based on the performance error $y(t) - y_m(t)$ such that the closed-loop system adjusts itself toward an operation condition at which the desired system performance is achieved asymptotically: $\lim_{t\to\infty}(y(t) - y_m(t)) = 0$.

There are two commonly used approaches for the design of an adaptive controller: a *direct* approach and an *indirect* approach. A direct adaptive control design employs a direct estimation of the controller parameters $\theta_1(t)$, $\theta_2(t)$, $\theta_{20}(t)$, $\theta_3(t)$, while an indirect adaptive control design first estimates the plant parameters (those in $G(s) = k_p \frac{Z(s)}{P(s)}$) and then maps the estimated plant parameters to the controller parameters from a design equation.

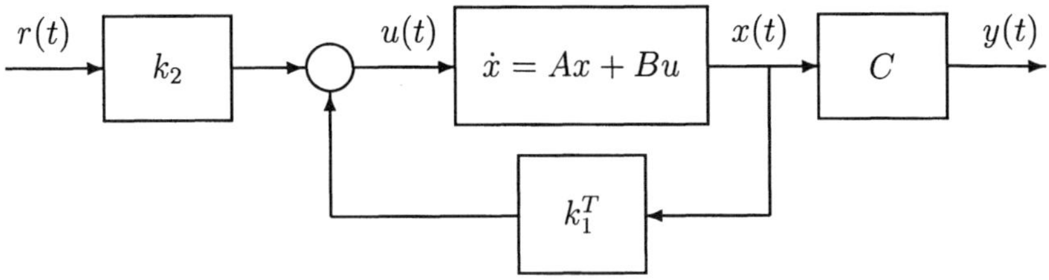

Figure 1.3: State feedback control system.

State Feedback Design

With the state variables in $x(t)$ available for feedback, the control objective is to design a state feedback control $u(t)$ such that all signals in the closed-loop system are bounded and either asymptotic state tracking or output tracking is achieved without knowledge of the system parameters.

For *state tracking*, the state vector $x(t)$ is required to track a given reference state vector $x_m(t)$, and for *output tracking*, the output $y(t)$ is required to track a given reference output $y_m(t)$. A state feedback controller is simpler than an output feedback controller. A typical state feedback controller structure is

$$u(t) = k_1^T(t)x(t) + k_2(t)r(t), \tag{1.32}$$

where $k_1(t)$ and $k_2(t)$ are the estimates of some ideal controller parameters $k_1^* \in R^n$ and $k_2^* \in R$ (which can be calculated from the system parameters for the controller (1.32) to achieve the desired control objective). The task of adaptive control is to generate the parameter estimates $k_1(t)$ and $k_2(t)$ without the knowledge of k_1^* and k_2^* to achieve the control objective.

Nonlinear Systems

In some cases, an output or state feedback adaptive controller can be developed for a nonlinear system of the general form $\dot{x} = f(x, u)$, $y = h(x, u)$, where f and h are some nonlinear functions, u is the system input, x is the system state vector, and y is the system output, all of appropriate dimensions. In this case, the linear dynamics and feedback blocks in Figures 1.2 and 1.3 are replaced by some nonlinear functions (see Section 1.5.4 and Chapter 10).

Actuator Nonlinearities

Actuators that generate control signals may also have dynamics (which may be included in the plant dynamics) and nonlinearities (which may be nonsmooth in nature, such as dead-zone, backlash, and hysteresis characteristics, and must be compensated in order to ensure desired performance of feedback control systems; see Chapter 10). Systems with actuator nonlinearities may be described as $\dot{x}(t) = Ax(t) + Bu(t)$, $u(t) = N(v(t))$, $y(t) = Cx(t)$, where $N(\cdot)$ represents an actuator nonlinearity and $v(t)$ is the applied control input, or $\dot{x} = f(x, u)$, $u = N(v)$, $y = h(x, u)$ if the system dynamics are nonlinear.

Discrete-Time Systems

To formulate an adaptive control system in discrete time, we consider a linear discrete-time time-invariant system described by a difference equation

$$P(z)[y](t) = k_p Z(z)[u](t), \; t \in \{0, 1, 2, \ldots\}, \tag{1.33}$$

where $u(t)$ and $y(t)$ are the system input and output, respectively, $P(z)$ and $Z(z)$ are monic polynomials of z, that is,

$$P(z) = z^n + p_{n-1} z^{n-1} + \cdots + p_1 z + p_0, \tag{1.34}$$

$$Z(z) = z^m + z_{m-1} z^{m-1} + \cdots + z_1 z + z_0, \tag{1.35}$$

with constant coefficients p_i and z_j, and k_p is a constant gain. The symbol z is used to denote the advance operator $z[x](t) = x(t+1)$.[1] For example, with $P(z) = z^2 + p_1 z + p_0$ and $Z(z) = z + z_0$, system (1.33) is

$$y(t+2) + p_1 y(t+1) + p_0 y(t) = k_p(u(t+1) + z_0 u(t)), \; t \in \{0, 1, 2, \ldots\}. \tag{1.36}$$

In this expression, the polynomials $P(z)$ and $Z(z)$ are also seen as operators on the signals $y(t)$ and $u(t)$ to generate the signals $P(z)[y](t)$ and $Z(z)[u](t)$, respectively. From its difference equation, one can obtain the transfer function of the system as $G(z) = P^{-1}(z) k_p Z(z)$, that is,

$$y(z) = G(z)u(z), \; G(z) = k_p \frac{Z(z)}{P(z)}, \tag{1.37}$$

where the symbol z is used to denote the z-transform variable and $y(z)$ and $u(z)$ are the z-transforms of $y(t)$ and $u(t)$, respectively, when the effect of the system initial conditions is ignored.

In an operator form, system (1.33) may be expressed as

$$y(t) = G(z)[u](t) \triangleq \mathcal{Z}^{-1}[G(z)u(z)], \tag{1.38}$$

[1] To simplify the notation, we denote a discrete-time signal $x(kT)$, $k = 0, 1, 2, \ldots$, as $x(t)$, $t = 0, 1, 2, \ldots$, and its advance value $x((k+1)T)$ as $x(t+1)$ throughout this book when discrete-time systems are studied; see Section 1.2.2.

where $\mathcal{Z}^{-1}[\cdot]$ is the inverse z-transform operator and $G(z)$ is the operator which maps $u(t)$ to $y(t)$.

A state-space form for a linear discrete-time time-invariant system is

$$x(t+1) = Ax(t) + Bu(t),\ y(t) = Cx(t),\ t \in \{0, 1, 2, \ldots\}, \tag{1.39}$$

where $A \in R^{n \times n}$, $B \in R^n$ are constant parameter matrices, $x(t) \in R^n$ is the state vector, $u(t) \in R$ is the control input, and $y(t) \in R$ is the output.

Either a state feedback or an output feedback design can be employed for adaptive output tracking control of a discrete-time system. The controller structure (1.31), as shown in Figure 1.2, can be modified by replacing the operator "s" with "z" under the stability definition for a discrete-time system: all zeros of $\Lambda(z)$ should be inside the unit circle of the complex z-plane: $|z| < 1$. Similarly, the controller structure (1.32), as shown in Figure 1.3, can be applied to system (1.39), while the block diagram in Figure 1.3 is modified by replacing "$\dot{x}(t)$" with "$x(t+1)$". Of course, different designs of adaptive laws for updating controller parameters are used in the discrete-time case.

Stochastic Systems

Adaptive control of stochastic systems is an important area of adaptive control which deals with systems with stochastic noise. Instead of treating a noise as a bounded disturbance, statistical properties of a noise are used in constructing an estimate (prediction) of the effect of the unknown noise to modify a controller structure designed for the noise-free case. With a noise predictor, certain optimality properties are ensured for the adaptive control system. Topics of stochastic adaptive control are not addressed in this book. There is a rich collection of literature for this area of adaptive control, for example, [23], [47], [50], [86], [87], [116], [125], [127], [208], [275], [446]. A self-tuning regulator is a popular adaptive controller for stochastic systems, which can be designed using a direct or indirect method [23], [125], [275].

Important issues in adaptive control theory include clarification of *a priori* plant information for adaptive control, parametrizations of the plant model and the controller, derivation of error models in terms of the tracking and parameter errors, development of adaptive laws for updating the controller parameters, and stability analysis for the closed-loop system. Simulations of adaptive control systems are often useful for performance evaluation.

More general cases of adaptive control include those with time-varying plants or nonlinear plants and those with structural modeling errors and external disturbances in a controlled system.

1.5 Simple Adaptive Control Systems

In this section some simple examples are used to illustrate basic concepts and design steps for adaptive control systems. Two different classes of adaptive control systems will be shown: those based on *direct* adaptive control and those based on *indirect* adaptive control.

1.5.1 Direct Adaptive Control

In direct adaptive control systems the controller parameters are *directly* updated from an adaptive law. There are two commonly used designs for direct adaptive control: a Lyapunov design and a gradient design.

Lyapunov Design Example

Example 1.1 Let us first consider a first-order linear time-invariant plant

$$\dot{y}(t) = a_p y(t) + u(t),\ t \geq 0, \tag{1.40}$$

where the constant a_p is the plant parameter, $y(t)$ is the plant output with initial value $y(0) = y_0$, and $u(t)$ is the control input.

The control objective is to design a feedback control $u(t)$ such that all closed-loop system signals are bounded[2] and the plant output $y(t)$ tracks, asymptotically, the output $y_m(t)$ of a chosen reference model

$$\dot{y}_m(t) = -a_m y_m(t) + r(t),\ t \geq 0, \tag{1.41}$$

with $y_m(0) = y_{m0}$, where $a_m > 0$ for stability and performance and $r(t)$ is a bounded external input which characterizes desired system response.

Design for a_p known. When the plant parameter a_p is known, we use the feedback controller

$$u(t) = k^* y(t) + r(t), \tag{1.42}$$

where

$$k^* = -a_p - a_m. \tag{1.43}$$

With this controller, the closed-loop system is

$$\dot{y}(t) = -a_m y(t) + r(t). \tag{1.44}$$

[2]A signal $x(t) \in R$ is bounded if $|x(t)| \leq c$, $\forall t \geq 0$, for some constant $c > 0$.

Define the tracking error as $e(t) = y(t) - y_m(t)$. Then from (1.41) and (1.44), we have the tracking error equation

$$\dot{e}(t) = -a_m e(t) \tag{1.45}$$

with $e(0) = y(0) - y_m(0)$. The solution to this equation is $e(t) = e^{-a_m t}e(0)$, $t \geq 0$, which has the desired property: $e(t)$ is bounded, and so are $y(t)$ and $u(t)$. Moreover, $\lim_{t\to\infty} e(t) = 0$. Hence, we have achieved the control objective.

Design for a_p unknown. When the plant parameter a_p is unknown, we cannot implement the control law (1.42) because k^* is unknown. Instead, we use an estimate $k(t)$ for k^* to implement the adaptive controller,

$$u(t) = k(t)y(t) + r(t). \tag{1.46}$$

In view of (1.42), this controller, when applied to the plant (1.40), results in the closed-loop system

$$\dot{y}(t) = -a_m y(t) + r(t) + (k(t) - k^*)y(t), \; t \geq 0. \tag{1.47}$$

In terms of the tracking error $e(t)$, we have

$$\dot{e}(t) = -a_m e(t) + \tilde{k}(t)y(t), \; t \geq 0, \tag{1.48}$$

where $\tilde{k}(t) = k(t) - k^*$ is the parameter error.

The design task is to choose an adaptive law to update the estimate $k(t)$ (i.e., to specify $\dot{k}(t)$, the time derivative of $k(t)$) so that the stated control objective is still achievable even if the plant parameter a_p is unknown.

Let us introduce a measure for the errors $e(t)$ and $\tilde{k}(t)$:

$$V(e, \tilde{k}) = e^2 + \tilde{k}^2, \tag{1.49}$$

which is positive whenever $e \neq 0$ and/or $\tilde{k} \neq 0$, and examine the time derivative $\dot{V} = \frac{dV}{dt}$ of $V(e, \tilde{k})|_{e=e(t), \tilde{k}=\tilde{k}(t)}$.

The time derivative of $V(e, \tilde{k})$ is

$$\begin{aligned} \dot{V} = \frac{d}{dt}V(e, \tilde{k}) &= \frac{\partial V(e, \tilde{k})}{\partial e}\dot{e}(t) + \frac{\partial V(e, \tilde{k})}{\partial \tilde{k}}\dot{\tilde{k}}(t) \\ &= 2(e(t)\dot{e}(t) + \tilde{k}(t)\dot{\tilde{k}}(t)). \end{aligned} \tag{1.50}$$

With (1.48) and the fact that $\dot{\tilde{k}}(t) = \dot{k}(t) - \dot{k}^* = \dot{k}(t)$ because k^* is a constant, equation (1.50) becomes

$$\dot{V} = -2a_m e^2(t) + 2e(t)y(t)\tilde{k}(t) + 2\tilde{k}(t)\dot{k}(t). \tag{1.51}$$

If an adaptive law $\dot{k}(t)$ ensures $\dot{V} \leq 0$, $\forall e(t)$, $\tilde{k}(t)$, then the errors $e(t)$, $\tilde{k}(t)$ will stay inside the circle centered at the origin with the radius equal to $\sqrt{V(e(0), \tilde{k}(0))}$. If, in addition, $\dot{V} < 0$ for any $e(t) \neq 0$, then the tracking error $e(t)$ may be forced to go to zero asymptotically. To make $\dot{V} \leq 0$, we choose the following adaptive law for $k(t)$:

$$\dot{k}(t) = -e(t)y(t), \; t \geq 0, \tag{1.52}$$

which leads (1.51) to

$$\dot{V} = -2a_m e^2(t) \leq 0. \tag{1.53}$$

The implementation of adaptive law (1.52) is simple:

$$k(t) = -\int_0^t e(\tau)y(\tau)\, d\tau + k(0), \tag{1.54}$$

where $k(0)$ is an initial estimate of the unknown parameter k^*.

With the adaptive law (1.52), $V(e(t), \tilde{k}(t))$ as a function of t does not increase, that is, $V(e(t), \tilde{k}(t)) \leq V(e(0), \tilde{k}(0))$, $\forall t \geq 0$. Therefore, both $e(t)$ and $\tilde{k}(t)$ are bounded signals, that is, there exists a finite positive constant γ_0 such that $|e(t)| \leq \gamma_0$, $|\tilde{k}(t)| \leq \gamma_0$, $\forall t \geq 0$, and so are the signals $y(t)$ and $k(t)$, because $y_m(t)$ from (1.41) is bounded and k^* is a constant. Furthermore, from (1.53), we have a finite energy error $e(t)$:

$$\int_0^\infty e^2(t)\, dt = \frac{1}{2a_m}\Big(V(e(0), \tilde{k}(0)) - V(e(\infty), \tilde{k}(\infty))\Big) < \infty, \tag{1.55}$$

and from (1.48), we have that $\dot{e}(t)$ is bounded. The boundness of $\dot{e}(t)$ and the property (1.55) ensure that $\lim_{t\to\infty} e(t)$ exists and is finite and such a limit is zero (see Lemma 2.14, Section 2.6). This means that the desired tracking performance $\lim_{t\to\infty}(y(t) - y_m(t)) = 0$ is achieved by the adaptive controller (1.46) updated from the adaptive law (1.52), despite the uncertainty of the plant parameter a_p (which is *unknown* to the controller (1.46)).

It should be noted that a measure of $e(t)$ and $\tilde{k}(t)$, useful for deriving an adaptive law, is not unique. For example, we can choose

$$V(e, \tilde{k}) = e^2 + \frac{1}{\gamma}\tilde{k}^2 \tag{1.56}$$

with $\gamma > 0$ being a constant. Then, the time derivative of $V(e, \tilde{k})$ is

$$\dot{V} = \frac{dV}{dt} = -2a_m e^2(t) + 2\tilde{k}(t)e(t)y(t) + \frac{2}{\gamma}\tilde{k}(t)\dot{\tilde{k}}(t). \tag{1.57}$$

The choice of the adaptive law

$$\dot{k}(t) = -\gamma e(t)y(t), \ t \geq 0, \tag{1.58}$$

also leads to (1.43): $\dot{V} = -2a_m e^2(t)$, from which desired closed-loop system properties follow. Such a measure $V(e, \tilde{k})$ is also called a *Lyapunov function*, which contains all error system states (in this example, they are $e(t)$ and $\tilde{k}(t)$)—that is why such an adaptive design is called a Lyapunov design. □

Gradient Design Example

Example 1.2 Consider the plant (1.40) with the controller structure (1.46) and reference model system (1.41). A different adaptive design can also be derived from the tracking error equation (1.48). Introduce the filtered signal

$$\zeta(t) = \frac{1}{s + a_m}[y](t), \tag{1.59}$$

where $\frac{1}{s+a_m}[y](t)$ denotes the output of the system with transfer function $\frac{1}{s+a_m}$ and input $y(t)$. We rewrite (1.48) as

$$e(t) = \frac{1}{s + a_m}[ky - k^*y](t) = \frac{1}{s + a_m}[ky](t) - k^*\zeta(t). \tag{1.60}$$

A desirable adaptive law for updating the controller parameter $k(t)$ should make both the parameter variation $\dot{k}(t)$ and the estimation error

$$\epsilon(t) = e(t) - \left(\frac{1}{s + a_m}[ky](t) - k(t)\zeta(t)\right) \tag{1.61}$$

"small" (see (1.68) below). With (1.60), this error can be expressed as

$$\epsilon(t) = \tilde{k}(t)\zeta(t), \ \tilde{k}(t) = k(t) - k^*. \tag{1.62}$$

We see that $\epsilon(t) = 0$ when $k(t) = k^*$. Introduce a measure of $\tilde{k}(t)$ as

$$V(\tilde{k}) = \tilde{k}^2\gamma^{-1}, \tag{1.63}$$

where $\gamma > 0$ is a constant scalar. The time derivative of $V(\tilde{k})$ is

$$\dot{V} = \frac{dV}{dt} = \frac{\partial V}{\partial k}\dot{k}(t) = 2\tilde{k}(t)\gamma^{-1}\dot{k}(t). \tag{1.64}$$

Choose the adaptive law for $k(t)$ as

$$\dot{k}(t) = -\frac{\gamma\epsilon(t)\zeta(t)}{m^2(t)}, \ t \geq 0, \tag{1.65}$$

where the normalizing signal to make $\dot{k}(t)$ bounded is

$$m(t) = \sqrt{1 + \zeta^2(t)}. \tag{1.66}$$

Then, from (1.62), (1.64), and (1.65), we have

$$\dot{V} = \frac{-2\epsilon^2(t)}{m^2(t)} \leq 0, \tag{1.67}$$

which implies that $V(\tilde{k})$ as a function of t does not increases, and it decreases whenever $\epsilon(t) \neq 0$, that is, $k(t)$ is bounded. It then follows from (1.62) that $\frac{\epsilon(t)}{m(t)}$ is bounded, from (1.65) and the boundedness of $\frac{\zeta(t)}{m(t)}$ that $\dot{k}(t)$ is bounded, and from (1.67), (1.65), and the boundedness of $\frac{\zeta(t)}{m(t)}$ that

$$\int_0^\infty \frac{\epsilon^2(t)}{m^2(t)}\, dt < \infty, \quad \int_0^\infty \dot{k}^2(t)\, dt < \infty. \tag{1.68}$$

From these desired properties, it can be proved that both $y(t)$ and $u(t)$ are bounded and $\lim_{t\to\infty}(y(t) - y_m(t)) = 0$ (see Theorem 5.4 of Chapter 5).

This design is called a gradient design because $\dot{k}$ is in the direction of the negative gradient of the normalized quadratic cost function $J(k) = \frac{\epsilon^2}{m^2}$, where $\epsilon = \tilde{k}\zeta$ as in (1.62). In this case, the gradient (see Section 3.3) of $J(k)$ is $\frac{\partial J}{\partial k} = \frac{\zeta\epsilon}{m^2}$. Based on optimization theory [325], the derivative of $k(t)$ should be in the negative gradient direction of $J(k)$. Therefore, with an adaptation gain $\gamma > 0$, the adaptive law is chosen as (1.65). □

1.5.2 Indirect Adaptive Control

In contrast to direct adaptive designs in which the controller parameters are updated directly from an adaptive law, in indirect adaptive control designs the plant parameters are adaptively estimated first, and the controller parameters are then calculated from some algebraic design equation using the estimates of the unknown plant parameters.

Example 1.3 Consider the plant (1.40) with the controller structure (1.46) and reference model system (1.41). Let $\widehat{a_p}(t)$ be an estimate of the unknown plant parameters a_p. Then an indirect adaptive controller is

$$u(t) = k(t)y(t) + r(t), \tag{1.69}$$

where $k(t)$ is *calculated* from the design equation

$$k(t) = -\widehat{a_p}(t) - a_m. \tag{1.70}$$

To generate such a desired parameter estimate $\widehat{a_p}(t)$, we choose a stable filter $\frac{1}{s+a_f}$ with $a_f > 0$ and express the plant (1.40) as

$$\dot{y} = -a_f y + (a_f + a_p)y + u, \tag{1.71}$$

which can be further written as

$$\begin{aligned} y(t) &= (a_f + a_p)\frac{1}{s+a_f}[y](t) + \frac{1}{s+a_f}[u](t) \\ &= \theta_p^* \phi(t) + \frac{1}{s+a_f}[u](t), \end{aligned} \tag{1.72}$$

where

$$\theta_p^* = a_f + a_p, \ \phi(t) = \frac{1}{s+a_f}[y](t). \tag{1.73}$$

Letting $\theta_p(t)$ be the estimate of the unknown parameter θ_p^*: $\theta_p(t) = a_f + \widehat{a_p}$ (i.e., $\widehat{a_p} = \theta_p(t) - a_f$), we introduce the estimation error

$$\epsilon(t) = \theta_p(t)\phi(t) - y(t) + \frac{1}{s+a_f}[u](t). \tag{1.74}$$

It follows from (1.72) and (1.74) that

$$\epsilon(t) = \tilde{\theta}_p(t)\phi(t), \ \tilde{\theta}_p(t) = \theta_p(t) - \theta_p^*. \tag{1.75}$$

We then choose the adaptive law for $\theta_p(t)$ as

$$\dot{\theta}_p(t) = -\frac{\gamma \epsilon(t)\phi(t)}{m^2(t)}, \ \gamma > 0, \ t \geq 0, \tag{1.76}$$

to ensure $\dot{\theta}_p(t)$ and $\epsilon(t)$ "small" (see (1.80) below), where

$$m(t) = \sqrt{1 + \phi^2(t)}. \tag{1.77}$$

Considering the measure of the parameter error $\tilde{\theta}_p$,

$$V(\tilde{\theta}_p) = \tilde{\theta}_p^2 \gamma^{-1}, \tag{1.78}$$

and its time derivative

$$\dot{V} = \frac{dV}{dt} = \frac{-2\epsilon^2(t)}{m^2(t)} \leq 0, \tag{1.79}$$

we conclude that $\theta_p(t)$ is bounded, and so are $\frac{\epsilon(t)}{m(t)}$ and $\dot{\theta}_p(t)$, and that

$$\int_0^\infty \frac{\epsilon^2(t)}{m^2(t)}\,dt < \infty, \ \int_0^\infty \dot{\theta}_p^2(t)\,dt < \infty. \tag{1.80}$$

From these properties of the parameter estimation algorithm (1.76), the desired closed-loop system properties that $y(t)$ and $u(t)$ are bounded and that $\lim_{t\to\infty}(y(t) - y_m(t)) = 0$ are ensured (see Section 7.1.1). □

1.5.3 Discrete-Time Designs

Adaptive control schemes for discrete-time systems can also be developed based on two approaches: *direct* or *indirect* adaptation of the controller parameters. Next we give two such discrete-time design examples.

Direct Adaptive Control Design Example

Example 1.4 Consider a first-order discrete-time time-invariant plant

$$y(t+1) = a_p y(t) + u(t),\ t \in \{0, 1, 2, \ldots\}, \tag{1.81}$$

where a_p is the unknown plant parameter, $y(t)$ is the plant output, and $u(t)$ is the control input. The control objective is to design a feedback control $u(t)$ such that all closed-loop system signals are bounded, and the plant output $y(t)$ asymptotically tracks the output $y_m(t)$ of a chosen reference model

$$y_m(t+1) = -a_m y_m(t) + r(t),\ t \in \{0, 1, 2, \ldots\}, \tag{1.82}$$

where $|a_m| < 1$ for the reference model system stability and performance and $r(t)$ is a bounded reference input. For a discrete-time design, it is common to choose $a_m = 0$ so that $y_m(t) = r(t-1)$.

The control structure and plant–model matching analysis are similar to that for a continuous-time case. When the plant parameter a_p is *known*, the above control objective can be met by using the the controller

$$u(t) = k^* y(t) + r(t), \tag{1.83}$$

where $k^* = -a_p - a_m$, which leads to the closed-loop system

$$y(t+1) = -a_m y(t) + r(t),\ t \in \{0, 1, 2, \ldots\}. \tag{1.84}$$

The tracking error $e(t) = y(t) - y_m(t)$ satisfies $e(t+1) = -a_m e(t)$, so that $e(t)$ is bounded, and so are $y(t)$ and $u(t)$. Moreover, $\lim_{t\to\infty} e(t) = 0$, exponentially. If $a_m = 0$ is chosen, $e(t) = 0$, $t \geq 1$.

When a_p is *unknown*, we use the adaptive version of (1.83), that is,

$$u(t) = k(t)y(t) + r(t), \tag{1.85}$$

where k is an estimate of the unknown parameter k^*. Substituting (1.85) into (1.80) gives the closed-loop system

$$y(t+1) = -a_m y(t) + r(t) + (k(t) - k^*)y(t),\ t \in \{0, 1, 2, \ldots\}, \tag{1.86}$$

from which, using (1.81), we have the tracking error equation

$$e(t+1) = -a_m e(t) + \tilde{k}(t)y(t), \tag{1.87}$$

where $e(t) = y(t) - y_m(t)$ and $\tilde{k}(t) = k(t) - k^*$.

A major difference between discrete-time adaptive control and continuous-time adaptive control is that a Lyapunov design such as that in Example 2.1 for the continuous-time case is not applicable to the discrete-time case.

To develop a gradient design, we introduce

$$\zeta(t) = \frac{1}{z+a_m}[y](t), \tag{1.88}$$

where $\frac{1}{z+a_m}[y](t)$ denotes the output of the system with transfer function $\frac{1}{z+a_m}$ and input $y(t)$. Then, we rewrite (1.87) as

$$e(t) = \frac{1}{z+a_m}[ky - k^*y](t) = \frac{1}{z+a_m}[ky](t) - k^*\zeta(t). \tag{1.89}$$

Based on this expression, we define the estimation error

$$\epsilon(t) = e(t) - \left(\frac{1}{z+a_m}[ky](t) - k(t)\zeta(t)\right). \tag{1.90}$$

With (1.89), this estimation error can be expressed as

$$\epsilon(t) = \tilde{k}(t)\zeta(t),\ \tilde{k}(t) = k(t) - k^*. \tag{1.91}$$

Choose the adaptive law for $k(t)$ as

$$k(t+1) = k(t) - \frac{\gamma\epsilon(t)\zeta(t)}{m^2(t)},\ 0 < \gamma < 2,\ t \in \{0, 1, 2, \ldots\}, \tag{1.92}$$

where

$$m(t) = \sqrt{1 + \zeta^2(t)}. \tag{1.93}$$

Introduce a measure of $\tilde{k}(t)$ as

$$V(\tilde{k}) = \tilde{k}^2\gamma^{-1}. \tag{1.94}$$

The time increment of $V(\tilde{k})$, along (1.92), is

$$V(\tilde{k}(t+1)) - V(\tilde{k}(t)) = -\left(2 - \frac{\zeta^2(t)\gamma}{m^2(t)}\right)\frac{\epsilon^2(t)}{m^2(t)} \le -\alpha_1\frac{\epsilon^2(t)}{m^2(t)} \tag{1.95}$$

for some constant $\alpha_1 > 0$. The last inequality follows from the condition $0 < \gamma < 2$. This result implies that $k(t)$ is bounded, and

$$\sum_{t=0}^{\infty} \frac{\epsilon(t)}{m(t)} < \infty, \ \sum_{t=0}^{\infty}(k(t+1) - k(t)) < \infty. \tag{1.96}$$

For $a_m = 0$, $\zeta(t) = y(t-1)$, we have from (1.90) and (1.93) that

$$\epsilon(t) = e(t) + (k(t) - k(t-1))y(t-1), \tag{1.97}$$

$$m(t+1) \leq 1 + |\zeta(t+1)| = 1 + |y(t)|. \tag{1.98}$$

At $t+1$, we can rewrite (1.90) with $a_m = 0$ as

$$y(t+1) = r(t) + \frac{\epsilon(t+1)}{m(t+1)} m(t+1) + (k(t+1) - k(t))y(t), \tag{1.99}$$

where $y_m(t) = r(t-1)$. Using (1.98) and (1.99), we have

$$|y(t+1)| \leq \left(\frac{|\epsilon(t+1)|}{m(t+1)} + |k(t+1) - k(t)| \right) |y(t)| + b_3 \tag{1.100}$$

for some constant $b_3 > 0$. From (1.96), we have that $\lim_{t\to\infty} \frac{\epsilon(t)}{m(t)} = 0$ and $\lim_{t\to\infty}(k(t+1) - k(t)) = 0$. Then, (1.100) implies that $y(t)$ is bounded, and so is $u(t)$ in (1.85). Therefore, $\lim_{t\to\infty} \epsilon(t) = 0$ and $\lim_{t\to\infty}(k(t) - k(t-1))y(t-1) = 0$, so that $\lim_{t\to\infty} e(t) = 0$, as from (1.97). □

Indirect Adaptive Control Design Example

Example 1.5 Consider the plant (1.81) with the controller structure (1.85) and reference model system (1.82). Let $\widehat{a_p}(t)$ be an estimate of the unknown plant parameter a_p. Then an indirect adaptive controller is

$$u(t) = k(t)y(t) + r(t), \tag{1.101}$$

where the controller parameter $k(t)$ is *calculated* from

$$k(t) = -\widehat{a_p}(t) - a_m. \tag{1.102}$$

To generate a desired parameter estimate, we choose a stable filter $\frac{1}{z+a_f}$ with $|a_f| < 1$ and express the plant (1.80) as

$$y(t+1) = -a_f y(t) + (a_f + a_p)y(t) + u(t), \tag{1.103}$$

which can be further written as

$$y(t) = (a_f + a_p)\frac{1}{z + a_f}[y](t) + \frac{1}{z + a_f}[u](t)$$
$$= \theta_p^{*T}\phi(t) + \frac{1}{z + a_f}[u](t), \tag{1.104}$$

where

$$\theta_p^* = a_f + a_p, \ \phi(t) = \frac{1}{z + a_f}[y](t). \tag{1.105}$$

Letting $\theta_p(t)$ be the estimate of θ_p^*, we introduce the estimation error

$$\epsilon(t) = \theta_p(t)\phi(t) - y(t) + \frac{1}{z + a_f}[u](t). \tag{1.106}$$

It follows from (1.104) and (1.106) that

$$\epsilon(t) = \tilde{\theta}_p(t)\phi(t), \ \tilde{\theta}_p(t) = \theta_p(t) - \theta_p^*. \tag{1.107}$$

Based on this error equation, we choose the adaptive law for $\theta_p(t)$ as

$$\theta_p(t+1) = \theta_p(t) - \frac{\gamma\epsilon(t)\phi(t)}{m^2(t)}, \ 0 < \gamma < 2, \ t \in \{0, 1, 2, \ldots\}, \tag{1.108}$$

where

$$m(t) = \sqrt{1 + \phi^2(t)}. \tag{1.109}$$

The time increment of the parameter error measure $V(\tilde{\theta}_p) = \tilde{\theta}_p^2\gamma^{-1}$, along the trajectories of (1.108), is

$$V(\tilde{\theta}_p(t+1)) - V(\tilde{\theta}_p(t)) = -\left(2 - \frac{\phi^2(t)\gamma\phi(t)}{m^2(t)}\right)\frac{\epsilon^2(t)}{m^2(t)} \leq -\alpha_1\frac{\epsilon^2(t)}{m^2(t)} \tag{1.110}$$

for some constant $\alpha_1 > 0$. From (1.110), we can derive the desired properties of (1.108): $\theta_p(t)$ is bounded, and so is $\frac{\epsilon(t)}{m(t)}$, and

$$\sum_{t=0}^{\infty}\frac{\epsilon(t)}{m(t)} < \infty, \ \sum_{t=0}^{\infty}(\theta_p(t+1) - \theta_p(t)) < \infty. \tag{1.111}$$

For $a_m = 0$ in (1.82) and $a_f = 0$ in (1.104), we have

$$y_m(t+1) = r(t), \ \theta_p^* = a_p, \ \phi(t) = y(t-1), \tag{1.112}$$

and, with (1.106), the controller (1.101) can be expressed as

$$r(t) = \theta_p(t)\phi(t+1) + u(t)$$
$$= \theta_p(t+1)\phi(t+1) - (\theta_p(t+1) - \theta_p(t))\phi(t+1) + u(t)$$
$$= y(t+1) + \epsilon(t+1) - (\theta_p(t+1) - \theta_p(t))\phi(t+1). \tag{1.113}$$

For $m(t)$ in (1.109), we have

$$m(t+1) \leq 1 + |y(t)| \tag{1.114}$$

and from (1.113) and (1.114), we obtain

$$\begin{aligned} |y(t+1)| &\leq \frac{|\epsilon(t+1)|}{m(t+1)} m(t+1) + |\theta_p(t+1) - \theta_p(t)||y(t)| + |r(t)| \\ &\leq \left(\frac{|\epsilon(t+1)|}{m(t+1)} + |\theta_p(t+1) - \theta_p(t)| \right) |y(t)| + b_0(t), \end{aligned} \tag{1.115}$$

where $b_0(t) = \frac{|\epsilon(t+1)|}{m(t+1)} + |r(t)|$. From (1.111), we have $\lim_{t\to\infty} \frac{\epsilon(t)}{m(t)} = 0$ and $\lim_{t\to\infty} |\theta_p(t+1) - \theta_p(t)| = 0$, which, together with the boundedness of $k(t)$ and $r(t)$, from (1.115), imply that $y(t)$ is bounded, and so is $u(t)$, as from (1.101). Then we have $\lim_{t\to\infty} \epsilon(t) = 0$ and $\lim_{t\to\infty}(\theta_p(t+1) - \theta_p(t))\phi(t+1) = 0$, so that $\lim_{t\to\infty} e(t) = 0$ for $e(t) = y(t) - y_m(t)$, as from (1.113). □

1.5.4 Backstepping Nonlinear Design

For the plant (1.40), the controller structure (1.46) is linear. To control a nonlinear system, a controller is also nonlinear. This can be illustrated by a backstepping design [206] for a second-order nonlinear system

$$\begin{aligned} \dot{x}_1 &= x_2 + \theta^{*T} \varphi_1(x_1), \\ \dot{x}_2 &= \varphi_0(x_1, x_2) + \theta^{*T} \varphi_2(x_1, x_2) + u, \quad y = x_1, \end{aligned} \tag{1.116}$$

where $\theta^* \in R^{n_\theta}$ is an unknown constant parameter vector and $\varphi_i, i = 0, 1, 2$, are known nonlinear functions which are differentiable with respect to x_1, x_2, the system state variables. This is called a *parametric-strict-feedback* system.

The control objective is to design a state feedback control $u(t)$ using x_1 and x_2 such that all closed-loop signals are bounded and the system output $y(t)$ tracks a given reference output $y_m(t)$ asymptotically, where $y_m(t)$, $\dot{y}_m(t)$, and $\ddot{y}_m(t)$ are all bounded functions of time t.

The adaptive control scheme [206] is formed in a *backstepping* procedure.

Step 1: Introducing $z_1 = x_1 - y_m$ and $z_2 = x_2 - \alpha_1$, where α_1 is a design function to be determined, from (1.116), we have

$$\dot{z}_1 = \dot{x}_1 - \dot{y}_m = z_2 + \alpha_1 + \theta^{*T} \varphi_1(x_1) - \dot{y}_m. \tag{1.117}$$

At this step, (1.117) is viewed as a first-order system to be stabilized by α_1 with respect to the error measure

$$V_1 = \frac{1}{2} z_1^2 + \frac{1}{2} (\theta - \theta^*)^T \Gamma^{-1} (\theta - \theta^*), \ \Gamma = \Gamma^T > 0, \tag{1.118}$$

where $\theta(t)$ is the estimate of θ^*. The time derivative of V_1 is

$$\dot{V}_1 = z_1(z_2 + \alpha_1 + \theta^T \varphi_1(x_1) - \dot{y}_m) + (\theta - \theta^*)^T \Gamma^{-1}(\dot{\theta} - \Gamma z_1 \varphi_1(x_1)). \quad (1.119)$$

Choosing the stabilizing function as

$$\alpha_1 = -c_1 z_1 - \theta^T \varphi_1(x_1) + \dot{y}_m, \ c_1 > 0, \quad (1.120)$$

and defining the tuning function as

$$\tau_1 = \Gamma z_1 \varphi_1(x_1), \quad (1.121)$$

from (1.117) and (1.119), we obtain

$$\dot{z}_1 = z_2 - c_1 z_1 - (\theta - \theta^*)^T \varphi_1(x_1), \quad (1.122)$$

$$\dot{V}_1 = -c_1 z_1^2 + z_1 z_2 + (\theta - \theta^*)^T \Gamma^{-1}(\dot{\theta} - \tau_1). \quad (1.123)$$

Step 2: Using (1.116) and $z_2 = x_2 - \alpha_1$, we express

$$\begin{aligned}\dot{z}_2 = \dot{x}_2 - \dot{\alpha}_1 = \varphi_0(x_1, x_2) + \theta^{*T} \varphi_2(x_1, x_2) + u - \frac{\partial \alpha_1}{\partial x_1}(x_2 + \theta^{*T} \varphi_1(x_1)) \\ - \frac{\partial \alpha_1}{\partial \theta}\dot{\theta} - \frac{\partial \alpha_1}{\partial y_m}\dot{y}_m - \frac{\partial \alpha_1}{\partial \dot{y}_m}\ddot{y}_m. \quad (1.124)\end{aligned}$$

Now, (1.117) and (1.124) can be viewed as a second-order system to be stabilized by α_1 given in (1.120) and the control input $u(t)$ with respect to the error measure $V_2 = V_1 + \frac{1}{2}z_2^2$. The time derivative of V_2 is

$$\begin{aligned}\dot{V}_2 = -c_1 z_1^2 + z_2 \bigg(z_1 + \varphi_0(x_1, x_2) + u - \frac{\partial \alpha_1}{\partial x_1} x_2 - \frac{\partial \alpha_1}{\partial \theta}\dot{\theta} - \frac{\partial \alpha_1}{\partial y_m}\dot{y}_m \\ - \frac{\partial \alpha_1}{\partial \dot{y}_m}\ddot{y}_m + \theta^T \left(\varphi_2(x_1, x_2) - \frac{\partial \alpha_1}{\partial x_1}\varphi_1(x_1) \right) \bigg) \\ + (\theta - \theta^*)^T \Gamma^{-1} \left(\dot{\theta} - \tau_1 - \Gamma z_2 \left(\varphi_2(x_1, x_2) - \frac{\partial \alpha_1}{\partial x_1}\varphi_1(x_1) \right) \right). \quad (1.125)\end{aligned}$$

Choose the adaptive law for $\theta(t)$ as

$$\dot{\theta} = \tau_2 = \tau_1 + \Gamma z_2 \left(\varphi_2 - \frac{\partial \alpha_1}{\partial x_1}\varphi_1 \right) \quad (1.126)$$

and the stabilizing control u as

$$\begin{aligned}u(t) = -z_1 - \varphi_0(x_1, x_2) - c_2 z_2 + \frac{\partial \alpha_1}{\partial x_1} x_2 + \frac{\partial \alpha_1}{\partial \theta}\tau_2 + \frac{\partial \alpha_1}{\partial y_m}\dot{y}_m \\ + \frac{\partial \alpha_1}{\partial \dot{y}_m}\ddot{y}_m - \theta^T \left(\varphi_2(x_1, x_2) - \frac{\partial \alpha_1}{\partial x_1}\varphi_1(x_1) \right), \ c_2 > 0. \quad (1.127)\end{aligned}$$

Then, we can rewrite $\dot{V}_2$ as

$$\dot{V}_2 = -c_1 z_1^2 - c_2 z_2^2. \tag{1.128}$$

Since $\dot{V}_2 \le 0$, we have that the error measure $V_2 = \frac{1}{2}z_1^2 + \frac{1}{2}z_2^2 + (\theta - \theta^*)^T \Gamma^{-1}(\theta - \theta^*)$ does not increase as a function of t, that is, $z_1(t) = x_1(t) - y_m(t)$, $z_2(t) = x_2(t) - \alpha_1(t)$ and $\theta(t) - \theta^*$ are bounded, and so are $x_1(t)$ and $\theta(t)$. Then, $\varphi_1(x_1)$ is bounded, and from (1.120), $\alpha_1(t)$ is bounded. From $z_2(t) = x_2(t) - \alpha_1(t)$, we have that $x_2(t)$ is bounded, and hence $\varphi_0(x_1, x_2)$ and $\varphi_2(x_1, x_2)$ are bounded. It follows from (1.127) that $u(t)$ is bounded so that all signals in the closed-loop system are bounded.

To analyze tracking performance, we have from (1.128) that

$$c_1 \int_0^t z_1^2(\tau)\, d\tau + c_2 \int_0^t z_2^2(\tau)\, d\tau = V_2(0) - V_2(t). \tag{1.129}$$

Since V_2 as a function of t is bounded, it follows that $\int_0^\infty z_1^2(\tau)\, d\tau < \infty$ and $\int_0^\infty z_2^2(\tau)\, d\tau < \infty$. From $\int_0^\infty z_1^2(\tau)\, d\tau < \infty$ and the boundedness of $\dot{z}_1(t)$ (see (1.117)), we can show that $\lim_{t\to\infty} z_1(t) = 0$, where $z_1(t) = y(t) - y_m(t)$.

This design procedure is called a *backstepping* design as it starts from the first equation for x_1 to the last equation for z_n ($n = 2$ in this case). It can be applied to an nth-order system structurally similar to that in (1.116), which is called a *parametric-strict-feedback* system [206].

1.5.5 Adaptive Control versus Fixed Control

It may be interesting to compare an adaptive control design with a fixed control design for a first-order linear time-invariant plant

$$\dot{y}(t) = (a_p + \Delta) y(t) + u(t), \tag{1.130}$$

where a_p is the nominal plant parameter and Δ is parameter uncertainty.

Given a reference model

$$\dot{y}_m(t) = -a_m y_m(t) + r(t), \; a_m > 0, \tag{1.131}$$

which in the frequency s-domain is $y_m(s) = \frac{1}{s+a_m} r(s)$, the nominal controller

$$u(t) = k^* y(t) + r(t), \; k^* = -a_p - a_m, \tag{1.132}$$

applied to the plant (1.130) with $\Delta = 0$, leads to the closed-loop system

$$\dot{y}(t) = -a_m y(t) + r(t), \tag{1.133}$$

which matches the reference system in the sense that $\dot{e}(t) = -a_m e(t)$, where $e(t) = y(t) - y_m(t)$, so that $\lim_{t\to\infty} e(t) = 0$ exponentially fast.

Fixed Controller

A fixed controller with certain robustness to Δ is designed to stabilize the closed-loop system, based on the knowledge of a_p and bounds of Δ, for example, $\Delta \in [-20, 10]$. The fixed controller is

$$u(t) = ky(t) + r(t), \tag{1.134}$$

where k is a constant to be specified such that the closed-loop system is stable (i.e., all signals are bounded) for any $\Delta \in [-20, 10]$.

In this case, the closed-loop system is

$$\dot{y}(t) = (a_p + \Delta + k)y(t) + r(t). \tag{1.135}$$

For stability, $a_p + \Delta + k < 0$. Hence the robust control gain k should satisfy

$$k < -a_p - 10 \tag{1.136}$$

in order to stabilize the system for any $\Delta \in [-20, 10]$.

To study the tracking performance of the controller (1.134) in the presence of parameter uncertainty, we express (1.135) as

$$\dot{y}(t) = -a_m y(t) + (a_p + \Delta + k + a_m)y(t) + r(t), \tag{1.137}$$

which in the frequency s-domain is

$$y(s) = \frac{1}{s + a_m} r(s) + \frac{a_p + \Delta + k + a_m}{s + a_m} y(s). \tag{1.138}$$

Using the s-domain expression of (1.135) $y(s) = \frac{1}{s - a_p - \Delta - k} r(s)$, we have

$$y(s) = \frac{1}{s + a_m} r(s) + \frac{a_p + \Delta + k + a_m}{s + a_m} \frac{1}{s - a_p - \Delta - k} r(s). \tag{1.139}$$

Therefore, the tracking error in the frequency s-domain is

$$e(s) = y(s) - y_m(s) = \frac{a_p + \Delta + k + a_m}{s + a_m} \frac{1}{s - a_p - \Delta - k} r(s), \tag{1.140}$$

which can be large in the time domain. To make the closed-loop system (1.135) as fast as the reference system (1.131), the condition on k is

$$k \leq -a_m - a_p - 10. \tag{1.141}$$

For $k = -a_m - a_p - 10$, from (1.140), the s-domain tracking error is

$$e(s) = \frac{\Delta - 10}{s + a_m} \frac{1}{s + a_m + 10 - \Delta} r(s). \tag{1.142}$$

For $r(t) = 1$ [i.e., $r(s) = \frac{1}{s}$] and $a_m = 1$,

$$e_{ss} = \lim_{t\to\infty} e(t) = \frac{\Delta - 10}{11 - \Delta}. \tag{1.143}$$

The maximum of $|e_{ss}|$ is reached when $\Delta = -20$, that is, $e_{ss} = -\frac{30}{31}$ for $\Delta = -20$, and the minimum of $|e_{ss}|$ is zero when $\Delta = 10$.

In summary, a fixed controller works only for a known range of parameter uncertainty Δ and cannot ensure asymptotic tracking for nonzero $r(t)$.

Adaptive Controller

For any a_p unknown and any Δ uncertain, the adaptive controller

$$u(t) = k(t)y(t) + r(t), \tag{1.144}$$

where

$$\dot{k}(t) = -\gamma e(t)y(t),\ \gamma > 0, \tag{1.145}$$

ensures that both $u(t)$ and $y(t)$ are bounded and $\lim_{t\to\infty} e(t) = 0$.

As a comparison, in this simple example, an adaptive controller, with the help of a more complex control algorithm (see (1.145)) for implementation of parameter adaptation, admits a much larger class of plant parameter uncertainty than a fixed controller does, while ensuring closed-loop signal boundedness. In addition, an adaptive controller is able to meet desired system tracking performance despite large parameter uncertainty.

1.5.6 Summary

Adaptive control schemes can be developed using the controller structures shown in Figures 1.2 and 1.3, based on two types of designs: a direct design and an indirect design. In a *direct* (model reference) adaptive control system whose block diagram is shown in Figure 1.4, the controller parameters $\theta(t)$ (which are estimates of some unknown ideal plant–model controller matching parameters) are directly updated from an adaptive law (which specifies $\dot{\theta}(t)$ or $\theta(t+1) - \theta(t)$, based on the tracking error $y(t) - y_m(t)$). In an *indirect* adaptive control system whose block diagram is shown in Figure 1.5, the controller parameters $\theta_c(t)$ (which are also estimates of some unknown ideal plant–model matching parameters) are simultaneously calculated from a design equation using the on-line estimates $\theta_p(t)$ of the unknown plant parameters θ_p^*, updated from a parameter estimator (which specifies $\dot{\theta}_p(t)$ or $\theta_p(t+1) - \theta_p(t)$ based

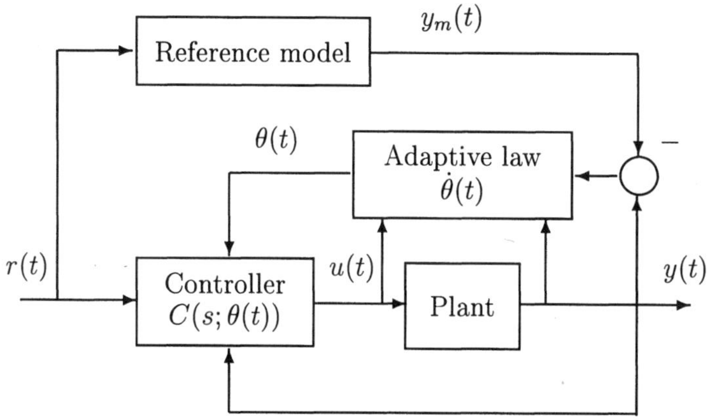

Figure 1.4: Direct model reference adaptive control system.

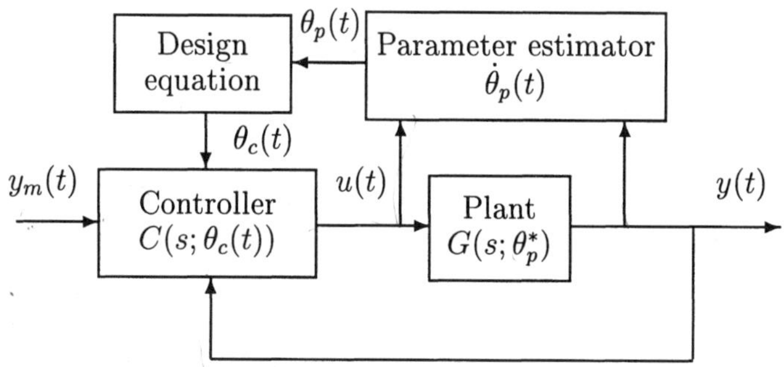

Figure 1.5: Indirect adaptive control system.

on an estimation error presenting the mismatch between the plant output $y(t)$ and its estimated version generated from the parameter estimate $\theta_p(t)$).

In either a direct or an indirect adaptive linear control design or an adaptive nonlinear backstepping control design, the key idea is to use the estimates of some unknown parameters of an ideal controller (in a *direct* design as shown in Figure 1.4) or the controlled plant (in an *indirect* design as shown in Figure 1.5) as if they were the true values of those unknown parameters. The parameter estimates are obtained from an adaptive update law driven by system performance error (tracking error), and used in implementing an adaptive controller either directly or through a design equation to map the plant parameter estimates to the controller parameters. This idea is called the *certainty equivalence principle.* For such a principle to work efficiently, the used parameter estimates should satisfy some desired properties: smallness of estimation errors (see (1.80), (1.111)), boundedness of parameter estimates, and slowness of parameter variations (see (1.80), (1.111)). With such properties of parameter estimates, an adaptive control system is stable and has desired tracking performance. For a backstepping nonlinear design, the situation is special: the parameter vector $\theta(t)$ may be seen as an estimate of the plant parameter vector θ^* or it may be considered as an estimate of a controller parameter vector. (What is the feedback controller when θ^* is known?)

For control system analysis, important concepts include system modeling, signal measure, feedback structure, feedback gain, signal boundedness, system stability, and error convergence. For adaptive control, some important issues are the choice of controller structures and their parametrization, derivation of error systems, choice of estimation errors, expression of error models, development of adaptive update laws, analysis of stability and robustness, and evaluation of system performance, and most of them are unique for achieving both *stabilization* and *tracking* in the presence of system parameter uncertainties. *Robustness* is important as it ensures that an adaptive control system remains stable in the presence of unmodeled dynamics and external disturbances (see Chapters 5 and 6). Disturbances whose function components are known can be rejected by adaptive control designs (see Chapter 4).

This book emphasizes the design and analysis of different adaptive control systems. The direct adaptive control approach is presented in Chapters 4–6, 8, and 9, while the indirect adaptive control approach is addressed in Chapter 7. The adaptive nonlinear backstepping design method is investigated for more general classes of nonlinear systems (see Chapter 10). To help understand how adaptive control systems are developed, some fundamental theoretical results for dynamic systems are presented in Chapter 2, and two commonly used adaptive parameter estimation algorithms—gradient and least-squares algorithms—are studied in Chapter 3.

Problems

1.1 In Figure 1.2, let $Z(s) = 1$ and $P(s) = s^2 + p_1 s + p_0$; k_p, p_1, and p_0 be constant parameters; $\Lambda(s) = s + 3$, $a(s) = 1$; and θ_1^*, θ_2^*, θ_{20}^*, and θ_3^* be scalar parameters. Find θ_1^*, θ_2^*, θ_{20}^*, and θ_3^* such that the closed-loop transfer function from r to y equals $\frac{1}{s^2+a_1 s+a_0}$ with $a_1 > 0$, $a_0 > 0$.

1.2 In Figure 1.3, let $A = \begin{bmatrix} 0 & 1 \\ -p_0 & -p_1 \end{bmatrix}$, $b = \begin{bmatrix} 0 \\ 1 \end{bmatrix}$, and $c = \begin{bmatrix} 1 & 0 \end{bmatrix}$. Find $k_1 = k_1^* \in R^2$ and $k_2 = k_2^* \in R$ such that the closed-loop transfer function from r to y equals $\frac{1}{s^2+a_1s+a_0}$ with $a_1 > 0$, $a_0 > 0$.

1.3 Simulate the adaptive control system consisting of (1.40), (1.41), (1.46), and (1.52), with $y_m(0) = 0$, $a_m = 2$, $r(t) = 11.3\sin(1.3t)$, $y(0) = 1.5$, $a = -5$, for (i) $k(0) = 0$, (ii) $k(0) = -4$, and (iii) $k(0) = -10$. Plot $u(t)$, $y(t)$, and $e(t)$ versus time and $e(t)$ versus $k(t) - k^*$. Comment on the simulation results.

1.4 For the adaptive law (1.65), prove that $\frac{\epsilon(t)}{m(t)}$ and $\dot{\theta}(t)$ are bounded and have finite energy, that is, they satisfy (1.68).

1.5 Simulate the direct gradient adaptive control system consisting of (1.40), (1.41), (1.46), and (1.65), with $a_p = 2$, $b_p = 1$, $a_m = 1$, for two sets of design parameters, initial conditions and reference signal $r(t)$ ($r(t) = 1$ and $r(t) = \sin 1.7t$), and plot your simulation results (the tracking errors and parameter errors, etc.) to verify the stated analytical properties.

1.6 Simulate the indirect adaptive control system consisting of (1.40), (1.41), (1.46), (1.70), and (1.76) for $r(t) = 1$ and $r(t) = \sin 1.7t$.

1.7 Verify (1.95), (1.96), and (1.100).

1.8 Simulate the adaptive control system consisting of (1.81), (1.82), (1.85) and (1.92), with $a_m = 0.5$, $a_m = 0.1$, and $a_m = 0$ respectively.

1.9 Verify (1.111), (1.113), and (1.115).

1.10 Simulate the adaptive control system consisting of (1.81), (1.82), (1.85), (1.102), and (1.108) for four different cases of $a_m = 0.5$, $a_m = 0$, $a_f = 0.5$, and $a_f = 0$, respectively, with two different adaptation gains γ.

1.11 Specify the range of the constant α_1 in (1.95) and (1.110).

1.12 Apply the controller (1.132) to the plant (1.1353) and analyze the closed-loop system stability and tracking properties.

1.13 Simulate the backstepping design example in Section 1.5.4, with $\theta^* = [2, -1.4]^T$, $\varphi_1(x_1) = [x_1^2, \sin x_1]^T$, $\varphi_0(x_1, x_2) = [\cos x_2, x_1^2 \sin x_1]^T$, $\varphi_2(x_1, x_2) = [x_2^2, \cos(x_1 + x_2)]^T$, for $y_m(t) = 1$ and $y_m(t) = \sin 0.5t$.

1.14 Specify the class of second-order linear systems to which the backstepping design in Section 1.5.4 can be applied and develop such a design.

1.15 Compare the adaptive control system structures in Figure 1.4 and 1.5 with the adaptive control algorithms in Sections 1.5.2 and 1.5.3 and identify the corresponding blocks in those adaptive control systems.

Chapter 2

Systems Theory

This chapter presents some fundamentals of dynamic system theory, including system models, system characterizations, signal measures, system stability (Lyapunov stability and input–output stability) theory, signal convergence lemmas, operator norms, and their discrete-time counterparts, which are needed in developing and analyzing different adaptive control systems.

2.1 Dynamic System Models

Dynamic systems are described by differential equations. While most dynamic systems are nonlinear in nature, study of their linearized models and linear systems has played a crucial role in understanding dynamic system behaviors.

We first introduce some notation. A real n-dimensional vector x is denoted as $x \in R^n$, and a real $n \times m$ matrix A is denoted as $A \in R^{n \times m}$ (both x and A can be functions of t, the time variable, in study of dynamic systems).

2.1.1 Nonlinear Systems

A dynamic system can be described by a set of differential equations

$$F_i(y^{(p)}(t), \ldots, y^{(1)}(t), y(t), u^{(p)}(t), \ldots, u^{(1)}(t), u(t), t) = 0, \; t \geq t_0, \tag{2.1}$$

$i = 1, 2, \ldots, l$, where t is the time variable, t_0 is the initial time of system operation, and $y(t) \in R^q$ is the system output and $u(t) \in R^M$ is the system input.[1] A specific form of the function F_i depends on a specific system under consideration. For a single-input, single-output system, that is, when $q = M = 1$, both y and u are scalar signals, denoted as $y \in R$ and $u \in R$.

[1] Here $y^{(i)}(t)$ and $u^{(i)}(t)$ denote the ith time derivatives $\frac{d^i y(t)}{dt^i}$ and $\frac{d^i u(t)}{dt^i}$ of $y(t)$ and $u(t)$, with a common notation $\dot{y}(t) = y^{(1)}(t)$, $\dot{u}(t) = u^{(1)}(t)$ and $\ddot{y}(t) = y^{(2)}(t)$, $\ddot{u}(t) = u^{(2)}(t)$.

An nth-order dynamic system can be described by a group of n interactive first-order differential equations

$$\dot{x} = f_0(x, u, t), \ y = h_0(x, u, t), \ t \geq t_0, \tag{2.2}$$

for some functions $f_0 \in R^n$ and $h_0 \in R^q$ with $q \geq 1$, $n \geq 1$, where $x = [x_1, \ldots, x_n]^T \in R^n$ is the system state vector with state variables x_i, $i = 1, 2, \ldots, n$, physical or artificial, to completely define the system behavior, $u(t) \in R^M$ with $M \geq 1$ is the system input, and $y(t)$ is the system output.

The system behavior depends on the control $u(t)$. When a feedback control law $u(t) = \alpha(x, t)$ is used, system (2.2) becomes

$$\dot{x} = f(x, t), \ y = h(x, t), \ t \geq t_0, \tag{2.3}$$

for some functions $f \in R^n$, $h \in R^q$, whose solution is denoted as $x(t) = x(t; t_0, x_0)$, where $x(t_0) = x_0$ is the initial state vector. For a vector function $x(t)$, a measure of its "magnitude" is its norm $\|x(t)\|$ (see Section 2.3.1).

The first question about system (2.3) would be: Does the solution $x(t)$ exist and is the solution $x(t)$ unique?

While the existence of the solution to system (2.3) is ensured if the function $f(x, t)$ is continuous in both t and x, the existence and uniqueness of the solution are ensured by an additional so-called Lipschitz condition [179].

Local existence and uniqueness. If $f(x, t)$ is continuous in t such that

$$\|f(x, t) - f(y, t)\| \leq L\|x - y\|, \tag{2.4}$$

$\forall x, y \in B(h) = \{x \in R^n : \|x - x_0\| \leq h\}$, $\forall t \in [t_0, t_0 + T]$, and $\|f(x_0, t)\| \leq g$, $\forall t \in [t_0, t_0 + T]$, for some positive constant T, h, g, and L, then system (2.3) has a unique solution over $[t_0, t_0 + \delta]$ for any $\delta > 0$ which satisfies $g\delta e^{L\delta} \leq h$ and $\delta \leq \min\{T, \frac{\rho}{L}, \frac{h}{g+Lh}\}$ for some constant $\rho < 1$.

Global existence and uniqueness. If $f(x, t)$ is continuous in t and for each $T \in [0, \infty)$ there exist finite constants L_T and g_T such that

$$\|f(x, t) - f(y, t)\| \leq L_T\|x - y\|, \tag{2.5}$$

$\forall x, y \in R^n$, $\forall t \in [t_0, t_0 + T]$, and $\|f(x_0, t)\| \leq g_T$, $\forall t \in [t_0, t_0 + T]$, then system (2.3) has a unique solution over $[t_0, \infty)$.

These and more results can be found in the literature (see, e.g., [179]). Condition (2.4) or (2.5) is called the *Lipschitz condition.*

2.1.2 Linear Systems

A linear time-varying dynamic system is described as

$$\dot{x} = A(t)x + B(t)u, \; y = C(t)x + D(t)u, \; t \geq t_0, \tag{2.6}$$

with $x(t_0) = x_0$, where $x \in R^n$, $u \in R^M$, $y \in R^q$ are the state, input, and output vectors, respectively, and $A(t) \in R^{n \times n}$, $B(t) \in R^{n \times M}$, $C(t) \in R^{q \times n}$, and $D(t) \in R^{q \times M}$ are continuous functions of t.

It is crucial to understand the behavior of the dynamic part of system (2.6), that is, $\dot{x} = A(t)x$, in order to describe the solution of (2.6).

Proposition 2.1 [335] *The solution $x(t; x_0, t_0)$ of $\dot{x} = A(t)x$ has the form*

$$x(t; x_0, t_0) = \Phi(t, t_0)x_0, \tag{2.7}$$

where $\Phi(t, t_0)$ is the state transition matrix of $A(t)$ such that

(i) $\Phi(t, \tau)$ is the unique solution of $\frac{\partial \Phi(t,\tau)}{\partial t} = A(t)\Phi(t, \tau)$ with $\Phi(\tau, \tau) = I_n$;
(ii) $\Phi(t, \tau) = \Phi(t, \sigma)\Phi(\sigma, \tau)$, $\forall t, \sigma, \tau$;
(iii) $\Phi(t, \tau)$ is nonsingular, $\forall t, \tau$; and
(iv) $[\Phi(t, \tau)]^{-1} = \Phi(\tau, t)$, $\forall t, \tau$.

Formally, the state transition matrix $\Phi(t, t_0)$ is

$$\begin{aligned}\Phi(t, t_0) = I &+ \int_{t_0}^{t} A(\sigma_1)\, d\sigma_1 + \int_{t_0}^{t} A(\sigma_1) \int_{t_0}^{\sigma_1} A(\sigma_2)\, d\sigma_2\, d\sigma_1 + \cdots \\ &+ \int_{t_0}^{t} A(\sigma_1) \int_{t_0}^{\sigma_1} A(\sigma_2) \cdots \int_{t_0}^{\sigma_{k-1}} A(\sigma_k)\, d\sigma_k \cdots d\sigma_2\, d\sigma_1 + \cdots .\end{aligned}$$

The matrix exponential property

$$\Phi(t, t_0) = \sum_{k=0}^{\infty} \frac{1}{k!} \left[\int_{t_0}^{t} A(\sigma)\, d\sigma \right]^k \triangleq e^{\int_{t_0}^{t} A(\sigma)\, d\sigma} \tag{2.8}$$

holds if for every t and t_0,

$$A(t) \int_{t_0}^{t} A(\sigma)\, d\sigma = \int_{t_0}^{t} A(\sigma)\, d\sigma A(t). \tag{2.9}$$

This condition is satisfied when $A(t) = A$, a constant matrix. In this case,

$$\Phi(t, t_0) = e^{A(t-t_0)} = \mathcal{L}^{-1}\left[(sI - A)^{-1} \right], \tag{2.10}$$

which can be calculated for any given constant matrix $A \in R^{n \times n}$.

With the help of the system transition matrix $\Phi(t, t_0)$, the state solution of system (2.6) can be formally expressed as

$$x(t) = \Phi(t, t_0)x_0 + \int_{t_0}^{t} \Phi(t, \tau)B(\tau)u(\tau)\, d\tau. \tag{2.11}$$

In the time-invariant case, system (2.6) becomes

$$\dot{x} = Ax + Bu, \ y = Cx + Du, \tag{2.12}$$

where $A \in R^{n\times n}$, $B \in R^{n\times M}$, $C \in R^{q\times n}$, and $D \in R^{q\times M}$ are all constant system matrices. In this case, the state solution can be explicitly expressed as

$$x(t) = e^{A(t-t_0)}x_0 + \int_{t_0}^{t} e^{A(t-\tau)}Bu(\tau)\, d\tau. \tag{2.13}$$

2.2 System Characterizations

Many physical properties of dynamic systems can be mathematically characterized. In the linear case, system characterizations can be rigorously described, and their mathematical descriptions are useful for system study.

Laplace transform. The (unilateral) Laplace transform of a continuous-time scalar function (signal) $f(t) \in R$ is defined as

$$F(s) = \mathcal{L}[f(t)] = \int_{0}^{\infty} f(t)e^{-st}\, dt, \tag{2.14}$$

where $s = \sigma + j\omega$ is the complex variable. If $f(t)$ is such that $|f(t)| \le a\, e^{bt}$ for some constants a and b, $t \ge 0$, then $F(s) = \mathcal{L}[f(t)]$ exists (for $\sigma > b$).[2]

Laplace transforms of vector or matrix functions are vector or matrix functions with the Laplace transforms of their component functions (signals).

Impulse response functions. The impulse response function (matrix) of system (2.6) is defined as

$$g(t, \tau) = \begin{cases} C(t)\Phi(t,\tau)B(\tau) + D(t)\delta(t-\tau) & \text{for } t \ge \tau, \\ 0 & \text{for } t < \tau, \end{cases} \tag{2.15}$$

where $\delta(\cdot)$ is the Dirac δ-function,[3] which has the property $\int_{t_0}^{t} D(t)\delta(t-\tau)u(\tau)\, d\tau = D(t)u(t)$ for $t \ge t_0$. Then, the system output is expressed as

$$y(t) = \int_{t_0}^{t} g(t, \tau)u(\tau)\, d\tau. \tag{2.16}$$

[2]For example, the transform of $f(t) = 1$, or $\sin \omega t$, $t \ge 0$, is $F(s) = \frac{1}{s}$ or $\frac{\omega}{s^2+\omega^2}$.

[3]It is defined with $\int_{-\infty}^{\infty} \delta(t)\, dt = 1$, $\delta(t) = 0$ for $t \ne 0$ and $\delta(t) = \infty$ for $t = 0$.

In the time-invariant case (2.12), we have

$$g(t,\tau) = Ce^{A(t-\tau)}B + D\delta(t-\tau) = g(t-\tau). \tag{2.17}$$

Transfer functions. The transfer function (matrix) $G(s)$ of the *linear time-invariant* system (2.12) is defined from $y(s) = G(s)u(s)$ for $x(0) = 0$, where $y(s) = \mathcal{L}[y(t)]$ and $u(s) = \mathcal{L}[u(t)]$,[4] that is,

$$G(s) = C(sI-A)^{-1}B + D. \tag{2.18}$$

It follows that $G(s) = \mathcal{L}[g(t)]$, where $g(t)$ is the impulse response function of system (2.12), defined in (2.17). A transfer function of the form (2.18) is a proper rational function, and it is called strictly proper if $D = 0$.[5]

When $q = M = 1$ in system (2.12), its transfer function is

$$G(s) = C(sI-A)^{-1}B + D = \frac{Z(s)}{P(s)} + D, \tag{2.19}$$

where

$$P(s) = \det(sI-A), \ Z(s) = \det\begin{bmatrix} sI-A & B \\ -C & 0 \end{bmatrix} \tag{2.20}$$

and both $P(s)$ and $Z(s)$ are polynomials of real coefficients, in the forms

$$P(s) = s^n + p_{n-1}s^{n-1} + \cdots + p_1 s + p_0, \tag{2.21}$$

$$Z(s) = k_p(s^m + z_{m-1}s^{m-1} + \cdots + z_1 s + z_0) \tag{2.22}$$

for some constants $k_p \neq 0$, p_i, z_j, $i = 0,1,2,\ldots,n-1$, $j = 0,1,2,\ldots,m-1$, with $m \leq n-1$, where n and m are the degrees of $P(s)$ and $Z(s)$, respectively. In particular, k_p is called the system *high frequency gain*.

A polynomial $P(s)$ (matrix A) is said to be *stable* if all its zeros (eigenvalues) are in the stable region of the complex plane, which in the continuous-time case is the open left-half of the complex s-plane: $\mathrm{Re}[s] < 0$.[6] For example, $P_1(s) = (s+2)(s+3)$ is stable, and so is $P_2(s) = s^2+2s+2 = (s+j+1)(s-j+1)$ with $j = \sqrt{-1}$, while $P_3(s) = s-1$ is not stable.

A polynomial is *monic* if its leading coefficient is 1; for example, s^2+s+1 is monic, while $5s^2+s+1$ is not monic. Two polynomials are *coprime* if they do not have any common factor of degree larger than zero; for example, $P_1(s) = s^2-s-12 = (s+3)(s-4)$ and $P_2(s) = s+3$ are not coprime as they have

[4] The same symbol y or u is used to denote either $y(t)$ or $u(t)$, a function of t, or its Laplace transform $y(s)$ or $u(s)$, a function of s, as the case may be.

[5] A rational function $R(s)$ is called improper if $\lim_{s\to\infty} R(s) = \infty$, e.g., $R(s) = \frac{s^3+1}{s^2+s+1}$.

[6] In discrete time, it is inside the unit circle of the complex z-plane (see Section 2.7).

the common factor $s+3$ of degree 1, while $P_3(s) = 4(s^2-s-12) = 4(s+3)(s-4)$ and $P_4(s) = 4(s+5)$ are coprime.

When $q > 1$ and/or $M > 1$ in system (2.12), its transfer function $G(s)$ can be expressed in either a left polynomial matrix decomposition form

$$G(s) = C(sI - A)^{-1}B + D = P_l^{-1}(s)Z_l(s) + D \tag{2.23}$$

for some $q \times q$ polynomial matrix $P_l(s)$ and $q \times M$ polynomial matrix $Z_l(s)$ [7] or a right polynomial matrix decomposition form

$$G(s) = C(sI - A)^{-1}B + D = Z_r(s)P_r^{-1}(s) + D \tag{2.24}$$

for some $P_r(s) \in R^{M\times M}(s)$ and $Z_r(s) \in R^{q\times M}(s)$.

In (2.23) and (2.24), $\det[P_l(s)] \not\equiv 0$ and $\det[P_l(s)] \not\equiv 0$ for all s.

As we will see later on, the right polynomial matrix decomposition form is useful for direct adaptive control, whereas the left polynomial matrix decomposition form is useful for indirect adaptive control (see Chapter 9).

A square polynomial matrix $U(s)$ is *unimodular* if $\det[U(s)]$ is a nonzero real number (so that $U^{-1}(s)$ is also a polynomial matrix).

A $q\times q$ polynomial matrix $C_l(s)$ is a *common left divisor* of $Z_l(s)$ and $P_l(s)$ if $Z_l(s) = C_l(s)\hat{Z}_l(s)$ and $P_l(s) = C_l(s)\hat{P}_l(s)$ for some $\hat{P}_l(s) \in R^{q\times q}(s)$ and $\hat{Z}_l(s) \in R^{q\times M}(s)$ (in this case, $P_l^{-1}(s)Z_l(s) = \hat{P}_l^{-1}(s)\hat{Z}_l(s)$). A $q\times q$ polynomial matrix $C_l(s)$ is the *greatest common left divisor* of $Z_l(s)$ and $P_l(s)$ if $Z_l(s) = \bar{C}_l(s)\bar{Z}_l(s)$ and $P_l(s) = \bar{C}_l(s)\bar{P}_l(s)$ for some $\bar{C}_l(s) \in R^{q\times q}(s)$, $\bar{P}_l(s) \in R^{q\times q}(s)$ and $\bar{Z}_l(s) \in R^{q\times M}(s)$ implies that $C_l(s) = \bar{C}_l(s)\hat{C}_l(s)$ for some $\hat{C}_l(s) \in R^{q\times q}(s)$. Two polynomial matrices $Z_l(s)$ and $P_l(s)$ are *left coprime* if their greatest common left divisor is a unimodular matrix.

Similarly, we can define the *right coprimeness* of the two polynomial matrices $Z_r(s)$ and $P_r(s)$ in the decomposition (2.24).

The left polynomial matrix decomposition (2.23) can be written as

$$G(s) = P_l^{-1}(s)Z_l(s) + D = P_l^{-1}(s)(Z_l(s) + P_l(s)D). \tag{2.25}$$

Assume that the two polynomial matrices $Z_l(s) + P_l(s)D$ and $P_l(s)$ in (2.25) are left coprime. Then, the system order n = degree of $\det[P_l(s)]$. System zeros and poles can be defined for a general case of $q > 0$ and $M > 0$ [335]. For the case when $q = M$ and $\det[G(s)] \not\equiv 0$ for all s, the zeros of $G(s)$ are the zeros of $\det[Z_l(s) + P_l(s)D]$, and the poles of $G(s)$ are the zeros of $\det[P_l(s)]$. This is also valid for the right polynomial matrix decomposition $\{Z_r(s), P_r(s)\}$.

[7] A $q\times M$ polynomial matrix $P(s)$ is a $q\times M$ matrix whose components are all polynomials (with real coefficients), denoted as $P(s) \in R^{q\times m}(s)$.

Controllability. System controllability is a physical concept. By definition [335], the system $\dot{x} = A(t)x + B(t)u$ is controllable on $[t_0, t_f]$ if for any given initial state $x(t_0) = x_0$ there exists a continuous input signal $u(t)$ such that $x(t_f) = x(t_f; t_0, x_0) = 0$. The necessary and sufficient condition is that

$$W_c(t_0, t_f) = \int_{t_0}^{t_f} \Phi(t_0, t)B(t)B^T(t)\Phi^T(t_0, t)\, dt \ \text{ is nonsingular.} \tag{2.26}$$

For the time-invariant system (2.12) with A and B constant, the condition is

$$\text{rank}[B|AB|A^2B|\cdots|A^{n-1}B] = n \tag{2.27}$$

(when $M = 1$, it becomes $\det[B|AB|A^2B|\cdots|A^{n-1}B] \neq 0$).

Given a controllable (A, B), for any set of n chosen real or complex conjugate numbers s_i, $i = 1, 2, \ldots, n$, there exists a constant matrix $K \in R^{M\times n}$ such that the ith eigenvalue of $A + BK$ is $\lambda_i[A + BK] = s_i$, $i = 1, 2, \ldots, n$.

A special case of uncontrollable systems is when all eigenvalues of $A+BK$ can be made stable (i.e., in $\text{Re}[s] < 0$) by some constant matrix $K \in R^{M\times n}$. In this case, system (2.12) is called *stabilizable.*

Observability. A dual concept to controllability is the system observability. By definition [335], the system $\dot{x} = A(t)x + B(t)u$, $y = C(t)x + D(t)u$ (or equivalently, $\dot{x} = A(t)x$, $y = C(t)x$) is observable on $[t_0, t_f]$ if any initial state $x(t_0) = x_0$ is uniquely determined by the output $y(t)$ for $t \in [t_0, t_f]$. The necessary and sufficient condition for observability is that

$$M(t_0, t_f) = \int_{t_0}^{t_f} \Phi^T(t, t_0)C^T(t)C(t)\Phi(t, t_0)\, dt \ \text{ is nonsingular.} \tag{2.28}$$

For the time-invariant system (2.12) with A and C constant, the condition is

$$\text{rank}[C^T|A^TC^T|(A^T)^2C^T|\cdots|(A^T)^{n-1}C^T] = n \tag{2.29}$$

(when $q = 1$, it becomes $\det[C^T|A^TC^T|(A^T)^2C^T|\cdots|(A^T)^{n-1}C^T] \neq 0$).

Given an observable (A, B), for any set of n chosen real or complex conjugate numbers s_i, $i = 1, 2, \ldots, n$, there exists a constant matrix $L \in R^{n\times q}$ such that the ith eigenvalue of $A + LC$ is $\lambda_i[A + LC] = s_i$, $i = 1, 2, \ldots, n$.

A special case of unobservable systems is when all eigenvalues of $A + LC$ can be made stable (i.e., in $\text{Re}[s] < 0$) by some constant matrix $L \in R^{n\times q}$. In this case, system (2.12) is called *detectable.*

An important property is that any control input based on output feedback does not change system observability [171]. Physically, this is easy to visualize, because observability assessment can be done, independent of any known input signal whose effect can be taken into account in determining the initial state x_0, and an output feedback signal is known (while a state feedback control, which does not change system controllability, may change observability). However, a rigorous mathematical proof for the case of uniform observability (i.e., when, with (2.28) for a bounded $C(t)$, the condition that $M(t_1, t_1+\gamma) \geq \beta I$ for some $\gamma > 0$ and $\beta > 0$, $\forall t_1 \geq 0$, is required) is needed [151].

The system observability index ν is the least integer such that

$$\text{rank}[C^T|A^TC^T|(A^T)^2C^T|\cdots|(A^T)^{\nu-1}C^T] = n. \tag{2.30}$$

We should note that $\nu = n$ when $q = 1$.

Interactor matrix and high frequency gain matrix. For model reference control of a multi-input, multi-output system (2.12), an interactor matrix [443] plays an important role. Consider the case when $q = M$.

Proposition 2.2 [443] *For any $M \times M$ proper rational full rank transfer matrix $G(s)$ there exists a unique lower triangular polynomial matrix $\xi(s)$, defined as the left interactor matrix of $G(s)$, of the form*

$$\xi(s) = \begin{bmatrix} s^{l_1} & 0 & \cdots & \cdots & 0 \\ s^{l_1}h_{21}(s) & s^{l_2} & 0 & \cdots & 0 \\ \vdots & \vdots & \vdots & \vdots & \vdots \\ s^{l_1}h_{M1}(s) & s^{l_2}h_{M2}(s) & \cdots & s^{l_{M-1}}h_{M\,M-1}(s) & s^{l_M} \end{bmatrix}, \tag{2.31}$$

where $h_{ij}(s)$, $j = 1, \ldots, M-1$, $i = 2, \ldots, M$, are polynomials divisible by s and $l_i \geq 0$, $i = 1, \ldots, M$, are integers such that

$$\lim_{s\to\infty} \xi(s)G(s) = K_p^0 \tag{2.32}$$

is finite and nonsingular.

It is easy to see that the inverse of the above left interactor matrix $\xi(s)$ is not stable. Since in the design of model reference control (see Chapter 9) $\xi^{-1}(s)$ is required to be stable, we need to modify $\xi(s)$ so that a modified left interactor matrix has a stable inverse. Such a modification is given below.

Proposition 2.3 [377] *For any $M \times M$ strictly proper rational full rank transfer matrix $G(s)$ there exists a lower triangular polynomial matrix $\xi_m(s)$, defined as the modified left interactor (MLI) matrix of $G(s)$, of the form*

$$\xi_m(s) = \begin{bmatrix} d_1(s) & 0 & \cdots & \cdots & 0 \\ h_{21}^m(s) & d_2(s) & 0 & \cdots & 0 \\ \vdots & \vdots & \vdots & \vdots & \vdots \\ h_{M1}^m(s) & h_{M2}^m(s) & \cdots & h_{M\,M-1}^m(s) & d_M(s) \end{bmatrix}, \tag{2.33}$$

where $h_{ij}^m(s)$, $j = 1, \ldots, M-1$, $i = 2, \ldots, M$, *are some polynomials and* $d_i(s) = s^{l_i} + a_1^i s^{l_i - 1} + \cdots + a_{l_i}^i$, $i = 1, \ldots, M$, *are any chosen monic stable polynomials such that*

$$\lim_{s \to \infty} \xi_m(s) G(s) = K_p \tag{2.34}$$

is finite and nonsingular. Furthermore $h_{ij}^m(s)$ *can be chosen as*

$$h_{ij}^m(s) = \text{polynomial part of } s^{l_j}(h_{ij}(s) + a_1^i s^{-1} h_{ij}(s) + \cdots + a_{l_i}^i s^{-l_i} h_{ij}(s)) \tag{2.35}$$

for $j = 1, \ldots, M-1$, $i = 2, \ldots, M$. *For this choice of* $h_{ij}^m(s)$, *we have*

$$\lim_{s \to \infty} \xi_m(s) G_0(s) = K_p^0. \tag{2.36}$$

An equivalent concept is that of a Hermite matrix $H(s)$ of an $M \times M$ transfer matrix $G(s)$ [266], [282] such that $\lim_{s\to\infty} H^{-1}(s)G(s) = K_p$.

The matrix K_p^0 or K_p in (2.32) or (2.34) is called the *high frequency gain matrix* of the system transfer function (matrix) $G(s)$. In the case when $q = M = 1$, the interactor matrix is $\xi_l(s) = s^{n-m}$ for $G(s)$ in (2.19), and $\xi_m(s)$ is any stable polynomial of degree $n - m$. The gain k_p is a scalar whose sign information is only needed for adaptive control. For $q = M > 1$, K_p is a matrix and a different knowledge about K_p may be used in a different adaptive control scheme. An important issue is to use less information about K_p for adaptive control. For this purpose, the following decompositions specify the essential knowledge of K_p needed for adaptive control designs.

Gain matrix decompositions. Assume that all leading principal minors [126] of the gain matrix K_p are nonzero.[8] Then, several decompositions of K_p exist and can be employed for adaptive control of the plant (2.12).

Let Δ_i, $i = 1, 2, \ldots, M$, be the leading principal minors of K_p, and define the sgn function as sign$[x] = 1$ for $x > 0$ and sign$[x] = -1$ for $x < 0$. The following proposition gives the key result for such decompositions.

Proposition 2.4 (LDU decomposition) *A matrix* $K_p \in R^{M \times M}$ *with all its leading principal minors being nonzero has a unique decomposition:*

$$K_p = LD^*U \tag{2.37}$$

[8] The kth leading principal minor of a matrix $A = \{a_{ij}\} \in R^{M\times M}$ is

$$\det \begin{bmatrix} a_{11} & \cdots & a_{1k} \\ \vdots & & \vdots \\ a_{k1} & \cdots & a_{kk} \end{bmatrix}, \; k = 1, 2, \ldots, M.$$

for some $M \times M$ unit (i.e., with all diagonal elements being 1) lower triangular matrix L and unit upper triangular matrix U, and

$$D^* = \operatorname{diag}\{d_1^*, d_2^*, \ldots, d_M^*\} = \operatorname{diag}\left\{\Delta_1, \frac{\Delta_2}{\Delta_1}, \ldots, \frac{\Delta_M}{\Delta_{M-1}}\right\}. \tag{2.38}$$

This is the well-known LDU decomposition of a nonsingular matrix with nonzero leading principal minors (see, e.g., [126]).

From Proposition 2.4, one can derive the following results.

Corollary 2.1 (SDU decomposition) *A matrix $K_p \in R^{M \times M}$ with all its leading principal minors being nonzero has a nonunique decomposition:*

$$K_p = S D_s U_s, \tag{2.39}$$

where S is an $M \times M$ real symmetric and positive definite matrix,[9] *U_s is an $M \times M$ unit upper triangular matrix, and*

$$\begin{aligned} D_s &= \operatorname{diag}\{s_1^*, s_2^*, \ldots, s_M^*\} \\ &= \operatorname{diag}\left\{\operatorname{sign}\left[\Delta_1\right]\gamma_1, \operatorname{sign}\left[\frac{\Delta_2}{\Delta_1}\right]\gamma_2, \ldots, \operatorname{sign}\left[\frac{\Delta_M}{\Delta_{M-1}}\right]\gamma_M\right\} \end{aligned} \tag{2.40}$$

such that $\gamma_i > 0$, $i = 1, \ldots, M$, may be arbitrary.

This SDU decomposition of K_p follows from the LDU decomposition, with $S = LD^*D_s^{-1}L^T$ and $U_s = D_s^{-1}L^{-T}D_sU$, which has been used for multivariable model reference adaptive control [270] (where $\gamma_i = 1$, $i = 1, 2, \ldots, M$)[137].

Corollary 2.2 (LDS decomposition) *A matrix $K_p \in R^{M \times M}$ with all its leading principal minors being nonzero has a nonunique decomposition:*

$$K_p = L_s D_s S, \tag{2.41}$$

where S is an $M \times M$ real symmetric and positive definite matrix, L_s is an $M \times M$ unit lower triangular matrix, and

$$\begin{aligned} D_s &= \operatorname{diag}\{s_1^*, s_2^*, \ldots, s_M^*\} \\ &= \operatorname{diag}\left\{\operatorname{sign}\left[\Delta_1\right]\gamma_1, \operatorname{sign}\left[\frac{\Delta_2}{\Delta_1}\right]\gamma_2, \ldots, \operatorname{sign}\left[\frac{\Delta_M}{\Delta_{M-1}}\right]\gamma_M\right\} \end{aligned} \tag{2.42}$$

such that $\gamma_i > 0$, $i = 1, \ldots, M$, may be arbitrary.

[9] A symmetric matrix $S \in R^{M \times M}$ is positive definite matrix if and only if the M eigenvalues of S, which are real for S being symmetric, are all positive.

Note that if some of the leading principal minors of K_p are zero but K_p is nonsingular, additional information is needed for such decompositions. Such information is characterized by a permutation matrix P such that PK_p has the above LDU decomposition. The knowledge of this permutation matrix, if known, can be incorporated in the design of a multivariable model reference adaptive control scheme based on the decompositions of PK_p.

2.3 Signal Measures

A linear space is a collection of elements (vectors) in which two operations are defined: addition of elements (vectors) and multiplication of an element (vector) by a scalar, satisfying the commutative, associative, and distributive laws. Typical linear spaces are the space of $n \times m$ complex matrices, denoted as $C^{n\times m}$ (C^n for $m = 1$), and the space of all $n \times m$ real matrices, denoted as $R^{n\times m}$ (R^n for $m = 1$), which is more relevant in adaptive control literature.

Consider a vector signal $x(t) = [x_1(t), \ldots, x_n(t)]^T \in R^n$. At any t, $x(t) = [x_1(t), \ldots, x_n(t)]^T$ is a vector, and as t changes, $x(t)$ is a vector function. A vector x can be measured with a vector norm, while a function can be measured with a function norm; both norms are presented in this section.

Definition 2.1 *A real-valued function $\|\cdot\|$ on a linear space $\mathcal{S}$ is a norm if*

(i) $\|x\| \geq 0$ for all $x \in \mathcal{S}$ and $\|x\| = 0$ only if $x = 0$;
(ii) $\|\alpha x\| = |\alpha|\,\|x\|$ for all $x \in \mathcal{S}$ and any scalar α; and
(iii) $\|x + y\| \leq \|x\| + \|y\|$ for all $x, y \in \mathcal{S}$.

2.3.1 Vector and Matrix Norms

There are different ways to define a vector or matrix norm.

Vector norms on R^n. The l^1 norm of a constant vector $x = [x_1, \ldots, x_n]^T \in R^n$ is defined as

$$\|x\|_1 = |x_1| + \cdots + |x_n| \tag{2.43}$$

($|x_i|$ = the absolute value of x_i), the l^2 (Euclidean) norm of x is defined as

$$\|x\|_2 = \sqrt{x_1^2 + \cdots + x_n^2}, \tag{2.44}$$

and the l^∞ norm of x is defined as

$$\|x\|_\infty = \max_{1\leq i\leq n} |x_i|. \tag{2.45}$$

These vector norms have the following properties.

Proposition 2.5 [79] *For any $x \in R^n$,*

$$\|x\|_2 \leq \|x\|_1 \leq \sqrt{n}\|x\|_2, \tag{2.46}$$

$$\|x\|_\infty \leq \|x\|_1 \leq n\|x\|_\infty, \tag{2.47}$$

$$\|x\|_\infty \leq \|x\|_2 \leq \sqrt{n}\|x\|_\infty. \tag{2.48}$$

Each of the equalities holds for some $x \neq 0$.

This implies that $\|\cdot\|_1$, $\|\cdot\|_2$, and $\|\cdot\|_\infty$ are equivalent norms in R^n. In view of this, the notation $\|\cdot\|$ is frequently used to represent any of $\|\cdot\|_1$, $\|\cdot\|_2$, and $\|\cdot\|_\infty$. From (2.46) and (2.48), it follows that $\|x\|_\infty \leq \|x\|_2 \leq \|x\|_1$.

In general, the l^p norm of x is defined as

$$\|x\|_p = (|x_1|^p + \cdots + |x_n|^p)^{\frac{1}{p}} \tag{2.49}$$

for $p \in [1, \infty)$. (How about $p \in (0, 1)$?)

It can be verified [78] that for any $p \in [1, \infty)$ and $q \in [1, \infty)$ there exist constants $c_1 > 0$ and $c_2 > 0$ such that

$$k_1\|x\|_p \leq \|x\|_q \leq k_2\|x\|_p, \ \forall x \in R^n. \tag{2.50}$$

Vector norms on C^n. For $x = [x_1, x_2, \ldots, x_n]^T \in C^n$, that is, x_i, $i = 1, 2, \ldots, n$, are complex numbers of the form $x_i = \sigma_i + j\omega_i$, for real numbers σ_i and ω_i, the l^p norm of x is also defined by (2.49) with $|x_i| = \sqrt{\sigma_i^2 + \omega_i^2}$, the magnitude of x_i, $i = 1, 2, \ldots, n$, for $p \in [1, \infty)$, such as $p = 1, 2, \infty$.

Vector norms on R^n, as functions of t. For a vector signal $x(t) = [x_1(t), \ldots, x_n(t)]^T \in R^n$, we denote its vector norms as

$$\|x(t)\|_1 = |x_1(t)| + \cdots + |x_n(t)|, \tag{2.51}$$

$$\|x(t)\|_2 = \sqrt{x_1^2(t) + \cdots + x_n^2(t)}, \tag{2.52}$$

$$\|x(t)\|_\infty = \max_{1 \leq i \leq n} |x_i(t)|, \tag{2.53}$$

$$\|x(t)\|_p = (|x_1(t)|^p + \cdots + |x_n(t)|^p)^{\frac{1}{p}}, \ p \in [1, \infty). \tag{2.54}$$

Induced matrix norms. Matrix norms are defined (induced) from vector norms. The induced norm of a constant matrix $A \in R^{m \times n}$ is defined as

$$\|A\| = \sup_{x \in R^n, \|x\|=1} \|Ax\|, \tag{2.55}$$

where $\|\cdot\|$ is any of the vector norms $\|\cdot\|_1$, $\|\cdot\|_2$, $\|\cdot\|_\infty$, or $\|\cdot\|_p$ ($p \in [1,\infty)$). Similarly, the norm of a matrix function $A(t) \in R^{n\times n}$ is defined as

$$\|A(t)\| = \sup_{x\in R^n, \|x\|=1} \|A(t)x\|. \tag{2.56}$$

For different vector norms, the corresponding induced matrix norms are [10]

$$\|A\|_\infty = \max_{1\le i\le m} \sum_{j=1}^{n} |a_{ij}|, \tag{2.57}$$

$$\|A\|_1 = \max_{1\le j\le n} \sum_{i=1}^{m} |a_{ij}|, \tag{2.58}$$

$$\|A\|_2 = \sqrt{\lambda_{\max}(A^T A)}. \tag{2.59}$$

Frobenius norm. The frequently used Frobenius norm is defined as

$$\|A\|_F = \sqrt{\mathrm{trace}[A^T A]} = \sqrt{\sum_{i=1}^{m}\sum_{j=1}^{n} |a_{ij}|^2}, \tag{2.60}$$

but it is not an induced norm as $\|I_n\| = \sqrt{n} \neq 1$ for $n > 1$. This norm is useful for stability analysis of multivariable adaptive control systems (see Chapter 9) where a proper measure of some gain matrices is needed.

Proposition 2.6 [126] *For any $A \in R^{m\times n}$ and $B \in R^{n\times p}$,*

$$\max_{ij} |a_{ij}| \le \|A\|_2 \le \sqrt{mn}\, \max_{ij} |a_{ij}|, \tag{2.61}$$

$$\frac{1}{\sqrt{n}} \|A\|_\infty \le \|A\|_2 \le \sqrt{m}\, \|A\|_\infty, \tag{2.62}$$

$$\frac{1}{\sqrt{m}} \|A\|_1 \le \|A\|_2 \le \sqrt{n}\, \|A\|_1, \tag{2.63}$$

$$\|A\,B\|_p \le \|A\|_p\, \|B\|_p,\ p \in [1,\infty], \tag{2.64}$$

$$\|A\|_2 \le \|A\|_F \le \sqrt{n}\, \|A\|_2. \tag{2.65}$$

For any $D(t) \in R^{m\times n}$, $x(t) \in R^n$, and $t \ge t_0$, we also have

$$\left\| \int_{t_0}^{t} D(\tau)x(\tau)\, d\tau \right\|_p \le \int_{t_0}^{t} \|D(\tau)x(\tau)\|_p\, d\tau \le \int_{t_0}^{t} \|D(\tau)\|_p\, \|x(\tau)\|_p\, d\tau. \tag{2.66}$$

This result is from the norm property: $\|x(t_1) + x(t_2)\|_p \le \|x(t_1)\|_p + \|x(t_2)\|_p$, applied to the sum approximating the integral, and from (2.64) as well.

[10] For $A \in C^{m\times n}$, the results (2.57)–(2.59) also hold with absolute value $|a_{ij}|$ being replaced by magnitude $|a_{ij}|$ (the same notation $|\cdot|$) and A^T being replaced by A^{*T}, the complex conjugate transpose of $A = \{a_{ij}\}$: $A^{*T} = \{a^*_{ji}\}$, with $(\sigma + j\omega)^* = \sigma - j\omega$.

2.3.2 Signal Norms

For a *continuous-time* vector signal $x(t) \in R^n$, its L^1 signal norm is defined as

$$\|x(\cdot)\|_1 = \int_0^\infty \|x(t)\|_1 \, dt = \int_0^\infty (|x_1(t)| + \cdots + |x_n(t)|) \, dt, \tag{2.67}$$

its L^2 signal norm is defined as

$$\|x(\cdot)\|_2 = \sqrt{\int_0^\infty \|x(t)\|_2^2 \, dt} = \sqrt{\int_0^\infty (x_1^2(t) + \cdots + x_n^2(t)) \, dt}, \tag{2.68}$$

and its L^∞ signal norm is defined as

$$\|x(\cdot)\|_\infty = \sup_{t \geq 0} \|x(t)\|_\infty = \sup_{t \geq 0} \max_{1 \leq i \leq n} |x_i(t)|. \tag{2.69}$$

In general, the L^p signal norm, $p \in [1, \infty)$, is defined as

$$\|x(\cdot)\|_p = \left(\int_0^\infty \|x(t)\|_p^p \, dt\right) = \left(\int_0^\infty (|x_1(t)|^p + \cdots + |x_n(t)|^p) \, dt\right)^{\frac{1}{p}}. \tag{2.70}$$

Because all vector norms are equivalent (see Proposition 2.5 and (2.50)), the L^p signal norm, $p \in [1, \infty)$, can be also defined as

$$\|x(\cdot)\|_p = \left(\int_0^\infty \|x(t)\|^p \, dt\right)^{\frac{1}{p}}, \tag{2.71}$$

where $\|\cdot\|$ is any vector norm on R^n.

Important inequalities. The following inequalities are useful in adaptive control system analysis where signal bounding is needed.

Hölder inequality [333]: For any $p \in [1, \infty)$, $q \in [1, \infty)$ such that $\frac{1}{p} + \frac{1}{q} = 1$, any $\sigma > \tau \geq 0$, and integrable scalar functions $f(t)$ and $g(t)$, we have

$$\int_\tau^\sigma |f(t)g(t)| \, dt \leq \left(\int_\tau^\sigma |f(t)|^p \, dt\right)^{\frac{1}{p}} \left(\int_\tau^\sigma |g(t)|^q \, dt\right)^{\frac{1}{q}}. \tag{2.72}$$

Hence, if $f(t) \in L^p$ and $g(t) \in L^q$, then $f(t)g(t) \in L^1$.

Schwarz inequality: For $p = q = 2$ in the Hölder inequality, we have

$$\int_\tau^\sigma |f(t)g(t)| \, dt \leq \sqrt{\int_\tau^\sigma f^2(t) \, dt} \sqrt{\int_\tau^\sigma g^2(t) \, dt}. \tag{2.73}$$

Minkovski inequality: For any $p \in [1, \infty)$, any $b > a \geq 0$, and two integrable scalar functions $f(t)$ and $g(t)$, we have

$$\left(\int_\tau^\sigma |f(t) + g(t)|^p \, dt\right)^{\frac{1}{p}} \leq \left(\int_\tau^\sigma |f(t)|^p \, dt\right)^{\frac{1}{p}} + \left(\int_\tau^\sigma |g(t)|^p \, dt\right)^{\frac{1}{p}}. \tag{2.74}$$

Hence, if $f(t) \in L^p$ and $g(t) \in L^p$, then $f(t) + g(t) \in L^p$.

The proofs of these inequalities are based on the following result [333]: For any $p \in [1, \infty)$, $q \in [1, \infty)$ such that $\frac{1}{p} + \frac{1}{q} = 1$ and any $a \geq 0$, $b \geq 0$, we have

$$ab \leq \frac{a^p}{p} + \frac{b^q}{q}. \tag{2.75}$$

Signal spaces. The signal space L^1 is defined as

$$L^1 = \{x(t) \in R^n : \|x(\cdot)\|_1 < \infty\}, \tag{2.76}$$

the signal space L^2 is defined as

$$L^2 = \{x(t) \in R^n : \|x(\cdot)\|_2 < \infty\}, \tag{2.77}$$

the signal space L^∞ is defined as

$$L^\infty = \{x(t) \in R^n : \|x(\cdot)\|_\infty < \infty\}, \tag{2.78}$$

and the signal space L^p is defined as

$$L^p = \{x(t) \in R^n : \|x(\cdot)\|_p < \infty\}. \tag{2.79}$$

These are linear spaces of elements with finite norms of respective types.

A signal $x(t)$ is (uniformly) bounded if $x(t) \in L^\infty$, that is, for some $\beta > 0$, $\|x(t)\| \leq \beta$, $\forall t \geq 0$ (here any vector norm $\|\cdot\|$ is applicable).

For the signal norms L^1, L^2, and L^∞, in the *continuous-time* case, we have
(i) $x(t) \in L^1 \cap L^\infty \Rightarrow x(t) \in L^p$, $p \in (1, \infty)$; and
(ii) $x(t) \in L^1$ or $x(t) \in L^2$ may not imply $x(t) \in L^\infty$ or $\lim_{t\to\infty} x(t) = 0$.

2.4 Lyapunov Stability

System stability is a measure of the actual system response in relation to a nominal response, which depends on the differential equations describing the system, that is, on their parameters and structures. A system's response is influenced by the system's initial states and the input to the system. Input–output stability (see Section 2.5) characterizes system response to input signals while the system's initial states are zero. Internal stability or Lyapunov stability, to be studied in this section, characterizes system response to a system's initial states while the system's input is zero.

Consider a system described by the nonlinear differential equation

$$\dot{x} = f(x,t), \ x(t) \in R^n \tag{2.80}$$

with $x(t_0) = x_0$ and $f(0,t) = 0$, $\forall t \geq t_0$ (i.e., the initial time is t_0 and the initial state is x_0), and assume that system (2.80) has a solution (state trajectory) $x(t) = x(t; x_0, t_0)$ for a given pair of $\{x_0, t_0\}$. For example, the solution to $\dot{x}(t) = (\cos(t) - 2t)x(t)$ is $x(t) = x(t_0)e^{\sin(t) - t^2 - \sin(t_0) + t_0^2}$.

System (2.80) is called *autonomous* if $f(x,t)$ does not explicitly depend on t, that is, $f(x,t) = f(x)$, for example, $f(x) = x^2$. Otherwise, the system is called *nonautonomous*, for example, $f(x,t) = \sin(t)x$.

The condition that $f(0,t) = 0$, $\forall t \geq t_0$, means that the origin $x_e = 0$ is an *equilibrium state* of the system, that is, $x(t_0) = 0 \Rightarrow x(t) = x(t; x_0, t_0) \equiv 0$, $\forall t \geq t_0$. We assume that $x_e = 0$ is an isolated equilibrium state: there exists an $h > 0$ such that the ball $B(h) = \{x \in R^n : \|x\| \leq h\}$ contains no equilibrium state other than $x_e = 0$. In general, a constant vector $x_e \in R^n$ is called an equilibrium of system (2.80) if $f(x_e, t) = 0$, $\forall t \geq t_0$.

2.4.1 Stability Definitions

The following stability definitions specify what is called *Lyapunov stability* of a dynamic system in the form (2.80), which has been extensively studied in the literature (see, e.g., [151], [179], [259], [426]). Lyapunov stability study of the system $\dot{x} = f(x,t)$ is important for designing a desired feedback control law $u = \alpha(x,t)$ for a control system of the form $\dot{x} = f_0(x,u,t)$.

Definition 2.2 *The equilibrium state $x_e = 0$ of system (2.80) is stable if for every $\epsilon > 0$ and any $t_0 \geq 0$ there exists a $\delta(\epsilon, t_0) > 0$ such that $\|x_0\| < \delta$ implies that $\|x(t; x_0, t_0)\| < \epsilon$, $\forall t \geq t_0$.*

Such a stability concept means that the state trajectory of a system stable at $x_e = 0$ is close to $x = 0$ if it starts near $x_e = 0$.

Definition 2.3 *The equilibrium state $x_e = 0$ of system (2.80) is unstable if it is not stable.*

For example, $x_e = 0$ is unstable for $\dot{x} = ax$, $a > 0$: $x(t) = x_0 e^{a(t-t_0)}$, $\forall t \geq t_0$, such that $|x(t)| = |x_0|e^{a(t-t_0)} \to \infty$, as $t \to \infty$, for any $x_0 \neq 0$. $x_e = 0$ is unstable for $\dot{x} = t\sin(t)x$: $x(t) = e^{-t\cos(t) + \sin(t) + t_0\cos(t_0) - \sin(t_0)}x_0$.

Definition 2.4 *The equilibrium state $x_e = 0$ of system (2.80) is attractive if for every $t_0 \geq 0$ there exists a $\rho(t_0) > 0$ such that $\|x_0\| < \rho$ implies that $\lim_{t\to\infty} x(t; x_0, t_0) = 0$.*

For example, $x_e = 0$ is attractive for $\dot{x} = \frac{-1}{t+1}x$: $x(t) = x_0\frac{t_0+1}{t+1}$, with $\rho \in (0,\infty)$, and also for $\dot{x} = ax$, $a < 0$. There are systems which have an attractive but not stable equilibrium state $x_e = 0$.

Definition 2.5 *The equilibrium state $x_e = 0$ of system (2.80) is asymptotically stable if it is stable and attractive.*

The system state trajectory which starts near $x_e = 0$ is close and tends to $x_e = 0$ if $x_e = 0$ is an asymptotically stable equilibrium state. For example, $x_e = 0$ is asymptotically stable for $\dot{x} = \frac{-1}{t+1}x$ because, in addition to attractivity, $x_e = 0$ is stable: $|x(t)| \leq |x_0|$, $\forall t \geq t_0$, with $\delta = \epsilon$.

Definition 2.6 *The equilibrium state $x_e = 0$ of system (2.80) is uniformly stable if $\delta(\epsilon, t_0) = \delta(\epsilon)$ in Definition 2.2.*

For example, for $\dot{x} = \frac{-1}{t+1}x$, $x_e = 0$ is uniformly stable: $\delta(\epsilon, t_0) = \delta(\epsilon) = \epsilon$. For $\dot{x} = (t\sin(t) - 2)x$, $x_e = 0$ is asymptotically stable but not uniformly stable: $x(t) = e^{\int_{t_0}^{t}(\tau\sin(\tau)-2)\,d\tau}x(t_0) = e^{-t\cos(t)+\sin(t)+t_0\cos(t_0)-\sin(t_0)-2(t-t_0)}x(t_0)$, and for $t = (2k+1)\pi$, $t_0 = 2k\pi$, $x(t) = e^{(4k-1)\pi}x(t_0)$, that is, there is no $\delta(\epsilon, t_0) = \delta(\epsilon)$, independent of t_0, to meet Definition 2.6.

$x_e = 0$ is stable for $\dot{x} = (t\sin(t) - 1)x$: $|x(t)| \leq e^{t_0(1+\cos(t_0))+2}|x_0|$, as $x(t) = e^{-t\cos(t)+\sin(t)+t_0\cos(t_0)-\sin(t_0)-t+t_0}x_0$, but it is neither uniformly stable (for $t = (2k+1)\pi$ and $t_0 = 2k\pi$, $x(t) = e^{4k\pi}x(t_0)$, that is, $\delta(\epsilon, t_0) = \delta(\epsilon)$ does not hold), nor asymptotically stable (as $e^{-t\cos(t)+\sin(t)-t} \not\to 0$ as $t \to \infty$). $x_e = 0$ is uniformly stable for $\dot{x} = \sin(t)x$: $x(t) = e^{-\cos(t)+\cos(t_0)}x_0$, with $\delta = e^{-2}\epsilon$.

Definition 2.7 *The equilibrium state $x_e = 0$ of system (2.80) is uniformly asymptotically stable if it is uniformly stable and if for some $\delta_1 > 0$, every $\epsilon_1 > 0$, and any $t_0 \geq 0$ there exists a $T(\epsilon_1, \delta_1) > 0$ such that $\|x_0\| < \delta_1$ implies that $\|x(t; x_0, t_0)\| < \epsilon_1$, $\forall t \geq t_0 + T$.*

For example, $x_e = 0$ is not uniformly asymptotically stable for $\dot{x} = \frac{-1}{t+1}x$: $|x(t)| < \epsilon_1$ for $t \geq t_0 + T$ requires $T > T_1(t_0) = \frac{(|x_0|-\epsilon_1)(1+t_0)}{\epsilon_1}$ for $|x_0| > \epsilon_1$ but $T_1(t_0) \to \infty$ as $t_0 \to \infty$. However, $x_e = 0$ is uniformly asymptotically stable for $\dot{x} = -x^3$: $x(t) = \frac{x_0}{\sqrt{2x_0^2(t-t_0)+1}}$, and $|x(t)| < \epsilon_1$ for $t \geq t_0 + T$ is satisfied if $T > T_1(\epsilon_1, x_0) = \frac{x_0^2-\epsilon_1^2}{2x_0^2\epsilon_1^2}$ (for $x_0^2 - \epsilon_1^2 > 0$).

Definition 2.8 *The equilibrium state $x_e = 0$ of system (2.80) is exponentially stable if for some $\alpha > 0$, every $\epsilon > 0$, and any $t_0 \geq 0$ there exists a $\delta(\epsilon) > 0$ such that $\|x_0\| < \delta$ implies that $\|x(t; x_0, t_0)\| < \epsilon e^{-\alpha(t-t_0)}$, $\forall t \geq t_0$.*

For example, $x_e = 0$ is exponentially stable for $\dot{x} = -ax$, $a > 0$: $x(t) = x_0 e^{-a(t-t_0)}$, $\forall t \geq t_0$, with $\alpha = a$, $\delta = \epsilon$.

Exponential stability implies uniform asymptotic stability, and the converse is true for linear systems $\dot{x} = A(t)x$, but not for a general nonlinear system.

Definition 2.9 *The equilibrium state $x_e = 0$ of system (2.80) is globally asymptotically stable if it is stable and if every solution of (2.80) tends to zero as $t \to \infty$.*

Definition 2.10 *The equilibrium state $x_e = 0$ of system (2.80) is globally uniformly asymptotically stable if it is uniformly stable and if for every $\delta_1 > 0$, every $\epsilon_1 > 0$, and any $t_0 \geq 0$ there exists a $T(\delta_1, \epsilon_1) > 0$ such that $\|x_0\| < \delta_1$ implies that $\|x(t; x_0, t_0)\| < \epsilon_1$, $\forall t \geq t_0 + T$.*

Definition 2.11 *The equilibrium state $x_e = 0$ of system (2.80) is globally exponentially stable if for some $\alpha > 0$, every $\beta > 0$, and any $t_0 \geq 0$ there exists a $k(\beta) > 0$ such that $\|x_0\| < \beta$ implies $\|x(t; x_0, t_0)\| < k(\beta)e^{-\alpha(t-t_0)}$, $\forall t \geq t_0$.*

For example, for $\dot{x} = \frac{-1}{t+1}x$, $x_e = 0$ is globally asymptotically stable: $x(t) = x_0 \frac{t_0+1}{t+1} \to 0$ as $t \to \infty$, $\forall t_0 \geq 0$, $\forall x_0 \in R$; for $\dot{x} = -x^3$, $x_e = 0$ is globally uniformly asymptotically stable; but both are not exponentially stable. For $\dot{x} = -ax$ with $a > 0$, $x_e = 0$ is globally exponentially stable.

Another concept is the boundedness of the solutions of system (2.80).

Definition 2.12 *A solution $x(t; x_0, t_0)$ of system (2.80) is bounded if there exists a $\beta > 0$ such that $\|x(t; x_0, t_0)\| < \beta$, $\forall t \geq t_0$, where β may depend on each solution, that is, on x_0 and t_0.*

A bounded solution trajectory $x(t; x_0, t_0)$ of system (2.80) is a (uniformly) bounded signal when t_0 and x_0 are both fixed.

Definition 2.13 *System (2.80) possesses Lagrange stability if for each $t_0 > 0$ and each x_0 the solution $x(t; x_0, t_0)$ is bounded.*

Definition 2.14 *The solutions of system (2.80) are uniformly bounded if for every $\alpha > 0$ and any $t_0 > 0$ there exists a $\beta(\alpha) > 0$ such that $\|x(t; x_0, t_0)\| < \beta$, $\forall t \geq t_0$, whenever $\|x_0\| < \alpha$.*

Definition 2.15 *The solutions of (2.80) are uniformly ultimately bounded (with bound B) if for some $B > 0$, every $\alpha > 0$, and any $t_0 > 0$ there exists a $T(\alpha) > 0$ such that $\|x_0\| < \alpha$ implies that $\|x(t; x_0, t_0)\| < B$, $\forall t \geq t_0 + T$.*

2.4.2 Positive Definite Functions

To examine the stability of system (2.80), one can use the Lyapunov direct method (Section 2.4.3), which employs the following concepts of positive definite (semidefinite) and negative definite (semidefinite) functions.

Definition 2.16 *A real scalar continuous function $W(x)$ with $W(0) = 0$ is positive definite (semidefinite) in $B(h) = \{x \in R^n : \|x\| \leq h, h > 0\}$, if $W(x) > 0$ ($W(x) \geq 0$), $\forall x \in B(h)$, $x \neq 0$; and $W(x)$ is negative definite (semidefinite), if $-W(x)$ is positive definite (semidefinite).*

For example, for $x = [x_1, x_2]^T \in R^2$, $W_1(x) = x_1^2 + 2x_2^2$ is positive definite; $W_2(x) = x_1^2 + 2\sin^2(x_2)$ is positive definite in $B(h) = \{x \in R^2 : \|x\| \leq h < \frac{\pi}{2}\}$; while $W_3(x) = x_1^2$ is only positive semidefinite.

Definition 2.17 *A real scalar continuous function $V(x,t)$ with $V(0,t) = 0$, $\forall t \geq 0$, is positive definite in $B(h)$, if $V(x,t) \geq W(x)$, for some positive definite $W(x)$, $\forall t \geq 0$, $\forall x \in B(h)$; $V(x,t)$ with $V(0,t) = 0$, $\forall t \geq 0$, is positive semidefinite in $B(h)$, if $V(x,t) \geq 0$, $\forall t \geq 0$, $\forall x \in B(h)$; and $V(x,t)$ is negative definite (semidefinite), if $-V(x,t)$ is positive definite (semidefinite).*

For $x = [x_1, x_2]^T \in R^2$, $V(x,t) = (2t+1)(x_1^2 + 2x_2^2)$ is positive definite, while $V(x,t) = \frac{x_1^2 + 2x_2^2}{t+1}$ is only positive semidefinite.

Definition 2.18 *A real scalar continuous function $W(x)$ with $W(0) = 0$ is radially unbounded if $W(x) > 0$, $\forall x \in R^n$, $x \neq 0$, and $\lim_{\|x\| \to \infty} W(x) = \infty$.*

For example, $W(x) = x_1^2 + 2x_2^2$ is radially unbounded in R^2.

Definition 2.19 *A real scalar continuous function $V(x,t)$ with $V(0,t) = 0$, $\forall t \geq 0$, is radially unbounded if there exists a continuous and radially unbounded function $W(x)$ in R^n such that $V(x,t) \geq W(x)$, $\forall x \in R^n$, $\forall t \geq 0$.*

For $x = [x_1, x_2]^T \in R^2$, $V(x,t) = x_1^2 + 2(t+1)x_2^2$ is radially unbounded.

Definition 2.20 *A real scalar continuous function $V(x,t)$ is decrescent in $B(h)$ if there exists a positive definite function $W(x)$ such that $|V(x,t)| \leq W(x)$, $\forall x \in B(h)$, $\forall t \geq 0$.*

For example, $V(x,t) = x_1^2 + 2(t+1)x_2^2$ is not decrescent in any $B(h)$.

Equivalent definitions for the positive definite functions can be given in terms of the so-called class-$\mathcal{K}$ functions (see, e.g., [151], [259], [426]).[11]

[11] A real scalar continuous and single variable function $\phi(r) \geq 0$ for $r \geq 0$, belongs to the class-$\mathcal{K}$ if (i) $\phi(0) = 0$ and $\phi(r) > 0$, $\forall r > 0$; (ii) $\phi(r_1) < \phi(r_2)$, $\forall r_1 < r_2$ (i.e., $\phi(\cdot)$

2.4.3 Lyapunov Direct Method

For a nonlinear vector function $f(x(t), t)$, it may be difficult to find the solution $x(t; x_0, t_0)$ of system (2.80) to check the system's stability. Instead, the Lyapunov direct method employs a measure of the system's state variables, that is, a positive definite function $V(x, t)$ (a Lyapunov function candidate) which is positive for any x and t and $V(0, t) = 0$ for any $t \geq 0$, and evaluates the time derivative of $V(x, t)$ along the trajectory of $\dot{x}(t) = f(x(t), t)$:

$$\dot{V} = \frac{dV(x(t), t)}{dt} = \left(\frac{\partial V(x, t)}{\partial x}\right)^T f(x, t) + \frac{\partial V(x, t)}{\partial t}. \tag{2.81}$$

If such a derivative, as a function of the time t, does not increase as t increases, the equilibrium state $x_e = 0$ of the system has certain stability properties defined above (then $V(x, t)$ becomes a Lyapunov function). This method is useful not only for analyzing a system's stability properties but also for checking how robust those properties are with respect to uncertainties in the system dynamics, that is, the stability of the system $\dot{x}(t) = f(x(t), t) + \Delta(x(t), t)$, where $\Delta(x(t), t)$ denotes uncertain system dynamics modeling errors.

In the following study, we assume that $\frac{\partial V(x,t)}{\partial x}$ and $\frac{\partial V(x,t)}{\partial t}$ exist and are continuous (i.e., $V(x, t)$ is continuously differentiable).

Theorem 2.1 ((Uniform) stability) *If in some ball $B(h) = \{x \in R^n : \|x\| \leq$*

is a strictly monotonically increasing function). A real scalar continuous function $\phi(r)$, for $r \geq 0$, belongs to the class-$\mathcal{KR}$ if $\phi(\cdot) \in \mathcal{K}$ and (iii) $\lim_{r\to\infty} \phi(r) = \infty$.

Definition 2.17e *A real scalar continuous function $V(x, t)$ with $V(0, t) = 0$, $\forall t \geq 0$, is positive definite in $B(h)$ if there exists a continuous class-$\mathcal{K}$ function $\phi(\cdot)$ such that $V(x, t) \geq \phi(\|x\|)$, $\forall t \geq 0$, $\forall x \in B(h)$.*

The equivalence of Definition 2.17 and Definition 2.17e follows from the choice $\phi(r) = \inf_{r \leq \|x\| \leq h} W(x)$ which belongs to the class-$\mathcal{K}$ (i.e., $\phi(\|x\|) \leq W(x)$, $\forall x \in B(h)$).

Definition 2.19e *A real scalar continuous function $V(x, t)$ with $V(0, t) = 0$, $\forall t \geq 0$, is radially unbounded if there exists a continuous class-$\mathcal{KR}$ function $\phi(\cdot)$ such that $V(x, t) \geq \phi(\|x\|)$, $\forall t \geq 0$, $\forall x \in R^n$.*

The equivalence of Definition 2.19 and Definition 2.19e follows from the choice $\phi(r) = \inf_{r \leq \|x\|} W(x)$ which belongs to the class-$\mathcal{KR}$.

Definition 2.20e *A real scalar continuous function $V(x, t)$ with $V(0, t) = 0$, $\forall t \geq 0$, is decrescent in $B(h)$ if there exists a continuous class-$\mathcal{K}$ function $\phi(\cdot)$ such that $|V(x, t)| \leq \phi(\|x\|)$, $\forall t \geq 0$, $\forall x \in B(h)$.*

The equivalence of Definition 2.20 and Definition 2.20e follows from the choice $\phi(r) = \sup_{\|x\| \leq r} W(x)$ which also belongs to the class-$\mathcal{K}$ (i.e., $\phi(\|x\|) \geq W(x)$, $\forall x \in B(h)$).

$h, h > 0\}$ there exists a positive definite function $V(x,t)$ such that $\dot{V} \le 0$, then the equilibrium state $x_e = 0$ of system (2.80) is stable. If, in addition, $V(x,t)$ is decrescent, then it is uniformly stable.

For example, $x_e = 0$ is stable for $\dot{x} = -e^{-t}x$: $V(x,t) = x^2$ is positive definite, $\dot{V} = -2e^{-t}x^2 \le 0$; it is also uniformly stable: $V(x,t)$ is decrescent. Since $\dot{V}$ is only negative semidefinite, that is, $\dot{V} = -2e^{-t}x^2 \to 0$ as $t \to \infty$ for any $x \in R$, we cannot conclude asymptotic stability of $x_e = 0$. In fact, $x_e = 0$ is not asymptotically stable for $\dot{x} = -e^{-t}x$: $x(t) = x_0 e^{e^{-t}-e^{-t_0}} \not\to 0$ as $t \to \infty$.

For example, $x_e = [0,0]^T$ is uniformly stable for $\dot{x}_1 = x_2$, $\dot{x}_2 = -k\sin(x_1)$, $k > 0$: $V(x) = x_2^2 + 2k(1 - \cos(x_1))$ is positive definite in $B(h) = \{x \in R^2 : \|x\| \le h < 2\pi\}$, and $\dot{V} = 2\dot{x}_2 x_2 + 2k\sin(x_1)\dot{x}_1 = 0$.

Theorem 2.2 ((Global) uniform asymptotic stability) *If in some $B(h)$ there exists a positive definite and decrescent function $V(x,t)$ such that $\dot{V}$ is negative definite, then the equilibrium state $x_e = 0$ of system (2.80) is uniformly asymptotically stable. If $V(x,t)$ is radially unbounded and decrescent in R^n such that $\dot{V}$ is negative definite in R^n, then the equilibrium state $x_e = 0$ of system (2.80) is globally uniformly asymptotically stable.*

For example, $x_e = [0,0]^T$ is uniformly asymptotically stable for $\dot{x}_1 = (x_1 - \gamma_2 x_2)(x_1^2 + x_2^2 - 1)$, $\dot{x}_2 = (\gamma_1 x_1 + x_2)(x_1^2 + x_2^2 - 1)$, $\gamma_1 > 0$, $\gamma_2 > 0$: $V(x) = \gamma_1 x_1^2 + \gamma_2 x_2^2$ is positive definite and decrescent, $\dot{V} = 2(\gamma_1 x_1^2 + \gamma_2 x_2^2)(x_1^2 + x_2^2 - 1)$ is negative definite in $B(h) = \{x \in R^2 : \|x\|_2 \le h < 1\}$. $x_e = [0,0]^T$ is globally uniformly asymptotically stable for $\dot{x}_1 = x_2 + cx_1(x_1^2 + x_2^2)$, $\dot{x}_2 = -x_1 + cx_2(x_1^2 + x_2^2)$, $c < 0$: $V(x) = x_1^2 + x_2^2$ is radially unbounded and decrescent, and $\dot{V} = 2c(x_1^2 + x_2^2)^2$ is negative definite in R^2.

$x_e = 0$ is uniformly stable and asymptotically stable for $\dot{x} = \frac{-1}{t+1}x$ but not uniformly asymptotically stable: $\dot{V} = -\frac{2x^2}{t+1}$ is only negative semidefinite for $V = x^2$ positive definite and decrescent; or $\dot{V} = -x^2$ is negative definite for $V = (t+1)x^2$ positive definite but not decrescent.

$x_e = 0$ is uniformly stable and asymptotically stable for $\dot{x} = \frac{-1}{\sqrt{t+1}}x$ but not uniformly asymptotically stable (the solution is $x(t) = x_0 e^{-2(\sqrt{t+1}-\sqrt{t_0+1})}$; for $\|x_0\| > \epsilon_1 > 0$, to make $\|x(t)\| < \epsilon_1$ for $t \ge t_0 + T$, one needs $\sqrt{t_0 + T + 1} > \sqrt{t_0 + 1} + \frac{1}{2}\ln(\frac{\|x_0\|}{\epsilon_1})$, that is, $T > T(t_0) = \ln(\frac{\|x_0\|}{\epsilon_1})\sqrt{t_0+1} + \frac{1}{4}(\ln(\frac{\|x_0\|}{\epsilon_1}))^2 \to \infty$ as $t_0 \to \infty$): $\dot{V} = -\frac{2x^2}{\sqrt{t+1}}$ is only negative semidefinite for $V = x^2$ positive definite and decrescent; or $\dot{V} = -(2 - \frac{1}{2\sqrt{t+1}})x^2$ is negative definite for $V = \sqrt{t+1}x^2$ positive definite but not decrescent; or $\dot{V} = -(2\sqrt{t+1} - 1)x^2$ is negative definite for $V = (t+1)x^2$ positive definite but not decrescent.

$x_e = 0$ is uniformly stable for $\dot{x} = \frac{-1}{(t+1)^2}x$ but not asymptotically stable (as $x(t) = x_0 e^{\frac{1}{t+1} - \frac{1}{t_0+1}}$): $\dot{V} = -\frac{2}{(t+1)^2}x^2$ is only negative semidefinite for $V = x^2$.

Remark 2.1 In general, the condition that in some $B(h)$ there exists a positive definite function $V(x,t)$ such that $\dot{V}$ is negative definite is not sufficient for asymptotic stability (see p. 111 of [351], for such an example).

However, under some additional condition on the system, that is, the nonlinear function $f(x,t)$, asymptotical stability can be ensured. One such condition is that $f(x,t)$ is bounded for any $x \in B(h)$ and $\forall t \geq 0$ [324].

Asymptotic stability. If in some $B(h)$ there exists a positive definite function $V(x,t)$ such that $\dot{V}$ is negative definite, then the equilibrium state $x_e = 0$ of the system $\dot{x}(t) = f(x(t),t)$, with $f(x,t)$ being bounded for any $x \in B(h)$ and $\forall t \geq 0$, is asymptotically stable.

To prove this result, we first show that the equilibrium state $x_e = 0$ is stable, under the condition that $V(x,t)$ is positive definite and $\dot{V} \leq 0$. Since $V(x,t)$ is positive definite, by definition, there is some function $W(x) > 0$, $\forall x \in B(h)$ and $x \neq 0$, with $W(0) = 0$, such that $V(x,t) \geq W(x)$, $\forall t \geq 0$ and $\forall x \in B(h)$. Introduce the function $\phi(r) = \inf_{r \leq \|x\| \leq h} W(x)$ which satisfies: (i) $\phi(0) = 0$ and $\phi(r) > 0$, $\forall r \in [0,h]$; (ii) $\phi(r_1) < \phi(r_2)$, for any $0 \leq r_1 < r_2 \leq h$; and $\phi(\|x\|) \leq W(x)$, $\forall x \in B(h)$. For any $\epsilon \in (0,h)$, let $\delta > 0$ such that $V(x_0,t_0) < \phi(\epsilon)$ if $\|x_0\| < \delta$. Since $\dot{V} \leq 0$, it follows that $\phi(\|x(t)\|) \leq V(x(t),t) \leq V(x_0,t_0) < \phi(\epsilon)$, $\forall t \geq t_0$, which implies that $\|x(t)\| < \epsilon$.

It remains to show that for every $t_0 \geq 0$, there exists a $\rho(t_0) > 0$ such that $\|x_0\| < \rho$ implies that $\lim_{t\to\infty} x(t;x_0,t_0) = 0$. For any $x_0 \in B(h)$, that is, $\|x_0\| \leq h$, we show $\lim_{t\to\infty} x(t;x_0,t_0) = 0$ by contradiction. If $\lim_{t\to\infty} x(t;x_0,t_0) \neq 0$, then there exists a sequence of t_k, $k = 0,1,2,\ldots$, with $\lim_{k\to\infty} t_k = \infty$, such that $\|x(t_k;x_0,t_0)\| \geq \eta$, $k = 0,1,2,\ldots$, for some $\eta > 0$. Since $f(x,t)$ is bounded, $\forall x \in B(h)$, $\forall t \geq 0$, it follows from $x(t;x_0,t_0) - x(t_k;x_0,t_0) = \int_{t_k}^{t} f(x(\tau),\tau)\,d\tau$ that $\|x(t;x_0,t_0)\| - \|x(t_k;x_0,t_0)\| \leq b_f(t-t_k)$ for $t \geq t_k$, where b_f is the upper bound of $\|f(x,t)\|$: $\|f(x,t)\| \leq b_f$, $\forall x \in B(h)$, $\forall t \geq 0$, and that $\|x(t_k;x_0,t_0)\| - \|x(t;x_0,t_0)\| \leq b_f(t_k - t)$ for $t \leq t_k$. Therefore,

$$\|x(t;x_0,t_0)\| \geq \frac{\eta}{2},\ \forall t \in I_k = (t_k - \frac{\eta}{2b_f}, t_k + \frac{\eta}{2b_f}), \tag{2.82}$$

The condition that $\dot{V}$ is negative definite implies that $\dot{V} \leq -W(x(t))$ for some positive definite function $W(x)$. For this $W(x)$, it follows from (2.82) that $W(x(t)) \geq \gamma$, $\forall t \in I_k = (t_k - \frac{\eta}{2b_f}, t_k + \frac{\eta}{2b_f})$, for some $\gamma > 0$. Then,

$$0 \leq V(x(t;x_0,t_0),t) \leq V(x_0,t_0) - \int_{t_0}^{t} W(x(\tau))\,d\tau$$

$$\leq V(x_0, t_0) - \frac{\gamma\eta}{b_f} k \to -\infty \text{ as } k \to \infty \qquad (2.83)$$

(here we assumed that the intervals I_k, $k = 0, 1, 2, \ldots$, are disjoint, which can be ensured by a selection of the sequence t_k, $k = 0, 1, 2, \ldots$). This leads to a contradiction. Hence, $\lim_{t\to\infty} x(t; x_0, t_0) = 0$. A simple proof also follows from Barbălat lemma in Section 2.6: the first inequality of (2.83) implies that $\int_{t_0}^{t} W(x(\tau))\, d\tau < \infty$, the boundedness of $f(x, t)$ with respect to t implies that $\dot{W} = \frac{dW(x(t)}{dt} = \left(\frac{\partial W(x)}{\partial x}\right)^T f(x, t)$ is bounded, that is, $W(x(t))$ is uniformly continuous, so that $\lim_{t\to\infty} W(x(t)) = 0$, that is, $\lim_{t\to\infty} x(t; x_0, t_0) = 0$.

This result is extensible to global asymptotic stability: if there exists a radially unbounded $V(x, t)$ such that $\dot{V}$ is negative definite in R^n, then the equilibrium state $x_e = 0$ of the system $\dot{x}(t) = f(x(t), t)$, with $f(x, t)$ bounded for x in any bounded $B(h)$ and $\forall t \geq 0$, is globally asymptotically stable.

The boundedness of $f(x, t)$, $\forall x \in B(h)$, $\forall t \geq 0$, is ensured if a stronger Lipschitz condition is satisfied: $\|f(x, t) - f(y, t)\| \leq L_h \|x - y\|$, $\forall x, y \in B(h) = \{x \in R^n : \|x\| \leq h\}$, $\forall t \in [0, \infty)$, that is, the parameter L_h is defined for all $t \in [0, \infty)$, unlike that in the original Lipschitz condition (2.4) or (2.5), for $t \in [t_0, t_0 + T]$, with only a finite $T > 0$. □

Theorem 2.3 ((Global) exponential stability) *If in some $B(h)$ there exists a real scalar function $V(x, t)$ such that $\alpha_1 \|x\|^2 \leq V(x, t) \leq \alpha_2 \|x\|^2$, $\dot{V} \leq -\alpha_3 \|x\|^2$, for some constant $\alpha_i > 0$, $i = 1, 2, 3$, $\forall t \geq 0$, then the equilibrium state $x_e = 0$ of system (2.80) is exponentially stable. If the above inequalities hold $\forall x \in R^n$, $\forall t \geq 0$, then the equilibrium state $x_e = 0$ of system (2.80) is globally exponentially stable.*

For example, $x_e = [0, 0]^T$ is exponentially stable for $\dot{x}_1 = (x_1 - \gamma_2 x_2)(x_1^2 + x_2^2 - 1)$, $\dot{x}_2 = (\gamma_1 x_1 + x_2)(x_1^2 + x_2^2 - 1)$, $\gamma_1 > 0$, $\gamma_2 > 0$: $\min\{\gamma_1, \gamma_2\}\|x\|_2^2 \leq V(x) = \gamma_1 x_1^2 + \gamma_2 x_2^2 \leq \max\{\gamma_1, \gamma_2\}\|x\|_2^2$, $\dot{V} = 2(\gamma_1 x_1^2 + \gamma_2 x_2^2)(x_1^2 + x_2^2 - 1) \leq -2\min\{\gamma_1, \gamma_2\}(1 - h^2)\|x\|_2^2$ in $B(h) = \{x \in R^2 : \|x\|_2 \leq h < 1$. $x_e = [0, 0]^T$ is globally exponentially stable for $\dot{x}_1 = -a(t)x_1 - bx_2$, $\dot{x}_2 = bx_1 - c(t)x_2$, $a(t) \geq \alpha > 0$, $c(t) \geq \alpha > 0$, $\forall t \geq 0$, $b \in R$: $V(x) = x_1^2 + x_2^2 = \|x\|_2^2$, $\dot{V} = -2a(t)x_1^2 - 2c(t)x_2^2 \leq -2\alpha(x_1^2 + x_2^2) = -2\alpha\|x\|_2^2$ in R^2.

For an autonomous system, f does not explicitly depend on t:

$$\dot{x} = f(x),\ x(t_0) = x_0,\ f(0) = 0. \qquad (2.84)$$

For system (2.84), the equilibrium state $x_e = 0$ is uniformly ((globally) asymptotically) stable if it is ((globally) asymptotically) stable.

Theorem 2.4 [Lasalle] ((Global) asymptotic stability) *If in some $B(h)$ there exists a positive definite function $V(x)$ with $\dot{V} \leq 0$ but $\dot{V}$ is not identically zero along any state trajectory of (2.84) except for $x(t) \equiv 0$, then the equilibrium state $x_e = 0$ of system (2.84) is asymptotically stable. If, in addition, $V(x)$ is radially unbounded with $\dot{V} \leq 0$ in R^n, then the equilibrium state $x_e = 0$ of (2.84) is globally asymptotically stable.*

Note that $\dot{V}$ is not identically zero along any state trajectory of (2.84) except for $x(t) \equiv 0$ means that $\dot{V} \equiv 0$ implies $x(t) \equiv 0$.

For example, $x_e = [0,0]^T$ is globally uniformly asymptotically stable for $\dot{x}_1 = -x_1 + x_2(x_1 + c)$, $\dot{x}_2 = -x_1(x_1 + c)$, $c \neq 0$: $V(x) = x_1^2 + x_2^2$, $\dot{V} = -2x_1^2$, and $\dot{V} \equiv 0$ implies $x_1(t) \equiv 0$ (i.e., $\dot{x}_1(t) \equiv 0$), and thus $x_2 \equiv 0$.

Theorem 2.5 (Uniform boundedness) *The solutions of system (2.80) are uniformly bounded if in $\|x\| \geq R$ for some $R \geq 0$ there exists a $V(x,t)$ such that $\phi_1(\|x\|) \leq V(x,t) \leq \phi_2(\|x\|)$, $\dot{V} \leq 0$, $\forall t \geq 0$, where $\phi_i(\cdot)$, $i = 1,2$, are continuous scalar functions such that $\phi_i(0) = 0$, $\phi_i(r_1) > \phi_i(r_2) > 0$, $\forall r_1 > r_2 > 0$, and $\lim_{r\to\infty} \phi_i(r) = \infty$, that is, the class-$\mathcal{KR}$ functions.*

Theorem 2.6 (Uniform ultimate boundedness) *If, in addition to the conditions of Theorem 2.5, $\dot{V}$ is negative definite for $\|x\| \geq R$, then the solutions of system (2.80) are uniformly ultimately bounded.*

For example, all solutions of $\dot{x}_1 = -x_1 - x_2$, $\dot{x}_2 = -x_2 + x_1 - x_2(x_2^2 - 6)$ are uniformly bounded: $V(x) = x_1^2 + x_2^2$, $\phi_i(\|x\|_2) = \|x\|_2^2$, $i = 1,2$, $\dot{V} = -2(x_1^2 + x_2^2) + 2(6x_2^2 - x_2^4)$, $\max_{x_2}(6x_2^2 - x_2^4) = 9$, $\dot{V} \leq 0$ for $\|x\|_2 \geq R$ with $R > 3$. In fact, all the solutions are uniformly ultimately bounded.

2.4.4 Linear Systems

A linear system has the form

$$\dot{x} = A(t)x, \ A(t) \in R^{n\times n}, \ x \in R^n, \tag{2.85}$$

with $x(t_0) = x_0$, where $A(t)$ is continuous. The solution $x(t;x_0,t_0)$ of system (2.85) has the form: $x(t;x_0,t_0) = \Phi(t,t_0)x_0$, where $\Phi(t,t_0)$ is the state transition matrix of system (2.85) (see Section 2.1.2).

Theorem 2.7 ((Uniform) stability) *The equilibrium state $x_e = 0$ of system (2.85) is stable if and only if $\sup_{t\geq t_0} \|\Phi(t,t_0)\| = c(t_0) < \infty$. The equilibrium state $x_e = 0$ of system (2.85) is uniformly stable if and only if $\sup_{t_0\geq 0} c(t_0) = \sup_{t_0\geq 0}(\sup_{t\geq t_0} \|\Phi(t,t_0)\|) = \gamma_0 < \infty$.*

Equivalently, $x_e = 0$ is stable if and only if the solutions of (2.85) are bounded (see Definition 2.12); $x_e = 0$ is uniformly stable if and only if the solutions of (2.85) are uniformly bounded (see Definition 2.14).

Theorem 2.8 ((Global, uniform) asymptotic stability) *The equilibrium state $x_e = 0$ of system (2.85) is asymptotically stable iff $\lim_{t\to\infty} \|\Phi(t,t_0)\| = 0$, is uniformly asymptotically stable if and only if $\sup_{t_0\geq 0}(\sup_{t\geq t_0} \|\Phi(t,t_0)\|) = \gamma_0 < \infty$ and $\lim_{t\to\infty} \|\Phi(t,t_0)\| = 0$ uniformly in t_0, and is (uniformly) asymptotically stable if and only if it is globally (uniformly) asymptotically stable.*

Note that $\lim_{t\to\infty} \|\Phi(t,t_0)\| = 0$ uniformly in t_0 means that for every $\epsilon > 0$ there exists a $T(\epsilon) > 0$ such that $\|\Phi(t,t_0)\| < \epsilon$, $\forall t \geq t_0 + T$.

To construct a positive definite function for analyzing stability of linear systems, we first review symmetric and positive definite matrices.

Definition 2.21 *A symmetric matrix $Q(t) = Q^T(t) \in R^{n\times n}$ is positive definite, denoted as $Q(t) = Q^T(t) > 0$, if, for each $t \geq 0$, $x^T Q(t) x > 0$, $\forall x \in R^n$ with $x \neq 0$. A symmetric matrix $Q(t) = Q^T(t) \in R^{n\times n}$ is uniformly positive definite if $Q(t) - \alpha I$ is positive definite for some constant $\alpha > 0$. A symmetric matrix $Q(t) = Q^T(t) \in R^{n\times n}$ is positive semidefinite, denoted as $Q(t) = Q^T(t) \geq 0$, if, for each $t \geq 0$, $x^T Q(t) x \geq 0$, $\forall x \in R^n$.*

Definition 2.22 *Two (symmetric positive definite) matrices $Q_i(t) = Q_i^T(t) \in R^{n\times n}$ with $Q_i(t) > 0$, $i = 1,2$, have the relation $Q_1(t) \geq Q_2(t)$ or $Q_2(t) \leq Q_1(t)$ if $Q_1(t) - Q_2(t) \geq 0$. A symmetric matrix $Q(t) = Q^T(t) \in R^{n\times n}$ is bounded and uniformly positive definite if there exist constant $\alpha_i > 0$, $i = 1,2$, such that $\alpha_1 I_n \leq Q(t) \leq \alpha_2 I_n$, $\forall t \geq 0$.*

Proposition 2.7 *A symmetric matrix $Q(t) = Q^T(t) \in R^{n\times n}$ is positive definite (positive semidefinite) if and only if for each $t \geq 0$ all eigenvalues of $Q(t)$ are positive (nonnegative).*

A symmetric positive definite or semidefinite matrix $Q(t) \in R^{n\times n}$ can be expressed as $Q(t) = Q_s^T(t) Q_s(t)$ for some matrix $Q_s(t)$ ($Q_s(t)$ can be chosen to be symmetric: $Q_s^T = Q_s$). This is because for a real symmetric matrix Q, there is an orthogonal matrix M (i.e., $MM^T = M^T M = I$ [298]) such that $Q = M\Lambda M^T$, where $\Lambda = \mathrm{diag}\{\lambda_1, \lambda_2, \ldots, \lambda_n\}$ contains the eigenvalues λ_i, $i = 1,2,\ldots,n$, of Q, which are real numbers. For a positive definite (semidefinite) matrix Q, $\lambda_i > 0$ ($\lambda_i \geq 0$), $i = 1,2,\ldots,n$, so that $Q = M\Lambda^{\frac{1}{2}}\Lambda^{\frac{1}{2}}M^T = M\Lambda^{\frac{1}{2}}M^T M\Lambda^{\frac{1}{2}}M^T$, for $\Lambda^{\frac{1}{2}} = \mathrm{diag}\{\sqrt{\lambda_1}, \sqrt{\lambda_2}, \ldots, \sqrt{\lambda_n}\}$.

A symmetric matrix Q is positive definite if and only if all its leading principal minors are positive. A symmetric matrix Q is positive semidefinite if and only if all its principal minors are nonnegative.

Theorem 2.9 *Suppose that $Q(t) = Q^T(t) \in R^{n\times n}$ is continuous, bounded, and uniformly positive definite, $\forall t \geq 0$, $A(t) \in R^{n\times n}$ is continuous and bounded, $\forall t \geq 0$. Then the equilibrium state $x_e = 0$ of system (2.85) is uniformly asymptotically stable if and only if $P(t) = \int_t^\infty \Phi^T(\tau,t)Q(\tau)\Phi(\tau,t)\,d\tau$ is bounded and uniformly positive definite, $\forall t \geq 0$.*

As $\frac{\partial \Phi^T(\tau,t)Q(\tau)\Phi(\tau,t)}{\partial t} = -A^T(t)\Phi^T(\tau,t)Q(\tau)\Phi(\tau,t) - \Phi^T(\tau,t)Q(\tau)\Phi(\tau,t)A(t)$ and $\frac{\partial \Phi(\tau,t)}{\partial t} = -\Phi(\tau,t)A(t)$, we have that $P(t) = \int_t^\infty \Phi^T(\tau,t)Q(\tau)\Phi(\tau,t)\,d\tau$ satisfies $\dot{P}(t) = -A^T(t)P(t) - P(t)A(t) - Q(t)$. On the other hand, if there is a symmetric, uniformly positive definite and bounded matrix $P(t)$ to satisfy this equation for a bounded and uniformly positive definite matrix $Q(t) = Q^T(t)$, then the equilibrium state $x_e = 0$ is is uniformly asymptotically stable. In this case, $V(x,t) = x^T P(t) x$ is a Lyapunov function.

Theorem 2.10 ((Global) exponential stability) *The equilibrium state $x_e = 0$ of system (2.85) is exponentially stable if and only if it is uniformly asymptotically stable and if and only if there exist $\beta > 0$, $\alpha > 0$ such that $\|\Phi(t,t_0)\| \leq \beta e^{-\alpha(t-t_0)}$, $\forall t \geq t_0$, $\forall t_0 \geq 0$. The equilibrium state $x_e = 0$ of system (2.85) is exponentially stable only if it is globally exponentially stable.*

When the linear system is time invariant, system (2.85) becomes

$$\dot{x} = Ax, \; x \in R^n, \tag{2.86}$$

with $x(0) = x_0$, for a constant matrix $A \in R^{n\times n}$.

Theorem 2.11 *The equilibrium state $x_e = 0$ of system (2.86) is asymptotically stable (if and only if it is exponentially stable) if and only if all eigenvalues of A have negative real parts and is stable if and only if all eigenvalues of A have nonpositive real parts and every eigenvalue having a zero real part is a simple zero of the minimal polynomial of A.*

The minimal polynomial of A is the minimum-degree polynomial $(\lambda - \lambda_1)^{n_1}(\lambda - \lambda_2)^{n_2}\cdots(\lambda - \lambda_m)^{n_m}$ such that $(A - \lambda_1 I_n)^{n_1}(A - \lambda_2 I_n)^{n_2}\cdots(A - \lambda_m I_n)^{n_m} = 0$, where λ_i, $i = 1,2,\ldots,m$, are all the distinct eigenvalues of A. These integers n_i, $i = 1,2,\ldots,m$, are also the sizes of the largest Jordan blocks associated with the eigenvalues λ_i, $i = 1,2,\ldots,m$.

If $A(t)$ is time varying, system (2.85) may be even unstable at $x_e = 0$ even if all eigenvalues of $A(t)$ have negative real parts (and even are also constant) [151], [426]. An additional condition for stability is $\|\dot{A}(t)\|$ small.

Theorem 2.12 *The equilibrium state $x_e = 0$ of system (2.86) is asymptotically stable if and only if for every positive definite matrix $Q = Q^T \in R^{n\times n}$ the Lyapunov equation $A^T P + PA = -Q$ has a unique and positive definite solution $P = P^T \in R^{n\times n}$.*

In this case, $V(x) = x^T P x$ is a Lyapunov function.

2.4.5 Lyapunov Indirect Method

Lyapunov indirect method employs linearization of $f(x,t)$ at an equilibrium x_e of $\dot{x}(t) = f(x(t),t)$ to obtain a linear system whose stability is related to the stability of the original system $\dot{x}(t) = f(x(t),t)$. Consider the system $\dot{x} = f(x,t)$, $x(t) \in R^n$ with the origin $x_e = 0$ being an equilibrium state. Using Taylor series expansion, we express

$$f(x,t) = A(t)x + g(x,t),\ A(t) = \left.\frac{\partial f(x,t)}{\partial x}\right|_{x=0} \tag{2.87}$$

and assume that

$$\lim_{\|x\|\to 0} \frac{\|g(x,t)\|}{\|x\|} = 0 \tag{2.88}$$

uniformly in t. The stability properties of the linearized system

$$\dot{z} = A(t)z \tag{2.89}$$

are useful for examining the stability properties of the original nonlinear system: (i) if $z_e = 0$ is uniformly asymptotically stable, then $x_e = 0$ is uniformly asymptotically stable; (ii) if $z_e = 0$ is unstable, then $x_e = 0$ is unstable; and (iii) otherwise no conclusion can be made on the stability of $x_e = 0$.

2.5 Input–Output Stability

Adaptive control systems are feedback control systems whose control inputs depend on system outputs, and they are nonlinear and time varying in nature. Rigorous analysis is needed to show the desired performance of an adaptive control system. There are two main features of a stable adaptive control system: a feedback structure involving some critical system signals and an equivalent feedback loop gain which is small so that the closed-loop system is stable. Another important feature is that an error system used in adaptive control is dissipative so that a closed-loop error system is stable despite the fact that the dynamics of an adaptive law is an integration action (not asymptotically stable). In this section, we present some fundamental tools for analyzing the stability of adaptive control systems: the Bellman-Gronwall lemma (for feedback structure equivalence), a small-gain lemma (for feedback stability), some operator stability properties (for system analysis), and the Lefschetz-Kalman-Yakubovich lemma (for passivity and system matrices).

2.5.1 Bellman-Gronwall Lemma

There are different versions of the Bellman-Gronwall lemma [78], [151], [426]. Two of them are presented here with proofs, which are useful in stability analysis for different adaptive control schemes in the next chapters.

Lemma 2.1 *Let $f(t)$, $g(t)$, and $k(t)$ be continuous and $g(t) \geq 0$, $k(t) \geq 0$, $\forall t \geq t_0$ and for some $t_0 \geq 0$. If a continuous function $z(t)$ satisfies*

$$z(t) \leq f(t) + g(t) \int_{t_0}^{t} k(\tau) z(\tau)\, d\tau, \ \forall t \geq t_0, \tag{2.90}$$

then

$$z(t) \leq f(t) + g(t) \int_{t_0}^{t} k(\tau) f(\tau) e^{\int_\tau^t k(\sigma) g(\sigma)\, d\sigma}\, d\tau, \ \forall t \geq t_0. \tag{2.91}$$

Proof: Introducing $w(t) = \int_{t_0}^t k(\tau) z(\tau)\, d\tau$, and using (2.90) and the fact that $k(t) \geq 0$, we obtain

$$\dot{w}(t) = k(t) z(t) \leq k(t) f(t) + k(t) g(t) w(t). \tag{2.92}$$

Defining $\lambda(t) = w(t) e^{-\int_{t_0}^t k(\sigma) g(\sigma)\, d\sigma}$, and using (2.92), we have

$$\begin{aligned} \dot{\lambda}(t) &= \dot{w}(t) e^{-\int_{t_0}^t k(\sigma) g(\sigma)\, d\sigma} - w(t) k(t) g(t) e^{-\int_{t_0}^t k(\sigma) g(\sigma)\, d\sigma} \\ &\leq k(t) f(t) e^{-\int_{t_0}^t k(\sigma) g(\sigma)\, d\sigma}. \end{aligned} \tag{2.93}$$

Since $\lambda(t_0) = \omega(t_0) = 0$, from (2.93), it follows that

$$\lambda(t) \leq \int_{t_0}^{t} k(\tau) f(\tau) e^{-\int_{t_0}^\tau k(\sigma) g(\sigma)\, d\sigma}\, d\tau. \tag{2.94}$$

With the definition of $\lambda(t)$, (2.94) leads to

$$w(t) = \lambda(t) e^{\int_{t_0}^t k(\sigma) g(\sigma)\, d\sigma} \leq \int_{t_0}^{t} k(\tau) f(\tau) e^{\int_\tau^t k(\sigma) g(\sigma)\, d\sigma}\, d\tau, \tag{2.95}$$

which, with (2.90) for $g(t) \geq 0$ and the definition of $w(t)$, implies (2.91). ∇

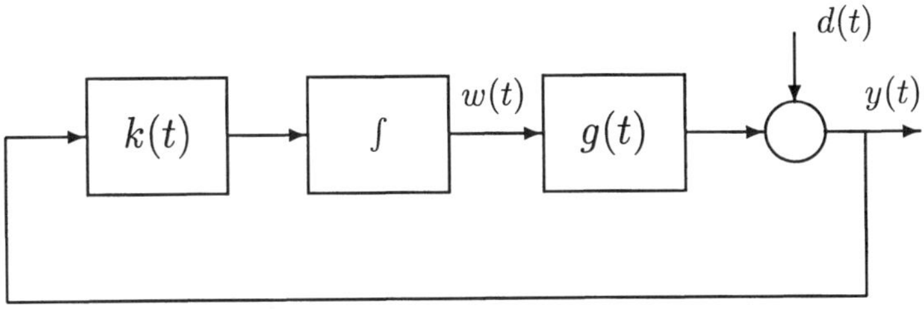

Figure 2.1: Feedback system diagram of Bellman-Gronwall lemma.

A control system block diagram can be used to illustrate Lemma 2.1, as shown in Figure 2.1, where $g(t) \in R$ and $k(t) \in R$, from which we have

$$\begin{aligned} y(t) &= d(t) + g(t)\left(w(t_0) + \int_{t_0}^{t} k(\tau)y(\tau)\,d\tau\right) \\ &= f(t) + g(t)\int_{t_0}^{t} k(\tau)y(\tau)\,d\tau \end{aligned} \tag{2.96}$$

for $f(t) = d(t) + g(t)w(t_0)$. From (2.96), similar to (2.90), we have

$$|y(t)| \leq |f(t)| + |g(t)| \int_{t_0}^{t} |k(\tau)|\,|y(\tau)|\,d\tau. \tag{2.97}$$

Lemma 2.2 *If $k(t)$ is continuous and $k(t) \geq 0$, $\forall t \geq t_0 \geq 0$, and a continuous function $z(t)$ satisfies*

$$z(t) \leq c + \int_{t_0}^{t} k(\tau)z(\tau)\,d\tau,\ \forall t \geq t_0, \tag{2.98}$$

where c is a constant, then

$$z(t) \leq c e^{\int_{t_0}^{t} k(\tau)\,d\tau},\ \forall t \geq t_0. \tag{2.99}$$

Proof: Introducing $w(t) = c + \int_{t_0}^{t} k(\tau)z(\tau)\,d\tau$ and $\lambda(t) = w(t)e^{-\int_{t_0}^{t} k(\sigma)\,d\sigma}$, with $\lambda(t_0) = w(t_0) = c$, from (2.98), we derive

$$\dot{\omega}(t) = k(t)z(t) \leq k(t)\omega(t), \tag{2.100}$$

$$\dot{\lambda}(t) = \dot{w}(t)e^{-\int_{t_0}^{t} k(\sigma)\,d\sigma} - w(t)k(t)e^{-\int_{t_0}^{t} k(\sigma)\,d\sigma} \leq 0. \tag{2.101}$$

Hence, $\lambda(t) \leq \lambda(t_0) = c$, and $z(t) \leq \omega(t) \leq c e^{\int_{t_0}^{t} k(\sigma)\,d\sigma}$. ∇

2.5.2 Small-Gain Lemma

From the Bellman-Gronwall lemma (Lemma 2.1), we have the following result.

Lemma 2.3 *If a continuous signal $z_1(t)$ satisfies*

$$z_1(t) \le b_1 + b_2 \int_0^t e^{-\alpha(t-\tau)} x_1(\tau) z_1(\tau)\, d\tau \tag{2.102}$$

for some $\alpha > 0$, $b_1 > 0$, $b_2 > 0$ and some signal $x_1(t) \ge 0$, $\forall t \ge 0$, then there exists a $\nu > 0$ such that $z_1(t)$ is bounded for

$$\int_{t_1}^{t_2} x_1^2(t)\, dt \le \nu(t_2 - t_1) + \gamma_0, \ \forall t_2 > t_1 \ge 0, \tag{2.103}$$

where $\gamma_0 > 0$ is any constant.

Proof [366]: Rewriting (2.102) as

$$e^{\alpha t} z_1(t) \le e^{\alpha t} b_1 + b_2 \int_0^t x_1(\tau) e^{\alpha \tau} z_1(\tau)\, d\tau, \tag{2.104}$$

setting $f(t) = e^{\alpha t} b_1$, $g(t) = b_2$, $k(t) = x_1(t)$, and $z(t) = e^{\alpha t} z_1(t)$, and using the Bellman-Gronwall lemma (Lemma 2.1), we obtain

$$e^{\alpha t} z_1(t) \le e^{\alpha t} b_1 + b_2 \int_0^t x_1(\tau) b_1 e^{\alpha \tau} e^{b_2 \int_\tau^t x_1(\sigma)\, d\sigma}\, d\tau, \tag{2.105}$$

which is equivalent to

$$z_1(t) \le b_1 + b_1 b_2 \int_0^t e^{-\alpha(t-\tau) + b_2 \int_\tau^t x_1(\sigma)\, d\sigma} x_1(\tau)\, d\tau. \tag{2.106}$$

Using the Schwarz inequality (see Section 2.3.2)

$$\int_\tau^t f(\sigma) g(\sigma)\, d\sigma \le \sqrt{\int_\tau^t f^2(\sigma)\, d\sigma} \sqrt{\int_\tau^t g^2(\sigma)\, d\sigma}, \tag{2.107}$$

with $f(t) = x_1(t)$ and $g(t) = 1$, and (2.103), we have

$$\begin{aligned} \int_\tau^t x_1(\sigma)\, d\sigma &\le \sqrt{\int_\tau^t d\sigma} \sqrt{\int_\tau^t x_1^2(\sigma)\, d\sigma} \le \sqrt{t-\tau} \sqrt{\nu(t-\tau) + \gamma_0} \\ &\le \sqrt{t - \tau + \frac{\gamma_0}{\nu}} \sqrt{\nu(t-\tau) + \gamma_0} = \sqrt{\nu}(t-\tau) + \frac{\gamma_0}{\sqrt{\nu}}. \end{aligned} \tag{2.108}$$

It follows from (2.106) and (2.108) that

$$z_1(t) \le b_1 + b_1 b_2 \int_0^t e^{-\alpha(t-\tau) + b_2 \sqrt{\nu}(t-\tau) + \frac{b_2 \gamma_0}{\sqrt{\nu}}} x_1(\tau)\, d\tau. \tag{2.109}$$

Choose $\nu \in (0, \left(\frac{\alpha}{b_2}\right)^2)$. Then $\alpha_1 = \alpha - b_2\sqrt{\nu} > 0$ and $e^{\frac{b_2\gamma_0}{\sqrt{\nu}}}$ is finite. Let $N \leq t$ be the nearest integer to t. Then, for such choice of ν,

$$\int_0^t e^{-\alpha(t-\tau)+b_2\sqrt{\nu}(t-\tau)+\frac{b_2\gamma_0}{\sqrt{\nu}}} x_1(\tau)\,d\tau = c_\nu z_2(t), \tag{2.110}$$

where $c_\nu = e^{\frac{b_2\gamma_0}{\sqrt{\nu}}}$ is a constant and

$$\begin{aligned} z_2(t) &= \int_0^t e^{-\alpha_1(t-\tau)} x_1(\tau)\,d\tau = \sum_{i=0}^{N-1}\int_i^{i+1} e^{-\alpha_1(t-\tau)} x_1(\tau)\,d\tau + \int_N^t e^{-\alpha_1(t-\tau)} x_1(\tau)\,d\tau \\ &\leq \sum_{i=0}^{N-1}\sqrt{\int_i^{i+1} e^{-2\alpha_1(t-\tau)}d\tau}\sqrt{\int_i^{i+1} x_1^2(\tau)\,d\tau} + \sqrt{\int_N^t e^{-2\alpha_1(t-\tau)}\,d\tau}\sqrt{\int_N^t x_1^2(\tau)\,d\tau} \\ &\leq \sum_{i=0}^{N-1} e^{-\alpha_1(t-i)}\sqrt{\frac{e^{2\alpha_1}-1}{2\alpha_1}}\sqrt{\nu+\gamma_0} + \sqrt{\frac{1-e^{-2\alpha_1(t-N)}}{2\alpha_1}}\sqrt{\nu(t-N)+\gamma_0} \\ &\leq \frac{1}{e^{\alpha_1}-1}\sqrt{\frac{e^{2\alpha_1}-1}{2\alpha_1}}\sqrt{\nu+\gamma_0} + \sqrt{\frac{1}{2\alpha_1}}\sqrt{\nu+\gamma_0}, \end{aligned} \tag{2.111}$$

that is, $z_2(t)$ is bounded. In view of (2.109), $z_1(t)$ is bounded. ∇

A feedback system can be used to illustrate Lemma 2.3, as shown in Figure 2.2, where $x_1(t) \in R$, $b_0 \in R$, $b_2 \in R$, and $\alpha > 0$, from which we have

$$y(t) = b_0 + e^{-\alpha t}x(0) + b_2\int_0^t e^{-\alpha(t-\tau)} x_1(\tau) y(\tau)\,d\tau, \tag{2.112}$$

$$|y(t)| \leq |b_0| + |x(0)| + |b_2|\int_0^t e^{-\alpha(t-\tau)} |x_1(\tau)|\,|y(\tau)|\,d\tau. \tag{2.113}$$

In view of (2.102), with $z_1(t) = |y(t)|$, we have

$$z_1(t) \leq |b_0| + |x(0)| + |b_2|\int_0^t e^{-\alpha(t-\tau)} |x_1(\tau)|\, z_1(\tau)\,d\tau. \tag{2.114}$$

A similar structure can be shown for a general system $\dot{x} = Ax + Bu$, $w = Cx$, replacing the block $\frac{1}{s+\alpha}$ in Figure 2.2, where, with $x_1(t)$, b_0, b_2 as before, $A \in R^{n\times n}$, $B \in R^n$, and $C \in R^{1\times n}$ are constant matrices with all eigenvalues of A being in $\mathrm{Re}[s] < 0$. In this case, we have

$$\begin{aligned} y(t) &= b_0 + w(t) = b_0 + Ce^{At}x(0) + C\int_0^t e^{A(t-\tau)} Bu(\tau)\,d\tau \\ &= b_0 + Ce^{At}x(0) + C\int_0^t e^{A(t-\tau)} Bb_2 x_1(\tau) y(\tau)\,d\tau \end{aligned} \tag{2.115}$$

$$\begin{aligned} |y(t)| &\leq |b_0| + \|C\|\, e^{-\alpha t}\, \|x(0)\| + \|C\|\int_0^t e^{-\alpha(t-\tau)} \|B\|\,|b_2|\,|x_1(\tau)|\,|y(\tau)|\,d\tau \\ &\leq \bar{b}_1 + \bar{b}_2\int_0^t e^{-\alpha(t-\tau)} |x_1(\tau)|\,|y(\tau)|\,d\tau \end{aligned} \tag{2.116}$$

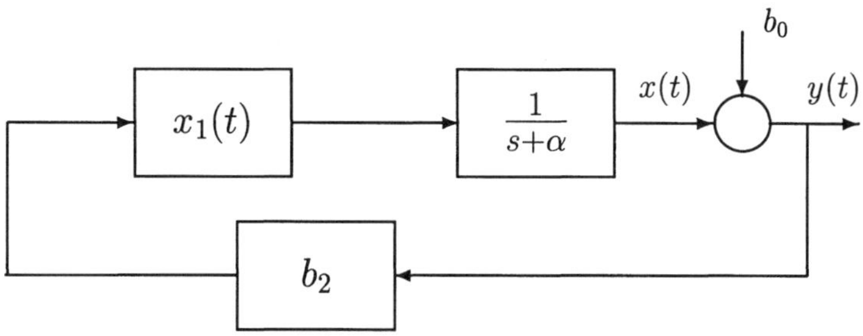

Figure 2.2: Feedback system diagram of Lemma 2.3 for (2.112).

for some $\alpha > 0$, where $\bar{b}_1 = |b_0| + \|C\| \, \|x(0)\|$ and $\bar{b}_2 = \|C\| \, \|B\| \, |b_2|$.

A special case of Lemma 2.2 is when $x_1(t) \in L^2$, that is, $\int_0^\infty x_1^2(t)\, dt \leq \gamma_0$. In this case, (2.103) is satisfied for any $\nu > 0$, so $z_1(t)$ is bounded.

Corollary 2.3 *If $x_1(t) \in L^2$, then the signal $z_1(t)$ is bounded.*

Corollary 2.4 *If $x_1(t) \in L^2$, then $\int_0^t e^{-\alpha(t-\tau)} x_1(\tau)\, d\tau$ is bounded for $\alpha > 0$, and so is $H(s)[x_1](t)$ for any strictly proper rational transfer function $H(s)$ whose poles are all in* $\mathrm{Re}[s] < 0$.

In fact, for $x_1(t) \in L^2$, both the signals $\int_0^t e^{-\alpha(t-\tau)} x_1(\tau)\, d\tau$ and $H(s)[x_1](t)$ converge to zero as $t \to \infty$ (see Lemmas 2.18 and 2.19 in Section 2.6).

The inequality (2.102) presents a feedback structure for the signal $z_1(t)$, with an equivalent loop gain $\frac{b_2}{\alpha}\sqrt{\nu} < 1$ for $\nu \in (0, (\frac{\alpha}{b_2})^2)$, so Lemma 2.3 (or Lemma 2.4) is essentially a small-gain theorem, which unifies some main stability properties of almost all continuous-time adaptive control systems.

Lemma 2.4 *If a continuous signal $z_1(t)$ satisfies*

$$z_1(t) \leq b_1 + b_2 \int_0^t e^{-\alpha(t-\tau)} x_1(\tau) z_1(\tau)\, d\tau \tag{2.117}$$

for some $\alpha > 0$, $b_1 > 0$, $b_2 > 0$ and some signal $x_1(t) \geq 0$, $\forall t \geq 0$, then there exists a $\nu > 0$ such that $z_1(t)$ is bounded for

$$\int_{t_1}^{t_2} |x_1(t)|^p\, dt \leq \nu(t_2 - t_1) + \gamma_0, \ \forall t_2 > t_1 \geq 0, \tag{2.118}$$

where $p \in (1, \infty)$ and $\gamma_0 > 0$ is any constant.

Proof: As from the proof of Lemma 2.3, we also have (2.106). For $p \in (1,\infty)$, using the Hölder inequality (see Section 2.3.2)

$$\int_{\tau}^{t} f(\sigma)g(\sigma)\,d\sigma \leq \left(\int_{\tau}^{t} |f(\sigma)|^p\,d\sigma\right)^{\frac{1}{p}} \left(\int_{\tau}^{t} |g(\sigma)|^q\,d\sigma\right)^{\frac{1}{q}} \tag{2.119}$$

for $\frac{1}{p}+\frac{1}{q}=1$, with $f(t)=x_1(t)$ and $g(t)=1$, and (2.118), we have

$$\begin{aligned}\int_{\tau}^{t} x_1(\sigma)\,d\sigma &\leq (t-\tau)^{\frac{1}{q}}(\nu(t-\tau)+\gamma_0)^{\frac{1}{p}} \\ &\leq \left(t-\tau+\frac{\gamma_0}{\nu}\right)^{\frac{1}{q}}(\nu(t-\tau)+\gamma_0)^{\frac{1}{p}} \\ &\leq \nu^{\frac{1}{p}}\left(t-\tau+\frac{\gamma_0}{\nu}\right) = \nu^{\frac{1}{p}}(t-\tau)+\gamma_0\nu^{-\frac{1}{q}}.\end{aligned} \tag{2.120}$$

From (2.106) and (2.120), we obtain

$$z_1(t) \leq b_1 + b_1 b_2 \int_0^t e^{-\alpha(t-\tau)+b_2\nu^{\frac{1}{p}}(t-\tau)+b_2\gamma_0\nu^{-\frac{1}{q}}} x_1(\tau)\,d\tau. \tag{2.121}$$

Choose $\nu \in (0, \left(\frac{\alpha}{b_2}\right)^p)$. Then, $\alpha_1 = \alpha - b_2\nu^{\frac{1}{p}} > 0$ and $e^{b_2\gamma_0\nu^{-\frac{1}{q}}}$ is finite.

Similar to (2.111), using the Hölder inequality, we derive

$$\begin{aligned}\int_i^{i+1} e^{-\alpha_1(t-\tau)}x_1(\tau)\,d\tau &\leq e^{-\alpha_1 t}\left(\int_i^{i+1} e^{q\alpha_1\tau}\,d\tau\right)^{\frac{1}{q}}\left(\int_i^{i+1}|x_1(\tau)|^p\,d\tau\right)^{\frac{1}{p}} \\ &\leq e^{-\alpha_1 t}e^{\alpha_1 i}\left(\frac{e^{q\alpha_1}-1}{q\alpha_1}\right)^{\frac{1}{q}}(\nu+\gamma_0)^{\frac{1}{p}}\end{aligned} \tag{2.122}$$

for any $i \leq t$, based on which we can prove that $z_2(t) = \int_0^t e^{-\alpha_1(t-\tau)}x_1(\tau)\,d\tau$ is bounded so that $z_1(t)$ in (2.121) is bounded. ∇

A special case of Lemma 2.4 is when $x_1(t) \in L^p$, $p \in (1,\infty)$, that is, $\int_0^\infty |x_1(t)|^p\,dt \leq \gamma_0$, that is, (2.103) is met for any $\nu > 0$, so $z_1(t)$ is bounded.

Corollary 2.5 *If $x_1(t) \in L^p$, $p \in (1,\infty)$, then the signal $z_1(t)$ is bounded.*

Corollary 2.6 *If $x_1(t) \in L^p$, $p \in (1,\infty)$, then $\int_0^t e^{-\alpha(t-\tau)}x_1(\tau)\,d\tau$ is bounded for $\alpha > 0$, and so is $H(s)[x_1](t)$ for any strictly proper rational transfer function $H(s)$ whose poles are all in* $\mathrm{Re}[s] < 0$.

When $x_1(t) \in L^1$, that is, $\int_0^\infty |x_1(t)|\,dt \leq \gamma_0$, from (2.106), $z_1(t)$ is bounded.

In fact, for $x_1(t) \in L^p$, $p \in [1,\infty)$, both the signals $\int_0^t e^{-\alpha(t-\tau)}x_1(\tau)\,d\tau$ and $H(s)[x_1](t)$ converge to zero as $t \to \infty$ (see Section 2.6).

Questions: How about the case when $\int_{t_1}^{t_2} |x_1(t)|\,dt \leq \nu(t_2-t_1)+\gamma_0$ with $\nu > 0$, in (2.118)? How about the case when $p \in (0,1)$ in (2.118)?

2.5.3 Operator Stability

An operator concept is the extension of that of a transfer function for linear time-invariant systems to linear time-varying systems. A linear time-invariant system has an input–output representation $y(s) = G(s)u(s)$, where $G(s)$ is the system transfer function and $u(s)$ and $y(s)$ are the Laplace transforms of the system input signal $u(t)$ and output signal $y(t)$. Let $g(t) = \mathcal{L}^{-1}[G(s)]$, the inverse Laplace transform of $G(s)$, or the impulse response function of the system represented by $G(s)$. Then, the system output is the convolution

$$y(t) = \int_0^t g(t-\tau)u(\tau)\,d\tau. \tag{2.123}$$

If we consider $G(s)$ as an operator whose operation is defined by (2.123), we can conveniently express the system $y(s) = G(s)u(s)$ as

$$y(t) = G(s)[u](t), \tag{2.124}$$

that is, $G(s)[u](t)$ denotes the output of a system whose operator representation (transfer function) is $G(s)$ and input is $u(t)$.

While a transfer function is for a linear time-invariant system, an operator for a linear time-varying system (2.6) can be defined by (2.16), that is,[12]

$$y(t) = \int_{t_0}^t g(t,\tau)u(\tau)\,d\tau, \tag{2.125}$$

where $g(t,\tau)$ is the system impulse response function. The linear time-varying systems we first consider are those in the form

$$y(t) = \sum_{i=1}^{p} \phi_i(t)H_i(s)[u](t), \tag{2.126}$$

where ϕ_i, $i = 1, 2, \ldots, p$, are time-varying parameters and $H_i(s)$, $i = 1, \ldots, p$, are operators (transfer functions) of linear time-invariant systems. Clearly, system (2.126) can be written as

$$y(t) = \sum_{i=1}^{p} \phi_i(t)H_i(s)[u](t) = \left(\sum_{i=1}^{p} \phi_i(t)H_i(s)\right)[u](t) \triangleq H(s,t)[u](t), \tag{2.127}$$

where $H(s,t) = \sum_{i=1}^{p} \phi_i(t)H_i(s)$ is the system operator. Now if the signal $y(t)$ is considered as the input to another linear time-invariant system with transfer function $G(s)$, then the output of this system is

$$z(t) = G(s)[y](t) = G(s)[H(s,\cdot)[u]](t) \triangleq T(s,\cdot)[u](t). \tag{2.128}$$

This operator $T(s,t)$ is well defined and it is clearly a linear operator.

[12]For operator representations, the system's initial conditions are assumed to be zero.

Example 2.1 Consider $y(t) = \phi(t)H(s)[u](t)$ and $z(t) = G(s)[y](t)$. With $h(t) = \mathcal{L}^{-1}[H(s)]$ and $g(t) = \mathcal{L}^{-1}[G(s)]$, we have $y(t) = \phi(t)\int_0^t h(t-\tau)u(\tau)\,d\tau$ and $z(t) = \int_0^t g(t-\tau)y(\tau)\,d\tau$. Hence, it follows that

$$z(t) = \int_0^t g(t-\tau)\phi(\tau)\int_0^\tau h(\tau-\sigma)u(\sigma)\,d\sigma\,d\tau. \tag{2.129}$$

Clearly, this expression defines a linear operator. □

Bounded-input–bounded-output stability. A system is said to be bounded-input–bounded-output (BIBO) stable if every bounded input results in a bounded output. A system represented by a transfer function $G(s)$ is BIBO stable if and only if all poles (i.e., complex numbers s_i, $i = 1, 2, \ldots, n$, that make $G(s_i) = \infty$) are in the open left-half of the s-plane: $\mathrm{Re}[s] < 0$.

Consider the case when $q = M = 1$ in (2.12), that is, both $y(t)$ and $u(t)$ are scalar functions, and denote $G(s) = C(sI - A)^{-1}B + D$ and $G_0(s) = C(sI - A)^{-1}B$. If $G(s)$ is BIBO stable (and so is $G_0(s)$), then

$$|g_0(t)| = |\mathcal{L}^{-1}[G_0(s)| \leq \beta e^{-\alpha t} \tag{2.130}$$

for some constants $\beta > 0$ and $\alpha > 0$, and

$$|y(t)| = \int_0^t \beta e^{-\alpha(t-\tau)}|u(\tau)|\,d\tau + \gamma|u(t)|, \tag{2.131}$$

where $\gamma = |D|$. This motivates us to define the following operator stability.

Definition 2.23 *An operator $T(s,t)$ is linear if, for any constants α_1, α_2 and signals $x_1(t)$, $x_2(t)$, $T(s,\cdot)[\alpha_1 x_1 + \alpha_2 x_2](t) = \alpha_1 T(s,\cdot)[x_1](t) + \alpha_2 T(s,\cdot)[x_2](t)$. A linear operator $T(s,t)$ is stable and proper if for some constants $\beta \geq 0$, $\alpha > 0$, and $\gamma > 0$, all $t \geq 0$, $|T(s,\cdot)[x](t)| \leq \beta\int_0^t e^{-\alpha(t-\tau)}|x(\tau)|\,d\tau + \gamma|x(t)|$, where $T(s,\cdot)[x](t)$ denotes the convolution of the impulse response of $T(s,\cdot)$ with $x(\cdot)$ at t. A linear operator $T(s,t)$ is stable and strictly proper if it is stable with $\gamma = 0$.*

Stability of cascade systems. For two BIBO stable transfer functions $H_1(s)$ and $H_2(s)$, their cascade systems $H_1(s)H_2(s)$ and $H_2(s)H_1(s)$ are also BIBO stable. For two linear time-varying operators, we have the following result, which can be easily verified by definition.

Proposition 2.8 *For two stable linear time-varying operators $T_1(s,t)$ and $T_2(s,t)$, their cascade operators $T_1(s,t) \circ T_2(s,t)$, $T_2(s,t) \circ T_1(s,t)$ are stable. The cascade operators are strictly proper if one of them is strictly proper.*

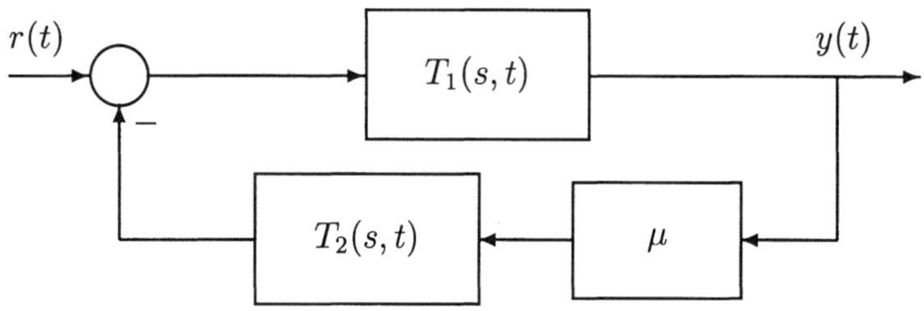

Figure 2.3: Feedback control system with linear operators.

Stability of feedback systems. For two BIBO stable transfer functions $H_1(s)$ and $H_2(s)$, their feedback system $(1+\mu H_1(s)H_2(s))^{-1}H_1(s)$ is also BIBO stable for $\mu > 0$ small. For two linear time-varying operators in a feedback loop shown in Figure 2.3, we have the following result.

Proposition 2.9 *For two stable linear time-varying operators $T_1(s,t)$ and $T_2(s,t)$, their feedback connection in Figure 2.3 is stable for $\mu > 0$ small.*

Next we present a stability result for time-varying operators in a feedback structure, which are some key components of an adaptive control system. We first introduce two filters $K_1(s)$ and $K(s)$ as

$$sK_1(s) = 1 - K(s),\ K(s) = \frac{a^{n^*}}{(s+a)^{n^*}}, \tag{2.132}$$

where $a > 0$ is a real number and $n^* > 0$ is an integer, and a bounded time-varying parameter vector $\theta_1(t) \in R^{\bar{n}}$ for some integer $\bar{n} > 0$, and let

$$\Lambda(s) = s^{\bar{n}-1} + \lambda_{\bar{n}-2}s^{n-2} + \cdots + \lambda_1 s + \lambda_0 \tag{2.133}$$

be a stable polynomial and $a(s)$ be a polynomial vector defined as

$$a(s) = [1, s, \ldots, s^{\bar{n}-2}]^T. \tag{2.134}$$

Lemma 2.5 *There exists $a^0 > 0$ such that for any fixed $a > a^0$ the operator*

$$T_0(s,t) \triangleq \left(1 + K_1(s)\left(a_0 - \theta_1^T(t)\frac{a(s)}{\Lambda(s)}(s+a_0)\right)\right)^{-1} \tag{2.135}$$

is stable and proper, where $a_0 > 0$ is a constant.

To prove this lemma, we need the following results.

Proposition 2.10 *The filter $K_1(s)$ defined in (2.132) has the expression*

$$K_1(s) = \frac{1}{s+a}\left(1 + \frac{a}{s+a} + \cdots + \frac{a^{n^*-1}}{(s+a)^{n^*-1}}\right). \tag{2.136}$$

Proof: Using the geometric sum $\sum_{i=0}^{k} q^i = \frac{1-q^{k+1}}{1-q}$, we have

$$\begin{aligned} &1 + \frac{a}{s+a} + \cdots + \frac{a^{n^*-1}}{(s+a)^{n^*-1}} \\ &= \frac{1 - \frac{a^{n^*}}{(s+a)^{n^*}}}{1 - \frac{a}{s+a}} = \frac{s+a}{s}\,\frac{(s+a)^{n^*} - a^{n^*}}{(s+a)^{n^*}} = (s+a)K_1(s), \end{aligned} \tag{2.137}$$

where the last equality follows from the definition (2.132).[13] ∇

Lemma 2.6 *If $x_1(t) = H(s)\frac{1}{s+\delta}[x_2](t)$, with $\delta > 0$, and the impulse response function $h_\delta(t)$ of $H(s-\delta)$ satisfies $\|h_\delta(\cdot)\|_1 = \int_0^\infty |h_\delta(t)|\,dt < \infty$, then*

$$|x_1(t)| \le \|h_\delta(\cdot)\|_1 \frac{1}{s+\delta}[|x_2|](t). \tag{2.138}$$

Proof: Introducing $x_3(t) = e^{\delta t}x_2(t)$, we have

$$\begin{aligned} |x_1(t)| &\le \frac{1}{s+\delta}[|H(s)[x_2]|](t) = \int_0^t e^{-\delta(t-\tau)}|H(s)[x_2](\tau)|\,d\tau \\ &= e^{-\delta t}\int_0^t |e^{\delta\tau}H(s)[x_2](\tau)|\,d\tau = e^{-\delta t}\int_0^t |H(s-\delta)[x_3](\tau)|\,d\tau \\ &\le e^{-\delta t}\|h_\delta(\cdot)\|_1 \int_0^t e^{\delta t}|x_2(\tau)|\,d\tau = \|h_\delta(\cdot)\|_1 \int_0^t e^{-\delta(t-\tau)}|x_2(\tau)|\,d\tau, \end{aligned} \tag{2.139}$$

which is (2.138) (in the third equality, we used the Laplace transform property $\mathcal{L}[e^{\delta t}h(t)] = H(s-\delta)$ and the expression $H(s-\delta)[x_3](\tau) = \int_0^\tau e^{\delta(\tau-\sigma)}h(\tau-\sigma)x_3(\sigma)\,d\sigma = e^{\delta\tau}\int_0^\tau h(\tau-\sigma)x_2(\sigma)\,d\sigma = e^{\delta\tau}H(s)[x_2](\tau)$; for the second inequality, see Section 2.8 for an operator norm $\|h_\delta(\cdot)\|_1$). ∇

Proposition 2.11 *For $k = 0, 1, 2, \ldots, n^*-1$,*

$$\left|\frac{1}{s+a}\,\frac{a^k}{(s+a)^k}[x](t)\right| \le 2^k \frac{1}{s+\frac{a}{2}}[|x|](t). \tag{2.140}$$

[13] Using $\int_0^\infty e^{-at}t^{n^*-i}\,dt = \frac{(n^*-i)!}{a^{n^*-i}a}$ $\left(\int e^{-at}t^{n^*-i}\,dt = \frac{t^{n^*-i}}{-a}e^{-at} + \frac{n^*-i}{a}\int e^{-at}t^{n^*-i-1}\,dt\right)$ and $k_1(t) = \mathcal{L}^{-1}[K_1(s)] = e^{-at}\sum_{i=1}^{n^*}\frac{a^{n^*-i}}{(n^*-i)!}t^{n^*-i}$, we have $\|k_1(\cdot)\|_1 = \int_0^\infty |k_1(t)|\,dt = \frac{n^*}{a}$.

Proof: Applying Lemma 2.6, we have

$$\begin{aligned}
&\left|\frac{1}{s+a}\frac{a^k}{(s+a)^k}[x](t)\right| \\
&\leq \frac{1}{s+a}\frac{a^k}{(s+a)^k}[|x|](t) \leq \frac{1}{s+\frac{a}{2}}\frac{a^k}{(s+a)^k}[|x|](t) \\
&\leq \left\|\mathcal{L}^{-1}\left[\frac{a^k}{(s-\frac{a}{2}+a)^k}\right](\cdot)\right\|_1 \frac{1}{s+\frac{a}{2}}[|x|](t) \leq 2^k\frac{1}{s+\frac{a}{2}}[|x|](t), \quad (2.141)
\end{aligned}$$

where $\|\mathcal{L}^{-1}[\frac{a^k}{(s+\frac{a}{2})^k}](\cdot)\|_1 = \frac{a^k}{(\frac{a}{2})^k} = 2^k$. ∇

Proposition 2.12 *For $k = 0, 1, 2, \ldots, n^* - 1$ and a constant $\alpha_1 > 0$,*

$$\left|\frac{1}{s+\alpha_1}\frac{1}{s+a}\frac{a^k}{(s+a)^k}[x](t)\right| \leq 2^k\frac{1}{s+\alpha_1}\frac{1}{s+\frac{a}{2}}[|x|](t). \quad (2.142)$$

Proof: Applying Lemma 2.6, with $v(t) = \frac{1}{s+\alpha_1}[|x|](t)$, we have

$$\begin{aligned}
&\left|\frac{1}{s+\alpha_1}\frac{1}{s+a}\frac{a^k}{(s+a)^k}[x](t)\right| \\
&\leq \frac{1}{s+\alpha_1}\frac{1}{s+a}\frac{a^k}{(s+a)^k}[|x|](t) \leq \frac{1}{s+\frac{a}{2}}\frac{a^k}{(s+a)^k}\frac{1}{s+\alpha_1}[|x|](t) \\
&\leq \left\|\mathcal{L}^{-1}\left[\frac{a^k}{(s-\frac{a}{2}+a)^k}\right](\cdot)\right\|_1 \frac{1}{s+\frac{a}{2}}[v](t) \\
&\leq 2^k\frac{1}{s+\frac{a}{2}}[v](t) = 2^k\frac{1}{s+\frac{a}{2}}\frac{1}{s+\alpha_1}[|x|](t). \quad (2.143)
\end{aligned}$$

Hence, we have proved the proposition. ∇

Proposition 2.13 *There exist constants $\bar{c}_1 > 0$, $\bar{c}_2 > 0$ such that*

$$|K_1(s)[x](t)| \leq \frac{\bar{c}_1}{s+\frac{a}{2}}[|x|](t), \quad (2.144)$$

$$\left|K_1(s)\frac{1}{s+\alpha_1}[x](t)\right| \leq \frac{\bar{c}_2}{(s+\frac{a}{2})(s+\alpha_1)}[|x|](t). \quad (2.145)$$

Proof: Applying Propositions 2.11 and 2.12, we have

$$|K_1(s)[x](t)| \leq (1 + 2 + 2^2 + \cdots + 2^{n^*-1})\frac{1}{s+\frac{a}{2}}[|x|](t). \quad (2.146)$$

Hence, with $\bar{c}_1 = 1 + 2 + 2^2 + \cdots + 2^{n^*-1} = 2^{n^*} - 1$, we have (2.144). In a similar way, we can derive (2.145) for $\bar{c}_1 = \bar{c}_2$. ∇

Proof of Lemma 2.4: Letting $z_0(t)$ and $x(t)$ satisfy

$$\left(1 + K_1(s)\left(a_0 - \theta_1^T(t)\frac{a(s)}{\Lambda(s)}(s + a_0)\right)\right)[z_0](t) = x(t), \tag{2.147}$$

denoting parameter vectors

$$\theta_1(t) = [\theta_{11}(t), \theta_{12}(t), \ldots, \theta_{1\,\bar{n}-1}(t)]^T \in R^{\bar{n}-1}, \tag{2.148}$$

$$\theta_1^s(t) = [0, \theta_{11}(t), \theta_{12}(t), \ldots, \theta_{1\,\bar{n}-2}(t)]^T \in R^{\bar{n}-1}, \tag{2.149}$$

and $\lambda = [\lambda_0, \lambda_1, \ldots, \lambda_{\bar{n}-2}]^T \in R^{\bar{n}-1}$, and introducing new parameters

$$\bar{\theta}_1(t) = \theta_{1\,\bar{n}-1}(t)\lambda - a_0\theta_1(t) - \theta_1^s(t),\ \bar{a}_0(t) = a_0 - \theta_{1\,\bar{n}-1}(t) \tag{2.150}$$

(it follows that $\bar{\theta}_1(t) \in L^\infty$, $\bar{a}_0(t) \in L^\infty$), we rewrite (2.147) as

$$\left(1 + K_1(s)\bar{a}_0(\cdot) + K_1(s)\bar{\theta}_1^T(\cdot)\frac{a(s)}{\Lambda(s)}\right)[z_0](t) = x(t), \tag{2.151}$$

from which it follows that

$$|z_0(t)| \le c_1K_1(s)[|z_0|](t) + c_2K_1(s)\frac{1}{s + \alpha_1}[|z_0|](t) + |x(t)| \tag{2.152}$$

for some constants $c_1 > 0$, $c_2 > 0$, $\alpha_1 > 0$. Using Proposition 2.13, we have

$$|z_0(t)| \le \frac{c_1\bar{c}_1}{s + \frac{a}{2}}[|z_0|](t) + \frac{c_2\bar{c}_2}{(s + \frac{a}{2})(s + \alpha_1)}[|z_0|](t) + |x(t)|. \tag{2.153}$$

Defining the signal $\bar{z}_0(t)$ from

$$\bar{z}_0(t) = \frac{c_1\bar{c}_1}{s + \frac{a}{2}}[\bar{z}_0](t) + \frac{c_2\bar{c}_2}{(s + \frac{a}{2})(s + \alpha_1)}[\bar{z}_0](t) + |x(t)|, \tag{2.154}$$

it follows that

$$|z_0(t)| - \bar{z}_0(t) \le \frac{c_1\bar{c}_1}{s + \frac{a}{2}}[|z_0| - \bar{z}_0](t) + \frac{c_2\bar{c}_2}{(s + \frac{a}{2})(s + \alpha_1)}[|z_0| - \bar{z}_0](t). \tag{2.155}$$

Let the impulse response function of $\frac{c_1\bar{c}_1}{s+\frac{a}{2}} + \frac{c_2\bar{c}_2}{(s+\frac{a}{2})(s+\alpha_1)}$ be bounded by $\gamma e^{-\beta t}$; from Lemma 2.2 with $u(t) = e^{\beta t}(|z_0(t)| - \bar{z}_0(t))$ and $c = 0$, we have

$$u(t) = e^{\beta t}(|z_0(t)| - \bar{z}_0(t)) \le 0 \ \Rightarrow |z_0(t)| \le \bar{z}_0(t). \tag{2.156}$$

Now system (2.154) can be written as

$$\bar{z}_0(s) = \left(1 - \frac{c_1\bar{c}_1}{s + \frac{a}{2}} - \frac{c_2\bar{c}_2}{(s + \frac{a}{2})(s + \alpha_1)}\right)^{-1}|x|(s), \tag{2.157}$$

where $|x|(s)$ denotes the Laplace transform of $|x(t)|$. This system is linear time invariant, and there exists a constant $a^0 > 0$ such that for any fixed $a > a^0$ the system is BIBO stable (i.e., all poles of the transfer function $\left(1 - \frac{c_1\bar{c}_1}{s+\frac{a}{2}} - \frac{c_2\bar{c}_2}{(s+\frac{a}{2})(s+\alpha_1)}\right)^{-1}$ are in $\mathrm{Re}[s] < 0$) so that

$$|\bar{z}_0(t)| \leq \beta \int_0^t e^{-\alpha(t-\tau)}|x(\tau)|\,d\tau + \gamma|x(t)| \tag{2.158}$$

for some constants $\beta \geq 0$, $\alpha > 0$, and $\gamma > 0$. Finally, from (2.156), we have $|z_0(t)| \leq \bar{z}_0(t) = |\bar{z}_0(t)|$, which, together with (2.158), implies that the operator in (2.135), defined in the sense of (2.147), is stable (see Definition 2.22) with the parameter $a > a^0$ for some $a^0 > 0$. ∇

The following input–output L^p stability properties of a stable operator are useful in stability analysis of adaptive control systems.

Lemma 2.7 *If $u(t) \in L^p$, $p \in [1, \infty]$, then, for $\alpha > 0$, $\frac{1}{s+\alpha}[u](t) \in L^p$.*

Proof: Express $y(t) \triangleq \frac{1}{s+\alpha}[u](t)$ in the convolution form

$$y(t) = \int_0^t e^{-\alpha(t-\tau)}u(\tau)\,d\tau = \int_0^\infty e^{-\alpha(t-\tau)}u_s(t-\tau)u(\tau)\,d\tau, \tag{2.159}$$

where $u_s(t) = 1$ for $t \geq 0$ and $u_s(t) = 0$ for $t < 0$, the unit step function.

For $p = 1$, we have

$$\begin{aligned}
\|y(\cdot)\|_1 &= \int_0^\infty |y(t)|\,dt = \int_0^\infty \left|\int_0^\infty e^{-\alpha(t-\tau)}u_s(t-\tau)u(\tau)\,d\tau\right| dt \\
&\leq \int_0^\infty \int_0^\infty e^{-\alpha(t-\tau)}u_s(t-\tau)|u(\tau)|\,d\tau\,dt \\
&= \int_0^\infty \int_0^\infty e^{-\alpha(t-\tau)}u_s(t-\tau)\,dt\,|u(\tau)|\,d\tau = \int_0^\infty \int_\tau^\infty e^{-\alpha\sigma}u_s(\sigma)\,d\sigma|u(\tau)|\,d\tau \\
&= \int_0^\infty \int_0^\infty e^{-\alpha\sigma}\,d\sigma|u(\tau)|\,d\tau = \int_0^\infty \frac{1}{\alpha}|u(\tau)|\,d\tau = \frac{1}{\alpha}\|u(\cdot)\|_1
\end{aligned} \tag{2.160}$$

(the iterated integral equals the double integral as the iterated integral exists, see pp. 233 and 240 of [78]), so $y(t) \in L^1$ if $u(t) \in L^1$.

For $p \in (1, \infty)$, we have

$$\begin{aligned}
|y(t)| &\leq \int_0^\infty e^{-\alpha(t-\tau)}u_s(t-\tau)|u(\tau)|\,d\tau \\
&= \int_0^\infty \left(e^{-\alpha(t-\tau)}u_s(t-\tau)\right)^{\frac{1}{p}}|u(\tau)|\left(e^{-\alpha(t-\tau)}u_s(t-\tau)\right)^{\frac{1}{q}}\,d\tau
\end{aligned}$$

$$
\begin{aligned}
&\le \left(\int_0^\infty e^{-\alpha(t-\tau)}u_s(t-\tau)|u(\tau)|^p\,d\tau\right)^{\frac{1}{p}}\left(\int_0^\infty e^{-\alpha(t-\tau)}u_s(t-\tau)\,d\tau\right)^{\frac{1}{q}}\\
&\le \left(\int_0^\infty e^{-\alpha(t-\tau)}u_s(t-\tau)|u(\tau)|^p\,d\tau\right)^{\frac{1}{p}}\left(\frac{1}{\alpha}\right)^{\frac{1}{q}},\\
\|y(\cdot)\|_p &= \left(\int_0^\infty |y(t)|^p\,dt\right)^{\frac{1}{p}}\\
&\le \left(\int_0^\infty\int_0^\infty e^{-\alpha(t-\tau)}u_s(t-\tau)|u(\tau)|^p\,d\tau\,dt\right)^{\frac{1}{p}}\left(\frac{1}{\alpha}\right)^{\frac{1}{q}}\\
&= \left(\int_0^\infty\int_0^\infty e^{-\alpha(t-\tau)}u_s(t-\tau)dt\,|u(\tau)|^p\,d\tau\right)^{\frac{1}{p}}\left(\frac{1}{\alpha}\right)^{\frac{1}{q}}\\
&= \left(\frac{1}{\alpha}\right)^{\frac{1}{p}}\left(\int_0^\infty |u(\tau)|^p\,d\tau\right)^{\frac{1}{p}}\left(\frac{1}{\alpha}\right)^{\frac{1}{q}} = \frac{1}{\alpha}\|u(\cdot)\|_p, \qquad (2.161)
\end{aligned}
$$

so $y(t) \in L^p$ if $u(t) \in L^p$ for $p \in (1, \infty)$.

For $p = \infty$, we have

$$
\begin{aligned}
\|y(\cdot)\|_\infty &= \sup_{t\ge 0}|y(t)| = \sup_{t\ge 0}\int_0^t e^{-\alpha(t-\tau)}|u(\tau)|\,d\tau\\
&\le \sup_{t\ge 0}\int_0^t e^{-\alpha(t-\tau)}\,d\tau\ \sup_{t\ge 0}|u(t)| = \frac{1}{\alpha}\|u(\cdot)\|_\infty, \qquad (2.162)
\end{aligned}
$$

so $y(t) \in L^\infty$ if $u(t) \in L^\infty$. ∇

Corollary 2.7 *If $u(t) \in L^p$, $p \in [1, \infty]$, then $y(t) = H(s)[u](t) \in L^p$ for any stable and proper transfer function $H(s)$ or $y(t) = T(s,t)[u](t) \in L^p$ for any stable and proper operator $T(s,t)$.*

Proof: Under the condition on $H(s)$ or $T(s,t)$, we have

$$|y(t)| \le \frac{\beta}{s+\alpha}[|u|](t) + \gamma|u(t)| \qquad (2.163)$$

for some constants $\alpha > 0$, $\beta > 0$, and $\gamma > 0$. Using Lemma 2.7, we have $y(t) \in L^p$ for $p = 1$ or $p = \infty$ directly and $y(t) \in L^p$ for $p \in (1, \infty)$, with the help of the Minkovski inequality (see Section 2.3.2). ∇

Similar results are given in [78] for the more general case, when

$$y(t) = \int_0^\infty w(t-\tau)u(\tau)\,d\tau, \qquad (2.164)$$

where $w(t) \in L^1$: $\|w(\cdot)\|_1 \triangleq \int_0^\infty |w(t)|\,dt < \infty$. In this case, if $u(t) \in L^p$, $p \in [1, \infty]$, then $y(t) \in L^p$. This result can be proved by replacing $e^{-\alpha(t-\tau)}u_s(t-\tau)$ with $w(t-\tau)$ in the proof of Lemma 2.7.

The following result is useful in robust adaptive control analysis.

Lemma 2.8 *If* $y(t) = \frac{1}{s+\alpha}[u](t)$, *with* $\alpha > 0$, *and*

$$\int_{t_1}^{t_2} u^2(t)\, dt \le \nu(t_2 - t_1) + \gamma_0 \tag{2.165}$$

for any $t_2 > t_1 \ge 0$ *and some* $\nu > 0$ *and* $\gamma_0 > 0$, *then*

$$\int_{t_1}^{t_2} y^2(t)\, dt \le c_0\nu(t_2 - t_1) + \gamma_1 \tag{2.166}$$

for some constants $c_0 > 0$ *and* $\gamma_1 > 0$.

Proof: From $y(t) = \int_0^t e^{-\alpha(t-\tau)}u(\tau)\, d\tau$, we have

$$\int_{t_1}^{t_2} y^2(t)\, dt \le \int_{t_1}^{t_2} \left(\int_0^t e^{-\alpha(t-\tau)}|u(\tau)|\, d\tau\right)^2 dt$$
$$\le 2\int_{t_1}^{t_2} \left(\int_{t_1}^t e^{-\alpha(t-\tau)}|u(\tau)|\, d\tau\right)^2 dt + 2\int_{t_1}^{t_2} \left(\int_0^{t_1} e^{-\alpha(t-\tau)}|u(\tau)|\, d\tau\right)^2 dt. \tag{2.167}$$

The first term in (2.167), with Schwarz inequality, satisfies

$$\int_{t_1}^{t_2} \left(\int_{t_1}^t e^{-\alpha(t-\tau)}|u(\tau)|\, d\tau\right)^2 dt$$
$$= \int_{t_1}^{t_2} \left(\int_{t_1}^{t_2} e^{-\alpha(t-\tau)}u_s(t-\tau)|u(\tau)|\, d\tau\right)^2 dt$$
$$= \int_{t_1}^{t_2} \left(\left(\int_{t_1}^{t_2} e^{-\alpha(t-\tau)}u_s(t-\tau)\, d\tau\right)^{\frac{1}{2}} \left(\int_{t_1}^{t_2} e^{-\alpha(t-\tau)}u_s(t-\tau)|u(\tau)|^2\, d\tau\right)^{\frac{1}{2}}\right)^2 dt$$
$$= \int_{t_1}^{t_2} \frac{1}{\alpha}\left(1 - e^{-\alpha(t-t_1)}\right)\left(\int_{t_1}^{t_2} e^{-\alpha(t-\tau)}u_s(t-\tau)|u(\tau)|^2\, d\tau\right) dt$$
$$= \int_{t_1}^{t_2} \left(\int_{t_1}^{t_2} \frac{1}{\alpha}\left(1 - e^{-\alpha(t-t_1)}\right) e^{-\alpha(t-\tau)}u_s(t-\tau)u_s(\tau-t_1)\, dt\right) |u(\tau)|^2\, d\tau$$
$$\le \int_{t_1}^{t_2} \left(\int_{t_1}^{t_2} \frac{1}{\alpha}e^{-\alpha(t-\tau)}u_s(t-\tau)u_s(\tau-t_1)\, dt\right) |u(\tau)|^2\, d\tau$$
$$= \int_{t_1}^{t_2} \left(\int_{\tau}^{t_2} \frac{1}{\alpha}e^{-\alpha(t-\tau)}dt\right) |u(\tau)|^2\, d\tau = \int_{t_1}^{t_2} \frac{1}{\alpha^2}\left(1 - e^{-\alpha(t_2-\tau)}\right) |u(\tau)|^2\, d\tau$$
$$\le \frac{1}{\alpha^2}\int_{t_1}^{t_2} |u(\tau)|^2\, d\tau \le \frac{1}{\alpha^2}\left(\nu(t_2 - t_1) + \gamma_0\right), \tag{2.168}$$

and the second term in (2.167) satisfies

$$\int_{t_1}^{t_2} \left(\int_0^{t_1} e^{-\alpha(t-\tau)}|u(\tau)|\, d\tau\right)^2 dt = \int_{t_1}^{t_2} e^{-2\alpha(t-t_1)} \left(\int_0^{t_1} e^{-\alpha(t_1-\tau)}|u(\tau)|\, d\tau\right)^2 dt$$
$$= \int_{t_1}^{t_2} e^{-2\alpha(t-t_1)}y^2(t_1)\, dt. \tag{2.169}$$

From (2.111) in the proof of Lemma 2.3, we have

$$|y(t)| = |\int_0^t e^{-\alpha(t-\tau)} u(\tau)\, d\tau| \leq \frac{1}{e^\alpha - 1}\sqrt{\frac{e^{2\alpha}-1}{2\alpha}}\sqrt{\nu+\gamma_0} + \sqrt{\frac{1}{2\alpha}}\sqrt{\nu+\gamma_0},\ t \geq 0. \tag{2.170}$$

Using (2.167)–(2.170), we can derive (2.166) for some constants c_0 and γ_1. ∇

The inequality (2.166) also holds if $\frac{1}{s+\alpha}$ in Lemma 2.8 is replaced by a stable and proper transfer function $H(s)$ [372] or a stable and proper operator $T(s,t)$. For a stable and strictly proper transfer function $H_0(s)$ whose impulse response function is $h(t) = \mathcal{L}^{-1}[H_0(s)]$, we have $y(t) = \int_0^t h(t-\tau)u(\tau)\, d\tau$ and

$$\int_{t_1}^{t_2} y^2(t)\, dt = \int_{t_1}^{t_2}\left(\int_0^t h(t-\tau)u(\tau)\, d\tau\right)^2 dt$$
$$\leq 2\int_{t_1}^{t_2}\left(\int_{t_1}^{t} |h(t-\tau)|\,|u(\tau)| d\tau\right)^2 dt + 2\int_{t_1}^{t_2}\left(\int_0^{t_1} |h(t-\tau)|\,|u(\tau)|\, d\tau\right)^2 dt$$
$$\leq 2\|h(\cdot)\|_1^2 \int_{t_1}^{t_2} |u(t)|^2\, dt + 2\int_{t_1}^{t_2} |h(t-t_1)|^2\, |y(t_1)|^2\, dt. \tag{2.171}$$

For $H(s)$ stable and proper, with $H_0(s) = H(s) - h_0$ (for $h_0 = H(\infty)$) being strictly proper, we have $y(t) = \int_0^t h(t-\tau)u(\tau)\, d\tau + h_0 u(t)$ and $y^2(t) \leq 2\left(\int_0^t h(t-\tau)u(\tau)\, d\tau\right)^2 + 2h_0^2 u^2(t)$, and so we can also have (2.166).

2.5.4 Strictly Positive Real Systems

The concept of positive realness of a system transfer function is related to the passivity of the system. An error system which is made to be strictly positive real is essential for stability of an adaptive control system. The definition of (strictly) positive real functions is motivated from circuit network theory. A positive real (strictly positive real) function corresponds to the driving point impedance of a passive (dissipative) network. A passive element is a resistor, a capacitor, or an inductor. A dissipative element is either a resistor or a capacitor *in parallel* with a resistor or an inductor *in series* with a resistor. For example, a series RC circuit has an impedance which is only positive real but not strictly positive real, and so is a parallel LR circuit.

Definition 2.24 [288] *A function $h(s)$ of the complex variable $s = \sigma + j\omega$ is positive real (PR) if*

(i) $h(s)$ is real for real s; and

(ii) Re$[h(s)] \geq 0$ for all s such that Re$[s] > 0$.

Lemma 2.9 [10] *A rational function $h(s)$ of the complex variable $s = \sigma + j\omega$ is positive real if and only if*

(i) all poles of $h(s)$ are in $\mathrm{Re}[s] \leq 0$*;*

(ii) $\mathrm{Re}[h(j\omega)] \geq 0$ *for any real ω for which $j\omega$ is not a pole of $h(s)$; and*

(iii) poles of $h(s)$ on the $j\omega$-axis are simple and the associated residues are positive: $k_0 = \lim_{s \to j\omega_0}(s - j\omega_0)h(s) > 0$ for ω_0 finite or $k_\infty = \lim_{\omega \to \infty} \frac{h(j\omega)}{j\omega} > 0$ for ω_0 infinite, where $j\omega_0$ is a pole of $h(s)$.

It can be verified that if $h(s)$ is positive real, $h^{-1}(s)$ is also positive real.

Definition 2.25 [288] *A rational function $h(s)$ is strictly positive real (SPR) if $h(s - \epsilon)$ is positive real for some $\epsilon > 0$.*

For example, $h_1(s) = \frac{1}{s}$ is positive real but not strictly positive real, while $h_2(s) = \frac{1}{s+1}$ is strictly positive real. A series RC circuit has an impedance $h_s(s) = \frac{R+sC}{sCR}$, while a parallel LR circuit has an impedance $h_p(s) = \frac{sLR}{sL+R}$. Both impedance functions $h_s(t)$ and $h_p(s)$ are only positive real but not strictly positive real. However, a parallel RC circuit has an impedance $H_s(s) = \frac{R}{sC+R}$, while a series LR circuit has an impedance $H_p(s) = R + sL$. Both impedance functions $H_s(s)$ and $H_p(s)$ are strictly positive real.

Let n^* be the relative degree of a rational function $h(s) = \frac{n(s)}{d(s)}$: $n^* =$ degree of $d(s)$ − degree of $n(s)$. The next lemma gives necessary and sufficient conditions for a SPR $h(s)$ in the frequency domain.

Lemma 2.10 [152] *A rational function $h(s)$ is SPR if and only if*

(i) $h(s)$ does not have any poles $\mathrm{Re}[s] \geq 0$*;*

(ii) $\mathrm{Re}[h(j\omega)] > 0$ *for all $\omega \in (-\infty, +\infty)$; and*

(iii) $\lim_{\omega^2 \to \infty} \omega^2 \mathrm{Re}[h(j\omega)] > 0$ if $n^* = 1$, or $\lim_{\omega \to \infty} \mathrm{Re}[h(j\omega)] > 0$ if $n^* = 0$, or $\lim_{\omega \to \infty} \frac{h(j\omega)}{j\omega} > 0$ if $n^* = -1$.

Lemma 2.10 means that if $h(s)$ is strictly positive real and strictly proper, then $\mathrm{Re}[h(j\omega)]$ can go to zero as $|\omega| \to \infty$ not faster than ω^{-2}. It can be verified that for $h(s)$ to be positive real it is necessary that $|n^*| \leq 1$.

The positive realness or strictly positive realness of a rational function $h(s)$ can also be characterized in terms of its state-space model

$$h(s) = c(sI - A)^{-1}b + d, \tag{2.172}$$

where $c^T \in R^n$, $b \in R^n$, $d \in R$, $A \in R^{n \times n}$, $n > 0$, by the following lemmas.

Lemma 2.11 (Positive real) [10] *Assume (A, b, c, d) is controllable and observable. Then, $h(s)$ is positive real if and only if a real matrix $P = P^T > 0$ and a vector q exist such that*

$$A^T P + PA = -qq^T, \quad Pb - c^T = \sqrt{2d}q. \tag{2.173}$$

Lemma 2.12 [10] *Assume (A, b, c, d) is controllable and observable. Then, $h(s - \epsilon)$ with $\epsilon > 0$ is positive real (i.e., $h(s)$ in (2.172) is strictly positive real) if and only if a real matrix $P = P^T > 0$ and a vector q exist such that*

$$A^T P + PA = -qq^T - 2\epsilon P, \quad Pb - c^T = \sqrt{2d}q. \tag{2.174}$$

Lemma 2.13 (Lefschetz-Kalman-Yakubovich) [215] *Given a scalar $\delta > 0$, a real matrix A such that all zeros of $\det(sI - A)$ are in the open left-half plane, a real vector b such that (A, b) is completely controllable, a real vector c, a real scalar d, and an arbitrary real matrix $L = L^T > 0$. Then a real vector q and a real matrix $P = P^T > 0$ exist such that*

$$A^T P + PA = -qq^T - \delta L, \quad Pb - c^T = \sqrt{2d}q \tag{2.175}$$

if and only if $h(s)$ in (2.172) is strictly positive real and δ is sufficiently small.

Construction of a strictly positive real system. To develop a Lyapunov method based adaptive control scheme (see Section 5.3.4), we need to choose a (monic) stable polynomial $L(s)$ of degree $n^* - 1$ such that $\frac{L(s)}{P_m(s)}$ is strictly positive real for a given monic stable polynomial $P_m(s)$ of degree n^*. Such a choice of $L(s)$ can be made by using the following procedure:

(i) with $P_m(s) = s^{n^*} + a_{n^*-1}s^{n^*-1} + \cdots + a_1 s + a_0$ being stable, define

$$A = \begin{bmatrix} 0 & 1 & 0 & \cdots & 0 & 0 \\ 0 & 0 & 1 & 0 & \cdots & 0 \\ \vdots & \vdots & \vdots & \vdots & \vdots & \vdots \\ 0 & 0 & \cdots & \cdots & 0 & 1 \\ -a_0 & -a_1 & \cdots & \cdots & -a_{n^*-2} & -a_{n^*-1} \end{bmatrix}, \; b = \begin{bmatrix} 0 \\ \vdots \\ 0 \\ 1 \end{bmatrix} \in R^{n^*}; \tag{2.176}$$

(ii) choose $q \in R^{n^*}$ and $\delta_0 > 0$ such that $(A + \delta_0 I, q)$ is observable and $A + \delta_0 I$ is stable, and solve the Lyapunov equation

$$P(A + \delta_0 I) + (A^T + \delta_0 I)P = -qq^T \tag{2.177}$$

for a symmetric and positive definite $P \in R^{n^* \times n^*}$; and

(iii) let $c = b^T P = [c_1, c_2, \ldots, c_{n^*-1}, c_{n^*}]$ and set $L(s) = c[1, s, \ldots, s^{n^*-1}]^T$ or $L(s) = s^{n^*-1} + \frac{c_{n^*-1}}{c_{n^*}} s^{n^*-2} + \cdots + \frac{c_2}{c_{n^*}} s + \frac{c_1}{c_{n^*}}$ for being monic.

This construction ensures that $\frac{L(s-\delta_0)}{P_m(s-\delta_0)}$ is positive real (see Lemma 2.12) with $\delta_0 > 0$, so that $\frac{L(s)}{P_m(s)}$ is strictly positive real (see Definition 2.25).

2.6 Signal Convergence Lemmas

A key task in adaptive control system analysis is to show that certain error signals, such as a tracking error, converge to zero as time t goes to infinity. Such a convergence property is based on the properties that the error signal belongs to L^2 and its derivative belongs to L^∞, as derived from an adaptive control system. The following lemmas establish the link between the stated L^2 and L^∞ properties and the convergence property of an error signal.

Lemma 2.14 *If $\dot{f}(t) \in L^\infty$ and $f(t) \in L^2$, then $\lim_{t\to\infty} f(t) = 0$.*

Proof: Because $\int_{t_0}^{t} |f^2(\tau)\dot{f}(\tau)|\, d\tau \le \|\dot{f}(\cdot)\|_\infty \|f(\cdot)\|_2^2 < \infty$, $\forall t \ge t_0$, that is, $\lim_{t\to\infty} \int_{t_0}^{t} |f^2(\tau)\dot{f}(\tau)|\, d\tau$ exists and is finite, we have that

$$f^3(t) = 3\int_{t_0}^{t} f^2(\tau)\dot{f}(\tau)\, d\tau + f^3(t_0) \tag{2.178}$$

has a finite limit as $t \to \infty$ (an absolutely integrable function $f(t)$ is also integrable because both the negative and positive parts of such an $f(t)$ are integrable), and so does $f(t)$. Since $f(t) \in L^2$, we have $\lim_{t\to\infty} f(t) = 0$. ∇

If $\lim_{t\to\infty} f(t) = f_0 \neq 0$, then $\lim_{t\to\infty} \int_{t_0}^{t} f^2(\tau)\, d\tau = \infty$. If $\dot{f}(t) \in L^\infty$, $f(t) \in L^2$, then $f(t) \in L^\infty$: $|f^3(t)| \le 3\int_{t_0}^{t} |f^2(\tau)\dot{f}(\tau)|\, d\tau + |f^3(t_0)|$.

A more general version of Lemma 2.14 can be stated.

Lemma 2.15 *If $\dot{f}(t) \in L^\infty$ and $f(t) \in L^p$, $0 < p < \infty$, then $\lim_{t\to\infty} f(t) = 0$.*

Proof: Because $\int_{t_0}^{t} |f^p(\tau)\dot{f}(\tau)|\, d\tau \le \|\dot{f}(\cdot)\|_\infty \|f(\cdot)\|_p^p < \infty$, $\forall t \ge t_0$, that is, $\lim_{t\to\infty} \int_{t_0}^{t} |f^p(\tau)\dot{f}(\tau)|\, d\tau$ exists and is finite, we have that

$$f^{p+1}(t) = (p+1)\int_{t_0}^{t} f^p(\tau)\dot{f}(\tau)\, d\tau + f^{p+1}(t_0) \tag{2.179}$$

has a finite limit as $t \to \infty$, and so does $f(t)$. Since $f(t) \in L^p$ and $\lim_{t\to\infty} f(t)$ exists, we have $\lim_{t\to\infty} f(t) = 0$. ∇

The following lemma is known as the Barbălat lemma, commonly used in the adaptive control literature, based on the uniform continuity concept.

Definition 2.26 *$f(t)$ is uniformly continuous if for every $\epsilon > 0$ there exists a $\delta(\epsilon)$ such that $|t - \tau| < \delta$ implies $|f(t) - f(\tau)| < \epsilon$, $\forall t, \tau \ge 0$.*

As a comparison, a function $f(t)$ is continuous at t if for every $\epsilon > 0$ there exists a $\delta(\epsilon)$ such that $|t - \tau| < \delta$ implies $|f(t) - f(\tau)| < \epsilon$. For example, $f(t) = \sin(t)$ is uniformly continuous, while $f(t) = \sin(t^2)$ is continuous but not uniformly continuous. Uniform continuity implies continuity.

Lemma 2.16 [Barbălat] *If a scalar function $f(t)$ is uniformly continuous such that $\lim_{t\to\infty} \int_0^t f(\tau)\, d\tau$ exists and is finite, then $\lim_{t\to\infty} f(t) = 0$.*

The proof of this lemma can be found in [151], [206], [318].

In an adaptive control system, a certain error signal $e(t)$ has the desired properties $e(t) \in L^2 \cap L^\infty$, $\dot{e}(t) \in L^\infty$. Either Lemma 2.14 or Lemma 2.16 can be applied to conclude the desired convergence property: $\lim_{t\to\infty} e(t) = 0$. To apply Lemma 2.16, the following results are needed: (i) $\lim_{t\to\infty} \int_0^t f(\tau)\, d\tau$ exists and is finite if $\lim_{t\to\infty} \int_0^t |f(\tau)|\, d\tau$ exists and is finite and (ii) if $\dot{f}(t)$ is bounded, that is, $\dot{f}(t) \in L^\infty$, then $f(t)$ is uniformly continuous.

From Lemma 2.16, if $f(t) \in L^1$ and $\dot{f}(t) \in L^\infty$, then $\lim_{t\to\infty} f(t) = 0$, as $f(t) \in L^1$ implies $\lim_{t\to\infty} \int_0^t f(\tau)\, d\tau$ exists and is finite.

Corollary 2.8 *If $f(t) \in L^2 \cap L^\infty$, $\dot{f}(t) \in L^\infty$, then $\lim_{t\to\infty} f(t) = 0$.*

Proof: The condition $f(t) \in L^2$ implies $\lim_{t\to\infty} \int_0^t f^2(\tau)\, d\tau$ exists and is finite, and $f(t) \in L^\infty$ and $\dot{f}(t) \in L^\infty$ imply $\frac{d}{dt}(f^2(t)) = 2f(t)\dot{f}(t) \in L^\infty$ (i.e., $f^2(t)$ is uniformly continuous), and so $\lim_{t\to\infty} f^2(t) = 0$, by Lemma 2.16. ∇

Using Lemma 2.16, one can verify that if $f(t) \in L^p \cap L^\infty$, $1 \le p < \infty$, and $\dot{f}(t) \in L^\infty$, then $\lim_{t\to\infty} f(t) = 0$. (How about $f(t) \in L^p \cap L^\infty$, $0 < p < 1$?)

We now give an alternative proof of Corollary 2.8. Let $a > 0$ and write

$$f^2(t) = \frac{s}{s+a}[f^2](t) + \frac{a}{s+a}[f^2](t) = \frac{2}{s+a}[f\dot{f}](t) + \frac{a}{s+a}[f^2](t). \tag{2.180}$$

Since $f(t) \in L^\infty$ and $\dot{f}(t) \in L^\infty$, it follows that

$$\left|\frac{2}{s+a}[f\,\dot{f}](t)\right| \le c_0 \int_0^t e^{-a(t-\tau)}\, d\tau \le \frac{c_0}{a} \tag{2.181}$$

for some constant $c_0 > 0$. Let $g(t) = \frac{a}{s+a}[f^2](t)$, which belongs to L^∞ as $f(t) \in L^\infty$. Since $f^2(t) \in L^1$, from Lemma 2.7, we have $g(t) \in L^1$, and $\dot{g}(t) = a(f^2(t) - g(t)) \in L^1$, from which we have that $\lim_{t\to\infty} \int_0^t \dot{g}(\tau)\, d\tau = \lim_{t\to\infty}(g(t) - g(0))$ exists and is finite as $\lim_{t\to\infty} \int_0^t |\dot{g}(\tau)|\, d\tau$ exists and is finite. Since $\lim_{t\to\infty} g(t)$ exists and is finite and $g(t) \in L^1$, it follows that $\lim_{t\to\infty} g(t) = 0$.

For any $\epsilon > 0$, we let $a = \frac{2}{c_0 \epsilon}$. Then, there is a $T > 0$ such that $\frac{a}{s+a}[f^2](t) < \frac{\epsilon}{2}$, $\forall t \geq T$. Hence, from (2.180) and (2.181), we have $f^2(t) \leq \frac{c_0}{a} + \frac{a}{s+a}[f^2](t) < \epsilon$, $\forall t \geq T$, which implies that $\lim_{t\to\infty} f^2(t) = 0$. ∇

Following this proof of Corollary 2.8, we can derive the following result.

Lemma 2.17 *If $f(t) \in L^1$, then $\frac{b}{s+a}[f](t) \in L^\infty$ and $\lim_{t\to\infty} \frac{b}{s+a}[f](t) = 0$ for any constants $a > 0$ and $b > 0$.*

Proof: The boundedness of $\frac{b}{s+a}[f](t)$ follows from

$$\left|\frac{b}{s+a}[f](t)\right| \leq b \int_0^t e^{-a(t-\tau)} |f(\tau)| \, d\tau \leq b \int_0^t |f(\tau)| \, d\tau < \infty. \tag{2.182}$$

The convergence property follows from the facts that $g(t) = \frac{b}{s+a}[f](t) \in L^1$ and $\dot{g}(t) \in L^1$ so that $g(t)$ has a limit which must be zero. ∇

The same idea can be used to prove the following general result.

Lemma 2.18 *If $f(t) \in L^p$, $1 \leq p < \infty$, then $g(t) = \frac{b}{s+a}[f](t) \in L^\infty$ and $\lim_{t\to\infty} g(t) = 0$ for any constants $a > 0$ and $b > 0$.*

Proof: For $1 < p < \infty$, we write

$$\begin{aligned} \left|\frac{b}{s+a}[f](t)\right| &\leq b \int_0^t e^{-a(t-\tau)} |f(\tau)| \, d\tau = b \int_0^t \left(e^{-a(t-\tau)}\right)^{\frac{p-1}{p}} \left(e^{-a(t-\tau)}\right)^{\frac{1}{p}} |f(\tau)| \, d\tau \\ &\leq b \left(\int_0^t e^{-a(t-\tau)} \, d\tau\right)^{\frac{p-1}{p}} \left(\int_0^t e^{-a(t-\tau)} |f(\tau)|^p \, d\tau\right)^{\frac{1}{p}} \\ &\leq b \, a^{\frac{p}{p-1}} \left(\int_0^t e^{-a(t-\tau)} |f(\tau)|^p \, d\tau\right)^{\frac{1}{p}} \leq b \, a^{\frac{p}{p-1}} \|f(\cdot)\|_p \end{aligned} \tag{2.183}$$

(we used Hölder's inequality (see Section 2.3.2), with $q = p/(p-1) > 1$ for $p > 1$, for the second inequality in (2.183)). Hence, we have $g(t) \in L^\infty$ as $f(t) \in L^p$. Since $|f(t)|^p \in L^1$, from Lemma 2.17, we have $\lim_{t\to\infty} \int_0^t e^{-a(t-\tau)} |f(\tau)|^p \, d\tau = 0$, and from (2.183) that $\lim_{t\to\infty} g(t) = 0$. ∇

How about $f(t) \in L^p$ with $0 < p < 1$? Hölder's inequality is not applicable. However if, in additon, $f(t) \in L^\infty$, then $f(t) \in L^1$, so that from Lemma 2.17 we also have $\lim_{t\to\infty} \frac{b}{s+a}[f](t) = 0$.

These convergence properties can also be verified using a Laplace transform method. For example, to prove that $f(t) \in L^1$ implies that $\lim_{t\to\infty} \frac{1}{s+a}[f](t) =$

0, we use the results that $g(t) = \frac{1}{s+a}[f](t) \in L^1$ and $\dot{g}(t) \in L^1$ to show that for $s = \sigma + j\omega$, $\sigma \geq 0$, the transform $G(s) = \int_0^\infty g(t)e^{-st}\,dt$ satisfies

$$|G(s)| \leq \int_0^\infty |g(t)|e^{-\sigma t}\,dt \leq \int_0^\infty |g(t)|\,dt < \infty, \tag{2.184}$$

that is, $G(s)$ is analytic [14] in $\mathrm{Re}[s] > 0$ [78] and converges in $\mathrm{Re}[s] \geq 0$ and, similarly, that the transform $G_d(s) = \int_0^\infty \dot{g}(t)e^{-st}\,dt$ satisfies

$$|G_d(s)| \leq \int_0^\infty |\dot{g}(t)|\,|e^{-(\sigma+j\omega)t}|\,dt = \int_0^\infty |\dot{g}(t)|e^{-\sigma t}\,dt < \infty, \tag{2.185}$$

that is, $G_d(s)$ is analytic in $\mathrm{Re}[s] > 0$ and converges in $\mathrm{Re}[s] \geq 0$.

Since $G(s)$ and $G_d(s)$ are well defined for s in $\mathrm{Re}[s] \geq 0$, we have

$$\lim_{s\to 0} G_d(s) = \lim_{s\to 0}\int_0^\infty \dot{g}(t)e^{-st}\,dt = \lim_{t\to\infty} g(t) - g(0), \tag{2.186}$$

and for s in $\mathrm{Re}[s] > 0$, we can derive

$$G_d(s) = \int_0^\infty \dot{g}(t)e^{-st}\,dt = e^{-st}g(t)|_0^\infty + s\int_0^\infty g(t)e^{-st}\,dt = sG(s) - f(0) \tag{2.187}$$

(note that $e^{-st}g(t)|_{t=\infty} = 0$ as $g(t) \in L^\infty$ and $\mathrm{Re}[s] > 0$). Hence, we have

$$\lim_{t\to\infty} g(t) = \lim_{s\to 0} sG(s) = 0. \tag{2.188}$$

A more general result is with a stable, strictly proper system $H(s)$.

Lemma 2.19 *If $f(t) \in L^p$, $1 \leq p < \infty$, then $g(t) = H(s)[f](t) \in L^\infty$ and $\lim_{t\to\infty} f(t) = 0$ for any stable and strictly proper transfer function $H(s)$.*

Proof: For a stable and strictly proper transfer function $H(s)$, we have

$$|H(s)[f](t)| \leq \frac{b}{s+a}[|f|](t) \tag{2.189}$$

for some constants $b > 0$, $a > 0$. From Lemma 2.18, we have the lemma. ∇

The following results are useful in analyzing convergence of an adaptive parameter vector estimate and its vector norm (see Chapter 3).

Proposition 2.14 *If a function $f(t)$ is monotonically increasing (i.e., $f(t_1) \leq f(t_2)$ for $t_1 < t_2$) and bounded from above (that is, $f(t) \leq c$, $\forall t \geq 0$, for some constant c), then $\lim_{t\to\infty} f(t)$ exists and is finite.*

Proposition 2.15 *If a function $f(t)$ is monotonically decreasing and bounded from below, then $\lim_{t\to\infty} f(t)$ exists and is finite.*

The proofs of these results can be found from a calculus textbook.

[14] A function $F(s)$ of the complex variable s is analytic at $s = s_0$ if $F(s)$ is differentiable in a neighborhood of s_0 [274]. A rational function $F(s)$ is analytic at s_0 if $F(s_0) \neq \infty$.

2.7 Discrete-Time Systems

Although most dynamic systems are operating in continuous time, many control systems are designed and implemented based on discrete-time control theory using discretized models of continuous-time systems.

2.7.1 System Modeling

For a continuous-time linear time-invariant system

$$\dot{x}(t) = A_c x(t) + B_c u(t), \ y(t) = C_c x(t) + D_c u(t), \tag{2.190}$$

where $A_c \in R^{n\times n}$, $B_c \in R^{n\times M}$, $C_c \in R^{q\times n}$, and $D_c \in R^{q\times M}$ are all constant system matrices, if the control signal $u(t)$ is applied in the special form

$$u(t) = u(kT), \ \forall t \in [kT, (k+1)T), \tag{2.191}$$

with $T > 0$ as the sampling period, the state vector $x(kT)$ can be described by the discrete-time state-space equation

$$\begin{aligned} x((k+1)T) &= A_d x(kT) + B_d u(kT), \\ y(kT) &= C_d x(kT) + D_d u(kT), \ k \in \{0, 1, 2, \ldots\}, \end{aligned} \tag{2.192}$$

where $C_d = C_c$, $D_d = D_c$, and

$$A_d = e^{A_c T} \in R^{n\times n}, \ B_d = e^{A_c T} \int_0^T e^{-A_c \sigma} B_c \, d\sigma \in R^{n\times M}, \tag{2.193}$$

with $e^{A_c t} = \mathcal{L}^{-1}[(sI - A_c)^{-1}]$ (see Section 2.1.2). This procedure is the so-called discretization of a linear time-invariant continuous-time system [209].

Since $T > 0$ is a constant, we can simplify the presentation of (2.192) as

$$\begin{aligned} x(t+1) &= Ax(t) + Bu(t), \\ y(t) &= Cx(t) + Du(t), \ t \in \{0, 1, 2, \ldots\}, \end{aligned} \tag{2.194}$$

for $A \in R^{n\times n}$, $B \in R^{n\times M}$, $C \in R^{q\times n}$, and $D \in R^{q\times p}$ constant whenever there is no confusion of notation, especially, noting that the time variable t should take nonnegative integers $\{0, 1, 2, \ldots\}$ for *discrete-time* systems (as compared with nonnegative real numbers $t \geq 0$ for *continuous-time* systems).

The z-transform of a discrete-time signal $f(t)$, $t \in \{0, 1, \ldots\}$, is defined as

$$F(z) = \mathcal{Z}[f(t)] = \sum_{t=0}^{\infty} f(t) z^{-t}. \tag{2.195}$$

If $|f(t)| \leq ab^t$ for some $b \geq 0$ and $a > 0$, then $F(z)$ exists and $\sum_{t=0}^{\infty} f(t)z^{-t}$ is convergent for $|z| > b$. For example, the z-transform of a constant or sinusoidal signal $f(t) = f_0$ or $\sin \omega t$, $t \in \{0, 1, 2, \ldots\}$, is $\frac{f_0 z}{z-1}$ or $\frac{z \sin \omega}{z^2 - 2\cos \omega z + 1}$.

A useful property of the z-transform is

$$\mathcal{Z}[f(t+1)u_s(t)] = zF(z) - zf(0), \tag{2.196}$$

where $u_s(t)$ is the discrete-time unit step function such that $u_s(t) = 1$, $t \in \{0, 1, 2, \ldots\}$, and $u_s(t) = 0$ otherwise. Using this property, we can express system (2.194) in the frequency z-domain as

$$\begin{aligned} x(z) &= (zI - A)^{-1} z x(0) + (zI - A)^{-1} B u(z), \\ y(z) &= C(zI - A)^{-1} z x(0) + C(zI - A)^{-1} B u(z) + D u(z), \end{aligned} \tag{2.197}$$

where $y(z) = \mathcal{Z}[y(t)]$, $u(z) = \mathcal{Z}[u(t)]$, and $x(z) = \mathcal{Z}[x(t)]$.

The system transfer function (matrix) is

$$G(z) = C(zI - A)^{-1}B + D \tag{2.198}$$

such as $y(z) = G(z)u(z)$ when $x(0) = 0$. With the inverse z-transform operation $\mathcal{Z}^{-1}[\cdot]$, it can be verified that $A^t = \mathcal{Z}^{-1}[(zI - A)^{-1}z]$ and that the time-domain solution of (2.194) is

$$\begin{aligned} x(t) &= A^t x(0) + \sum_{i=0}^{t-1} A^{t-i-1} B u(i), \\ y(t) &= C A^t x(0) + C \sum_{i=0}^{t-1} A^{t-i-1} B u(i) + D u(t), \end{aligned} \tag{2.199}$$

from which we can define the system transition matrix as

$$\Phi(t, t_0) = A^{t-t_0} = \underbrace{A\,A \ldots A}_{i-i_0},\ t \geq t_0. \tag{2.200}$$

Time-varying systems. We can generalize the above time-domain solution description to the discrete-time linear time-varying system

$$\begin{aligned} x(t+1) &= A(t)x(t) + B(t)u(t), \\ y(t) &= C(t)x(t) + D(t)u(t),\ t \in \{0, 1, 2, \ldots\}, \end{aligned} \tag{2.201}$$

for $A(t) \in R^{n \times n}$ and $B(t) \in R^{n \times M}$, $C(t) \in R^{q \times n}$, $D(t) \in R^{q \times p}$ being time-dependent matrices. Defining the discrete-time state transition matrix

$$\Phi(t, i) = \begin{cases} A(t-1)A(t-2) \cdots A(i) & \text{for } t \geq i+1 \\ I & \text{for } t = i, \end{cases} \tag{2.202}$$

we can derive the solution to system (2.201) as

$$x(t) = \Phi(t, t_0)x(t_0) + \sum_{i=t_0}^{t-1} \Phi(t, i+1)B(i)u(i),$$

$$y(t) = C(t)\Phi(t, t_0)x(t_0) + C(t)\sum_{i=t_0}^{t-1} \Phi(t, i+1)B(i)u(i) + D(t)u(t) \quad (2.203)$$

for $t \geq t_0 + 1$. The state transition matrix has the following properties [335]:
(i) $\Phi(k+1, j) = A(k)\Phi(k, j)$, $\Phi(k, j-1) = \Phi(k, j)A(j-1)$, $k \geq j$;
(ii) $\Phi(k, i) = \Phi(k, j)\Phi(j, i)$, $i \leq j \leq k$; and
(iii) $\Phi^{-1}(k, j) = \Phi(j, k)$ if $A(k)$ is invertible for all k.

For system (2.201) or (2.194), the concepts of reachability [15] and observability can be physically defined and mathematically characterized in a similar way to that in the continuous-time case (see Section 2.2), especially the concept of uniform observability [125], [151], in addition to impulse response functions (matrices), coprimeness of polynomials (polynomial matrices) and interactor matrices, and associated gain matrix decompositions.

2.7.2 Norms and Signal Spaces

The vector and matrix norms defined in Section 2.3.1 and their properties are applicable to the discrete-time case; in particular, as the counterparts of (2.65) and (2.66), for $D(t) \in R^{m\times n}$ and $x(t) \in R^n$, we have

$$\left\| \sum_{\tau=t_0}^{t} D(\tau) \right\| \leq \sum_{\tau=t_0}^{t} \|D(\tau)\|, \ \forall t \geq t_0 \quad (2.204)$$

$$\left\| \sum_{\tau=t_0}^{t} D(\tau)x(\tau) \right\| \leq \sum_{\tau=t_0}^{t} \|D(\tau)\| \, \|x(\tau)\|, \ \forall t \geq t_0. \quad (2.205)$$

[15]The system $x(t+1) = A(t)x + B(t)u$ is reachable on $\{t_0, t_0+1, \ldots, t_f\}$ if for any given final state $x(t_f) = x_f$ there exists an input signal $u(t)$ such that $x(t_f) = x(t_f; t_0, x_0) = x_f$ for $x(t_0) = x_0 = 0$. A necessary and sufficient condition for the system $x(t+1) = A(t)x + B(t)u$ to be reachable on $\{t_0, t_0+1, \ldots, t_f\}$ is that

$$W_r(t_0, t_f) = \sum_{t=t_0}^{t_f-1} \Phi(t_f, t+1)B(t)B^T(t)\Phi^T(t_f, i+1) \ \text{ is nonsingular.}$$

For time-invariant systems with $A(t) = A$ and $B(t) = B$ constant, this condition is equivalent to (2.27) under which arbitrary pole placement by state feedback $u(t) = Kx(t)$ is ensured. Similar to the continuous-time case, the system $x(t+1) = Ax + Bu$ is stabilizable if there is a constant matrix $K \in R^{M\times n}$ such that all eigenvalues of $A + BK$ are stable (i.e., all are inside the unit circle of the complex z-plane).

For a discrete-time vector signal $x(t) \in R^n$, its L^1 signal norm is defined as

$$\|x(\cdot)\|_1 = \sum_{t=0}^{\infty} \|x(t)\|_1 = \sum_{t=0}^{\infty} (|x_1(t)| + \cdots + |x_n(t)|), \tag{2.206}$$

its L^2 signal norm is defined as

$$\|x(\cdot)\|_2 = \sqrt{\sum_{t=0}^{\infty} \|x(t)\|_2^2} = \sqrt{\sum_{t=0}^{\infty} (x_1^2(t) + \cdots + x_n^2(t))}, \tag{2.207}$$

its L^∞ signal norm is defined as

$$\|x(\cdot)\|_\infty = \sup_{t\geq 0} \|x(t)\|_\infty = \sup_{t\geq 0} \max_{1\leq i\leq n} |x_i(t)|, \tag{2.208}$$

and its L^p ($p \in [1, \infty)$) signal norm is defined as

$$\|x(\cdot)\|_p = \left(\sum_{t=0}^{\infty} \|x(t)\|_p^p\right)^{\frac{1}{p}} = \left(\sum_{t=0}^{\infty} (|x_1(t)|^p + \cdots + |x_n(t)|^p)\right)^{\frac{1}{p}}. \tag{2.209}$$

The L^p signal norm can also be defined as

$$\|x(\cdot)\|_p = \left(\sum_{t=0}^{\infty} \|x(t)\|^p dt\right)^{\frac{1}{p}}, \tag{2.210}$$

where $\|\cdot\|$ is any vector norm on R^n.

The signal space L^p ($p \in [1, \infty]$) is defined as

$$L^p = \{x(t) \in R^n : \|x(\cdot)\|_p < \infty\}. \tag{2.211}$$

In the *discrete-time* case, we have that

(i) $x(t) \in L^p$, $p \in [1, \infty)$, $\Rightarrow x(t) \in L^\infty$, and $\lim_{t\to\infty} x(t) = 0$; and

(ii) $x(t) \in L^1 \Rightarrow x(t) \in L^p$, $p \in (1, \infty)$.

Important inequalities. The following counterparts of the inequalities (2.72)–(2.74) are useful for discrete-time adaptive control system analysis.

Hölder inequality [78]: For any $p \in [1, \infty)$, $q \in [1, \infty)$ such that $\frac{1}{p} + \frac{1}{q} = 1$, any integers $\sigma > \tau \geq 0$, and scalar functions $f(t)$ and $g(t)$, we have

$$\sum_{t=\tau}^{\sigma} |f(t)g(t)| \leq \left(\sum_{t=\tau}^{\sigma} |f(t)|^p\right)^{\frac{1}{p}} \left(\sum_{t=\tau}^{\sigma} |g(t)|^q\right)^{\frac{1}{q}}. \tag{2.212}$$

Hence, if $f(t) \in L^p$ and $g(t) \in L^q$, then $f(t)g(t) \in L^1$.

Schwarz inequality: For $p = q = 2$ in the Hölder inequality, we have

$$\sum_{t=\tau}^{\sigma} |f(t)g(t)| \leq \sqrt{\sum_{t=\tau}^{\sigma} f^2(t)} \sqrt{\sum_{t=\tau}^{\sigma} g^2(t)}. \tag{2.213}$$

Minkovski inequality: For any $p \in [1, \infty)$, any $b > a \geq 0$, and two integrable scalar functions $f(t)$ and $g(t)$, we have

$$\left(\sum_{t=\tau}^{\sigma} |f(t) + g(t)|^p\right)^{\frac{1}{p}} \leq \left(\sum_{t=\tau}^{\sigma} |f(t)|^p\right)^{\frac{1}{p}} + \left(\sum_{t=\tau}^{\sigma} |g(t)|^p\right)^{\frac{1}{p}}. \tag{2.214}$$

Hence, if $f(t) \in L^p$ and $g(t) \in L^p$, then $f(t) + g(t) \in L^p$.

The proofs of these inequalities are also based on the inequality (2.75).

2.7.3 Stability

The Lyapunov stability concepts presented in Section 2.4 for continuous-time systems, such as Lyapunov stability, asymptotic, uniform, uniformly asymptotic, or exponential stability, are also applicable to discrete-time systems. The Lyapunov direct method for examining system stability also works in the discrete-time case, with the difference that the time increment $\Delta V \triangleq V(x(t+1), t+1) - V(x(t), t)$ (instead of the time derivative $\dot{V}$ for the continuous-time case) of a positive definite function $V(x(t), t)$ is used for stability analysis.

For a linear system of the form

$$x(t+1) = A(t)x(t) \tag{2.215}$$

with a state transition matrix $\Phi(t, t_0)$, stability results, analogous to that in Section 2.4.4 for a continuous-time linear system of the form $\dot{x} = A(t)x$ (see Theorems 2.7 and 2.8), can be developed in terms of $\Phi(t, t_0)$.

To use the Lyapunov direct method, we can employ a quadratic positive definite function $V(x, t) = x^T P(t)x$ as a Lyapunov function candidate.

Theorem 2.13 [335] *The equilibrium state $x_e = 0$ of system (2.215) is stable if there exist a $P(t) = P^T(t) \in R^{n\times n}$ and a constant $\eta > 0$ such that*

$$\eta I \leq P(t),\ A^T(t)P(t+1)A(t) - P(t) \leq 0,\ \forall t \geq t_0, \tag{2.216}$$

is uniformly stable if, for some constant $\rho > 0$,

$$\eta I \leq P(t) \leq \rho I,\ A^T(t)P(t+1)A(t) - P(t) \leq 0,\ \forall t \geq t_0, \tag{2.217}$$

and is uniformly asymptotically stable if, for some constant $\nu > 0$,

$$\eta I \leq P(t) \leq \rho I,\ A^T(t)P(t+1)A(t) - P(t) \leq -\nu I,\ \forall t \geq t_0. \tag{2.218}$$

In this case, $V(x,t) = x^T P(t) x$ is a Lyapunov function (a desired positive definite function for proving the Lyapunov stability).

Theorem 2.14 *The equilibrium state $x_e = 0$ of system (2.215) with $A(t) = A$ constant is stable if and only if all eigenvalues of A have magnitudes less than or equal to 1, and for those eigenvalues whose magnitude are equal to 1, the associated Jordan blocks all have size 1×1. It is asymptotically stable if and only all eigenvalues of A have magnitudes less than 1.*

When all eigenvalues of A have magnitudes less than 1, the following equation always has a solution $P = P^T > 0$ for any given $Q = Q^T > 0$,

$$A^T P A - P = -Q, \tag{2.219}$$

so that $V(x) = x^T P x$ is a Lyapunov function.

The following results are useful in stability and convergence analysis for discrete-time adaptive control and parameter estimation system.

An inequality. For any function $x_s(t) \geq 0$, any $j > 0$, $c_1 \geq 0$, and $a_0 \geq 0$,

$$\prod_{t=t_0}^{t_0+j} (a_0 + c_1 x_s(t)) \leq \left(a_0 + \frac{c_1}{j+1} \sum_{t=t_0}^{t_0+j} x_s(t) \right)^{j+1}. \tag{2.220}$$

Proof: We use the well-known inequality

$$(a_1 a_2 \cdots a_N)^{\frac{1}{N}} \leq \frac{1}{N}(a_1 + a_2 + \cdots + a_N),\ a_i \geq 0,\ i = 1, \ldots, N,\ N \geq 1, \tag{2.221}$$

with $a_i = a_0 + c_1 x_s(i)$, to obtain (2.220). ∇

A special matrix norm. From Section 2.3.1, for a given vector norm on C^n, and a matrix $A \in C^{n \times n}$, an induced matrix norm for A is $\|A\| = \sup_{\|x\|=1} \|Ax\|$. For a constant matrix $A \in R^{n \times n}$ whose eigenvalues all have magnitudes less than 1, there is a norm $\|\cdot\|$ on C^n such that

$$\|A\| < 1. \tag{2.222}$$

Proof [298]: Denote $\lambda = |\lambda_M|$, where λ_M is the eigenvalue of A which has the maximum magnitude, and let $\epsilon \in (0, 1-\lambda)$. For $A \in R^{n \times n}$, there is a matrix $P \in C^{n \times n}$ such that $PAP^{-1} = J$, where J is the Jordan canonical form of A [298]. Let $D = \text{diag}\{1, \frac{1}{\epsilon}, \frac{1}{\epsilon^2}, \ldots, \frac{1}{\epsilon^{n-1}}\}$ and $Q = DP$. Then, $QAQ^{-1} = J_\epsilon$, where J_ϵ is exactly the same as J except that each 1 in a superdiagonal position of J is replaced by ϵ in J_ϵ. Denoting the eigenvalues of A as λ_i, $i = 1, 2, \ldots, n$, setting ϵ_i as ϵ or 0 in a superdiagonal position of J_ϵ, and letting x_i^* be the complex conjugate of x_i, for $x \in C^n$, we calculate

$$\begin{aligned} \|J_\epsilon x\|_2^2 &= \sum_{i=1}^{n-1} \left(|\lambda_i|^2 |x_i|^2 + \epsilon_i \lambda_i x_{i+1}^* x_i + \epsilon_i \lambda_i^* x_i^* x_{i+1} + \epsilon_i^2 |x_{i+1}|^2 \right) + |\lambda_n|^2 |x_n|^2 \\ &\leq \left(\lambda^2 + 2\lambda\epsilon + \epsilon^2 \right) \|x\|_2^2 = (\lambda + \epsilon)^2 \|x\|_2^2. \end{aligned} \tag{2.223}$$

Then, it follows that

$$\|QAQ^{-1}\|_2 = \|J_\epsilon\|_2 = \sup_{\|x\|=1} \|J_\epsilon x\|_2 \leq \lambda + \epsilon < 1. \tag{2.224}$$

This motivates us to define the norm on C^n:

$$\|x\| = \|Qx\|_2. \tag{2.225}$$

For this norm, we have

$$\begin{aligned} \|A\| &= \sup_{\|x\|=1} \|Ax\| = \sup_{\|Qx\|_2=1} \|QAx\|_2 \\ &= \sup_{\|y\|_2=1} \|QAQ^{-1}y\|_2 = \|QAQ^{-1}\|_2 \leq \lambda + \epsilon < 1. \end{aligned} \tag{2.226}$$

The first and last equalities are just the norm definition. ∇

A stability analysis example. Consider the time-invariant system $x(t+1) = Ax(t)$, where $A \in R^{n \times n}$ whose eigenvalues all have magnitudes less than 1. In view of (2.219), we introduce $V(x) = x^T P x$ and calculate

$$\begin{aligned} V(x(t+1)) - V(x(t)) &= x^T(t)(A^T P A - P)x(t) \\ &= -x^T(t) Q x(t) \leq -\gamma V(x(t)) \end{aligned} \tag{2.227}$$

for some $\gamma > 0$, that is, $V(x(t+1)) \leq (1-\gamma)V(x(t))$, where $1 - \gamma \in (0, 1)$ as $\gamma > 0$ and $V(x(t)) > 0$ for $x(t) \neq 0$. We then conclude that $V(x(t)) \leq (1-\gamma)^t V(x(0))$ so that $\lim_{t\to\infty} V(x(t)) = 0$ exponentially, and so does $x(t)$.

We can derive this result using the norm of A defined in (2.226): $\|A\| = \|QAQ^{-1}\|_2$, with $QAQ^{-1} = J_\epsilon$, $\lambda + 0 < \epsilon < \lambda - 1$. Expressing the system $x(t+1) = Ax(t)$ as $\bar{x}(t+1) = QAQ^{-1}\bar{x}(t)$, where $\bar{x}(t) = Qx(t)$, we have

$$\|\bar{x}(t+1)\|_2 = \|QAQ^{-1}\bar{x}(t)\|_2 \leq \|QAQ^{-1}\|_2 \, \|\bar{x}(t)\|_2 \leq (\lambda + \epsilon) \, \|\bar{x}(t)\|_2 \tag{2.228}$$

so that $\|\bar{x}(t)\|_2 \leq (\lambda + \epsilon)^t \, \|\bar{x}(0)\|_2$, which implies that $\lim_{t\to\infty} \|\bar{x}(t)\|_2 = 0$ exponentially, and so does $x(t)$.

A robust stability example. Let us now consider the time-invariant system $x(t+1) = Ax(t) + g(x(t), t)$, where $A \in R^{n \times n}$ whose eigenvalues all have magnitudes less than 1, $g(x(t), t)$ satisfies $g(0, t) = 0$, $\forall t \geq 0$, and

$$\|g(x,t)\|_2 \leq x_s(t)\|x\|_2, \ \forall t \geq t_0, \ \forall x \in R^n, \tag{2.229}$$

for some scalar function $x_s(t) \geq 0$ such that

$$\sum_{t=t_1}^{t_2} x_s(t) \leq \nu(t_2 - t_1 + 1) + \gamma_0, \ \forall t_2 > t_1 \geq t_0, \tag{2.230}$$

with $\gamma_0 \geq 0$, $\nu \geq 0$. With the above notation, we have

$$\begin{aligned} \|\bar{x}(t+1)\|_2 &\leq \|QAQ^{-1}\|_2 \, \|\bar{x}(t)\|_2 + \|Q\|_2 \, \|Q^{-1}\|_2 \, x_s(t) \, \|\bar{x}(t)\|_2 \\ &\leq (a_0 + c_1 x_s(t)) \, \|\bar{x}(t)\|_2, \end{aligned} \tag{2.231}$$

where $a_0 = \lambda + \epsilon \in (0,1)$, $c_1 = \|Q\|_2 \, \|Q^{-1}\|_2$, and $\|x\|_2 \leq \|Q^{-1}\|_2 \|\bar{x}(t)\|_2$ for the third term in (2.231). Using (2.220), we have

$$\begin{aligned} \|\bar{x}(t_0+j)\|_2 &\leq \prod_{\tau=t_0}^{t_0+j-1} (a_0 + c_1 x_s(\tau)) \|\bar{x}(t_0)\|_2 \leq \left(a_0 + \frac{c_1 \sum_{\tau,=t_0}^{t_0+j-1} x_s(\tau)}{j}\right)^j \|\bar{x}(t_0)\|_2 \\ &\leq \left(a_0 + c_1\nu + \frac{c_1\gamma_0}{j}\right)^j \|\bar{x}(t_0)\|_2, \ j \geq 1. \end{aligned} \tag{2.232}$$

From this expression, we can conclude that there exists a $\nu^* > 0$ such that, for any $\nu \in [0, \nu^*)$, $\lim_{t\to\infty} \|\bar{x}(t)\|_2 = 0$, so that $\lim_{t\to\infty} \|x(t)\|_2 = 0$.

This result is useful for analysis of discrete-time adaptive control systems.

Convergence of monotonic functions. A function $f(t)$ is monotonically increasing (or decreasing) if $f(t_1) \leq f(t_2)$ (or $f(t_1) \geq f(t_2)$ for $t_1 < t_2$).

Proposition 2.16 *If a function $f(t)$ is monotonically increasing and bounded from above or $f(t)$ is monotonically decreasing and bounded from below, then $\lim_{t\to\infty} f(t)$ exists and is finite.*

Discrete-time Bellman-Gronwall lemma. Let $z(t)$, $f(t)$, $g(t)$ and $k(t)$ be real-valued functions with $g(t) \geq 0$, $k(t) \geq 0$, $t \in \{0, 1, 2, \ldots\}$. If

$$z(t) \leq f(t) + g(t) \sum_{i=0}^{t-1} k(i) z(i), \ \forall t \in \{0, 1, 2, \ldots\}, \tag{2.233}$$

then

$$z(t) \leq f(t) + g(t) \sum_{i=0}^{t-1} \prod_{j=i+1}^{t-1} (1 + k(j)g(j)) k(i) f(i), \ \forall t \in \{0, 1, 2, \ldots\}, \tag{2.234}$$

where $\prod_{j=i+1}^{t-1}(1 + k(j)g(j)) = 1$ for $i = k - 1$.

This result can be proved by induction ([78], p. 254; for $g(t) = 1$ [151]).

2.8 Operator Norms

Consider a linear time-invariant system

$$y(t) = H(D)[u](t), \tag{2.235}$$

where $y(t) \in R^M$ and $u(t) \in R^M$ are the system output and input and $H(D)$ is an $M \times M$ proper rational transfer matrix in D which denotes the Laplace transform variable or the time differentiation operator $D[x](t) = \dot{x}(t)$, $t \in [0, \infty)$, in the continuous-time case; or the z-transform variable or the time advance operator $D[x](t) = x(t+1)$, $t \in \{0, 1, 2, 3, \ldots\}$, in the discrete-time case. Assume that the system is BIBO stable, that is, all poles p_i of $H(D)$ are stable: $\mathrm{Re}[p_i] < 0$ in continuous time or $|p_i| < 1$ in discrete time.

Recall from Sections 2.3.3 and 2.7.2 the signal norms: for a *continuous-time* vector signal $x(t) \in R^M$, its L^p ($p \in [1, \infty)$) signal norm is defined as

$$\|x(\cdot)\|_p = \left(\int_0^\infty \|x(t)\|_p^p \, dt\right)^{\frac{1}{p}} = \left(\int_0^\infty (|x_1(t)|^p + \cdots + |x_M(t)|^p) \, dt\right)^{\frac{1}{p}} \tag{2.236}$$

and its L^∞ signal norm is defined as

$$\|x(\cdot)\|_\infty = \sup_{t \geq 0} \|x(t)\|_\infty = \sup_{t \geq 0} \max_{1 \leq i \leq M} |x_i(t)|; \tag{2.237}$$

for a *discrete-time* signal $x(t) \in R^M$, its L^p ($p \in [1, \infty)$) signal norm is

$$\|x(\cdot)\|_p = \left(\sum_{t=0}^{\infty} \|x(t)\|_p^p\right) = \left(\sum_{t=0}^{\infty} (|x_1(t)|^p + \cdots + |x_M(t)|^p)\right)^{\frac{1}{p}} \tag{2.238}$$

and its L^∞ signal norm is

$$\|x(\cdot)\|_\infty = \sup_{t \geq 0} \|x(t)\|_\infty = \sup_{t \geq 0} \max_{1 \leq i \leq M} |x_i(t)|. \tag{2.239}$$

Definition of Operator Norms

The input–output mapping of a system can be considered as an operator (see Section 2.5.3). For a linear time-invariant case, the system transfer matrix $H(D)$ is represented by the system impulse response $h(t) \in R^{M \times M}$ (which is the inverse Laplace transform or z-transform of $H(D)$):

$$\begin{aligned} y(t) &= H(D)[u](t) = h(t) * u(t) \\ &= \begin{cases} \int_0^t h(t-\tau) u(\tau) \, d\tau & \text{in continuous time,} \\ \sum_{\tau=0}^t h(t-\tau) u(\tau) & \text{in discrete time,} \end{cases} \end{aligned} \tag{2.240}$$

where, with $\delta(t)$ given in (2.15), $\delta_d(t)$ defined as $\delta_d(t) = 0$ for $t \neq 0$ and $\delta_d(0) = 1$, and $u_d(t)$ defined as $u_d(t) = 1$ for $t \geq 0$ and $u_d(t) = 0$ for $t < 0$,

$$g(t) = \begin{cases} Ce^{At}B + D\delta(t) & \text{in continuous time,} \\ CA^{t-1}Bu_d(t-1) + D\delta_d(t) & \text{in discrete time,} \end{cases} \tag{2.241}$$

for ($A \in R^{n\times n}$, $B \in R^{n\times M}$, $C \in R^{M\times n}$, $D \in R^{M\times M}$) being a minimal realization of $H(D)$, that is, $H(D) = C(DI - A)^{-1}B + D$.

The (induced) operator norm of H is defined as

$$\|H\| = \sup_{\|u(\cdot)\|=1} \|y(\cdot)\| = \sup_{\|u(\cdot)\|=1} \|h * u(\cdot)\|, \tag{2.242}$$

where $\|u(\cdot)\|$ and $\|y(\cdot)\|$ are a specific L^1, L^2, or L^∞ signal norm.

The following results [78] are useful for adaptive control design and analysis.

L^1 Operator Norms

Let $u(t) \in L^1$. Then, for the *continuous-time* case, the L^1 norm of H is

$$\|H\|_1 = \|h(\cdot)\|_1 \triangleq \max_{1\leq j\leq M} \int_0^\infty \sum_{i=1}^M |h_{ij}(\tau)|\, d\tau, \tag{2.243}$$

where $h_{ij}(t)$ is the (i,j)th element of $h(t)$, which is the impulse response of $H_{ij}(s)$, the (i,j)th element of $H(s)$; and for the *discrete-time* case, the L^1 norm of H is

$$\|H\|_1 = \|h(\cdot)\|_1 \triangleq \max_{1\leq j\leq M} \sum_{\tau=0}^\infty \sum_{i=1}^M |h_{ij}(\tau)|, \tag{2.244}$$

where $h_{ij}(t)$ is the impulse response of $H_{ij}(z)$, the (i,j)th element of $H(z)$.

L^2 Operator Norms

Let $u(t) \in L^2$. Then, for the *continuous-time* case, the L^2 norm of H is

$$\|H\|_2 = \|h(\cdot)\|_2 \triangleq \sqrt{\sup_{\omega\geq 0} \max_{1\leq i\leq M} \lambda_i[H^T(-j\omega)H(j\omega)]}, \tag{2.245}$$

where $\lambda_i[\cdot]$ denotes the ith eigenvalue of a matrix, and $H(j\omega) = H(s)|_{s=j\omega}$; for the *discrete-time* case, the L^2 norm of H is

$$\|H\|_2 = \|h(\cdot)\|_2 \triangleq \sqrt{\sup_{\Omega\in[0,2\pi]} \max_{1\leq i\leq M} \lambda_i[H^T(e^{-j\Omega})H(e^{j\Omega})]}, \tag{2.246}$$

where $H(e^{j\Omega}) = H(z)|_{z=e^{j\Omega}}$. The norms defined in (2.245) and (2.246) are also called the H^∞ norms of the (stable) transfer matrices $H(s)$ and $H(z)$ (or other notation $G(s)$, $G(z)$), denoted as $\|H(s)\|_\infty$ and $\|H(z)\|_\infty$, respectively.

L^∞ Operator Norms

Let $u(t) \in L^\infty$. Then, for the *continuous-time* case, the L^∞ norm of H is

$$\|H\|_\infty = \|h(\cdot)\|_\infty \triangleq \max_{1\le i\le M} \int_0^\infty \sum_{j=1}^{M} |h_{ij}(\tau)|\, d\tau; \tag{2.247}$$

for the *discrete-time* case, the L^∞ norm of H is

$$\|H\|_\infty = \|h(\cdot)\|_\infty \triangleq \max_{1\le i\le M} \sum_{\tau=0}^{\infty} \sum_{j=1}^{M} |h_{ij}(\tau)|. \tag{2.248}$$

Remark 2.2 The above operator norms are equivalent on the linear space of $M \times M$ stable and proper rational transfer matrices $H(D)$ of the same minimal order n, in the sense that, for example, there exist constants $c_1 > 0$ and $c_1 > 0$, independent of $H(D)$ with given M and n, such that $c_1\|H\|_\infty \le \|H\|_1 \le c_2\|H\|_\infty$, similar to that in the vector and matrix norms case.

When $H(D)$ is a scalar transfer function ($M = 1$), it follows that $\|H\|_\infty = \|H\|_1 = \|h(\cdot)\|_1$ and that $\|H\|_2 \le \|H\|_1$ [78]. Note that in this case $\|H\|_2 = \sup_{\omega\ge 0} |H(j\omega)| = \|H(s)\|_\infty$ or $\|H\|_2 = \sup_{\Omega\in[0,2\pi]} |H(e^{j\Omega})| = \|H(z)\|_\infty$. □

System Properties Based Operator Norms

For system (2.235), $y(t) = H(D)[u](t)$, the following properties hold:

$$\|y(\cdot)\|_p \le \|H\|_p\, \|u(\cdot)\|_p,\ p = 1, 2, \infty, \tag{2.249}$$

$$\sup_{\tau\in[0,t]} \|y(\tau)\|_\infty \le \|H\|_\infty \sup_{\tau\in[0,t]} \|u(\tau)\|_\infty, \tag{2.250}$$

$$\left(\int_0^t \|y(\tau)\|_p^p\, d\tau\right)^{\frac{1}{p}} \le \|H\|_p \left(\int_0^t \|u(\tau)\|_p^p\, d\tau\right)^{\frac{1}{p}},\ p = 1, 2, \tag{2.251}$$

$$\left(\sum_{\tau=0}^{t} \|y(\tau)\|_p^p\right)^{\frac{1}{p}} \le \|H\|_p \left(\sum_{\tau=0}^{t} \|u(\tau)\|_p^p\right)^{\frac{1}{p}},\ p = 1, 2. \tag{2.252}$$

The properties (2.250)–(2.252) are for the truncated signals $y(\tau)$, $u(\tau)$, $\tau \in [0, t]$, and can be proved by considering an input $u(\cdot)$ with $u(\tau) = 0$ for $\tau > t$.

2.9 Pole Placement

For a system in continuous time: $\dot{x}(t) = Ax(t) + Bu(t)$, $y = Cx(t)$ (or in discrete time: $x(t+1) = Ax(t) + Bu(t)$, $y(t) = Cx(t)$), with $x \in R^n$, $u \in R$, $y \in R$, $A \in R^{n\times n}$, $B \in R^n$ and $C \in R^{1\times n}$, the system transfer function is $G(s) = C(sI - A)^{-1}B$ (or $G(z) = C(zI - A)^{-1}B$), so that the system poles are the eigenvalues of the matrix A. Using a state feedback control $u(t) = -Kx(t) + r$, where $K \in R^{1\times n}$ is a constant gain vector, and $r(t)$ is a reference input. The resulting closed-loop poles are the eigenvalues of the matrix $A - BK$. If (A, B) is controllable, that is, the $n \times n$ matrix $M_c = [B|AB|A^2B|\cdots|A^{n-1}B]$ is nonsingular, then for any given n poles p_i, $i = 1, 2, \ldots, n$, there is a gain vector K such that the eigenvalues of the matrix $A - BK$ are equal to p_i, $i = 1, 2, \ldots, n$. This is the so-called *pole placement* design, for which the key is to find such a gain vector K such that

$$\begin{aligned}\det(\lambda I - A + BK) &= (s - p_1)(s - p_2)\cdots(s - p_n) \\ &\triangleq \lambda^n + \alpha_{n-1}\lambda^{n-1} + \cdots + \alpha_1\lambda + \alpha_0 = \alpha(\lambda). \quad (2.253)\end{aligned}$$

Canonical form design. For A, B in the controllable canonical form:

$$A = A^c = \begin{bmatrix} 0 & 1 & 0 & \cdots & 0 & 0 \\ 0 & 0 & 1 & 0 & \cdots & 0 \\ \vdots & \vdots & \vdots & \vdots & \vdots & \vdots \\ 0 & 0 & \cdots & \cdots & 0 & 1 \\ -a_0 & -a_1 & \cdots & \cdots & -a_{n-2} & -a_{n-1} \end{bmatrix}, B = B^c = \begin{bmatrix} 0 \\ 0 \\ \vdots \\ 0 \\ 1 \end{bmatrix} \quad (2.254)$$

such that $\det(\lambda I - A^c) = \lambda^n + a_{n-1}\lambda^{n-1} + \ldots + a_1\lambda + a_0$, the design is

$$K = [k_1, k_2, \ldots, k_n] = K^c = [\alpha_0 - a_0, \alpha_1 - a_1, \ldots, \alpha_{n-1} - a_{n-1}]. \quad (2.255)$$

General form design. For a general pair of (A, B), the design is based on a transformation matrix T_c such that $T_c A T_c^{-1} = A^c$ and $T_c B = B^c$, where (A^c, B^c) is in the canonical form (2.254), and T_c is defined from its inverse:

$$T_c^{-1} = [B|AB|A^2B|\cdots|A^{n-1}B] \begin{bmatrix} a_1 & a_2 & \cdots & a_{n-1} & 1 \\ a_2 & \cdots & a_{n-1} & 1 & 0 \\ \vdots & \vdots & \vdots & \vdots & \vdots \\ a_{n-1} & 1 & 0 & \cdots & 0 \\ 1 & 0 & 0 & \cdots & 0 \end{bmatrix}. \quad (2.256)$$

This transformation can be directly verified from $AT_c^{-1} = T_c^{-1}A^c$, using the Cayley-Hamilton equality: $A^n + a_{n-1}A^{n-1} + \cdots + a_1 A + a_0 I = 0$, and from $T_c^{-1}B^c = B$. Then, the desired feedback gain vector is

$$K = K^c T_c,\ K^c = [\alpha_0 - a_0, \alpha_1 - a_1, \ldots, \alpha_{n-1} - a_{n-1}], \quad (2.257)$$

such that $\det(\lambda I - A + BK) = \det(\lambda I - A^c + B^c K^c) = \alpha(\lambda)$.

Problems

2.1 Prove the inequality (2.75) (*Hint*: In the (x, y)-plane, consider the curve $y = x^{p-1}$, or equivalently, $x = y^{q-1}$, for $p \in [1, \infty)$, $\frac{1}{p} + \frac{1}{q} = 1$, and for $a \geq 0$ and $b \geq 0$, evaluate $\int_0^a x^{p-1}\,dx$, $\int_0^b y^{q-1}\,dy$, and relate them to ab).

2.2 Prove Hölder inequality (2.72). (*Hint*: Use the inequality (2.75), with

$$a = \frac{|f(t)|}{\left(\int_c^d |f(t)|^p\,dt\right)^{\frac{1}{p}}},\ b = \frac{|g(t)|}{\left(\int_c^d |g(t)|^q\,dt\right)^{\frac{1}{q}}},$$

and integrate it.)

2.3 Prove Minkovski inequality (2.74). (*Hint*: Use Hölder inequality (2.72) and $(|f(t)| + |g(t)|)^p = (|f(t)| + |g(t)|)^{p-1}|f(t)| + (|f(t)| + |g(t)|)^{p-1}|g(t)|$.)

2.4 Derive Hölder and Minkovski inequalities for $p = 1$ and $q = \infty$.

2.5 Analyze the stability of the system $\dot{x}_1 = -x_1^3 + x_2^2 x_1$, $\dot{x}_2 = -2x_1^2 x_2 - x_2$. (*Hint*: Use the Lyapunov function candidate $V(x) = 2x_1^2 + x_2^2$.)

2.6 (i) Find the explicit solution of the system $\dot{x} = (6t\cos(t) - t)x$, $x(t_0) = x_0$ and analyze its stability. (ii) Examine the stability of the equilibrium state $x = 0$ of the system $\dot{x} = -x^5$ with $x(0) = x_0$.

2.7 Analyze the stability of the system $\dot{x} = -\zeta(t)\zeta^T(t)x$, where $\zeta \in R^n$ is a unformly bounded and continuous vector signal. (*Hint*: Use $V(x, t) = x^T x$.)

2.8 Consider the system $\dot{x} = Ax + bw$, where $x(t) \in R^n$, $b \in R^n$ is a constant vector, and $A \in R^{n\times n}$ is a constant matrix with all eigenvalues having negative real parts, and $w(t) \in R$ is a bounded noise. Show that the solutions of this system are uniformly bounded and uniformly ultimately bounded.

2.9 Consider each $x(t)$ in Problems 2.5–2.8. To which of the spaces L^1, L^2, L^∞ does the signal $x(t)$ belong?

2.10 Consider the system

$$\dot{e} = -2e - \theta^T\omega,\ e \in R, \theta \in R^n, \omega \in R^n$$

$$\dot{\theta} = \Gamma e\omega,\ 0 < \Gamma = \Gamma^T \in R^{n\times n}.$$

(i) Prove that $\theta \in L^\infty$ and $e \in L^\infty$ for any $\omega(t)$.
(ii) Show that $e \in L^2$ for any $\omega(t)$.
(iii) Prove that $\lim_{t\to\infty} e(t) = 0$ if $\omega(t) \in L^\infty$.
(iv) Prove that $\lim_{t\to\infty} \theta(t) = 0$ if $\omega = 1$ and $n = 1$.
(v) Study the convergence of $\theta = [\theta_1, \theta_2]^T$ for $\Gamma = I_n$, $\omega = [1, -1]^T$.

2.11 Consider the system

$$\dot{x} = -x - z(t)x, \; x \in R, z(t) \in R, x(0) = 1.$$

(i) Find the range of a constant $z(t) = z_0$ for exponential stability.
(ii) Prove that $\lim_{t\to\infty} x(t) = 0$ exponentially for $z(t) \in L^1$.
(iii) Prove that $\lim_{t\to\infty} x(t) = 0$ exponentially for $z(t) = a \sin t$, $a \in R$.
(iv) Find the range of a such that $\lim_{t\to\infty} x(t) = 0$ for $z(t) = a|\sin t|$.
(v) Show that $\lim_{t\to\infty} x(t) = 0$ for $z(t) \in L^p$, $1 < p < \infty$. (*Hint:* With $f(t) = |z(t)|$ and $g(t) = 1$, use Hölder's inequality.)

2.12 Consider the system

$$\dot{x}_1 = -x_1 + x_1^3 - x_2, \; \dot{x}_2 = -x_2 + x_2^3 + x_1,$$

where $x_1 \in R$, $x_2 \in R$. Examine the stability of $x_e = [0, 0]^T$.

2.13 (i) Apply Lemma 2.1 to the case

$$u(t) \leq c + \int_{t_0}^{t} k(\tau)u(\tau)\,d\tau, \; \forall t \geq t_0,$$

where $k(t)$ is continuous and $k(t) \geq 0$, $\forall t \geq t_0$, and c is a constant, and compare your result with Lemma 2.2.

(ii) Show that if

$$0 \leq u(t) \leq \int_{t_0}^{t} k(\tau)u(\tau)\,d\tau, \; \forall t \geq t_0,$$

where $k(t)$ is continuous and $k(t) \geq 0$, then $u(t) \equiv 0$, $\forall t \geq t_0$.

2.14 Use Lemma 2.2 to study the stability of the system

$$\dot{x} = A(t)x + g(x, t), \; x(t_0) = x_0,$$

where $A(t) \in R^{n\times n}$, $g(x,t) \in R^n$ are continuous with $g(0,t) = 0$, $\forall t \geq 0$. Show that if the equilibrium state $x_e = 0$ of the unperturbed linear system $\dot{x} = A(t)x$ is uniformly asymptotically stable and

$$\|g(x,t)\| \leq \gamma(t)\|x\|, \; \forall t \geq t_0, \; \forall x \in R^n,$$

for some continuous function $\gamma(t) \geq 0$, $\forall t \geq t_0$, such that

$$\int_{t_1}^{t_2} \gamma(t)\, dt \leq \nu(t_2 - t_1) + \gamma_0, \ \forall t_2 > t_1 \geq t_0,$$

where $\gamma_0 \geq 0$, $\nu \geq 0$, then there exists a $\nu^* > 0$ such that for any $\nu \in [0, \nu^*)$ the equilibrium state $x_e = 0$ of the perturbed system is uniformly asymptotically stable. (*Hint*: Use $x(t) = \Phi(t, t_0)x_0 + \int_{t_0}^{t} \Phi(t, \tau)g(x(\tau), \tau)\, d\tau$ and $\|\Phi(t, \tau)\| \leq \beta e^{-\alpha(t-\tau)}$ for some $\beta > 0$, $\alpha > 0$, $\forall t \geq \tau \geq t_0$.)

2.15 Redo Problem 2.14 for

$$\int_{t_1}^{t_2} \gamma^p(t)\, dt \leq \nu(t_2 - t_1) + \gamma_0, \ \forall t_2 > t_1 \geq t_0.$$

2.16 Investigate the question stated at the end of Section 2.5.2.

2.17 Consider the rational function $h(s) = k\frac{s+a}{s^2+b_1 s+b_2}$.
(i) Find the conditions on k, a, b_1, b_2 for $h(s)$ to be strictly positive real.
(ii) With $h(s)$ being strictly positive real, $k = 1$, $a = 1$, $b_1 = 3$, $b_2 = 1$, solve the equations in the Lefschetz-Kalman-Yakubovich lemma for

$$A = \begin{bmatrix} 0 & 1 \\ -b_2 & -b_1 \end{bmatrix}, \ c = [a, 1], \ b = [0, 1]^T, \ d = 0, \ L = \begin{bmatrix} 1 & 0 \\ 0 & 1 \end{bmatrix}.$$

(iii) Show that for $a = b_1$, $h(s)$ can be positive real but not strictly positive real and the Lefschetz-Kalman-Yakubovich equations cannot be satisfied.

2.18 Show that $h(s) = \frac{k}{s^2+b_1 s+b_2}$ cannot be PR for any $k \neq 0$, b_1, and b_2.

2.19 Prove that if $h_1(s)$ and $h_2(s)$ is (strictly) positive real, then $h_1(s)+h_2(s)$ are (strictly) positive real. How about $h_1(s)h_2(s)$? How about $\frac{h_1(s)}{1+h_1(s)h_2(s)}$? How about $\frac{h_1(s)h_2(s)}{1+h_1(s)h_2(s)}$? How about $\frac{h_1(s)h_2(s)}{1+h_1(s)}$?

2.20 Given that $h(s) = k\frac{s+a}{s^2+b_1 s+b_2}$ is PR, show that $\frac{1}{h(s)}$ is also PR.

2.21 Use one of the signal convergence lemmas in Section 2.6 to show $\lim_{t\to\infty} e(t) = 0$ for the adaptive control system $\dot{y}(t) = -ay(t) + u(t)$, $u(t) = k(t)y(t) + r(t)$, $\dot{k}(t) = -e(t)y(t)$, $e(t) = y(t) - y_m(t)$, $\dot{y}_m(t) = -a_m y_m(t) + r(t)$, where $r(t)$ is bounded.

2.22 Derive the discrete-time versions of the Hölder inequality, Schwarz inequality, and Minkovski inequality by using (2.75).

2.23 Prove Theorem 2.13.

2.24 Verify (2.223).

Chapter 3

Adaptive Parameter Estimation

Dynamic systems are characterized by their structures and parameters. When input signals are applied to a system, response signals are generated by the system. Parameter estimation is to use a collection of available system signals, based on certain system structure information and optimality criteria, to produce estimates of the system parameters. Adaptive parameter estimation is a dynamic estimation procedure that makes use of updated system signals to produce updated parameter estimates on-line. Dynamic behavior of adaptive parameter estimates can be rigorously analyzed. Adaptive parameter estimation is crucial for adaptive control designs where controller parameters need to be updated on-line as the controlled system is operating.

Key issues in adaptive parameter estimation include linear parametrization of system models, linear representation of parametric error models, stable design of adaptive estimation algorithms, analytical proof of system stability, parameter convergence, and robustness of adaptive estimation.

3.1 A Parametrized System Model

Consider a linear time-invariant system described by the differential equation

$$P(s)[y](t) = Z(s)[u](t), \tag{3.1}$$

where $y(t) \in R$ and $u(t) \in R$ are the measured system input and output;

$$P(s) = s^n + p_{n-1}s^{n-1} + \cdots + p_1 s + p_0, \tag{3.2}$$

$$Z(s) = z_m s^m + z_{m-1}s^{m-1} + \cdots + z_1 s + z_0 \tag{3.3}$$

are polynomials in s with s being the time differentiation operator $s[x](t) = \dot{x}(t)$; and p_i, $i = 0, 1, \ldots, n-1$, z_i, $i = 0, 1, \ldots, m$, with $n > m$, are the unknown but constant system parameters to be estimated.

Parametrization

Choose a stable polynomial $\Lambda(s) = s^n + \lambda_{n-1}s^{n-1} + \cdots + \lambda_1 s + \lambda_0$ (its zeros all in $\mathrm{Re}[s] < 0$). Operating both sides of (3.1) by the stable filter $\frac{1}{\Lambda(s)}$, we have

$$y(t) = \frac{Z(s)}{\Lambda(s)}[u](t) + \frac{\Lambda(s) - P(s)}{\Lambda(s)}[y](t). \tag{3.4}$$

Introducing the parameter and regressor vectors

$$\theta^* = [z_0, z_1, \ldots, z_{m-1}, z_m, \\ \lambda_0 - p_0, \lambda_1 - p_1, \ldots, \lambda_{n-2} - p_{n-2}, \lambda_{n-1} - p_{n-1}]^T \in R^{n+m+1}, \tag{3.5}$$

$$\phi(t) = \left[\frac{1}{\Lambda(s)}[u](t), \frac{s}{\Lambda(s)}[u](t), \ldots, \frac{s^{m-1}}{\Lambda(s)}[u](t), \frac{s^m}{\Lambda(s)}[u](t), \right. \\ \left. \frac{1}{\Lambda(s)}[y](t), \frac{s}{\Lambda(s)}[y](t), \ldots, \frac{s^{n-2}}{\Lambda(s)}[y](t), \frac{s^{n-1}}{\Lambda(s)}[y](t)\right]^T \in R^{n+m+1}, \tag{3.6}$$

we can express (3.4) as

$$y(t) = \theta^{*T}\phi(t). \tag{3.7}$$

We can also introduce $\Lambda_{n-1}(s) = \lambda_{n-1}s^{n-1} + \cdots + \lambda_1 s + \lambda_0$ and

$$\theta_p^* = [z_0, z_1, \ldots, z_{m-1}, z_m, -p_0, -p_1, \ldots, -p_{n-2}, -p_{n-1}]^T \in R^{n+m+1} \tag{3.8}$$

to parametrize (3.4) as

$$y(t) - \frac{\Lambda_{n-1}(s)}{\Lambda(s)}[y](t) = \theta_p^{*T}\phi(t). \tag{3.9}$$

When some of the system parameters are known, the dimension of the parameter vector can be reduced. For example, if p_1 is known, then the unknown parameter vector does not contain the known parameter $\lambda_1 - p_1$, that is,

$$\bar{\theta}^* = [z_0, z_1, \ldots, z_{m-1}, z_m, \\ \lambda_0 - p_0, \lambda_2 - p_2, \ldots, \lambda_{n-2} - p_{n-2}, \lambda_{n-1} - p_{n-1}]^T \in R^{n+m} \tag{3.10}$$

and the associated regressor vector does not contain $\frac{s}{\Lambda(s)}[y](t)$, that is,

$$\bar{\phi}(t) = \left[\frac{1}{\Lambda(s)}[u](t), \frac{s}{\Lambda(s)}[u](t), \ldots, \frac{s^{m-1}}{\Lambda(s)}[u](t), \frac{s^m}{\Lambda(s)}[u](t), \right. \\ \left. \frac{1}{\Lambda(s)}[y](t), \frac{s^2}{\Lambda(s)}[y](t), \ldots, \frac{s^{n-2}}{\Lambda(s)}[y](t), \frac{s^{n-1}}{\Lambda(s)}[y](t)\right]^T \in R^{n+m}. \tag{3.11}$$

Then, we can express (3.4) as

$$y(t) - (\lambda_1 - p_1)\frac{s}{\Lambda(s)}[y](t) = \bar{\theta}^{*T}\bar{\phi}(t),\ t \geq t_0. \tag{3.12}$$

Implementation

To generate the vector signal $\phi(t)$, instead of implementing the individual filters $\frac{s^i}{\Lambda(s)}$, $i = 0, 1, \ldots, n-1$, in (3.6), for a chosen stable polynomial $\Lambda(s) = s^n + \lambda_{n-1}s^{n-1} + \cdots + \lambda_1 s + \lambda_0$, we construct two dynamic systems

$$\dot{\omega}_1(t) = A_\lambda \omega_1(t) + bu(t) \tag{3.13}$$

$$\dot{\omega}_2(t) = A_\lambda \omega_2(t) + by(t), \tag{3.14}$$

where $\omega_1(t) \in R^n$, $\omega_1(t) \in R^n$, and

$$A_\lambda = \begin{bmatrix} 0 & 1 & 0 & \cdots & 0 & 0 \\ 0 & 0 & 1 & 0 & \cdots & 0 \\ \vdots & \vdots & \vdots & \vdots & \vdots & \vdots \\ 0 & 0 & \cdots & \cdots & 0 & 1 \\ -\lambda_0 & -\lambda_1 & \cdots & \cdots & -\lambda_{n-2} & -\lambda_{n-1} \end{bmatrix}, \; b = \begin{bmatrix} 0 \\ \vdots \\ 0 \\ 1 \end{bmatrix} \in R^n. \tag{3.15}$$

Then, we generate the regressor vector $\phi(t)$ from

$$\phi(t) = \left[(C_m \omega_1(t))^T, \omega_2^T(t) \right]^T, \tag{3.16}$$

where, with I_{m+1} being the $(m+1)\times(m+1)$ identity matrix and $0_{(m+1)\times(n-m-1)}$ being the $(m+1) \times (n-m-1)$ zero matrix,

$$C_m = [I_{m+1}, 0_{(m+1)\times(n-m-1)}] \in R^{(m+1)\times n}. \tag{3.17}$$

The representation (3.16) means that $C_m\omega_1(t)$ contains the first $m+1$ components of $\omega_1(t)$. In particular, $\phi(t) = [\omega_1^T(t), \omega_2^T(t)]^T$ for $m = n-1$.

3.2 Linear Parametric Models

A linear parametric model has the form

$$y(t) = \theta^{*T}\phi(t), \; t \geq t_0, \tag{3.18}$$

where $\theta^* \in R^{n_\theta}$ is an unknown parameter vector, $y(t) \in R$ is a known (measured) signal and $\phi(t) \in R^{n_\theta}$ is a known vector signal (called a *regressor* vector), and $n_\theta > 0$ is the dimension of the parametric model ($n_\theta = n+m+1$ for (3.7)). The model (3.18) is commonly seen in system modeling when unknown system parameters can be separated from known system signals. While, in general, the components of $\phi(t)$ may contain nonlinear functions of some system signals, the parametrized models (3.7), (3.9), and (3.12) have linear components of filtered $u(t)$ and $y(t)$.

To develop an adaptive algorithm, based on the available signals $y(t)$ and $\phi(t)$, to estimate the unknown parameter vector θ^* in (3.18) on-line, a parametric error model is needed. Letting $\theta(t)$ be the estimate of θ^* to be obtained from an adaptive update law, we define the estimation error

$$\epsilon(t) = \theta^T(t)\phi(t) - y(t), \; t \geq t_0. \tag{3.19}$$

Using (3.7), we have

$$\epsilon(t) = \tilde{\theta}^T(t)\phi(t), \; \tilde{\theta}(t) = \theta(t) - \theta^*, \; t \geq t_0, \tag{3.20}$$

which is a linear parametric error model (linear in the parameter error $\tilde{\theta}(t)$).

For the parametric model (3.9) or (3.12), we define the estimation error

$$\epsilon(t) = \begin{cases} \theta_p^T(t)\phi(t) - \left(y(t) - \frac{\Lambda_{n-1}(s)}{\Lambda(s)}[y](t)\right) & \text{for (3.9)}, \\ \bar{\theta}^T(t)\bar{\phi}(t) - \left(y(t) - (\lambda_1 - p_1)\frac{s}{\Lambda(s)}[y](t)\right) & \text{for (3.12)}, \end{cases} \tag{3.21}$$

where $\theta_p(t)$ and $\bar{\theta}(t)$ are the estimates of θ_p^* and $\bar{\theta}^*$, respectively.

The linear error model (3.20) is common in adaptive control and system identification for which the key task is to develop an adaptive law to update the parameter estimate $\theta(t)$, based on the knowledge of the measured signals $y(t)$ and $\phi(t)$, to ensure some desired properties for $\theta(t)$. Next, we present two of the most commonly used adaptive parameter update laws: a normalized gradient algorithm and a normalized least-squares algorithm.

3.3 Normalized Gradient Algorithm

A normalized gradient algorithm for updating the parameter estimate $\theta(t)$ is to choose the derivative of $\theta(t)$, in a steepest descent direction, to successively generate $\theta(t)$ to minimize a normalized quadratic cost function

$$J(\theta) = \frac{\epsilon^2}{2m^2} = \frac{(\theta - \theta^*)^T\phi\phi^T(\theta - \theta^*)}{2m^2}, \tag{3.22}$$

where $\epsilon = \theta^T\phi - y$ as in (3.19) and $m(t)$ is a so-called normalizing signal which does not explicitly depend on $\theta(t)$. (A preferred choice of $m(t)$ is the one which ensures the boundedness of $\frac{\phi\phi^T}{m^2}$; see Lemma 3.1 below.)

The steepest descent direction of $J(\theta)$ is $-\frac{\partial J}{\partial \theta}$ [325]. For the above chosen $J(\theta)$, $\frac{\partial J}{\partial \theta} = \frac{\phi\epsilon}{m^2}$, which suggests the adaptive update law for $\theta(t)$,

$$\dot{\theta}(t) = -\frac{\Gamma\phi(t)\epsilon(t)}{m^2(t)}, \; \theta(t_0) = \theta_0, \; t \geq t_0, \tag{3.23}$$

where $\Gamma = \Gamma^T > 0$ is a gain matrix, θ_0 is an initial estimate of θ^*, and

$$m(t) = \sqrt{1 + \kappa \phi^T(t)\phi(t)} \tag{3.24}$$

with $\kappa > 0$ being a design parameter.

Lemma 3.1 *The adaptive algorithm (3.23) guarantees*

(i) $\theta(t)$, $\dot{\theta}(t)$, $\frac{\epsilon(t)}{m(t)}$ *are bounded; and*

(ii) $\frac{\epsilon(t)}{m(t)}$ *and* $\dot{\theta}(t)$ *belong to* L^2.

Proof: (i) Introducing the positive definite function $V(\tilde{\theta}) = \tilde{\theta}^T \Gamma^{-1} \tilde{\theta}$, we have the time derivative of $V = V(\tilde{\theta}(t))$, along (3.23), as

$$\dot{V} = \frac{dV}{dt} = -\frac{2\tilde{\theta}^T(t)\phi(t)\phi^T(t)\tilde{\theta}(t)}{m^2(t)} = -\frac{2\epsilon^2(t)}{m^2(t)},\ t \geq t_0, \tag{3.25}$$

where $\dot{\tilde{\theta}}(t) = \dot{\theta}(t)$ as θ^* is constant. Hence $V(\tilde{\theta}(t))$, $\tilde{\theta}(t)$, and $\theta(t)$ are bounded.

From the boundedness of $\tilde{\theta}(t)$ and the inequality

$$\frac{|\epsilon(t)|}{m(t)} \leq \frac{\|\tilde{\theta}(t)\|_2 \|\phi(t)\|_2}{\sqrt{1 + \kappa \|\phi(t)\|^2}}, \tag{3.26}$$

we have the boundedness of $\frac{\epsilon(t)}{m(t)}$, which, together with the inequality

$$\|\dot{\theta}(t)\|_2 \leq \frac{\|\Gamma\|_2 \|\phi(t)\|_2}{\sqrt{1 + \kappa \|\phi(t)\|_2^2}} \frac{|\epsilon(t)|}{m(t)}, \tag{3.27}$$

implies that $\dot{\theta}(t)$ is bounded.

(ii) It follows from (3.25) and the boundedness of $V(\tilde{\theta}(t))$ that

$$2\int_0^t \frac{\epsilon^2(\tau)}{m^2(\tau)}\, d\tau = V(\tilde{\theta}(0)) - V(\tilde{\theta}(t)) \leq V(\tilde{\theta}(0)),\ t \geq 0, \tag{3.28}$$

that is, $\frac{\epsilon(t)}{m(t)} \in L^2$, and then from (3.27) that $\dot{\theta}(t) \in L^2$. ∇

Equation (3.25) shows that system (3.23), which has the form

$$\dot{\tilde{\theta}} = -\frac{\Gamma \phi(t) \phi^T(t)}{m^2(t)} \tilde{\theta}, \tag{3.29}$$

has a uniformly stable equilibrium point at $\tilde{\theta} = 0$. If we consider $V(\tilde{\theta}) = \tilde{\theta}^T \Gamma^{-1} \tilde{\theta}$ as a measure of the parameter error $\tilde{\theta} = \theta(t) - \theta^*$, (3.25) indicates that the choice of the gradient adaptive law (3.23) makes $V(\tilde{\theta})$ nonincreasing always and decreasing whenever $\epsilon(t)$ is nonzero. The parameter estimation objective is to make $V(\tilde{\theta}(t))$ converge to zero, which implies that the signal $\epsilon(t)$ will converge to zero. For $\epsilon(t) = 0$, it is sufficient but not necessary that $\tilde{\theta} = \theta(t) - \theta^* = 0$. Therefore, from (3.25), the gradient algorithm (3.23) is to make $\frac{\epsilon(t)}{m(t)}$ small but not $\tilde{\theta} = \theta(t) - \theta^*$ small, in general. From (3.20), to make $\tilde{\theta} = \theta(t) - \theta^*$ small when $\epsilon(t)$ is small, certain conditions on the vector signal $\phi(t)$ are needed (see Section 3.5). From (3.25), we conclude that $\lim_{t\to\infty} V(\tilde{\theta}(t)) = \lim_{t\to\infty} \tilde{\theta}^T(t)\Gamma^{-1}\tilde{\theta}(t) = V_0$ for some constant $V_0 \geq 0$ as $V(\tilde{\theta}(t))$ is nonincreasing and bounded from below ($V(\tilde{\theta}(t)) \geq 0$). If $V_0 = 0$, then $\lim_{t\to\infty} \theta(t) = \theta^*$. However, for $n_\theta > 1$, if $V_0 > 0$, we cannot conclude $\lim_{t\to\infty} \theta(t) = \theta_\infty$ for some constant vector $\theta_\infty \in R^{n_\theta}$.

If $\ddot{\theta}(t) \in L^\infty$, then $\dot{\theta}(t) \in L^2 \cap L^\infty$ implies that $\lim_{t\to\infty} \dot{\theta}(t) = 0 \in R^{n_\theta}$. However, this does not mean $\lim_{t\to\infty} \theta(t) = \theta_\infty$ for some constant vector θ_∞.

A simple example is $\theta(t) = [\sin((t+1)^{\frac{1}{4}}) - 1, \cos((t+1)^{\frac{1}{4}}) + 1]^T \in R^2$, for which $\dot{\theta}(t) = [\frac{\cos((t+1)^{\frac{1}{4}})}{4(t+1)^{\frac{3}{4}}}, \frac{-\sin((t+1)^{\frac{1}{4}})}{4(t+1)^{\frac{3}{4}}}]^T \in L^2 \cap L^\infty$, and $\lim_{t\to\infty} \dot{\theta}(t) = 0$, but $\lim_{t\to\infty} \theta(t)$ does not exist. For $\theta^* = [-1, 1]^T$ and $\Gamma = I_2$, $V(\tilde{\theta}(t)) = V_0 = 1$.

When applied to the parametric model (3.7), the adaptive law (3.23) is called an adaptive estimator (identifier) for the system parameter vector θ^*.

3.4 Normalized Least-Squares Algorithm

Another commonly used parameter estimation algorithm for updating $\theta(t)$ in the error model (3.20) is the normalized least-squares algorithm:

$$\dot{\theta}(t) = -\frac{P(t)\phi(t)\epsilon(t)}{m^2(t)}, \; \theta(t_0) = \theta_0, \; t \geq t_0, \tag{3.30}$$

$$\dot{P}(t) = -\frac{P(t)\phi(t)\phi^T(t)P(t)}{m^2(t)}, \; P(t_0) = P_0 = P_0^T > 0, \; t \geq t_0, \tag{3.31}$$

$$m(t) = \sqrt{1 + \kappa\phi^T(t)P(t)\phi(t)}, \; \kappa > 0, \tag{3.32}$$

where θ_0 is the initial estimate of $\theta^* \in R^{n_\theta}$, P_0 is the initial value of the gain matrix $P(t) \in R^{n_\theta \times n_\theta}$ for $\theta \in R^{n_\theta}$, and κ is a design parameter.

Lemma 3.2 *The adaptive algorithm (3.30)–(3.31) guarantees*

(i) $P(t) = P^T(t) > 0$, $\forall t \geq t_0$, *and* $P(t)$, $\dot{P}(t)$ *are bounded;*

(ii) $\theta(t)$ *and* $\frac{\epsilon(t)}{\bar{m}(t)}$ *are bounded, where* $\bar{m}(t) = \sqrt{1 + \phi^T(t)\phi(t)}$;

(iii) $\frac{\epsilon(t)}{m(t)}$, $\frac{\epsilon(t)}{\bar{m}(t)}$, *and* $\dot{\theta}(t)$ *belong to* L^2; *and*

(iv) there exist a constant matrix $P_\infty \in R^{n_\theta \times n_\theta}$, *and a constant vector* $\theta_\infty \in R^{n_\theta}$ *such that* $\lim_{t\to\infty} P(t) = P_\infty$ *and* $\lim_{t\to\infty} \theta(t) = \theta_\infty$.

Proof: (i) Since $\det(P(t_0)) \neq 0$, at $t = t_0$, the identity $P(t)P^{-1}(t) = I$ implies

$$\frac{d}{dt}(P^{-1}(t)) = -P^{-1}(t)\dot{P}(t)P^{-1}(t) = \frac{\phi(t)\phi^T(t)}{m^2(t)}. \tag{3.33}$$

This equality means that $P^{-1}(t)$ is nondecreasing in the sense

$$P^{-1}(t) = P^{-1}(t_0) + \int_{t_0}^{t} \frac{\phi(\tau)\phi^T(\tau)}{m^2(\tau)}\, d\tau,\ t \geq t_0, \tag{3.34}$$

that is, $P^{-1}(t) \geq P^{-1}(t_0) > 0$, and so $P(t) > 0$, $\forall t \geq t_0$, and $P(t)$ is bounded. From (3.31) and (3.32), it follows that $\dot{P}(t)$ is bounded.

(ii) Define the positive definite function $V(\tilde{\theta}, t) = \tilde{\theta}^T P^{-1}(t)\tilde{\theta}$ (as $P^{-1}(t) \geq P^{-1}(t_0) > 0$). Then, the time derivative of $V = V(\tilde{\theta}(t), t)$, along (3.30), is

$$\dot{V} = \frac{dV}{dt} = -\frac{\tilde{\theta}^T(t)\phi(t)\phi^T(t)\tilde{\theta}(t)}{m^2(t)} = -\frac{\epsilon^2(t)}{m^2(t)},\ t \geq t_0. \tag{3.35}$$

Hence $V(\tilde{\theta}(t), t)$ is bounded. Using (3.34), we have

$$V(\tilde{\theta}(t), t) = \tilde{\theta}^T(t)P^{-1}(t_0)\tilde{\theta}(t) + \tilde{\theta}^T(t)\left(\int_{t_0}^{t} \frac{\phi(\tau)\phi^T(\tau)}{m^2(\tau)}\, d\tau\right)\tilde{\theta}(t), \tag{3.36}$$

which implies that $\tilde{\theta}^T(t)P^{-1}(t_0)\tilde{\theta}(t)$ is bounded, and so are $\tilde{\theta}(t)$ and $\theta(t)$.

From the boundedness of $\tilde{\theta}(t)$ and the inequality

$$\frac{|\epsilon(t)|}{\bar{m}(t)} \leq \frac{\|\tilde{\theta}(t)\|_2 \|\phi(t)\|_2}{\sqrt{1 + \|\phi(t)\|_2^2}}, \tag{3.37}$$

where $\epsilon(t) = \tilde{\theta}^T(t)\phi(t)$ as in (3.20), we have the boundedness of $\frac{\epsilon(t)}{\bar{m}(t)}$.

(iii) It follows from (3.35) and the boundedness of $V(\tilde{\theta}(t), t)$ that $\frac{\epsilon(t)}{m(t)} \in L^2$ and that $\frac{\epsilon(t)}{\bar{m}(t)} = \frac{\epsilon(t)}{m(t)}\frac{m(t)}{\bar{m}(t)} \in L^2$ as $\frac{m(t)}{\bar{m}(t)} \in L^\infty$. Then, from the boundedness of $P(t)$ and the inequality

$$\|\dot{\theta}(t)\|_2 \leq \frac{\|P_s(t)\|_2 \|P_s(t)\phi(t)\|_2}{\sqrt{1 + \kappa\|P_s(t)\phi(t)\|_2^2}} \frac{|\epsilon(t)|}{m(t)}, \tag{3.38}$$

where $P(t) \triangleq P_s(t)P_s(t)$, it follows that $\dot{\theta}(t) \in L^2$.

(iv) The adaptive law (3.31) and the property that $P(t) > 0$ imply that

$$P(t) = P(t_0) - \int_{t_0}^{t} \frac{P(\tau)\phi(\tau)\phi^T(\tau)P(\tau)}{m^2(\tau)} d\tau > 0, \ t \geq t_0. \tag{3.39}$$

For any constant $z \in R^{n_\theta}$, we have

$$z^T P(t) z = z^T P(t_0) z - \int_{t_0}^{t} \frac{z^T P(\tau)\phi(\tau)\phi^T(\tau)P(\tau)z}{m^2(t)} d\tau \geq 0, \ t \geq t_0. \tag{3.40}$$

From (3.40), $f_z(t) \triangleq \int_{t_0}^{t} \frac{z^T P(\tau)\phi(\tau)\phi^T(\tau)P(\tau)z}{m^2(t)} d\tau$, a nondecreasing function of t, is bounded from above by $z^T P(t_0) z \geq$. From Proposition 2.14, we have that $\lim_{t\to\infty} f_z(t)$ exists and is finite. Hence there exists a constant $\gamma_z \geq 0$ such that $\lim_{t\to\infty} z^T P(t) z = \gamma_z$. Since $z \in R^{n_\theta}$ can be arbitrary, there exists a constant matrix $P_\infty \in R^{n_\theta \times n_\theta}$ such that $\lim_{t\to\infty} P(t) = P_\infty$.[1]

To prove the convergence of $\theta(t)$, we note that (3.30) is equivalent to

$$\dot{\tilde{\theta}}(t) = \dot{P}(t)P^{-1}(t)\tilde{\theta}(t), \tag{3.41}$$

whose solution is

$$\tilde{\theta}(t) = P(t)P^{-1}(t_0)\tilde{\theta}(t_0), \ t \geq t_0. \tag{3.42}$$

Since $\lim_{t\to\infty} P(t) = P_\infty$, from (3.42) we have

$$\begin{aligned} \lim_{t\to\infty} \theta(t) &= \theta^* + \lim_{t\to\infty} P(t)P^{-1}(t_0)\tilde{\theta}(t_0) \\ &= \theta^* + P_\infty P^{-1}(t_0)\tilde{\theta}(t_0) \triangleq \theta_\infty \in R^{n_\theta}. \end{aligned} \tag{3.43}$$

This completes the lemma's proof. ∇

In view of (3.34), system (3.30) can be expressed as

$$\dot{\tilde{\theta}} = -\frac{P(t)\phi(t)\phi^T(t)}{m^2(t)}\tilde{\theta} \triangleq A(t, t_0)\tilde{\theta}, \tag{3.44}$$

where the dependence $A(t, t_0)$ on t_0 is due to the dependence of $P(t)$ on t_0 in the integral (3.34) (not in $P(t_0) = P_0$). This situation is different from (3.29) where Γ is a constant matrix, and the adaptive law (3.30)–(3.31) is a nonlinear system in nature. From (3.42), we have $\|\tilde{\theta}(t)\|_2 \leq \|P(t)\|_2 \, \|P^{-1}(t_0)\| \, \|\tilde{\theta}(t_0)\|_2 \leq$

[1] As an illustrative example, let us consider $P = \begin{bmatrix} p_{11} & p_{21} \\ p_{21} & p_{22} \end{bmatrix}$. With $z = \begin{bmatrix} 1 & 0 \end{bmatrix}^T$, $z^T P z = p_{11}$, with $z = \begin{bmatrix} 0 & 1 \end{bmatrix}^T$, $z^T P z = p_{22}$, and with $z = \begin{bmatrix} 1 & 1 \end{bmatrix}^T$, $z^T P z = p_{11} + p_{22} + 2p_{21}$. Applying the property $\lim_{t\to\infty} z^T P(t) z = \gamma_z$ for z equal to the above choices, respectively, we see that the components p_{11}, p_{22}, and p_{21} all have limits as $t \to \infty$.

$c_0\|\tilde{\theta}(t_0)\|_2$ for some positive constant c_0 independent of t_0 as $0 < P(t) \leq P(t_0)$ from (3.39) and $0 < P(t_0) = P_0 = P_0^T$ independent of t_0.

The least-squares algorithm (3.30)–(3.31) minimizes a cost function which is an integral of squared errors at many time instants with a penalty on the initial estimate $\theta(t_0) = \theta_0$ of θ^*. Such a cost function is

$$J(\theta) = \frac{1}{2}\int_{t_0}^{t} \frac{(\theta^T(t)\phi(\tau) - y(\tau))^2}{m^2(\tau)}\, d\tau + \frac{1}{2}(\theta(t) - \theta_0)^T P_0^{-1}(\theta(t) - \theta_0),\ P_0 = P_0^T > 0. \tag{3.45}$$

An estimation error $\hat{\epsilon}(\tau) = \theta^T(t)\phi(\tau) - y(\tau)$ is defined for all $\tau \in [0, t]$. The least-squares choice of $\theta(t)$ is to make $J(\theta)$ minimized: $\frac{\partial J(\theta)}{\partial \theta} = 0 \Rightarrow$ (3.30)–(3.31) [23], [151], [340]. More precisely, with $J(\theta)$ in (3.45), it follows that

$$\frac{\partial J(\theta)}{\partial \theta} = \int_{t_0}^{t} \frac{(\theta^T(t)\phi(\tau) - y(\tau))}{m^2(\tau)}\phi(\tau)\, d\tau + P_0^{-1}(\theta(t) - \theta_0). \tag{3.46}$$

Setting $\frac{\partial J(\theta)}{\partial \theta} = 0$, we have

$$\theta(t) = P(t)\left(P_0^{-1}\theta_0 + \int_{t_0}^{t} \frac{y(\tau)\phi(\tau)}{m^2(\tau)}\, d\tau\right), \tag{3.47}$$

where $\theta(t_0) = \theta_0$ and $P(t_0) = P_0$ and

$$P(t) = \left(P_0^{-1} + \int_{t_0}^{t} \frac{\phi(\tau)\phi^T(\tau)}{m^2(\tau)}\, d\tau\right)^{-1}. \tag{3.48}$$

Using the definition of $P(t)$ and the identity

$$0 = \frac{d}{dt}I = \frac{d}{dt}(P(t)P^{-1}(t)) = \frac{d}{dt}(P(t))P^{-1}(t) + P(t)\frac{d}{dt}(P^{-1}(t)), \tag{3.49}$$

we have

$$\dot{P}(t) = \frac{d}{dt}(P(t)) = -\frac{P(t)\phi(t)\phi^T(t)P(t)}{m^2(t)},\ P(t_0) = P_0, \tag{3.50}$$

which is (3.31). Using the above expression of $\theta(t)$, we have

$$\begin{aligned}\dot{\theta}(t) &= \dot{P}(t)\left(P_0^{-1}\theta_0 + \int_{t_0}^{t} \frac{y(\tau)\phi(\tau)}{m^2(\tau)}\, d\tau\right) + P(t)\frac{y(t)\phi(t)}{m^2(t)}\\ &= \dot{P}(t)\left(P_0^{-1}\theta_0 + \int_{t_0}^{t} \frac{y(\tau)\phi(\tau)}{m^2(\tau)}\, d\tau\right) + P(t)\frac{\phi(t)\phi^T(t)}{m^2(t)}\theta(t) - \frac{P(t)\epsilon(t)\phi(t)}{m^2(t)}\\ &= -\frac{P(t)\epsilon(t)\phi(t)}{m^2(t)},\end{aligned} \tag{3.51}$$

which is (3.30) (here we used the expression $\epsilon(t) = \theta^T(t)\phi(t)$ for the second equality and (3.47), (3.50) of $\theta(t)$, $\dot{P}(t)$ for the third equality).

3.5 Parameter Convergence

In this section we address the issue of parameter convergence of both the normalized gradient algorithm and the normalized least-squares algorithm.

Corollary 3.1 *For the gradient algorithm (3.23) or the least-squares algorithm (3.30), if* $m(t) \in L^\infty$ *and* $\dot{\phi}(t) \in L^\infty$, *then* $\lim_{t\to\infty} \epsilon(t) = 0$.

Proof: Lemma 3.1 or 3.2 shows that $\frac{\epsilon(t)}{m(t)} \in L^2$ and $\theta(t)$, $\dot{\theta}(t)$, $\frac{\epsilon(t)}{m(t)} \in L^\infty$. From (3.20), that is, $\epsilon(t) = (\theta(t) - \theta^*)^T\phi(t)$, we have that $\dot{\epsilon}(t) = \dot{\theta}^T(t)\phi(t) + (\theta(t) - \theta^*)^T\dot{\phi}(t)$. Hence, if $m(t) \in L^\infty$ and $\dot{\phi}(t) \in L^\infty$, then $\epsilon(t) \in L^2$ and $\dot{\epsilon}(t) \in L^\infty$, which, by Lemma 2.14, implies that $\lim_{t\to\infty} \epsilon(t) = 0$. ∇

In view of Corollary 4.1, we examine the simplified limit situation $\epsilon(t) = (\theta(t) - \theta^*)^T\phi(t) = 0$ for $\theta \in R^2$ by the following introductory example.

Example 3.1 Consider the limit situation $\epsilon(t) = (\theta(t) - \theta^*)^T\phi(t) = 0$, with $\theta = \begin{bmatrix} \theta_1 & \theta_2 \end{bmatrix}^T$, $\theta^* = \begin{bmatrix} \theta_1^* & \theta_2^* \end{bmatrix}^T$, and $\phi(t) = \begin{bmatrix} \phi_1(t) & \phi_2(t) \end{bmatrix}^T$.

If $\phi(t) = \begin{bmatrix} 1 & 0 \end{bmatrix}^T$, then $(\theta(t) - \theta^*)^T\phi(t) = 0$ implies that $\theta_1 - \theta_1^* = 0$; if $\phi(t) = \begin{bmatrix} 1 & 1 \end{bmatrix}^T$, then $(\theta(t) - \theta^*)^T\phi(t) = 0$ implies that $\theta_1 + \theta_2 - \theta_1^* - \theta_2^* = 0$; and if $\phi(t) = \begin{bmatrix} 1 & \sin(t) \end{bmatrix}^T$, then $(\theta(t) - \theta^*)^T\phi(t) = 0$ implies that $\theta_1 - \theta_1^* + \sin(t)(\theta_2 - \theta_2^*) = 0$, which in turn implies that $\theta_1 - \theta_1^* = 0$ and $\theta_2 - \theta_2^* = 0$. □

As seen from this example, for the parameter estimate $\theta(t)$ to converge to its true value θ^*, the regressor $\phi(t)$ should have enough signal components.

3.5.1 Persistency of Excitation

We now introduce the concept of excitation for the regressor $\phi(t)$ related to the richness of some system input signals, which is crucial for ensuring parameter convergence with the gradient and least-squares algorithms.

Definition 3.1 *A bounded vector signal* $x(t) \in R^q$, $q \geq 1$, *is exciting over the finite time interval* $[\sigma_0, \sigma_0 + \delta_0]$, $\delta_0 > 0$, $\sigma_0 \geq t_0$, *if for some* $\alpha_0 > 0$

$$\int_{\sigma_0}^{\sigma_0+\delta_0} x(\tau)x^T(\tau)\, d\tau \geq \alpha_0 I. \tag{3.52}$$

Definition 3.2 *A bounded vector signal* $x(t) \in R^q$, $q \geq 1$, *is persistently exciting (PE) if there exist* $\delta > 0$ *and* $\alpha_0 > 0$ *such that*

$$\int_{\sigma}^{\sigma+\delta} x(t)x^T(t)\, dt \geq \alpha_0 I, \ \forall \sigma \geq t_0. \tag{3.53}$$

It can be verified that $x(t) = \begin{bmatrix} 1 & 1 \end{bmatrix}^T$ is not exciting over any interval, $x(t) = \begin{bmatrix} 1 & e^{-t} \end{bmatrix}^T$ is exciting over some finite intervals but not persistently exciting, and $x(t) = \begin{bmatrix} 1 & \sin t \end{bmatrix}^T$ is persistently exciting.

For a general linear model (3.20), whether the vector signal $\phi(t)$ meets Definition 3.1 or 3.2 depends on the original system associated to the model (3.20). For the parametric system model (3.7), the following result holds.

Theorem 3.1 [151], [282], [340] *Consider system (3.1), $P(s)[y](t) = Z(s)[u](t)$, and the parametric model (3.7), $y(t) = \theta^{*T}\phi(t)$, with $\theta^* \in R^{n+m+1}$. Assume that all zeros of $P(s)$ are in the open left-half of the complex s-plane, and that for any set of $\{\omega_1, \omega_2, \ldots, \omega_{n+m+1}\}$, with $\omega_j \neq \omega_k$ for $j \neq k$, $H(j\omega_i)$, $i = 1, \ldots, n+m+1$, are linearly independent vectors over C^n, where $H(s) = [\frac{b^T(s)}{\Lambda(s)}, \frac{a^T(s)}{\Lambda(s)}\frac{Z(s)}{P(s)}]^T$ with $b(s) = [1, s, \ldots, s^m]^T$ and $a(s) = [1, s, \ldots, s^{n-2}, s^{n-1}]^T$. Then the signal $\phi(t)$ defined in (3.6) is persistently exciting if the system input $u(t)$ has at least $n+m+1$ frequencies.*

The richness of the input signal $u(t)$ is determined by the number of frequencies in $u(t)$. While there is a general definition for signal frequencies [340], the frequency concept can be illustrated with sinusoidal signals. For example, $u_1(t) = 10$ has one frequency at $\omega_1 = 0$. $u_2(t) = 10 - \sin(t)$ has three frequencies at $\omega_1 = 0$, $\omega_2 = 1$, $\omega_3 = -1$. $u_3(t) = \sin(t) + \cos(2t)$ or $u_4(t) = \sin(t) + \sin(2t)$ has four frequencies at $\omega_1 = -1$, $\omega_2 = 1$, $\omega_3 = 2$, and $\omega_4 = -2$. For the case in Theorem 3.1, if an input $u(t)$ has at least $n+m+1$ frequencies to make $\phi(t)$ PE, this $u(t)$ is said to be sufficiently rich [282], [340].

Note that if $\phi(t) \in L^\infty$ and $\phi(t)$ is PE, then $\frac{\phi(t)}{\sqrt{1+\kappa\phi^T(t)\phi(t)}}$ is PE.

The condition on the vector function $H(s)$ is also crucial for $\phi(t)$ to be PE.

Proposition 3.1 *$H(j\omega_i)$, $i = 1, \ldots, n+m+1$, with $\omega_j \neq \omega_k$ for $j \neq k$, are linearly independent vectors over C^n if and only if $Z(s)$ and $P(s)$ are coprime, that is, they do not have common factors. If $Z(s)$ and $P(s)$ are not coprime, then $\phi(t)$ cannot be PE for any bounded input $u(t)$.*

When $Z(s)$ and $P(s)$ are not coprime, there is a nonzero constant vector $c \in R^{n_\theta}$ such that $c^T H(s) = 0$, $\forall s \in C$, which implies that the above set of $H(j\omega_i)$ cannot be linear independent. In the time domain, since $\phi(t) = H(s)[u](t)$, (3.35) cannot be met for $\phi(t)$, as $c^T(\int_\sigma^{\sigma+\delta} \phi(t)\phi^T(t)\,dt)c = 0$.

When a modified parametric model, for example, (3.12) for a reduced-order parametrization, is used, the persistent excitation property of a resulting regressor should be checked by definition, as shown in the next example.

Example 3.2 Consider the plant $(s^3 + 2s^2 + s + z_0)[y](t) = (2s^2 + z_0)[u](t)$, with an unknown parameter $0 < z_0 < 2$ (a stable plant). If we design a first-order gradient algorithm to estimate z_0, for $\Lambda(s) = (s+1)^3$, we can write $\frac{s^3+2s^2+s+z_0}{\Lambda(s)}[y](t) = \frac{2s^2+z_0}{\Lambda(s)}[u](t)$ as $\bar{y}(t) = z_0\phi(t)$ with $\phi(t) = \frac{1}{\Lambda(s)}[u-y](t)$ and $\bar{y}(t) = \frac{s^3+2s^2+s}{\Lambda(s)}[y](t) - \frac{2s^2}{\Lambda(s)}[u](t)$. Then, we can define $\epsilon(t) = \theta(t)\phi(t)$, where $\theta(t)$ is an estimate of z_0, and update $\theta(t)$ from $\dot{\theta}(t) = -\frac{\gamma\phi(t)\epsilon(t)}{1+\kappa\phi^2(t)}$, $\gamma > 0$, with any initial estimate $\theta(0)$ of θ^* and $\kappa > 0$ being a design parameter.

In this case, since $u(t) - y(t) = \frac{s(s^2+1)}{s^3+2s^2+s+z_0}[u](t)$, we have $\phi(s) = \frac{1}{\Lambda(s)}(u(s) - y(s)) = \frac{1}{\Lambda(s)}\frac{s(s^2+1)}{s^3+2s^2+s+z_0}u(s)$, which means that for $u(t) = 1$, $u(t) = \sin t$, or $u(t) = 1 + \sin t$, the signal $\phi(t)$ cannot be PE as $\lim_{t\to\infty}\phi(t) = 0$ exponentially. For example, for $u(t) = \sin t$, we have $u(s) = \frac{1}{s^2+1}$ so that $\phi(s) = \frac{1}{\Lambda(s)}\frac{s}{s^3+2s^2+s+z_0}$, that is, all signal components of $\phi(t)$ decay exponentially. However, for $u(t) = \sin 1.01t$, the signal $\phi(t)$ is PE.

3.5.2 Convergence of the Gradient Algorithm

Consider the normalized gradient algorithm (3.23) with $\Gamma = I$ and (3.20):

$$\dot{\tilde{\theta}}(t) = A(t)\tilde{\theta}(t),\ A(t) = -\frac{\phi(t)\phi^T(t)}{m^2(t)},\ t \geq t_0. \tag{3.54}$$

We can view the signal $\frac{\epsilon(t)}{m(t)}$ as the output of system (3.54), that is,

$$\bar{\epsilon}(t) \triangleq \frac{\epsilon(t)}{m(t)} = \psi^T(t)\tilde{\theta}(t),\ \psi(t) = \frac{\phi(t)}{m(t)}, \tag{3.55}$$

where $\frac{\phi(t)}{m(t)}$ is the observation vector (as $C(t)$ in $\dot{x} = A(t)x$, $y = C(t)x$).

Lemma 3.3 *If the signal $\psi(t)$ is exciting over the interval $[\sigma_0, \sigma_0+\delta_0]$, then $\|\tilde{\theta}(\sigma_0+\delta_0)\|_2 \leq \alpha_0\|\tilde{\theta}(\sigma_0)\|_2$ for some $\alpha_0 \in (0,1)$.*

Proof: Let $\Phi(t, t_0)$ be the state transition matrix of system (3.54), which is determined by the signal $\psi(t)$. If the signal $\psi(t)$ is exciting over the interval $[\sigma_0, \sigma_0+\delta_0]$, that is, the matrix $S_0 = \int_{\sigma_0}^{\sigma_0+\delta_0}\psi(\tau)\psi^T(\tau)\,d\tau$ is positive definite, then the matrix $S = \int_{\sigma_0}^{\sigma_0+\delta_0}\Phi^T(\tau,\sigma_0)\psi(\tau)\psi^T(\tau)\Phi(\tau,\sigma_0)\,d\tau$ is also positive definite (and vice verse).[2] For $V(\tilde{\theta}) = \frac{1}{2}\tilde{\theta}^T\tilde{\theta}$, we have

$$\dot{V} = \frac{dV}{dt} = -\tilde{\theta}^T(t)\psi(t)\psi^T(t)\tilde{\theta}(t), \tag{3.56}$$

[2]The matrix S is the observability Gramian of the system $\dot{\tilde{\theta}}(t) = A(t)\tilde{\theta}(t)$, $\bar{\epsilon}(t) = \psi^T(t)\tilde{\theta}(t)$ and S_0 is the observability Gramian of the system $\dot{\tilde{\theta}}(t) = A(t)\tilde{\theta}(t) + \psi(t)v(t) = A(t)\tilde{\theta}(t) + \psi(t)\bar{\epsilon}(t) = 0$, where $v(t) = \bar{\epsilon}(t) = \psi^T(t)\tilde{\theta}(t)$ is the input signal, and these two systems have the same observability property because the output feedback control $\bar{\epsilon}(t) = \psi^T(t)\tilde{\theta}(t)$ does not change system observability [171].

which, together with $\tilde{\theta}(\tau) = \Phi(\tau, \sigma_0)\tilde{\theta}(\sigma_0)$, leads to

$$\begin{aligned}
&V(\tilde{\theta}(\sigma_0 + \delta_0)) \\
&= V(\tilde{\theta}(\sigma_0)) - \int_{\sigma_0}^{\sigma_0+\delta_0} \tilde{\theta}^T(\tau)\psi(\tau)\psi^T(\tau)\tilde{\theta}(\tau)\, d\tau \\
&= V(\tilde{\theta}(\sigma_0)) - \tilde{\theta}^T(\sigma_0) \int_{\sigma_0}^{\sigma_0+\delta_0} \Phi^T(\tau, \sigma_0)\psi(\tau)\psi^T(\tau)\Phi(\tau, \sigma_0)\, d\tau\, \tilde{\theta}(\sigma_0) \\
&= V(\tilde{\theta}(\sigma_0)) - \tilde{\theta}^T(\sigma_0)S\tilde{\theta}(\sigma_0).
\end{aligned} \tag{3.57}$$

Since the matrix S is positive definite, we have

$$\tilde{\theta}^T(\sigma_0)S\tilde{\theta}(\sigma_0) \geq \lambda_m \tilde{\theta}^T(\sigma_0)\tilde{\theta}(\sigma_0) = \lambda_s V(\tilde{\theta}(\sigma_0)) \tag{3.58}$$

for $\lambda_s = 2\lambda_m$. Hence, from (3.57) and (3.58), we have

$$V(\tilde{\theta}(\sigma_0 + \delta_0)) \leq (1 - \lambda_s)V(\tilde{\theta}(\sigma_0)), \tag{3.59}$$

where $1 - \lambda_s \in (0, 1)$ as $V(\tilde{\theta}(t)) > 0$ for $\tilde{\theta}(t) \neq 0$. Hence, we have ∇

Lemma 3.4 *If the signal $\psi(t)$ is persistently exciting, then $\lim_{t\to\infty} \|\tilde{\theta}(t)\|_2 = 0$ exponentially (i.e., $\lim_{t\to\infty} \theta(t) = \theta^*$ exponentially).*

Proof: From Lemma 3.3, we have that $\|\tilde{\theta}(\sigma+\delta)\|_2 \leq \alpha_s \|\tilde{\theta}(\sigma)\|_2$ for some $\alpha_s \in (0, 1)$, some $\delta > 0$, and any $\sigma \geq t_0$. It follows from the persistent excitation property of $\psi(t)$ that $\|\tilde{\theta}(k\delta+t_0)\|_2 \leq \alpha_s^k \|\tilde{\theta}(t_0)\|_2$ so that $\lim_{k\to\infty} \|\tilde{\theta}(k\delta+t_0)\|_2 = 0$ exponentially, which implies that $\lim_{t\to\infty} \|\tilde{\theta}(t)\|_2 = 0$ exponentially because (3.25) means that, in general, $\lim_{t\to\infty} \|\tilde{\theta}(t)\|_2$ exists and is finite. ∇

Example 3.3 Consider the scalar case when $\theta^* \in R$ and let $\frac{\phi(t)}{m(t)} = \beta_0 > 0$ appromixately. From (3.54), we have $\tilde{\theta}(t) = e^{-\beta_0^2(t-t_0)}\tilde{\theta}(t_0)$, that is, $\tilde{\theta}(t)$ converges to zero exponentially. □

3.5.3 Convergence of the Least-Squares Algorithm

Consider the normalized least-squares algorithm (3.30)–(3.31). The parameter error $\tilde{\theta}(t)$ has the form (3.42). Using $\psi(t) = \frac{\phi(t)}{m(t)}$, we write (3.33) as

$$\frac{d}{dt}(P^{-1}(t)) = \psi(t)\psi^T(t),\ t \geq t_0. \tag{3.60}$$

Lemma 3.5 *If the signal $\psi(t)$ is exciting over the interval $[\sigma_0, \sigma_0 + \delta_0]$, then $(I - P(\sigma_0 + \delta_0)P^{-1}(\sigma_0))$ is nonsingular, and θ^* is identified as*

$$\theta^* = \left(I - P(\sigma_0 + \delta_0)P^{-1}(\sigma_0)\right)^{-1} (\theta(\sigma_0+\delta_0) - P(\sigma_0+\delta_0)P^{-1}(\sigma_0)\theta(\sigma_0)). \tag{3.61}$$

Proof: From (3.34), we have

$$I - P(\sigma_0 + \delta_0)P^{-1}(\sigma_0) = P(\sigma_0 + \delta_0) \int_{\sigma_0}^{\sigma_0+\delta_0} \psi(\tau)\psi^T(\tau)\, d\tau. \tag{3.62}$$

Hence, if the signal $\psi(t)$ is exciting over the interval $[\sigma_0, \sigma_0 + \delta_0]$, that is, $\int_{\sigma_0}^{\sigma_0+\delta_0} \psi(\tau)\psi^T(\tau)\, d\tau$ is nonsingular, then $I - P(\sigma_0 + \delta_0)P^{-1}(\sigma_0)$ is nonsingular as $P(\sigma_0 + \delta_0)$ is nonsingular. Finally, using (3.42), we obtain (3.61). ∇

From (3.31), we have

$$\begin{aligned} &I - P(\sigma_0 + \delta_0 + \gamma_0)P^{-1}(\sigma_0) \\ &= \left(P(\sigma_0) - P(\sigma_0 + \delta_0) + \int_{\sigma_0+\delta_0}^{\sigma_0+\delta_0+\gamma_0} \frac{P(\tau)\phi(\tau)\phi^T(\tau)P(\tau)}{m^2(\tau)}\, d\tau \right) P^{-1}(\sigma_0), \end{aligned} \tag{3.63}$$

which implies that if $I - P(\sigma_0+\delta_0)P^{-1}(\sigma_0)$ is nonsingular (i.e., $P(\sigma_0) - P(\sigma_0 + \delta_0) = \int_{\sigma_0}^{\sigma_0+\delta_0} \frac{P(\tau)\phi(\tau)\phi^T(\tau)P(\tau)}{m^2(\tau)}\, d\tau$ is nonsingular and positive definite as well), then $(I - P(\sigma_0+\delta_0+\gamma_0)P^{-1}(\sigma_0))$ is also nonsingular for any $\gamma_0 > 0$. Therefore, if θ^* is identifiable over $[\sigma_0, \sigma_0 + \delta_0]$, then θ^* is identifiable over $[\sigma_0, \sigma_0 + \delta_0 + \gamma_0]$ for any $\gamma_0 > 0$.

Lemma 3.6 *If the signal $\psi(t)$ is persistently exciting, then the least-squares algorithm (3.30)–(3.31) ensures that $\lim_{t\to\infty} \|\tilde{\theta}(t)\|_2 = 0$.*

Proof: If $\psi(t)$ is persistently exciting, then there exists a $\delta > 0$ such that $\int_{\sigma}^{\sigma+\delta} \psi(t)\psi^T(t)\, dt \geq \alpha_0 I$ for some $\alpha_0 > 0$ and $\forall \sigma \geq t_0$. Taking $\sigma = t_0 + i\delta$ and letting $i = 0, 1, \ldots,$ we have

$$\int_{t_0+i\delta}^{t_0+(i+1)\delta} \psi(t)\psi^T(t)\, dt \geq \alpha_0 I. \tag{3.64}$$

Using (3.60), we have

$$P^{-1}(k\delta + t_0) = P^{-1}(t_0) + \sum_{i=0}^{k-1} \int_{t_0+i\delta}^{t_0+(i+1)\delta} \psi(\tau)\psi^T(\tau)\, d\tau. \tag{3.65}$$

From (3.33), (3.64), and (3.65), we conclude that

$$\lim_{t\to\infty} \min_{1\leq i\leq n+m+1} \lambda_i(P^{-1}(t)) = \infty, \tag{3.66}$$

where $\lambda_i(P^{-1}(t))$ is the ith eigenvalue of $P^{-1}(t)$, which implies that

$$\begin{aligned} \lim_{t\to\infty} \|P(t)\|_2 &= \lim_{t\to\infty} \max_{1\leq j\leq n+m+1} \lambda_j(P(t)) \\ &= \lim_{t\to\infty} \left(\min_{1\leq i\leq n+m+1} \lambda_i(P^{-1}(t)) \right)^{-1} = 0. \end{aligned} \tag{3.67}$$

Therefore, from (3.42), we conclude that

$$\lim_{t\to\infty}\tilde{\theta}(t) = \lim_{t\to\infty} P(t)P^{-1}(t_0)\tilde{\theta}(t_0) = 0, \tag{3.68}$$

which completes the proof of the lemma. ∇

To evaluate the rate at which $\tilde{\theta}(t)$ converges to zero with the normalized least-squares algorithm (3.30)–(3.31), using (3.64) and (3.65), we obtain

$$P(T+t_0) \le \left(P^{-1}(t_0) + \frac{\alpha_0 T}{\delta} I_{n_\theta}\right)^{-1}, \tag{3.69}$$

where $T = k\delta$. For $P(t_0) = I_{n_\theta}$, from (3.42) and (3.69), we have

$$\|\tilde{\theta}(T+t_0)\|_2 \le \|P(T+t_0)\|_2\,\|P^{-1}(t_0)\|_2\,\|\tilde{\theta}(t_0)\|_2 \le \frac{1}{1+\frac{\alpha_0 T}{\delta}}\|\tilde{\theta}(t_0)\|_2. \tag{3.70}$$

Hence the convergence rate is $\frac{1}{t}$ (i.e., in general, it is not exponential).

Example 3.4 Consider the scalar case when $\theta(t)$, $\theta^* \in R$ and let $\frac{\phi(t)}{m(t)} = \beta_0 > 0$ (appromixately, with a small κ in (3.30) and in view of $P(t) \to 0$ as $t \to \infty$) and $P(t_0) = 1$. For $t \ge t_0$, from (3.34), we have $P(t) = \frac{1}{1+\beta_0^2(t-t_0)}$, and from (3.42), we have $\tilde{\theta}(t) = \frac{1}{1+\beta_0^2(t-t_0)}\tilde{\theta}(t_0)$, which satisfies (3.30): $\dot{\tilde{\theta}}(t) = -\frac{\beta_0^2}{1+\beta_0^2(t-t_0)}\tilde{\theta}(t)$. In this case, $\frac{\phi(t)}{m(t)} = \beta_0 > 0$ is persistently exciting and $\tilde{\theta}(t)$ converges to zero but the convergence rate is not exponential. $\square$

There are several ways to modify the least-squares algorithm (3.30) so that under a PE condition, the parameter error $\tilde{\theta}(t)$ converges to zero exponentially as $t \to \infty$. One way is the so-called normalized least-squares algorithm with covariance (the matrix $P(t)$ in (3.31)) resetting [151], [340]:

$$\dot{\theta}(t) = -\frac{P(t)\phi(t)\epsilon(t)}{m^2(t)},\ \theta(t_0) = \theta_0,\ t \ge t_0, \tag{3.71}$$

$$\dot{P}(t) = -\frac{P(t)\phi(t)\phi^T(t)P(t)}{m^2(t)},\ P(t_0) = \rho_0 I,$$

$$P(t_r^+) = \rho_0 I,\ \rho_0 > 0, \tag{3.72}$$

where t_r is the time at which $\lambda_{\min}(P(t_r)) = \rho_1 < \rho_0$ for some constant $\rho_1 \in (0, \rho_0)$. This algorithm has the desired properties as stated in Lemma 3.2. Moreover, with this algorithm, the parameter error $\tilde{\theta}(t)$ converges to zero exponentially fast if the signal $\psi(t) = \frac{\phi(t)}{m(t)}$ in (3.30) is persistently exciting.

3.6 Discrete-Time Algorithms

Adaptive parameter estimation algorithms can be developed based on discrete-time system models described by difference equations. System signals are defined on the time sequence $t \in \{t_0, t_0+1, \ldots\}$. For example, a system output signal is described as $y(t)$, $t = t_0, t_0+1, \ldots$, and an advance operator z is defined as $z[y](t) = y(t+1)$, while a delay operator is z^{-1}: $z^{-1}[y](t) = y(t-1)$.

3.6.1 Linear Parametric Models

A discrete-time linear parametric model has the same form as (3.18), that is,

$$y(t) = \theta^{*T}\phi(t), \; t \in \{t_0, t_0+1, \ldots\}, \tag{3.73}$$

where $\theta^* \in R^{n_\theta}$ is an unknown parameter vector, $y(t) \in R$ is a known (measured) signal, $\phi(t) \in R^{n_\theta}$ is a known regressor vector, and $n_\theta > 0$.

To develop an adaptive algorithm to estimate the unknown parameter vector θ^* in (3.73), we need a parametric error model. Letting $\theta(t)$ be the estimate of θ^*, we define the estimation error

$$\epsilon(t) = \theta^T(t)\phi(t) - y(t), \; \in \{t_0, t_0+1, \ldots\}, \tag{3.74}$$

which, in view of (3.73), can be expressed as

$$\epsilon(t) = \tilde{\theta}^T(t)\phi(t), \; \tilde{\theta}(t) = \theta(t) - \theta^*, \; t \in \{t_0, t_0+1, \ldots\}. \tag{3.75}$$

This error model is common in discrete-time adaptive control and system identification. The key task is to develop an adaptive law to update the parameter estimate $\theta(t)$, based on the measured signals $y(t)$ and $\phi(t)$.

One example of the linear parametric model (3.73) is a parametrized model, similar to its continuous-time counterpart in Section 3.1, of the discrete-time linear time-invariant system described by the difference equation

$$P(z)[y](t) = Z(z)[u](t), \tag{3.76}$$

where $y(t) \in R$ and $u(t) \in R$ are the measured system input and output,

$$P(z) = z^n + p_{n-1}z^{n-1} + \cdots + p_1 z + p_0, \tag{3.77}$$

$$Z(z) = z_m z^m + z_{m-1}z^{m-1} + \cdots + z_1 z + z_0 \tag{3.78}$$

are polynomials in z with z being the time advance operator $z[x](t) = x(t+1)$, and p_i, $i = 0, 1, \ldots, n-1$, and z_i, $i = 0, 1, \ldots, m$, with $n > m$, are the unknown but constant system parameters to be estimated.

We choose a stable polynomial $\Lambda(z) = z^n + \lambda_{n-1}z^{n-1} + \cdots + \lambda_1 z + \lambda_0$ whose zeros are in $|z| < 1$ and operate (3.76) by $\frac{1}{\Lambda(z)}$ to obtain (3.73) with

$$\theta^* = [z_0, z_1, \ldots, z_{m-1}, z_m, \\ \lambda_0 - p_0, \lambda_1 - p_1, \ldots, \lambda_{n-2} - p_{n-2}, \lambda_{n-1} - p_{n-1}]^T \in R^{n_\theta}, \tag{3.79}$$

$$\phi(t) = \left[\frac{1}{\Lambda(z)}[u](t), \frac{z}{\Lambda(z)}[u](t), \ldots, \frac{z^{m-1}}{\Lambda(z)}[u](t), \frac{z^m}{\Lambda(z)}[u](t), \right. \\ \left. \frac{1}{\Lambda(z)}[y](t), \frac{z}{\Lambda(z)}[y](t), \ldots, \frac{z^{n-2}}{\Lambda(z)}[y](t), \frac{z^{n-1}}{\Lambda(z)}[y](t)\right]^T \in R^{n_\theta}, \tag{3.80}$$

where $n_\theta = n+m+1$. The vector signal $\phi(t)$ can be generated in a way similar to (3.13)–(3.14), with $\dot{\omega}_1(t)$ and $\dot{\omega}_2(t)$ replaced by $\omega_1(t+1)$ and $\omega_2(t+1)$.

In discrete time, the polynomial $\Lambda(z)$ can be chosen as $\Lambda(z) = z^n$, so that the linear model (3.73) has the parameter vector and regressor as

$$\theta^* = [z_0, z_1, \ldots, z_{m-1}, z_m, -p_0, -p_1, \ldots, -p_{n-2}, -p_{n-1}]^T \in R^{n_\theta}, \tag{3.81}$$

$$\phi(t) = [u(t-n), u(t-n+1), \ldots, u(t-n+m-1), u(t-n+m), \\ y(t-n), y(t-n+1), \ldots, y(t-2), y(t-1)]^T \in R^{n_\theta}. \tag{3.82}$$

When some system parameters are known, the dimension of the parameter vector θ^* can be reduced by regrouping known signals and unknown parameters (similar to (3.10)–(3.12) for the continuous-time case).

3.6.2 Normalized Gradient Algorithm

The discrete-time version of the normalized gradient algorithm (3.23) for updating the parameter vector $\theta(t)$ is

$$\theta(t+1) = \theta(t) - \frac{\Gamma\phi(t)\epsilon(t)}{m^2(t)}, \ \theta(t_0) = \theta_0, \ t \in \{t_0, t_0+1, \ldots\}, \tag{3.83}$$

where $0 < \Gamma = \Gamma^T < 2I_{n_\theta}$ is a gain matrix, θ_0 is an initial estimate of θ^*, and

$$m(t) = \sqrt{\kappa + \phi^T(t)\phi(t)} \tag{3.84}$$

with $\kappa > 0$ being a design parameter.

Lemma 3.7 *The adaptive algorithm (3.83) guarantees that*

(i) $\theta(t)$ *and* $\frac{\epsilon(t)}{m(t)}$ *are bounded; and*

(ii) $\frac{\epsilon(t)}{m(t)}$ *and* $\theta(t+1) - \theta(t)$ *belong to* L^2.

Proof: (i) Introducing the positive definite function $V(\tilde{\theta}) = \tilde{\theta}^T \Gamma^{-1} \tilde{\theta}$, the time increment of $V = V(\tilde{\theta}(t))$, along (3.83), is

$$\begin{aligned} &V(\tilde{\theta}(t+1)) - V(\tilde{\theta}(t)) \\ &= -\left(2 - \frac{\phi^T(t)\Gamma\phi(t)}{m^2(t)}\right)\frac{\epsilon^2(t)}{m^2(t)} \le -\alpha_1 \frac{\epsilon^2(t)}{m^2(t)} \end{aligned} \tag{3.85}$$

for $\alpha_1 = 2 - \lambda_{\max}[\Gamma] > 0$, where $\lambda_{\max}[\Gamma] \in (0, 2)$ is the maximum eigenvalue of Γ which satisfies the condition $0 < \Gamma = \Gamma^T < 2I_{n_\theta}$. This implies that $\theta(t) \in L^\infty$ and, with (3.75), that $\frac{\epsilon(t)}{m(t)} \in L^2 \cap L^\infty$, and $\theta(t+1) - \theta(t) \in L^2$.

(ii) It follows from (3.85) and the boundedness of $V(\tilde{\theta}(t))$ that

$$\alpha_1 \sum_{\tau=0}^{t} \frac{\epsilon^2(\tau)}{m(\tau)} = V(\tilde{\theta}(0)) - V(\tilde{\theta}(t)) \le V(\tilde{\theta}(0)), \tag{3.86}$$

that is, $\frac{\epsilon(t)}{m(t)} \in L^2$, and then from (3.83) that $\theta(t+1) - \theta(t) \in L^2$. ∇

From (3.85), we have that $\lim_{t\to\infty} V(\tilde{\theta}(t)) = \lim_{t\to\infty} \tilde{\theta}^T(t)\Gamma^{-1}\tilde{\theta}(t) = V_0$ for some constant $V_0 \ge 0$ as $V(\tilde{\theta}(t))$ is nonincreasing and bounded from below ($V(\tilde{\theta}(t)) \ge 0$). If $V_0 = 0$, then $\lim_{t\to\infty} \theta(t) = \theta^*$. However, when $n_\theta > 1$, if $V_0 > 0$, we cannot conclude $\lim_{t\to\infty} \theta(t) = \theta_\infty$ for some constant vector $\theta_\infty \in R^{n_\theta}$. Moreover, since $\theta(t+1) - \theta(t) \in L^2$, we have that $\lim_{t\to\infty}(\theta(t+1) - \theta(t)) = 0 \in R^{n_\theta}$ (as $\lim_{t\to\infty} \|\theta(t+1) - \theta(t)\| = 0$). However, this also does not mean $\lim_{t\to\infty} \theta(t) = \theta_\infty$ for some constant vector θ_∞.

3.6.3 Normalized Least-Squares Algorithm

The discrete-time version of the normalized least-squares algorithm (3.30)–(3.31) for updating the parameter vector $\theta(t)$ is

$$\theta(t+1) = \theta(t) - \frac{P(t-1)\phi(t)\epsilon(t)}{m^2(t)}, \ \theta(t_0) = \theta_0, \ t \in \{t_0, t_0+1, \ldots\}, \tag{3.87}$$

$$\begin{aligned} &P(t) = P(t-1) - \frac{P(t-1)\phi(t)\phi^T(t)P(t-1)}{m^2(t)}, \\ &P(t_0 - 1) = P_0 = P_0^T > 0, \ t \in \{t_0, t_0+1, \ldots\}, \end{aligned} \tag{3.88}$$

$$m(t) = \sqrt{\kappa + \phi^T(t)P(t-1)\phi(t)}, \ \kappa > 0, \tag{3.89}$$

where θ_0 is the initial estimate of $\theta^* \in R^{n_\theta}$, P_0 is the initial value of the "gain" matrix $P(t) \in R^{n_\theta \times n_\theta}$ at $t = t_0 - 1$, and $\kappa > 0$ is a design parameter.

Lemma 3.8 *The adaptive algorithm (3.87)–(3.88) guarantees that*

(i) $P(t) = P^T(t) > 0$, $\forall t \in \{t_0, t_0+1, \ldots\}$, *and* $P(t)$ *is bounded;*

(ii) $\theta(t)$ *and* $\frac{\epsilon(t)}{\bar{m}(t)}$ *are bounded, where* $\bar{m}(t) = \sqrt{1+\phi^T(t)\phi(t)}$*;*

(iii) $\frac{\epsilon(t)}{m(t)}$, $\frac{\epsilon(t)}{\bar{m}(t)}$, *and* $\theta(t+1) - \theta(t)$ *belong to* L^2*; and*

(iv) there exist a constant matrix $P_\infty \in R^{n_\theta \times n_\theta}$, *and a constant vector* $\theta_\infty \in R^{n_\theta}$ *such that* $\lim_{t\to\infty} P(t) = P_\infty$ *and* $\lim_{t\to\infty} \theta(t) = \theta_\infty$.

Proof: (i) From (3.88), by direct calculation, we have

$$P^{-1}(t) = P^{-1}(t-1) + \frac{1}{\kappa}\phi(t)\phi^T(t),\ t \in \{t_0, t_0+1, \ldots\}. \tag{3.90}$$

This equality means that $P^{-1}(t)$ is nondecreasing in the sense that

$$P^{-1}(t) = P^{-1}(t_0-1) + \sum_{\tau=t_0}^{t} \frac{1}{\kappa}\phi(\tau)\phi^T(\tau),\ t \in \{t_0, t_0+1, \ldots\}, \tag{3.91}$$

that is, $P^{-1}(t) = (P^{-1}(t))^T \geq P^{-1}(t_0) > 0$, and so $P(t) = P^T(t) > 0$, $\forall t \in \{t_0, t_0+1, \ldots\}$, and $P(t)$ is bounded.

(ii) From (3.75), (3.87), and (3.88), it follows that

$$\begin{aligned}\tilde{\theta}(t+1) &= \tilde{\theta}(t) - \frac{P(t-1)\phi(t)\phi^T(t)\tilde{\theta}(t)}{m^2(t)} \\ &= \left(I - \frac{P(t-1)\phi(t)\phi^T(t)}{m^2(t)}\right)\tilde{\theta}(t) = P(t)P^{-1}(t-1)\tilde{\theta}(t).\end{aligned} \tag{3.92}$$

Define the positive definite function $V(\tilde{\theta}, t) = \tilde{\theta}^T P^{-1}(t-1)\tilde{\theta}$ (as $P^{-1}(t-1) \geq P^{-1}(t_0) > 0$). Then, the time increment of $V = V(\tilde{\theta}(t), t)$, along (3.87), is

$$\begin{aligned}&V(\tilde{\theta}(t+1), t+1) - V(\tilde{\theta}(t), t) \\ &= \tilde{\theta}^T(t+1)P^{-1}(t)\tilde{\theta}(t+1) - \tilde{\theta}^T(t)P^{-1}(t-1)\tilde{\theta}(t) \\ &= \tilde{\theta}^T(t+1)P^{-1}(t-1)\tilde{\theta}(t) - \tilde{\theta}^T(t)P^{-1}(t-1)\tilde{\theta}(t) \\ &= -\frac{\epsilon(t)\phi^T(t)\tilde{\theta}(t)}{m^2(t)} = -\frac{\epsilon^2(t)}{m^2(t)}.\end{aligned} \tag{3.93}$$

Hence $V(\tilde{\theta}(t), t)$ is bounded. Using (3.91), we have

$$V(\tilde{\theta}(t), t) = \tilde{\theta}^T(t)P^{-1}(t_0-1)\tilde{\theta}(t) + \tilde{\theta}^T(t)\left(\sum_{\tau=t_0}^{t}\frac{1}{\kappa}\phi(\tau)\phi^T(\tau)\right)\tilde{\theta}(t), \tag{3.94}$$

which implies that $\tilde{\theta}^T(t)P^{-1}(t_0-1)\tilde{\theta}(t)$ is bounded, and so are $\tilde{\theta}(t)$ and $\theta(t)$.

From the boundedness of $\tilde{\theta}(t)$ and the inequality

$$\frac{|\epsilon(t)|}{\bar{m}(t)} \leq \frac{\|\tilde{\theta}(t)\|_2 \|\phi(t)\|_2}{\sqrt{1+\|\phi(t)\|_2^2}}, \tag{3.95}$$

where $\epsilon(t) = \tilde{\theta}^T(t)\phi(t)$ as in (3.75), we have the boundedness of $\frac{\epsilon(t)}{\bar{m}(t)}$.

(iii) It follows from (3.93) and the boundedness of $V(\tilde{\theta}(t),t)$ that $\frac{\epsilon(t)}{m(t)} \in L^2$, $\frac{\epsilon(t)}{m(t)} \in L^2$, and $\frac{\epsilon(t)}{\bar{m}(t)} = \frac{\epsilon(t)}{m(t)}\frac{m(t)}{\bar{m}(t)} \in L^2$ as $\frac{m(t)}{\bar{m}(t)} \in L^\infty$. Then, from the boundedness of $P(t)$ and the inequality

$$\|\theta(t+1)-\theta(t)\|_2 \leq \frac{\|P_s^T(t-1)\|_2 \|P_s(t-1)\phi(t)\|_2}{\sqrt{1+\kappa\|P_s(t-1)\phi(t)\|_2^2}} \frac{|\epsilon(t)|}{m(t)}, \tag{3.96}$$

where $P(t-1) \triangleq P_s^T(t-1)P_s(t-1)$, it follows that $\theta(t+1)-\theta(t) \in L^2$.

(iv) The adaptive law (3.88) with $P(t)>0$ implies that

$$P(t) = P(t_0-1) - \sum_{\tau=t_0}^{t} \frac{P(\tau-1)\phi(\tau)\phi^T(\tau)P(\tau-1)}{m^2(\tau)} > 0. \tag{3.97}$$

For any constant vector $z \in R^{n_\theta}$, we have

$$z^T P(t) z = z^T P(t_0-1) z - \sum_{\tau=t_0}^{t} \frac{z^T P(\tau-1)\phi(\tau)\phi^T(\tau)P(\tau-1)z}{m^2(t)} \geq 0, \tag{3.98}$$

that is, $f_z(t) \triangleq \sum_{\tau=t_0}^{t} \frac{z^T P(\tau-1)\phi(\tau)\phi^T(\tau)P(\tau-1)z}{m^2(t)} \leq z^T P(t_0) z$ is nondecreasing and bounded from above. From Proposition 2.16, we have that $\lim_{t\to\infty} f_z(t)$ exists and is finite and $\lim_{t\to\infty} z^T P(t) z$ exists and is finite. Therefore, given that $z \in R^{n_\theta}$ is arbitrary, there exists a constant matrix $P_\infty \in R^{n_\theta \times n_\theta}$ such that $\lim_{t\to\infty} P(t) = P_\infty$ (see the footnote after (3.40)).

To prove the convergence of $\theta(t)$, we use (3.92) to obtain

$$\tilde{\theta}(t) = P(t-1)P^{-1}(t_0-1)\tilde{\theta}(t_0),\ t \in \{t_0+1, t_0+2, \ldots\}. \tag{3.99}$$

Since $\lim_{t\to\infty} P(t) = P_\infty$, from (3.99) we have

$$\begin{aligned}\lim_{t\to\infty}\theta(t) &= \theta^* + \lim_{t\to\infty} P(t-1)P^{-1}(t_0-1)\tilde{\theta}(t_0) \\ &= \theta^* + P_\infty P^{-1}(t_0-1)\tilde{\theta}(t_0) \triangleq \theta_\infty \in R^{n_\theta},\end{aligned} \tag{3.100}$$

where $P(t_0-1)=P_0$. This completes the lemma's proof. ∇

The least-squares algorithm (3.87)–(3.88) minimizes a cost function which is a sum of squared errors at many time instants with a penalty on the initial estimate $\theta(t_0) = \theta_0$ of θ^*. Such a cost function in discrete time is

$$J(\theta) = \frac{1}{2}\sum_{\tau=t_0}^{t-1}\frac{1}{\kappa}(\theta^T(t)\phi(\tau) - y(\tau))^2 + \frac{1}{2}(\theta(t) - \theta_0)^T P_0^{-1}(\theta(t) - \theta_0),\ P_0 = P_0^T > 0,\ \kappa > 0. \quad (3.101)$$

An estimation error $\hat{\epsilon}(\tau) = \theta^T(t)\phi(\tau) - y(\tau)$ is defined for all $\tau \in \{t_0, t_0 + 1, \ldots, t-1\}$. The least-squares choice of $\theta(t)$ is to make $J(\theta)$ minimized: $\frac{\partial J(\theta)}{\partial \theta} = 0 \Rightarrow$ (3.87)–(3.88) [23], [125]. With $J(\theta)$ in (3.101), it follows that

$$\frac{\partial J(\theta)}{\partial \theta} = \sum_{\tau=t_0}^{t-1}\frac{1}{\kappa}(\theta^T(t)\phi(\tau) - y(\tau))\,\phi(\tau) + P_0^{-1}(\theta(t) - \theta_0). \quad (3.102)$$

Setting $\frac{\partial J(\theta)}{\partial \theta} = 0$, we have

$$\theta(t) = P(t-1)\left(P_0^{-1}\theta_0 + \sum_{\tau=t_0}^{t-1}\frac{1}{\kappa}y(\tau)\phi(\tau)\right),\ t \in \{t_0+1, t_0+2, \ldots\}, \quad (3.103)$$

where $\theta(t_0) = \theta_0$ and $P(t_0 - 1) = P_0$, and

$$P(t-1) = \left(P_0^{-1} + \sum_{\tau=t_0}^{t-1}\frac{1}{\kappa}\,\phi(\tau)\phi^T(\tau)\right)^{-1},\ t \in \{t_0+1, t_0+2, \ldots\}. \quad (3.104)$$

From (3.104), we have (3.90), which leads to (3.88).

To derive (3.87), we use (3.103) and (3.88) to obtain

$$P^{-1}(t-1)\theta(t) = P_0^{-1}\theta_0 + \sum_{\tau=t_0}^{t-1}\frac{1}{\kappa}y(\tau)\phi(\tau), \quad (3.105)$$

$$P(t)\phi(t) = P(t-1)\phi(t) - \frac{P(t-1)\phi(t)\phi^T(t)P(t-1)\phi(t)}{m^2(t)} = \frac{\kappa P(t-1)\phi(t)}{m^2(t)}. \quad (3.106)$$

Based on (3.74), (3.90), (3.105), and (3.106), we use (3.103) to obtain

$$\begin{aligned}\theta(t+1) &= P(t)\left(P^{-1}(t-1)\theta(t) + \frac{1}{\kappa}y(t)\phi(t)\right)\\ &= P(t)\left(P^{-1}(t)\theta(t) - \frac{1}{\kappa}\phi(t)\phi^T(t)\theta(t) + \frac{1}{\kappa}y(t)\phi(t)\right)\\ &= \theta(t) - \frac{1}{\kappa}P(t)\phi(t)\epsilon(t) = \theta(t) - \frac{P(t-1)\phi(t)\epsilon(t)}{m^2(t)},\end{aligned} \quad (3.107)$$

which is the adaptive law (3.87). Note that a choice of $P_0 > \alpha I_{n_\theta}$ for a large $\alpha > 0$ is preferred, as the error $\theta(t) - \theta_0$ should be less penalized.

3.6.4 Parameter Convergence

We now address the issue of parameter convergence of the normalized gradient and least-squares algorithms for discrete-time systems by starting with the following excitation definitions for the regressor $\phi(t)$.

Definition 3.3 *A bounded vector signal $x(t) \in R^q$, $q \geq 1$, is exciting over the time sequence set $\{\sigma_0, \sigma_0 + 1, \ldots, \sigma_0 + \delta_0\}$, $\delta_0 > 0$, $\sigma_0 \geq t_0$, if for some $\alpha_0 > 0$*

$$\sum_{\tau=\sigma_0}^{\sigma_0+\delta_0} x(\tau)x^T(\tau) \geq \alpha_0 I. \tag{3.108}$$

Definition 3.4 *A bounded vector signal $x(t) \in R^q$, $q \geq 1$, is persistently exciting (PE) if there exist $\delta > 0$ and $\alpha_0 > 0$ such that*

$$\sum_{t=\sigma}^{\sigma+\delta} x(t)x^T(t) \geq \alpha_0 I, \ \forall \sigma \geq t_0. \tag{3.109}$$

The discrete-time version of Theorem 3.1 may be stated as follows.

Theorem 3.2 *Consider system (3.76), that is, $P(z)[y](t) = Z(z)[u](t)$, and the associated parametric model (3.73), that is, $y(t) = \theta^{*T}\phi(t)$ with $\theta^* \in R^{n+m+1}$. Assume that all zeros of $P(z)$ are inside the unit circle of the complex z-plane and for any set of $\{\omega_1, \omega_2, \ldots, \omega_{n+m+1}\}$, with $\omega_j \neq \omega_k + 2\pi l$ for $j \neq k$ and an integer l, $H(e^{j\omega_i})$, $i = 1, \ldots, n+m+1$, are linearly independent vectors over C^n, where $H(z) = [\frac{b^T(z)}{\Lambda(z)}, \frac{a^T(z)}{\Lambda(z)}\frac{Z(z)}{P(z)}]^T$ with $b(z) = [1, z, \cdots, z^m]^T$ and $a(z) = [1, z, \ldots, z^{n-2}, z^{n-1}]^T$. Then the signal $\phi(t)$ is persistently exciting (PE) if the system input $u(t)$ has at least $n+m+1$ frequencies.*

The frequency concept in discrete time has some difference from that in continuous time. For example, while $u_1(t) = 10$ has one frequency at $\omega_1 = 0$ and $u_2(t) = 10 - \sin(t)$ has three frequencies at $\omega_1 = 0$, $\omega_2 = 1$, $\omega_3 = -1$, the signal $u_3(t) = \sin(t) + \sin((1 + 2\pi l)t)$ has only two frequencies, 1 and -1, for any integer l, as $\sin((1 + 2\pi l)t) = \sin t$.

Similar to Proposition 3.1, there is a condition on $H(s)$ for $\phi(t)$ to be PE.

Proposition 3.2 *$H(je^{\omega_i})$, $i = 1, \ldots, n+m+1$, with $\omega_j \neq \omega_k + 2\pi l$ for $j \neq k$ and an integer l, are linearly independent vectors over C^n if and only if $Z(z)$ and $P(z)$ are coprime, that is, they do not have common factors. If $Z(z)$ and $P(z)$ are not coprime, then $\phi(t)$ cannot be PE for any bounded input $u(t)$.*

Convergence of the Gradient Algorithm

Consider the normalized gradient algorithm (3.83) with $\Gamma = I$. Using (3.75), we express system (3.83) as

$$\tilde{\theta}(t+1) = A(t)\tilde{\theta}(t),\ A(t) = I - \psi(t)\psi^T(t),\ t \in \{t_0, t_0+1, \ldots\}, \qquad (3.110)$$

where $\psi(t) = \frac{\phi(t)}{m(t)}$. Similar to their continuous-time counterparts in Lemma 3.3 and Lemma 3.4 for (3.53), for (3.110), we have the following results.

Proposition 3.3 *If the signal $\psi(t)$ is exciting over the time sequence set $\{\sigma_0, \sigma_0+1, \ldots, \sigma_0+\delta_0\}$, then $\|\tilde{\theta}(\sigma_0+\delta_0)\|_2 \leq \alpha_0\|\tilde{\theta}(\sigma_0)\|_2$ for some $\alpha_0 \in (0,1)$.*

Proposition 3.4 *If $\psi(t)$ is persistently exciting, then $\lim_{t\to\infty} \|\tilde{\theta}(t)\|_2 = 0$ exponentially (that is, $\lim_{t\to\infty} \theta(t) = \theta^*$ exponentially).*

Their proofs are based on the derivable inequality

$$V(\tilde{\theta}(\sigma_0+\delta_0)) \leq \alpha_0^2 V(\tilde{\theta}(\sigma_0)) \qquad (3.111)$$

for some $\alpha_0 \in (0,1)$, where $V(\tilde{\theta}) = \tilde{\theta}^T\tilde{\theta}$, as similar to (3.59).

Convergence of the Least-Squares Algorithm

Consider the normalized least-squares algorithm (3.87)–(3.88) for the error model (3.75). The parameter error $\tilde{\theta}(t) = \theta(t) - \theta^*$ has the form (3.99).

Lemma 3.9 *If the signal $\phi(t)$ is exciting over the time sequence set $\{\sigma_0, \sigma_0 + 1, \ldots, \sigma_0+\delta_0\}$, then $(I - P(\sigma_0+\delta_0)P^{-1}(\sigma_0-1))$ is nonsingular, and the unknown parameter vector θ^* is identified as*

$$\theta^* = \left(I - P(\sigma_0+\delta_0)P^{-1}(\sigma_0-1)\right)^{-1} \times \left(\theta(\sigma_0+\delta_0+1) - P(\sigma_0+\delta_0)P^{-1}(\sigma_0-1)\theta(\sigma_0)\right). \qquad (3.112)$$

Proof: From (3.91), we have

$$I - P(\sigma_0+\delta_0)P^{-1}(\sigma_0-1) = P(\sigma_0+\delta_0)\sum_{\tau=\sigma_0}^{\sigma_0+\delta_0}\frac{1}{\kappa}\phi(\tau)\phi^T(\tau), \qquad (3.113)$$

which is nonsingular under the lemma's condition. From (3.99), we obtain

$$\tilde{\theta}(\sigma_0+\delta_0+1) = P(\sigma_0+\delta_0)P^{-1}(\sigma_0-1)\tilde{\theta}(\sigma_0), \qquad (3.114)$$

which implies (3.112) under the nonsingularity condition. ∇

Lemma 3.10 *If the signal $\phi(t)$ is persistently exciting, then the least-squares algorithm (3.87)–(3.88) ensures that* $\lim_{t\to\infty}\|\tilde{\theta}(t)\|_2 = 0$.

Proof: If $\phi(t)$ is persistently exciting, then $\sum_{t=\sigma}^{\sigma+\delta}\phi(t)\phi^T(t) \geq \alpha_0 I$, $\forall\sigma \geq t_0$, for some $\alpha_0 > 0$ and $\delta > 0$. Letting $\sigma = t_0 + i(\delta+1)$, $i = 0, 1, \ldots$, we have

$$\sum_{t=t_0+i(\delta+1)}^{t_0+i(\delta+1)+\delta}\phi(t)\phi^T(t) \geq \alpha_0 I. \tag{3.115}$$

Using (3.91), we have

$$P^{-1}(k(\delta+1)+t_0-1) = P^{-1}(t_0-1) + \sum_{i=0}^{k-1}\sum_{t=t_0+i(\delta+1)}^{t_0+i(\delta+1)+\delta}\phi(t)\phi^T(t). \tag{3.116}$$

From (3.90), (3.115), and (3.116), we conclude that

$$\lim_{t\to\infty}\min_{1\leq i\leq n+m+1}\lambda_i(P^{-1}(t)) = \infty, \tag{3.117}$$

where $\lambda_i(P^{-1}(t))$ is the ith eigenvalue of $P^{-1}(t)$, which implies that

$$\begin{aligned}\lim_{t\to\infty}\|P(t)\|_2 &= \lim_{t\to\infty}\max_{1\leq j\leq n+m+1}\lambda_j(P(t))\\ &= \lim_{t\to\infty}\left(\min_{1\leq i\leq n+m+1}\lambda_i(P^{-1}(t))\right)^{-1} = 0.\end{aligned} \tag{3.118}$$

Therefore, from (3.99), we conclude that

$$\lim_{t\to\infty}\tilde{\theta}(t) = \lim_{t\to\infty}P(t-1)P^{-1}(t_0-1)\tilde{\theta}(t_0) = 0, \tag{3.119}$$

which completes the proof of the lemma. ∇

In this discrete-time case, the rate at which $\tilde{\theta}(t)$ converges to zero with the normalized least-squares algorithm (3.87)–(3.88) can also be evaluated as follows. Using (3.115) and (3.116), we obtain

$$P(T+t_0-1) \leq \left(P^{-1}(t_0-1) + \frac{\alpha_0 T}{\delta}I\right)^{-1}, \tag{3.120}$$

where $T = k(\delta+1)$. For $P(t_0) = I$, from (3.99) and (3.120), we have

$$\|\tilde{\theta}(T+t_0)\|_2 \leq \|P(T+t_0)\|_2\,\|P^{-1}(t_0)\|_2\,\|\tilde{\theta}(t_0)\|_2 \leq \frac{1}{1+\frac{\alpha_0 T}{\delta}}\|\tilde{\theta}(t_0)\|_2. \tag{3.121}$$

Hence the convergence rate is $\frac{1}{t}$ (i.e., in general, not exponential).

In the discrete-time case, it is the persistent excitation property of the vector signal $\phi(t)$ which plays a crucial role in the parameter convergence analysis for the normalized least-squares algorithm. For $\phi(t) \in L^\infty$ and $\kappa > 0$, $\phi(t)$ is PE if and only if $\psi(t) = \frac{\phi(t)}{\sqrt{\kappa+\phi^T(t)P(t-1)\phi(t)}}$.

Similar to the continuous-time case, there are several ways [125] to modify the least-squares algorithm (3.87)–(3.88) so that under a PE condition the parameter error $\tilde{\theta}(t)$ converges to zero exponentially as $t \to \infty$. One choice is the normalized least-squares algorithm with covariance resetting [125]:

$$\theta(t+1) = \theta(t) - \frac{P(t-1)\phi(t)\epsilon(t)}{m^2(t)}, \; \theta(t_0) = \theta_0, \tag{3.122}$$

$$P(t) = P(t-1) - \frac{P(t-1)\phi(t)\phi^T(t)P(t-1)}{m^2(t)}, \; P(t_{0-1}) = \rho_0 I,$$

$$P(t_r+1) = \rho_0 I, \; \rho_0 > 0, \tag{3.123}$$

where t_r is the time at which $\lambda_{\min}(P(t_r)) = \rho_1 < \rho_0$ for some constant $\rho_1 \in (0, \rho_0)$. This algorithm has the desired properties as stated in Lemma 3.4. Moreover, with this algorithm, the parameter error $\tilde{\theta}(t)$ converges to zero exponentially fast if the signal $\phi(t)$ is persistently exciting.

3.7 Robustness of Adaptive Algorithms

The gradient and least-squares algorithms for adaptive parameter estimation have the desired properties as stated in Lemmas 3.1, 3.2, 3.7, and 3.8 for the ideal parametric models (3.18) and (3.73): $y(t) = \theta^{*T}\phi(t)$ for some unknown parameter vector to be estimated and some available (known) signals $y(t)$ and $\phi(t)$. This model does not contain any modeling errors such as system disturbances (noises) and parameter variations. It is of both theoretical and practical interest to study the robustness properties of the gradient and least-squares algorithms in the presence of modeling errors. In this section, we will address a class of system modeling errors, the $L^\infty \cap L^{1+\alpha}$ $(0 < \alpha < \infty)$ errors, to study the inherent robustness properties of these adaptive algorithms.

3.7.1 Continuous-Time Algorithms

Consider a continuous-time linear parametric model with modeling errors:

$$y(t) = \theta^{*T}\phi(t) - \delta(t), \; t \geq t_0, \tag{3.124}$$

where $\theta^* \in R^{n_\theta}$ is an unknown parameter vector, $y(t) \in R$ is a measured signal, $\phi(t) \in R^{n_\theta}$ is a known vector signal, with $n_\theta > 0$, and $\delta(t)$ represents modeling errors in the system associated with the model (3.124). In this study, we consider modelings errors which satisfy

$$|\delta(t)| \le \delta_1(t)\|\phi(t)\|_2 + \delta_2(t), \tag{3.125}$$

where $\delta_1(t) \in L^\infty \cap L^{1+\alpha}$ and $\delta_2(t) \in L^\infty \cap L^{1+\alpha}$, $0 < \alpha < \infty$, that is,

$$\int_0^\infty |\delta_1(t)|^{1+\alpha}\, dt < \infty, \int_0^\infty |\delta_2(t)|^{1+\alpha}\, dt < \infty, \tag{3.126}$$

in addition to the boundedness of $\delta_1(t)$ and $\delta_2(t)$.

Letting $\theta(t)$ be the estimate of θ^*, we define the estimation error

$$\epsilon(t) = \theta^T(t)\phi(t) - y(t),\ t \ge 0. \tag{3.127}$$

Using (3.124), we have

$$\epsilon(t) = \tilde{\theta}^T(t)\phi(t) + \delta(t),\ \tilde{\theta}(t) = \theta(t) - \theta^*,\ t \ge t_0. \tag{3.128}$$

To handle the modeling error $\delta(t)$, we use the modified adaptive law for $\theta(t)$:

$$\dot{\theta}(t) = -\frac{\Gamma\phi(t)f(\epsilon(t))}{m^{1+\bar{\alpha}}(t)},\ \theta(t_0) = \theta_0, \tag{3.129}$$

where $\Gamma = \Gamma^T > 0$ is a gain matrix, the continuous function $f(\cdot)$ is

$$f(\epsilon(t)) = \begin{cases} (\epsilon(t))^{\bar{\alpha}} & \text{if } \epsilon(t) \ge 0, \\ -(-\epsilon(t))^{\bar{\alpha}} & \text{if } \epsilon(t) < 0 \end{cases} \tag{3.130}$$

for $\alpha \le \bar{\alpha} < \infty$, θ_0 is an initial estimate of θ^*, and

$$m(t) = \sqrt{1 + \kappa\phi^T(t)\phi(t)}, \tag{3.131}$$

with $\kappa > 0$ being a design parameter.

The time derivative of the positive definite function $V(\tilde{\theta}) = \tilde{\theta}^T\Gamma^{-1}\tilde{\theta}$ is

$$\dot{V} = \frac{-2f(\epsilon(t))\epsilon(t)}{m^{1+\bar{\alpha}}(t)} + \frac{2f(\epsilon(t))\delta(t)}{m^{1+\bar{\alpha}}(t)} \le \frac{-2|\epsilon(t)|^{1+\bar{\alpha}}}{m^{1+\bar{\alpha}}(t)} + \frac{2|\epsilon(t)|^{\bar{\alpha}}|\delta(t)|}{m^{1+\bar{\alpha}}(t)}, \tag{3.132}$$

from which it follows that

$$\int_0^t \frac{2|\epsilon(\tau)|^{1+\bar{\alpha}}}{m^{1+\bar{\alpha}}(\tau)}\, d\tau + V(\tilde{\theta}(t)) \le \int_0^t \frac{2|\epsilon(\tau)|^{\bar{\alpha}}|\delta(\tau)|}{m^{1+\bar{\alpha}}(\tau)}\, d\tau + V(\tilde{\theta}(0)). \tag{3.133}$$

Using the Hölder inequality (see Section 2.3.2), we have

$$\int_0^t \frac{|\epsilon(\tau)|^{\bar{\alpha}}|\delta(\tau)|}{m^{1+\bar{\alpha}}(\tau)}\,d\tau$$
$$\leq \left(\int_0^t \left(\frac{|\epsilon(\tau)|^{\bar{\alpha}}}{m^{\bar{\alpha}}(\tau)}\right)^{\frac{1+\bar{\alpha}}{\bar{\alpha}}} d\tau\right)^{\frac{\bar{\alpha}}{1+\bar{\alpha}}} \left(\int_0^t \left(\frac{|\delta(\tau)|}{m(\tau)}\right)^{1+\bar{\alpha}} d\tau\right)^{\frac{1}{1+\bar{\alpha}}}. \quad (3.134)$$

Since $\delta(t)$ satisfies (3.125), with $\delta_1(t), \delta_2(t) \in L^\infty \cap L^{1+\alpha}$, and $\bar{\alpha} \geq \alpha$, it follows that $(\frac{|\delta(t)|}{m(t)})^{\bar{\alpha}-\alpha}$ is bounded and $\frac{|\delta(t)|}{m(t)} \in L^{1+\alpha}$, that is,

$$\int_0^\infty \left(\frac{|\delta(\tau)|}{m(\tau)}\right)^{1+\alpha} d\tau < \infty, \quad (3.135)$$

by using Minkovski inequality (see Section 2.3.2). Therefore, we have

$$\left(\int_0^t \left(\frac{|\delta(\tau)|}{m(\tau)}\right)^{1+\bar{\alpha}} d\tau\right)^{\frac{1}{1+\bar{\alpha}}} = \left(\int_0^t \left(\frac{|\delta(\tau)|}{m(\tau)}\right)^{1+\alpha} \left(\frac{|\delta(\tau)|}{m(\tau)}\right)^{\bar{\alpha}-\alpha} d\tau\right)^{\frac{1}{1+\bar{\alpha}}} \leq k_d \quad (3.136)$$

for some constant $k_d > 0$. From (3.133)–(3.136), we obtain

$$\int_0^t \frac{2|\epsilon(\tau)|^{1+\bar{\alpha}}}{m^{1+\bar{\alpha}}(\tau)}\,d\tau + V(\tilde{\theta}(t)) \leq 2k_d \left(\int_0^t \frac{|\epsilon(\tau)|^{1+\bar{\alpha}}}{m^{1+\bar{\alpha}}(\tau)}\,d\tau\right)^{\frac{\bar{\alpha}}{1+\bar{\alpha}}} + V(\tilde{\theta}(0)). \quad (3.137)$$

Let $h(t) = \int_0^t \frac{|\epsilon(\tau)|^{1+\bar{\alpha}}}{m^{1+\bar{\alpha}}(\tau)}\,d\tau$. Then, using (3.137), we get

$$2h(t) \leq 2k_d(h(t))^{\frac{\bar{\alpha}}{1+\bar{\alpha}}} + V(\tilde{\theta}(0)), \quad (3.138)$$

which means that $h(t) \in L^\infty$, that is, $\frac{\epsilon(t)}{m(t)} \in L^{1+\bar{\alpha}}$. With this result, using (3.137) again, we have that $V(\tilde{\theta}(t)) \in L^\infty$, so that $\theta(t) \in L^\infty$. In view of (3.128) and (3.129), we have that $\frac{\epsilon(t)}{m(t)} \in L^\infty$, and $\dot{\theta}(t) \in L^\infty \cap L^{\frac{1+\bar{\alpha}}{\bar{\alpha}}}$.

In summary, we have the following result.

Lemma 3.11 *The modified gradient algorithm (3.129) is robust with respect to $L^\infty \cap L^{1+\alpha}$ modeling errors $\delta(t)$, $0 < \alpha < \infty$, in the sense that $\theta(t) \in L^\infty$, $\frac{\epsilon(t)}{m(t)} \in L^\infty \cap L^{1+\bar{\alpha}}$, and $\dot{\theta}(t) \in L^\infty \cap L^{\frac{1+\bar{\alpha}}{\bar{\alpha}}}$, $\alpha \leq \bar{\alpha} < \infty$.*

For adaptation robustness, the case of $1 < \alpha < \infty$ is of our main interest, as $L^{1+\beta} \cap L^\infty \subset L^{1+\alpha} \cap L^\infty$ for $0 < \beta < \alpha$; in particular, $L^{1+\beta} \cap L^\infty \subset L^2 \cap L^\infty$ for $0 < \beta < 1$. In the adaptive design (3.129), one may choose $\max\{1, \alpha\} \leq \bar{\alpha}$.

In particular, when $\alpha = \bar{\alpha} = 1$, we have that $\delta_1(t) \in L^\infty \cap L^2$ and $\delta_2(t) \in L^\infty \cap L^2$, and (3.129) becomes (3.23). The above result means that the properties in Lemma 3.1 are still true in the presence of $\delta(t)$ in the linear parametric model (3.124), that is, the standard gradient algorithm (3.23) is robust with respect to $L^\infty \cap L^2$ modeling errors $\delta(t)$ [342].

3.7.2 Discrete-Time Algorithms

A discrete-time linear parametric model with modeling errors is

$$y(t) = \theta^{*T}\phi(t) - \delta(t),\ t \in \{0, 1, 2, \ldots\}, \tag{3.139}$$

where $\theta^* \in R^{n_\theta}$ is an unknown parameter vector, $y(t) \in R$ is a measured signal, and $\phi(t) \in R^{n_\theta}$ is a known vector signal, with $n_\theta > 0$, and $\delta(t)$ represents modeling errors which, in this study, satisfy

$$|\delta(t)| \leq \delta_1(t)\|\phi(t)\|_2 + \delta_2(t), \tag{3.140}$$

where $\delta_1(t) \in L^{1+\alpha}$ and $\delta_2(t) \in L^{1+\alpha}$, $1 \leq \alpha < \infty$, that is,

$$\sum_{t=0}^{\infty} |\delta_1(t)|^{1+\alpha} < \infty,\ \sum_{t=0}^{\infty} |\delta_2(t)|^{1+\alpha} < \infty. \tag{3.141}$$

For a discrete-time signal $x(t)$, $x(t) \in L^p$, $p > 0$, implies that $x(t) \in L^\infty$.

Letting $\theta(t)$ be the estimate of θ^*, we define the estimation error

$$\epsilon(t) = \theta^T(t)\phi(t) - y(t),\ t \in \{0, 1, 2, \ldots\}, \tag{3.142}$$

and using (3.139), we have the error model

$$\epsilon(t) = (\theta(t) - \theta^*)^T\phi(t) + \delta(t). \tag{3.143}$$

We first study the robustness of the gradient algorithm (3.83) in the presence of L^2 modeling error $\delta(t)$ in the linear model (3.139), that is, for $\alpha = 1$ in (3.141). In this case, the time increment of the positive definite function $V(\tilde{\theta}) = \tilde{\theta}^T\Gamma^{-1}\tilde{\theta}$, along (3.83), that is, along $\theta(t+1) = \theta(t) - \frac{\Gamma\phi(t)\epsilon(t)}{m^2(t)}$ with $m(t) = \sqrt{\kappa + \phi^T(t)\phi(t)}$, $\kappa > 0$ and $0 < \Gamma = \Gamma^T < 2I_{n_\theta}$, is

$$\begin{aligned}
&V(\tilde{\theta}(t+1)) - V(\tilde{\theta}(t)) \\
&= -\left(2 - \frac{\phi^T(t)\Gamma\phi(t)}{m^2(t)}\right)\frac{\epsilon^2(t)}{m^2(t)} + 2\frac{\epsilon(t)\delta(t)}{m^2(t)} \\
&\leq -\alpha_1\frac{\epsilon^2(t)}{m^2(t)} + \frac{2\epsilon(t)\delta(t)}{m^2(t)} \\
&\leq -\frac{\alpha_1\epsilon^2(t)}{2m^2(t)} - \frac{\alpha_1}{2}\left(\frac{|\epsilon(t)|}{m(t)} - \frac{2|\delta(t)|}{\alpha_1 m(t)}\right)^2 + \frac{2\delta^2(t)}{\alpha_1 m^2(t)}
\end{aligned} \tag{3.144}$$

for some constant $\alpha_1 > 0$, as from the condition $0 < \Gamma = \Gamma^T < 2I_{n_\theta}$ in the adaptive law (3.83). From (3.140) and (3.141), we have

$$\sum_{t=0}^{\infty} \frac{2\delta^2(t)}{\alpha_1 m^2(t)} \leq k_d \tag{3.145}$$

for some constant $k_d > 0$. This, together with (3.144), implies that

$$V(\tilde{\theta}(t)) \leq V(\tilde{\theta}(0)) + \sum_{\tau=0}^{t} \frac{2\delta^2(\tau)}{\alpha_1 m^2(\tau)} \leq V(\tilde{\theta}(0)) + k_d, \ \forall t \geq 0. \tag{3.146}$$

Hence, $\theta(t) \in L^\infty$, and with (3.144) again, $\frac{\epsilon(t)}{m(t)} \in L^2 \cap L^\infty$, and $\theta(t+1) - \theta(t) \in L^2$. We conclude that the standard gradient algorithm (3.83) is robust with respect to L^2 modeling errors in a linear parametric model.

Now we consider the general case when $1 \leq \alpha < \infty$. To handle such modeling errors $\delta(t)$ in (3.139), we choose the adaptive law for $\theta(t)$ as

$$\theta(t+1) = \theta(t) - \frac{\Gamma \epsilon(t)\phi(t)}{m^2(t)} g(t), \ 0 < \Gamma = \Gamma^T < 2I_{n_\theta}, \tag{3.147}$$

where

$$m(t) = \sqrt{\kappa + \phi^T(t)\phi(t) + \epsilon^2(t)}, \ \kappa > 0, \tag{3.148}$$

$$g(t) = \left(\frac{|\epsilon(t)|}{m(t)}\right)^{\bar{\alpha}-1}, \ 1 \leq \alpha \leq \bar{\alpha} < \infty. \tag{3.149}$$

In this case, the time increment of $V(\tilde{\theta}) = \tilde{\theta}^T \Gamma^{-1} \tilde{\theta}$, along (3.147), is

$$\begin{aligned} & V(\tilde{\theta}(t+1)) - V(\tilde{\theta}(t)) \\ &= -\left(2 - \frac{\phi^T(t)\Gamma\phi(t)}{m^2(t)} g(t)\right) \frac{\epsilon^2(t)}{m^2(t)} g(t) + 2\frac{\epsilon(t)\delta(t)}{m^2(t)} g(t) \\ &\leq -\alpha_1 \left(\frac{|\epsilon(t)|}{m(t)}\right)^{1+\bar{\alpha}} + 2\left(\frac{|\epsilon(t)|}{m(t)}\right)^{\bar{\alpha}} \frac{|\delta(t)|}{m(t)} \end{aligned} \tag{3.150}$$

for some constant $\alpha_1 > 0$ as $0 < \Gamma = \Gamma^T < 2I_{n_\theta}$ and $0 \leq g(t) < 1$ in the modified adaptive law (3.147). From (3.150), it follows that

$$\alpha_0 \sum_{\tau=0}^{t} \frac{|\epsilon(\tau)|^{1+\bar{\alpha}}}{m^{1+\bar{\alpha}}(\tau)} + V(\tilde{\theta}(t+1)) \leq 2\sum_{\tau=0}^{t} \frac{|\epsilon(\tau)|^{\bar{\alpha}}|\delta(\tau)|}{m^{1+\bar{\alpha}}(\tau)} + V(\tilde{\theta}(0)). \tag{3.151}$$

Using the Hölder inequality (see Section 2.7), we have

$$\sum_{\tau=0}^{t} \frac{|\epsilon(\tau)|^{\bar{\alpha}}|\delta(\tau)|}{m^{1+\bar{\alpha}}(\tau)} \leq \left(\sum_{\tau=0}^{t} \left(\frac{|\epsilon(\tau)|^{\bar{\alpha}}}{m^{\bar{\alpha}}(\tau)}\right)^{\frac{1+\bar{\alpha}}{\bar{\alpha}}}\right)^{\frac{\bar{\alpha}}{1+\bar{\alpha}}} \left(\sum_{\tau=0}^{t} \left(\frac{|\delta(\tau)|}{m(\tau)}\right)^{1+\bar{\alpha}}\right)^{\frac{1}{1+\bar{\alpha}}}. \tag{3.152}$$

From (3.140) and (3.141), using Minkovski inequality (see Section 2.7), we have $\frac{|\delta(t)|}{m(t)} \in L^{1+\alpha}$, and so $\frac{|\delta(t)|}{m(t)} \in L^\infty$. Then,

$$\left(\sum_{\tau=0}^{t} \left(\frac{|\delta(\tau)|}{m(\tau)}\right)^{1+\bar{\alpha}}\right)^{\frac{1}{1+\bar{\alpha}}} = \left(\sum_{\tau=0}^{t} \left(\frac{|\delta(\tau)|}{m(\tau)}\right)^{1+\alpha} \left(\frac{|\delta(\tau)|}{m(\tau)}\right)^{\bar{\alpha}-\alpha}\right)^{\frac{1}{1+\bar{\alpha}}} \leq k_\delta \tag{3.153}$$

for some constant $k_\delta > 0$. From (3.150)–(3.153), we obtain

$$\alpha_0 \sum_{\tau=0}^{t} \frac{|\epsilon(\tau)|^{1+\bar{\alpha}}}{m^{1+\bar{\alpha}}(\tau)} + V(\tilde{\theta}(t+1)) \leq 2k_\delta \left(\sum_{\tau=0}^{t} \frac{|\epsilon(\tau)|^{1+\bar{\alpha}}}{m^{1+\bar{\alpha}}(\tau)} \right)^{\frac{\bar{\alpha}}{1+\bar{\alpha}}} + V(\tilde{\theta}(0)). \quad (3.154)$$

Define the auxiliary signal $h(t) = \sum_{\tau=0}^{t} \frac{|\epsilon(\tau)|^{1+\bar{\alpha}}}{m^{1+\bar{\alpha}}(\tau)}$. Then, using (3.154), we get

$$\alpha_0 h(t) \leq 2k_\delta (h(t))^{\frac{\bar{\alpha}}{1+\bar{\alpha}}} + V(\tilde{\theta}(0)), \quad (3.155)$$

which means that $h(t) \in L^\infty$, that is, $\frac{\epsilon(t)}{m(t)} \in L^{1+\bar{\alpha}}$. With this result, using (3.154) again, we have that $V(\tilde{\theta}(t)) \in L^\infty$, so that $\theta(t) \in L^\infty$. The fact that $\frac{\epsilon(t)}{m(t)} \in L^{1+\bar{\alpha}}$ implies that $\frac{\epsilon(t)}{m(t)} \in L^\infty$ (which also follows from (3.143)). In view of (3.147), we have that $\theta(t+1) - \theta(t) \in L^{\frac{1+\bar{\alpha}}{\bar{\alpha}}}$. In summary, we have:

Lemma 3.12 *The adaptive law (3.147) ensures that* $\theta(t) \in L^\infty$, $\frac{\epsilon(t)}{m(t)} \in L^{1+\bar{\alpha}}$ *(and* $\frac{\epsilon(t)}{m(t)} \in L^\infty$*),* $\theta(t+1) - \theta(t) \in L^{\frac{1+\bar{\alpha}}{\bar{\alpha}}}$.

The adaptive design (3.147) with $\bar{\alpha} \geq 1$ and analysis are also applicable to modeling errors which belong to $L^{1+\alpha}$, $0 < \alpha < 1$, as a signal $x(t) \in L^{1+\alpha}$ with $0 < \alpha < 1$, also belongs to $L^{1+\beta}$ with $\beta > \alpha$, in particular, for $\beta = 1$.

3.8 Robust Adaptive Laws

When the system modeling errors are bounded but do not belong to $L^{1+\alpha}$ for any finite $\alpha > 0$, a gradient algorithm (3.23) or its modified version (3.129) or a least-squares algorithm (3.50)–(3.51) (or their discrete-time counterparts) cannot ensure the desired properties of an adaptive parameter estimator, not even the boundedness of $\theta(t)$. Different types of modifications are needed to make those algorithms robust with respect to more general modeling errors. In this section, we present several such modifications (redesigns) for the gradient algorithm (3.23) or (3.83) to enhance their robustness.

3.8.1 Continuous-Time Algorithms

Consider a continuous-time linear parametric model with modeling errors:

$$y(t) = \theta^{*T}\phi(t) - \delta(t),\ t \geq t_0, \quad (3.156)$$

where $\theta^* \in R^{n_\theta}$ is an unknown parameter vector, $y(t) \in R$ is a measured signal, and $\phi(t) \in R^{n_\theta}$ is a known vector signal. The term $\delta(t)$ represents system modeling errors which satisfy

$$|\delta(t)| \leq \delta_1(t)\|\phi(t)\|_2 + \delta_2(t), \quad (3.157)$$

where, unlike their counterparts in (3.124), $\delta_1(t) \in L^\infty$ and $\delta_2(t) \in L^\infty$ only, that is, $\delta_1(t)$ and $\delta_2(t)$ are just bounded.

Let $\theta(t)$ be the estimate of θ^* and define the estimation error

$$\epsilon(t) = \theta^T(t)\phi(t) - y(t),\ t \geq 0, \tag{3.158}$$

which can be expressed as

$$\epsilon(t) = \tilde{\theta}^T(t)\phi(t) + \delta(t),\ \tilde{\theta}(t) = \theta(t) - \theta^*,\ t \geq t_0. \tag{3.159}$$

To handle such $\delta(t)$, the standard gradient algorithm (3.23) is modified as

$$\dot{\theta}(t) = -\frac{\Gamma\phi(t)\epsilon(t)}{m^2(t)} + f(t) \tag{3.160}$$

with an initial estimate $\theta(t_0) = \theta_0$ of θ^* and a gain matrix $\Gamma = \Gamma^T > 0$, where

$$m(t) = \sqrt{1 + \kappa\phi^T(t)\phi(t)},\ \kappa > 0, \tag{3.161}$$

is the normalization signal and the function $f(\cdot)$ is a modification term for robustness with respect to the modeling error $\delta(t)$ in (3.159).

The time derivative of the positive definite function $V(\tilde{\theta}) = \tilde{\theta}^T\Gamma^{-1}\tilde{\theta}$ is

$$\dot{V} = \frac{dV}{dt} = -\frac{2\epsilon^2(t)}{m^2(t)} + \frac{2\epsilon(t)\delta(t)}{m^2(t)} + 2\tilde{\theta}^T(t)\Gamma^{-1}f(t). \tag{3.162}$$

Our goal is to choose $f(t)$ such that $\theta(t) \in L^\infty$ and $\dot{\theta}(t)$ and $\frac{\epsilon(t)}{m(t)}$ have some desired properties close to that in Lemma 3.1 for the case when $\delta(t) = 0$.

Dead-Zone Modification

From (3.157) and (3.161), it follows that

$$\frac{|\delta(t)|}{m(t)} \leq c_1 + \frac{c_2}{m(t)} \leq d_1 + \frac{d_2}{m(t)} \tag{3.163}$$

for some constants $d_1 \geq c_1 > 0$ and $d_2 \geq c_2 > 0$. The use of constants $\{c_1, c_2\}$ is for characterizing the modeling error $\delta(t)$, while the use of constants $\{d_1, d_2\}$ is for designing the modification function $f(t)$.

We see from (3.162) and (3.163) that when $\frac{|\epsilon(t)|}{m(t)} \geq d_1 + \frac{d_2}{m(t)}$, we have $\dot{V} \leq 0$ for $\tilde{\theta}^T(t)\Gamma^{-1}f(t) \leq 0$. When $\frac{|\epsilon(t)|}{m(t)} < d_1 + \frac{d_2}{m(t)}$, if we set $\dot{\theta}(t) = 0$, then $\dot{V} = 0$. This motivates the following choice of $f(t)$:

$$f(t) = \frac{\Gamma\phi(t)f_0(\epsilon(t))}{m^2(t)},\ f_0(\epsilon(t)) = \begin{cases} \epsilon(t) & \text{if } \frac{|\epsilon(t)|}{m(t)} < d_1 + \frac{d_2}{m(t)}, \\ 0 & \text{if } \frac{|\epsilon(t)|}{m(t)} \geq d_1 + \frac{d_2}{m(t)}. \end{cases} \tag{3.164}$$

With $f(t)$ in (3.164), we have $\tilde{\theta}^T(t)\Gamma^{-1}f(t) = 0$ when $\frac{|\epsilon(t)|}{m(t)} \geq d_1 + \frac{d_2}{m(t)}$ and $\dot{\theta}(t) = 0$ when $\frac{|\epsilon(t)|}{m(t)} < d_1 + \frac{d_2}{m(t)}$. Hence, $\dot{V} \leq 0$, $\forall t \geq 0$, so that $\theta(t) \in L^\infty$. From (3.157) and (3.159), we have $\frac{\epsilon(t)}{m(t)} \in L^\infty$, and from (3.160) and (3.164), $\dot{\theta}(t) \in L^\infty$. Furthermore, from (3.162)–(3.164), with $\theta(t) \in L^\infty$, we have

$$\int_{t_1}^{t_2} \frac{\epsilon^2(t)}{m^2(t)}\, dt \leq c_3 \left(c_1^2(t_2 - t_1) + \int_{t_1}^{t_2} \frac{c_2^2}{m^2(t)}\, dt\right) + c_4 \left(d_1(t_2 - t_1) + \int_{t_1}^{t_2} \frac{d_2}{m(t)}\, dt\right) + c_5, \forall t_2 \geq t_1 \geq 0, \tag{3.165}$$

for some constants $c_i > 0$, $i = 3, 4, 5$. From (3.160) and (3.164), we have

$$\begin{aligned} \dot{\theta}^T(t)\dot{\theta}(t) &= \frac{\phi^T(t)\Gamma^T\Gamma\phi(t)}{m^2(t)} \frac{(\epsilon(t) - f_0(\epsilon(t)))^2}{m^2(t)} \\ &\leq \frac{\phi^T(t)\Gamma^T\Gamma\phi(t)}{m^2(t)} \frac{\epsilon^2(t)}{m^2(t)}. \end{aligned} \tag{3.166}$$

Since $\frac{\phi^T(t)\Gamma^T\Gamma\phi(t)}{m^2(t)} \in L^\infty$, from (3.165) and (3.166), we have

$$\int_{t_1}^{t_2} \dot{\theta}^T(t)\dot{\theta}(t)\, dt \leq c_6 \left(c_1^2(t_2 - t_1) + \int_{t_1}^{t_2} \frac{c_2^2}{m^2(t)}\, dt\right) + c_7 \left(d_1(t_2 - t_1) + \int_{t_1}^{t_2} \frac{d_2}{m(t)}\, dt\right) + c_8, \forall t_2 \geq t_1 \geq 0, \tag{3.167}$$

for some constants $c_i > 0$, $i = 6, 7, 8$.

A dead-zone modification design. A choice of a continuous $f(t)$ is

$$\begin{aligned} f(t) &= \frac{\Gamma\phi(t) f_s(\epsilon(t))}{m^2(t)}, \\ f_s(\epsilon(t)) &= \begin{cases} \epsilon(t) & \text{if } \frac{|\epsilon(t)|}{m(t)} < d_1 + \frac{d_2}{m(t)}, \\ d_1 m(t) + d_2 & \text{if } \frac{\epsilon(t)}{m(t)} \geq d_1 + \frac{d_2}{m(t)}, \\ -(d_1 m(t) + d_2) & \text{if } \frac{\epsilon(t)}{m(t)} \leq -\left(d_1 + \frac{d_2}{m(t)}\right). \end{cases} \end{aligned} \tag{3.168}$$

This $f(t)$ is called a *dead-zone* modification [151], [282] because the function

$$f_d(\epsilon) \stackrel{\triangle}{=} \epsilon - f_s(\epsilon) = \begin{cases} 0 & \text{if } \frac{|\epsilon|}{m} < d_1 + \frac{d_2}{m}, \\ \epsilon - (d_1 m + d_2) & \text{if } \frac{\epsilon}{m} \geq d_1 + \frac{d_2}{m}, \\ \epsilon + (d_1 m + d_2) & \text{if } \frac{\epsilon}{m} \leq -\left(d_1 + \frac{d_2}{m}\right) \end{cases} \tag{3.169}$$

is a dead-zone function of ϵ. With this $f(t)$, from (3.162), we have

$$\dot{V} = \begin{cases} -\frac{2}{m^2}\big((\epsilon - (d_1 m + d_2))^2 \\ \quad +(d_1 m + d_2 - \delta)(\epsilon - (d_1 m + d_2))\big) & \text{if } \frac{\epsilon}{m} \geq d_1 + \frac{d_2}{m}, \\ 0 & \text{if } \frac{|\epsilon|}{m} \leq d_1 + \frac{d_2}{m}, \\ -\frac{2}{m^2}\big((\epsilon + (d_1 m + d_2))^2 \\ \quad -(d_1 m + d_2 + \delta)(\epsilon + (d_1 m + d_2))\big) & \text{if } \frac{\epsilon}{m} \leq -(d_1 + \frac{d_2}{m}), \end{cases} \tag{3.170}$$

that is, $\dot{V} \leq 0$ so that $\theta(t) \in L^\infty$. From (3.157) and (3.159), we have $\frac{\epsilon(t)}{m(t)} \in L^\infty$, and from (3.160) and (3.168), $\dot{\theta}(t) \in L^\infty$. From (3.162), (3.163), and (3.168), we also have (3.165) and (3.167). Using (3.160) and (3.170), it can be shown that $\dot{\theta}(t) \in L^2$, $\dot{\theta}(t) \in L^1$, and $\lim_{t\to\infty} \theta(t) = \theta_\infty$ for some constant θ_∞.

σ-Modification

To examine the destabilizing effect of $\delta(t)$ on $\theta(t)$, we rewrite (3.162) as

$$\dot{V} = -\frac{\epsilon^2(t)}{m^2(t)} - \left(\frac{\epsilon(t)}{m(t)} - \frac{\delta(t)}{m(t)}\right)^2 + \frac{\delta^2(t)}{m^2(t)} + 2\tilde{\theta}^T(t)\Gamma^{-1}f(t). \tag{3.171}$$

In view of (3.163), $\frac{\delta^2(t)}{m^2(t)}$ is a bounded term. If we choose $f(t) = f_\theta(\theta(t))$ such that $2\tilde{\theta}^T(t)\Gamma^{-1}f(t) < 0$ for $\|\theta(t)\| \geq \theta^0$ for some constant $\theta^0 > 0$ and $\lim_{\|\theta\|\to\infty} |\tilde{\theta}^T\Gamma^{-1}f_\theta(\theta)| = \infty$, then the boundedness of $\theta(t)$ is ensured. With this motivation, the σ-modification [147] is given as

$$f(t) = f_\theta(\theta(t)) = -\sigma\Gamma\theta(t), \tag{3.172}$$

where $\sigma > 0$ is a constant design parameter.

With this choice of $f(t)$, $\dot{V} < 0$ whenever

$$\begin{aligned} \frac{\delta^2(t)}{m^2(t)} + 2\tilde{\theta}^T(t)\Gamma^{-1}f(t) &= \frac{\delta^2(t)}{m^2(t)} - 2\sigma(\theta^T(t)\theta(t) - \theta^{*T}\theta(t)) \\ &\leq \frac{\delta^2(t)}{m^2(t)} - 2\sigma\|\theta(t)\|_2^2 + 2\sigma\|\theta(t)\|_2\|\theta^*\|_2 < 0. \end{aligned} \tag{3.173}$$

This condition is satisfied when

$$\begin{aligned} \|\theta(t)\|_2 > \theta^0 &= \frac{1}{2}\left(\sqrt{\|\theta^*\|_2^2 + \frac{2(c_1+c_2)^2}{\sigma}} + \|\theta^*\|_2\right) \\ &\geq \frac{1}{2}\left(\sqrt{\|\theta^*\|_2^2 + \frac{2\delta^2(t)}{\sigma m^2(t)}} + \|\theta^*\|_2\right). \end{aligned} \tag{3.174}$$

Hence, $V(\tilde{\theta}(t))$ is bounded, and so is $\theta(t)$. Then, from (3.159), (3.160), and (3.172), we have $\frac{\epsilon(t)}{m(t)} \in L^\infty$ and $\dot{\theta}(t) \in L^\infty$. From (3.160), (3.163), (3.171), and (3.172), for some $c_i > 0$, $i = 3, \ldots, 8$, we have

$$\int_{t_1}^{t_2} \frac{\epsilon^2(t)}{m^2(t)}\,dt \le c_3 \left(c_1^2(t_2 - t_1) + \int_{t_1}^{t_2} \frac{c_2^2}{m^2(t)}\,dt \right), \\ + c_4\sigma(t_2 - t_1) + c_5, \forall t_2 \ge t_1 \ge 0 \tag{3.175}$$

$$\int_{t_1}^{t_2} \dot{\theta}^T(t)\dot{\theta}(t)\,dt \le c_6 \left(c_1^2(t_2 - t_1) + \int_{t_1}^{t_2} \frac{c_2^2}{m^2(t)}\,dt \right) \\ + c_7\sigma(t_2 - t_1) + c_8, \forall t_2 \ge t_1 \ge 0. \tag{3.176}$$

Switching σ-Modification

A switching σ-modification [151], [154] is given as

$$f(t) = f_\theta(\theta(t)) = -\sigma(t)\Gamma\theta(t),$$
$$\sigma(t) = \begin{cases} 0 & \text{if } \|\theta(t)\|_2 < M_\theta, \\ \sigma_0\left(\frac{\|\theta(t)\|_2}{M_\theta} - 1\right) & \text{if } M_\theta \le \|\theta(t)\|_2 < 2M_\theta, \\ \sigma_0 & \text{if } \|\theta(t)\|_2 \ge 2M_\theta, \end{cases} \tag{3.177}$$

where $M_\theta \ge \|\theta^*\|_2$ and $\sigma_0 > 0$ is any design parameter.

With this choice of $f(t)$, it follows that

$$\tilde{\theta}^T(t)\Gamma f(t) = -\sigma(t)(\theta^T(t)\theta(t) - \theta^{*T}\theta(t)) \\ \le -\sigma(t)\|\theta(t)\|_2(\|\theta(t)\|_2 - \|\theta^*\|_2) \le 0. \tag{3.178}$$

It can be shown that $\theta(t) \in L^\infty$, $\frac{\epsilon(t)}{m(t)} \in L^\infty$, and $\dot{\theta}(t) \in L^\infty$ and that (3.175) and (3.176) are satisfied with $c_4 = c_7 = 0$.

Remark 3.1 The functional difference between (3.172) and (3.177) is that when $\delta(t) = 0$ in (3.159), the switching σ-modification (3.177), employing the additional knowledge of the upper bound M_θ on the parameter norm $\|\theta^*\|$, ensures the ideal properties that $\dot{\theta}(t) \in L^2$ and $\frac{\epsilon(t)}{m(t)} \in L^2$ as from (3.175) and (3.176) with $c_1 = c_2 = 0$ plus $c_4 = c_7 = 0$. However, the fixed σ-modification (3.172) can only ensure $\int_{t_1}^{t_2} \frac{\epsilon^2(t)}{m^2(t)}\,dt \le +c_4\sigma(t_2 - t_1) + c_5$ and $\int_{t_1}^{t_2} \dot{\theta}^T(t)\dot{\theta}(t)\,dt \le c_7\sigma(t_2 - t_1) + c_8$, $\forall t_2 \ge t_1 \ge 0$, as from (3.175) and (3.176) with $c_1 = c_2 = 0$.

The dead-zone modification (3.168) has the similar feature to that of a fixed σ-modification: $\int_{t_1}^{t_2} \frac{\epsilon^2(t)}{m^2(t)}\,dt \le c_4 \left(d_1(t_2 - t_1) + \int_{t_1}^{t_2} \frac{d_2}{m(t)}\,dt\right) + c_5$, $\int_{t_1}^{t_2} \dot{\theta}^T(t)\dot{\theta}(t)\,dt \le c_7 \left(d_1(t_2 - t_1) + \int_{t_1}^{t_2} \frac{d_2}{m(t)}\,dt\right) + c_8$, as from (3.165) and (3.167) with $c_1 = c_2 = 0$ to indicate $\delta(t) = 0$, while $d_1 > 0$ and $d_2 > 0$ are fixed parameters. □

Parameter Projection

A parameter projection design for the modification term $f(t)$ in (3.160) uses the knowledge of the parameter region $[\theta_j^a, \theta_j^b]$, $j = 1, 2, \ldots, n_\theta$, such that $\theta_j^* \in [\theta_j^a, \theta_j^b]$, $j = 1, 2, \ldots, n_\theta$, for $\theta^* = [\theta_1^*, \theta_2^*, \ldots, \theta_{n_\theta}^*]^T$.

For parameter projection, a convenient choice of Γ in (3.16) is

$$\Gamma = \text{diag}\{\gamma_1, \gamma_2, \ldots, \gamma_{n_\theta}\},\ \gamma_i > 0, i = 1, 2, \ldots, n_\theta. \tag{3.179}$$

Denote $\theta_j(t)$, $f_j(t)$, and $g_j(t)$ as the jth components of $\theta(t)$, $f(t)$, and

$$g(t) = -\frac{\Gamma\phi(t)\epsilon(t)}{m^2(t)}, \tag{3.180}$$

respectively, for $j = 1, 2, \ldots, n_\theta$, and choose the initial estimates as

$$\theta_j(0) \in [\theta_j^a, \theta_j^b] \tag{3.181}$$

and set the projection function components as

$$f_j(t) = \begin{cases} 0 & \text{if } \theta_j(t) \in (\theta_j^a, \theta_j^b), \text{ or} \\ & \text{if } \theta_j(t) = \theta_j^a,\ g_j(t) \geq 0, \text{ or} \\ & \text{if } \theta_j(t) = \theta_j^b,\ g_j(t) \leq 0, \\ -g_j(t) & \text{otherwise.} \end{cases} \tag{3.182}$$

This choice of $f(t)$ ensures that $\theta_j(t) \in [\theta_j^a, \theta_j^b]$, $i = 1, 2, \ldots, n_\theta$, so that $\theta(t) \in L^\infty$. From (3.159) and (3.160), we have $\frac{\epsilon(t)}{m(t)} \in L^\infty$ and $\dot{\theta}(t) \in L^\infty$. Furthermore, it can be verified that

$$(\theta_j(t) - \theta_j^*) f_j(t) \leq 0,\ 1 \leq j \leq n_\theta, \tag{3.183}$$

so that with $\tilde{\theta}_j(t) = \theta_j(t) - \theta_j^*$, we have

$$\tilde{\theta}^T(t)\Gamma^{-1} f(t) = \sum_{j=1}^{n_\theta} \tilde{\theta}_j(t)\gamma_j^{-1} f_j(t) \leq 0. \tag{3.184}$$

Based on this property, using (3.160) and (3.171), we can also derive (3.175) and (3.176) with $c_4 = c_7 = 0$. Hence, the parameter projection modification $f(t)$ also has the feature that it leads to the ideal properties that $\dot{\theta}(t) \in L^2$ and $\frac{\epsilon(t)}{m(t)} \in L^2$, when the modeling error $\delta(t)$ is absent in the system model (3.159), similar to the case with a switching σ-modification (3.177).

ϵ-Modification

An ϵ-modification design [282] for $f(t)$ is given as

$$f(t) = -\Gamma\frac{\lambda|\epsilon(t)|}{m(t)}\theta(t)\left(d_1 + \frac{d_2}{m(t)}\right), \tag{3.185}$$

where $\lambda > 0$ is a constant parameter and d_1, d_2 are defined in (3.163).

The time derivative of $V(\tilde{\theta}) = \tilde{\theta}^T\Gamma^{-1}\tilde{\theta}$, along (3.160) with (3.185), is

$$\begin{aligned}\dot{V} &= -\frac{2\epsilon^2(t)}{m^2(t)} + \frac{2\epsilon(t)\delta(t)}{m^2(t)} - \frac{2\lambda|\epsilon(t)|}{m(t)}\tilde{\theta}^T(t)\theta(t)\left(d_1 + \frac{d_2}{m(t)}\right)\\ &\le -\frac{2\epsilon^2(t)}{m^2(t)} + \frac{2|\epsilon(t)|}{m(t)}\left(\frac{|\delta(t)|}{m(t)} - \lambda\tilde{\theta}^T(t)\theta(t)\left(d_1 + \frac{d_2}{m(t)}\right)\right)\\ &\le -\frac{2\epsilon^2(t)}{m^2(t)} + \frac{2|\epsilon(t)|}{m(t)}\left(d_1 + \frac{d_2}{m(t)}\right)\left(1 - \lambda\tilde{\theta}^T(t)\theta(t)\right).\end{aligned} \tag{3.186}$$

Hence, $\dot{V} \le 0$ whenever

$$1 - \lambda\tilde{\theta}^T(t)\theta(t) \le 0, \tag{3.187}$$

which is satisfied when

$$\|\theta(t)\|_2 > \theta^0 = \frac{1}{2}\left(\sqrt{\|\theta^*\|_2^2 + \frac{4}{\lambda}} + \|\theta^*\|_2\right). \tag{3.188}$$

Hence, $\theta(t)$ is bounded, and so are $\frac{\epsilon(t)}{m(t)} \in L^\infty$ and $\dot{\theta}(t) \in L^\infty$.

From (3.171) with $f(t)$ in (3.185), we can derive (3.175) and (3.176) with σ used in (3.172) replaced by λ used in (3.185). As in the fixed switching σ-modification case with (3.172), with this ϵ-modification $f(t)$ in (3.185), the corresponding c_4 and c_7 are nonzero (positive), while with the switching σ or parameter projection modification $f(t)$, $c_4 = c_7 = 0$.

3.8.2 Discrete-Time Algorithms

In the discrete-time case, we consider the linear parametric model

$$y(t) = \theta^{*T}\phi(t) - \delta(t),\ t \in \{0, 1, 2, \ldots,\}, \tag{3.189}$$

where $\theta^* \in R^{n_\theta}$ is an unknown constant parameter vector, $y(t) \in R$ is a measured signal, $\phi(t) \in R^{n_\theta}$ is a known vector signal, and $\delta(t)$ represents modeling errors which satisfy

$$|\delta(t)| \le \delta_1(t)\|\phi(t)\|_2 + \delta_2(t) \tag{3.190}$$

with $\delta_1(t) \in L^\infty$ and $\delta_2(t) \in L^\infty$.

Let $\theta(t)$ be the estimate of θ^* and define the estimation error

$$\epsilon(t) = \theta^T(t)\phi(t) - y(t),\ t \in \{0, 1, 2, \ldots\}. \tag{3.191}$$

Using (3.199), we have the error model

$$\epsilon(t) = \tilde{\theta}^T(t)\phi(t) + \delta(t),\ \tilde{\theta}(t) = \theta(t) - \theta^*. \tag{3.192}$$

To handle $\delta(t)$ in (3.189), we choose the adaptive law for $\theta(t)$ as

$$\theta(t+1) = \theta(t) - \frac{\Gamma\epsilon(t)\phi(t)}{m^2(t)} + f(t),\ 0 < \Gamma = \Gamma^T < 2I_{n_\theta}, \tag{3.193}$$

where

$$m(t) = \sqrt{\kappa + \phi^T(t)\phi(t)},\ \kappa > 0, \tag{3.194}$$

and $f(t)$ is a modification term for robustness with respect to $\delta(t)$.

In this case, the time increment of $V(\tilde{\theta}) = \tilde{\theta}^T\Gamma^{-1}\tilde{\theta}$, along (3.193), is

$$\begin{aligned} &V(\tilde{\theta}(t+1)) - V(\tilde{\theta}(t)) \\ &= -\left(2 - \frac{\phi^T(t)\Gamma\phi(t)}{m^2(t)}\right)\frac{\epsilon^2(t)}{m^2(t)} + \frac{2\epsilon(t)\delta(t)}{m^2(t)} \\ &\quad + 2f^T(t)\Gamma^{-1}\left(\tilde{\theta}(t) - \frac{\Gamma\epsilon(t)\phi(t)}{m^2(t)} + f(t)\right) - f^T(t)\Gamma^{-1}f(t). \end{aligned} \tag{3.195}$$

From the condition (3.190), we see that for some constants $c_1 > 0$ and $c_2 > 0$,

$$\frac{|\delta(t)|}{m(t)} < c_1 + \frac{c_2}{m(t)}. \tag{3.196}$$

We need to choose $f(t)$ to ensure that $V(\tilde{\theta}(t+1)) - V(\tilde{\theta}(t)) \leq 0$ whenever $\|\theta(t)\|$ becomes large and that $\frac{\epsilon(t)}{m(t)}$, $\theta(t+1) - \theta(t)$ satisfy some desired properties.

A suitable design of $f(t)$ will make use of the knowledge of c_1 and c_2 in (3.196), that of $\|\theta^*\|$ or individual parameter bounds.

The discrete-time counterparts of the dead-zone, fixed σ, switching σ, and parameter projection modification designs for $f(t)$ are given next.

Dead-Zone Modification

A dead-zone modification $f(t)$, similar to that in (3.164), is

$$f(t) = \frac{\Gamma\phi(t)f_0(\epsilon(t))}{m^2(t)},$$

$$f_0(\epsilon(t)) = \begin{cases} \epsilon(t) & \text{if } \frac{|\epsilon(t)|}{m(t)} < \frac{2}{\alpha_1}\left(d_1 + \frac{d_2}{m(t)}\right), \\ 0 & \text{if } \frac{|\epsilon(t)|}{m(t)} \geq \frac{2}{\alpha_1}\left(d_1 + \frac{d_2}{m(t)}\right) \end{cases} \tag{3.197}$$

for $d_1 \geq c_1$, $d_2 \geq c_2$, where $\alpha_1 = 2 - \lambda_{\max}[\Gamma] > 0$, with $\lambda_{\max}[\Gamma] \in (0, 2)$ the maximum eigenvalue of Γ which satisfies $0 < \Gamma = \Gamma^T < 2I_{n_\theta}$ (see (3.193)).

With this choice, $V(\tilde{\theta}(t+1)) - V(\tilde{\theta}(t)) = 0$ for $\frac{|\epsilon(t)|}{m(t)} < \frac{2}{\alpha_1}\left(d_1 + \frac{d_2}{m(t)}\right)$ and $V(\tilde{\theta}(t+1)) - V(\tilde{\theta}(t)) \leq \alpha_1 \frac{\epsilon^2(t)}{m^2(t)} + \frac{2\epsilon(t)\delta(t)}{m^2(t)} \leq 0$ for $\frac{|\epsilon(t)|}{m(t)} \geq \frac{2}{\alpha_1}\left(d_1 + \frac{d_2}{m(t)}\right)$. Hence, $\theta(t) \in L^\infty$ and $\frac{\epsilon(t)}{m(t)} \in L^\infty$. Using the inequality

$$
\begin{aligned}
& -\left(2 - \frac{\phi^T(t)\Gamma\phi(t)}{m^2(t)}\right)\frac{\epsilon^2(t)}{m^2(t)} + \frac{2\epsilon(t)\delta(t)}{m^2(t)} \\
& \leq -\frac{\alpha_1\epsilon^2(t)}{2m^2(t)} - \frac{\alpha_1}{2}\left(\frac{|\epsilon(t)|}{m(t)} - \frac{2|\delta(t)|}{\alpha_1 m(t)}\right)^2 + \frac{2\delta^2(t)}{\alpha_1 m^2(t)},
\end{aligned} \tag{3.198}
$$

from (3.193), (3.195), and (3.197), we have

$$
\begin{aligned}
\sum_{t=t_1}^{t_2} \frac{\epsilon^2(t)}{m^2(t)} \leq & \; c_3\left(c_1^2(t_2+1-t_1) + \sum_{t=t_1}^{t_2} \frac{c_2^2}{m^2(t)}\right) \\
& + c_4\left(d_1(t_2+1-t_1) + \sum_{t=t_1}^{t_2} \frac{d_2}{m^2(t)}\right) + c_5, \forall t_2 \geq t_1 \geq 0,
\end{aligned} \tag{3.199}
$$

for some constants $c_i > 0$, $i = 3, 4, 5$. From (3.193) and (3.197), we can derive

$$
\begin{aligned}
\sum_{t=t_1}^{t_2} \|\theta(t+1) - \theta(t)\|_2^2 \leq & \; c_6\left(c_1^2(t_2+1-t_1) + \sum_{t=t_1}^{t_2} \frac{c_2^2}{m^2(t)}\right) \\
& + c_7\left(d_1(t_2+1-t_1) + \sum_{t=t_1}^{t_2} \frac{d_2}{m^2(t)}\right) + c_8,
\end{aligned} \tag{3.200}
$$

$\forall t_2 \geq t_1 \geq 0$, for some constants $c_i > 0$, $i = 6, 7, 8$.

How about a dead-zone modification (3.168) for the discrete-time case?

σ-Modification

A discrete-time σ-modification can be chosen as

$$
f(t) = f_\theta(\theta(t)) = -\sigma\theta(t), \; 0 < \sigma < \frac{1}{2}(1-\gamma), \tag{3.201}
$$

where $\gamma \in (0, 1)$ defines the adaptation gain matrix $\Gamma = \gamma I_{n_\theta}$ in (3.193).

Then, we can express the time increment of $V(\tilde{\theta}) = \tilde{\theta}^T\Gamma^{-1}\tilde{\theta}$ as

$$
\begin{aligned}
& V(\tilde{\theta}(t+1)) - V(\tilde{\theta}(t)) \\
& \leq -\frac{\gamma}{2}\frac{\epsilon^2(t)}{m^2(t)} + \frac{2}{2-\gamma}\frac{\delta^2(t)}{m^2(t)} - 2\sigma\theta^T(t)\gamma^{-1}\tilde{\theta}(t) \\
& \quad + \frac{\sigma^2\theta^T(t)\theta(t)}{1-\gamma} + \sigma^2\theta^T(t)\gamma^{-1}\theta(t).
\end{aligned} \tag{3.202}
$$

When $\|\theta(t)\|_2 > 2\|\theta^*\|_2$, we have

$$-2\sigma\theta^T(t)\gamma^{-1}\tilde{\theta}(t) + \frac{\sigma^2\theta^T(t)\theta(t)}{1-\gamma} + \sigma^2\theta^T(t)\gamma^{-1}\theta(t) \leq -\frac{\sigma}{2}\theta^T(t)\theta(t), \quad (3.203)$$

as $0 < \gamma < 1$, $0 < \sigma \leq \frac{1-\gamma}{2}$, and $\theta^T(t)\tilde{\theta}(t) \geq \frac{1}{2}\theta^T(t)\theta(t)$ for $\|\theta(t)\|_2 > 2\|\theta^*\|_2$.

Based on (3.202) and (3.203), we see that since $\frac{\delta^2(t)}{m^2(t)}$ is bounded, there exist constants $\theta^0 > 0$ such that $V(\tilde{\theta}(t+1)) - V(\tilde{\theta}(t)) < 0$ whenever $\|\theta(t)\|_2 \geq \theta^0$. Hence, $V(\tilde{\theta}(t))$ is bounded, and so are $\theta(t)$ and $\frac{\epsilon(t)}{m(t)}$. From (3.201) and (3.193) with (3.203), we have (3.199) and (3.200) with $d_1 = \sigma$ and $d_2 = 0$.

Switching σ-Modification

A switching σ-modification [155] is given as

$$f(t) = f_\theta(\theta(t)) = -\sigma(t)\theta(t), \ \sigma(t) = \begin{cases} \sigma_0 & \text{if } \|\theta(t)\|_2 \geq 2M_\theta, \\ 0 & \text{if } \|\theta(t)\|_2 < 2M_\theta, \end{cases} \quad (3.204)$$

where $M_\theta \geq \|\theta^*\|_2$, and for $\Gamma = \gamma I_{n_\theta}$ in (3.193),

$$0 < \sigma_0 < \frac{1}{2}(1-\gamma), \ 0 < \gamma < 1. \quad (3.205)$$

With this choice of $f(t)$, we also have (3.202) but with $\sigma = \sigma(t)$:

$$\begin{aligned} & V(\tilde{\theta}(t+1)) - V(\tilde{\theta}(t)) \\ & \leq -\frac{\gamma}{2}\frac{\epsilon^2(t)}{m^2(t)} + \frac{2}{2-\gamma}\frac{\delta^2(t)}{m^2(t)} - 2\sigma(t)\theta^T(t)\gamma^{-1}\tilde{\theta}(t) \\ & \quad + \frac{\sigma^2(t)\theta^T(t)\theta(t)}{1-\gamma} + \sigma^2(t)\theta^T(t)\gamma^{-1}\theta(t). \end{aligned} \quad (3.206)$$

In this case, with $0 < \gamma < 1$, $0 < \sigma(t) = \sigma_0 \leq \frac{1-\gamma}{2}$ and $\theta^T(t)\tilde{\theta}(t) \geq \frac{1}{2}\theta^T(t)\theta(t)$ for $\|\theta(t)\|_2 > 2\|\theta^*\|_2$, and $\sigma(t) = 0$ for $\|\theta(t)\|_2 < 2M_\theta$, we have

$$\begin{aligned} & -2\sigma(t)\theta^T(t)\gamma^{-1}\tilde{\theta}(t) + \frac{\sigma^2(t)\theta^T(t)\theta(t)}{1-\gamma} + \sigma^2(t)\theta^T(t)\gamma^{-1}\theta(t) \\ & \leq -\frac{\sigma(t)}{2}\theta^T(t)\theta(t). \end{aligned} \quad (3.207)$$

Finally, using (3.206) and (3.207), we obtain

$$\begin{aligned} & V(\tilde{\theta}(t+1)) - V(\tilde{\theta}(t)) \\ & \leq -\frac{\gamma}{2}\frac{\epsilon^2(t)}{m^2(t)} + \frac{2}{2-\gamma}\frac{\delta^2(t)}{m^2(t)} - \frac{\sigma(t)}{2}\theta^T(t)\theta(t). \end{aligned} \quad (3.208)$$

From (3.208), we see that the term $\sigma(t)\theta^T(t)\theta(t)$ grows unbounded if $\theta(t)$ grows unbounded. Since $\frac{\delta^2(t)}{m^2(t)}$ is bounded, there exist constants $\theta^0 > 0$ such that $V(\tilde{\theta}(t+1)) - V(\tilde{\theta}(t)) < 0$ whenever $\|\theta(t)\|_2 \geq \theta^0$. Hence, $V(\tilde{\theta}(t))$ is bounded, and so are $\theta(t)$ and $\frac{\epsilon(t)}{m(t)}$. From (3.208), we have (3.199) with $d_1 = d_2 = 0$, and from (3.193) and (3.204), we get

$$\|\theta(t+1) - \theta(t)\|_2^2 \leq c_9 \frac{\epsilon^2(t)}{m^2(t)} + c_{10}\sigma(t)\theta^T(t)\theta(t) \tag{3.209}$$

for some constants $c_i > 0$, $i = 9, 10$, which, together with (3.208), leads to (3.200) with $d_1 = d_2 = 0$ (i.e., $c_7 = 0$).

Clearly this switching σ-modification also has the ideal property that it ensures that $\frac{\epsilon(t)}{m(t)} \in L^2$ and $\theta(t+1) - \theta(t) \in L^2$ when the modeling error $\delta(t) = 0$ (i.e., $c_1 = c_2 = 0$ in (3.199) and (3.200)).

Parameter Projection

For a projection design of $f(t)$, we assume that the jth component of $\theta^* \in R^{n_\theta}$ belongs to a known interval $\theta_j^* \in [\theta_j^a, \theta_j^b]$, $j = 1, \ldots, n_\theta$, and choose

$$\Gamma = \operatorname{diag}\{\gamma_1, \ldots, \gamma_{n_\theta}\},\ 0 < \gamma_j < 2,\ j = 1, 2, \ldots, n_\theta, \tag{3.210}$$

in (3.193). We then define the base adaptation vector as

$$g(t) = -\frac{\Gamma\epsilon(t)\phi(t)}{m^2(t)}. \tag{3.211}$$

Denoting the jth component of $\theta(t)$, $f(t)$, $g(t)$ as $\theta_j(t)$, $f_j(t)$, $g_j(t)$, $j = 1, 2, \ldots, n_\theta$, we choose the initial estimates as

$$\theta_j(0) \in [\theta_j^a, \theta_j^b] \tag{3.212}$$

and set the projection function components as

$$f_j(t) = \begin{cases} 0 & \text{if } \theta_j(t) + g_j(t) \in [\theta_j^a, \theta_j^b], \\ \theta_j^b - \theta_j(t) - g_j(t) & \text{if } \theta_j(t) + g_j(t) > \theta_j^b, \\ \theta_j^a - \theta_j(t) - g_j(t) & \text{if } \theta_j(t) + g_j(t) < \theta_j^a. \end{cases} \tag{3.213}$$

This algorithm has the key property

$$f_j(t)(\theta_j(t) - \theta_j^* + g_j(t) + f_j(t)) \leq 0,\ j = 1, 2, \ldots, n_\theta, \tag{3.214}$$

based on which we have

$$f^T(t)\Gamma^{-1}\left(\tilde{\theta}(t) - \frac{\Gamma\epsilon(t)\phi(t)}{m^2(t)} + f(t)\right)$$
$$= \sum_{j=1}^{n_\theta} \gamma_j^{-1} f_j(t)(\theta_j(t) - \theta_j^* + g_j(t) + f_j(t)) \leq 0. \tag{3.215}$$

Using this result, (3.195), and (3.198), we can also obtain (3.199) and (3.200) with $c_4 = c_7 = 0$ (with this result, the projection algorithm also has the same ideal property as that of a switching σ-modification).

3.8.3 Summary

The above redesigns of a standard gradient algorithm (3.23) or (3.83), for robustness with respect to the modeling errors $\delta(t) \in L^\infty$ in (3.156) or (3.189), have the desired properties that $\theta(t)$, $\dot{\theta}(t)$, and $\frac{\epsilon(t)}{m(t)}$ are bounded and $\frac{\epsilon(t)}{m(t)}$ and $\dot{\theta}(t)$ (or $\theta(t+1) - \theta(t)$) satisfy the mean boundedness properties (3.165) and (3.167) (or (3.199) and (3.200)), where $c_4 = c_7 = 0$ for the switching σ and parameter projection modifications which are able to restore the ideal properties of the gradient algorithm (3.23) or (3.83) when the modeling errors are absent, while the dead-zone, fixed σ, and ϵ modifications are not.

When the modeling error $\delta(t)$ is small, the mean boundedness becomes the mean smallness. We should note that the mean smallness of $\epsilon(t)$ (assume $m(t)$ is bounded) does not imply the absolute smallness of $\epsilon(t)$. Certain bursting phenomena may occur in an adaptive system when there is lack of persistency of excitation [5], [247], that is, even for very small nonzero modeling errors $\delta(t)$, the estimation error $\epsilon(t)$, while small most of time, may have large values for a short interval of time, and such a pattern continues in the steady state. This bursting phenomenon may also occur in the continuous-time case [134].

The choice of the design parameter σ or σ_0 in the fixed σ or switching σ modification is determined by the stability of the adaptive law (3.160) or (3.193). For (3.160), using (3.159) and (3.172) (or (3.177)), we have

$$\dot{\tilde{\theta}}(t) = -\left(\frac{\Gamma\phi(t)\phi^T(t)}{m^2(t)} + \Gamma\sigma\right)\tilde{\theta}(t) - \frac{\Gamma\phi(t)\delta(t)}{m^2(t)} - \Gamma\sigma\theta^*, \tag{3.216}$$

where $\tilde{\theta}(t) = \theta(t) - \theta^*$. This expression indicates that any choice of $\sigma > 0$ can make the homogeneous part of system (3.215) exponentially stable so that $\tilde{\theta}(t)$ remains bounded as both $\frac{\Gamma\phi(t)\delta(t)}{m^2(t)}$ and $\Gamma\sigma\theta^*$ are bounded.

For (3.193), using (3.192) and (3.201) (or (3.204)), we have

$$\tilde{\theta}(t+1) = \left(I - \frac{\Gamma\phi(t)\phi^T(t)}{m^2(t)} - \Gamma\sigma\right)\tilde{\theta}(t) - \frac{\Gamma\phi(t)\delta(t)}{m^2(t)} - \Gamma\sigma\theta^*. \tag{3.217}$$

For this system to be stable, there are certain conditions on Γ and σ. The choice of Γ and σ in (3.201) or (3.205) is only a sufficient condition for stability. There are other possible choices of Γ and σ to make (3.217) stable.

The parameter projection algorithm (3.182) or (3.213) not only has the desired robustness properties but also ensures that the components of the parameter vector $\theta(t)$ stay in the same intervals where their true values stay. This feature is important when some parameter estimates need to stay in certain regions for stability and performance of an adaptive control system (see Chapter 7 for indirect adaptive control which needs this feature).

3.9 Discussion

Parameter Convergence for Nonminimum Systems

For system (3.1), an adaptive parameter estimator, based on either the gradient algorithm (3.23) or the least-squares algorithm (3.30)–(3.31), uses only the measured system input and output signals $u(t)$, $y(t)$ in the parametric model (3.7) and the estimation error signal $\epsilon(t)$ in (3.19). A system in the form (3.1) is nonminimum if the polynomials $P(s) = A(s)F(s)$ and $Z(s) = B(s)F(s)$ in (3.1) have a common factor

$$F(s) = s^q + f_{q-1}s^{q-1} + \cdots + f_1 s + f_0,\ q > 0, \tag{3.218}$$

in which case, system (3.1) becomes

$$A(s)F(s)[y](t) = B(s)F(s)[u](t), \tag{3.219}$$

$$A(s) = s^{n-q} + a_{n-q-1}s^{n-q-1} + \cdots + a_1 s + a_0, \tag{3.220}$$

$$B(s) = b_{m-q}s^{m-q} + b_{m-q-1}s^{m-q-1} + \cdots + b_1 s + b_0. \tag{3.221}$$

Since an adaptive parameter estimator uses only the measured system input and output signals $u(t)$, $y(t)$, the parameter estimate $\theta(t)$ cannot uniquely reproduce the parameters of $F(s)$, that is, to the parameter estimator, the system parameter vector θ^* is not unique. Moreover, from Proposition 3.1, the persistency of excitation of the regressor $\phi(t)$ depends on the coprimeness of $P(s)$ and $Z(s)$. If $P(s)$ and $Z(s)$ are not coprime, that is, they have a common factor $F(s)$, then $\phi(t)$ cannot be persistently exciting for an input $u(t)$ with any number of frequencies, as the vectors $H(j\omega_i)$, $i = 1, \ldots, N$, cannot be linearly independent for any ω_i and any N.

In this case, what we may conclude is if, given that all poles of $P(s)$ are in $\mathrm{Re}[s] < 0$ and that $A(s)$ and $B(s)$ do not have any common factor, the system input $u(t)$ has at least $n+m-q+1$ frequencies, then

$$\lim_{t\to\infty} \text{dist}(\theta(t), \mathcal{L}) = 0, \tag{3.222}$$

where $\text{dist}(a,b)$ denotes the Euclidean distance between two vectors a and b and $\mathcal{L}$ is a set of vectors: a linear variety defined as

$$\mathcal{L} = \{\theta^* + \theta_h \,|\theta_h \in R^{n+m+1} : \theta_h^T H(s) = 0, \forall s \in C\} \tag{3.223}$$

for θ^* in (3.5) based on $P(s)$ and $Z(s)$ and $H(s) = [\frac{b^T(s)}{\Lambda(s)}, \frac{a^T(s)}{\Lambda(s)}\frac{Z(s)}{P(s)}]^T$ with $b(s) = [1, s, \ldots, s^m]^T$ and $a(s) = [1, s, \ldots, s^{n-2}, s^{n-1}]^T$. Note that the parameter vector θ^* used in (3.8) is uniquely determined by $P(s)$ and $Z(s)$ but θ_h is not unique, and that the dimension of the subspace

$$\mathcal{S}_h = \{\theta_h \,|\theta_h \in R^{n+m+1} : \theta_h^T H(s) = 0, \forall s \in C\} \tag{3.224}$$

is q, which is the order of the common factor $F(s)$ in $P(s)$ and $Z(s)$.

Robustness and Convergence

Consider the case when $1 \le \alpha \le \bar{\alpha} < \infty$ [3] in the modified adaptive law (3.129), which can be rewritten as

$$\dot{\theta}(t) = -\frac{\Gamma\phi(t)\epsilon(t)}{m^2(t)}\frac{|\epsilon(t)|^{\bar{\alpha}-1}}{m^{\bar{\alpha}-1}(t)}. \tag{3.225}$$

Assume $1 < \bar{\alpha} < \infty$ (as the case when $\bar{\alpha} = 1$ is with the standard gradient algorithm (3.23)). Then, since $\frac{\epsilon(t)}{m(t)} \in L^\infty \cap L^{1+\bar{\alpha}}$, under some additional conditions such that $\frac{d}{dt}\frac{\epsilon(t)}{m(t)} \in L^\infty$, we would have $\lim_{t\to\infty}\frac{|\epsilon(t)|^{\bar{\alpha}-1}}{m^{\bar{\alpha}-1}(t)} = 0$. This would imply that, as compared with the standard gradient algorithm (3.23), the rate of convergence of $\theta(t)$ from the modified adaptive algorithm (3.129), if it converges, is slower. While the exact difference in convergence rate, caused by the term $\frac{|\epsilon(t)|^{\bar{\alpha}-1}}{m^{\bar{\alpha}-1}(t)}$ in (3.226), needs further study, it is clear that there is a trade-off between the robustness (with respect to modeling errors) and fastness (of convergence) for the design of an adaptive parameter estimation scheme.

The class of modeling errors (3.125) belong to $L^\infty \cap L^{1+\alpha}$, with $0 < \alpha < \infty$, for which the adaptive law (3.129) is robust in the sense of results in Lemma 3.11. For modeling errors $\delta(t)$ which belong to L^∞ only, the modified adaptive law (3.129) in general is not powerful enough to handle these errors, and different modifications such as dead-zone, parameter projection, ϵ-, and σ-modifications (see Section 3.7.3) are needed, but they can only ensure that $\theta(t) \in L^\infty$ and $\int_{t_1}^{t_2} \frac{\epsilon^2(t)}{m^2(t)}\,dt \le c_0 \int_{t_1}^{t_2} \frac{\delta^2((t)}{m^2(t)}\,dt$ for some $c_0 > 0$, $\forall t_2 \ge t_1 \ge 0$.

[3] For robustness consideration, an adaptive design (3.129) with $1 \le \bar{\alpha} < \infty$ is also applicable to the case when $\delta_1(t) \in L^\infty \cap L^{1+\alpha}$ and $\delta_2(t) \in L^\infty \cap L^{1+\alpha}$ in (3.125), with $0 < \alpha < 1$, as in this case we have $\delta_1(t) \in L^\infty \cap L^2$ and $\delta_2(t) \in L^\infty \cap L^2$ (see Section 2.3.2).

Perspective for Adaptive Control

As in the following three chapters, a linear error model to be derived for adaptive control of system (3.1) with unknown parameters has the form

$$\epsilon(t) = \rho^*(\theta(t) - \theta^*)^T \zeta(t) + (\rho(t) - \rho^*)\xi(t), \tag{3.226}$$

where $\rho^* = z_m$, $\theta(t)$ is an estimate of a constant parameter vector θ^* (which contains some unknown ideal controller parameters), $\rho(t)$ is an estimate of a constant parameter ρ^*, and $\epsilon(t)$, $\zeta(t)$, $\xi(t)$ are some known signals. This error model is similar to that in (3.20) except for the additional parameter ρ^*.

Consider the quadratic cost function

$$J(\theta, \rho) = \frac{\epsilon^2}{2m^2}, \tag{3.227}$$

where $m(t)$ is a normalizing signal to be defined. As seen in Section 3.3, a gradient algorithm updates $\theta(t)$ and $\rho(t)$ (i.e., sets $\dot{\theta}(t)$ and $\dot{\rho}(t)$) in the steepest descent directions of $J(\theta, \rho)$ in terms of θ and ρ. As specified in [151], [325], the steepest descent direction of $J(\theta, \rho)$ for θ is the direction of

$$-\frac{\partial J}{\partial \theta} = -\frac{\rho^* \phi \epsilon}{m^2}. \tag{3.228}$$

It is clear that the sign of the unknown parameter ρ^* specifies the the direction of $-\frac{\partial J}{\partial \theta}$. This motivates us to choose the adaptive update law for $\theta(t)$:

$$\dot{\theta}(t) = -\frac{\text{sign}[\rho^*]\Gamma\phi(t)\epsilon(t)}{m^2(t)}, \; \theta(t_0) = \theta_0, \; t \geq t_0, \tag{3.229}$$

where $\Gamma = \Gamma^T > 0$ is a constant matrix and θ_0 is an initial estimate of θ^*.

For ρ, the steepest descent direction of $J(\theta, \rho)$ is

$$-\frac{\partial J}{\partial \rho} = -\frac{\xi \epsilon}{m^2}. \tag{3.230}$$

Based on this result, we choose the adaptive update law for $\rho(t)$ as

$$\dot{\rho}(t) = -\frac{\gamma \xi(t)\epsilon(t)}{m^2(t)}, \; \rho(t_0) = \rho_0, \; t \geq t_0, \tag{3.231}$$

where $\gamma > 0$ is a constant scalar and ρ_0 is an initial estimate of ρ^*.

To specify the normalizing signal $m(t)$, we need to ensure the boundedness of $\frac{\epsilon(t)}{m(t)}$ when $\theta(t)$ and $\rho(t)$ are bounded. This leads to the choice

$$m(t) = \sqrt{1 + \zeta^T(t)\zeta(t) + \xi^2(t)}. \tag{3.232}$$

Analysis of this adaptive scheme is similar to that for (3.23), by considering the positive definite function

$$V(\tilde{\theta}, \tilde{\rho}) = \frac{1}{2}(|\rho^*|\tilde{\theta}^T\Gamma^{-1}\tilde{\theta} + \gamma^{-1}\tilde{\rho}^2), \tag{3.233}$$

where

$$\tilde{\theta}(t) = \theta(t) - \theta^*, \ \tilde{\rho}(t) = \rho(t) - \rho^*. \tag{3.234}$$

The time derivative of $V(\tilde{\theta}, \tilde{\rho})$, along the trajectories of (3.229) and (3.231), is

$$\dot{V} = -\frac{\epsilon^2(t)}{m^2(t)}, \ t \geq 0. \tag{3.235}$$

From this result, we can conclude that the adaptive laws (3.229) and (3.231) ensure that $\theta(t) \in L^\infty$, $\rho(t) \in L^\infty$, $\frac{\epsilon(t)}{m(t)} \in L^2 \cap L^\infty$, $\dot{\theta}(t) \in L^2 \cap L^\infty$, and $\dot{\rho}(t) \in L^2 \cap L^\infty$. These desired properties are standard in an adaptive system (for either parameter estimation or adaptive control) and crucial for achieving desired closed-loop system stability (the boundedness of the signals $y(t)$ and $u(t)$ in a controlled plant such as (3.1)) and tracking performance in the adaptive control case. Note that in the parameter estimation case the boundedness of the signals $y(t)$ and $u(t)$, that is, that of $\phi(t)$ in (3.6), has not been required in the analysis of the developed adaptive laws. However, when in real-time implementation of a parameter estimation scheme, such a boundedness condition is needed to avoid any damage to a real-life system.

Discrete-time adaptive laws can also be developed for the error model (3.226) in discrete time. A normalized gradient algorithm is

$$\theta(t+1) = \theta(t) - \frac{\text{sign}[\rho^*]\Gamma\zeta(t)\epsilon(t)}{m^2(t)}, \ 0 < \Gamma = \Gamma^T < \frac{2}{\rho^0} I_{n_\theta}, \tag{3.236}$$

$$\rho(t+1) = \rho(t) - \frac{\gamma\xi(t)\epsilon(t)}{m^2(t)}, \ 0 < \gamma < 2, \tag{3.237}$$

for $t \in \{0, 1, 2, \ldots\}$, where $\rho^0 \geq |\rho^*|$ is a known upper bound of $|\rho^*|$, n_θ is the dimension of θ: $\theta \in R^{n_\theta}$, and $m(t) = \sqrt{1 + \zeta^T(t)\zeta(t) + \xi^2(t)}$.

This adaptive parameter update law has the following properties: $\theta(t) \in L^\infty$, $\rho(t) \in L^\infty$, $\frac{\epsilon(t)}{m(t)} \in L^2 \cap L^\infty$, $\theta(t+1) - \theta(t) \in L^2$, and $\rho(t+1) - \rho(t) \in L^2$.

In this case, the time increment of $V(\tilde{\theta}, \tilde{\rho})$ of (3.233) in discrete time, along the trajectories of (3.236) and (3.237), is derived as

$$\begin{aligned} & V(\tilde{\theta}(t+1), \tilde{\rho}(t+1)) - V(\tilde{\theta}(t), \tilde{\rho}(t)) \\ & = -\left(2 - \frac{|\rho^*|\zeta^T(t)\Gamma\zeta(t) + \gamma\xi^2(t)}{m^2(t)}\right)\frac{\epsilon^2(t)}{m^2(t)} \leq -\alpha_1\frac{\epsilon^2(t)}{m^2(t)}, \end{aligned} \tag{3.238}$$

for $\alpha_1 = 2 - \max\{|\rho^*|\lambda_{\max}[\Gamma], \gamma\} > 0$, where $\gamma \in (0, 2)$, and $\lambda_{\max}[\Gamma] \in (0, \frac{2}{\rho^0})$ is the maximum eigenvalue of Γ which satisfies $0 < \Gamma = \Gamma^T < \frac{2}{\rho^0} I_{n_\theta}$. From (3.238), we can derive the above properties of (3.236) and (3.237).

When the adaptive law (3.129) is used for adaptive control, it has stronger robustness properties (see Lemma 3.11) than that of the standard gradient algorithm (3.23), as well as ensures the desired output tracking property that $\lim_{t\to\infty}(y(t) - y_m(t)) = 0$, where $y(t)$ is the output of a controlled system and $y_m(t)$ is the output of a reference system. The difference is that with (3.129) the tracking error $e(t) = y(t) - y_m(t)$ can be shown to belong to $L^\infty \cap L^{1+\bar{\alpha}}$, instead of $L^\infty \cap L^2$ ensured by the gradient adaptive law (3.23) applied to the no-modeling-error case or the $L^\infty \cap L^2$ modeling error case (see Chapter 5).

The modified adaptive law (3.129) in general cannot handle modeling errors which belong to L^∞ only. For such modeling errors, different modifications such as dead-zone, parameter projection, ϵ-modification, and σ-modification are needed to ensure closed-loop signal boundedness, which is indeed important for adaptive control where closed-loop stability is crucial. However, for L^∞ but not $L^{1+\alpha}$ modeling errors, an adaptive controller with any of those modified adaptive laws cannot ensure asymptotical tracking.

Parameter Estimation for Multivariable Systems

A multivariable linear parametric model has the form

$$y(t) = \Theta^{*T}\phi(t), \tag{3.239}$$

where $\Theta^* \in R^{n_\theta \times m_\theta}$ is an unknown parameter matrix, $y(t) \in R^{m_\theta}$ is a known (measured) vector signal, $\phi(t) \in R^{n_\theta}$ is a known vector signal, and $m_\theta > 0$, $n_\theta > 0$. This parametric model is commonly seen in system identification for systems with multiple inputs and multiple outputs.

The design and analysis of adaptive laws for updating the estimate $\Theta(t)$ of the unknown parameter matrix Θ^* are similar to those for the case when $m_\theta = 1$. We can start with defining the estimation error

$$\epsilon(t) = \Theta^T(t)\phi(t) - y(t) \tag{3.240}$$

and obtaining its expression

$$\epsilon(t) = \tilde{\Theta}^T(t)\phi(t), \ \tilde{\Theta}(t) = \Theta(t) - \Theta^*. \tag{3.241}$$

Then, introducing the positive definite function

$$V(\tilde{\Theta}) = \mathrm{tr}[\tilde{\Theta}^T\Gamma^{-1}\tilde{\Theta}], \ 0 < \Gamma = \Gamma^T \in R^{n_\theta \times n_\theta}, \tag{3.242}$$

where tr[·] is the trace operator for a matrix, and choosing the adaptive law

$$\dot{\Theta}(t) = -\frac{\Gamma\phi(t)\epsilon^T(t)}{1+\kappa\phi^T(t)\phi(t)}, \tag{3.243}$$

where $\kappa > 0$ is a design parameter, we have

$$\dot{V} = -\frac{2\epsilon^T(t)\epsilon(t)}{1+\kappa\phi^T(t)\phi(t)}, \tag{3.244}$$

which implies that $\Theta(t) \in L^\infty$, $\dot{\Theta}(t) \in L^\infty$, $\frac{\epsilon(t)}{m(t)} \in L^\infty$, $\frac{\epsilon(t)}{m(t)} \in L^2$, and $\dot{\Theta}(t) \in L^2$, where $m(t) = \sqrt{1+\kappa\phi^T(t)\phi(t)}$. The related parameter convergence issues can be addressed [72], and so can the robustness issue, and a least-squares algorithm can also be derived and analyzed, similar to that in the previous sections. Adaptive parameter estimation is crucial for adaptive control of multivariable systems with known parameters (see Chapter 9).

A discrete-time version can also be formulated and solved.

Problems

3.1 Consider the system $(s^2 + p_1 s + p_0)[y](t) = (z_1 s + z_0)[u](t)$.

(i) Design a gradient estimator for the parameters z_1, z_0, p_1, and p_0.

(ii) Design a reduced-order estimator when $p_0 = 6$ is known.

3.2 Consider the system $(s + p_0)[y](t) = z_0 u(t)$ with $z_0 \neq 0$ and $p_0 > 0$.

(i) Design a least-squares estimator for the parameters z_0, p_0.

(ii) Design a gradient estimator for the parameters z_0, p_0.

(iii) Show that for $u(t) = \sin(t)$, $\forall t \geq 0$, the estimated parameters converge to the true ones as t goes to infinity for any initial estimates.

(iv) Simulate both the gradient and least-squares estimators with $z_0 = 1$, $p_0 = 2$ for (a) $u(t) = 1$ and (b) $u(t) = \sin(t)$, and plot the parameter errors for $\theta(0) = [1, 1]^T$ (and some other choices of yours).

(v) Comment on the simulation results.

3.3 Verify (3.35), (3.37), and (3.38).

3.4 Consider the case in Example 3.4: $\theta(t)$, $\theta(t_0) = \theta_0$, $\theta^* \in R$, $\frac{\phi(t)}{m(t)} = \beta_0 > 0$, $P(t) = \frac{1}{1+\beta_0^2(t-t_0)}$, and $\tilde{\theta}(t) = \frac{1}{1+\beta_0^2(t-t_0)}\tilde{\theta}(t_0)$, which is the solution of

$$\dot{\tilde{\theta}}(t) = -\frac{\beta_0^2}{1+\beta_0^2(t-t_0)}\tilde{\theta}(t).$$

(i) Show that $|\tilde{\theta}(t)| \leq c_0 |\tilde{\theta}(t_0)|$, $\forall t \geq t_0$, for some $c_0 > 0$ independent of t_0.

(ii) Show that for some $\delta_1 > 0$, every $\epsilon_1 > 0$, and any $t_0 \geq 0$, there exists a $T(\epsilon_1, \delta_1) > 0$ such that $|\theta_0| < \delta_1$ implies that $|\theta(t; \theta_0, t_0)| < \epsilon_1$, $\forall t \geq t_0 + T$.

(iii) Review the stability definitions 2.1, 2.5–2.7.

(iv) Comment on your results, in view of the expression of $\tilde{\theta}(t)$.

3.5 Show that for $u(t) = 1$, $u(t) = \sin t$, or $u(t) = 1 + \sin t$, the signal $\phi(t)$ defined in Example 3.2 is not PE and the parameter estimate $\theta(t)$ does not converge to $\theta^* = z_0$. How about $u(t) = \sin 1.01t$? Construct a similar case.

3.6 Study the solutions of the systems (a) $\dot{x} = -2x$, (b) $\dot{x} = -\sin^2 tx$, (c) $\dot{x} = -\frac{1}{t+1}x$, and (d) $\dot{x} = -e^{-0.1t}x$, $x \in R$, and compare them with parameter convergence with the gradient algorithm (3.54).

3.7 Derive the parametric model (3.73) for system (3.72), with θ^* and $\phi(t)$ given in (3.79) and (3.80), and verify (3.81) and (3.82).

3.8 Derive (3.85) and (3.86).

3.9 Verify (3.90)–(3.92).

3.10 Verify (3.106) and (3.107).

3.11 Prove Proposition 3.3 and Proposition 3.4.

3.12 Verify (3.134) and (3.135).

3.13 From (3.144), verify the desired robustness properties of the adaptive law (3.83): $\theta(t) \in L^\infty$, $\frac{\epsilon(t)}{m(t)} \in L^2 \cap L^\infty$, and $\theta(t+1) - \theta(t) \in L^2$, in the presence of L^2 modeling errors $\delta(t)$ in the linear model (3.139).

3.14 Verify (3.150) and the results of Lemma 3.12.

3.15 Verify (3.165). (*Hint*: Use $\tilde{\theta}^T(t)\Gamma^{-1}f(t) = \frac{(\epsilon(t)-\delta(t))f_d(\epsilon(t))}{m^2(t)}$ (see (3.159), (3.164)), $\frac{|\delta(t)|}{m(t)} \leq c_1 + \frac{c_2}{m(t)}$ (see (3.163)), $\frac{|f_d(\epsilon(t))|}{m(t)} < c_1 + \frac{c_2}{m(t)}$ (see (3.164)), and $-\frac{\epsilon^2(t)}{m^2(t)} + \frac{2\epsilon(t)\delta(t)}{m^2(t)} = -\frac{\epsilon^2(t)}{m^2(t)} - \left(\frac{\epsilon(t)}{m(t)} - \frac{\delta(t)}{m(t)}\right)^2 + \frac{\delta^2(t)}{m^2(t)}$.)

3.16 Derive (3.165) and (3.167) for the choice of $f(\epsilon(t))$ in (3.168).

3.17 Simulate the robust adaptive parameter estimation scheme with the dead-zone modification (3.164) (and (3.168) for comparison) for a second-order dynamic system (3.1) with $\delta(t)$ in (3.156) being a sinusoidal signal.

3.18 Verify (3.170).

3.19 Derive (3.174)–(3.176).

3.20 Prove the properties of the discrete-time fixed σ or switching σ modification design in (3.201) or (3.204) by deriving (3.202) and (3.203) or (3.208).

3.21 Make a different choice of the design parameters Γ and σ for the discrete-time fixed σ or switching σ modification design in (3.201) or (3.204), which should stabilize the system (3.217).

3.22 Prove the desired property (3.214) of the projection algorithm (3.213).

3.23 Develop a discrete-time ϵ-modification $f(t)$ for (3.193). *Hint*: Use $m(t) = \sqrt{\kappa + \phi^T(t)\phi(t) + \epsilon^2(t)}$, study the resulting system

$$\begin{aligned}\tilde{\theta}(t+1) = &\left(I - \frac{\Gamma\phi(t)\phi^T(t)}{m^2(t)} - \Gamma\frac{\lambda|\epsilon(t)|}{m(t)}\left(c_1 + \frac{c_2}{m(t)}\right)\right)\tilde{\theta}(t)\\ &- \frac{\Gamma\phi(t)\delta(t)}{m^2(t)} - \Gamma\frac{\lambda|\epsilon(t)|}{m(t)}\theta^*\left(c_1 + \frac{c_2}{m(t)}\right),\end{aligned}$$

and specify a design parameter $\lambda > 0$ to make the system stable. As a comparison, the corresponding system in the continuous-time case is

$$\begin{aligned}\dot{\tilde{\theta}}(t) = &\left(-\frac{\Gamma\phi(t)\phi^T(t)}{m^2(t)} - \Gamma\frac{\lambda|\epsilon(t)|}{m(t)}\left(c_1 + \frac{c_2}{m(t)}\right)\right)\tilde{\theta}(t)\\ &- \frac{\Gamma\phi(t)\delta(t)}{m^2(t)} - \Gamma\frac{\lambda|\epsilon(t)|}{m(t)}\theta^*\left(c_1 + \frac{c_2}{m(t)}\right),\end{aligned}$$

whose homogeneous part is stable for any $\lambda > 0$.

Advanced Topics

Overparametrization. An overparametrized version of the system (3.1), with an overestimated system order $\bar{n} > n$, can be modeled as

$$\begin{aligned}\bar{P}(s)[y](t) &= \bar{Z}(s)[u](t),\\ \bar{P}(s) &= s^{\bar{n}} + \bar{p}_{\bar{n}-1}s^{\bar{n}-1} + \cdots + \bar{p}_1 s + \bar{p}_0,\\ \bar{Z}(s) &= \bar{z}_{\bar{m}}s^{\bar{m}} + \bar{z}_{\bar{m}-1}s^{\bar{m}-1} + \cdots + \bar{z}_1 s + \bar{z}_0,\end{aligned}$$

where $\bar{n} > n$ and $\bar{m} = m + \bar{n} - n$, $\bar{p}_{i+\bar{n}-n} = p_i$, $i = 0, 1, \ldots, n$, $\bar{z}_{i+\bar{n}-n} = z_i$, $i = 0, 1, \ldots, m$, and $\bar{p}_i = 0$, $\bar{z}_i = 0$, $i = 0, 1, \ldots, \bar{n} - n - 1$, that is,

$\bar{P}(s) = s^{\bar{n}-n}P(s)$, $\bar{Z}(s) = s^{\bar{n}-n}Z(s)$. The resulting parametric model (3.18) has $n_\theta = \bar{n} + \bar{m} + 1 = 2\bar{n} - n + m + 1$, for which (3.23) or (3.30)–(3.31) can be designed. How about its parameter convergence properties?

Finite-time parameter estimation with gradient algorithms. Recall that a least-squares algorithm (3.30) or (3.87) has a finite-time parameter identifiability property as stated in Lemma 3.5 or Lemma 3.9. Consider the gradient algorithm (3.23) or the related system (3.29) with $\Gamma = I_{n_\theta}$ and $\kappa = 1$:

$$\dot{\tilde{\theta}}(t) = A(t)\tilde{\theta}(t),\ A(t) = -\psi(t)\psi^T(t).$$

Let $\Phi(t,\tau)$ be the state transition matrix of this system. Prove or disprove that if the signal $\psi(t)$ is exciting over the interval $[\sigma_0, \sigma_0 + \delta_0]$, then the matrix $I - \Phi(\sigma_0 + \delta_0, \sigma_0)$ is nonsingular so that the parameter vector θ^* is identified as $\theta^* = (I - \Phi(\sigma_) + \delta_0, \sigma_0))^{-1}(\theta(\sigma_0 + \delta_0) - \Phi(\sigma_0 + \delta_0, \sigma_0)\theta(\sigma_0))$, and that if the matrix $I - \Phi(\sigma_0 + \delta_0, \sigma_0)$ is nonsingular, then $I - \Phi(\sigma_0 + \delta_0 + \gamma_0, \sigma_0)$ is nonsingular for any $\gamma_0 > 0$, so that θ^* can be identified any time after $\sigma_0 + \delta_0$.

Formulate the discrete-time counterpart of this problem.

Convergence of a gradient algorithm. Consider the modeling error free linear model (3.18) and estimation error (3.19) for which the modified gradient algorithm (3.129) is used, with $0 < \bar{\alpha} < 1$. Defining $g(t) = \left(\frac{|\epsilon(t)|}{m(t)}\right)^{\bar{\alpha}-1}$, we can write (3.129) with $\Gamma = I_{n_\theta}$ and $\kappa = 1$ as

$$\dot{\tilde{\theta}}(t) = g(t)A(t)\tilde{\theta}(t),\ A(t) = -\psi(t)\psi^T(t).$$

Study the stability and convergence properties of this adaptive law.

As an example, let $n_\theta = 1$ and $\bar{\alpha} = 1/3$. Then, $\dot{\tilde{\theta}}(t) = -\psi^{4/3}(t)\tilde{\theta}^{1/3}(t)$,

$$0 \leq \tilde{\theta}^{2/3}(t) = \eta(t) \triangleq \tilde{\theta}^{2/3}(0) - \frac{2}{3}\int_0^t \psi^{4/3}(\tau)\,d\tau,$$

for all $t \in [0, T)$, where $T > 0$ is such that $\eta(T) > 0$. If for some $T_0 > 0$, $\eta(T_0) = 0$, then $\tilde{\theta}(t) = 0$, $\forall t \geq T_0$, which is a finite-time convergence faster than exponential but under a condition weaker than a PE condition.

How about its discrete-time formulation?

Multivariable bilinear models. Consider the multivariable version (for multivariable adaptive control, see Chapter 9) of the bilinear model (3.226):

$$\epsilon(t) = \rho^*(\theta(t) - \theta^*)^T\zeta(t) + (\rho(t) - \rho^*)\xi(t), \tag{3.245}$$

where $\epsilon \in R^M$, $\rho^* \in R^{M\times M}$, $\theta \in R^{n_\theta \times M}$, $\zeta \in R^{n_\theta}$, and $\xi \in R^M$ for some $M > 1$. What is the minimum knowledge of the parameter matrix ρ^* suitable for a stable parameter estimation scheme for $\theta(t)$ and $\rho(t)$?

Chapter 4

Adaptive State Feedback Control

State feedback control is commonly used in control systems, due to its simple structure and powerful functions. To develop an adaptive controller, it is necessary to solve the related nonadaptive control problem when the parameters of the controlled plant are known. A crucial issue is the existence of an ideal (nominal) controller for a given control objective, which is equivalent to a set of matching conditions. For adaptive control when the plant parameters are unknown, adaptive schemes for updating controller parameters can be derived to ensure closed-loop stability and desired tracking. Adaptive state feedback control is applicable to either state tracking or output tracking, under different matching conditions, based on different control objectives.

4.1 Design for State Tracking

For state tracking, a plant model for control, a reference model system to be tracked, and a plant–model matching condition are needed for both a non-adaptive control design and an adaptive control design. For adaptive control, adaptive laws are also derived for updating the controller parameters.

4.1.1 Design Example

We first consider the first-order plant

$$\dot{y}(t) = a_p y(t) + b_p u(t),\ t \geq 0, \tag{4.1}$$

with two parameters a_p and b_p and the reference model

$$\dot{y}_m(t) = -a_m y_m(t) + b_m r(t),\ t \geq 0, \tag{4.2}$$

with $a_m > 0$. When both plant parameters a_p and b_p are known, the above control objective can be met by using the the controller

$$u(t) = k_1^* y(t) + k_2^* r(t), \tag{4.3}$$

where

$$k_1^* = \frac{-a_p - a_m}{b_p}, \; k_2^* = \frac{b_m}{b_p}, \tag{4.4}$$

which leads to the closed-loop system

$$\dot{y}(t) = -a_m y(t) + b_m r(t). \tag{4.5}$$

With the tracking error $e(t) = y(t) - y_m(t)$, from (4.2) and (4.5), we have $\dot{e}(t) = -a_m e(t)$, so that $e(t)$ is bounded, and so are $y(t)$ and $u(t)$. Moreover, $\lim_{t\to\infty} e(t) = 0$, exponentially.

When the plant parameters a_p and b_p are unknown, we use the adaptive version of the fixed controller (4.3),

$$u(t) = k_1(t)y(t) + k_2(t)r(t), \tag{4.6}$$

where $k_1(t)$ and $k_2(t)$ are estimates of the unknown parameters k_1^* and k_2^* as the plant parameters a_p and b_p are unknown in (4.4), to be updated from an adaptive law. Substituting (4.6) into (4.1) gives the closed-loop system

$$\dot{y}(t) = -a_m y(t) + b_m r(t) + b_p(k_1(t) - k_1^*)y(t) + b_p(k_2(t) - k_2^*)r(t). \tag{4.7}$$

In this case, the tracking error equation becomes

$$\dot{e}(t) = -a_m e(t) + b_p \tilde{k}_1(t)y(t) + b_p \tilde{k}_2(t)r(t), \; t \geq 0, \tag{4.8}$$

where $\tilde{k}_1(t) = k_1(t) - k_1^*$ and $\tilde{k}_2(t) = k_2(t) - k_2^*$.

Consider the following measure for the errors $e(t)$, $\tilde{k}_1(t)$, and $\tilde{k}_2$:

$$V(e, \tilde{k}_1, \tilde{k}_2) = e^2 + \frac{|b_p|}{\gamma_1}\tilde{k}_1^2 + \frac{|b_p|}{\gamma_2}\tilde{k}_2^2 \tag{4.9}$$

with $\gamma_1 > 0$ and $\gamma_2 > 0$, which is positive when any of e, $\tilde{k}_1$, and $\tilde{k}_2$ is nonzero (so it is also called a positive definite function). The term $|b_p|$ is included in this measure V in order to take into account the scaling term b_p in the error equation (4.9). In this case, the time derivative of $V(e, \tilde{k}_1, \tilde{k}_2)$ is

$$\begin{aligned}
\dot{V} &= \frac{d}{dt}V(e, \tilde{k}_1, \tilde{k}_2) \\
&= \frac{\partial V(e, \tilde{k}_1, \tilde{k}_2)}{\partial e}\dot{e}(t) + \frac{\partial V(e, \tilde{k}_1, \tilde{k}_2)}{\partial \tilde{k}_1}\dot{\tilde{k}}_1(t) + \frac{\partial V(e, \tilde{k}_1, \tilde{k}_2)}{\partial \tilde{k}_2}\dot{\tilde{k}}_2(t) \\
&= 2e(t)\dot{e}(t) + \frac{2|b_p|}{\gamma_1}\tilde{k}_1(t)\dot{\tilde{k}}_1(t) + \frac{2|b_p|}{\gamma_2}\tilde{k}_2(t)\dot{\tilde{k}}_2(t).
\end{aligned} \tag{4.10}$$

Choose the adaptive laws for $k_1(t)$ and $k_2(t)$ as

$$\dot{k}_1(t) = -\text{sign}[b_p]\gamma_1 e(t)y(t),\ t \geq 0, \tag{4.11}$$

$$\dot{k}_2(t) = -\text{sign}[b_p]\gamma_2 e(t)r(t),\ t \geq 0, \tag{4.12}$$

where $\text{sign}[b_p]$ denotes the sign of b_p, which is assumed to be known for an adaptive control design. Then, using the fact that $\dot{\tilde{k}}_1(t) = \dot{k}_1(t)$ and $\dot{\tilde{k}}_2(t) = \dot{k}_2(t)$ and (4.8), we can express (4.10) as

$$\dot{V} = -2a_m e^2(t). \tag{4.13}$$

This implies that $V(e(t), \tilde{k}_1(t), \tilde{k}_2)$ as a function of t does not increase, that is, $e(t)$, $\tilde{k}_1(t)$, and $\tilde{k}_2(t)$ are bounded, and so is $u(t)$ in (4.6). Then, $\dot{e}(t)$ in (4.8) is bounded, and from (4.13), $\int_0^\infty e^2(t)\, dt < \infty$, that is, $e(t) \in L^2$. Finally, from Lemma 2.14, we conclude that $\lim_{t\to\infty}(y(t) - y_m(t)) = 0$.

4.1.2 Control System Structure

Consider a general linear time-invariant plant in the state-space form

$$\dot{x}(t) = Ax(t) + Bu(t),\ x(t) \in R^n,\ u(t) \in R,\ t \geq 0, \tag{4.14}$$

with $x(0) = x_0$, where $A \in R^{n\times n}$, $B \in R^n$ are unknown constant parameter matrices, assuming that the state vector $x(t)$ is available for measurement.

The control objective is to design a feedback control law for the plant input $u(t)$ such that all signals in the closed-loop system are bounded and the plant state vector $x(t)$ asymptotically tracks a reference state vector $x_m(t)$ generated from a chosen reference model system

$$\dot{x}_m(t) = A_m x_m(t) + B_m r(t),\ x_m(t) \in R^n,\ r(t) \in R,\ t \geq 0, \tag{4.15}$$

with $x_m(0) = x_{m0}$, where $A_m \in R^{n\times n}$, $B_m \in R^n$ are known constant matrices, all the eigenvalues of A_m are in the open left-half complex plane, and $r(t)$ is bounded and piecewise continuous for a stable well-defined reference system.

To meet this control objective, we assume

> (A4.1): There exist a constant vector $k_1^* \in R^n$ and a nonzero constant scalar $k_2^* \in R$ such that the following equations are satisfied:
>
> $$A + Bk_1^{*T} = A_m,\ Bk_2^* = B_m. \tag{4.16}$$
>
> (A4.2): The sign of k_2^*, $\text{sign}[k_2^*]$, is known.

Remark 4.1 Assumption (A4.1) is the so-called matching condition, with which if the parameters of A and B were known, the control law

$$u(t) = k_1^{*T}x(t) + k_2^* r(t) \tag{4.17}$$

would lead to the closed-loop system

$$\dot{x}(t) = Ax(t) + B(k_1^{*T}x(t) + k_2^* r(t)) = A_m x(t) + B_m r(t),\ t \geq 0, \tag{4.18}$$

whose state vector $x(t)$ belongs to L^∞ (i.e., $x(t)$ is bounded), and the tracking error $e(t) = x(t) - x_m(t)$ satisfies

$$\dot{e}(t) = A_m e(t),\ e(0) = x_0 - x_{m0}, \tag{4.19}$$

so that $\lim_{t\to\infty} e(t) = 0$ exponentially, achieving the control objective. It is clear that Assumption (A4.1) is also necessary for achieving the stated control objective even if the plant parameters A, B are known. □

In the adaptive control problem, the parameters of A and B are unknown and so (4.17) cannot be implemented for control. In this case, an adaptive controller which has the same structure as that in (4.17) is used:

$$u(t) = k_1^T(t)x(t) + k_2(t)r(t), \tag{4.20}$$

where $k_1(t)$ and $k_2(t)$ are the estimates of k_1^*, k_2^*, respectively. The design task now is to choose suitable adaptive laws to update these estimates so that the control objective is still achievable even if A and B are unknown.

4.1.3 Adaptive Laws

To develop adaptive laws for $k_1(t)$ and $k_2(t)$, we need an error equation in terms of the tracking error $x(t) - x_m(t)$ and the parameter errors

$$\tilde{k}_1(t) = k_1(t) - k_1^*,\ \tilde{k}_2(t) = k_2(t) - k_2^*. \tag{4.21}$$

Using (4.14) and (4.20), we obtain

$$\begin{aligned}
\dot{x}(t) &= Ax(t) + B\left(k_1^T(t)x(t) + k_2(t)r(t)\right) \\
&= A_m x(t) + B_m r(t) + B\left(\tilde{k}_1^T(t)x(t) + \tilde{k}_2(t)r(t)\right) \\
&= A_m x(t) + B_m r(t) + B_m\left(\frac{1}{k_2^*}\tilde{k}_1^T(t)x(t) + \frac{1}{k_2^*}\tilde{k}_2(t)r(t)\right).
\end{aligned} \tag{4.22}$$

Substituting (4.15) into (4.22), we have the tracking error equation

$$\dot{e}(t) = A_m e(t) + B_m \left(\frac{1}{k_2^*} \tilde{k}_1^T(t) x(t) + \frac{1}{k_2^*} \tilde{k}_2(t) r(t) \right). \tag{4.23}$$

Since $k_1(t)$ and $k_2(t)$ are generated from some dynamic adaptive laws, the state vector of the closed-loop error system is

$$e_c(t) = \left[e^T(t), \tilde{k}_1^T(t), \tilde{k}_2(t) \right]^T \in R^{2n+1}. \tag{4.24}$$

Consider the positive definite function

$$V(e_c) = e^T P e + \frac{1}{|k_2^*|} \tilde{k}_1^T \Gamma^{-1} \tilde{k}_1 + \frac{1}{|k_2^*|} \tilde{k}_2^2 \gamma^{-1} \tag{4.25}$$

as a measure of the system errors e, $\tilde{k}_1$ and $\tilde{k}_2$, where $P \in R^{n \times n}$ is a constant matrix such that $P = P^T > 0$ and satisfies

$$P A_m + A_m^T P = -Q \tag{4.26}$$

for any chosen constant matrix $Q \in R^{n \times n}$ such that $Q = Q^T > 0$, $\Gamma \in R^{n \times n}$ is a constant matrix such that $\Gamma = \Gamma^T > 0$, and $\gamma > 0$ is a constant scalar.

The time derivative of $V(e_c)$ is

$$\begin{aligned} \dot{V} &= \frac{d}{dt} V(e_c) \\ &= \left(\frac{\partial V(e_c)}{\partial e} \right)^T \dot{e}(t) + \left(\frac{\partial V(e_c)}{\partial \tilde{k}_1} \right)^T \dot{\tilde{k}}_1(t) + \frac{\partial V(e_c)}{\partial \tilde{k}_2} \dot{\tilde{k}}_2(t) \\ &= 2e^T(t) P \dot{e}(t) + \frac{2}{|k_2^*|} \tilde{k}_1^T(t) \Gamma^{-1} \dot{\tilde{k}}_1(t) + \frac{2}{|k_2^*|} \tilde{k}_2(t) \gamma^{-1} \dot{\tilde{k}}_2(t). \end{aligned} \tag{4.27}$$

Substituting (4.23) and (4.26) into (4.27), we have

$$\begin{aligned} \dot{V} = &-e^T(t) Q e(t) + e^T(t) P B_m \frac{2}{k_2^*} \tilde{k}_1^T(t) x(t) + e^T(t) P B_m \frac{2}{k_2^*} \tilde{k}_2(t) r(t) \\ &+ \frac{2}{|k_2^*|} \tilde{k}_1^T(t) \Gamma^{-1} \dot{\tilde{k}}_1(t) + \frac{2}{|k_2^*|} \tilde{k}_2(t) \gamma^{-1} \dot{\tilde{k}}_2(t). \end{aligned} \tag{4.28}$$

To make $\dot{V} \leq 0$, we choose the adaptive laws for $k_1(t)$ and $k_2(t)$ as

$$\dot{k}_1(t) = -\mathrm{sign}[k_2^*] \Gamma x(t) e^T(t) P B_m, \ t \geq 0, \tag{4.29}$$

$$\dot{k}_2(t) = -\mathrm{sign}[k_2^*] \gamma r(t) e^T(t) P B_m, \ t \geq 0, \tag{4.30}$$

with $k_1(0)$ and $k_2(0)$ being arbitrary initial estimates of the unknown and constant k_1^* and k_2^*, respectively.

Indeed, with this choice of $\dot{k}_1(t)$ and $\dot{k}_1(t)$, (4.28) becomes

$$\dot{V} = -e^T(t) Q e(t) \leq -\lambda_{\min}[Q] \|e(t)\|_2^2 \leq 0, \tag{4.31}$$

where $\lambda_{\min}[Q] > 0$ is the minimum eigenvalue of Q.

4.1.4 Adaptive System Properties

From (4.31), we can derive the desired properties of this adaptive control system. Using the stability theory of Section 2.4.3, a positive definite and decrescent V in (4.25) and a negative semidefinite $\dot{V}$ in (4.31) imply that the equilibrium state $e_c = 0$ of the closed-loop system consisting of (4.23), (4.29), and (4.30) is uniformly stable and its solution $e_c(t)$ is uniformly bounded, which gives the boundedness of $x(t)$, $k_1(t)$, and $k_2(t)$ and, in turn, the boundedness of $\dot{e}(t)$ in (4.23). Then, (4.31) implies that $e(t) = x(t) - x_m(t) \in L^2$. With $e(t) \in L^2 \cap L^\infty$ and $\dot{e}(t) \in L^\infty$, applying the convergence lemma (Lemma 2.14), we conclude that $\lim_{t\to\infty} e(t) = 0$.

In summary, we have the following results.

Theorem 4.1 *The controller (4.20) with the adaptive laws (4.29) and (4.30), applied to the plant (4.14), guarantees that all closed-loop signals are bounded and that the tracking error $e(t)$ belongs to L^2 and goes to zero as t goes to ∞.*

In addition to Theorem 4.1, we also have $\dot{\tilde{k}}_i(t) \in L^2 \cap L^\infty$, $\dot{k}_i(t) \in L^2 \cap L^\infty$, $\lim_{t\to\infty} \dot{\tilde{k}}_i(t) = 0$, $\lim_{t\to\infty} \dot{k}_i(t) = 0$, as k_1^* and k_2^* are constant. In general, we do not have $\lim_{t\to\infty} \tilde{k}_1(t) = 0$ and $\lim_{t\to\infty} \tilde{k}_2(t) = 0$, which, however, are not needed for achieving the stated control objective: $\lim_{t\to\infty} e(t) = 0$.

References [151], [212], [282] also address adaptive state tracking control.

4.2 Design for Output Tracking

For state tracking control, the matching condition (4.16) is needed. To satisfy such conditions, the system matrices A and B need to be in certain canonical restrictive forms. As it will be shown in this section, for output tracking, the plant–model matching conditions are much less restrictive than those for state tracking, based on which an adaptive output tracking controller can be developed without the knowledge of the matching parameters.

4.2.1 Introductory Example

Consider a two-body system with joint flexibility and damping described by

$$\begin{aligned} J_m\ddot{\theta}_m(t) + b_m\dot{\theta}_m(t) &= u(t) - u_k(t), \\ J_l\ddot{\theta}_l(t) + b_l\dot{\theta}_l(t) &= u_k(t), \end{aligned} \tag{4.32}$$

where $\theta_m(t)$, J_m, and b_m are the motor angular position, inertia, and viscous friction coefficient, $\theta_l(t)$, J_l, and b_l are the load inertia and viscous friction coefficient; $u(t)$ is the control input (motor voltage); and $u_k(t)$ is the joint force due to flexibility and damping:

$$u_k(t) = k\delta(t) + b_k\dot{\delta}(t), \;\; \delta(t) = \theta_m(t) - \theta_l(t), \tag{4.33}$$

where $k > 0$ is the joint stiffness constant and b_k is the damping coefficient.

The state-space form of this plant is

$$\begin{aligned} \dot{x}(t) &= Ax(t) + Bu(t), \\ y(t) &= Cx(t), \; t \geq 0, \end{aligned} \tag{4.34}$$

where $x(t) = [\theta_m(t), \dot{\theta}_m(t), \theta_l(t), \dot{\theta}_l(t)]^T$ and

$$A = \begin{bmatrix} 0 & 1 & 0 & 0 \\ -\frac{k}{J_m} & -\frac{b_m+b_k}{J_m} & \frac{k}{J_m} & \frac{b_k}{J_m} \\ 0 & 0 & 0 & 1 \\ \frac{k}{J_l} & \frac{b_k}{J_l} & -\frac{k}{J_l} & -\frac{b_l+b_k}{J_l} \end{bmatrix}, \; B = \begin{bmatrix} 0 \\ \frac{1}{J_m} \\ 0 \\ 0 \end{bmatrix},$$
$$C = \begin{bmatrix} 0 & 0 & 1 & 0 \end{bmatrix}. \tag{4.35}$$

In view of the expression

$$A + Bk_1^{*T} = \begin{bmatrix} 0 & 1 & 0 & 0 \\ \frac{k_{11}^*}{J_m} - \frac{k}{J_m} & \frac{k_{12}^*}{J_m} - \frac{b_m+b_k}{J_m} & \frac{k_{13}^*}{J_m} + \frac{k}{J_m} & \frac{k_{14}^*}{J_m} + \frac{b_k}{J_m} \\ 0 & 0 & 0 & 1 \\ \frac{k}{J_l} & \frac{b_k}{J_l} & -\frac{k}{J_l} & -\frac{b_l+b_k}{J_l} \end{bmatrix}, \tag{4.36}$$

we see that for any choice of the feedback gain $k_1^* = [k_{11}^*, k_{12}^*, k_{13}^*, k_{14}^*]^T$ the closed-loop system matrix $A + Bk_1^{*T}$ in (4.36) cannot be made as a (stable) matrix A_m independent of the plant parameters. This means that a state variable reference model of the form (4.15) is not available for designing an adaptive controller for the plant (4.34), based on the adaptive state feedback controller developed in Section 4.1 for state tracking.

However, if the control objective is to achieve closed-loop signal boundedness and tracking of a reference output by the plant output $y(t)$ (in this example, $y(t) = \theta_l(t)$), we can use an alternative design: adaptive state feedback for output tracking, to be developed in this section.

4.2.2 Control System Structure

Consider a linear time-invariant plant in the state-space form

$$\begin{aligned} \dot{x}(t) &= Ax(t) + Bu(t), \; x \in R^n, \; u \in R, \\ y(t) &= Cx(t), \; y \in R, \; t \geq 0, \end{aligned} \tag{4.37}$$

for some unknown constant matrices $A \in R^{n\times n}$, $B \in R^{n\times 1}$, and $C \in R^{1\times n}$, with $n > 0$. The input–output description of this plant is

$$y(s) = G(s)u(s), \tag{4.38}$$

where

$$G(s) = C(sI - A)^{-1}B = k_p \frac{Z(s)}{P(s)} \tag{4.39}$$

with $k_p \neq 0$ a constant and $P(s) = \det(sI - A)$ and $Z(s)$ being monic polynomials with constant coefficients and of degrees n and m, respectively.

To design an adaptive state feedback control law for $u(t)$ which ensures closed-loop signal boundedness and asymptotic tracking of a chosen reference signal $y_m(t)$ by the plant output $y(t)$, we need the following assumptions:

(A4.3): (A, B, C) is stabilizable and detectable.

(A4.4): $Z(s)$ is a stable polynomial.

(A4.5): The degree m of $Z(s)$ is known.

(A4.6): The sign of k_p, $\text{sign}[k_p]$, is known.

These assumptions have their meanings for the stated control problem: Assumption (A4.3) is needed for output matching and internal signal boundedness, Assumption (A4.4) is for model reference control, and Assumption (A4.5) is needed for choosing a reference model for adaptive control, while Assumption (A4.6) is used for designing an adaptive parameter update law.

Based on Assumption (A4.5), the reference model system, which is independent of the dynamics of the plant (4.37) or (4.38), is chosen as

$$y_m(t) = W_m(s)[r](t),\ t \geq 0,\ W_m(s) = \frac{1}{P_m(s)}, \tag{4.40}$$

where $P_m(s)$ is a stable monic polynomial of degree $n^* = n - m$ and $r(t)$ is an external reference input signal for defining a desired $y_m(t)$.

The model reference adaptive controller structure is

$$u(t) = k_1^T(t)x(t) + k_2(t)r(t), \tag{4.41}$$

where

$$k_1(t) = [k_{11}(t), k_{12}(t), \ldots, k_{1n}(t)]^T \in R^n,\ k_2(t) \in R \tag{4.42}$$

are the adaptive estimates of the unknown parameters

$$k_1^* = [k_{11}^*, k_{12}^*, \ldots, k_{1n}^*]^T \in R^n,\ k_2^* \in R, \tag{4.43}$$

which are defined to satisfy

$$\det(sI - A - Bk_1^{*T}) = P_m(s)Z(s)\frac{1}{k_p},\ k_2^* = \frac{1}{k_p}. \tag{4.44}$$

This implies that all zeros of $\det(sI - A - Bk_1^{*T})$ are stable. The existence of such k_1^* and k_2^* is guaranteed by Assumptions (A4.3) and (A4.4), that is, all the uncontrollable modes (which cannot be changed by k_1^*) of A are stable and equal to some zeros of $Z(s)$, and all the controllable modes of A are moved to the zeros of $P_m(s)$ and the other zeros of $Z(s)$ (any controllable but unobservable modes of A, which are stable due to detectability of (A, C), are left unchanged because they are equal to some of the other zeros of $Z(s)$).

With this definition of k_1^* and k_2^*, the ideal version of (4.41) is

$$u(t) = k_1^{*T}x(t) + k_2^*r(t), \tag{4.45}$$

which would lead to the desired closed-loop system: $y(s) = W_m(s)r(s)$. Since in adaptive control A, B and C are all unknown parameters, the ideal controller parameters k_1^* and k_2^* are also unknown. We need to develop an adaptive law to update the parameter estimates $k_1(t)$ and $k_2(t)$.

With the adaptive controller (4.41), the plant (4.37) becomes

$$\begin{aligned}\dot{x}(t) &= Ax(t) + B(k_1^T(t)x(t) + k_2(t)r(t))\\ &= (A + Bk_1^{*T})x(t) + Bk_2^*r(t) + B\left((k_1^T(t) - k_1^{*T})x(t) + (k_2(t) - k_2^*)r(t)\right),\\ y(t) &= Cx(t),\ t \geq 0.\end{aligned} \tag{4.46}$$

From (4.44) it follows that

$$C(sI - A - Bk_1^{*T})^{-1}Bk_2^* = \frac{Z(s)k_2^*}{\det(sI - A - Bk_1^{*T})} = \frac{1}{P_m(s)} = W_m(s). \tag{4.47}$$

In view of (4.40), (4.46), and (4.47), the tracking error equation is

$$\begin{aligned}e(t) &= y(t) - y_m(t)\\ &= \rho^* W_m(s)[(k_1^T - k_1^{*T})x + (k_2 - k_2^*)r](t) + Ce^{(A+Bk_1^{*T})t}x(0),\ t \geq 0,\end{aligned} \tag{4.48}$$

where $\rho^* = k_p$ and $Ce^{(A+Bk_1^{*T})t}x(0)$ converges to zero exponentially fast due to the stability of $A + Bk_1^{*T}$. The reference output $y_m(t) = W_m(s)[r](t)$ is generated from an $(n - m)$th-order state variable equation which depends on the parameters of $W_m(s)$ only and is defined to have zero initial conditions.

4.2.3 Adaptive Laws

To derive an estimation error equation, we define

$$\theta(t) = [k_1^T(t), k_2(t)]^T, \ \theta^* = [k_1^{*T}, k_2^*]^T \tag{4.49}$$

and introduce the auxiliary signals

$$\omega(t) = [x^T(t), r(t)]^T, \tag{4.50}$$

$$\zeta(t) = W_m(s)[\omega](t), \tag{4.51}$$

$$\xi(t) = \theta^T(t)\zeta(t) - W_m(s)[\theta^T\omega](t) \tag{4.52}$$

and the estimation error

$$\epsilon(t) = e(t) + \rho(t)\xi(t), \tag{4.53}$$

where $\rho(t)$ is an estimate of $\rho^* = k_p$. Then, substituting (4.48)–(4.52) into (4.53) and neglecting the exponentially decaying $Ce^{(A+Bk_1^{*T})t}x(0)$,[1] we have

$$\epsilon(t) = \rho^*(\theta(t) - \theta^*)^T\zeta(t) + (\rho(t) - \rho^*)\xi(t), \tag{4.54}$$

which is linear in the parameter errors $\theta(t) - \theta^*$ and $\rho(t) - \rho^*$.

Based on a gradient algorithm (see Section 3.3 and (3.226)–(3.228)), this error equation suggests the following adaptive laws for $\theta(t)$ and $\rho(t)$:

$$\dot{\theta}(t) = -\frac{\text{sign}[\rho^*]\Gamma\zeta(t)\epsilon(t)}{m^2(t)}, \ \Gamma = \Gamma^T > 0, \ t \geq 0, \tag{4.55}$$

$$\dot{\rho}(t) = -\frac{\gamma\xi(t)\epsilon(t)}{m^2(t)}, \ \gamma > 0, \ t \geq 0, \tag{4.56}$$

where sign$[\rho^*]$ is the sign of the parameter $\rho^* = k_p$, $\Gamma \in R^{(n+1)\times(n+1)}$ and γ are constant adaptation gains, and

$$m(t) = \sqrt{1 + \zeta^T(t)\zeta(t) + \xi^2(t)}. \tag{4.57}$$

This adaptive parameter update law has the following properties.

Lemma 4.1 *The adaptive laws (4.55) and (4.56) ensure that* $\theta(t) \in L^\infty$, $\rho(t) \in L^\infty$, $\frac{\epsilon(t)}{m(t)} \in L^2 \cap L^\infty$, $\dot{\theta}(t) \in L^2 \cap L^\infty$, *and* $\dot{\rho}(t) \in L^2 \cap L^\infty$.

[1] Since $\lim_{t\to\infty} Ce^{(A+Bk_1^{*T})t}x(0) = 0$ exponentially, it follows that $Ce^{(A+Bk_1^{*T})t}x(0) \in L^2$. A gradient adaptive law is robust with respect to such signals (see Section 3.7.1) in the sense that the desired L^2 and L^∞ properties obtained in the absence of such signals still hold.

Proof: Consider the positive definite function

$$V(\tilde{\theta}, \tilde{\rho}) = |\rho^*|\tilde{\theta}^T \Gamma^{-1} \tilde{\theta} + \gamma^{-1} \tilde{\rho}^2, \tag{4.58}$$

where

$$\tilde{\theta}(t) = \theta(t) - \theta^*, \ \tilde{\rho}(t) = \rho(t) - \rho^*. \tag{4.59}$$

The time derivative of $V(\tilde{\theta}, \tilde{\rho})$ along the trajectories of (4.55) and (4.56) is

$$\dot{V} = -\frac{2\epsilon^2(t)}{m^2(t)}, \ t \geq 0. \tag{4.60}$$

This, together with (4.54)–(4.57), proves the lemma. ∇

Using these desired properties of the adaptive laws (4.55) and (4.56), the following results can be established.

Theorem 4.2 *All signals in the closed-loop system with the plant (4.38), the reference model (4.40), and the controller (4.41) updated by the adaptive law (4.55)–(4.56) are bounded, and the tracking error* $e(t) = y(t) - y_m(t)$ *satisfies*

$$\lim_{t \to \infty} (y(t) - y_m(t)) = 0, \tag{4.61}$$

$$\int_0^\infty (y(t) - y_m(t))^2 \, dt < \infty. \tag{4.62}$$

The proof of this theorem is given in Section 5.7, with the needed background knowledge given in Chapter 5, especially the proof of Theorem 5.4.

4.3 Disturbance Rejection

The plant model (4.14) or (4.37) is disturbance free. In real life, there are disturbances acting on control systems. An adaptive control approach can be employed to reject the effect of unknown disturbances so that desired system performance (state tracking or output tracking, in addition to closed-loop signal boundedness) can be achieved, under certain matching conditions, using additional compensation introduced in the controller (4.20) or (4.41).

4.3.1 State Tracking

Consider a linear time-invariant plant,

$$\dot{x}(t) = Ax(t) + Bu(t) + Bd(t), \ t \geq 0, \tag{4.63}$$

where $A \in R^{n\times n}$, $B \in R^n$ are unknown constant matrices and

$$d(t) = d_0 + \sum_{j=1}^{q} d_j f_j(t) \tag{4.64}$$

for some *unknown* constants d_0, d_j and *known* bounded continuous functions $f_j(t)$, $j = 1, 2, \ldots, q$, and some $q > 0$. Assuming that the state vector $x(t)$ is available for measurement, the control objective is to design a feedback control $u(t)$ to ensure closed-loop signal boundedness and asymptotic state $x(t)$ tracking of the reference state vector $x_m(t)$ generated from (4.15), under Assumptions (A4.1) and (A4.2).

In this case, the ideal control law (4.17) is modified as

$$u(t) = k_1^{*T} x(t) + k_2^* r(t) + k_3^*(t), \tag{4.65}$$

where $k_3^*(t) = -d(t)$, which leads to the desired closed-loop system

$$\begin{aligned}\dot{x}(t) &= Ax(t) + B\left(k_1^{*T} x(t) + k_2^* r(t) + k_3^*(t) + d(t)\right)\\ &= A_m x(t) + B_m r(t).\end{aligned} \tag{4.66}$$

When the parameters of A and B and the value of $d(t)$ are unknown, the adaptive version of the controller (4.65) is used:

$$u(t) = k_1^T(t) x(t) + k_2(t) r(t) + k_3(t), \tag{4.67}$$

where $k_i(t)$, $i = 1, 2$, are the estimates of k_i^*, respectively, and

$$k_3(t) = k_{30}(t) + \sum_{j=1}^{q} k_{3j}(t) f_j(t) \tag{4.68}$$

with $k_{3j}(t)$ being the estimate of $k_{3j}^* = d_j$, $j = 0, 1, 2, \ldots, q$.

To develop adaptive laws for $k_i(t)$, $i = 1, 2$, and $k_{3j}(t)$, $j = 0, 1, 2, \ldots, q$, we first derive an error equation in terms of the tracking error $e(t) = x(t) - x_m(t)$ and the parameter errors

$$\tilde{k}_i(t) = k_i(t) - k_i^*,\ i = 1, 2,\ \tilde{k}_{3j}(t) = k_{3j}(t) - k_{3j}^*,\ j = 0, 1, 2, \ldots, q. \tag{4.69}$$

Let $\tilde{k}_3(t) = k_3(t) - k_3^*(t)$. Using (4.67)–(4.69), we obtain

$$\begin{aligned}\dot{x}(t) &= Ax(t) + B\left(k_1^T(t) x(t) + k_2(t) r(t) + k_3\right)\\ &= A_m x(t) + B_m r(t) + B_m \frac{1}{k_2^*}\left(\tilde{k}_1^T(t) x(t) + \tilde{k}_2(t) r(t) + \tilde{k}_3(t)\right).\end{aligned} \tag{4.70}$$

Substituting (4.15) into (4.70), we have the tracking error equation

$$\dot{e}(t) = A_m e(t) + B_m \left(\frac{1}{k_2^*} \tilde{k}_1^T(t) x(t) + \frac{1}{k_2^*} \tilde{k}_2(t) r(t) + \frac{1}{k_2^*} \tilde{k}_3(t) \right). \tag{4.71}$$

Based on this error equation, we choose the adaptive laws as

$$\dot{k}_1(t) = -\text{sign}[k_2^*]\Gamma_1 x(t) e^T(t) P B_m, \tag{4.72}$$

$$\dot{k}_2(t) = -\text{sign}[k_2^*]\gamma_2 r(t) e^T(t) P B_m, \tag{4.73}$$

$$\dot{k}_{30}(t) = -\text{sign}[k_2^*]\gamma_{30} e^T(t) P B_m, \tag{4.74}$$

$$\dot{k}_{3j}(t) = -\text{sign}[k_2^*]\gamma_{3j} f_j(t) e^T(t) P B_m, \; j = 1, 2, \ldots, q, \tag{4.75}$$

where $\Gamma_1 = \Gamma_1^T > 0$, $\gamma_2 > 0$ and $\gamma_{3j} > 0$, $j = 0, 1, 2, \ldots, q$, are adaptation gains, and $P = P^T > 0$ satisfies (4.26). The positive definite function

$$V = e^T P e + \frac{1}{|k_2^*|} \tilde{k}_1^T \Gamma_1^{-1} \tilde{k}_1 + \frac{1}{|k_2^*|} \tilde{k}_2^2 \gamma_2^{-1} + \frac{1}{|k_2^*|} \sum_{j=0}^{q} \tilde{k}_{3j}^2 \gamma_{3j}^{-1} \tag{4.76}$$

has a negative semidefinite time derivative

$$\dot{V} = -e^T(t) Q e(t), \; Q = Q^T > 0, \tag{4.77}$$

from which we can also have the results of Theorem 4.1.

Remark 4.2 The disturbance $d(t)$ in (4.63) is matched to the control input $u(t)$, that is, both act on the plant dynamics through the same vector B. If they are not matched, for example, for the plant

$$\dot{x}(t) = A x(t) + B u(t) + B_d (d_0 + d_1 f(t)), \tag{4.78}$$

where B and $B_d \in R^n$ are linear independent, the above adaptive design may not be able to ensure asymptotic state tracking. □

4.3.2 Output Tracking

Consider a linear time-invariant plant with disturbances, that is,

$$\dot{x}(t) = A x(t) + B u(t) + B_d d_u(t),$$
$$y(t) = C x(t) + d_y(t), \; t \geq 0, \tag{4.79}$$

where $B_d \in R^n$ is an unknown constant vector, $d_u(t) \in R$ represents the actuation disturbance, $d_y(t) \in R$ represents the output disturbance, and $A \in R^{n \times n}$,

$B \in R^{n\times 1}$, and $C \in R^{1\times n}$, with $n > 0$, are some unknown constant matrices which satisfy Assumptions (A4.3)–(A4.6) in Section 4.2.2. The control objective is to design an adaptive state feedback model reference control law for $u(t)$ to ensure closed-loop signal boundedness and asymptotic tracking of a chosen reference signal $y_m(t)$ by the plant output $y(t)$, despite the uncertainties in A, B, B_d, C, and $d_u(t)$, $d_y(t)$.

We consider the reference model (4.40) and use the controller structure

$$u(t) = k_1^T(t)x(t) + k_2(t)r(t) + k_3(t), \tag{4.80}$$

where $k_3(t) \in R$ is an adaptive compensation term for cancelling the effect of the actuation disturbance $B_d d_u$ and output disturbance d_y on the tracking error $e(t) = y(t) - y_m(t)$ and the gains $k_1(t)$ and $k_2(t)$ are adaptive gains whose ideal values k_1^* and k_2^* are defined in Section 4.2.2. Next we will first develop ideal parametrizations and the associated error equations for the compensation term $k_3(t)$ for three cases of disturbances—constant, sinusoidal, and nonsinusoidal—and then develop the adaptive scheme for updating $k_3(t)$.

Rejection of Constant Disturbances

For $d_u(t) = d_{u0}$ and $d_y(t) = d_{y0}$ being some unknown constants, $k_3(t)$ is the estimate of a constant parameter k_3^*. To derive this ideal value k_3^* of $k_3(t)$, we apply the ideal form of (4.80) for this case, $u(t) = k_1^{*T}x(t) + k_2^*r(t) + k_3^*$, to the plant (4.79) to get the closed-loop system in the frequency s-domain:

$$\begin{aligned} y(s) &= C(sI - A - Bk_1^{*T})^{-1}Bk_2^*r(s) + C(sI - A - Bk_1^{*T})^{-1}B\frac{k_3^*}{s} \\ &\quad + C(sI - A - Bk_1^{*T})^{-1}B_d\frac{d_{u0}}{s} + \frac{d_{y0}}{s} \\ &= y_m(s) + \Delta(s), \end{aligned} \tag{4.81}$$

where $y_m(s) = C(sI - A - Bk_1^{*T})^{-1}Bk_2^*r(s)$ from (4.40) and (4.47) and

$$\Delta(s) = C(sI - A - Bk_1^{*T})^{-1}B\frac{k_3^*}{s} + C(sI - A - Bk_1^{*T})^{-1}B_d\frac{d_{u0}}{s} + \frac{d_{y0}}{s}. \tag{4.82}$$

From Assumption (A4.4), all zeros and poles of $C(sI - A - Bk_1^{*T})^{-1}B$ are in the open left-half of the complex plane, so that

$$\lim_{s\to 0} s\,\Delta(s) = c_1 k_3^* + c_2 + d_{y0} \tag{4.83}$$

for some constants $c_1 \neq 0$ and c_2 which depends on $B_d d_{u0}$. We then define

$$k_3^* = -\frac{c_2 + d_{y0}}{c_1}. \tag{4.84}$$

With this k_3^*, from (4.82)–(4.84), in the time domain, we have

$$\lim_{t\to\infty}(y(t) - y_m(t)) = \lim_{t\to\infty}\delta(t) = 0 \tag{4.85}$$

exponentially fast, where $\delta(t) = \mathcal{L}^{-1}[\Delta(s)]$.

To develop adaptive laws for updating the controller parameters k_1, k_2, and k_3 in (4.80), we express the closed-loop system with (4.80) as

$$\begin{aligned}
\dot{x}(t) &= Ax(t) + B(k_1^T(t)x(t) + k_2(t)r(t) + k_3(t)) + B_d d_{u0} \\
&= (A + Bk_1^{*T})x(t) + Bk_2^* r(t) + Bk_3^* + B_d d_{u0} \\
&\quad + B\left((k_1^T(t) - k_1^{*T})x(t) + (k_2(t) - k_2^*)r(t) + (k_3(t) - k_3^*)\right), \\
y(t) &= Cx(t) + d_{y0}.
\end{aligned} \tag{4.86}$$

Using (4.47), (4.82), (4.86), and $\delta(t) = \mathcal{L}^{-1}[\Delta(s)]$, similar to (4.48), we obtain the tracking error equation

$$\begin{aligned}
e(t) &= y(t) - y_m(t) \\
&= \rho^* W_m(s)[(k_1^T - k_1^{*T})x + (k_2 - k_2^*)r + (k_3 - k_3^*)](t) \\
&\quad + Ce^{(A+Bk_1^{*T})t}x(0) + \delta(t) \\
&= \rho^* W_m(s)[(\theta - \theta^*)^T\omega](t) + Ce^{(A+Bk_1^{*T})t}x(0) + \delta(t),
\end{aligned} \tag{4.87}$$

in which $Ce^{(A+Bk_1^{*T})t}x(0)$ and $\delta(t)$ converge to zero exponentially and

$$\theta(t) = [k_1^T(t), k_2(t), k_3(t)]^T,\ \theta^* = [k_1^{*T}, k_2^*, k_3^*]^T \tag{4.88}$$

$$\omega(t) = [x^T(t), r(t), 1]^T. \tag{4.89}$$

Rejection of Sinusoidal Disturbances

Sinusoidal disturbances $d_u(t)$ and $d_y(t)$ can be expressed as

$$d_u(t) = d_{u0} + \sum_{j=1}^{q_u}(\alpha_{uj}\sin\omega_{uj}t + \beta_{uj}\cos\omega_{uj}t), \tag{4.90}$$

$$d_y(t) = d_{y0} + \sum_{j=1}^{q_y}(\alpha_{yj}\sin\omega_{yj}t + \beta_{yj}\cos\omega_{yj}t), \tag{4.91}$$

where d_{u0}, α_{uj}, β_{uj}, $j = 1, 2, \ldots, q_u$, d_{y0}, α_{yj}, β_{yj}, $j = 1, 2, \ldots, q_y$, are some *unknown* constants and ω_{uj}, $j = 1, 2, \ldots, q_u$, ω_{yj}, $j = 1, 2, \ldots, q_y$, are some *known* constants (i.e., the disturbance frequencies are assumed to be known).

In this case, we use the controller structure (4.80) with

$$\begin{aligned} k_3(t) = k_{30}(t) &+ \sum_{j=1}^{q_u} (k_{3\alpha j}(t) \sin \omega_{uj} t + k_{3\beta j}(t) \cos \omega_{uj} t) \\ &+ \sum_{j=1}^{q_y} (k_{3\alpha(q_u+j)}(t) \sin \omega_{yj} t + k_{3\beta(q_u+j)}(t) \cos \omega_{yj} t), \end{aligned} \tag{4.92}$$

where $k_{30}(t)$, $k_{3\alpha j}(t)$, and $k_{3\beta j}(t)$, $j = 1, 2, \ldots, q_u + q_y$, are estimates of some ideal constant parameters k_{30}^*, $k_{3\alpha j}^*$, and $k_{3\beta j}^*$. To derive these ideal values, we express the ideal form of $k_3(t)$ as

$$\begin{aligned} k_3^*(t) = k_{30}^* &+ \sum_{j=1}^{q_u} (k_{3\alpha j}^* \sin \omega_{uj} t + k_{3\beta j}^* \cos \omega_{uj} t) \\ &+ \sum_{j=1}^{q_y} (k_{3\alpha(q_u+j)}^* \sin \omega_{yj} t + k_{3\beta(q_u+j)}^* \cos \omega_{yj} t) \end{aligned} \tag{4.93}$$

and apply the ideal form of (4.80) for this case, $u(t) = k_1^{*T} x(t) + k_2^* r(t) + k_3^*(t)$, to the plant (4.79) with disturbances described in (4.90) and (4.91), to get the closed-loop system in the frequency s-domain:

$$\begin{aligned} y(s) &= C(sI - A - Bk_1^{*T})^{-1} B k_2^* r(s) + C(sI - A - Bk_1^{*T})^{-1} B k_3^*(s) \\ &\quad + C(sI - A - Bk_1^{*T})^{-1} B_d d_u(s) + d_y(s) \\ &= y_m(s) + \Delta(s), \end{aligned} \tag{4.94}$$

where $y_m(s) = C(sI - A - Bk_1^{*T})^{-1} B k_2^* r(s)$ and

$$\begin{aligned} \Delta(s) &= C(sI - A - Bk_1^{*T})^{-1} B k_3^*(s) \\ &\quad + C(sI - A - Bk_1^{*T})^{-1} B_d d_u(s) + d_y(s). \end{aligned} \tag{4.95}$$

From Assumption (A4.4) and (4.44), all zeros and poles of $C(sI - A - Bk_1^{*T})^{-1} B$ are in the open left-half of the complex plane. Given the sinusoidal forms of $d_u(t)$, $d_y(t)$, and $k_3^*(t)$ in (4.90), (4.91), and (4.93), it can be verified by a partial fraction expansion method that there exist constant parameters k_{30}^*, $k_{3\alpha j}^*$, and $k_{3\beta j}^*$ such that $s\Delta(s)$ does not have any pole in $\mathrm{Re}[s] \geq 0$ and

$$\lim_{s \to 0} s\,\Delta(s) = 0, \tag{4.96}$$

so that from the final-value theorem of the Laplace transform the time-domain function $\delta(t)$ of $\Delta(s)$ converges to zero exponentially fast as time t goes to infinity, that is, $\lim_{t\to\infty}(y(t) - y_m(t)) = \lim_{t\to\infty} \delta(t) = 0$ exponentially fast.

With the adaptive controller (4.80) and (4.92), the closed-loop system is

$$\begin{aligned}\dot{x}(t) &= Ax(t) + B(k_1^T(t)x(t) + k_2(t)r(t) + k_3(t)) + B_d d_u(t)\\ &= (A + Bk_1^{*T})x(t) + Bk_2^* r(t) + Bk_3^*(t) + B_d d_u(t)\\ &\quad + B\left((k_1^T(t) - k_1^{*T})x(t) + (k_2(t) - k_2^*)r(t) + (k_3(t) - k_3^*(t))\right),\\ y(t) &= Cx(t) + d_y(t). \end{aligned} \tag{4.97}$$

Using (4.47), (4.95), and (4.97), we obtain the tracking error equation

$$\begin{aligned} e(t) &= y(t) - y_m(t)\\ &= \rho^* W_m(s)[(k_1^T - k_1^{*T})x + (k_2 - k_2^*)r + (k_3 - k_3^*)](t)\\ &\quad + Ce^{(A+Bk_1^{*T})t}x(0) + \delta(t)\\ &= \rho^* W_m(s)[(\theta - \theta^*)^T\omega](t) + Ce^{(A+Bk_1^{*T})t}x(0) + \delta(t), \end{aligned} \tag{4.98}$$

where $Ce^{(A+Bk_1^{*T})t}x(0)$ and $\delta(t)$ converge to zero exponentially and

$$\begin{aligned} \theta(t) &= [k_1^T(t), k_2(t), k_{30}(t), k_{3\alpha 1}(t), \ldots, k_{3\alpha q_u}(t), k_{3\beta 1}(t), \ldots, k_{3\beta q_u}(t),\\ &\qquad k_{3\alpha(q_u+1)}(t), \ldots, k_{3\alpha(q_u+q_y)}(t), k_{3\beta(q_u+1)}(t), \ldots, k_{3\beta(q_u+q_y)}(t)]^T,\\ \theta^* &= [k_1^{*T}, k_2^*, k_{30}^*, k_{3\alpha 1}^*, \ldots, k_{3\alpha q_u}^*, k_{3\beta 1}^*, \ldots, k_{3\beta q_u}^*,\\ &\qquad k_{3\alpha(q_u+1)}^*, \ldots, k_{3\alpha(q_u+q_y)}^*, k_{3\beta(q_u+1)}^*, \ldots, k_{3\beta(q_u+q_y)}^*]^T, \end{aligned} \tag{4.99}$$

$$\begin{aligned} \omega(t) &= [x^T(t), r(t), 1, \sin\omega_{u1}, \ldots, \sin\omega_{uq_u}, \cos\omega_{u1}, \ldots, \cos\omega_{uq_u},\\ &\qquad \sin\omega_{y1}, \ldots, \sin\omega_{yq_y}, \cos\omega_{y1}, \ldots, \cos\omega_{yq_y}]^T. \end{aligned} \tag{4.100}$$

Rejection of Nonsinusoidal Disturbances

We now consider the disturbances $d_u(t)$ and $d_y(t)$ in (4.79) of the form (4.64) for some nonsinusoidal functions $f_j(t)$, $j = 1, 2, \ldots, q$. For an ideal form of (4.80), $u(t) = k_1^{*T}x(t) + k_2^* r(t) + k_3^*(t)$, the closed-loop system is also given by (4.94) and (4.95) with $C(sI - A - Bk_1^{*T})^{-1}B = \frac{1}{k_2^* P_m(s)}$ (see (4.40)).

To cancel a nonsinusoidal $d_y(t)$ by $k_3^*(t)$, one may have to choose

$$k_3^*(s) = -k_2^* P_m(s) d_y(s), \tag{4.101}$$

whose time-domain function may be parametrizable if the derivatives (up to n^*th order) of $d_y(t)$ are parametrizable (for which sinusoidal components (4.91) are a special case when such derivatives are available). An adaptive scheme is to estimate the unknown parameters in (4.101).

However, to cancel a nonsinusoidal $d_u(t)$ by $k_3^*(t)$, a similar choice of

$$k_3^*(s) = -k_2^* P_m(s) C(sI - A - Bk_1^{*T})^{-1} B_d d_u(s) \tag{4.102}$$

is not parametrizable when (C, A, B) has finite zeros (i.e., when $n^* < n$; in this case, the zeros of (C, A, B) are in the denominator of (4.102)).

What is of interest here is to derive a parametrizable compensation scheme for an adaptive control design to reject the effect of the disturbance $d_u(t)$ in the plant (4.79) with $d_y(t) = 0$ on the plant output $y(t)$.

Let us consider the linear time-invariant plant

$$\dot{x}(t) = Ax(t) + Bu(t) + B_d d_u(t),$$
$$y(t) = Cx(t), \tag{4.103}$$

where the disturbance $d_u(t)$ has the form

$$d_u(t) = d_0 + \sum_{j=1}^{q} d_j f_j(t) \tag{4.104}$$

for some *unknown* constants d_0, d_j and *known* bounded continuous functions $f_j(t)$, $j = 1, 2, \ldots, q$, and some $q > 0$. Since $f_j(t)$, $j = 1, 2, \ldots, q$, are arbitrary, the existence of a $k_3^*(t)$ of the same components $f_j(t)$ to meet the final-value theorem condition (4.96) (with $d_y(s) = 0$) may be impossible unless the following transfer function matching is ensured:

$$\frac{1}{k_2^*} W_m(s) k_3^* + C(sI - A - Bk_1^{*T})^{-1} B_d = 0 \tag{4.105}$$

for some constant $k_3^* \in R$. Under this condition, the choice of $k_3^*(t) = k_3^* d_u(t)$ (i.e., $k_3^*(s) = k_3^* d_u(s)$), which leads to $\Delta(s) = 0$, ensures the desired plant–model output matching: $y(s) = y_m(s)$.

Clearly, if $B_d = \alpha_d B$ for some constant $\alpha_d \in R$ (that is, the case of *matched* input disturbances), then k_3^* exists to satisfy (4.105) which is crucial for adaptive rejection of the disturbances $d_u(t)$ when the parameters d_0, d_j, $j = 1, 2 \ldots, q$, are unknown. What if B and B_d are not parallel? This is the question to be answered next: What class of (C, A, B, B_d) can satisfy the matching equations (4.47) and (4.105) simultaneously?

Proposition 4.1 *Assume that (A, B) is controllable. There exist constant $k_1^* \in R^n$, $k_2^*, k_3^* \in R$ such that*

$$C(sI - A - Bk_1^{*T})^{-1} Bk_2^* = W_m(s) = \frac{1}{P_m(s)}, \tag{4.106}$$

$$\frac{1}{k_2^*} W_m(s) k_3^* + C(sI - A - Bk_1^{*T})^{-1} B_d = 0, \tag{4.107}$$

where $P_m(s)$ is a monic polynomial of degree n^* if and only if the two systems (C, A, B) and (C, A, B_d) have the same relative degree n^*.

The zeros of (C, A, B) are required to be stable (see Assumption (A4.4)) for $A + Bk_1^{*T}$ to have all its eigenvalues in $\mathrm{Re}[s] < 0$, so that the pole-zero cancellation in (4.106) is stable, which is crucial for closed-loop stability.

The complete proof of this proposition is given in [399], using an expansion of $(sI - \bar{A})^{-1} = \frac{N(s)}{\det(sI-\bar{A})}$, by the resolvent formula [103]. Here, we give a third-order example to illustrate such a basic property of linear systems.

Example 4.1 If (A, B) is controllable and (C, A, B) has relative degree n^*, then (4.106) follows from state feedback pole placement. To show how to use (4.106) to derive (4.107), for $n = 3$, we present the resolvent formula

$$(sI - \bar{A})^{-1} = \frac{N(s)}{\det(sI - \bar{A})},\ \bar{A} = A + Bk_1^{*T} \in R^{3\times 3}, \tag{4.108}$$

$$\det(sI - \bar{A}) = s^3 + a_3 s^2 + a_2 s + a_1, \tag{4.109}$$

$$N(s) = s^2 I + s(\bar{A} + a_3 I) + (\bar{A}^2 + a_3\bar{A} + a_2 I). \tag{4.110}$$

For (C, A, B) controllable, without loss of generality, we set $\bar{A}$, C, and B as

$$\bar{A} = A + Bk_1^{*T} = \begin{bmatrix} 0 & 1 & 0 \\ 0 & 0 & 1 \\ -a_1 & -a_2 & -a_3 \end{bmatrix},$$

$$C = [c_1, c_2, c_3],\ B = [0, 0, 1]^T. \tag{4.111}$$

If $n^* = 1$, then $k_2^* = c_3 = CB \neq 0$ and $CB_d \neq 0$. Let $c_3 = 1$. From

$$C(sI - A - Bk_1^{*T})^{-1}B = \frac{s^2 + c_2 s + c_1}{s^3 + a_3 s^2 + a_2 s + a_1}, \tag{4.112}$$

$P_m(s) = s + a_1^*$, and (4.111), it follows that $C(a_1^* I + \bar{A}) = 0$. Set $k_3^* = CB_d$ for (4.107) so that $C(B - \frac{1}{k_3^*}B_d) = 0$ as $CB = 1$. Then, using (4.111), we have

$$\begin{aligned} & C(sI - \bar{A})^{-1}\left(B - \frac{1}{k_3^*}B_d\right) \\ & = \frac{C\bar{A}(B - \frac{1}{k_3^*}B_d)s + C(\bar{A}^2 + a_3\bar{A})(B - \frac{1}{k_3^*}B_d)}{\det(sI - \bar{A})}. \end{aligned} \tag{4.113}$$

Combining (4.106) and (4.108), we have

$$(s^2 I + s(C\bar{A}B + a_3) + C\bar{A}^2 B + Ca_3\bar{A}B + a_2)(s + a_1^*) = s^3 + a_3 s^2 + a_2 s + a_1 \tag{4.114}$$

as $CB = 1$, which implies that

$$\begin{aligned} C\bar{A}B &= -a_1^*, \\ C\bar{A}^2 B &= -a_1^* C\bar{A}B - a_n(C\bar{A}B - a_1^*) = -a_1^* C\bar{A}B. \end{aligned} \tag{4.115}$$

Using this result, we write

$$C\bar{A}\left(B - \frac{1}{k_3^*}B_d\right) = -\frac{1}{k_3^*}C(a_1^* + \bar{A})B_d = 0,$$

$$C\bar{A}^2\left(B - \frac{1}{k_3^*}B_d\right) = -a_1^*C\bar{A}B - C\bar{A}^2\frac{1}{k_3^*}B_d$$

$$= -\frac{1}{k_3^*}C(a_1^*I + \bar{A})\bar{A}B_d = 0. \tag{4.116}$$

Substituting (4.116) into (4.113) leads to (4.107): $C(sI-\bar{A})^{-1}(B-\frac{1}{k_3^*}B_d) = 0$.

If $n^* = 2$, then $c_3 = CB = 0 = CB_d$, $k_2^* = c_2 = CAB \neq 0$, and $CAB_d \neq 0$. Let $c_2 = 1$ so that $C\bar{A}B = 1$. From

$$C(sI - A - Bk_1^{*T})^{-1}B = \frac{s + c_1}{s^3 + a_3s^2 + a_2s + a_1}, \tag{4.117}$$

$P_m(s) = s^2 + a_2^*s + a_1^*$, and (4.111), it follows that $C(a_1^*I + a_2^*\bar{A} + \bar{A}^2) = 0$, so that $C(a_2^*\bar{A} + \bar{A}^2)B_d = 0$ as $CB_d = 0$. Set $k_3^* = C\bar{A}B_d$ for (4.107), so that $C\bar{A}(B - \frac{1}{k_3^*}B_d) = 0$ as $C\bar{A}B = 1$. Then, using (4.108), we have

$$C(sI - \bar{A})^{-1}\left(B - \frac{1}{k_3^*}B_d\right) = \frac{C\bar{A}^2(B - \frac{1}{k_3^*}B_d)}{\det(sI - \bar{A})}. \tag{4.118}$$

Combining (4.106) and (4.108), we have

$$(sI + C\bar{A}^2B + a_3)(s^2 + a_2^*s + a_1^*) = s^3 + a_3s^2 + a_2s + a_1 \tag{4.119}$$

as $CB = 0$ and $C\bar{A}B = 1$, which implies that

$$C\bar{A}^2B = -a_2^*. \tag{4.120}$$

Using this result, we write

$$C\bar{A}^2\left(B - \frac{1}{k_3^*}B_d\right) = a_2^* - C\bar{A}^2\frac{1}{k_3^*}B_d = -\frac{1}{k_3^*}C(a_2^*\bar{A} + \bar{A}^2)B_d = 0. \tag{4.121}$$

Substituting (4.121) into (4.118) leads to (4.107).

If $n^* = 3$, then $c_{3-i} = CA^iB = 0 = CA^iB_d$, $i = 0, 1$, $k_2^* = c_1 = CA^2B \neq 0$, and $CA^2B_d \neq 0$. Let $c_1 = 1$ so that $C\bar{A}^2B = 1$. Set $k_3^* = C\bar{A}^2B_d$ in (4.107) so that $C\bar{A}^2(B - \frac{1}{k_3^*}B_d) = 0$ as $C\bar{A}^2B = 1$. Then, using (4.108), we have (4.107). In this case, from $CB_d = CAB_d = 0$, $B_d = \alpha_b B$ for some $\alpha_b \in R$.

For the general case, the parameter k_3^* is taken as $k_3^* = CA^{n^*-1}B_d$ (with $CA^{n^*-1}B = 1$), and the equality $C(\bar{A}^{n^*} + a_{n^*}^*\bar{A}^{n^*-1} + a_{n^*-1}^*\bar{A}^{n^*-2} + \cdots + a_2^*\bar{A} + a_1^*I) = 0$ can be established. Expressions of $C\bar{A}^kB$, $k = n^*, n^*+1, \ldots, n-1$, in terms of the parameters a_i^*, $i = 1, 2, \ldots, n^*$, of $P_m(s)$, and $C\bar{A}^jB$, $j < k$, can be derived, similar to those in (4.115) and (4.120). □

Assuming that (C, A, B) and (C, A, B_d) in (4.103) have the relative degree n^* as that of $W_m(s)$ in (4.40), we use the controller structure (4.80) for rejecting the effect of $d_u(t)$ on $y(t)$, with the compensation term

$$k_3(t) = k_{30}(t) + \sum_{j=1}^{q} k_{3j}(t) f_j(t) \tag{4.122}$$

with $k_{3j}(t)$ being the estimate of $k_{3j}^* = k_3^* d_j$, $j = 0, 1, 2 \ldots, q$.

From (4.40), (4.80), (4.103), (4.106), (4.107), and (4.122), we obtain

$$e(t) = \rho^* W_m(s)[(\theta - \theta^*)^T \omega](t) + C e^{(A+Bk_1^{*T})t} x(0), \tag{4.123}$$

where $C e^{(A+Bk_1^{*T})t} x(0)$ converges to zero exponentially and

$$\theta(t) = [k_1^T(t), k_2(t), k_{30}(t), k_{31}(t), \ldots, k_{3q}(t)]^T,$$

$$\theta^* = [k_1^{*T}, k_2^*, k_{30}^*, k_{31}^*, \ldots, k_{3q}^*]^T, \tag{4.124}$$

$$\omega(t) = [x^T(t), r(t), 1, f_1(t), \ldots, f_q(t)]^T. \tag{4.125}$$

Adaptive Scheme

For three types of disturbances—constant, sinusoidal, and nonsinusoidal—we have developed a unified error equation (4.87), (4.98), or (4.123) in terms of the tracking error $e(t) = y(t) - y_m(t)$ and parameter error $\theta(t) - \theta^*$, where $\theta(t)$ is the controller parameter vector. To develop an adaptive law for updating $\theta(t)$, we introduce the auxiliary signals

$$\zeta(t) = W_m(s)[\omega](t), \tag{4.126}$$

$$\xi(t) = \theta^T(t)\zeta(t) - W_m(s)[\theta^T \omega](t) \tag{4.127}$$

and the estimation error

$$\epsilon(t) = e(t) + \rho(t)\xi(t), \tag{4.128}$$

where $\rho(t)$ is an estimate of ρ^*. Then, neglecting the exponentially decaying terms $C e^{(A+Bk_1^{*T})t} x(0)$ and $\delta(t)$, we have

$$\epsilon(t) = \rho^*(\theta(t) - \theta^*)^T \zeta(t) + (\rho(t) - \rho^*)\xi(t), \tag{4.129}$$

which suggests the following adaptive laws for $\theta(t)$ and $\rho(t)$:

$$\dot{\theta}(t) = -\frac{\text{sign}[\rho^*]\Gamma\zeta(t)\epsilon(t)}{m^2(t)}, \quad \Gamma = \Gamma^T > 0, \ t \geq 0, \tag{4.130}$$

$$\dot{\rho}(t) = -\frac{\gamma\xi(t)\epsilon(t)}{m^2(t)}, \quad \gamma > 0, \ t \geq 0, \tag{4.131}$$

where $m(t) = \sqrt{1 + \zeta^T(t)\zeta(t) + \xi^2(t)}$.

This adaptive parameter update law has the following properties.

Lemma 4.2 *The adaptive laws (4.93) and (4.94) ensure that* $\theta(t) \in L^\infty$, $\rho(t) \in L^\infty$, $\frac{\epsilon(t)}{m(t)} \in L^2 \cap L^\infty$, $\dot{\theta}(t) \in L^2 \cap L^\infty$, *and* $\dot{\rho}(t) \in L^2 \cap L^\infty$.

Based on this result, the following desired properties can be established: closed-loop signal boundedness, asymptotic tracking $\lim_{t\to\infty}(y(t) - y_m(t)) = 0$, and $y(t) - y_m(t) \in L^2$, which are the same as those in Theorem 4.2.

4.4 Parametrization of State Feedback

We have shown that when the state vector $x(t)$ of the plant (4.37) is available for measurement, a state feedback controller, adaptive with $u(t) = k_1^T(t)x(t) + k_2(t)r(t)$ (see (4.41)) or ideal with $u(t) = k_1^{*T}x(t) + k_2^*r(t)$ (see (4.45)), can be developed to ensure closed-loop signal boundedness and plant–model output tracking. In this case, with the knowledge of $x(t)$, the adaptive term $k_1^T(t)x(t)$ is an estimate of the ideal term $k_1^{*T}x(t)$, simply by replacing k_1^* with $k_1(t)$ updated from an adaptive law.

When the plant state vector $x(t)$ is not available for measurement, an output feedback controller is to be employed using an estimate of $k_1^{*T}x(t)$ which needs to be re-parametrized. This re-parametrization is based on a state observer which can be designed in a full order or in a reduced order.

4.4.1 Parametrization with Full-Order Observer

Consider the linear time-invariant plant

$$\dot{x} = Ax + Bu, \ x \in R^n, \ y = Cx \in R, \ t \geq 0, \tag{4.132}$$

for some constant matrices $A \in R^{n\times n}$, $B \in R^n$, and $C \in R^{1\times n}$, with $n > 0$, where $x(t) = [x_1(t), \ldots, x_n(t)]^T$ is the plant state vector, $u(t) \in R$ is the control input, and $y(t) \in R$ is the plant output.

A state observer is a dynamic system which provides an estimate $\hat{x}(t)$ of the state variable $x(t)$ such that $\hat{x}$ converges to the true state x so that we could use $\hat{x}$ for state feedback control if the plant parameters were known.

When the initial plant state vector $x(0)$ is unknown, assuming that the plant (4.132) is observable, we generate an estimate $\hat{x}(t)$ for the unknown state vector $x(t)$ from the full-order state observer (estimator), that is,

$$\dot{\hat{x}}(t) = A\hat{x}(t) + Bu(t) + L(y(t) - C\hat{x}(t)), \ \hat{x}(0) = \hat{x}_0, \ t \geq 0, \tag{4.133}$$

where $L = [L_1, L_2, \ldots, L_n]^T \in R^n$ is a constant-gain vector and $\hat{x}_0 \in R^n$ is an estimate of $x(0) = x_0$. For the estimation error $\tilde{x}(t) = x(t) - \hat{x}(t)$, from (4.132) and (4.133), we obtain the estimation error equation

$$\dot{\tilde{x}}(t) = (A - LC)\tilde{x}(t),\ t \geq 0, \tag{4.134}$$

from which it is clear that the task of designing an estimator is equivalent to the task of choosing a vector gain L to make the eigenvalues of the $n \times n$ matrix $A - LC$ equal to some given stable ones. Then, it follows that $\lim_{t\to\infty} \tilde{x}(t) = 0$ at an exponential rate determined by the eigenvalues of $A - LC$.

Rewriting the observer (4.133) as

$$\dot{\hat{x}}(t) = (A - LC)\hat{x}(t) + Bu(t) + Ly(t),\ t \geq 0, \tag{4.135}$$

we express $\hat{x}(t)$ as

$$\begin{aligned}\hat{x}(t) &= (sI - A + LC)^{-1}B[u](t) + (sI - A + LC)^{-1}L[y](t) + e^{(A-LC)t}\hat{x}(0)\\ &= \frac{G_u(s)}{\Lambda(s)}[u](t) + \frac{G_y(s)}{\Lambda(s)}[y](t) + e^{(A-LC)t}\hat{x}(0),\end{aligned} \tag{4.136}$$

where $\Lambda(s) = \det(sI - A + LC)$, $G_u(s)$ and $G_y(s)$ are some n-dimensional polynomial vectors (operators), and $\lim_{t\to\infty} e^{(A-LC)t}\hat{x}(0) = 0$ exponentially.

Instead of using $k_1^{*T}x(t)$ for feedback control, we now use $k_1^{*T}\hat{x}(t)$, which, having the desired property that $\lim_{t\to\infty}(k_1^{*T}\hat{x}(t) - k_1^{*T}x(t)) = 0$, in view of (4.136), can be parametrized as

$$k_1^{*T}\hat{x}(t) = \theta_1^{*T}\omega_1(t) + \theta_2^{*T}\omega_2(t), \tag{4.137}$$

where

$$\omega_1(t) = \frac{a(s)}{\Lambda(s)}[u](t),\ \omega_2(t) = \frac{a(s)}{\Lambda(s)}[y](t), \tag{4.138}$$

$$a(s) = [1, s, \ldots, s^{n-1}]^T, \tag{4.139}$$

$\Lambda(s)$ is a chosen monic stable polynomial of degree n (whose zeros are the eigenvalues of $A - LC$, which can be made as any desired ones), and $\theta_1^* \in R^n$, $\theta_2^* \in R^n$ are some constant parameters which could be calculated if the plant parameters were known. For this parametrization, the effect of the exponentially decaying term $e^{(A-LC)t}\hat{x}(0)$ in (4.136) can be neglected.

For *adaptive control*, the term $k_1^T\hat{x}(t)$ is parametrized as

$$k_1^T\hat{x}(t) = \theta_1^T\omega_1(t) + \theta_2^T\omega_2(t), \tag{4.140}$$

where $\theta_1(t) \in R^n$, $\theta_2(t) \in R^n$ are estimates of the unknown constant parameters $\theta_1^* \in R^n$, $\theta_2^* \in R^n$, to be updated from an adaptive law.

4.4.2 Parametrization with Reduced-Order Observer

To derive a reduced-order observer, we express the plant (4.132) as

$$\begin{bmatrix} \dot{x}_1(t) \\ \dot{x}_2(t) \end{bmatrix} = \begin{bmatrix} A_{11} & A_{12} \\ A_{21} & A_{22} \end{bmatrix} \begin{bmatrix} x_1(t) \\ x_2(t) \end{bmatrix} + \begin{bmatrix} B_1 \\ B_2 \end{bmatrix} u(t),$$

$$y(t) = x_1(t) = \begin{bmatrix} 1 & 0 & 0 & \cdots & 0 \end{bmatrix} \begin{bmatrix} x_1(t) \\ x_2(t) \end{bmatrix} \in R,\ t \geq 0, \quad (4.141)$$

where $x_2(t) \in R^{n-1}$ is to be estimated.[2]

Assuming that $(A,\ C)$ is observable, which implies that $(A_{22},\ A_{12})$ is observable [13], we generate an estimate $\hat{x}(t)$ for the unknown state vector $x(t)$ from the reduced-order observer with a dynamic system for $\hat{x}_2(t)$:

$$\hat{x}(t) = \begin{bmatrix} x_1(t) \\ \hat{x}_2(t) \end{bmatrix} = \begin{bmatrix} y(t) \\ w(t) + L_r y(t) \end{bmatrix}, \quad (4.142)$$

where $L_r = [L_1, L_2, \ldots, L_{n-1}]^T \in R^{n-1}$ is a constant-gain vector such that the eigenvalues of the $(n-1) \times (n-1)$ matrix $A_{22} - L_r A_{12}$ are equal to some given stable ones and $w(t) \in R^{n-1}$ is generated from the system

$$\begin{aligned} \dot{w}(t) &= (A_{22} - L_r A_{12}) w(t) + (B_2 - L_r B_1) u(t) \\ &\quad + ((A_{22} - L_r A_{12}) L_r + A_{21} - L_r A_{11}) y(t),\ w(0) = w_0,\ t \geq 0, \quad (4.143) \end{aligned}$$

where $w_0 \in R^{n-1}$ is an estimate of $L_r y(0) - x_2(0)$. From (4.142) and (4.143), the dynamic equation for $\hat{x}_2(t)$ can be expressed as

$$\begin{aligned} \dot{\hat{x}}_2(t) &= A_{22}\hat{x}_2(t) + \begin{bmatrix} A_{21} & B_2 \end{bmatrix} \begin{bmatrix} x_1(t) \\ u(t) \end{bmatrix} \\ &\quad + L_r(\dot{x}_1(t) - A_{11}x_1(t) - B_1 u(t) - A_{12}\hat{x}_2(t)). \quad (4.144) \end{aligned}$$

For the estimation error $\tilde{x}_2(t) = x_2(t) - \hat{x}_2(t)$, from (4.141) and (4.144), we obtain the estimation error equation

$$\dot{\tilde{x}}_2(t) = (A_{22} - L_r A_{12}) \tilde{x}_2(t), \quad (4.145)$$

that is, $\lim_{t\to\infty} \tilde{x}_2(t) = 0$ exponentially, as determined by $A_{22} - L_r A_{12}$.

---

[2]If $y(t) = Cx(t) \neq x_1(t)$, then the state transformation $\bar{x}(t) = Px(t)$, where $P = \begin{bmatrix} C \\ \hat{C} \end{bmatrix}$ and $\hat{C}$ is such that $P \in R^{n \times n}$ is nonsingular, results in $y(t) = [1, 0, 0, \ldots, 0]\bar{x}(t)$ [13]. In this case, the estimate of $x(t)$ is $P^{-1}\hat{x}(t)$, where $\hat{x}(t)$ is the estimate of $\bar{x}(t)$.

To proceed, we express $w(t)$ in (4.143) as

$$\begin{aligned} w(t) &= e^{(A_{22}-L_rA_{12})t}w(0) + (sI - A_{22} + L_rA_{12})^{-1}(B_2 - L_rB_1)[u](t) \\ &\quad + (sI - A_{22} + L_rA_{12})^{-1}((A_{22} - L_rA_{12})L_r + A_{21} - L_rA_{11})[y](t) \\ &= \frac{G_u(s)}{\Lambda(s)}[u](t) + \frac{G_y(s)}{\Lambda(s)}[y](t) + e^{(A_{22}-L_rA_{12})t}w(0), \end{aligned} \tag{4.146}$$

where $\Lambda(s) = \det(sI - A_{22} + L_rA_{12})$, $G_u(s)$ and $G_y(s)$ are some $(n-1)$-dimensional polynomial vectors (operators), and $\lim_{t\to\infty} e^{(A_{22}-L_rA_{12})t}w(0) = 0$.

Instead of using $k_1^{*T}x(t)$ for feedback control, we now use $k_1^{*T}\hat{x}(t)$, where $\hat{x}_2(t) = w(t) + L_ry(t)$ and $x_1(t) = y(t)$ as in (4.142), both available for measurement. In view of (4.146), $k_1^{*T}\hat{x}(t)$ can be parametrized as

$$k_1^{*T}\hat{x}(t) = \theta_1^{*T}\omega_1(t) + \theta_2^{*T}\omega_2(t) + \theta_{20}^*y(t), \tag{4.147}$$

where

$$\omega_1(t) = \frac{a(s)}{\Lambda(s)}[u](t),\ \omega_2(t) = \frac{a(s)}{\Lambda(s)}[y](t), \tag{4.148}$$

$$a(s) = [1, s, \dots, s^{n-2}]^T, \tag{4.149}$$

$\Lambda(s)$ is a chosen monic stable polynomial of degree $n-1$ (whose zeros are the desired eigenvalues of $A_{22} - L_rA_{12}$), and $\theta_1^* \in R^{n-1}$, $\theta_2^* \in R^{n-1}$, $\theta_{20}^* \in R$ are some constant parameters which are related to the plant and observer parameters. For this parametrization, the effect of the exponentially decaying term $e^{(A_{22}-L_rA_{12})t}w(0)$ in (4.142) can be neglected (without making a control system unstable; see Section 5.4).

For *adaptive control*, the term $k_1^T\hat{x}(t)$ is parametrized as

$$k_1^T\hat{x}(t) = \theta_1^T\omega_1(t) + \theta_2^T\omega_2(t) + \theta_{20}(t)y(t), \tag{4.150}$$

where $\theta_1(t) \in R^{n-1}$, $\theta_2(t) \in R^{n-1}$, $\theta_{20}(t)$ are estimates of the unknown parameters $\theta_1^* \in R^{n-1}$, $\theta_2^* \in R^{n-1}$, $\theta_{20}^* \in R$, to be adaptively updated.

Based on the parametrization (4.140) or (4.150), which depends on the plant input $u(t)$ and output $y(t)$ only, adaptive laws can be developed for updating the parameters θ_1, θ_2, and θ_{20} to implement output feedback adaptive controllers for achieving output tracking (see Chapter 5).

Remark 4.3 When (A, C) is only detectable, the choice of $\Lambda(s)$ in (4.136) is limited to one which contains the unobservable modes (eigenvalues which are stable due to detectability of (A, C)) of A. For the reduced-order observer case, the condition that (A, C) is detectable implies that (A_{22}, A_{12}) is detectable (which can be verified using the rank test for observability [13]). Therefore, $\Lambda(s)$ in (4.146) also contains those stable unobservable modes of A. In this case, since some modes (eigenvalues) of A are not observable, the state feedback $k_1^{*T}x(t)$ cannot be parametrized using an arbitrary $\Lambda(s)$. However, the existence of θ_1^*, θ_2^* (and θ_{20}^*) in (4.137) and (4.147) with an arbitrary stable $\Lambda(s)$ of a specified degree is guaranteed such that the feedback control law $u(t) = \theta_1^{*T}\omega_1(t) + \theta_2^{*T}\omega_2(t) + \theta_{20}^* y(t) + k_2^* r(t)$ is capable of ensuring the plant–model input–output matching: $y(t) = y_m(t)$ (see Chapter 5). □

4.5 Discrete-Time Adaptive Control

Stable adaptive control designs can be derived for discrete-time systems, based on a gradient algorithm for controller parameter adaptation. The continuous-time results of Section 4.2 (design for output tracking), Section 4.3.2 (output disturbance rejection), and Section 4.4 (parametrization of state feedback) can be in parallel developed for discrete-time linear time-invariant systems, as shown in this section, starting with a design example.

4.5.1 Design Example

Consider a first-order discrete-time time-invariant plant

$$y(t+1) = a_p y(t) + b_p u(t),\ t \in \{0, 1, 2, \ldots\}, \tag{4.151}$$

where a_p and b_p are the plant parameters, $y(t)$ is the plant output, and $u(t)$ is the control input. The control objective is to design a feedback control $u(t)$ such that all closed-loop system signals are bounded and the plant output $y(t)$ asymptotically tracks the output $y_m(t)$ of a chosen reference model

$$y_m(t+1) = -a_m y_m(t) + b_m r(t),\ t \in \{0, 1, 2, \ldots\}, \tag{4.152}$$

where $|a_m| < 1$ for the model system stability and performance and $r(t)$ is a bounded reference input. For a discrete-time design, it is common to choose $a_m = 0$ and $b_m = 1$ so that $y_m(t) = r(t-1)$.

The control structure and plant–model matching analysis are similar to that for a continuous-time case. When both plant parameters a_p and b_p are known, the above control objective can be met by using the controller

$$u(t) = k_1^* y(t) + k_2^* r(t), \tag{4.153}$$

where

$$k_1^* = \frac{-a_p - a_m}{b_p},\ k_2^* = \frac{b_m}{b_p}, \tag{4.154}$$

which leads to the closed-loop system

$$y(t+1) = -a_m y(t) + b_m r(t). \tag{4.155}$$

The tracking error $e(t) = y(t) - y_m(t)$ satisfies $e(t+1) = -a_m e(t)$, so that $e(t)$ is bounded, and so are $y(t)$ and $u(t)$. Moreover, $\lim_{t\to\infty} e(t) = 0$ exponentially. If $a_m = 0$ is chosen, $e(t) = 0$, $t \geq 1$.

When the plant parameters a_p and b_p are unknown, we use the adaptive version of the fixed controller (4.153),

$$u(t) = k_1(t)y(t) + k_2(t)r(t), \tag{4.156}$$

where $k_1(t)$ and $k_2(t)$ are estimates of the unknown parameters k_1^* and k_2^* in (4.154). Substituting (4.156) into (4.151) gives the closed-loop system

$$y(t+1) = -a_m y(t) + b_m r(t) + b_p((k_1(t) - k_1^*)y(t) + (k_2(t) - k_2^*)r(t)), \tag{4.157}$$

from which, using (4.155), we have the tracking error equation

$$e(t+1) = -a_m e(t) + b_p \tilde{k}_1(t)y(t) + b_p \tilde{k}_2(t)r(t), \; t \in \{0, 1, 2, \ldots\}, \tag{4.158}$$

where $e(t) = y(t) - y_m(t)$, $\tilde{k}_1(t) = k_1(t) - k_1^*$, and $\tilde{k}_2(t) = k_2(t) - k_2^*$.

To develop a gradient design, we define

$$\rho^* = b_p, \; \theta^* = [k_2^*, k_2^*]^T, \; \theta = [k_2, k_2]^T, \tag{4.159}$$

$$\omega(t) = [y(t), r(t)]^T \tag{4.160}$$

and introduce the filtered vector signal

$$\zeta(t) = \frac{1}{z+a_m}[\omega](t) = \left[\frac{1}{z+a_m}[y](t), \frac{1}{z+a_m}[r](t)\right]^T, \tag{4.161}$$

where $\frac{1}{z+a_m}[y](t)$ denotes the output of the system with transfer function $\frac{1}{z+a_m}$ and input $y(t)$. Then, we rewrite (4.158) as

$$\begin{aligned} e(t) &= \frac{\rho^*}{z+a_m}[\theta^T\omega - \theta^{*T}\omega](t) \\ &= \rho^* \left(\frac{1}{z+a_m}[\theta^T\omega](t) - \theta^{*T}\zeta(t)\right). \end{aligned} \tag{4.162}$$

Based on this expression, we define the estimation error

$$\epsilon(t) = e(t) - \rho(t)\left(\frac{1}{z+a_m}[\theta^T\omega](t) - \theta^T(t)\zeta(t)\right). \tag{4.163}$$

With (4.162), this estimation error can be expressed as

$$\epsilon(t) = \rho^* \tilde{\theta}^T(t)\zeta(t) + \tilde{\rho}(t)\xi(t), \tag{4.164}$$

where

$$\xi(t) = \theta^T(t)\zeta(t) - \frac{1}{z+a_m}[\theta^T \omega](t), \tag{4.165}$$

$$\tilde{\theta}(t) = \theta(t) - \theta^*, \ \tilde{\rho}(t) = \rho(t) - \rho^*. \tag{4.166}$$

Choose the adaptive laws for $\theta(t)$ and $\rho(t)$ as

$$\theta(t+1) = \theta(t) - \frac{\text{sign}[b_p]\Gamma\epsilon(t)\zeta(t)}{m^2(t)}, \ 0 < \Gamma = \Gamma^T < \frac{2}{|b_p|}I_2, \tag{4.167}$$

$$\rho(t+1) = \rho(t) - \frac{\gamma\epsilon(t)\xi(t)}{m^2(t)}, \ 0 < \gamma < 2, \tag{4.168}$$

for $t \in \{0, 1, 2, \ldots\}$, where I_2 is the 2×2 identity matrix and

$$m(t) = \sqrt{1 + \zeta^T(t)\zeta(t) + \xi^2(t)}. \tag{4.169}$$

Introduce a measure of $\tilde{\theta}(t)$ and $\tilde{\rho}(t)$ as

$$V(\tilde{\theta}, \tilde{\rho}) = |\rho^*|\tilde{\theta}^T\Gamma^{-1}\tilde{\theta} + \gamma^{-1}\tilde{\rho}^2. \tag{4.170}$$

The time increment of $V(\tilde{\theta}, \tilde{\rho})$, along (4.167) and (4.168), is

$$\begin{aligned} &V(\tilde{\theta}(t+1), \tilde{\rho}(t+1)) - V(\tilde{\theta}(t), \tilde{\rho}(t)) \\ &= -\left(2 - \frac{|b_p|\zeta^T(t)\Gamma\zeta(t) + \gamma\xi^2(t)}{m^2(t)}\right)\frac{\epsilon^2(t)}{m^2(t)} \leq -\alpha_1 \frac{\epsilon^2(t)}{m^2(t)} \end{aligned} \tag{4.171}$$

for some constant $\alpha_1 > 0$. The last inequality follows from the conditions $0 < \Gamma = \Gamma^T < \frac{2}{|b_p|}I_2$ and $0 < \gamma < 2$. This result implies that $\theta(t) \in L^\infty$, $\rho(t) \in L^\infty$, $\frac{\epsilon(t)}{m(t)} \in L^2 \cap L^\infty$, and $\theta(t+1) - \theta(t) \in L^2$.

For $a_m = 0$, $\zeta(t) = [y(t-1), r(t-1)]^T$, we have $\epsilon(t) = \rho^*\tilde{\theta}^T(t)\omega(t-1) + \tilde{\rho}(t)\xi(t)$, $\xi(t) = (\theta(t) - \theta(t-1))^T\omega(t-1)$, with $\omega(t) = [y(t), r(t)]^T$, and

$$\begin{aligned} m^2(t+1) &= 1 + \zeta^T(t+1)\zeta(t+1) + \xi^2(t+1) \\ &= 1 + y^2(t) + r^2(t) + (\theta_1(t+1) - \theta_1(t))^2 y^2(t) \\ &\quad + (\theta_2(t+1) - \theta_2(t))^2 r^2(t). \end{aligned} \tag{4.172}$$

Since $\theta(t) = [\theta_1(t), \theta_2(t)]^T \in L^\infty$ and $r(t) \in L^\infty$, it follows from (4.169) that

$$m(t+1) \leq b_0(t) + b_1(t)|y(t)| \tag{4.173}$$

for some $b_0(t), b_1(t) \in L^\infty$. We can rewrite (4.163) with $a_m = 0$ as

$$y(t+1) = r(t) + \frac{\epsilon(t+1)}{m(t+1)}m(t+1) - \rho(t+1)(\theta(t+1) - \theta(t))^T\omega(t). \quad (4.174)$$

Using (4.173) and (4.174), we have

$$|y(t+1)| \leq \left(\frac{|\epsilon(t+1)|}{m(t+1)}b_1(t) + |\rho(t+1)|\,|\theta_1(t+1) - \theta_1(t)|\right)|y(t)| + b_3 \quad (4.175)$$

for some constant $b_3 > 0$. Since $\frac{\epsilon(t)}{m(t)} \in L^2 \cap L^\infty$ and $\theta(t+1) - \theta(t) \in L^2$, we have that $\lim_{t\to\infty} \frac{\epsilon(t)}{m(t)} = 0$ and $\lim_{t\to\infty}(\theta_i(t+1) - \theta_i(t)) = 0$, $i = 1, 2$. With this fact, (4.175) implies that $y(t) \in L^\infty$, and all signals are bounded. Therefore, $\lim_{t\to\infty} \epsilon(t) = 0$ and $\lim_{t\to\infty} \xi(t) = 0$, so that $\lim_{t\to\infty} e(t) = 0$. □

4.5.2 Output Tracking Design

We now derive an adaptive state feedback control scheme for an nth-order system described by a state variable equation which ensures closed-loop signal boundedness and asymptotic output tracking in the presence of system parameter uncertainties and is based on a set of assumptions for stable plant–model matching and controller parameter adaptation.

Control System Structure

Consider the linear discrete-time time-invariant plant

$$\begin{aligned} x(t+1) &= Ax(t) + Bu(t),\ x \in R^n,\ u \in R, \\ y(t) &= Cx(t),\ y \in R,\ t \in \{0, 1, 2, \ldots\}, \end{aligned} \quad (4.176)$$

for some unknown constant matrices $A \in R^{n\times n}$, $B \in R^{n\times 1}$, and $C \in R^{1\times n}$, with $n > 0$. The input–output description of this plant is

$$y(z) = G(z)u(z), \quad (4.177)$$

where

$$G(z) = C(zI - A)^{-1}B = k_p\frac{Z(z)}{P(z)} \quad (4.178)$$

with $k_p \neq 0$ a constant and $P(z) = \det(zI - A)$ and $Z(z)$ being monic polynomials with constant coefficients and of degrees n and $n - n^*$, respectively, for some integer $n^* > 0$, which is called the *plant relative degree.* The symbol z denotes, as the case may be, the z-transform variable (in this case, $G(z)$ is the plant transfer function) or the time advance operator $z[x](t) = x(t+1)$ (in this case, $G(z)$ is the plant operator and the notation $y(t) = G(z)[u](t)$ denotes the output of a plant with transfer function $G(z)$ and input $u(t)$).

To design an adaptive state feedback model reference controller for generating the plant input $u(t)$, which ensures closed-loop signal boundedness and asymptotic tracking of a chosen reference signal $y_m(t)$ by the plant output $y(t)$, we need the following assumptions on the plant (4.176):

(A4.3-D): (A, B, C) is stabilizable and detectable.

(A4.4-D): $Z(z)$ is a stable polynomial (i.e., its zeros are in $|z| < 1$).

(A4.5-D): The degree $n - n^*$ of $Z(z)$ is known.

(A4.6-D): The sign of k_p, $\text{sign}[k_p]$, is known, and so is the upper bound of $|k_p|$, $k_p^0 \geq |k_p|$.

Assumption (A4.3-D) is for output matching by pole placement and for internal signal boundedness, Assumption (A4.4-D) is for stable model reference control, and Assumption (A4.5-D) is needed for choosing a reference model for adaptive control, while Assumption (A4.6-D) is used for designing an adaptive parameter update law (as compared with (A4.6) in Section 4.2.2 for a continuous-time adaptive design, the additional knowledge of the upper bound k_p^0 on $|k_p|$ is needed for an adaptive law in discrete time).

Based on Assumption (A4.5-D), the reference model system, which is independent of the dynamics of the plant (4.176) or (4.177), is chosen as

$$y_m(t) = W_m(z)[r](t),\ t \in \{0, 1, 2, \ldots\},\ W_m(z) = \frac{1}{P_m(z)}, \tag{4.179}$$

where $P_m(z)$ is a stable monic polynomial of degree n^* and $r(t)$ is an external reference input signal. In discrete-time adaptive control, it is common to choose $P_m(z) = z^{n^*}$ so that $y_m(t) = r(t - n^*)$ is the reference output.

The model reference adaptive controller structure is

$$u(t) = k_1^T(t)x(t) + k_2(t)r(t), \tag{4.180}$$

where $r(t)$ is the input signal to the reference system (4.179) and

$$k_1(t) = [k_{11}(t), k_{12}(t), \ldots, k_{1n}(t)]^T \in R^n,\ k_2(t) \in R, \tag{4.181}$$

are the adaptive estimates of the unknown parameters

$$k_1^* = [k_{11}^*, k_{12}^*, \ldots, k_{1n}^*]^T \in R^n,\ k_2^* \in R, \tag{4.182}$$

which are defined to satisfy

$$\det(zI - A - Bk_1^{*T}) = P_m(z)Z(z)\frac{1}{k_p}, \; k_2^* = \frac{1}{k_p}, \tag{4.183}$$

that is, all zeros of $\det(zI - A - Bk_1^{*T})$ are stable. The existence of such parameters k_1^* and k_2^* is guaranteed by Assumptions (A4.3-D) and (A4.4-D), that is, all the uncontrollable modes (which cannot be changed by k_1^*) of A are stable and equal to some zeros of $Z(z)$ and all the controllable modes of A are moved to the zeros of $P_m(z)$ and the other zeros of $Z(z)$ (any controllable but unobservable (stable) modes of A are left unchanged because they are equal to some of the other zeros of $Z(z)$), as in the continuous-time case.

With this definition of k_1^* and k_2^*, the ideal version of (4.180) is

$$u(t) = k_1^{*T}x(t) + k_2^*r(t), \tag{4.184}$$

which would lead to the desired closed-loop system $y(z) = W_m(z)r(z)$. When A, B, and C are all unknown parameters, the ideal control gains k_1^* and k_2^* are also unknown. An adaptive law is needed to update the parameter estimates $k_1(t)$ and $k_2(t)$, and to derive a stable adaptive law, the existence of the stable plant–model matching k_1^* and k_2^*, not their values, is crucial.

With the adaptive controller (4.180), the plant (4.176) becomes

$$\begin{aligned} x(t+1) &= Ax(t) + B(k_1^T(t)x(t) + k_2(t)r(t)) \\ &= (A + Bk_1^{*T})x(t) + Bk_2^*r(t) \\ &\quad + B\left((k_1^T(t) - k_1^{*T})x(t) + (k_2(t) - k_2^*)r(t)\right), \\ y(t) &= Cx(t). \end{aligned} \tag{4.185}$$

From (4.183) it follows that

$$\begin{aligned} &C(zI - A - Bk_1^{*T})^{-1}Bk_2^* \\ &= \frac{Z(z)k_2^*}{\det(zI - A - Bk_1^{*T})} = \frac{1}{P_m(z)} = W_m(z). \end{aligned} \tag{4.186}$$

In view of (4.179), (4.185), and (4.186), the tracking error equation is

$$\begin{aligned} e(t) &= y(t) - y_m(t) \\ &= \rho^* W_m(z)[(k_1^T - k_1^{*T})x + (k_2 - k_2^*)r](t) + C(A + Bk_1^{*T})^t x(0), \end{aligned} \tag{4.187}$$

where $\rho^* = k_p$ and $C(A + Bk_1^{*T})^t x(0)$ converges to zero exponentially fast due to the stability of $A + Bk_1^{*T}$ (whose eigenvalues are all in $|z| < 1$).

Adaptive Laws

To derive an estimation error equation, we define

$$\theta(t) = [k_1^T(t), k_2(t)]^T, \ \theta^* = [k_1^{*T}, k_2^*]^T, \tag{4.188}$$

and introduce the auxiliary signals

$$\omega(t) = [x^T(t), r(t)]^T, \tag{4.189}$$

$$\zeta(t) = W_m(z)[\omega](t), \tag{4.190}$$

$$\xi(t) = \theta^T(t)\zeta(t) - W_m(z)[\theta^T\omega](t) \tag{4.191}$$

and the estimation error

$$\epsilon(t) = e(t) + \rho(t)\xi(t), \tag{4.192}$$

where $\rho(t)$ is an estimate of $\rho^* = k_p$. Then, substituting (4.187)–(4.191) into (4.192) and neglecting the exponentially decaying $C(A + Bk_1^{*T})^t x(0)$, we have

$$\epsilon(t) = \rho^*(\theta(t) - \theta^*)^T\zeta(t) + (\rho(t) - \rho^*)\xi(t), \tag{4.193}$$

which is linear in the parameter errors $\theta(t) - \theta^*$ and $\rho(t) - \rho^*$.

Using a gradient method (see Section 3.6.2), this error equation suggests the following adaptive laws for $\theta(t)$ and $\rho(t)$:

$$\theta(t+1) = \theta(t) - \frac{\text{sign}[\rho^*]\Gamma\zeta(t)\epsilon(t)}{m^2(t)}, \ 0 < \Gamma = \Gamma^T < \frac{2}{k_p^0} I_{n+1}, \tag{4.194}$$

$$\rho(t+1) = \rho(t) - \frac{\gamma\xi(t)\epsilon(t)}{m^2(t)}, \ 0 < \gamma < 2, \tag{4.195}$$

$t \in \{0, 1, 2, \ldots\}$, where $\text{sign}[\rho^*]$ is the sign of the parameter $\rho^* = k_p$, $k_p^0 \geq |k_p|$ as in (A4.6-D), $\Gamma \in R^{(n+1)\times(n+1)}$ and γ are constant adaptation gains, and

$$m(t) = \sqrt{1 + \zeta^T(t)\zeta(t) + \xi^2(t)}. \tag{4.196}$$

This adaptive parameter update law has the following properties.

Lemma 4.3 *The adaptive laws (4.194) and (4.195) ensure that* $\theta(t) \in L^\infty$, $\rho(t) \in L^\infty$, $\frac{\epsilon(t)}{m(t)} \in L^2 \cap L^\infty$, $\theta(t+1) - \theta(t) \in L^2$, *and* $\rho(t+1) - \rho(t) \in L^2$.

Proof: Consider the positive definite function

$$V(\tilde{\theta}, \tilde{\rho}) = |\rho^*|\tilde{\theta}^T\Gamma^{-1}\tilde{\theta} + \gamma^{-1}\tilde{\rho}^2, \tag{4.197}$$

where

$$\tilde{\theta}(t) = \theta(t) - \theta^*, \ \tilde{\rho}(t) = \rho(t) - \rho^*. \tag{4.198}$$

The time increment of $V(\tilde{\theta}, \tilde{\rho})$ along the trajectories of (4.194) and (4.195) is

$$\begin{aligned} &V(\tilde{\theta}(t+1), \tilde{\rho}(t+1)) - V(\tilde{\theta}(t), \tilde{\rho}(t)) \\ &= -\left(2 - \frac{|k_p|\zeta^T(t)\Gamma\zeta(t) + \gamma\xi^2(t)}{m^2(t)}\right)\frac{\epsilon^2(t)}{m^2(t)} \leq -\alpha_1 \frac{\epsilon^2(t)}{m^2(t)} \end{aligned} \tag{4.199}$$

for some constant $\alpha_1 > 0$. The last inequality in (4.199) follows from the conditions $0 < \Gamma = \Gamma^T < \frac{2}{k_p^0} I_{n+1}$, $b^0 \geq |k_p| = |\rho^*|$, and $0 < \gamma < 2$. This result implies that $\theta(t) \in L^\infty$, $\rho(t) \in L^\infty$, and, with (4.193) and (4.194), that $\frac{\epsilon(t)}{m(t)} \in L^2 \cap L^\infty$ and $\theta(t+1) - \theta(t) \in L^2$. ∇

Using these desired properties of the adaptive laws (4.194) and (4.195), the following results can be established.

Theorem 4.3 *All signals in the closed-loop system, which consists of the plant (4.176), the reference model (4.179), and the controller (4.180) updated by the adaptive law (4.194)–(4.195), are bounded, and the output tracking error $e(t) = y(t) - y_m(t)$ satisfies*

$$\lim_{t \to \infty} (y(t) - y_m(t)) = 0, \ \sum_{t=0}^{\infty} (y(t) - y_m(t))^2 < \infty. \tag{4.200}$$

The proof of this theorem is similar to that for Theorem 6.2, using the parametrization (4.258), and can also be derived from Theorem 9.2.

4.5.3 Disturbance Rejection

An adaptive control approach can be employed to reject the effect of unknown disturbances in the plant (4.176) so that desired system performance output tracking, in addition to closed-loop signal boundedness, can be achieved by using additional compensation in a feedback control law.

Consider a linear time-invariant plant with disturbances:

$$\begin{aligned} x(t+1) &= Ax(t) + Bu(t) + B_d d_u(t), \\ y(t) &= Cx(t) + d_y(t), \ t \in \{0, 1, 2, \ldots\}, \end{aligned} \tag{4.201}$$

where $B_d \in R^n$ is an unknown constant vector, $d_u(t) \in R$ represents the actuation disturbance, $d_y(t) \in R$ represents the output disturbance, and $A \in R^{n \times n}$, $B \in R^{n \times 1}$, and $C \in R^{1 \times n}$, with $n > 0$, are some unknown constant matrices which satisfy Assumptions (A4.3-D)–(A4.6-D) in Section 4.5.2. The control objective is to design an adaptive state feedback model reference control law for $u(t)$ to ensure closed-loop signal boundedness and asymptotic tracking of a chosen reference signal $y_m(t)$ by the plant output $y(t)$, despite the uncertainties in A, B, B_d, C, and $d_u(t)$, $d_y(t)$.

We consider the reference model (4.179) and use the controller structure

$$u(t) = k_1^T(t)x(t) + k_2(t)r(t) + k_3(t), \tag{4.202}$$

where $k_3(t) \in R$ is an adaptive compensation term for cancelling the effect of the actuation disturbance $B_d d_u$ and output disturbance d_y on the tracking error $e(t) = y(t) - y_m(t)$ and the gains $k_1(t)$ and $k_2(t)$ are adaptive gains whose ideal values k_1^* and k_2^* are defined in Section 4.5.2. Similar to that in Section 4.3.2 for the continuous-time case, next we will first develop ideal parametrizations and the associated error equations for the compensation term $k_3(t)$ for three types of disturbances—constant, sinusoidal, and nonsinusoidal—and then develop the adaptive scheme for updating $k_3(t)$.

Rejection of Constant Disturbances

For $d_u(t) = d_{u0}$ and $d_y(t) = d_{y0}$ being some unknown constants, $k_3(t)$ is the estimate of a constant parameter k_3^*. To derive this ideal value k_3^* of $k_3(t)$, we apply the ideal form of (4.202) for this case, $u(t) = k_1^{*T}x(t) + k_2^*r(t) + k_3^*$, to the plant (4.201) to get the closed-loop system in the frequency z-domain:

$$y(z) = y_m(z) + \Delta(z), \tag{4.203}$$

where $y_m(z) = C(zI - A - Bk_1^{*T})^{-1}Bk_2^*r(z)$ and

$$\begin{aligned}\Delta(z) &= C(zI - A - Bk_1^{*T})^{-1}B\frac{k_3^* z}{z-1} \\ &\quad + C(zI - A - Bk_1^{*T})^{-1}B_d\frac{d_{u0}z}{z-1} + \frac{d_{y0}z}{z-1}.\end{aligned} \tag{4.204}$$

From Assumption (A4.4-D), all zeros and poles of $C(zI - A - Bk_1^{*T})^{-1}B$ are inside the unit circle of the complex plane, so that

$$\lim_{z\to 1}(z-1)\,\Delta(z) = c_1 k_3^* + c_2 + d_{y0} \tag{4.205}$$

for some constants $c_1 \neq 0$ and c_2 (which depends on $B_d d_{u0}$). We then set

$$k_3^* = -\frac{c_2 + d_{y0}}{c_1}. \tag{4.206}$$

With this k_3^*, from (4.203)–(4.205), in the time domain, we have

$$\lim_{t\to\infty}(y(t) - y_m(t)) = \lim_{t\to\infty}\delta(t) = 0, \tag{4.207}$$

exponentially fast, where $\delta(t) = \mathcal{Z}^{-1}[\Delta(z)]$.

To develop adaptive laws to update the controller parameters $k_1(t)$, $k_2(t)$, and $k_3(t)$ in (4.202), we express the closed-loop system with (4.202) as

$$\begin{aligned} x(t+1) &= (A + Bk_1^{*T})x(t) + Bk_2^* r(t) + Bk_3^* + B_d d_{u0} \\ &\quad + B\left((k_1^T(t) - k_1^{*T})x(t) + (k_2(t) - k_2^*)r(t) + (k_3(t) - k_3^*)\right), \\ y(t) &= Cx(t) + d_{y0}. \end{aligned} \tag{4.208}$$

Using (4.186), (4.204), and (4.208), with $\delta(t) = \mathcal{Z}^{-1}[\Delta(z)]$, similar to (4.187), we obtain the tracking error equation

$$\begin{aligned} e(t) &= y(t) - y_m(t) \\ &= \rho^* W_m(z)[(k_1^T - k_1^{*T})x + (k_2 - k_2^*)r + (k_3 - k_3^*)](t) \\ &\quad + C(A + Bk_1^{*T})^t x(0) + \delta(t) \\ &= \rho^* W_m(z)[(\theta - \theta^*)^T\omega](t) + C(A + Bk_1^{*T})^t x(0) + \delta(t), \end{aligned} \tag{4.209}$$

where $C(A + Bk_1^{*T})^t x(0)$ and $\delta(t)$ converge to zero exponentially and

$$\theta(t) = [k_1^T(t), k_2(t), k_3(t)]^T,\ \theta^* = [k_1^{*T}, k_2^*, k_3^*]^T, \tag{4.210}$$

$$\omega(t) = [x^T(t), r(t), 1]^T. \tag{4.211}$$

Rejection of Sinusoidal Disturbances

Consider the sinusoidal disturbances $d_u(t)$ and $d_y(t)$:

$$d_u(t) = d_{u0} + \sum_{j=1}^{q_u}(\alpha_{uj}\sin\omega_{uj}t + \beta_{uj}\cos\omega_{uj}t), \tag{4.212}$$

$$d_y(t) = d_{y0} + \sum_{j=1}^{q_y}(\alpha_{yj}\sin\omega_{yj}t + \beta_{yj}\cos\omega_{yj}t), \tag{4.213}$$

where d_{u0}, α_{uj}, β_{uj}, $j = 1, 2, \ldots, q_u$, d_{y0}, α_{yj}, β_{yj}, $j = 1, 2, \ldots, q_y$, are some *unknown* constants and ω_{uj}, $j = 1, 2, \ldots, q_u$, ω_{yj}, $j = 1, 2, \ldots, q_y$, are some *known* constants (i.e., we assume that the disturbance frequencies are known).

For disturbance rejection, we use the controller structure (4.202) with

$$\begin{aligned} k_3(t) &= k_{30}(t) + \sum_{j=1}^{q_u}(k_{3\alpha j}(t)\sin\omega_{uj}t + k_{3\beta j}(t)\cos\omega_{uj}t) \\ &\quad + \sum_{j=1}^{q_y}(k_{3\alpha(q_u+j)}(t)\sin\omega_{yj}t + k_{3\beta(q_u+j)}(t)\cos\omega_{yj}t), \end{aligned} \tag{4.214}$$

where $k_{30}(t)$, $k_{3\alpha j}(t)$, and $k_{3\beta j}(t)$, $j = 1, 2, \ldots, q_u + q_y$, are adaptive estimates of some ideal constant parameters k_{30}^*, $k_{3\alpha j}^*$, and $k_{3\beta j}^*$. To derive these ideal values, we choose the ideal version of $k_3(t)$ as

$$\begin{aligned} k_3^*(t) = k_{30}^* &+ \sum_{j=1}^{q_u}(k_{3\alpha j}^* \sin\omega_{uj}t + k_{3\beta j}^* \cos\omega_{uj}t) \\ &+ \sum_{j=1}^{q_y}(k_{3\alpha(q_u+j)}^* \sin\omega_{yj}t + k_{3\beta(q_u+j)}^* \cos\omega_{yj}t) \end{aligned} \tag{4.215}$$

and apply the ideal version of (4.202), that is, $u(t) = k_1^{*T}x(t)+k_2^*r(t)+k_3^*(t)$, to the plant (4.201) with disturbances described in (4.212) and (4.213) to obtain the closed-loop system in the frequency z-domain as

$$y(z) = y_m(z) + \Delta(z), \tag{4.216}$$

where $y_m(z) = C(zI - A - Bk_1^{*T})^{-1}Bk_2^*r(z)$ and

$$\begin{aligned} \Delta(z) = &\, C(zI - A - Bk_1^{*T})^{-1}Bk_3^*(z) \\ &+ C(zI - A - Bk_1^{*T})^{-1}B_d d_u(z) + d_y(z). \end{aligned} \tag{4.217}$$

Similar to the continuous-time case, from Assumption (A4.4-D) and (4.183), all zeros and poles of $C(zI - A - Bk_1^{*T})^{-1}B$ are inside the unit circle of the complex z-plane: $|z| < 1$. Given the sinusoidal forms of $d_u(t)$, $d_y(t)$, and $k_3^*(t)$ in (4.212), (4.213), and (4.215), it can be verified by a partial fraction expansion method that there exist constant parameters k_{30}^*, $k_{3\alpha j}^*$, and $k_{3\beta j}^*$ such that $(z-1)\Delta(z)$ does not have any pole in $|z| \geq 1$ and that

$$\lim_{z\to 1}(z-1)\,\Delta(z) = 0, \tag{4.218}$$

so that from the final-value theorem of the z-transform, the time-domain function $\delta(t)$ of $\Delta(z)$ converges to zero exponentially fast as time t goes to infinity, that is, $\lim_{t\to\infty}(y(t) - y_m(t)) = \lim_{t\to\infty}\delta(t) = 0$ exponentially fast.

For the adaptive controller (4.202) with (4.214), the closed-loop system is

$$\begin{aligned} x(t+1) &= (A + Bk_1^{*T})x(t) + Bk_2^*r(t) + Bk_3^*(t) + B_d d_u(t) \\ &\quad + B\left((k_1^T(t) - k_1^{*T})x(t) + (k_2(t) - k_2^*)r(t) + (k_3(t) - k_3^*(t))\right), \\ y(t) &= Cx(t) + d_y(t). \end{aligned} \tag{4.219}$$

Using (4.186), (4.217), and (4.219), we obtain the tracking error equation

$$\begin{aligned} e(t) &= y(t) - y_m(t) \\ &= \rho^*W_m(z)[(k_1^T - k_1^{*T})x + (k_2 - k_2^*)r + (k_3 - k_3^*)](t) \\ &\quad + C(A + Bk_1^{*T})^t x(0) + \delta(t) \\ &= \rho^*W_m(z)[(\theta - \theta^*)^T\omega](t) + C(A + Bk_1^{*T})^t x(0) + \delta(t), \end{aligned} \tag{4.220}$$

where $C(A+Bk_1^{*T})^t x(0)$ and $\delta(t)$ converge to zero exponentially and

$$\begin{aligned}\theta(t) &= [k_1^T(t), k_2(t), k_{30}(t), k_{3\alpha 1}(t), \ldots, k_{3\alpha q_u}(t), k_{3\beta 1}(t), \ldots, k_{3\beta q_u}(t),\\ &\quad k_{3\alpha(q_u+1)}(t), \ldots, k_{3\alpha(q_u+q_y)}(t), k_{3\beta(q_u+1)}(t), \ldots, k_{3\beta(q_u+q_y)}(t)]^T,\\ \theta^* &= [k_1^{*T}, k_2^*, k_{30}^*, k_{3\alpha 1}^*, \ldots, k_{3\alpha q_u}^*, k_{3\beta 1}^*, \ldots, k_{3\beta q_u}^*,\\ &\quad k_{3\alpha(q_u+1)}^*, \ldots, k_{3\alpha(q_u+q_y)}^*, k_{3\beta(q_u+1)}^*, \ldots, k_{3\beta(q_u+q_y)}^*]^T, \end{aligned} \tag{4.221}$$

$$\begin{aligned}\omega(t) &= [x^T(t), r(t), 1, \sin\omega_{u1}, \ldots, \sin\omega_{uq_u}, \cos\omega_{u1}, \ldots, \cos\omega_{uq_u},\\ &\quad \sin\omega_{y1}, \ldots, \sin\omega_{yq_y}, \cos\omega_{y1}, \ldots, \cos\omega_{yq_y}]^T. \end{aligned} \tag{4.222}$$

Rejection of Nonsinusoidal Disturbances

We now consider the disturbances $d_u(t)$ and $d_y(t)$ in (4.201) of the form

$$d(t) = d_0 + \sum_{j=1}^{q} d_j f_j(t) \tag{4.223}$$

for some nonsinusoidal functions $f_j(t)$, $j = 1, 2, \ldots, q$. For an ideal version of (4.201), $u(t) = k_1^{*T}x(t) + k_2^* r(t) + k_3^*(t)$, the closed-loop system is also given by (4.216) and (4.217) with $C(zI - A - Bk_1^{*T})^{-1}B = \frac{1}{k_2^* P_m(z)}$ (see (4.186)).

To cancel a nonsinusoidal $d_y(t)$ by $k_3^*(t)$, one chooses

$$k_3^*(z) = -k_2^* P_m(z) d_y(z), \tag{4.224}$$

whose time-domain function may be parametrizable if the derivatives (up to n^*th order) of $d_y(t)$ are parametrizable (for which sinusoidal components (4.213) are a special case where such derivatives are indeed parametrizable). An adaptive scheme is to estimate the unknown parameters in (4.224).

However, to cancel a nonsinusoidal $d_u(t)$ by $k_3^*(t)$, an analogous choice,

$$k_3^*(z) = -k_2^* P_m(z) C(zI - A - Bk_1^{*T})^{-1} B_d d_u(z), \tag{4.225}$$

is not parametrizable using $f_j(t)$ when (C, A, B) has finite zeros (i.e., when $n^* < n$, as the zeros of (C, A, B) are in the denominator of (4.225)).

What is of interest here is to derive a parametrizable compensation scheme and the associated condition for an adaptive control design to reject the effect of the disturbance $d_u(t)$ in the plant (4.201) on the plant output $y(t)$.

Let us consider the linear time-invariant plant

$$\begin{aligned} x(t+1) &= Ax(t) + Bu(t) + B_d d_u(t),\\ y(t) &= Cx(t), \end{aligned} \tag{4.226}$$

where the disturbance $d_u(t) = d(t)$ given in (4.223) for some *unknown* constants d_0, d_j and *known* bounded continuous functions $f_j(t)$, $j = 1, 2, \ldots, q$, and some $q > 0$. Similar to the continuous-time case, since $f_j(t)$, $j = 1, 2, \ldots, q$, can be arbitrary, the existence of $k_3^*(t)$ of the same components $f_j(t)$ to meet the final-value theorem condition (4.218) (with $d_y(z) = 0$) requires that the transfer function matching

$$\frac{1}{k_2^*} W_m(z) k_3^* + C(zI - A - Bk_1^{*T})^{-1} B_d = 0, \tag{4.227}$$

is satisfied for some constant $k_3^* \in R$. Under this condition, the choice of $k_3^*(t) = k_3^* d_u(t)$ (i.e., $k_3^*(z) = k_3^* d_u(z)$), which leads to $\Delta(z) = 0$, is able to ensure the desired plant–model output matching $y(z) = y_m(z)$.

If $B_d = \alpha_d B$ for some constant $\alpha_d \in R$ (that is, the case of *matched* input disturbances), then k_3^* exists to satisfy (4.227), which is crucial for adaptive rejection of the disturbances $d_u(t)$ when the parameters d_0, d_j, $j = 1, 2 \ldots, q$, are unknown. When B and B_d are not parallel, the following discrete-time version of Proposition 4.1 specifies the class of (C, A, B, B_d) which satisfy the matching equations (4.186) and (4.227) simultaneously.

Proposition 4.2 *Assume that (A, B) is controllable. There exist constants $k_1^* \in R^n$, $k_2^*, k_3^* \in R$ such that*

$$C(zI - A - Bk_1^{*T})^{-1} B k_2^* = W_m(z) = \frac{1}{P_m(z)}, \tag{4.228}$$

$$\frac{1}{k_2^*} W_m(z) k_3^* + C(zI - A - Bk_1^{*T})^{-1} B_d = 0, \tag{4.229}$$

where $P_m(z)$ is a monic polynomial of degree n^* if and only if the two systems (C, A, B) and (C, A, B_d) have the same relative degree n^*.

The zeros of (C, A, B) are required to be stable (see Assumption (A4.4-D)) for $A + Bk_1^{*T}$ to have all its eigenvalues in $|z| < 1$, so that the pole-zero cancellation in (4.228) is stable, which is crucial for closed-loop stability. The proof of this proposition is the same as that for Proposition 4.1, and Example 4.1 is also meaningful in discrete time, with s replaced by z.

Now, assuming that (C, A, B) and (C, A, B_d) in (4.225) have relative degree n^*, we use the controller structure (4.202) to reject the effect of $d_u(t)$ on $y(t)$, with the compensation term being

$$k_3(t) = k_{30}(t) + \sum_{j=1}^{q} k_{3j}(t) f_j(t) \tag{4.230}$$

with $k_{3j}(t)$ being the estimate of $k_{3j}^* = k_3^* d_j$, $j = 0, 1, 2 \ldots, q$.

From (4.179), (4.202), (4.226), and (4.228)–(4.230), we obtain

$$e(t) = \rho^* W_m(z)[(\theta - \theta^*)^T \omega](t) + C(A + Bk_1^{*T})^t x(0), \tag{4.231}$$

where $C(A + Bk_1^{*T})^t x(0)$ converges to zero exponentially and

$$\theta(t) = [k_1^T(t), k_2(t), k_{30}(t), k_{31}(t), \ldots, k_{3q}(t)]^T,$$
$$\theta^* = [k_1^{*T}, k_2^*, k_{30}^*, k_{31}^*, \ldots, k_{3q}^*]^T, \tag{4.232}$$
$$\omega(t) = [x^T(t), r(t), 1, f_1(t), \ldots, f_q(t)]^T. \tag{4.233}$$

Adaptive Scheme

For the above three types of disturbances: constant, sinusoidal, and nonsinusoidal, we have developed a unified error equation (4.209), (4.220), or (4.231), in terms of the tracking error $e(t) = y(t) - y_m(t)$ and parameter error $\theta(t) - \theta^*$, where $\theta(t)$ is the controller parameter vector. To develop an adaptive law for updating $\theta(t)$, we introduce the auxiliary signals

$$\zeta(t) = W_m(z)[\omega](t), \tag{4.234}$$
$$\xi(t) = \theta^T(t)\zeta(t) - W_m(z)[\theta^T \omega](t) \tag{4.235}$$

and the estimation error

$$\epsilon(t) = e(t) + \rho(t)\xi(t), \tag{4.236}$$

where $\rho(t)$ is an estimate of ρ^*. Then, neglecting the exponentially decaying terms $C(A + Bk_1^{*T})^t x(0)$ and $\delta(t)$, we have

$$\epsilon(t) = \rho^*(\theta(t) - \theta^*)^T \zeta(t) + (\rho(t) - \rho^*)\xi(t), \tag{4.237}$$

which suggests the following adaptive laws for $\theta(t)$ and $\rho(t)$:

$$\theta(t+1) = \theta(t) - \frac{\text{sign}[\rho^*]\Gamma\zeta(t)\epsilon(t)}{m^2(t)}, \ 0 < \Gamma = \Gamma^T < \frac{2}{k_p^0} I_{n+1}, \tag{4.238}$$

$$\rho(t+1) = \rho(t) - \frac{\gamma\xi(t)\epsilon(t)}{m^2(t)}, \ 0 < \gamma < 2, \tag{4.239}$$

$t \in \{0, 1, 2, \ldots\}$, where $m(t) = \sqrt{1 + \zeta^T(t)\zeta(t) + \xi^2(t)}$.

This adaptive parameter update law has the following properties.

Lemma 4.4 *The adaptive laws (4.238) and (4.239) ensure that* $\theta(t) \in L^\infty$, $\rho(t) \in L^\infty$, $\frac{\epsilon(t)}{m(t)} \in L^2 \cap L^\infty$, $\theta(t+1) - \theta(t) \in L^2$, *and* $\rho(t+1) - \rho(t) \in L^2$.

Based on this result, the following desired system properties can also be established: closed-loop signal boundedness, asymptotic tracking $\lim_{t\to\infty}(y(t) - y_m(t)) = 0$, and $y(t) - y_m(t) \in L^2$, the same as those stated in Theorem 4.3.

4.5.4 Parametrizations of State Feedback

Our results in this section so far indicate that when the state vector $x(t)$ of the plant (4.175) is available for measurement, a state feedback controller, adaptive with $u(t) = k_1^T(t)x(t) + k_2(t)r(t)$ (see (4.179)) or ideal with $u(t) = k_1^{*T}x(t) + k_2^* r(t)$ (see (4.183)), can be developed to ensure closed-loop signal boundedness and plant–model output tracking. In this case, with the knowledge of $x(t)$, the adaptive term $k_1^T(t)x(t)$ is an estimate of the ideal term $k_1^{*T}x(t)$, simply by replacing k_1^* with $k_1(t)$ updated from an adaptive law.

When the plant state vector $x(t)$ is not available for measurement, an output feedback controller is to be used, using an estimate of $k_1^{*T}x(t)$ which needs to be re-parametrized. Such a re-parametrization can be developed based on either a full-order or a reduced-order state observer.

Parametrization with Full-Order Observer

Consider the discrete-time linear time-invariant plant

$$x(t+1) = Ax + Bu,\ x \in R^n,\ y = Cx \in R,\ t \in \{0, 1, 2, \ldots\}, \tag{4.240}$$

for some constant matrices $A \in R^{n\times n}$, $B \in R^n$, and $C \in R^{1\times n}$, with $n > 0$, where $x(t) = [x_1(t), \ldots, x_n(t)]^T$ is the plant state vector, $u(t) \in R$ is the applied control input, and $y(t) \in R$ is the measured plant output.

Assuming that the plant (4.240) is observable, we generate an estimate $\hat{x}(t)$ for the unknown state vector $x(t)$ from the full-order state observer:

$$\hat{x}(t+1) = A\hat{x}(t) + Bu(t) + L(y(t) - C\hat{x}(t)),\ \hat{x}(0) = \hat{x}_0, \tag{4.241}$$

where $L = [L_1, L_2, \ldots, L_n]^T \in R^n$ is a constant-gain vector and $\hat{x}_0 \in R^n$ is an estimate of $x(0) = x_0$. For the estimation error $\tilde{x}(t) = x(t) - \hat{x}(t)$, from (4.240) and (4.241), we obtain the state estimation error equation

$$\tilde{x}(t+1) = (A - LC)\tilde{x}(t). \tag{4.242}$$

To design a desired state observer, we choose a vector gain L to make the eigenvalues of the $n \times n$ matrix $A - LC$ equal to some given stable ones.

Then, it follows that $\lim_{t\to\infty} \tilde{x}(t) = 0$ at an exponential rate determined by the eigenvalues of $A - LC$, which can be arbitrarily assigned by L.

Rewriting the observer (4.241) as

$$\hat{x}(t+1) = (A - LC)\hat{x}(t) + Bu(t) + Ly(t), \tag{4.243}$$

we express $\hat{x}(t)$ as

$$\begin{aligned}\hat{x}(t) &= (zI - A + LC)^{-1}B[u](t) + (zI - A + LC)^{-1}L[y](t) + (A - LC)^t\hat{x}(0)\\ &= \frac{G_u(z)}{\Lambda(z)}[u](t) + \frac{G_y(z)}{\Lambda(z)}[y](t) + (A - LC)^t\hat{x}(0),\end{aligned} \tag{4.244}$$

where $\Lambda(z) = \det(zI - A + LC)$, $G_u(z)$ and $G_y(z)$ are some n-dimensional polynomial vectors, and $\lim_{t\to\infty}(A - LC)^t\hat{x}(0) = 0$.

Instead of using $k_1^{*T}x(t)$ for feedback control, we now use $k_1^{*T}\hat{x}(t)$, which, having the desired property that $\lim_{t\to\infty}(k_1^{*T}\hat{x}(t) - k_1^{*T}x(t)) = 0$ exponentially, in view of (4.244), can be parametrized in terms of $u(t)$ and $y(t)$ as

$$k_1^{*T}\hat{x}(t) = \theta_1^{*T}\omega_1(t) + \theta_2^{*T}\omega_2(t), \tag{4.245}$$

where

$$\omega_1(t) = \frac{a(z)}{\Lambda(z)}[u](t),\ \omega_2(t) = \frac{a(z)}{\Lambda(z)}[y](t), \tag{4.246}$$

$$a(z) = [1, z, \ldots, z^{n-1}]^T, \tag{4.247}$$

$\Lambda(z)$ is a chosen monic stable polynomial of degree n (whose zeros are the desired eigenvalues of $A-LC$ inside $|z| < 1$, that is, $\Lambda(z) = \det(zI-A+LC)$), and $\theta_1^* \in R^n$, $\theta_2^* \in R^n$ are some constant parameters which could be calculated if the plant parameters were known. For this parametrization, the effect of the exponentially decaying term $(A - LC)^t\hat{x}(0)$ in (4.244) can be neglected.

For *adaptive control*, the term $k_1^T\hat{x}(t)$ is parametrized as

$$k_1^T\hat{x}(t) = \theta_1^T\omega_1(t) + \theta_2^T\omega_2(t), \tag{4.248}$$

where $\theta_1(t) \in R^n$, $\theta_2(t) \in R^n$ are estimates of the unknown constant parameters $\theta_1^* \in R^n$, $\theta_2^* \in R^n$, to be updated from an adaptive law.

For discrete-time designs, it is common to choose $\Lambda(z) = z^n$, that is, all n eigenvalues of $A - LC$ are placed at the origin of the z-plane.

Parametrization with Reduced-Order Observer

To derive a reduced-order observer, we express the plant (4.240) as

$$\begin{bmatrix} x_1(t+1) \\ x_2(t+1) \end{bmatrix} = \begin{bmatrix} A_{11} & A_{12} \\ A_{21} & A_{22} \end{bmatrix} \begin{bmatrix} x_1(t) \\ x_2(t) \end{bmatrix} + \begin{bmatrix} B_1 \\ B_2 \end{bmatrix} u(t),$$

$$y(t) = x_1(t) = \begin{bmatrix} 1 & 0 & 0 & \cdots & 0 \end{bmatrix} \begin{bmatrix} x_1(t) \\ x_2(t) \end{bmatrix} \in R, \tag{4.249}$$

where $x_2(t) \in R^{n-1}$ is to be estimated (for $y(t) = Cx(t) \neq x_1(t)$, the state transformation $\bar{x}(t) = Px(t)$, where $P = \begin{bmatrix} C \\ \hat{C} \end{bmatrix} \in R^{n\times n}$ is nonsingular for some $\hat{C}$, results in $y(t) = [1, 0, 0, \ldots, 0]\bar{x}(t)$, as in the continuous-time case).

Assuming that $(A,\ C)$ is observable, which implies that $(A_{22},\ A_{12})$ is observable [13], we generate an estimate $\hat{x}(t)$ for the unknown state vector $x(t)$ from the reduced-order observer with a dynamic system for $\hat{x}_2(t)$:

$$\hat{x}(t) = \begin{bmatrix} x_1(t) \\ \hat{x}_2(t) \end{bmatrix} = \begin{bmatrix} y(t) \\ w(t) + L_r y(t) \end{bmatrix}, \tag{4.250}$$

where $L_r = [L_1, L_2, \ldots, L_{n-1}]^T \in R^{n-1}$ is a constant-gain vector such that the eigenvalues of the $(n-1)\times(n-1)$ matrix $A_{22} - L_r A_{12}$ equal some given stable ones and $w(t) \in R^{n-1}$ is generated from the system

$$\begin{aligned} w(t+1) = {} & (A_{22} - L_r A_{12})w(t) + (B_2 - L_r B_1)u(t) \\ & + ((A_{22} - L_r A_{12})L_r + A_{21} - L_r A_{11})y(t),\ w(0) = w_0, \end{aligned} \tag{4.251}$$

$t \in \{0, 1, 2, \ldots\}$, where $w_0 \in R^{n-1}$ is an estimate of $L_r y(0) - x_2(0)$. From (4.249) and (4.250), the dynamic equation for $\hat{x}_2(t)$ can be expressed as

$$\begin{aligned} \hat{x}_2(t+1) = {} & A_{22}\hat{x}_2(t) + \begin{bmatrix} A_{21} & B_2 \end{bmatrix} \begin{bmatrix} x_1(t) \\ u(t) \end{bmatrix} \\ & + L_r(x_1(t+1) - A_{11}x_1(t) - B_1 u(t) - A_{12}\hat{x}_2(t)). \end{aligned} \tag{4.252}$$

For the estimation error $\tilde{x}_2(t) = x_2(t) - \hat{x}_2(t)$, from (4.249) and (4.252), we obtain the estimation error equation

$$\tilde{x}_2(t+1) = (A_{22} - L_r A_{12})\tilde{x}_2(t), \tag{4.253}$$

that is, $\lim_{t\to\infty} \tilde{x}_2(t) = 0$ exponentially, as determined by $A_{22} - L_r A_{12}$.

To proceed, we express $w(t)$ in (4.251) as

$$\begin{aligned} w(t) = {} & (A_{22} - L_r A_{12})^t w(0) + (zI - A_{22} + L_r A_{12})^{-1}(B_2 - L_r B_1)[u](t) \\ & + (zI - A_{22} + L_r A_{12})^{-1}((A_{22} - L_r A_{12})L_r + A_{21} - L_r A_{11})[y](t) \\ = {} & \frac{G_u(z)}{\Lambda(z)}[u](t) + \frac{G_y(z)}{\Lambda(z)}[y](t) + (A_{22} - L_r A_{12})^t w(0), \end{aligned} \tag{4.254}$$

where $\Lambda(z) = \det(zI - A_{22} + L_r A_{12})$, $G_u(z)$ and $G_y(z)$ are some $(n-1)$-dimensional polynomial vectors, and $\lim_{t\to\infty}(A_{22} - L_r A_{12})^t w(0) = 0$.

Instead of using $k_1^{*T} x(t)$ for feedback control, we now use $k_1^{*T}\hat{x}(t)$, where $\hat{x}_2(t) = w(t) + L_r y(t)$ and $x_1(t) = y(t)$ (see (4.250)), both available for measurement. In view of (4.254), $k_1^{*T}\hat{x}(t)$ can be parametrized as

$$k_1^{*T}\hat{x}(t) = \theta_1^{*T}\omega_1(t) + \theta_2^{*T}\omega_2(t) + \theta_{20} y(t), \tag{4.255}$$

where

$$\omega_1(t) = \frac{a(z)}{\Lambda(z)}[u](t),\ \omega_2(t) = \frac{a(z)}{\Lambda(z)}[y](t), \tag{4.256}$$

$$a(z) = [1, z, \ldots, z^{n-2}]^T, \tag{4.257}$$

$\Lambda(z)$ is a chosen monic stable polynomial of degree $n-1$ (whose zeros are the desired eigenvalues of $A_{22} - L_r A_{12}$), and $\theta_1^* \in R^{n-1}$, $\theta_2^* \in R^{n-1}$, $\theta_{20}^* \in R$ are some constant parameters which depend on the plant and observer parameters. For this parametrization, the effect of the exponentially decaying term $(A_{22} - L_r A_{12})^t w(0)$ in (4.254) can be neglected.

For *adaptive control*, the term $k_1^T\hat{x}(t)$ is parametrized as

$$k_1^T\hat{x}(t) = \theta_1^T\omega_1(t) + \theta_2^T\omega_2(t) + \theta_{20}(t) y(t), \tag{4.258}$$

where $\theta_1(t) \in R^{n-1}$, $\theta_2(t) \in R^{n-1}$, $\theta_{20}(t) \in R$ are estimates of the unknown parameters $\theta_1^* \in R^{n-1}$, $\theta_2^* \in R^{n-1}$, $\theta_{20}^* \in R$, to be adaptively updated.

For discrete-time designs, it is common to choose $\Lambda(z) = z^{n-1}$, that is, all $n-1$ eigenvalues of $A_{22} - L_r A_{12}$ are placed at the origin of the z-plane.

Based on the parametrization (4.248) or (4.258), which depends on the plant input $u(t)$ and output $y(t)$ only, adaptive laws can be developed for updating the parameters $\theta_1(t)$, $\theta_2(t)$, and $\theta_{20}(t)$ to implement output feedback adaptive controllers for achieving output tracking (see Chapter 6; also see Remark 4.3 for the case when (A, C) is only detectable).

To conclude this chapter, we note that the key to adaptive control is a stable adaptive law for updating the parameters of a controller applied to a plant with unknown parameters. Such an adaptive law should have the desired properties such as bounded parameter estimates and small estimation errors. This idea is applicable not only to state feedback designs (see this chapter) but also to output feedback model reference designs (see Chapters 5 and 6), pole placement using output feedback (see Chapter 7), multi-input and multi-output cases (see Chapter 9), and nonlinear cases (see Chapter 10).

Problems

4.1 Consider the plant

$$(s^2 + p_1 s + p_0)[y](t) = k_p u(t)$$

with $\dot{y}(0) = y(0) = 0$ and $k_p \neq 0$, where $y(t)$, $\dot{y}(t)$ are available for measurement. Let $x(t) = (y(t), \dot{y}(t))^T$ be the plant state variable.

(i) Find (A, b) such that $\dot{x}(t) = Ax(t) + bu(t)$.

(ii) Choose a reference model $\dot{x}_m(t) = A_m x_m(t) + b_m r(t)$ such that the equilibrium state $x_m = 0$ of $\dot{x}_m(t) = A_m x_m(t)$ is exponentially stable and A_m and b_m are independent of the system parameters k_p, p_1, and p_0 but the matching equations $A_m = A + bk_1^{*T}$, $b_m = bk_2^*$ are satisfied by some $k_1^* \in R^2$ and $k_2^* \in R$ which are dependent on k_p, p_1, and p_0.

4.2 Consider the plant in Problem 4.1.

(i) For k_p, p_1, p_0 unknown but sign$[k_p]$ known, formulate the adaptive control problem, and design an adaptive control law $u(t) = k_1^T(t)x(t) + k_2(t)r(t)$, that is, specify $\dot{k}_1(t)$ and $\dot{k}_2(t)$ in terms of measured signals.

(ii) For your design, analyze the closed-loop system by deriving that $k_1(t)$, $k_2(t)$, $x(t)$ are bounded and $\lim_{t\to\infty} e(t) = 0$.

(iii) Simulate the adaptive control system with $p_1 = -1$, $p_0 = 2$, $k_p = 2$ for $r(t) = 10$, $r(t) = 10\sin(t)$, and $r(t) = 10\sin(t) + 13\sin(4.3t)$, respectively, with $k_1(0) = (1, 1)^T$, $k_2(0) = 1$, and plot the control $u(t)$, the tracking error $e(t)$, and the parameter errors $k_1(t) - k_1^*$, $k_2(t) - k_2^*$.

(iii) Make a few more simulations with your own choice of initial conditions, and comment on the system responses.

4.3 Consider the plant

$$\dot{x}(t) = Ax(t) + Bu(t),$$
$$y(t) = Cx(t),$$

with

$$A = \begin{bmatrix} 0 & -1 & 3 \\ 1 & 0 & -1 \\ 0 & 1 & -1 \end{bmatrix}, \quad B = \begin{bmatrix} 1 \\ 0 \\ 0 \end{bmatrix}, \quad C = \begin{bmatrix} 1 & 0 & 0 \end{bmatrix},$$

whose transfer function is $\frac{s^2+s+1}{s^3+s^2+2s-2}$. Assume that the plant parameters in (A, B, C) are all unknown.

(i) Verify the conditions for an adaptive controller in Section 4.2.

(ii) Design an adaptive controller for the given plant.

(iii) Simulate the adaptive control system with $r(t) = 1$, $\sin(t)$, and $\sin(t) + 2\sin(2t)$, respectively, for different choices of initial conditions and plot the control signal, the tracking error, and parameter errors.

(iv) Discuss your simulation results.

4.4 Verify (4.70) and (4.77).

4.5 Verify (4.83) and (4.87).

4.6 Verify (4.96) for $q_u = q_y = 1$, $n = 2$, and $n^* = 1$.

4.7 Verify Proposition 4.1 for a few $n = 4$ examples.

4.8 Find k_1^*, θ_1^*, and θ_2^* in (4.137) for an $n = 3$ case.

4.9 Find k_1^*, θ_1^*, θ_2^*, and θ_{20}^* in (4.147) for an $n = 3$ case.

4.10 Conduct a simulation study of the design example in Section 4.5.1.

4.11 Derive (4.199).

4.12 Present the simplfied version of the adaptive control scheme in Section 4.5.2 with $P_m(z) = z^{n^*}$ and work out a simulation example with $n = 3$.

4.13 Conduct a case study for adaptive disturbance rejection of different cases addressed in Section 4.5.3.

4.14 Present the parametrization schemes of Section 4.5.4 with $\Lambda(z) = z^n$ and $\Lambda(z) = z^{n-1}$, respectively.

Advanced Topics

Actuator failure compensation. Consider the linear plant

$$\dot{x}(t) = Ax(t) + Bu(t),\ y(t) = Cx(t),$$

where $A \in R^{n\times n}$, $B \in R^{n\times m}$, $C \in R^{1\times n}$ are unknown constant parameter matrices, the state vector $x(t) \in R^n$ is available for measurement, $u(t) = [u_1, \ldots, u_m]^T \in R^m$ is the input vector whose components may fail during system operation, and $y(t) \in R$ is the plant output.

The type of actuator failure is

$$u_j(t) = \bar{u}_j,\ t \geq t_j,\ j \in \{1, 2, \ldots, m\},$$

where the constant value $\bar{u}_j$ and the failure time instant t_j are unknown.

The basic assumption for the actuator failure compensation problem is:

(A0): The plant matrices (A, B, C) are so designed that for any up to $m-1$ actuator failures the remaining actuators can still achieve a desired control objective when implemented with known plant and failure parameters.

The task of adaptive control is to adjust the remaining controls to achieve the desired system performance when there are up to $m-1$ actuator failures whose parameters are unknown. The key issues are to specify plant conditions on (A, B, C) to satisfy the basic assumption (A0), to choose controller structures to achieve plant–model matching with known plant parameters and failure parameters, and to develop adaptive schemes to update controller parameters to ensure closed-loop stability and asymptotic plant–model tracking in the presence of uncertainties in actuator failures as well as plant parameters.

There are three adaptive state feedback problems in the presence of actuator failures: (i) state tracking in continuous time, (ii) output tracking in continuous time, and (iii) output tracking in discrete time.

The challenge of these control problems is that the failure values $\bar{u}_j$, failure time instants t_j, and failure actuators j are all uncertain, in addition to the uncertainty of plant parameters. An adaptive controller to be developed should work for all cases of up to $m-1$ uncertain actuator failures.

How about compensation of up to $m-q$ actuator failures with $1 < q < m$? How about time-varying failures with $\bar{u}_j = \bar{u}_j(t)$?

Adaptive tracking control of nonminimum phase systems. One of the design conditions for the output tracking adaptive control schemes developed in this chapter is that all zeros of the controlled plant are stable. With this condition, a stable nominal controller $u(t) = k_1^{*T}x(t) + k_2^*r(t)$ exists and leads the closed-loop system $\dot{x}(t) = Ax(t) + Bu(t)$, $y(t) = Cx(t)$, to $y(t) = y_m(t) = W_m(s)[r](t)$ for a chosen model transfer function $W_m(s)$ (see Section 4.2), and a stable adaptive controller can be developed to ensure this desired output tracking property in the asymptotic sense.

Using a pole placement design, with (A, B) being controllable, the same nominal controller structure is capable of assigning the closed-loop poles as zeros of an arbitrary stable polynomial $P_d(s)$, that is, $y(t) = y_d(t) = \frac{Z(s)}{P_d(s)}[r](t)$, where $Z(s)$ is the zero polynomial of (A, B, C). Since no pole-zero cancellations occur in this feedback control, zeros of $Z(s)$ do not need to be stable.

The problem is to design a stable adaptive controller which ensures that $\lim_{t\to\infty}(y(t) - y_d(t)) = 0$ (even if the unstable zeros of $Z(s)$ are known).

Chapter 5

Continuous-Time Model Reference Adaptive Control

Model reference adaptive control is a major adaptive control design method which has a rigorous and systematic theoretical foundation, an attractive and promising application perspective, and an easy and concise design procedure. The objective of model reference control is to ensure the output of a controlled system (a plant) to track the output of a given reference model system, in addition to closed-loop stability. While model reference control can be designed using either state feedback (see Chapter 4) or output feedback (see this chapter), output feedback model reference control is more challenging in theory and has more potential for applications. When the plant parameters are unknown, adaptive laws are needed to update the parameters of a model reference controller. The main issues in model reference adaptive control include controller parametrization, error model derivation, *a priori* plant knowledge specification, adaptive law design, and stability, tracking, and robustness analysis, which are addressed in this chapter for continuous-time systems.

5.1 Control System Structure

Consider a continuous-time linear time-invariant plant with transfer function $G_0(s) = k_p \frac{Z(s)}{P(s)}$, described by the differential equation

$$P(s)[y](t) = k_p Z(s)[u](t), \tag{5.1}$$

where $y(t) \in R$ is the measured output, $u(t) \in R$ is the applied input,

$$P(s) = s^n + p_{n-1}s^{n-1} + \cdots + p_1 s + p_0, \tag{5.2}$$

$$Z(s) = s^m + z_{m-1}s^{m-1} + \cdots + z_1 s + z_0 \tag{5.3}$$

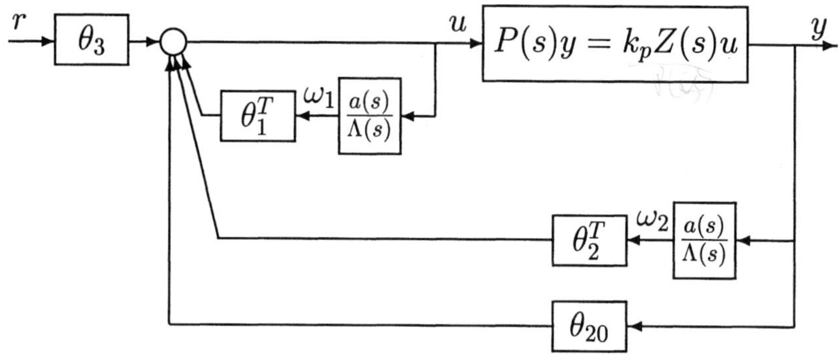

Figure 5.1: Model reference control system.

are polynomials in s, which denotes the time differentiation operator $s[x](t) = \dot{x}(t)$ or the Laplace transform variable, and p_i, $i = 0, 1, \ldots, n-1$, z_j, $j = 0, 1, \ldots, m-1$, $n > m$, and $k_p \neq 0$ are unknown constant parameters.

The control objective is: given a reference model characterized by

$$P_m(s)[y_m](t) = r(t), \tag{5.4}$$

where $P_m(s)$ is a monic stable polynomial and $r(t)$ is a bounded and piecewise continuous reference input signal,[1] find an output feedback control $u(t)$ such that all closed-loop system signals are bounded and the plant output $y(t)$ tracks the reference model output $y_m(t)$ asymptotically with time.

To meet the control objective, we make the following assumptions:

(A5.1): $Z(s)$ is a stable polynomial.[2]

(A5.2): An upper bound $\bar{n} \geq n$ for the degree n of $P(s)$ is known.

(A5.3): The sign of the plant high frequency gain k_p is known.

(A5.4): The degree of $P_m(s)$ is $n^* = n - m > 0$.

The model reference controller structure, as shown in Figure 5.1, is

$$u(t) = \theta_1^T \omega_1(t) + \theta_2^T \omega_2(t) + \theta_{20} y(t) + \theta_3 r(t), \tag{5.5}$$

[1]The general case when $\bar{P}_m(s)[y_m](t) = k_m \bar{Z}_m(s)[\bar{r}](t)$ with $\bar{P}_m(s)$ being a monic stable polynomial can be rewritten in the form (5.4) for some $P_m(s)$ and $r(t)$, provided that $P_m(s)\bar{Z}_m(s)$ and $\bar{P}_m(s)$ have the same degree: $P_m(s)[y_m](t) = \frac{k_m P_m(s) \bar{Z}_m(s)}{\bar{P}_m(s)}[\bar{r}](t) \triangleq r(t)$.

[2]This assumption implies that the state variable form of the plant (5.1),

$$\dot{x}(t) = Ax(t) + Bu(t),\ y(t) = Cx(t),$$

with $x(t) \in R^n$, $u(t) \in R$, $y(t) \in R$, is both stabilizable and detectable.

where

$$\omega_1(t) = \frac{a(s)}{\Lambda(s)}[u](t), \ \omega_2(t) = \frac{a(s)}{\Lambda(s)}[y](t), \tag{5.6}$$

$$a(s) = [1, s, \ldots, s^{\bar{n}-2}]^T, \tag{5.7}$$

$\theta_1 \in R^{\bar{n}-1}$, $\theta_2 \in R^{\bar{n}-1}$, $\theta_{20} \in R$, and $\theta_3 \in R$ are parameters, and $\Lambda(s)$ is a monic stable polynomial of degree $\bar{n} - 1$.

The controller (5.5) uses the measured plant output $y(t)$ for feedback control, based on the parametrization scheme (4.150) of a state feedback control law used for adaptive control using system internal state variables.

With $\Lambda(s) = s^{\bar{n}-1} + \lambda_{\bar{n}-2}s^{\bar{n}-2} + \cdots + \lambda_1 s + \lambda_0$ being stable, the signals $\omega_1(t)$ and $\omega_2(t)$ can be obtained as the states of the auxiliary systems

$$\dot{\omega}_1(t) = A_\lambda \omega_1(t) + bu(t), \tag{5.8}$$

$$\dot{\omega}_2(t) = A_\lambda \omega_2(t) + by(t), \tag{5.9}$$

where $\omega_1(t) \in R^{\bar{n}-1}$, $\omega_2(t) \in R^{\bar{n}-1}$, and

$$A_\lambda = \begin{bmatrix} 0 & 1 & 0 & \cdots & 0 & 0 \\ 0 & 0 & 1 & 0 & \cdots & 0 \\ & \cdots & \cdots & \cdots & \cdots & \\ 0 & 0 & \cdots & \cdots & 0 & 1 \\ -\lambda_0 & -\lambda_1 & \cdots & \cdots & -\lambda_{\bar{n}-3} & -\lambda_{\bar{n}-2} \end{bmatrix}, \ b = \begin{bmatrix} 0 \\ \vdots \\ 0 \\ 1 \end{bmatrix} \in R^{\bar{n}-1}. \tag{5.10}$$

As seen next, if the parameters of $G_0(s)$ are known, then the desired controller parameters θ_1, θ_2, θ_{20}, and θ_3 can be calculated from a design equation, with which the control objective can be met. In the adaptive control problem, the parameters of $G_0(s)$ are unknown so that θ_1, θ_2, θ_{20}, and θ_3 cannot be determined *a priori* and have to be updated from some adaptive laws.

5.2 Model Reference Control

Model reference control solves the stated control problem when the parameters of $G_0(s)$ are known, that is, the nonadaptive control problem, for whose solution the following matching equation (5.11) is the key.

Lemma 5.1 *Constants θ_1^*, θ_2^*, θ_{20}^*, θ_3^* exist such that*

$$\begin{aligned} &\theta_1^{*T} a(s)P(s) + (\theta_2^{*T} a(s) + \theta_{20}^* \Lambda(s))k_p Z(s) \\ &= \Lambda(s)(P(s) - k_p \theta_3^* Z(s) P_m(s)). \end{aligned} \tag{5.11}$$

Proof: Choosing the feedforward gain θ_3^* as

$$\theta_3^* = k_p^{-1} \tag{5.12}$$

and defining the following polynomial, vectors, and matrices

$$\begin{aligned} B(s) &= \Lambda(s)(P(s) - k_p\theta_3^* Z(s)P_m(s)) = \Lambda(s)(P(s) - Z(s)P_m(s)) \\ &\triangleq \beta_{n+\bar{n}-2}s^{n+\bar{n}-2} + \beta_{n+\bar{n}-3}s^{n+\bar{n}-3} + \cdots + \beta_1 s + \beta_0, \end{aligned} \tag{5.13}$$

$$\beta = [\beta_{n+\bar{n}-2}, \beta_{n+\bar{n}-3}, \ldots, \beta_1, \beta_0]^T \in R^{n+\bar{n}-1}, \tag{5.14}$$

$$\theta_1^* = [\theta_{11}^*, \theta_{12}^*, \ldots, \theta_{1\bar{n}-2}^*, \theta_{1\bar{n}-1}^*]^T \in R^{\bar{n}-1}, \tag{5.15}$$

$$\theta_2^* = [\theta_{21}^*, \theta_{22}^*, \ldots, \theta_{2\bar{n}-2}^*, \theta_{2\bar{n}-1}^*]^T \in R^{\bar{n}-1}, \tag{5.16}$$

$$\begin{aligned} \bar{\theta}^* = \Big[&\theta_{1\bar{n}-1}^*, \theta_{1\bar{n}-2}^*, \ldots, \theta_{12}^*, \theta_{11}^*, \theta_{20}^*[1, \lambda_{\bar{n}-2}, \ldots, \lambda_1, \lambda_0] \\ &+ [0, \theta_{2\bar{n}-1}^*, \theta_{2\bar{n}-2}^*, \ldots, \theta_{22}^*, \theta_{21}^*]\Big]^T \in R^{2\bar{n}-1}, \end{aligned} \tag{5.17}$$

$$p^* = [1, p_{n-1}, p_{n-2}, \ldots, p_1, p_0]^T \in R^{n+1}, \tag{5.18}$$

$$z^* = k_p[0, \ldots, 0, 1, z_{m-1}, z_{m-2}, \cdots, z_1, z_0]^T \in R^n, \tag{5.19}$$

$$S^* = [S_1^* | S_2^*] = \left[\begin{array}{ccccc|ccccc} p^* & 0 & \cdots & 0 & 0 & z^* & 0 & \cdots & 0 & 0 \\ 0 & p^* & \cdots & \vdots & 0 & 0 & z^* & \cdots & \vdots & 0 \\ \vdots & 0 & \cdots & 0 & \vdots & \vdots & 0 & \cdots & 0 & \vdots \\ 0 & \vdots & \cdots & 0 & 0 & 0 & \vdots & \cdots & 0 & 0 \\ 0 & 0 & \cdots & 0 & 0 & 0 & 0 & \cdots & 0 & 0 \\ 0 & 0 & \cdots & p^* & 0 & 0 & 0 & \cdots & z^* & 0 \\ 0 & 0 & \cdots & 0 & p^* & 0 & 0 & \cdots & 0 & z^* \end{array}\right] \tag{5.20}$$

with $S_1^* \in R^{(n+\bar{n}-1)\times(\bar{n}-1)}$, $S_2^* \in R^{(n+\bar{n}-1)\times\bar{n}}$, so that $S^* \in R^{(n+\bar{n}-1)\times(2\bar{n}-1)}$, we examine the existence of the solution $\bar{\theta}^*$ to the linear equation

$$S^*\bar{\theta}^* = \beta. \tag{5.21}$$

Such existence is equivalent to the existence of the solution $\{\theta_1^*, \theta_2^*, \theta_{20}^*, \theta_3^*\}$ to (5.11) which has the form $\bar{a}(s)S^*\bar{\theta}^* = \bar{a}(s)\beta$ for $\bar{a}(s) = [s^{n+\bar{n}-2}, \ldots, s, 1]$.

If $P(s)$ and $Z(s)$ are coprime, then $\text{rank}[S^*] = n + \bar{n} - 1$,[3] so that (5.21) has a solution $\bar{\theta}^*$. This solution is unique if and only if $\bar{n} = n$, as in this case the matrix $S^* \in R^{(2n-1)\times(2n-1)}$ is nonsingular.

[3]If $P(s)$ and $Z(s)$ are coprime, then the submatrix of S^* consisting of the first $n-1$ columns of S_1^* plus the first n columns of S_2^* has rank $2n-1$ and one of p_0, z_0 is nonzero. If $p_0 \neq 0$, then the submatrix of S^* consisting of S_1^* plus the first n columns of S_2^* has rank $n+\bar{n}-1$; if $z_0 \neq 0$, then the submatrix of S^* consisting of the first $n-1$ columns of S_1^* plus S_2^* has rank $n+\bar{n}-1$. Therefore, in this case $\text{rank}[S^*] = n + \bar{n} - 1$.

If $P(s)$ and $Z(s)$ are not coprime, that is, $P(s) = \bar{P}(s)C(s)$, $Z(s) = \bar{Z}(s)C(s)$ for some polynomial $C(s)$ of degree $n_c > 0$ and some $\bar{P}(s)$, $\bar{Z}(s)$ coprime, then, with (5.12), (5.11) becomes

$$\begin{aligned} &\theta_1^{*T} a(s)\bar{P}(s) + (\theta_2^{*T} a(s) + \theta_{20}^* \Lambda(s)) k_p \bar{Z}(s) \\ &= \Lambda(s)(\bar{P}(s) - \bar{Z}(s)P_m(s)), \end{aligned} \tag{5.22}$$

from which we have

$$\bar{S}^* \bar{\theta}^* = \bar{\beta} \tag{5.23}$$

for some $\bar{S}^* \in R^{(n+\bar{n}-n_c-1)\times(2\bar{n}-1)}$ (which has a structure similar to S^* in (5.21)) and some $\bar{\beta} \in R^{n+\bar{n}-n_c-1}$. Because $\bar{P}(s)$ and $\bar{Z}(s)$ are coprime, the rank of $\bar{S}^*$ is $n + \bar{n} - n_c - 1$ so that (5.23) has a solution $\bar{\theta}^*$. In this case, the solution is not unique because $\bar{n} \geq n > n - n_c$.

With the solution $\bar{\theta}^* \in R^{2\bar{n}-1}$ and the matrices $C_1 = [I_{\bar{n}-1}, 0_{(\bar{n}-1)\times\bar{n}}] \in R^{(\bar{n}-1)\times(2\bar{n}-1)}$, $C_2 = [0_{(\bar{n}-1)\times\bar{n}}, I_{\bar{n}-1}] \in R^{(\bar{n}-1)\times(2\bar{n}-1)}$, we have

$$[\theta_{1\bar{n}-1}^*, \theta_{1\bar{n}-2}^*, \ldots, \theta_{12}^*, \theta_{11}]^T = C_1 \bar{\theta}^* \in R^{\bar{n}-1}, \tag{5.24}$$

$$\theta_{20}^* = [0_{1\times(\bar{n}-1)}, 1, 0_{1\times(\bar{n}-1)}]\bar{\theta}^* \in R, \tag{5.25}$$

$$\begin{aligned} &[\theta_{2\bar{n}-1}^*, \theta_{2\bar{n}-2}^*, \ldots, \theta_{22}^*, \theta_{21}]^T \\ &= C_2\bar{\theta}^* - \theta_{20}^*[\lambda_{\bar{n}-2}, \lambda_{\bar{n}-2}, \ldots, \lambda_1, \lambda_0]^T \in R^{\bar{n}-1}. \end{aligned} \tag{5.26}$$

Hence, there exist θ_1^*, θ_2^*, θ_{20}^*, θ_3^* such that (5.11) is satisfied. ∇

The parameters θ_1^*, θ_2^*, θ_{20}^*, θ_3^* are the so-called matching parameters because with these parameters the controller (5.5) results in the matching of the closed-loop system to the reference model (5.4). Using Lemma 5.1, we have the following solution to the model reference control problem.

Theorem 5.1 *With θ_1^*, θ_2^*, θ_{20}^*, θ_3^* satisfying (5.11) and $\theta_1 = \theta_1^*$, $\theta_2 = \theta_2^*$, $\theta_{20} = \theta_{20}^*$, and $\theta_3 = \theta_3^*$, the controller (5.5) ensures that all signals in the closed-loop system are bounded and*

$$y(t) - y_m(t) = \epsilon_0(t) \tag{5.27}$$

for some initial condition-related exponentially decaying $\epsilon_0(t)$.

Proof: Operating both sides of (5.11) on $y(t)$, we have

$$\begin{aligned} &\theta_1^{*T} a(s)P(s)[y](t) + (\theta_2^{*T} a(s) + \theta_{20}^* \Lambda(s)) k_p Z(s)[y](t) \\ &= \Lambda(s)(P(s) - k_p\theta_3^* Z(s)P_m(s))[y](t). \end{aligned} \tag{5.28}$$

Substituting the plant (5.1), $P(s)[y](t) = k_pZ(s)[u](t)$, into (5.28), we have

$$\begin{aligned} &\theta_1^{*T}a(s)k_pZ(s)[u](t) + (\theta_2^{*T}a(s) + \theta_{20}^*\Lambda(s))k_pZ(s)[y](t) \\ &= \Lambda(s)k_pZ(s)[u](t) - \Lambda(s)k_p\theta_3^*Z(s)P_m(s)[y](t). \end{aligned} \tag{5.29}$$

Because $\Lambda(s)$ and $Z(s)$ are stable, (5.29) can be expressed as

$$\begin{aligned} u(t) = &\ \theta_1^{*T}\frac{a(s)}{\Lambda(s)}[u](t) + \theta_2^{*T}\frac{a(s)}{\Lambda(s)}[y](t) \\ &+ \theta_{20}^*y(t) + \theta_3^*P_m(s)[y](t) + \epsilon_1(t) \end{aligned} \tag{5.30}$$

for some initial condition-related exponentially decaying $\epsilon_1(t)$.

With $\theta_1 = \theta_1^*$, $\theta_2 = \theta_2^*$, $\theta_{20} = \theta_{20}^*$, and $\theta_3 = \theta_3^*$, the controller (5.5) becomes

$$u(t) = \theta_1^{*T}\omega_1(t) + \theta_2^{*T}\omega_2(t) + \theta_{20}^*y(t) + \theta_3^*r(t). \tag{5.31}$$

In view of (5.4), (5.6), (5.30), and (5.31), we have

$$\theta_3^*P_m(s)[y - y_m](t) + \epsilon_1(t) = 0. \tag{5.32}$$

Since $P_m(s)$ is stable, from (5.32) we have (5.27), which also implies that $y(t) \in L^\infty$. Using (5.1) and (5.32), we have

$$k_pP_m(s)Z(s)[u](t) = P(s)P_m(s)[y](t) = P(s)\frac{1}{\theta_3^*}(\theta_3^*r(t) - \epsilon_1(t)), \tag{5.33}$$

which implies that $u(t) \in L^\infty$ because $P_m(s)Z(s)$ is stable and has degree $n^* + m = n$. Therefore, both $\omega_1(t)$ and $\omega_2(t)$ are bounded. ∇

Lemma 5.1 and Theorem 5.1 show that when the plant parameters are known, the model reference control problem has a solution: the controller (5.5) with fixed parameters θ_1^*, θ_2^*, θ_{20}^*, and θ_3^* in (5.11). As we will see in the rest of this chapter, it is the existence of such plant–model matching parameters that is sufficient for the development of adaptive control schemes for the case when the plant parameters in $P(s)$ and $Z(s)$ are unknown.

5.3 Adaptive Control

Adaptive control is needed when the parameters p_i, $i = 0, 1, \ldots, n-1$, z_j, $j = 0, 1, \ldots, m-1$, and k_p in the plant (5.1) are unknown. The key issue is to adjust controller parameters adaptively to ensure desired performance.

5.3.1 Tracking Error Equation

When the parameters of $G_0(s) = k_p \frac{Z(s)}{P(s)}$ are unknown, we cannot solve (5.11) for θ_1^*, θ_2^*, θ_{20}^*, θ_3^* to implement an ideal fixed controller (5.5). We need to design an adaptive law to update the parameters of the controller (5.5) in order to achieve the control objective stated in Section 5.1.1.

With $\theta_1(t)$, $\theta_2(t)$, $\theta_{20}(t)$ and $\theta_3(t)$ being the time-varying estimates of θ_1^*, θ_2^*, θ_{20}^*, θ_3^*, we use the adaptive version of the controller (5.5):

$$u(t) = \theta_1^T(t)\omega_1(t) + \theta_2^T(t)\omega_2(t) + \theta_{20}(t)y(t) + \theta_3(t)r(t). \tag{5.34}$$

To derive a tracking error equation, we define

$$\theta^* = [\theta_1^{*T}, \theta_2^{*T}, \theta_{20}^*, \theta_3^*]^T \in R^{2\bar{n}}, \tag{5.35}$$

$$\theta(t) = [\theta_1^T(t), \theta_2^T(t), \theta_{20}(t), \theta_3(t)]^T \in R^{2\bar{n}}, \tag{5.36}$$

$$\tilde{\theta}(t) = \theta(t) - \theta^*, \ e(t) = y(t) - y_m(t), \tag{5.37}$$

$$\omega(t) = [\omega_1^T(t), \omega_2^T(t), y(t), r(t)]^T \in R^{2\bar{n}}. \tag{5.38}$$

Substituting (5.30) into (5.34) and ignoring the exponentially decaying effect of the initial conditions, we have the tracking error equation

$$e(t) = \frac{k_p}{P_m(s)}[\tilde{\theta}^T\omega](t) = -k_p\left(\theta^{*T}\frac{1}{P_m(s)}[\omega](t) - \frac{1}{P_m(s)}[\theta^T\omega](t)\right). \tag{5.39}$$

This equation is the basis for developing an adaptive law for updating the controller parameters $\theta_1(t)$, $\theta_2(t)$, $\theta_{20}(t)$, and $\theta_3(t)$.

5.3.2 Lyapunov Design for Relative Degree 1

When the plant transfer function $G_0(s) = k_p \frac{Z(s)}{P(s)}$ has a relative degree $n^* = 1$, the characteristic polynomial of the reference model is

$$P_m(s) = s + a_m, \ a_m > 0, \tag{5.40}$$

and the error equation (5.39) becomes

$$\dot{e}(t) = -a_m e(t) + k_p\tilde{\theta}^T(t)\omega(t). \tag{5.41}$$

Defining a partial state vector of the closed-loop system as

$$e_p(t) = [e(t), \tilde{\theta}^T(t)]^T \in R^{2\bar{n}+1}, \tag{5.42}$$

we choose the positive definite function

$$V(e_p) = e^2 + |k_p|\tilde{\theta}^T\Gamma^{-1}\tilde{\theta} \tag{5.43}$$

as a measure of $e_p(t)$. The time derivative of $V(e_p)$ is

$$\begin{aligned}\dot{V} &= 2e(t)\dot{e}(t) + 2|k_p|\tilde{\theta}^T(t)\Gamma^{-1}\dot{\tilde{\theta}}(t)\\ &= -2a_m e^2(t) + 2e(t)k_p\tilde{\theta}^T(t)\omega(t) + 2|k_p|\tilde{\theta}^T(t)\Gamma^{-1}\dot{\tilde{\theta}}(t).\end{aligned} \tag{5.44}$$

To make $\dot{V} \leq 0$, we choose the adaptive law for $\theta(t)$ as

$$\boxed{\dot{\theta}(t) = -\text{sign}[k_p]\Gamma\omega(t)e(t)} \tag{5.45}$$

with $\Gamma = \Gamma^T > 0$ and $\theta(0)$ being an (arbitrary) initial estimate of θ^*.

Since $\dot{\theta}(t) = \dot{\tilde{\theta}}(t)$ as θ^* is constant, with (5.45), (5.43) becomes

$$\dot{V} = -2a_m e^2(t) \leq 0. \tag{5.46}$$

Hence we have that $e(t) \in L^\infty$, $\theta(t) \in L^\infty$, and so $y(t) \in L^\infty$, $\omega_2(t) \in L^\infty$.

Using the plant (5.1), $P(s)[y](t) = k_p Z(s)[u](t)$, and ignoring the exponentially decaying effect of the initial conditions, we have

$$\frac{s^i}{\Lambda(s)}[u](t) = \frac{P(s)}{k_p Z(s)}\frac{s^i}{\Lambda(s)}[y](t), \tag{5.47}$$

which is bounded for $i = 0, 1, \ldots, \bar{n}-2$, because $\frac{P(s)}{k_p Z(s)}\frac{s^i}{\Lambda(s)}$ is stable and proper and $y(t) \in L^\infty$. This implies that $\omega_1(t) \in L^\infty$, and so does $u(t)$. From (5.41), we have that $\dot{e}(t) \in L^\infty$, and from (5.46), we have $e(t) \in L^2$: $\int_0^\infty e^2(t)\,dt < \infty$. Therefore, using Lemma 2.14, we conclude that $\lim_{t\to\infty} e(t) = 0$. ∇

In summary, we have the following results.

Theorem 5.2 *The controller (5.34) with the adaptive law (5.44) applied to the plant (5.1) with relative degree $n^* = 1$ guarantees that all closed-loop signals are bounded and that the tracking error $e(t) = y(t) - y_m(t)$ belongs to L^2 and goes to zero asymptotically as t goes to ∞.*

In addition, we also have $\dot{\tilde{\theta}}(t) = \dot{\theta}(t) \in L^2 \cap L^\infty$, $\lim_{t\to\infty} \dot{\tilde{\theta}}(t) = 0$.

In the above analysis we used a Lyapunov method, that is, we chose a positive definite function $V(e_p)$ which contains both the parameter error $\tilde{\theta}(t)$ and the error system state $e(t)$ in (5.41). This adaptive design is also called a Lyapunov design. Other Lyapunov-type designs are that in Example 1.1, the example in Section 4.1.1, and the general cases in Sections 4.1 and 9.1.

5.3.3 Alternative Design for Relative Degree 1

For the case when $n^* = 1$, a redesign of the adaptive scheme in Section 5.3.2 can ensure that the tracking error $e(t) = y(t) - y_m(t)$ satisfies

$$\int_0^\infty |e(t)|^{1+\alpha}\, dt < \infty,\ 0 < \alpha < \infty, \tag{5.48}$$

as compared with the result $\int_0^\infty |e(t)|^2\, dt < \infty$.

From the error equation (5.41), we consider the positive definite function

$$V(e, \tilde{\theta}) = \frac{|e(t)|^{1+\alpha}}{1+\alpha} + \frac{|k_p|\tilde{\theta}^T \Gamma^{-1} \tilde{\theta}}{2},\ \Gamma = \Gamma^T > 0, \tag{5.49}$$

where $\alpha > 0$ is a design parameter, and introduce the function

$$f(e(t)) = \begin{cases} (e(t))^\alpha & \text{if } e(t) \geq 0, \\ -(-e(t))^\alpha & \text{if } e(t) < 0 \end{cases} \tag{5.50}$$

for $0 < \alpha < \infty$. Since $\alpha > 0$, $|e(t)|^{1+\alpha}$ is differentiable with respect to t, and its derivative is $(1+\alpha)f(e(t))\dot{e}(t)$. Then, the time derivative of V is

$$\begin{aligned} \dot{V} &= f(e(t))\dot{e}(t) + |k_p|\tilde{\theta}^T(t)\Gamma^{-1}\dot{\tilde{\theta}}(t) \\ &= -a_m|e(t)|^{1+\alpha} + f(e(t))k_p\tilde{\theta}^T(t)\omega(t) + |k_p|\tilde{\theta}^T(t)\Gamma^{-1}\dot{\tilde{\theta}}(t). \end{aligned} \tag{5.51}$$

If we choose the adaptive law for $\theta(t)$ as

$$\dot{\theta}(t) = -\text{sign}[k_p]\Gamma\omega(t)f(e(t)), \tag{5.52}$$

then it follows from (5.51) that

$$\dot{V} = -a_m|e(t)|^{1+\alpha}. \tag{5.53}$$

Hence, we have $\dot{V} \leq 0$, which implies $e(t) \in L^\infty$, $\theta(t) \in L^\infty$, and so $y(t) \in L^\infty$, $\omega_2(t) \in L^\infty$. From (5.47), we have $\omega_1(t) \in L^\infty$, and so does $u(t)$. From (5.41), we have $\dot{e}(t) \in L^\infty$, and from (5.53), we have $e(t) \in L^{1+\alpha}$, that is, (5.48) holds. Therefore, from Lemma 2.14, we conclude $\lim_{t\to\infty} e(t) = 0$.

It can also be verified that $\dot{\theta}(t) \in L^{\frac{1+\alpha}{\alpha}}$, that is, $\int_0^\infty \|\dot{\theta}(t)\|^{\frac{1+\alpha}{\alpha}}\, dt < \infty$, $\dot{\theta}(t) \in L^\infty$, and $\lim_{t\to\infty} \dot{\theta}(t) = 0$.

For tracking performance, what may be interesting is the design (5.52) with $0 < \alpha < 1$. An $L^{1+\alpha} \cap L^\infty$ signal, with $1 < 1+\alpha < 2$, is also an $L^2 \cap L^\infty$ signal, and the converse is not true. The result that $e(t) \in L^{1+\alpha}$ means that the new adaptive law (5.52) ensures that the tracking error $e(t)$ belongs to the nontrivial subset $L^{1+\alpha} \cap L^\infty$ of $L^2 \cap L^\infty$ signals for $1 < 1+\alpha < 2$.

For $0 < \alpha < 1$, we can define the signal

$$e_n(t) = |e(t)|^{\frac{1+\alpha}{2}} = |e(t)| g_e(t), \; g_e(t) = |e(t)|^{\frac{\alpha-1}{2}}. \tag{5.54}$$

We see that $e_n(t)$ is an "enlarged" $|e(t)|$ for $|e(t)| < 1$ with the enlargement factor $g_e(t)$. Since $\lim_{t\to\infty} e(t) = 0$, we have that $\lim_{t\to\infty} g_e(t) = \infty$. Hence, if $e_n(t)$ is small, then $e(t)$ is "smaller." Using the signal energy as the measure of smallness, we see that a favorable adaptive design is the one like (5.52) with $0 < \alpha < 1$, which ensures that $e_n(t)$, an enlarged $|e(t)|$ with the enlargement factor $g_e(t)$ such that $\lim_{t\to\infty} g_e(t) = 0$, has finite energy.

Example 5.1 As an illustrative example, we simulated the adaptive control system consisting of (5.1) with $P(s) = s^2 + s - 2$, $Z(s) = s + 1$, (5.4) with $P_m(s) = s + 1$, (5.5) with $\Lambda(s) = s + 3$, and (5.52) with $\alpha = \frac{1}{15}$ or (5.45) for $\Gamma = 0.2I$. The controller structure is $u(t) = \theta_1 \frac{1}{s+3}[u](t) + \theta_2 \frac{1}{s+3}[y](t) + \theta_{20} y(t) + \theta_3 r(t)$. The nominal values of $\theta_1, \theta_2, \theta_{20}, \theta_3$ are $\theta_1^* = 2$, $\theta_2^* = 4$, $\theta_{20}^* = -3$, $\theta_3^* = 1$, and their initial estimates are $\theta_1(0) = 1$, $\theta_2(0) = 2$, $\theta_{20}(0) = -1$, $\theta_3(0) = 1.5$.

The simulation results are shown in Figures 5.2 and 5.3 for $r(t) = \sin(0.9t)$ and $r(t) = \sin(0.9t) + \sin(1.9t)$, respectively, which indicate that the modified adaptive control scheme (5.52) with $\alpha = \frac{1}{15}$ leads to faster convergence of the tracking error $e(t) = y(t) - y_m(t)$ to zero than the standard model reference adaptive control scheme (5.45) with $\alpha = 2$. □

5.3.4 Lyapunov Design for Arbitrary Relative Degrees

When the plant transfer function $G_0(s) = k_p \frac{Z(s)}{P(s)}$ has a relative degree $n^* \geq 1$, there are two types of adaptive schemes which can be used for updating the parameters of the controller (5.34) applied to the plant (5.1): one is a Lyapunov-like design and the other is a gradient design (see Section 5.3.4).

For a Lyapunov design, we choose $L(s)$ as a (monic) stable polynomial of degree $n^* - 1$ such that $\frac{L(s)}{P_m(s)}$ is strictly positive real (see Section 2.5.4).

From the controller (5.34) and the tracking error equation (5.39), we have

$$e(t) = \frac{L(s)}{P_m(s)}[\rho^* L^{-1}(s)[u] - \rho^* \theta^{*T} L^{-1}(s)[\omega]](t), \; \rho^* = k_p. \tag{5.55}$$

Let $\theta(t)$, $\rho(t)$ be the estimates of θ^*, ρ^*, respectively. Define the estimation error $\epsilon(t)$ from the dynamic system

$$\epsilon(t) = e(t) + \frac{L(s)}{P_m(s)}[\rho\xi - \kappa\epsilon m_0^2](t), \tag{5.56}$$

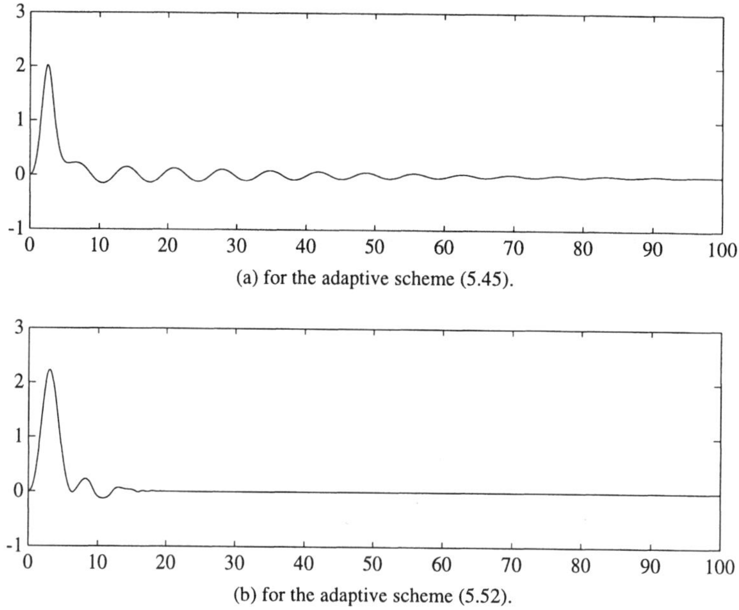

(a) for the adaptive scheme (5.45).

(b) for the adaptive scheme (5.52).

Figure 5.2: Tracking errors for $r(t) = \sin(0.9t)$ (t in sec).

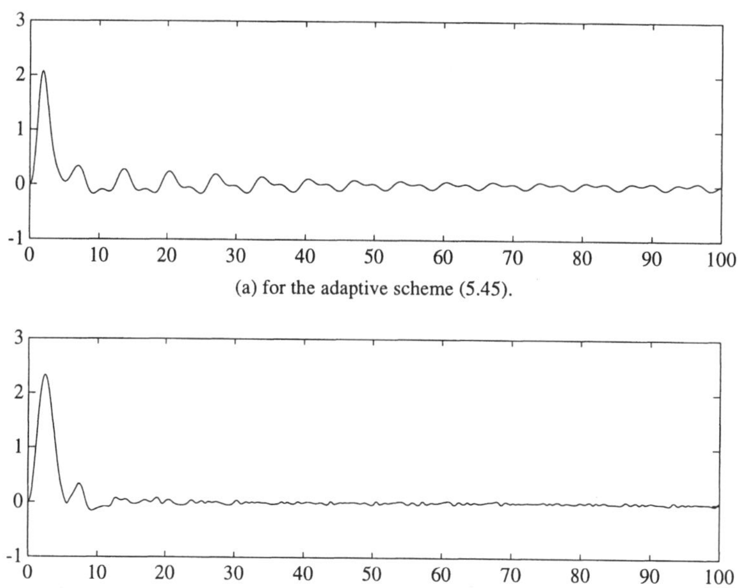

(a) for the adaptive scheme (5.45).

(b) for the adaptive scheme (5.52).

Figure 5.3: Tracking errors for $r(t) = \sin(0.9t) + \sin(1.9t)$ (t in sec).

where $\kappa > 0$ is an arbitrary constant, and

$$\xi(t) = \theta^T(t)\zeta(t) - L^{-1}(s)[\theta^T\omega](t), \tag{5.57}$$

$$\zeta(t) = L^{-1}(s)[\omega](t), \tag{5.58}$$

$$m_0(t) = \sqrt{\zeta^T(t)\zeta(t) + \xi^2(t)}. \tag{5.59}$$

We choose to design the adaptive laws for updating $\theta(t)$ and $\rho(t)$ as

$$\dot{\theta}(t) = -\text{sign}[k_p]\Gamma\zeta(t)\epsilon(t),\ \Gamma = \Gamma^T > 0,\ \theta(0) = \theta_0, \tag{5.60}$$

$$\dot{\rho}(t) = -\gamma\xi(t)\epsilon(t),\ \gamma > 0,\ \rho(0) = \rho_0, \tag{5.61}$$

where θ_0, ρ_0 are initial estimates of θ^*, ρ^*, respectively.

We have the following stability properties of this adaptive scheme.

Lemma 5.2 *The adaptive law (5.60)–(5.61) guarantees that $\epsilon(t) \in L^2 \cap L^\infty$, $\epsilon(t)m_0(t) \in L^2$, $\theta(t) \in L^\infty$, $\rho(t) \in L^\infty$, and $\dot{\theta}(t) \in L^2$, $\dot{\rho}(t) \in L^2$.*

Proof: From (5.34) and (5.56) we have the estimation error equation

$$\epsilon(t) = \frac{L(s)}{P_m(s)}[\rho^*\tilde{\theta}^T\zeta + \tilde{\rho}\xi - \kappa\epsilon m_0^2](t), \tag{5.62}$$

where $\tilde{\rho}(t) = \rho(t) - \rho^*$. Let $\frac{L(s)}{P_m(s)}$ in (5.62) have a controllable realization (A, b, c) and $e_\epsilon(t)$ be its state variable. Then, from (5.62), we have

$$\dot{e}_\epsilon(t) = Ae_\epsilon(t) + bv_0(t),\ \epsilon(t) = ce_\epsilon(t), \tag{5.63}$$

where

$$v_0(t) = \rho^*\tilde{\theta}^T(t)\zeta(t) + \tilde{\rho}(t)\xi(t) - \kappa\epsilon(t)m_0^2(t). \tag{5.64}$$

Since $\frac{L(s)}{P_m(s)}$ is strictly positive real, it follows from the Lefschetz-Kalman-Yakubovich lemma (Lemma 2.13) that real constant matrices $P = P^T > 0$, $Q = Q^T > 0$ exist such that

$$A^TP + PA = -Q,\ Pb = c^T. \tag{5.65}$$

Define the partial state vector of the adaptive system as

$$e_p(t) = [e_\epsilon^T(t), \tilde{\theta}^T(t), \tilde{\rho}(t)]^T \tag{5.66}$$

and choose the positive definite function as

$$V(e_p) = e_\epsilon^T P e_\epsilon + |\rho^*|\tilde{\theta}^T\Gamma^{-1}\tilde{\theta} + \gamma^{-1}\tilde{\rho}^2. \tag{5.67}$$

From (5.63)–(5.65), and (5.67), we have

$$\begin{aligned}\dot{V} &= -e_\epsilon^T(t)Qe_\epsilon(t) - 2\kappa\epsilon^2(t)m_0^2(t) \\ &\quad + 2k_p\tilde{\theta}^T(t)(\zeta(t)\epsilon(t) + \text{sign}[k_p]\Gamma^{-1}\dot{\tilde{\theta}}(t)) \\ &\quad + 2\tilde{\rho}(t)(\xi(t)\epsilon(t) + \gamma^{-1}\dot{\tilde{\rho}}(t)).\end{aligned} \tag{5.68}$$

Substituting (5.60) and (5.61) into (5.681), we have

$$\dot{V} = -e_\epsilon^T(t)Qe_\epsilon(t) - 2\kappa\epsilon^2(t)m_0^2(t), \tag{5.69}$$

from which we can derive that $\epsilon(t) \in L^2 \cap L^\infty$, $\epsilon(t)m_0(t) \in L^2$, $\theta(t) \in L^\infty$, $\rho(t) \in L^\infty$. Finally, from (5.59), it follows that

$$\frac{\|\zeta(t)\|}{m_0(t)} \leq b_1, \ \frac{|\xi(t)|}{m_0(t)} \leq b_2 \tag{5.70}$$

for some $b_1 > 0$, $b_2 > 0$, which, together with (5.60) and (5.61), implies that $\dot{\theta}(t) \in L^2$ and $\dot{\rho}(t) \in L^2$ (we have not shown $\dot{\theta}(t) \in L^\infty$, $\dot{\rho}(t) \in L^\infty$). ∇

In the proof of Lemma 5.2 we used $V(e_p)$, which contains not only the parameter errors $\tilde{\theta}(t)$, $\tilde{\rho}(t)$ but also the state variable $e_\epsilon(t)$ of the error system (5.63) for the estimation error $\epsilon(t)$ generated from the dynamic system (5.56). This $V(e_p)$ is a partial Lyapunov function, which does not contain the states of the tracking error system (5.39). We also note that for the special case when the relative degree is $n^* = 1$, letting $L(s) = 0$ in (5.56), we have $\epsilon(t) = e(t)$ and obtain the same adaptive law as that in (5.45) for $\theta(t)$.

Theorem 5.3 *All signals in the closed-loop system consisting of the plant (5.1), reference model (5.4), controller (5.34), and adaptive law (5.60)–(5.61) are bounded, and the tracking error $e(t) = y(t) - y_m(t)$ belongs to L^2 and converges to zero asymptotically with time.*

Proof: Introduce a fictitious signal

$$z_0(t) = \frac{1}{s + a_0}[u](t) \tag{5.71}$$

and two fictitious filters $K_1(s)$ and $K(s)$ as

$$sK_1(s) = 1 - K(s), \ K(s) = \frac{a^{n^*}}{(s+a)^{n^*}}, \tag{5.72}$$

where $a_0 > 0$ is arbitrary and $a > 0$ is to be specified. With $G_0(s) = k_p \frac{Z(s)}{P(s)}$, using the equality $-a_0 K_1(s) + (s + a_0)K_1(s) = 1 - K(s)$, we obtain

$$z_0(t) + a_0 K_1(s)[z_0](t) - K_1(s)[u](t) = K(s)G_0^{-1}(s)\frac{1}{s + a_0}[y](t). \tag{5.73}$$

Hence (5.34), operated by $K_1(s)$ on both sides, with $z_0(t)$ in (5.71), gives

$$\begin{aligned} K_1(s)[u](t) = K_1(s)\theta_1^T(\cdot)\frac{a(s)}{\Lambda(s)}(s + a_0)[z_0](t) + K_1(s)\theta_2^T(\cdot)\frac{a(s)}{\Lambda(s)}[y](t) \\ + K_1(s)\theta_{20}(\cdot)[y](t) + K_1(s)[\theta_3 r](t), \end{aligned} \tag{5.74}$$

where, with $a(s) = [1, s, \ldots, s^{\bar{n}-2}]^T$ as in (5.7),

$$K_1(s)\theta_1^T(\cdot)\frac{a(s)}{\Lambda(s)}(s + a_0)[z_0](t) \triangleq K_1(s)\left[\theta_1^T \frac{a(s)}{\Lambda(s)}(s + a_0)[z_0]\right](t), \tag{5.75}$$

$$K_1(s)(\theta_2^T(\cdot)\frac{a(s)}{\Lambda(s)}[y](t) \triangleq K_1(s)\left[\theta_2^T \frac{a(s)}{\Lambda(s)}[y]\right](t), \tag{5.76}$$

$$K_1(s)\theta_{20}(\cdot)[y](t) \triangleq K_1(s)\left[\theta_{20}[y]\right](t). \tag{5.77}$$

Substituting (5.74) into (5.73), we have

$$\begin{aligned} \left(1 + K_1(s)\left(a_0 - \theta_1^T(\cdot)\frac{a(s)}{\Lambda(s)}(s + a_0)\right)\right)[z_0](t) = K_1(s)[\theta_3 r](t) \\ + \left(K(s)G_0^{-1}(s)\frac{1}{s + a_0} + K_1(s)\theta_2^T(\cdot)\frac{a(s)}{\Lambda(s)} + K_1(s)\theta_{20}(\cdot)\right)[y](t). \end{aligned} \tag{5.78}$$

From Proposition 2.10, the impulse response $k_1(t)$ of $K_1(s)$ is

$$k_1(t) = \mathcal{L}^{-1}[K_1(s)] = e^{-at}\sum_{i=1}^{n^*}\frac{a^{n^*-i}}{(n^* - i)!}t^{n^*-i}, \tag{5.79}$$

where $\mathcal{L}^{-1}[\cdot]$ is the inverse Laplace transform operator, which satisfies

$$\|k_1(\cdot)\|_1 = \int_0^\infty |k_1(t)|\, dt = \frac{n^*}{a}. \tag{5.80}$$

Hence there exists $a^0 > 0$ such that for any finite $a > a^0$ the operator

$$T_0(s, t) \triangleq \left(1 + K_1(s)\left(a_0 - \theta_1^T(t)\frac{a(s)}{\Lambda(s)}(s + a_0)\right)\right)^{-1} \tag{5.81}$$

is stable and proper (see Lemma 2.5).

Let $a > a^0$ in $K(s)$ and $K_1(s)$ defined in (5.72). Then, (5.78) implies that

$$z_0(t) = T_1(s,\cdot)[y](t) + b_0(t), \tag{5.82}$$

where $T_1(s,t)$ is a stable and strictly proper operator and $b_0(t) \in L^\infty$ due to the bounded reference input $r(t)$.

Denoting the polynomial $L(s)$ in (5.56) as $L(s) = s^{n^*-1} + l_{n^*-2}s^{n^*-2} + \cdots + l_1 s + l_0$, we can express $\xi(t)$ in (5.57) as[4]

$$\begin{aligned}\xi(t) = {} & \frac{s^{n^*-2} + l_{n^*-2}s^{n^*-3} + \cdots + l_2 s + l_1}{L(s)}\left[\dot{\theta}^T \frac{1}{L(s)}[\omega]\right](t) \\ & + \frac{s^{n^*-3} + l_{n^*-2}s^{n^*-4} + \cdots + l_3 s + l_2}{L(s)}\left[\dot{\theta}^T \frac{s}{L(s)}[\omega]\right](t) + \cdots \\ & + \frac{s + l_{n^*-2}}{L(s)}\left[\dot{\theta}^T \frac{s^{n^*-3}}{L(s)}[\omega]\right](t) + \frac{1}{L(s)}\left[\dot{\theta}^T \frac{s^{n^*-2}}{L(s)}[\omega]\right](t)\end{aligned} \tag{5.83}$$

(see Problem 5.8). From (5.56) we have

$$y(t) = y_m(t) + \epsilon(t) - \frac{L(s)}{P_m(s)}[\rho\xi](t) + \kappa\frac{L(s)}{P_m(s)}[\epsilon\, m_0^2](t). \tag{5.84}$$

The inequality

$$m_0(t) \le \|\zeta(t)\|_1 + |\xi(t)|, \tag{5.85}$$

Lemma 5.2, and (5.82)–(5.84) imply that

$$|y(t)| \le x_0(t) + T_2(s,\cdot)[x_1 T_3(s,\cdot)[|y|]](t) \tag{5.86}$$

---

[4] As an illustrative example, we consider $L(s) = s^2 + l_1 s + l_0$ and derive

$$\begin{aligned} & L(s)[\theta^T\zeta](t) \\ & = (s^2 + l_1 s + l_0)[\theta^T\zeta](t) \\ & = (s + l_1)[\dot{\theta}^T\zeta + \theta^T\dot{\zeta}](t) + l_0\theta^T(t)\zeta(t) \\ & = (s + l_1)[\dot{\theta}^T\zeta](t) + \dot{\theta}^T(t)\dot{\zeta}(t) + \theta^T(t)\ddot{\zeta}(t) + l_1\theta^T(t)\dot{\zeta}(t) + l_0\theta^T(t)\zeta(t) \\ & = (s + l_1)[\dot{\theta}^T\zeta](t) + \dot{\theta}^T(t)s[\zeta](t) + \theta^T(t)(s^2 + l_1 s + l_0)[\zeta](t) \\ & = (s + l_1)\left[\dot{\theta}^T\frac{1}{L(s)}[\omega]\right](t) + \dot{\theta}^T(t)\frac{s}{L(s)}[\omega](t) + \theta^T(t)\omega(t), \end{aligned}$$

where $\zeta(t) = \frac{1}{L(s)}[\omega](t)$. Hence, from this equality and (5.57), we have

$$\xi(t) = \theta^T(t)\zeta(t) - \frac{1}{L(s)}[\theta^T\omega](t) = \frac{s + l_1}{L(s)}\left[\dot{\theta}^T\frac{1}{L(s)}[\omega]\right](t) + \frac{1}{L(s)}\left[\dot{\theta}^T\frac{s}{L(s)}[\omega]\right](t).$$

for some $x_0(t) \in L^\infty$, $x_1(t) \in L^2$ with $x_1(t) \geq 0$, $T_2(s,t)$ being a stable and strictly proper operator and $T_3(s,t)$ being a stable and proper operator, both with a nonnegative impulse response.

Introducing $z_1(t) = T_3(s,\cdot)[|y|](t)$, operating $T_3(s,t)$ on both sides of (5.86), and noting that $T_3(s,t)$ has a nonnegative impulse response, we have

$$z_1(t) \leq b_1 + b_2 \int_0^t e^{-\alpha(t-\tau)} x_1(\tau) z_1(\tau)\, d\tau \tag{5.87}$$

for some α, b_1, $b_2 > 0$ (such a feedback structure is shown in Figure 5.4, which satisfies the small-gain condition of Lemma 2.3 as $x_1(t) \in L^2$).[5]

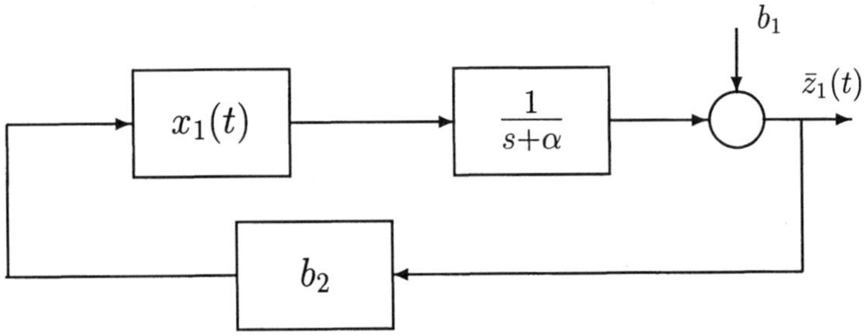

Figure 5.4: Feedback loop representation of (5.87).

Applying Lemma 2.3 to (5.87) with $x_1(t) \in L^2$ satisfying (2.103), we conclude that $z_1(t) \in L^\infty$, so $y(t) \in L^\infty$, and $z_0(t) \in L^\infty$ in (5.82). This implies that $u(t) \in L^\infty$ in (5.34). Thus all signals in the closed-loop system are bounded. From (5.62), we have $\dot{\epsilon}(t) \in L^\infty$, which, with $\epsilon(t) \in L^2 \cap L^\infty$, from Lemma 2.14, implies $\lim_{t\to\infty} \epsilon(t) = 0$. Hence $\lim_{t\to\infty} \dot{\theta}(t) = 0$ in (5.60), and from Corollary 2.7 and Lemma 2.19, $\xi(t) \in L^2$ and $\lim_{t\to\infty} \xi(t) = 0$ in (5.83). Finally, from (5.84), we have $e(t) = y(t) - y_m(t) \in L^2$ and $\lim_{t\to\infty} e(t) = 0$. (After proving signal boundedness, we now have $\dot{\theta}(t) \in L^\infty$, $\dot{\rho}(t) \in L^\infty$.) ∇

---

[5] For the signal $\bar{z}_1(t)$ defined from

$$\bar{z}_1(t) = b_1 + b_2 \int_0^t e^{-\alpha(t-\tau)} x_1(\tau) \bar{z}_1(\tau)\, d\tau,$$

we have

$$e^{\alpha t}(z_1(t) - \bar{z}_1(t)) \leq b_2 \int_0^t e^{\alpha\tau} x_1(\tau)(z_1(\tau) - \bar{z}_1(\tau))\, d\tau.$$

Using Lemma 2.2, we conclude that $z_1(t) \leq \bar{z}_1(t)$.

5.3.5 Gradient Design for Arbitrary Relative Degrees

In view of the second equality of the tracking error equation (5.39), to develop an adaptive scheme for updating $\theta(t)$ and $\rho(t)$, the estimates of θ^* and $\rho^* = k_p$, we can also define the estimation error (see Section 4.5.3) as

$$\epsilon(t) = e(t) + \rho(t)\xi(t), \tag{5.88}$$

where

$$\xi(t) = \theta^T(t)\zeta(t) - \frac{1}{P_m(s)}[\theta^T\omega](t), \tag{5.89}$$

$$\zeta(t) = \frac{1}{P_m(s)}[\omega](t). \tag{5.90}$$

Then we choose the gradient adaptive update law for $\theta(t)$ and $\rho(t)$ as

$$\dot{\theta}(t) = -\frac{\text{sign}[k_p]\Gamma\epsilon(t)\zeta(t)}{m^2(t)}, \ \Gamma = \Gamma^T > 0, \ \theta(0) = \theta_0, \tag{5.91}$$

$$\dot{\rho}(t) = -\frac{\gamma\epsilon(t)\xi(t)}{m^2(t)}, \ \gamma > 0, \ \rho(0) = \rho_0, \tag{5.92}$$

where θ_0 and ρ_0 are initial estimates of θ^* and ρ^* and

$$m(t) = \sqrt{1 + \zeta^T(t)\zeta(t) + \xi^2(t)}. \tag{5.93}$$

This design is called a gradient design because both $\dot{\theta}$ and $\dot{\rho}$ are in the direction of the negative gradient of the normalized quadratic cost function $J(\theta, \rho) = \frac{\epsilon^2}{m^2}$, where $\epsilon = \rho^*\tilde{\theta}^T\zeta + \tilde{\rho}\xi$ (see (5.94)). In this case, the gradients [151] of $J(\theta, \rho)$ are $\frac{\partial J}{\partial \theta} = \frac{\rho^*\zeta\epsilon}{m^2}$ with $\rho^* = k_p$, and $\frac{\partial J}{\partial \rho} = \frac{\xi\epsilon}{m^2}$, whose directions are the directions of $\frac{\text{sign}[k_p]\zeta\epsilon}{m^2}$, and $\frac{\xi\epsilon}{m^2}$. Therefore, with adaptation gains $\Gamma = \Gamma^T > 0$, $\gamma > 0$, the adaptive laws are (5.91) and (5.92).

This adaptive update law has the following properties.

Lemma 5.3 *The adaptive law (5.91)–(5.92) guarantees that $\theta(t), \rho(t) \in L^\infty$ and $\frac{\epsilon(t)}{m(t)} \in L^2 \cap L^\infty$, $\dot{\theta}(t) \in L^2 \cap L^\infty$, and $\dot{\rho}(t) \in L^2 \cap L^\infty$.*

Proof: Substituting (5.39) into (5.88), we have the linear model

$$\epsilon(t) = \rho^*\tilde{\theta}^T(t)\zeta(t) + \tilde{\rho}(t)\xi(t). \tag{5.94}$$

Then, the time derivative of the positive definite function

$$V(\tilde{\theta}, \tilde{\rho}) = |\rho^*|\tilde{\theta}^T\Gamma^{-1}\tilde{\theta} + \gamma^{-1}\tilde{\rho}^2 \tag{5.95}$$

along the trajectories of (5.91)–(5.92) satisfies

$$\dot{V} = -\frac{2\epsilon^2(t)}{m^2(t)} \leq 0. \tag{5.96}$$

Hence $\theta(t) \in L^\infty$, $\rho(t) \in L^\infty$, and $\frac{\epsilon(t)}{m(t)} \in L^2$, which, with (5.94), (5.91), and (5.92), in turn, imply $\frac{\epsilon(t)}{m(t)} \in L^\infty$, $\dot{\theta}(t) \in L^2 \cap L^\infty$, and $\dot{\rho}(t) \in L^2 \cap L^\infty$. ∇

If we choose $L(s) = P_m(s)$ and $\alpha = 1$ in (5.56), then the adaptive law (5.60)–(5.61) is identical to (5.91)–(5.92). Moreover, the signal $m(t)$ in (5.91)–(5.92) normalizes the adaptive law and leads to the additional properties that $\dot{\theta}(t) \in L^\infty$ and $\dot{\rho}(t) \in L^\infty$, as compared with (5.60)–(5.61).

Before presenting the proof for closed-loop stability and tracking performance for this gradient design, we first give two illustrative examples.

Example 5.2 This example shows the closed-loop stability and tracking properties of the simple adaptive control system in Example 1.2, with the plant

$$\dot{y}(t) = a_p y(t) + u(t) \tag{5.97}$$

with a_p unknown, the reference model system

$$\dot{y}_m(t) = -a_m y_m(t) + r(t) \tag{5.98}$$

with $a_m > 0$ and $r(t)$ bounded, and the controller structure

$$u(t) = k(t)y(t) + r(t), \tag{5.99}$$

where $k(t)$ is the estimate of $k^* = -a_p - a_m$. The tracking error equation is

$$\dot{e}(t) = -a_m e(t) + \tilde{k}(t)y(t), \tag{5.100}$$

where $e(t) = y(t) - y_m(t)$ and $\tilde{k}(t) = k(t) - k^*$.

Introducing the estimation error

$$\epsilon(t) = e(t) + \xi(t),\ \xi(t) = k(t)\zeta(t) - \frac{1}{s+a_m}[ky](t), \tag{5.101}$$

we choose the adaptive law for $k(t)$ as

$$\dot{k}(t) = -\frac{\gamma\epsilon(t)\zeta(t)}{m^2(t)},\ m(t) = \sqrt{1+\zeta^2(t)},\ \gamma > 0. \tag{5.102}$$

From Lemma 5.3, we have $k(t) \in L^\infty$, $\frac{\epsilon(t)}{m(t)} \in L^2$, $\frac{\epsilon(t)}{m(t)} \in L^\infty$, and $\dot{k}(t) \in L^\infty \cap L^2$. To prove that both $y(t)$ and $u(t)$ are bounded, we express

$$\xi(t) = \frac{1}{s+a_m}\left[\dot{k}\frac{1}{s+a_m}[y]\right](t), \tag{5.103}$$

which can be verified by using $\xi(t)$ in (5.101) and

$$\begin{aligned}
&(s+a_m)[\xi](t) \\
&= (s+a_m)[k\zeta](t) - k(t)y(t) \\
&= \dot{k}(t)\zeta(t) + k(t)\dot{\zeta}(t) + a_m k(t)\zeta(t) - k(t)y(t) \\
&= \dot{k}(t)\frac{1}{s+a_m}[y](t) + k(t)\frac{s}{s+a_m}[y](t) + k(t)\frac{a_m}{s+a_m}[y](t) - k(t)y(t) \\
&= \dot{k}(t)\frac{1}{s+a_m}[y](t).
\end{aligned} \tag{5.104}$$

Then, from the first equation in (5.101), we have

$$\begin{aligned}
y(t) &= y_m(t) + \epsilon(t) + \xi(t) \\
&= y_m(t) + \frac{\epsilon(t)}{m(t)}m(t) + \frac{1}{s+a_m}\left[\dot{k}\frac{1}{s+a_m}[y]\right](t).
\end{aligned} \tag{5.105}$$

From the definition of $m(t)$, it follows that

$$m(t) \le 1 + |\zeta(t)| = 1 + \left|\frac{1}{s+a_m}[y](t)\right| \le 1 + \frac{1}{s+a_m}[|y|](t). \tag{5.106}$$

From (5.105) and (5.106), we obtain

$$\begin{aligned}
|y(t)| \le |y_m(t)| &+ \frac{|\epsilon(t)|}{m(t)} + \frac{|\epsilon(t)|}{m(t)}\frac{1}{s+a_m}[|y|](t) \\
&+ \frac{1}{s+a_m}\left[|\dot{k}|\frac{1}{s+a_m}[|y|]\right](t),
\end{aligned} \tag{5.107}$$

$$\begin{aligned}
\frac{1}{s+a_m}[|y|](t) \le &\frac{1}{s+a_m}\left[|y_m| + \frac{|\epsilon|}{m}\right](t) + \frac{1}{s+a_m}\left[\frac{|\epsilon|}{m}\frac{1}{s+a_m}[|y|]\right](t) \\
&+ \frac{1}{s+a_m}\left[\frac{1}{s+a_m}\left[|\dot{k}|\frac{1}{s+a_m}[|y|]\right]\right](t).
\end{aligned} \tag{5.108}$$

Here we used the expression $\frac{1}{s+a_m}[f](t) = \int_0^t e^{-a_m(t-\tau)}f(\tau)\,d\tau$. Denoting $\bar{y}(t) = \frac{1}{s+a_m}[|y|](t)$, we rewrite (5.108) as

$$\bar{y}(t) \le b_0(t) + \frac{1}{s+a_m}\left[\frac{|\epsilon|}{m}\bar{y}\right](t) + \frac{1}{s+a_m}\left[\frac{1}{s+a_m}[|\dot{k}|\bar{y}]\right](t), \tag{5.109}$$

where $b_0(t) = \frac{1}{s+a_m}\left[|y_m| + \frac{|\epsilon|}{m}\right](t)$ is bounded as both $y_m(t)$ and $\frac{|\epsilon(t)|}{m(t)}$ are.

Let $\alpha > 0$ and $\beta > 0$ be such that

$$\alpha \leq a_m(1 + \ln(a_m\beta)), \tag{5.110}$$

which ensures that

$$\beta e^{-\alpha t} \geq te^{-a_m t},\ t \geq 0. \tag{5.111}$$

Using (5.111) and the expression $\frac{1}{s+a_m}[\frac{1}{s+a_m}[f]](t) = \int_0^t (t-\tau)e^{-a_m(t-\tau)} f(\tau)\, d\tau$, we express (5.109) as

$$\begin{aligned}
\bar{y}(t) &\leq b_0(t) + \int_0^t e^{-a_m(t-\tau)}\frac{|\epsilon(\tau)|}{m(\tau)}\bar{y}(\tau)\, d\tau + \int_0^t (t-\tau)e^{-a_m(t-\tau)}|\dot{k}(\tau)|\bar{y}(\tau)\, d\tau \\
&\leq b_0(t) + \int_0^t e^{-a_m(t-\tau)}\frac{|\epsilon(\tau)|}{m(\tau)}\bar{y}(\tau)\, d\tau + \int_0^t \beta e^{-\alpha(t-\tau)}|\dot{k}(\tau)|\bar{y}(\tau)\, d\tau \\
&\leq b_0(t) + \int_0^t e^{-a_m(t-\tau)}\frac{|\epsilon(\tau)|}{m(\tau)}\bar{y}(\tau)\, d\tau + \int_0^t \beta e^{-\alpha(t-\tau)}|\dot{k}(\tau)|\bar{y}(\tau)\, d\tau \\
&\leq b_0(t) + \bar{\beta}\int_0^t e^{-\bar{\alpha}(t-\tau)}(\frac{|\epsilon(\tau)|}{m(\tau)} + |\dot{k}(\tau)|)\bar{y}(\tau)\, d\tau,
\end{aligned} \tag{5.112}$$

where

$$\bar{\beta} = \max\{1, \beta\},\ \bar{\alpha} = \min\{a_m, \alpha\}. \tag{5.113}$$

Using Lemma 2.3, with $z_1(t) = \bar{y}(t)$, we conclude that $\bar{y}(t) \in L^\infty$. Then, from (5.107), $y(t) \in L^\infty$, and from (5.99), $u(t) \in L^\infty$. From (5.101), it can be verified that $\dot{\epsilon}(t) \in L^\infty$ so that $\lim_{t\to\infty} \epsilon(t) = 0$ and $\lim_{t\to\infty} \dot{k}(t) = 0$. From (5.101) and (5.104), it follows that $\lim_{t\to\infty} e(t) = 0$. □

Example 5.3 We consider the first-order plant

$$\dot{y}(t) = a_p y(t) + b_p u(t) \tag{5.114}$$

with two unknown parameters a_p and b_p and the reference model

$$\dot{y}_m(t) = -a_m y_m(t) + b_m r(t) \tag{5.115}$$

with $a_m > 0$. We use the adaptive controller structure

$$u(t) = k_1(t)y(t) + k_2(t)r(t), \tag{5.116}$$

where $k_1(t)$ and $k_2(t)$ are estimates of the unknown parameters

$$k_1^* = \frac{-a_p - a_m}{b_p},\ k_2^* = \frac{b_m}{b_p}. \tag{5.117}$$

In this case the tracking error equation becomes

$$\dot{e}(t) = -a_m e(t) + b_p \tilde{k}_1(t) y(t) + b_p \tilde{k}_2(t) r(t), \tag{5.118}$$

where $\tilde{k}_1(t) = k_1(t) - k_1^*$ and $\tilde{k}_2(t) = k_2(t) - k_2^*$.

Defining $\rho^* = b_p$ and

$$\theta^* = [k_1^*, k_2^*]^T, \ \theta = [k_1, k_2]^T, \tag{5.119}$$

$$\omega(t) = [y(t), r(t)]^T \tag{5.120}$$

and introducing the filtered vector signal

$$\zeta(t) = \frac{1}{s + a_m}[\omega](t) = \left[\frac{1}{s + a_m}[y](t), \frac{1}{s + a_m}[r](t)\right]^T, \tag{5.121}$$

where $\frac{1}{s+a_m}[y](t)$ denotes the output of the system with transfer function $\frac{1}{s+a_m}$ and input $y(t)$, we rewrite (5.119) as

$$\begin{aligned} e(t) &= \frac{\rho^*}{s + a_m}[\theta^T \omega - \theta^{*T} \omega](t) \\ &= \rho^* \left(\frac{1}{s + a_m}[\theta^T \omega](t) - \theta^{*T} \zeta(t)\right). \end{aligned} \tag{5.122}$$

We look for adaptive laws to update the parameter estimates $\theta(t)$ and $\rho(t)$ (which is an estimate of ρ^*) such that the estimation error

$$\epsilon(t) = e(t) - \rho(t) \left(\frac{1}{s + a_m}[\theta^T \omega](t) - \theta^T(t) \zeta(t)\right) \tag{5.123}$$

is small in some sense. With (5.122), this error can be expressed as

$$\epsilon(t) = \rho^* \tilde{\theta}^T(t) \zeta(t) + \tilde{\rho}(t) \xi(t), \tag{5.124}$$

where, following (5.103), with $b_1(t) = k_2(t)\frac{1}{s+a_m}[r](t) - \frac{1}{s+a_m}[k_2 r](t) \in L^\infty$,

$$\xi(t) \triangleq \theta^T(t)\zeta(t) - \frac{1}{s + a_m}[\theta^T \omega](t) = \frac{1}{s + a_m}\left[\dot{k}_1 \frac{1}{s + a_m}[y]\right](t) + b_1(t), \tag{5.125}$$

$$\tilde{\theta}(t) = \theta(t) - \theta^*, \ \tilde{\rho}(t) = \rho(t) - \rho^*. \tag{5.126}$$

Then, we can use the adaptive scheme (5.91)–(5.92) to update the estimates $\theta(t)$ and $\rho(t)$, and from Lemma 5.3, we have that $\theta(t)$, $\rho(t)$, $\frac{\epsilon(t)}{m(t)}$, $\dot{\theta}(t)$, and $\dot{\rho}(t)$ are all bounded and $\frac{\epsilon(t)}{m(t)} \in L^2$, $\dot{\theta}(t) \in L^2$. To prove closed-loop signal boundedness and asymptotic tracking, using (5.125), we write (5.123) as

$$\begin{aligned} y(t) &= y_m(t) + \epsilon(t) + \rho(t)\xi(t) \\ &= y_m(t) + \frac{\epsilon(t)}{m(t)} m(t) + \rho(t)\left(\frac{1}{s + a_m}\left[\dot{k}\frac{1}{s + a_m}[y]\right](t) + b_1(t)\right). \end{aligned} \tag{5.127}$$

From the definition of $m(t)$, it follows that

$$
\begin{aligned}
m(t) &\leq 1 + \|\zeta(t)\|_1 + |\xi(t)| \\
&\leq b_2(t) + \left|\frac{1}{s+a_m}[y](t)\right| + \left|\frac{1}{s+a_m}\left[\dot{k}_1 \frac{1}{s+a_m}[y]\right](t)\right|,
\end{aligned}
\tag{5.128}
$$

where $b_2(t) = 1 + |\frac{1}{s+a_m}[r](t)| + |b_1(t)|$. From (5.127) and (5.128), we obtain

$$
\begin{aligned}
|y(t)| \leq |y_m(t)| &+ \frac{|\epsilon(t)|}{m(t)} b_2(t) + |\rho(t) b_1(t)| + \frac{|\epsilon(t)|}{m(t)} \frac{1}{s+a_m}[|y|](t) \\
&+ \left(\frac{|\epsilon(t)|}{m(t)} + |\rho(t)|\right) \frac{1}{s+a_m}\left[|\dot{k}_1| \frac{1}{s+a_m}[|y|]\right](t).
\end{aligned}
\tag{5.129}
$$

Denoting $\bar{y}(t) = \frac{1}{s+a_m}[|y|](t)$, from (5.129), we obtain

$$
\bar{y}(t) \leq c_1 + \frac{1}{s+a_m}\left[\frac{|\epsilon|}{m}\bar{y}\right](t) + \frac{c_2}{s+a_m}\left[\frac{1}{s+a_m}[|\dot{k}_1|\bar{y}]\right](t),
\tag{5.130}
$$

where $\frac{1}{s+a_m}[|y_m| + \frac{|\epsilon|}{m} b_2 + |\rho b_1|](t) \leq c_1 < \infty$ and $\frac{|\epsilon(t)|}{m(t)} + |\rho(t)| \leq c_2 < \infty$. Then, following an analysis similar to that after (5.110) in Example 5.1, we can conclude that $\bar{y}(t) \in L^\infty$, and so $y(t) \in L^\infty$, $u(t) \in L^\infty$, and $\dot{e}(t) \in L^\infty$, and that $e(t) \in L^2$ so that $\lim_{t\to\infty} e(t) = 0$. □

Theorem 5.4 *All signals in the closed-loop system consisting of the plant (5.1), reference model (5.4), controller (5.34), and adaptive law (5.91)–(5.92) are bounded, and the tracking error $e(t) = y(t) - y_m(t)$ belongs to L^2 and reduces to zero asymptotically with time.*

Proof: Introducing a fictitious signal

$$
z(t) = \frac{1}{s+a_0}[y](t),
\tag{5.131}
$$

where $a_0 > 0$, and using the fictitious signal and filters $z_0(t)$, $K_1(s)$, and $K(s)$ defined in (5.71) and (5.72), we obtain

$$
z_0(t) + a_0 K_1(s)[z_0](t) - K_1(s)[u](t) = K(s) G_0^{-1}(s)[z](t).
\tag{5.132}
$$

Hence (5.34), operated by $K_1(s)$ on both sides, with the substitution of (5.71) and (5.131), gives the identity

$$
\begin{aligned}
K_1(s)[u](t) = K_1(s)&\left[\theta_1^T \frac{a(s)}{\Lambda(s)}(s+a_0)[z_0]\right](t) + K_1(s)\left[\theta_2^T \frac{a(s)}{\Lambda(s)}(s+a_0)[z]\right](t) \\
&+ K_1(s)[\theta_{20}(s+a_0)[z]](t) + K_1(s)[\theta_3 r](t).
\end{aligned}
\tag{5.133}
$$

Using the identity

$$\theta_{20}(t)(s+a_0)[z]](t) = \frac{d}{dt}(\theta_{20}(t)z(t)) - z(t)\dot{\theta}_{20} + \theta_{20}a_0 z(t) \tag{5.134}$$

and substituting (5.133) into (5.132), we see

$$\begin{aligned}&\left(1 + K_1(s)\left(a_0 - \theta_1^T(\cdot)\frac{a(s)}{\Lambda(s)}(s+a_0)\right)\right)[z_0](t)\\ &= \left(K(s)G_0^{-1}(s) + K_1(s)\theta_2^T(\cdot)\frac{a(s)}{\Lambda(s)}(s+a_0)\right)[z](t)\\ &\quad + K_1(s)[s[\theta_{20}z] - zs[\theta_{20}] + a_0\theta_{20}z](t) + K_1(s)[\theta_3 r](t).\end{aligned} \tag{5.135}$$

From (5.80) there exists $a^0 > 0$ such that for any finite $a > a^0$ the operator

$$T_0(s,t) \triangleq \left(1 + K_1(s)\left(a_0 - \theta_1^T(t)\frac{a(s)}{\Lambda(s)}(s+a_0)\right)\right)^{-1} \tag{5.136}$$

is stable and proper (see Lemma 2.5). Let $a > a^0$ be finite and fixed in $K(s)$ and $K_1(s)$. Then (5.135) implies that

$$z_0(t) = T_1(s,\cdot)[z](t) + b_0(t), \tag{5.137}$$

where $T_1(s,t)$ is a stable and proper operator and $b_0(t) \in L^\infty$ due to $r(t)$.

Denoting $P_m(s) = s^{n^*} + a_{n^*-1}s^{n^*-1} + \cdots + a_1 s + a_0$, we express $\xi(t)$ as

$$\begin{aligned}\xi(t) &= \frac{s^{n^*-1} + a_{n^*-1}s^{n^*-2} + \cdots + a_2 s + a_1}{P_m(s)}\left[\dot{\theta}^T\frac{1}{P_m(s)}[\omega]\right](t)\\ &+ \frac{s^{n^*-2} + a_{n^*-1}s^{n^*-3} + \cdots + a_2}{P_m(s)}\left[\dot{\theta}^T\frac{s}{P_m(s)}[\omega]\right](t) + \cdots\\ &+ \frac{s + a_{n^*-1}}{P_m(s)}\left[\dot{\theta}^T\frac{s^{n^*-2}}{P_m(s)}[\omega]\right](t) + \frac{1}{P_m(s)}\left[\dot{\theta}^T\frac{s^{n^*-1}}{P_m(s)}[\omega]\right](t).\end{aligned} \tag{5.138}$$

Filtering both sides of (5.88) by $\frac{1}{s+a_0}$, we obtain

$$z(t) = \frac{1}{s+a_0}[y_m](t) + \frac{1}{s+a_0}[\epsilon - \rho\xi](t). \tag{5.139}$$

Then, the inequality

$$|\epsilon(t)| \le \frac{|\epsilon(t)|}{m_0(t)}(1 + \|\zeta(t)\|_1 + |\xi(t)|), \tag{5.140}$$

Lemma 5.3, (5.89), and (5.137)–(5.139) imply that

$$|z(t)| \le x_0(t) + T_2(s,\cdot)[x_1 T_3(s,\cdot)[|z|]](t) \tag{5.141}$$

for some $x_0(t) \in L^\infty$, $x_1(t) \in L^\infty \cap L^2$ with $x_1(t) \geq 0$, $T_2(s,t)$ being a stable and strictly proper operator, $T_3(s,t)$ being a stable and proper operator with a nonnegative impulse response.

Introducing $z_1(t) = T_3(s,\cdot)[|z|](t)$ and operating $T_3(s,t)$ on both sides of (5.141), and noting that $T_3(s,t)$ has a nonnegative impulse response, we have

$$z_1(t) \leq b_1 + b_2 \int_0^t e^{-\alpha(t-\tau)} x_1(\tau) z_1(\tau)\, d\tau \tag{5.142}$$

for some α, b_1, $b_2 > 0$. Applying Lemma 2.3 to (5.142) with $x_1(t) \in L^2 \cap L^\infty$ satisfying (2.103), we conclude that $z_1(t) \in L^\infty$, and so $z(t) \in L^\infty$. That is, $z_0(t) \in L^\infty$ in (5.137), $\xi(t) \in L^\infty$ in (5.89), $\zeta(t) \in L^\infty$ in (5.90), $\epsilon(t) \in L^\infty$ in (5.94), $y(t) \in L^\infty$ in (5.88), and $u(t) \in L^\infty$ in (5.34). From (5.94) we have $\dot{\epsilon}(t) \in L^\infty$, which, with $\epsilon(t) \in L^2 \cap L^\infty$, implies $\lim_{t\to\infty} \epsilon(t) = 0$ in (5.94) and $\lim_{t\to\infty} \dot{\theta}(t) = 0$ in (5.91). From this and $\dot{\theta}(t) \in L^2$, using Corollary 2.7 and Lemma 2.19, it follows that $\xi(t) \in L^2$ and $\lim_{t\to\infty} \xi(t) = 0$ in (5.138). Finally, from (5.88), we have $e(t) = y(t) - y_m(t) \in L^2$ and $\lim_{t\to\infty} e(t) = 0$. ∇

5.3.6 Summary

There are several key steps in developing an adaptive control scheme. The first step is to build a controller structure which, when implemented with knowledge of the plant parameters, is able to achieve a desired control objective. Then, the derivation of a tracking error equation like (5.39) is important. For the relative degree $n^* = 1$ case, an adaptive control scheme based on a complete Lyapunov design leads to the desired stability and tracking performance (see Theorem 5.2). For higher relative degree cases, the use of an estimation error like that in (5.56) or (5.88) is crucial, based on which an adaptive law like (5.60)–(5.61) or (5.91)–(5.92) is developed to have certain desired properties as in Lemmas 5.2 and 5.3. The key feature of the developed Lyapunov and gradient adaptive control schemes is that the controller (5.34) with bounded parameters leads to the closed-loop inequality (5.86) or (5.141) (see Figure 5.4), while the adaptive law (5.60)–(5.61) or (5.91)–(5.92), through the L^2 property of $\dot{\theta}(t)$, $\epsilon(t)m_0(t)$ or $\frac{\epsilon(t)}{m_0(t)}$, ensures that the loop gain of (5.87) or (5.142) is small so that the signal boundedness is guaranteed.

An adaptive controller is able to handle parameter uncertainties of the plant model (5.1), $P(s)[y](t) = k_p Z(s)[u](t)$, that is, it is applicable to the unknown parameters k_p and p_i, $i = 0, 1, \ldots, n-1$, z_j, $j = 0, 1, \ldots, m-1$, in $P(s) = s^n + p_{n-1}s^{n-1} + \cdots + p_1 s + p_0$, $Z(s) = s^m + z_{m-1}s^{m-1} + \cdots + z_1 s + z_0$. Next, we will investigate adaptive control schemes for plant models with modeling errors such as disturbances $d(t)$ and unmodeled dynamics $\Delta(s)[u](t)$.

5.4 Robustness of MRAC

An important question is: Is an adaptive controller robust with respect to modelings errors? In this section, we show that model reference adaptive control (MRAC) schemes are inherently robust with respect to $L^2 \cap L^\infty$ modeling errors $d(t)$ (or $L^{1+\alpha} \cap L^\infty$ modeling errors $d(t)$, $1 < \alpha < \infty$, with a modified adaptive law similar to that used in Section 3.7.1 for the parameter estimation case), in the sense that closed-loop signal boundedness and asymptotic tracking are ensured in the presence of such modeling errors.

We should note that such robustness properties *no longer* hold for more general modeling errors $d(t)$ which belong to L^∞ only (see Section 5.5, where different modifications of adaptive laws are needed for ensuring robustness of stability and tracking, which is an important issue for adaptive control).

For modeling errors $d(t)$ such that $\lim_{t\to\infty} d(t) = 0$, see Section 5.7.

5.4.1 Lyapunov Designs for Relative Degree 1

We examine the inherent robustness of the Lyapunov designs in Sections 5.3.2 and 5.3.3, for the plant (5.1) with relative degree $n^* = 1$, to modeling errors. Consider the following version of the error system (5.41):

$$\dot{e}(t) = -a_m e(t) + k_p \tilde{\theta}^T(t)\omega(t) + d(t), \tag{5.143}$$

where $d(t)$ represents modeling errors which belong to $L^{1+\alpha} \cap L^\infty$, $\alpha \geq 1$.

Robustness to $L^2 \cap L^\infty$ Errors

We first consider the case when $d(t) \in L^2 \cap L^\infty$ and the adaptive law (5.45) is used. The time derivative of the positive definite function $V(e, \tilde{\theta}) = e^2 + |k_p|\tilde{\theta}^T\Gamma^{-1}\tilde{\theta}$, along (5.41) and (5.45), is

$$\begin{aligned}\dot{V} &= -2a_m e^2(t) + 2e(t)d(t) \\ &= -a_m e^2(t) - a_m\left(e(t) - \frac{d(t)}{a_m}\right)^2 + \frac{d^2(t)}{a_m},\end{aligned} \tag{5.144}$$

from which, using $\int_0^\infty d^2(t)\,dt < \infty$ as $d(t) \in L^2$, we derive

$$V(e(t), \tilde{\theta}(t)) + 2a_m \int_0^t e^2(t)\,d\tau \leq V(e(0), \tilde{\theta}(0)) + \frac{1}{a_m}\int_0^t d^2(t)\,d\tau < \infty. \tag{5.145}$$

Hence, $V(e(t), \tilde{\theta}(t))$ is bounded, and so are $e(t)$ and $\theta(t)$, and $e(t) \in L^2$. Then $u(t) \in L^\infty$ and $\dot{e}(t) \in L^\infty$ as $d(t) \in L^\infty$, so that $\lim_{t\to\infty} e(t) = 0$.

Robustness to $L^{1+\alpha} \cap L^\infty$ Errors

When $d(t) \in L^{1+\alpha} \cap L^\infty$ for $0 < \alpha < \infty$ (for adaptation robustness investigation, the case of $1 < \alpha < \infty$ is of our main interest, as $L^{1+\beta} \cap L^\infty \subset L^{1+\alpha} \cap L^\infty$ for $0 < \beta < \alpha$; in particular, $L^{1+\beta} \cap L^\infty \subset L^2 \cap L^\infty$ for $0 < \beta < 1$), we use a modified adaptive parameter update law similar to that in (5.52):

$$\dot{\theta}(t) = -\mathrm{sign}[k_p]\Gamma\omega(t)\,\mathrm{sign}[e(t)]\,|e(t)|^\alpha. \tag{5.146}$$

In this case, the time derivative of $V(e, \tilde{\theta}) = \frac{|e(t)|^{1+\alpha}}{1+\alpha} + \frac{|k_p|\tilde{\theta}^T\Gamma^{-1}\tilde{\theta}}{2}$ is

$$\dot{V} = -a_m|e(t)|^{1+\alpha} + f(e(t))d(t) \leq -a_m|e(t)|^{1+\alpha} + |e(t)|^\alpha\,|d(t)|, \tag{5.147}$$

from which we obtain

$$\begin{aligned} & V(e(t), \tilde{\theta}(t)) + a_m \int_0^t |e(t)|^{1+\alpha}\,d\tau \\ & \leq V(e(0), \tilde{\theta}(0)) + \int_0^t |e(t)|^\alpha\,|d(t)|\,d\tau, \\ & \leq V(e(0), \tilde{\theta}(0)) + \left(\int_0^t |e(t)|^{1+\alpha}\,d\tau\right)^{\frac{\alpha}{1+\alpha}} \left(\int_0^t |d(t)|^{1+\alpha}\,d\tau\right)^{\frac{1}{1+\alpha}}. \end{aligned} \tag{5.148}$$

Since $\int_0^t |d(t)|^{1+\alpha}\,d\tau \leq d_0 < \infty$ for some constant $d_0 > 0$, as $d(t) \in L^{1+\alpha}$, from (5.148), we have $a_m \int_0^t |e(t)|^{1+\alpha}\,d\tau \leq \left(\int_0^t |e(t)|^{1+\alpha}\,d\tau\right)^{\frac{\alpha}{1+\alpha}} d_0$, which implies that $\int_0^t |e(t)|^{1+\alpha}\,d\tau \in L^\infty$, so that $V(e(t), \tilde{\theta}(t)) \in L^\infty$, that is, $e(t) \in L^\infty$, $\theta(t) \in L^\infty$ and $e(t) \in L^{1+\alpha}$. Then $u(t) \in L^\infty$ and $\dot{e}(t) \in L^\infty$, so that $\lim_{t\to\infty} e(t) = 0$.

Remark 5.1 We can replace α by $\bar{\alpha} \in (\alpha, \infty)$ in the adaptive law (5.146) to derive the signal boundedness but with $e(t) \in L^{1+\bar{\alpha}}$ instead. □

An Illustrative Example

Let us reexamine the parameter drift example of [148]:

$$\dot{y}(t) = ay(t) + u(t) + d(t), \tag{5.149}$$

where the bounded decaying disturbance is

$$d(t) = \frac{1}{(1+t)^{\frac{1}{8}}}\left(1 - \frac{a}{(1+t)^{\frac{1}{4}}} - \frac{3}{8(1+t)^{\frac{5}{4}}}\right) \tag{5.150}$$

and the adaptive controller is

$$u(t) = -k(t)y(t),\ \dot{k}(t) = \gamma y^2(t),\ \gamma > 0. \tag{5.151}$$

As in [148], with $y(0) = 1$, $k(0) = 1$, $\gamma = 0.25$, the solution to this system is $y(t) = \frac{1}{(1+t)^{\frac{3}{8}}}$, $k(t) = (1+t)^{\frac{1}{4}}$, from which we see that the asymptotic tracking of $y_m(t) = 0$ by $y(t)$ is achieved, and, however, the parameter $k(t)$ drifts to ∞ as $t \to \infty$. This example shows that the adaptive control scheme (5.150) is not robust with respect to the disturbance $d(t)$, which decays to zero at $t \to \infty$.

Since this $d(t)$ is an $L^{1+\alpha}$ signal with $\alpha > 7$, we now take $\bar{\alpha} = 9 > \alpha$ to design the adaptive law (5.52) with $f(y(t)) = y^9(t)$ to update $k(t)$:

$$\dot{k}(t) = \gamma y^{10}(t), \ \gamma > 0. \tag{5.152}$$

Then, the time derivative of the positive definite function

$$V(y, \tilde{k}) = \frac{y^{10}}{10} + \frac{\tilde{k}^2}{2\gamma}, \ \tilde{k} = k - k^*, \ k^* > a, \tag{5.153}$$

along the trajectories of (5.149) and (5.152) is

$$\dot{V} = -(k^* - a)y^{10}(t) + y^9(t)d(t), \tag{5.154}$$

from which we obtain

$$\begin{aligned} &(k^* - a)\int_0^t y^{10}(\tau)\,d\tau + V(y(t), \tilde{k}(t)) \\ &\leq \int_0^t |y(\tau)|^9|d(\tau)|\,d\tau + V(y(0), \tilde{k}(0)), \ \forall t \geq 0. \end{aligned} \tag{5.155}$$

Since $d(t) \in L^{10}$, it follows that

$$\begin{aligned} \int_0^t |y(\tau)|^9|d(\tau)|\,d\tau &\leq \left(\int_0^t y^{10}(\tau)\,d\tau\right)^{\frac{9}{10}} \left(\int_0^t d^{10}(\tau)\,d\tau\right)^{\frac{1}{10}} \\ &\leq c_d \left(\int_0^t y^{10}(\tau)\,d\tau\right)^{\frac{9}{10}} \end{aligned} \tag{5.156}$$

for some constant $c_d > 0$. From (5.155) and (5.156), we conclude that both $\int_0^t y^{10}(\tau)\,d\tau$ and $V(y(t), \tilde{k}(t))$ are bounded, that is, $y(t) \in L^{10}$, $k(t) \in L^\infty$, $y(t) \in L^\infty$. It then follows from $u(t) = -k(t)y(t)$ and (5.149) that $u(t) \in L^\infty$, $\dot{y}(t) \in L^\infty$, and so we have $\lim_{t\to\infty} y(t) = 0$.

5.4.2 Gradient Algorithms

In this section we prove that a model reference adaptive control (MRAC) system with a standard gradient adaptive law is inherently robust with respect to $L^\infty \cap L^2$ disturbances in the sense that all closed-loop signals remain bounded and the tracking error $e(t)$ belongs to $L^\infty \cap L^2$. We also show that a MRAC scheme with a modified adaptive law based on a nonquadratic cost function is inherently robust with respect to disturbances in $L^\infty \cap L^{1+\alpha}$, $0 < \alpha < \infty$, with an $L^{1+\beta}$ tracking error, for $\beta = \max\{\bar{\alpha}, \frac{1}{\bar{\alpha}}\}$, $\alpha \leq \bar{\alpha} < \infty$.

System with Disturbances

In the presence of an input disturbance $d_u(t)$ and an output disturbance $d_y(t)$, the plant model (5.1) is expanded as

$$P(s)[y - d_y](t) = k_p Z(s)[u + d_u](t). \tag{5.157}$$

An important issue in adaptive control has been the robustness of the signal boundedness and tracking properties with respect to these disturbances. As illustrated in [23], [147], [151], [282], [340], without modification of an adaptive law, a model reference adaptive controller applied to the plant (5.157) with general disturbances $d_u(t)$ and $d_y(t)$ may not ensure signal boundedness.

When the controller (5.34), that is,

$$u(t) = \theta_1^T(t)\omega_1(t) + \theta_2^T(t)\omega_2(t) + \theta_{20}(t)y(t) + \theta_3(t)r(t), \tag{5.158}$$

is applied to the plant (5.157), the tracking error equation becomes

$$e(t) = y(t) - y_m(t) = \frac{k_p}{P_m(s)}[\tilde{\theta}^T\omega](t) + d(t) + \epsilon_0(t), \tag{5.159}$$

where $\tilde{\theta}(t)$ and $\omega(t)$ are defined in (5.35)–(5.38),

$$\begin{aligned} d(t) = &\left(1 + \frac{k_p}{P_m(s)}\left(\theta_2^{*T}\frac{a(s)}{\Lambda(s)} + \theta_{20}^*\right)\right)[d_y](t) \\ &+ \frac{k_p}{P_m(s)}\left(1 - \theta_1^{*T}\frac{a(s)}{\Lambda(s)}\right)[d_u](t), \end{aligned} \tag{5.160}$$

and $\epsilon_0(t)$ is an exponentially decaying term due to initial conditions. From (5.159), we express the estimation error $\epsilon(t) = e(t) + \rho(t)\xi(t)$ in (5.88) as

$$\epsilon(t) = k_p\tilde{\theta}^T(t)\zeta(t) + \tilde{\rho}(t)\xi(t) + d(t) + \epsilon_0(t), \tag{5.161}$$

where $\xi(t) = \theta^T(t)\zeta(t) - \frac{1}{P_m(s)}[\theta^T\omega](t)$, $\zeta(t) = \frac{1}{P_m(s)}[\omega](t)$, as in (5.89) and (5.90), and $\tilde{\rho}(t) = \rho(t) - \rho^*$ with $\rho^* = k_p$.

L^2 Robustness of MRAC

Consider the plant (5.157) with $d_u(t) \in L^\infty \cap L^2$ and $d_y(t) \in L^\infty \cap L^2$. From the definition of $d(t)$ in (5.160) and L^2 stability theory (see Lemma 2.19), it follows that $d(t) \in L^\infty \cap L^2$. Since $\epsilon_0(t) \in L^\infty \cap L^2$, we can ignore it in the presence of $d(t)$. With the adaptive law (5.91)–(5.92), the time derivative of the positive definite function $V(\tilde{\theta}, \tilde{\rho}) = |k_p|\tilde{\theta}^T\Gamma^{-1}\tilde{\theta} + \gamma^{-1}\tilde{\rho}^2$, along (5.91)–(5.92), is

$$\dot{V} = -\frac{2\epsilon^2(t)}{m^2(t)} + \frac{2\epsilon(t)d(t)}{m^2(t)} \leq \frac{-\epsilon^2(t)}{m^2(t)} + \frac{d^2(t)}{m^2(t)}. \tag{5.162}$$

From this expression, it follows that

$$\int_0^t \frac{\epsilon^2(\tau)}{m^2(\tau)}\,d\tau + 2V(\tilde{\theta}(t),\tilde{\rho}(t)) \le \int_0^t \frac{d^2(\tau)}{m^2(\tau)}\,d\tau + 2V(\tilde{\theta}(0),\tilde{\rho}(0)) \le c_1, \quad (5.163)$$

for some constant $c_1 > 0$, as $d(t) \in L^2$. From (5.163), we conclude that $\frac{\epsilon(t)}{m(t)} \in L^2$ and $V(\tilde{\theta}(t),\tilde{\rho}(t)) \in L^\infty$, that is, $\theta(t), \rho(t) \in L^\infty$. It can then be shown that $\frac{\epsilon(t)}{m(t)} \in L^\infty$ and $\dot{\theta}(t), \dot{\rho}(t) \in L^\infty \cap L^2$.

Following the stability proof of Theorem 5.4, it can be shown that all closed-loop signals are bounded and $e(t) \in L^2$. If, in addition, $\lim_{t\to\infty} d_y(t) = 0$, then $\lim_{t\to\infty} d(t) = 0$ so that $\lim_{t\to\infty} e(t) = 0$. Note that for $d_u(t) \in L^2$ the signal

$$\bar{d}_u(t) \triangleq \frac{k_p}{P_m(s)}\left(1 - \theta_1^{*T}\frac{a(s)}{\Lambda(s)}\right)[d_u](t) \quad (5.164)$$

is such that $\bar{d}_u(t) \in L^\infty$ and $\lim_{t\to\infty} \bar{d}_u(t) = 0$, because $\frac{k_p}{P_m(s)}(1 - \theta_1^{*T}\frac{a(s)}{\Lambda(s)})$ is stable and strictly proper (see Lemma 2.19).

In summary, we have the following robustness result.

Theorem 5.5 *The controller (5.34) updated from the adaptive law (5.91)–(5.92), applied to the plant (5.157) with $d_u(t) \in L^\infty \cap L^2$, $d_y(t) \in L^\infty \cap L^2$, ensures that all closed-loop system signals are bounded and the tracking error $e(t) \in L^2$. If, in addition, $\lim_{t\to\infty} d_y(t) = 0$, then $\lim_{t\to\infty} e(t) = 0$.*

$L^{1+\alpha}$ Robustness of MRAC

Now we consider the plant (5.157) with $d_u(t) \in L^\infty \cap L^{1+\alpha}$ and $d_y(t) \in L^\infty \cap L^{1+\alpha}$ for $0 < \alpha < \infty$ (especially, $1 \le \alpha < \infty$). Similar to the L^2 case, it follows that $d(t)$ in (5.160) is such that $d(t) \in L^\infty \cap L^{1+\alpha}$ (see Lemma 2.7).

In this case, the parameters of the controller (5.34) (that is, (5.158)) are updated from the modified gradient algorithm

$$\dot{\theta}(t) = -\frac{\mathrm{sign}[k_p]\Gamma f(\epsilon(t))\zeta(t)}{m^{1+\bar{\alpha}}(t)}, \quad \Gamma = \Gamma^T > 0 \quad (5.165)$$

$$\dot{\rho}(t) = -\frac{\gamma f(\epsilon(t))\xi(t)}{m^{1+\bar{\alpha}}(t)}, \quad \gamma > 0, \quad (5.166)$$

for $m(t)$ in (5.93), where the continuous function $f(\cdot)$ is, for $\alpha \le \bar{\alpha} < \infty$,

$$f(\epsilon(t)) = \begin{cases} (\epsilon(t))^{\bar{\alpha}} & \text{if } \epsilon(t) \ge 0, \\ -(-\epsilon(t))^{\bar{\alpha}} & \text{if } \epsilon(t) < 0. \end{cases} \quad (5.167)$$

This gradient algorithm is based on the normalized nonquadratic cost function $J(\theta, \rho) = \frac{|\epsilon|^{\bar{\alpha}+1}}{m^2}$, where $\epsilon = k_p^* \tilde{\theta}^T \zeta + \tilde{\rho}\xi + d$ as in (5.161). In this case, the gradients of $J(\theta, \rho)$ are $\frac{\partial J}{\partial \theta} = \frac{\rho^* \zeta f(\epsilon)}{m^2}$ with $\rho^* = k_p$ and $\frac{\partial J}{\partial \rho} = \frac{\xi f(\epsilon)}{m^2}$, whose directions are the directions of $\frac{\text{sign}[k_p]\zeta\epsilon}{m^2}$ and $\frac{\xi\epsilon}{m^2}$. Therefore, with adaptation gains $\Gamma = \Gamma^T > 0$, $\gamma > 0$, the adaptive laws are those in (5.165) and (5.166).

The time derivative of the positive definite function $V(\tilde{\theta}, \tilde{\rho}) = |k_p|\tilde{\theta}^T \Gamma^{-1} \tilde{\theta} + \gamma^{-1}\tilde{\rho}^2$ along the trajectories of (5.165) and (5.166) is

$$\dot{V} = \frac{-f(\epsilon(t))\epsilon(t)}{m^{1+\bar{\alpha}}(t)} + \frac{f(\epsilon(t))d(t)}{m^{1+\bar{\alpha}}(t)} \leq \frac{-|\epsilon(t)|^{1+\bar{\alpha}}}{m^{1+\bar{\alpha}}(t)} + \frac{|\epsilon(t)|^{\bar{\alpha}}|d(t)|}{m^{1+\bar{\alpha}}(t)}, \tag{5.168}$$

from which it follows that

$$\int_0^t \frac{|\epsilon(\tau)|^{1+\bar{\alpha}}}{m^{1+\bar{\alpha}}(\tau)}\, d\tau + V(\tilde{\theta}(t), \tilde{\rho}(t)) \leq \int_0^t \frac{|\epsilon(\tau)|^{\bar{\alpha}}|d(\tau)|}{m^{1+\bar{\alpha}}(\tau)}\, d\tau + V(\tilde{\theta}(0), \tilde{\rho}(0)). \tag{5.169}$$

Using the Hölder inequality (see Section 2.3.2), we have

$$\int_0^t \frac{|\epsilon(\tau)|^{\bar{\alpha}}|d(\tau)|}{m^{1+\bar{\alpha}}(\tau)}\, d\tau \leq \left(\int_0^t \left(\frac{|\epsilon(\tau)|^{\bar{\alpha}}}{m^{\bar{\alpha}}(\tau)}\right)^{\frac{1+\bar{\alpha}}{\bar{\alpha}}} d\tau\right)^{\frac{\bar{\alpha}}{1+\bar{\alpha}}} \left(\int_0^t \left(\frac{|d(\tau)|}{m(\tau)}\right)^{1+\bar{\alpha}} d\tau\right)^{\frac{1}{1+\bar{\alpha}}}. \tag{5.170}$$

Since $d(t) \in L^\infty \cap L^{1+\alpha}$ and $\bar{\alpha} \geq \alpha$, it follows that

$$\left(\int_0^t \left(\frac{|d(\tau)|}{m(\tau)}\right)^{1+\bar{\alpha}} d\tau\right)^{\frac{1}{1+\bar{\alpha}}} = \left(\int_0^t \left(\frac{|d(\tau)|}{m(\tau)}\right)^{1+\alpha}\left(\frac{|d(\tau)|}{m(\tau)}\right)^{\bar{\alpha}-\alpha} d\tau\right)^{\frac{1}{1+\bar{\alpha}}} \leq c_d \tag{5.171}$$

for some constant $c_d > 0$. From (5.169)–(5.171), we obtain

$$\int_0^t \frac{|\epsilon(\tau)|^{1+\bar{\alpha}}}{m^{1+\bar{\alpha}}(\tau)}\, d\tau + V(\tilde{\theta}(t), \tilde{\rho}(t)) \leq c_d \left(\int_0^t \frac{|\epsilon(\tau)|^{1+\bar{\alpha}}}{m^{1+\bar{\alpha}}(\tau)}\, d\tau\right)^{\frac{\bar{\alpha}}{1+\bar{\alpha}}} + c_0, \tag{5.172}$$

where $c_0 = V(\tilde{\theta}(0), \tilde{\rho}(0))$. Define $h(t) = \int_0^t \frac{|\epsilon(\tau)|^{1+\bar{\alpha}}}{m^{1+\bar{\alpha}}(\tau)}\, d\tau$. From (5.172), we get

$$h(t) \leq c_d (h(t))^{\frac{\bar{\alpha}}{1+\bar{\alpha}}} + V(\tilde{\theta}(0), \tilde{\rho}(0)), \tag{5.173}$$

which means that $h(t) \in L^\infty$, that is, $\frac{\epsilon(t)}{m(t)} \in L^{1+\bar{\alpha}}$. With this result, using (5.172) again, we have $V(\tilde{\theta}(t), \tilde{\rho}(t)) \in L^\infty$, so that $\theta(t), \rho(t) \in L^\infty$. In view of (5.171), (5.179), and (5.180), we have $\frac{\epsilon(t)}{m(t)} \in L^\infty$ and $\dot{\theta}(t), \dot{\rho}(t) \in L^\infty \cap L^{\frac{1+\bar{\alpha}}{\bar{\alpha}}}$.

Then, using the analysis method in Section 5.3.5, it can be shown that the controller (5.34) with $\theta(t) \in L^\infty$ and the estimation error (5.88) with $\rho(t) \in L^\infty$ lead to the closed-loop system

$$|y(t)| \leq c_1 + c_2 \int_0^t e^{-a_1(t-\tau)} z(\tau) \int_0^\tau e^{-a_2(\tau-\sigma)} |y(\sigma)|\, d\sigma\, d\tau \tag{5.174}$$

$$|u(t)| \leq c_3 + c_4 \int_0^t e^{-a_3(t-\tau)} |y(\tau)|\, d\tau, \tag{5.175}$$

where $0 \leq z(t) \leq \frac{|\epsilon(t)|}{m(t)} + c_5\|\dot{\theta}(t)\|_2$ with $\|\cdot\|_2$ being the l^2 (Euclidean) vector norm for some constant $c_i > 0$, $i = 1, \ldots, 5$, and $a_j > 0$, $j = 1, 2, 3$ (in particular, c_1 and c_3 depend on the bounded $r(t)$ and $d(t)$). Since $\frac{|\epsilon(t)|}{m(t)}, \dot{\theta}(t) \in L^\infty \cap L^{1+\beta}$, $\beta = \max\{\bar{\alpha}, \frac{1}{\bar{\alpha}}\}$, it follows that $z(t) \in L^\infty \cap L^{1+\beta}$.

Defining $y_0(t) = \int_0^t e^{-a_2(t-\tau)}|y(\tau)|\,d\tau$ and using (5.174), we obtain

$$y_0(t) \leq b_1 + b_2 \int_0^t e^{-a_4(t-\tau)} z(\tau) y_0(\tau)\,d\tau \tag{5.176}$$

for some constant $a_4 > 0$, $b_1 > 0$ and $b_2 > 0$. Since $z(t) \in L^\infty \cap L^{1+\beta}$, $\beta \geq 1$, using Lemma 2.3, we conclude that $y_0(t) \in L^\infty$ (as $y_0(t) \leq b_1 + b_1 b_2 \int_0^t e^{-a_4(t-\tau)+b_3(t-\tau)^{\frac{\beta}{\beta+1}}} z(\tau)\,d\tau$ for some constant $b_3 > 0$), so that $y(t) \in L^\infty$ because of (5.174): $|y(t)| \leq c_1 + c_2 \int_0^t e^{-a_1(t-\tau)} z(\tau) y_0(\tau)\,d\tau$. Then, from (5.175), we have $u(t) \in L^\infty$ so that all closed-loop signals are bounded.

Using the expressions of $\xi(t)$ in (5.138) and $\epsilon(t)$ in (5.88), $\epsilon(t) = e(t) + \rho(t)\xi(t)$, and the facts that $\frac{|\epsilon(t)|}{m(t)}, \dot{\theta}(t) \in L^{1+\beta}$, we conclude that $e(t) \in L^{1+\beta}$.

If $\lim_{t\to\infty} d_y(t) = 0$, then $\lim_{t\to\infty} d(t) = 0$ so that $\lim_{t\to\infty} e(t) = 0$. Note that, with $d_u(t) \in L^\infty \cap L^{1+\alpha}$, $\bar{d}_u(t) \triangleq \frac{k_p}{P_m(s)}(1 - \theta_1^{*T}\frac{a(s)}{\Lambda(s)})[d_u](t)$ has the property that $\lim_{t\to\infty} \bar{d}_u(t) = 0$ because $\bar{d}_u(t) \in L^{1+\alpha}$ and $\frac{d}{dt}(\bar{d}_u(t)) \in L^\infty$ so that the signal convergence lemmas (Lemmas 2.14 and 2.16) are applicable.

In summary, we have the following result.

Theorem 5.6 *The controller (5.34) updated from the adaptive law (5.165)–(5.166), applied to the plant (5.157) with $d_u(t) \in L^\infty \cap L^{1+\alpha}$ and $d_y(t) \in L^\infty \cap L^{1+\alpha}$, $0 < \alpha < \infty$, ensures that all closed-loop system signals are bounded and the tracking error $e(t) \in L^{1+\beta}$ for $\beta = \max\{\frac{1}{\bar{\alpha}}, \bar{\alpha}\}$, $\bar{\alpha} \geq \alpha$. If, in addition, $\lim_{t\to\infty} d_y(t) = 0$, then $\lim_{t\to\infty} e(t) = 0$.*

In summary, stability and tracking performance of model reference adaptive control systems is inherently robust with respect to modeling errors $\frac{d(t)}{m(t)} \in L^2 \cap L^\infty$ (or $\frac{d(t)}{m(t)} \in L^{1+\alpha} \cap L^\infty$, with a modified parameter update law). Such modeling errors may depend on certain system signals, for example, $d(t) = f(y(t), u(t)) d_0(t)$, with some $d_0(t) \subset L^2 \cap L^\infty$ and some $\frac{f(y(t),u(t))}{m(t)} \in L^\infty$.

However, in the presence of more general modeling errors such as $d(t) \in L^\infty$ or unmodeled dynamics (see (5.177) in the next section), such inherent properties are not strong enough for ensuring system stability. The need of adaptive control schemes for such strong robustness led to an active area of research: robust adaptive control (see Section 5.7 for literature review). Next we present a comprehensive study of issues in robust adaptive control.

5.5 Robust MRAC

In this section we address the redesign of adaptive laws for model reference adaptive control to ensure closed-loop stability and small tracking errors in the presence of modeling errors, including unmodeled dynamics and bounded disturbances. We present the plant model with unmodeled dynamics and output disturbances, modify the adaptive laws presented in Sections 5.3.2, 5.3.4, and 5.3.5, and analyze their robustness properties.

5.5.1 Modeling Errors

In the presence of the output disturbance $d(t)$ and the additive and multiplicative unmodeled dynamics $\mu\Delta_a(s)$, $\mu\Delta_m(s)$, the plant is described by

$$\begin{aligned} &y(t) = G(s)[u](t) + d(t), \\ &G(s) = G_0(s)(1+\mu\Delta_m(s)) + \mu\Delta_a(s),\ G_0(s) = k_p\frac{Z(s)}{P(s)}, \end{aligned} \tag{5.177}$$

as shown in Figure 5.5. The adaptive control schemes of Section 5.3 were designed for $\mu = 0$ and $d(t) = 0$ and may not ensure the closed-loop stability for $\mu \neq 0$ and/or $d(t) \neq 0$ [23], [90], [147], [148], [334]. Modifications of those adaptive laws are needed for robustness of stability and tracking performance of adaptive control systems with respect to $\mu\Delta_m(s)$, $\mu\Delta_a(s)$, and/or $d(t)$.

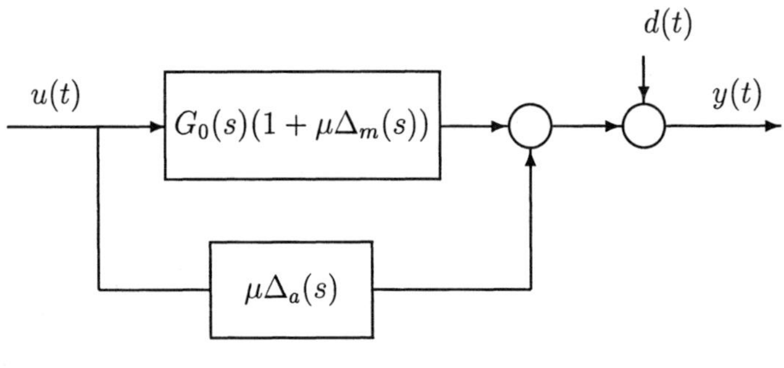

Figure 5.5: Plant with unmodeled dynamics and disturbance.

For a robust adaptive control scheme, the reference model is (5.4), that is,

$$P_m(s)[y_m](t) = r(t), \tag{5.178}$$

where $P_m(s)$ is a monic stable polynomial and $r(t)$ is a bounded reference input, and the controller structure is (5.34), that is,

$$u(t) = \theta_1^T\omega_1(t) + \theta_2^T\omega_2(t) + \theta_{20}y(t) + \theta_3 r(t), \tag{5.179}$$

where $\theta_1 \in R^{\bar{n}-1}$, $\theta_2 \in R^{\bar{n}-1}$, $\theta_{20} \in R$, and $\theta_3 \in R$ are parameters,

$$\omega_1(t) = \frac{a(s)}{\Lambda(s)}[u](t),\ \omega_2(t) = \frac{a(s)}{\Lambda(s)}[y](t), \tag{5.180}$$

$$a(s) = [1, s, \ldots, s^{\bar{n}-2}]^T, \tag{5.181}$$

and $\Lambda(s)$ is a monic stable polynomial of degree $\bar{n} - 1$. The modeled part of the plant (5.177), $G_0(s)$, is assumed to satisfy Assumptions (A5.1)–(A5.3), and $P_m(s)$ in (5.4) to satisfy Assumption (A5.4) (see Section 5.1).

For robust adaptive control, we make the following assumptions on the unmodeled dynamics $\mu\Delta_a(s)$, $\mu\Delta_m(s)$ and the output disturbance $d(t)$:

(A5.5a): $\Delta_a(s)$, $\frac{1}{P_m(s)}\Delta_m(s)$ are strictly proper rational functions.

(A5.6a): For $d_m = \lim_{s\to\infty} \frac{s}{P_m(s)}\Delta_m(s)$, $d_a = \lim_{s\to\infty} s\Delta_a(s)$, there exists a known constant $q_\delta > 0$ independent of μ and a constant $a > q_\delta$ such that the impulse response functions $h_m(t)$, $h_a(t)$ of $(\frac{s+a-q_\delta}{P_m(s-q_\delta)}\Delta_m(s-q_\delta) - d_m)s$, $((s+a-q_\delta)\Delta_a(s-q_\delta) - d_a)s$ satisfy

$$\|h_m(\cdot)\|_1 = \int_0^\infty |h_m(t)|\,dt \le c,\ \|h_a(\cdot)\|_1 = \int_0^\infty |h_a(t)|\,dt \le c \tag{5.182}$$

and $|d_m| \le c$, $|d_a| \le c$, for some constant $c > 0$ independent of μ.[6]

(A5.7a): $\dot{d}(t) \in L^\infty$, $d(t) \in L^\infty$.

With (A5.5a), Assumption (A5.6a) implies

(A5.6b): The impulse response functions $\bar{h}_m(t)$, $\bar{h}_a(t)$ of the proper rational functions $\frac{s}{P_m(s-q_\delta)}\Delta_m(s-q_\delta)$, $s\Delta_a(s-q_\delta)$ satisfy

$$\|\bar{h}_m(\cdot)\|_1 = \int_0^\infty |\bar{h}_m(t)|\,dt \le c,\ \|\bar{h}_a(\cdot)\|_1 = \int_0^\infty |\bar{h}_a(t)|\,dt \le c \tag{5.183}$$

for some constant $c > 0$ independent of μ.[7]

For a robust nonadaptive controller for $G_0(s)$ known, we assume

[6] For a proper rational function $H(s)$ and its impulse response function $h(t) = \mathcal{L}^{-1}[H(s)]$, if all poles of $H(s)$ are in $\mathrm{Re}[s] < 0$, then $\|h(\cdot)\|_1 = \int_0^\infty |h(t)|\,dt = c_h\|H(s)\|_\infty \triangleq \sup_{\omega\in R}|H(j\omega)|$ for some constant $c_h > 0$.

[7] To verify, for example, the second inequality of (5.183), let the impulse response function $s\Delta_a(s-q_\delta) - d_a$ be $\hat{h}_a(t)$. Then, we have

$$\|\bar{h}_a(\cdot)\|_1 \le \int_0^\infty |\hat{h}_a(t)|\,dt + |d_a|.$$

(A5.5b): $\Delta_a(s)$ and $\frac{1}{P_m(s)}\Delta_m(s)$ are proper rational functions.

(A5.6c): The impulse response functions $h_{m0}(t)$, $h_{a0}(t)$ of the proper rational functions $\frac{1}{P_m(s)}\Delta_m(s)$, $\Delta_a(s)$ satisfy

$$\|h_{m0}(\cdot)\|_1 = \int_0^\infty |h_{m0}(t)|\,dt \le c,\ \|h_{a0}(\cdot)\|_1 = \int_0^\infty |h_{a0}(t)|\,dt \le c \quad (5.184)$$

for some constant $c > 0$ independent of μ.

(A5.7b): $d(t) \in L^\infty$.

As examples of $\Delta_a(s)$, $\frac{1}{s+1}$ satisfies (A5.6a), (A5.6b), (A5.6c); $\frac{1}{(\mu s+1)(s+1)}$ satisfies (A5.6b), (A5.6c), not (A5.6a); $\frac{1}{\mu s+1}$ satisfies (A5.6c), not (A5.6a), (A5.6b); while $\frac{1}{s+\mu}$ satisfies none of (A5.6a), (A5.6b), and (A5.6c).

An implication of Assumptions (A5.5a), (A5.5b), and (A5.6a)–(A5.6c) is that Assumptions (A5.1) and (A5.4) can be relaxed, that is, the controlled plant may have unstable zeros and its actual relative degree may be unknown for stable model reference adaptive or nonadaptive control.

5.5.2 Robustness of MRC

We first have the following robustness results for a nonadaptive model reference control (MRC) system with the controller (5.179) designed based on the knowledge of the modeled part $G_0(s)$ of the plant (5.177).

Theorem 5.7 *Under Assumptions (A5.5b), (A5.6c), and (A5.7b), there exists a $\mu_0 > 0$ such that for any $\mu \in [0, \mu_0)$ the controller (5.179) with $\theta_1 = \theta_1^*$, $\theta_2 = \theta_2^*$, $\theta_{20} = \theta_{20}^*$ and $\theta_3 = \theta_3^*$, where θ_1^*, θ_2^*, θ_{20}^*, θ_3^* satisfy (5.11), ensures that all signals in the closed-loop system are bounded and the tracking error $e(t) = y(t) - y_m(t)$ converges exponentially to the residual set*

$$S_0 = \{e : \ |e| \le b_1\mu\bar{r} + b_2\bar{d}\} \quad (5.185)$$

for some constants $b_1, b_2 > 0$, where $\bar{r}$, $\bar{d}$ are the upper bounds for $|r(t)|$, $|d(t)|$.

To show that $\int_0^\infty |\hat{h}_a(t)|\,dt$ is less than a constant independent of μ, we use the equality

$$s\Delta_a(s-q_\delta) - d_a = \frac{1}{s+a-q_\delta}(((s+a-q_\delta)\Delta_a(s-q_\delta) - d_a)s - (a-q_\delta)d_a)$$

whose left-hand side has a finite H^∞ norm less than $\|\frac{1}{s+a-q_\delta}\|_\infty(\|((s+a-q_\delta)\Delta_a(s-q_\delta) - d_a)s\|_\infty + (a-q_\delta)|d_a|)$ which is less than a constant independent of μ, because $\|((s+a-q_\delta)\Delta_a(s-q_\delta) - d_a)s\|_\infty$ is, from (A5.6a), and $\|\frac{1}{s+a-q_\delta}\|_\infty = \frac{1}{a-q_\delta}$.

Proof: Ignoring the exponentially decaying effect of initial conditions, defining

$$F_1(s) = \theta_1^{*T} \frac{a(s)}{\Lambda(s)}, \; F_2(s) = \theta_2^{*T} \frac{a(s)}{\Lambda(s)} + \theta_{20}^* \tag{5.186}$$

from (5.11) and (5.179) with θ_1^*, θ_2^*, θ_{20}^*, θ_3^*, we have

$$\begin{aligned} P_m(s)[y - y_m](t) = {} & k_p(1 - F_1(s))\mu(\Delta_m(s) + G_0^{-1}(s)\Delta_a(s))[u](t) \\ & + k_p(1 - F_1(s))G_0^{-1}(s)[d](t), \end{aligned} \tag{5.187}$$

$$(1 - F_1(s))[u](t) = F_2(s)[y - y_m](t) + \left(\frac{F_2(s)}{P_m(s)} + \theta_3^*\right)[r](t). \tag{5.188}$$

Hence it follows from (5.11), (5.177), (5.187), and (5.188) that

$$\begin{aligned} & \left(1 - \mu\frac{k_p}{P_m(s)}(\Delta_m(s) + G_0^{-1}(s)\Delta_a(s))F_2(s)\right)[y - y_m](t) \\ & = \mu\frac{k_p}{P_m(s)}(\Delta_m(s) + G_0^{-1}(s)\Delta_a(s))\left(\frac{F_2(s)}{P_m(s)} + \theta_3^*\right)[r](t) \\ & \quad + \frac{k_p}{P_m(s)}(1 - F_1(s))G_0^{-1}(s)[d](t) \end{aligned} \tag{5.189}$$

$$\begin{aligned} & \left(1 - \mu\frac{k_p}{P_m(s)}(\Delta_m(s) + G_0^{-1}(s)\Delta_a(s))F_2(s)\right)[u](t) \\ & = G_0^{-1}(s)\frac{1}{P_m(s)}(r + k_pF_2(s)[d])(t). \end{aligned} \tag{5.190}$$

Since $\frac{1}{P_m(s)}(\Delta_m(s) + G_0^{-1}(s)\Delta_a(s))$ is stable and proper, for $\forall \mu \in [0, \mu_0)$, with

$$\mu_0 = \frac{1}{\sup_{\omega \in R} |\frac{k_p}{P_m(s)}(\Delta_m(s) + G_0^{-1}(s)\Delta_a(s))F_2(s)|_{s=j\omega}} > 0 \tag{5.191}$$

as ensured by the condition (5.184), from a well-known small-gain theorem (see, e.g., [459]), $\left(1 - \mu\frac{k_p}{P_m(s)}(\Delta_m(s) + G_0^{-1}(s)\Delta_a(s))F_2(s)\right)^{-1}$ is well-defined, proper, and stable, so the theorem's results hold. ∇

5.5.3 Robust Adaptive Laws

To design a robust adaptive controller for $G_0(s)$ unknown, we need to update the parameters of (5.179) with an adaptive law which is robust with respect to the unmodeled dynamics $\mu\Delta_m(s)$, $\mu\Delta_a(s)$ as well as to the output disturbance $d(t)$. In this case the tracking error equation becomes

$$e(t) = \frac{k_p}{P_m(s)}[\tilde{\theta}^T\omega](t) + \mu\Delta(s)[u](t) + H(s)[d](t), \tag{5.192}$$

where, with $\rho^* = k_p$,

$$\Delta(s) = \rho^*(1 - F_1(s))\frac{1}{P_m(s)}\Delta_m(s) + \left(1 + \frac{\rho^*}{P_m(s)}F_2(s)\right)\Delta_a(s), \tag{5.193}$$

$$H(s) = 1 + \frac{\rho^*}{P_m(s)}F_2(s). \tag{5.194}$$

For the case of unmodeled dynamics $\mu\Delta_m(s)$ and $\mu\Delta_a(s)$, we choose $\Lambda(s)$ in the adaptive controller structure (5.34), $P_m(s)$ in the reference model (5.4), and $L(s)$ in the estimation error (5.49) such that all zeros of $\Lambda(s)$, $P_m(s)$, and $L(s)$ are in $\mathrm{Re}[s] < -q_\delta$, where $q_\delta > 0$ is in (A5.6a) or (A5.6b). We also generate a new normalizing signal $m(t)$ [154], [320] from

$$\dot{m}(t) = -\delta_0 m(t) + \delta_1(|u(t)| + |y(t)|) + \delta_2, \ m(0) > 0, \tag{5.195}$$

where $\delta_1 > 0$, $\delta_2 > 0$, and $0 < \delta_0 < q_\delta$, with $q_\delta > 0$ being the stability margin of $\Delta_m(s)$ and $\Delta_a(s)$ as defined in Assumptions (A5.6a) and (A5.6b).[8]

To handle the unmodeled dynamics $\mu\Delta_m(s)$, $\mu\Delta_a(s)$ and disturbance $d(t)$, we need the switching signals $\sigma_1(t)$, $\sigma_2(t)$ [154], [156] (see Section 3.8.1):

$$\sigma_1(t) = \begin{cases} 0 & \text{if } \|\theta(t)\|_2 < M_1, \\ \sigma_{01}(\frac{\|\theta(t)\|_2}{M_1} - 1) & \text{if } M_1 \le \|\theta(t)\|_2 < 2M_1, \\ \sigma_{01} & \text{if } \|\theta(t)\|_2 \ge 2M_1, \end{cases} \tag{5.196}$$

$$\sigma_2(t) = \begin{cases} 0 & \text{if } |\rho(t)| < M_2, \\ \sigma_{02}(\frac{|\rho(t)|}{M_2} - 1) & \text{if } M_2 \le |\rho(t)| < 2M_2, \\ \sigma_{02} & \text{if } |\rho(t)| \ge 2M_2, \end{cases} \tag{5.197}$$

where $M_1 > \|\theta^*\|_2$, $\sigma_{01} > 0$, $M_2 > |\rho^*|$, $\sigma_{02} > 0$. To implement $\sigma_1(t)$ and $\sigma_2(t)$, we need the knowledge of parameter bounds M_1, M_2.

For robustness, we propose the following modifications to the adaptive laws (5.45), (5.60)–(5.61), and (5.91)–(5.92).

[8] We can also choose $\Lambda(s)$, $P_m(s)$, and $L(s)$ whose zeros are in $\mathrm{Re}[s] < -q_0$ for any $q_0 > 0$ and then use a normalizing signal $m(t)$ from (5.195) but with $0 < \delta_0 < \min\{q_\delta, q_0\}$. Then, the following robustness analysis is still valid if, in (5.209), (5.210), (5.216), and the proof of Lemma 5.5, the constant q_δ is replaced by a constant $q_1 \in (\delta_0, \min\{q_\delta, q_0\})$. We note that $\Delta_m(s)$ and $\Delta_a(s)$ satisfying Assumptions (A5.6a) and (A5.6b) with q_δ also satisfy the assumptions of the same form but with q_δ replaced by any $q_1 \in (0, q_\delta)$. For example, if $\Delta_a(s - q_\delta)s$ has a finite L^1 operator norm bound independent of μ, then, for $q_1 \in (0, q_\delta)$,

$$\Delta_a(s - q_1)s = \Delta_a(s - q_\delta + q_\delta - q_1)(s + q_\delta - q_1)\left(1 - \frac{q_\delta - q_1}{s + q_\delta - q_1}\right)$$

also has a finite L^1 operator norm bound independent of μ, as $\mathcal{L}^{-1}[\Delta_a(s - q_\delta + q_\delta - q_1)(s + q_\delta - q_1)] = e^{-(q_\delta - q_1)t}\mathcal{L}^{-1}[\Delta_a(s - q_\delta)s]$ and $1 - \frac{q_\delta - q_1}{s + q_\delta - q_1}$ does not depend on μ.

Lyapunov Design for Relative Degree 1

We first modify the adaptive law (5.45) for the plant (5.177) with $G_0(s)$ of relative degree $n^* = 1$ and output disturbance $d(t)$ satisfying (A5.7a) and without unmodeled dynamics, that is, with $\mu = 0$:

$$\dot{\tilde{\theta}}(t) = \dot{\theta}(t) = -\text{sign}[k_p]\Gamma\omega(t)e(t) - \Gamma\sigma_1(t)\theta(t). \tag{5.198}$$

Lyapunov Design with (A5.5a), (A5.6a), (A5.7a)

We then modify the adaptive law (5.60)–(5.61) (for a Lyapunov design) for the plant (5.177) with $G_0(s)$ of relative degree $n^* \geq 1$ and output disturbance $d(t)$ as well as unmodeled dynamics $\mu\Delta_m(s)$, $\mu\Delta_a(s)$:

$$\dot{\theta}(t) = -\text{sign}[k_p]\Gamma\zeta(t)\epsilon(t) - \Gamma\sigma_1(t)\theta(t), \ \Gamma = \Gamma^T > 0, \tag{5.199}$$

$$\dot{\rho}(t) = -\gamma\xi(t)\epsilon(t) - \gamma\sigma_2(t)\rho(t), \ \gamma > 0, \tag{5.200}$$

where $\epsilon(t) = e(t) + \frac{L(s)}{P_m(s)}(\rho\xi - \kappa\epsilon m^2)$, $\kappa > 0$, $\xi(t) = \theta^T(t)\zeta(t) - L^{-1}(s)[\theta^T\omega](t)$, $\zeta(t) = L^{-1}(s)[\omega](t)$, as in (5.56)–(5.59), and $L(s)$ is a stable polynomial of degree $n^* - 1$ such that $\frac{L(s)}{P_m(s)}$ is strictly positive real.

Gradient Design with (A5.5a), (A5.6b), (A5.7b)

We also modify the adaptive law (5.91)–(5.92) (for a gradient design) for the plant (5.177) with $G_0(s)$ of relative degree $n^* \geq 1$ and output disturbance $d(t)$ as well as unmodeled dynamics $\mu\Delta_m(s)$, $\mu\Delta_a(s)$:

$$\dot{\theta}(t) = -\frac{\text{sign}[k_p]\Gamma\epsilon(t)\zeta(t)}{m^2(t)} - \Gamma\sigma_1(t)\theta(t), \ \Gamma = \Gamma^T > 0, \tag{5.201}$$

$$\dot{\rho}(t) = -\frac{\gamma\epsilon(t)\xi(t)}{m^2(t)} - \gamma\sigma_2(t)\rho(t), \ \gamma > 0, \tag{5.202}$$

where $\epsilon(t) = e(t) + \rho(t)\xi(t)$, $\xi(t) = \theta^T(t)\zeta(t) - \frac{1}{P_m(s)}[\theta^T\omega](t)$, and $\zeta(t) = \frac{1}{P_m(s)}[\omega](t)$, as in (5.88)–(5.93).

5.5.4 Robust Stability Analysis

We introduce the following definition to be used in the stability analysis of these modified adaptive control schemes.

Definition 5.1 *A signal $x(t) \in B^{2,\nu}$ for some $\nu \geq 0$ if, for any $t_2 > t_1 \geq 0$ and some $\gamma_0 > 0$, $\int_{t_1}^{t_2} \|x(t)\|_2^2\, dt \leq \nu(t_2 - t_1) + \gamma_0$.*

Next we will analyze three adaptive control systems:

(i) The controller (5.179) with adaptive law (5.198) applied to the plant (5.177) with $n^* = 1$ and $d(t)$ satisfying Assumption (A5.7a) for $\mu = 0$.

(ii) The controller (5.179) with adaptive law (5.199)–(5.200) applied to the plant (5.177) with unmodeled dynamics $\mu\Delta_m(s)$, $\mu\Delta_a(s)$ satisfying Assumptions (A5.5a), (A5.6a) for $d(t) = 0$.

(iii) The controller (5.179) with adaptive law (5.201)–(5.202) applied to the plant (5.177) with unmodeled dynamics $\mu\Delta_m(s)$, $\mu\Delta_a(s)$ satisfying Assumptions (A5.5a), (A5.6b) for $d(t) = 0$.

Note that the controller (5.179) with adaptive law (5.199)–(5.200) also applies to the plant (5.177) with $\mu\Delta_m(s)$, $\mu\Delta_a(s)$ as well as $d(t)$ satisfying (A5.5a), (A5.6a), and (A5.7a), while the controller (5.179) with adaptive law (5.201)–(5.202) applies to the plant (5.177) with $\mu\Delta_m(s)$, $\mu\Delta_a(s)$ as well as $d(t)$ satisfying (A5.5a), (A5.6b), and (A5.7b).

Lyapunov Design for Relative Degree 1

With the adaptive law (5.198), the controller (5.179) has the following robustness properties when applied to the plant (5.179) with output disturbance $d(t)$ which satisfies Assumption (A5.7a).

Theorem 5.8 *All signals in the closed-loop system consisting of the plant (5.179) with $\mu = 0$, reference model (5.178), controller (5.179), and adaptive law (5.198) are bounded, and the tracking error $e(t) \in B^{2,k_0\bar{d}_0^2}$ for some $k_0 > 0$, where $\bar{d}_0 > 0$ is such that $\max\{|\dot{d}(t)|, |d(t)|\} \le \bar{d}_0$.*

Proof: With $\mu = 0$, $P_m(s) = s + a_m$ and $a_m > 0$, (5.192) becomes

$$\dot{e}(t) = -a_m e(t) + k_p \tilde{\theta}^T(t)\omega(t) + \bar{d}(t), \tag{5.203}$$

where, from (A5.7a) and (5.194), $\bar{d}(t) = (s + a_m)H(s)[d](t) \in L^\infty$.

With (5.198), the time derivative of $V(e_p)$ defined in (5.43) is

$$\begin{aligned}\dot{V} &= -2a_m e^2(t) + 2e(t)\bar{d}(t) - 2|k_p|\sigma_1(t)\tilde{\theta}^T(t)\theta(t)\\ &\le -a_m e^2(t) - a_m\left(|e(t)| - \frac{|\bar{d}(t)|}{a_m}\right)^2 + \frac{\bar{d}^2(t)}{a_m} - 2|k_p|\sigma_1(t)\tilde{\theta}^T(t)\theta(t).\end{aligned} \tag{5.204}$$

From (5.196), we have

$$\sigma_1(t)\tilde{\theta}^T(t)\theta(t) = \frac{\sigma_1(t)}{2}(\theta^T(t)\theta(t) + (\theta(t) - \theta^*)^T(\theta(t) - \theta^*) - \theta^{*T}\theta^*) \ge 0. \tag{5.205}$$

Hence $e(t) \in L^\infty$, $\theta(t) \in L^\infty$ and so $y(t) \in L^\infty$, $\omega_2(t) \in L^\infty$.

Using (5.179) with $\mu = 0$, that is, $P(s)[y - d](t) = k_p Z(s)[u](t)$, and ignoring the exponentially decaying effect of initial conditions, we obtain

$$\frac{s^i}{\Lambda(s)}[u](t) = \frac{P(s)}{k_p Z(s)} \frac{s^i}{\Lambda(s)}[y - d](t) \in L^\infty \tag{5.206}$$

for $i = 0, 1, \ldots, \bar{n} - 2$ because $\frac{P(s)}{k_p Z(s)} \frac{s^i}{\Lambda(s)}$ is stable and proper and $y(t) \in L^\infty$, $d(t) \in L^\infty$. Then we have $\omega_1(t) \in L^\infty$ and $u(t) \in L^\infty$ and from (5.204),

$$\begin{aligned} \int_{t_1}^{t_2} e^2(t)\, dt &\leq \frac{1}{a_m}(V(e_p(t_1)) - V(e_p(t_2))) + \int_{t_1}^{t_2} \frac{\bar{d}^2(t)}{a_m}\, dt \\ &\leq \gamma_0 + k_0(t_2 - t_1)\bar{d}_0^2 \end{aligned} \tag{5.207}$$

for some $k_0 > 0$, where $\gamma_0 = \frac{1}{a_m}\|V(e_p(\cdot))\|_\infty$, that is, $e(t) \in B^{2,k_0\bar{d}_0^2}$. ∇

Remark 5.2 We should note that with $\sigma_1(t) = 0$, from (5.204), we may not conclude that $\tilde{\theta}(t) \in L^\infty$, as $\dot{V}$ may be positive when $|e(t)|$ is small, due a nonzero disturbance $\bar{d}(t)$, which cannot prevent $\tilde{\theta}(t)$, that is, $\theta(t)$, from growing unbounded (see the illustrative example in (5.149)–(5.151) for such parameter drift instability). However, with $\sigma_1(t)$ in action, there exist positive constants e_0 and θ_0 such that $|e(t)| > e_0$ and/or $\|\tilde{\theta}(t)\|_2 > \theta_0$ implies that $\dot{V} < 0$ so that both $e(t)$ and $\tilde{\theta}(t)$ are ensured to be bounded.

One can study the simple system in Problem 5.21 for further understanding of robustness of this Lyapunov design to unmodeled dynamics. □

To study the robustness properties of the adaptive control designs (5.199)–(5.200) and (5.201)–(5.202), we only consider the case of unmodeled dynamics. We should note that in the following development Lemmas 5.4 and 5.5 can be readily modified for the bounded disturbance case, while Theorems 5.9 and 5.10 also hold for bounded disturbances $d(t)$ [151], [282], [395].

Lyapunov Design with (A5.5a), (A5.6a), (A5.7a)

For $d(t) = 0$, the adaptive law (5.199)–(5.200) has the following properties.

Lemma 5.4 *The adaptive law (5.199)–(5.200) guarantees that there exists a $\mu_0^* > 0$ such that, for any $\mu \in [0, \mu_0^*)$, $\theta(t)$, $\rho(t)$, $\epsilon(t)$ are bounded and $\epsilon(t)$, $\epsilon(t)m(t)$, $\dot{\theta}(t)$, $\dot{\rho}(t)$ all belong to $B^{2,k_0\mu^2}$ for some constant $k_0 > 0$.*

To prove Lemma 5.4, we recall the following lemma.

Lemma 2.6 *If $x_1(t) = H(s)\frac{1}{s+\delta}[x_2](t)$ with $\delta > 0$, and the impulse response function $h_\delta(t)$ of $H(s-\delta)$ satisfies $\|h_\delta(\cdot)\|_1 = \int_0^\infty |h_\delta(t)|\, dt < \infty$, then*

$$|x_1(t)| \leq \|h_\delta(\cdot)\|_1 \frac{1}{s+\delta}[|x_2|](t). \tag{5.208}$$

Using this lemma (whose proof is given in Section 2.5.3, see (2.139)), we can establish the following auxiliary results.

Corollary 5.1 *If (A5.6b) holds, then $\frac{\Delta(s)[u](t)}{m(t)} \in L^\infty$.*

Proof: From (5.193), we have

$$\begin{aligned}\Delta(s)[u](t) = {} & \rho^*(1 - F_1(s))\frac{1}{P_m(s)}\Delta_m(s)(s+q_\delta)\frac{1}{s+q_\delta}[u](t) \\ & + \left(1 + \frac{\rho^*}{P_m(s)}F_2(s)\right)\Delta_a(s)(s+q_\delta)\frac{1}{s+q_\delta}[u](t).\end{aligned} \tag{5.209}$$

With (A5.6b), applying Lemma 2.6, we have

$$|\Delta(s)[u](t)| \leq c_m \frac{1}{s+q_\delta}[|u|](t) + c_a \frac{1}{s+q_\delta}[|u|](t) \tag{5.210}$$

for some constants $c_m > 0$, $c_a > 0$, independent of μ. Since $q_\delta > \delta_0$ and $m(0) > 0$, it follows that $\frac{1}{s+q_\delta}[|u|](t) \leq \frac{1}{s+\delta_0}[|u|](t)$, that is, $\frac{1}{s+q_\delta}[|u|](t) \leq c_0 m(t)$ for some constant $c_0 > 0$, so that $\frac{\Delta(s)[u](t)}{m(t)} \leq b_0$ for some constant $b_0 > 0$. ∇

Corollary 5.2 *If (A5.6b) holds, then $\frac{\omega(t)}{m(t)} \in L^\infty$.*

Proof: From (5.11) and (5.192), we have

$$y(t) = \frac{k_p}{P_m(s)}(1 - F_1(s))[u](t) - \frac{k_p}{P_m(s)}F_2(s)[y](t) + \mu\Delta(s)[u](t). \tag{5.211}$$

For $\Delta_m(s)$, $\Delta_m(s)$ satisfying (A5.6b), Corollary 5.1 implies that there exists a constant $b_0 > 0$ such that $\frac{|\Delta(s)[u](t)|}{m(t)} \leq b_0$ and in turn $\frac{y(t)}{m(t)} \in L^\infty$. From Lemma 2.6 again, we have $\frac{\omega_1(t)}{m(t)} \in L^\infty$, $\frac{\omega_2(t)}{m(t)} \in L^\infty$. ∇

Proof of Lemma 5.4: In the presence of $\mu\Delta_m(s)$, $\mu\Delta_a(s)$, the estimation error equation (5.62) becomes

$$\epsilon(t) = \frac{L(s)}{P_m(s)}[\rho^*\tilde{\theta}^T\zeta + \tilde{\rho}\xi - \kappa\epsilon m^2 + \mu\bar{\eta}](t), \tag{5.212}$$

where, with $\Delta(s)$ in (5.193),

$$\bar{\eta}(t) = \bar{\Delta}(s)[u](t),\ \bar{\Delta}(s) = \frac{P_m(s)}{L(s)}\Delta(s). \tag{5.213}$$

Let (5.212) have the controllable state-space form

$$\dot{e}_\epsilon(t) = Ae_\epsilon(t) + b\nu(t),\ \epsilon(t) = ce_\epsilon(t), \tag{5.214}$$

where $\nu(t) = \rho^*\tilde{\theta}^T(t)\zeta(t) + \tilde{\rho}(t)\xi(t) - \kappa\epsilon(t)m^2(t) + \mu\bar{\eta}(t)$. With (5.199), (5.200), and (5.214), the time derivative of $V(e_p)$ in (5.67) is

$$\begin{aligned}\dot{V} &= -e_\epsilon^T(t)Qe_\epsilon(t) - 2\kappa\epsilon^2(t)m^2(t) + 2\mu\epsilon(t)\bar{\eta}(t)\\ &\quad - 2|\rho^*|\sigma_1(t)\tilde{\theta}^T(t)\theta(t) - 2\sigma_2(t)\tilde{\rho}(t)\rho(t).\end{aligned} \tag{5.215}$$

Using (5.193), (5.213), and Assumptions (A5.5a), (A5.6a), we express

$$\bar{\eta}(t) = \bar{\eta}_0(t) + \bar{\eta}_1 u(t),\ \bar{\eta}_0(t) = \bar{\Delta}_0(s)\frac{1}{s+q_\delta}[u](t), \tag{5.216}$$

where $|\bar{\eta}_1| \leq b_1$ and the impulse response function $h_0(t)$ of $\bar{\Delta}_0(s - q_\delta)$ satisfies $\|h_0(\cdot)\|_1 = \int_0^\infty |h_0(t)|\,dt \leq b_2$ for some constant $b_1 > 0$, $b_2 > 0$ independent of μ.[9] It then follows from Lemma 2.6, Corollary 5.2, and (5.195) that $\frac{|\bar{\eta}_0(t)|}{m(t)} \leq b_3$

[9] From (5.193) and (5.213), for $a > q_\delta > 0$, we have

$$\begin{aligned}\bar{\eta}(t) &= \rho^*\frac{P_m(s)}{L(s)(s+a)}(1 - F_1(s))\frac{s+a}{P_m(s)}\Delta_m(s)[u](t)\\ &\quad + \frac{P_m(s)}{L(s)(s+a)}\left(1 + \frac{\rho^*}{P_m(s)}F_2(s)\right)(s+a)\Delta_a(s)[u](t)\\ &\triangleq G_1(s)\frac{s+a}{P_m(s)}\Delta_m(s)[u](t) + G_2(s)(s+a)\Delta_a(s)[u](t).\end{aligned}$$

Considering $G_2(s)(s+a)\Delta_a(s)[u](t)$, $d_a = \lim_{s\to\infty} s\Delta_a(s)$, $\bar{d}_a = \lim_{s\to\infty} G_2(s)d_a$, we have

$$\begin{aligned}G_2(s)(s+a)\Delta_a(s)[u](t) &= G_2(s)((s+a)\Delta_a(s) - d_a)[u](t) + G_2(s)d_a[u](t)\\ &= G_2(s)((s+a)\Delta_a(s) - d_a)(s+q_\delta)\frac{1}{s+q_\delta}[u](t)\\ &\quad + (G_2(s)d_a - \bar{d}_a)(s+q_\delta)\frac{1}{s+q_\delta}[u](t) + \bar{d}_a u(t).\end{aligned}$$

Since $\|G_2(s-q_\delta)((s+a-q_\delta)\Delta_a(s-q_\delta) - d_a)s\|_\infty$ is less than a constant independent of μ (as a result of Assumption (A5.6a) and stability of $\Lambda(s-q_\delta)$, $L(s-q_\delta)$ and $P_m(s-q_\delta)$), and so is $\|(G_2(s-q_\delta)d_a - \bar{d}_a)s\|_\infty$ (as $G_2(s)$ does not depend on μ), with $\bar{d}_a$ independent of μ too, we have the result of (5.216) for the part related to $\Delta_a(s)$. The result for the part of $\bar{\eta}(t)$ related to $\Delta_m(s)$ can be similarly established.

and $\frac{|u(t)|}{m(t)} \leq b_3 \|\theta(t)\|_2$ for some constant $b_3 > 0$. Then we have

$$\begin{aligned} \dot{V} \leq & -e_\epsilon^T(t) Q e_\epsilon(t) - \kappa \epsilon^2(t) m^2(t) + \mu^2 \frac{2b_3^2}{\kappa} + \mu^2 \frac{2b_3^2}{\kappa} \bar{\eta}_1^{-2} \|\theta(t)\|_2^2 \\ & - 2|k_p| \sigma_1(t) \tilde{\theta}^T(t) \theta(t) - 2\sigma_2(t) \tilde{\rho}(t) \rho(t). \end{aligned} \tag{5.217}$$

From (5.217), we see that for any $\mu \in (0, \sqrt{\frac{|k_p|\sigma_{01}\kappa}{b_3^2 \bar{\eta}_1^2}})$ the existence of constant $\theta_0 > 0$, $\rho_0 > 0$ is guaranteed so that $\dot{V} \leq 0$ whenever $\|\theta(t)\| > \theta_0$, $|\rho(t)| > \rho_0$. Therefore, $\theta(t) \in L^\infty$, $\rho(t) \in L^\infty$, $e_\epsilon(t) \in L^\infty$, $\epsilon(t) \in L^\infty$, and $V(e_p(t)) \in L^\infty$. Using (5.196), (5.197), (5.199), (5.200), and (5.217), we can show that $\epsilon(t)$, $\epsilon(t)m(t)$, $\dot{\theta}(t)$, $\dot{\rho}(t)$ all belong to $B^{2,k_0\mu^2}$ for some constant $k_0 > 0$. ∇

As in Remark 5.2, we note that with $\sigma_1(t) = \sigma_2(t) = 0$, from (5.215), we may not conclude that $\tilde{\theta}(t) \in L^\infty$ and $\tilde{\rho}(t) \in L^\infty$, as $\dot{V}$ may be positive when both $|e_\epsilon(t)|_2$ and $|\epsilon(t)m(t)|$ are small, due to a nonzero modeling error $\bar{\eta}(t)$, which cannot prevent $\tilde{\theta}(t)$ and $\tilde{\rho}(t)$ from growing unbounded.

For the adaptive control scheme with the adaptive law (5.199)–(5.200), we have the following robust closed-loop stability and tracking performance.

Theorem 5.9 *There exists a $\mu^* > 0$ such that for any $\mu \in [0, \mu^*)$ all signals in the closed-loop system are bounded and $e(t) \in B^{2,\gamma_0\mu^2}$ for some $\gamma_0 > 0$. Moreover, $\lim_{t\to\infty} e(t) = 0$ when $\mu = 0$.*

Proof: In the presence of $\mu\Delta_a(s)$, $\mu\Delta_m(s)$, (5.73) becomes

$$\begin{aligned} & z_0(t) + a_0 K_1(s)[z_0](t) - K_1(s)[u](t) \\ & = K(s) G_0^{-1}(s) \frac{1}{s + a_0}[y](t) - \mu K(s) \left(\Delta_m(s) + G_0^{-1}(s)\Delta_a(s)\right)[z_0](t). \end{aligned} \tag{5.218}$$

Substituting (5.74) into (5.218) gives

$$\begin{aligned} & \left(1 + K_1(s)(a_0 - \theta_1^T(\cdot)\frac{a(s)}{\Lambda(s)}(s + a_0)) + \mu K(s)(\Delta_m(s) + G_0^{-1}(s)\Delta_a(s))\right)[z_0](t) \\ & = \left(K(s) G^{-1}(s) \frac{1}{s + a_0} + K_1(s)\theta_2^T(\cdot)\frac{a(s)}{\Lambda(s)} + K_1(s)\theta_{20}(\cdot)\right)[y](t) \\ & \quad + K_1(s)[\theta_3 r](t). \end{aligned} \tag{5.219}$$

There exists $a^0 > 0$ such that for any fixed finite $a > a^0$ the existence of $\mu_1^* > 0$ is guaranteed such that for any $\mu \in (0, \mu_1^*)$ the operator

$$\begin{aligned} T_0(s,t) \triangleq & \Big(1 + K_1(s)(a_0 - \theta_1^T(t)\,\frac{a(s)}{\Lambda(s)}(s + a_0)) \\ & + \mu K(s)(\Delta_m(s) + G_0^{-1}(s)\Delta_a(s))\Big)^{-1} \end{aligned} \tag{5.220}$$

is stable and proper, and so (5.82) holds (see Section 2.5.3).

Now from (5.56), we have

$$y(t) = y_m(t) + \epsilon(t) - \frac{L(s)}{P_m(s)}[\rho\xi](t) + \kappa\frac{L(s)}{P_m(s)}[\epsilon\, m^2](t), \tag{5.221}$$

where, from (5.195),

$$m(t) = m(0)e^{-\delta_0 t} + \frac{\delta_1}{s+\delta_0}[|u| + |y|](t) + \frac{\delta_2}{\delta_0}\left(1 - e^{-\delta_0 t}\right). \tag{5.222}$$

Using (5.83), (5.221), (5.222), and Lemmas 5.4, 2.6, with $x_1(t) \in B^{2,\bar{k}_0\mu^2}$ for some $\bar{k}_0 > 0$, we also obtain (5.87), from which, using Lemma 2.3, we conclude that there exists $\mu^* > 0$ such that, for any $\mu \in [0, \mu^*)$, $z_1(t) \in L^\infty$, and so $y(t) \in L^\infty$ because of (5.86). This implies that $z_0(t) \in L^\infty$ in (5.82). Hence $\omega(t) \in L^\infty$ in (5.38), $u(t) \in L^\infty$ in (5.34) ((5.179)), $\xi(t) \in L^\infty$ in (5.57), $\zeta(t) \in L^\infty$ in (5.58), $\epsilon(t) \in L^\infty$ in (5.56). That is, all closed-loop signals are bounded. Then, in view of (5.83), (5.56), from Lemma 5.4, we have $e(t) \in B^{2,\gamma_0\mu^2}$ for some $\gamma_0 > 0$. When $\mu = 0$, the adaptive law (5.99)–(5.200) ensures that all conditions of Theorem 5.3 hold, so that $\lim_{t\to\infty} e(t) = 0$. ∇

Gradient Design with (A5.5a), (A5.6b), (A5.7b)

The adaptive system consisting of the adaptive law (5.201)–(5.202), controller (5.179), and plant (5.177) with $\mu\Delta_m(s)$, $\mu\Delta_a(s)$ satisfying (A5.5a), (A5.6b) has the following robustness properties of parameter adaptation.

Lemma 5.5 *The adaptive law (5.201)–(5.202) guarantees that* $\theta(t)$, $\rho(t)$, $\frac{\epsilon(t)}{m(t)}$, $\dot\theta(t)$, $\dot\rho(t) \in L^\infty$ *and* $\frac{\epsilon(t)}{m(t)}$, $\dot\theta(t) \in B^{2,k_0\mu^2}$ *for some* $k_0 > 0$.

Proof: The estimation error equation now is

$$\epsilon(t) = \rho^*\tilde\theta^T(t)\zeta(t) + \tilde\rho(t)\xi(t) + \mu\eta(t),\ \eta(t) = \Delta(s)[u](t) \tag{5.223}$$

and the time derivative of $V(\tilde\theta, \tilde\rho)$ in (5.95), along (5.201)–(5.202), satisfies

$$\dot V \le -\frac{\epsilon^2(t)}{m^2(t)} + \mu^2\frac{\eta^2(t)}{m^2(t)} - 2\sigma_1(t)|\rho^*|\tilde\theta^T(t)\theta(t) - 2\sigma_2(t)\tilde\rho(t)\rho(t). \tag{5.224}$$

From Lemma 2.6 and Assumptions (A5.5a), (A5.6b), we have $\frac{|\eta(t)|}{m(t)} \le b_0$ for some constant $b_0 > 0$ (see Corollary 5.1). Hence $\theta(t) \in L^\infty$, $\rho(t) \in L^\infty$. Using (5.196), (5.197), (5.201), (5.202), and (5.224), we can show that $\frac{\epsilon(t)}{m(t)} \in L^\infty$ and $\frac{\epsilon(t)}{m(t)}$, $\dot\theta(t) \in B^{2,k_0\mu^2}$ for some $k_0 > 0$. ∇

We also note that with $\sigma_1(t)$ and $\sigma_2(t)$ in action the boundedness of $\theta(t)$ and $\rho(t)$ is ensured. Without $\sigma_1(t)$ and $\sigma_2(t)$, that is, when $\sigma_1(t) = \sigma_2(t) = 0$, the boundedness of $\theta(t)$ and $\rho(t)$ cannot be concluded from (5.224), as $\dot{V}$ may be positive for small $|\epsilon(t)|$, which may lead to unbounded $\theta(t)$ and $\rho(t)$.

With Lemma 5.5, we have the following robust stability result.

Theorem 5.10 *There exists a $\mu^* > 0$ such that for any $\mu \in [0, \mu^*)$ all signals in the closed-loop system are bounded and $e(t) \in B^{2,\gamma_0\mu^2}$ for some $\gamma_0 > 0$. Moreover, $\lim_{t\to\infty} e(t) = 0$ when $\mu = 0$.*

Proof: In the presence of $\mu\Delta_a(s)$, $\mu\Delta_m(s)$, (5.135) becomes

$$\left(1 + K_1(s)(a_0 - \theta_1^T(\cdot)\frac{a(s)}{\Lambda(s)}(s + a_0)) + \mu K(s)(\Delta_m(s) + G_0^{-1}(s)\Delta_a(s))\right)[z_0](t)$$

$$= \left(K(s)G_0^{-1}(s) + K_1(s)\theta_2^T(\cdot)\frac{a(s)}{\Lambda(s)}(s + a_0)\right)[z](t)$$

$$+ K_1(s)[s[\theta_{20}z] - zs[\theta_{20}] + a_0\theta_{20}z](t) + K_1(s)[\theta_3 r](t). \tag{5.225}$$

Then, using the same arguments as that for (5.136)–(5.139) and noticing that

$$|\epsilon(t)| \leq \frac{|\epsilon(t)|}{m(t)}\frac{\delta_1}{s + \delta_0}[|u| + |y| + 1](t), \tag{5.226}$$

we can conclude that there exists a $\mu^* > 0$ such that, for any $\mu \in [0, \mu^*)$, (5.141) holds and $z_1(t) \in L^\infty$ in (5.142), that is, $z(t) \in L^\infty$ in (5.132) because of (5.141), which then leads to the closed-loop signal boundedness. In view of (5.138), (5.88), from Lemma 5.5, we have $e(t) \in B^{2,\gamma_0\mu^2}$ for some $\gamma_0 > 0$. When $\mu = 0$, we have $\epsilon(t) \in L^2$, and so $e(t) \in L^2$ and $\lim_{t\to\infty} e(t) = 0$. ∇

5.5.5 Summary

To deal with modeling errors such as unmodeled dynamics $\mu\Delta_m(s)$, $\mu\Delta_a(s)$ and bounded disturbances $d(t)$ in the plant (5.177), adaptive control schemes need to be modified for robustness of the desired parameter adaptation properties. Unlike the case of $L^{1+\alpha} \cap L^\infty$ modeling errors $d(t)$ (see Section 5.4), adaptive law modifications for the case of unmodeled dynamics $\mu\Delta_m(s)$, $\mu\Delta_a(s)$ and bounded disturbances $d(t)$ are done by adding terms $-\Gamma\sigma_1(t)\theta(t)$ and $-\gamma\sigma_2(t)\rho(t)$ to the adaptive laws for $\theta(t)$ and $\rho(t)$ to ensure robust parameter adaptation. The controller (5.179) with bounded parameters still leads to the closed-loop inequality (5.86) or (5.141), while the adaptive law (5.199)–(5.200) or (5.201)–(5.202), through the $B^{2,k_0\mu^2}$ property (see Definition 5.1 and Lemmas 5.4 and 5.5) of $\dot{\theta}(t)$, $\epsilon(t)m_0(t)$ or $\frac{\epsilon(t)}{m_0(t)}$, for some constant $k_0 > 0$, ensures that the loop gain of (5.87) or (5.142) is small enough for a nonzero range of $\mu \geq 0$ so that the closed-loop signal boundedness is guaranteed.

In addition to the switching-σ modification with $\sigma_1(t)$ and $\sigma_2(t)$ in (5.196)–(5.202), other modifications such as the dead-zone, the fixed-σ, the parameter projection, and the ϵ modifications used in Section 3.8.1 for robust parameter estimation can also be used for robust adaptive control [151], [282].

5.6 Design for Unknown High Frequency Gain

For model reference adaptive control, a key assumption is that the sign of the high frequency gain k_p of the plant (5.1) is known (see Assumption (A5.3)). In [267], [293], adaptive control schemes without such sign information are developed for some classes of systems. The main technique used is the Nussbaum gain [267], [293], illustrated in the following examples and employed for a more general case in Section 5.6.2, which relaxes the sign assumption for the high frequency gain k_p for model reference adaptive control.

5.6.1 Adaptive Control Designs Using Nussbaum Gain

Consider the first-order linear time-invariant plant

$$\dot{y}(t) = a_p y(t) + b_p u(t), \; t \geq 0, \tag{5.227}$$

with unknown constant parameters a_p and b_p and the reference model

$$\dot{y}_m(t) = -a_m y_m(t) + b_m r(t), \; t \geq 0, \tag{5.228}$$

with $a_m > 0$. The control objective is to find a feedback control law $u(t)$ to ensure that $\lim_{t\to\infty}(y(t) - y_m(t)) = 0$, in addition to closed-loop signal boundedness, without any knowledge of the parameters a_p and b_p.

Recall from the example in Section 4.1.1 and Example 5.2 of Section 5.3.5 that a suitable adaptive controller structure is

$$u(t) = k_1(t)y(t) + k_2(t)r(t), \tag{5.229}$$

where $k_1(t)$ and $k_2(t)$ are estimates of the unknown parameters

$$k_1^* = \frac{-a_p - a_m}{b_p}, \; k_2^* = \frac{b_m}{b_p} \tag{5.230}$$

if it is updated from an adaptive law using the knowledge of sign$[b_p]$, the sign of the unknown parameter b_p. The question now is how to design an adaptive controller without using such knowledge.

Lyapunov Design Example

To solve the adaptive control problem without using the knowledge of sign$[b_p]$, we first derive an adaptive control scheme different from that with (5.229).

Introducing the parameter vectors

$$\theta(t) = [k_1(t), k_2(t)]^T, \; \theta^* = [k_1^*, k_2^*]^T \tag{5.231}$$

and the regressor vector $\omega(t)$ and tracking error $e(t)$ as

$$\omega(t) = [y(t), r(t)]^T, \; e(t) = y(t) - y_m(t), \tag{5.232}$$

we can express the tracking error system as

$$\dot{e}(t) = -a_m e(t) + b_p(u(t) - \theta^{*T}\omega(t)). \tag{5.233}$$

For (5.229), we have $u(t) = \theta^T(t)\omega(t)$. This controller leads (5.233) to

$$\dot{e}(t) = -a_m e(t) + b_p(\theta(t) - \theta^*)^T\omega(t), \tag{5.234}$$

based on which, in the previous designs, we used the adaptive law

$$\dot{\theta}(t) = -\text{sign}[b_p]\Gamma\omega(t)e(t), \; \Gamma = \Gamma^T > 0, \tag{5.235}$$

to ensure closed-loop signal boundedness and tracking: $\lim_{t\to\infty} e(t) = 0$.

A different control scheme is to use the control law

$$u(t) = \text{sign}[b_p]\left(k_2(t)y(t) + k_2 r(t)\right), \tag{5.236}$$

where $k_1(t)$ and $k_2(t)$ are estimates of sign$[k_p]k_1^*$ and sign$[b_p]k_2^*$ for k_1^* and k_2^* in (5.230). With (5.236), we have that $u(t) = \text{sign}[b_p]\theta^T(t)\omega(t)$, where $\theta(t)$ is the estimate of $\bar{\theta}^* = \text{sign}[b_p]\theta^*$ for θ^* in (5.231) and that

$$\dot{e}(t) = -a_m e(t) + |b_p|(\theta(t) - \text{sign}[b_p]\theta^*)^T\omega(t), \tag{5.237}$$

based on which we now use the adaptive law

$$\dot{\theta}(t) = -\Gamma\omega(t)e(t), \; \Gamma = \Gamma^T > 0, \tag{5.238}$$

to ensure closed-loop signal boundedness and tracking: $\lim_{t\to\infty} e(t) = 0$. This property can be easily verified by considering the positive definite function

$$V(e, \tilde{\theta}) = e^2 + |b_p|\tilde{\theta}^T\Gamma^{-1}\tilde{\theta}, \; \tilde{\theta} = \theta - \text{sign}[b_p]\theta^*, \tag{5.239}$$

whose time derivative is $\dot{V} = -2a_m e^2(t)$.

Design with Nussbaum gain. When the knowledge of sign$[b_p]$ is not available, we use the control scheme

$$u(t) = N(z(t))\theta^T(t)\omega(t), \tag{5.240}$$

$$\dot{z}(t) = \theta^T(t)\omega(t)e(t), \;\; z(0) = z_0 \in R, \tag{5.241}$$

$$N(z(t)) = z^2(t)\sin z(t), \tag{5.242}$$

for which the parameter vector θ is updated from the adaptive law

$$\dot{\theta}(t) = -\Gamma\omega(t)e(t), \;\; \Gamma = \Gamma^T > 0. \tag{5.243}$$

The function $N(z(t))$ is the *Nussbaum gain* [293], used to replace sign$[b_p]$.

With the control law (5.240), the error system (5.223) becomes

$$\dot{e}(t) = -a_m e(t) + b_p(N(z(t))\theta^T(t)\omega(t) - \theta^{*T}\omega(t)). \tag{5.244}$$

To analyze this adaptive control scheme, we consider the positive definite function $V(e, \tilde{\theta})$ in (5.239) whose time derivative now is

$$\dot{V} = -2a_m e^2(t) + 2b_p N(z(t))\theta^T(t)\omega(t)e(t) - 2|b_p|\theta^T(t)\omega(t)e(t). \tag{5.245}$$

With (5.241), integrating (5.245) gives

$$\begin{aligned} &V(e(t), \tilde{\theta}(t)) + 2a_m \int_0^t e^2(\tau)\,d\tau \\ &= V(e(0), \tilde{\theta}(0)) + 2b_p \int_{z_0}^{z} \xi^2 \sin\xi\,d\xi - 2|b_p| \int_{z_0}^{z} d\xi \triangleq \pi(z(t)), \end{aligned} \tag{5.246}$$

where

$$\begin{aligned} \pi(z(t)) = {} & V(e(0), \tilde{\theta}(0)) + z(t)(|b_p| + 2b_p \sin z(t)) + 2b_p \cos z(t) - b_p z^2(t) \cos z(t) \\ & - z_0(|b_p| + 2b_p \sin z_0) - 2b_p \cos z_0 + b_p z_0^2 \cos z_0 \end{aligned} \tag{5.247}$$

with $\pi(z(0)) = \pi(z_0) = V(e(0), \tilde{\theta}(0)) \geq 0$.

Examination of $\pi(z(t))$ shows that there exists a closed, bounded interval $[z^-, z^+]$ containing $\pi(z_0)$ for which both $\pi(z^-)$ and $\pi(z^+)$ are negative. However (5.246) implies that $z(t)$ cannot pass through either z^- or z^+, that is, $z(t)$ is bounded. From (5.247), $\pi(z(t))$ is bounded, and from (5.246), $V(e(t), \tilde{\theta}(t))$ is bounded and $e(t) \in L^2$. Then, we have that $y(t) \in L^\infty$ and $\dot{e}(t) \in L^\infty$ in (5.237) so that $u(t) \in L^\infty$ and $\lim_{t\to\infty} e(t) = 0$.

Gradient Design Example

In this example, we show how a Nussbaum gain can be used for an adaptive control scheme based on a gradient design.

We express the error system (5.233) as

$$e(t) = \frac{b_p}{s+a_m}[u - \theta^{*T}\omega](t). \tag{5.248}$$

As in Example 5.2 of Section 5.3.5, if we use the controller structure (5.229), that is, $u(t) = \theta^T(t)\omega(t)$, we have the tracking error equation

$$e(t) = \frac{b_p}{s+a_m}[(\theta - \theta^*)^T\omega](t), \tag{5.249}$$

based on which we define the estimation error

$$\epsilon(t) = e(t) + \rho(t)\xi(t), \tag{5.250}$$

where $\rho(t)$ is the estimate of $\rho^* = b_p$ and

$$\xi(t) = \theta^T(t)\zeta(t) - \frac{1}{s+a_m}[\theta^T\omega](t), \tag{5.251}$$

$$\zeta(t) = \frac{1}{s+a_m}[\omega](t). \tag{5.252}$$

From (5.249)–(5.252), we have the error equation

$$\epsilon(t) = \rho^*\tilde{\theta}^T(t)\zeta(t) + \tilde{\rho}(t)\xi(t), \tag{5.253}$$

where $\tilde{\theta}(t) = \theta(t) - \theta^*$, $\tilde{\rho}(t) = \rho(t) - \rho^*$. Then, the adaptive laws

$$\dot{\theta}(t) = -\frac{\mathrm{sign}[b_p]\Gamma\epsilon(t)\zeta(t)}{m^2(t)},\ \Gamma = \Gamma^T > 0, \tag{5.254}$$

$$\dot{\rho}(t) = -\frac{\gamma\epsilon(t)\xi(t)}{m^2(t)},\ \gamma > 0, \tag{5.255}$$

where $m(t) = \sqrt{1 + \zeta^T(t)\zeta(t) + \xi^2(t)}$, ensure closed-loop signal boundedness and $\lim_{t\to\infty} e(t) = 0$ (see Section 5.3.5).

A different control scheme is to use the control law

$$u(t) = \mathrm{sign}[b_p]\left(k_1(t)y(t) + k_2(t)r(t)\right), \tag{5.256}$$

where $k_1(t)$ and $k_2(t)$ are estimates of $\text{sign}[k_p]k_1^*$ and $\text{sign}[b_p]k_2^*$ for k_1^* and k_2^* in (5.230). With (5.256), we have that $u(t) = \text{sign}[b_p]\theta^T(t)\omega(t)$, where $\theta(t)$ is the estimate of $\bar{\theta}^* = \text{sign}[b_p]\theta^*$ for θ^* in (5.231), and from (5.248) that

$$e(t) = \frac{b_p}{s + a_m}[(\text{sign}[b_p]\theta - \theta^*)^T\omega](t), \tag{5.257}$$

based on which we define the estimation error

$$\epsilon(t) = e(t) + \rho(t)\xi(t), \tag{5.258}$$

where $\rho(t)$ is the estimate of $\rho^* = b_p$ and

$$\xi(t) = \text{sign}[b_p]\theta^T(t)\zeta(t) - \frac{1}{s + a_m}[\text{sign}[b_p]\theta^T\omega](t), \tag{5.259}$$

$$\zeta(t) = \frac{1}{s + a_m}[\omega](t). \tag{5.260}$$

From (5.257)–(5.260), we have the error equation

$$\begin{aligned}\epsilon(t) &= \rho^*(\text{sign}[b_p]\theta^T(t) - \theta^*)^T\zeta(t) + \tilde{\rho}(t)\xi(t)\\ &= |\rho^*|(\theta^T(t) - \text{sign}[b_p]\theta^*)^T\zeta(t) + \tilde{\rho}(t)\xi(t),\end{aligned} \tag{5.261}$$

where $\tilde{\rho}(t) = \rho(t) - \rho^*$. Then, the adaptive laws

$$\dot{\theta}(t) = -\frac{\Gamma\epsilon(t)\zeta(t)}{m^2(t)}, \quad \Gamma = \Gamma^T > 0, \tag{5.262}$$

$$\dot{\rho}(t) = -\frac{\gamma\epsilon(t)\xi(t)}{m^2(t)}, \quad \gamma > 0, \tag{5.263}$$

where $m(t) = \sqrt{1 + \zeta^T(t)\zeta(t) + \xi^2(t)}$, ensure closed-loop signal boundedness and $\lim_{t\to\infty} e(t) = 0$, by considering the positive definite function

$$V(\tilde{\theta}, \tilde{\rho}) = |b_p|\tilde{\theta}^T\Gamma^{-1}\tilde{\theta} + \tilde{\rho}^2\gamma^{-1}, \ \tilde{\theta} = \theta - \text{sign}[b_p]\theta^*, \tag{5.264}$$

whose time derivative is $\dot{V} = -\frac{2\epsilon^2(t)}{m^2(t)}$, which implies that $\frac{\epsilon(t)}{m(t)} \in L^2$, and from (5.261)–(5.263) that $\dot{\theta}(t) \in L^\infty \cap L^2$, $\dot{\rho}(t) \in L^\infty \cap L^2$, and $\frac{\epsilon(t)}{m(t)} \in L^\infty$.

Design with Nussbaum gain. When the knowledge of $\text{sign}[b_p]$ is not available, we use the control scheme

$$u(t) = N(z(t))\theta^T(t)\omega(t), \tag{5.265}$$

$$N(z(t)) = z^2(t)\sin z(t). \tag{5.266}$$

To design the signal $z(t)$ and to develop an adaptive law for $\theta(t)$, with (5.265), we write the tracking error system (2.48) as

$$e(t) = \frac{b_p}{s + a_m}[N(z)\theta^T \omega - \theta^{*T} \omega](t) \tag{5.267}$$

and, motivated by (5.259) with sign$[b_p]$ replaced by Nussbaum gain $N(z(t))$, we define the estimation error

$$\epsilon(t) = e(t) + \rho(t)\xi(t), \tag{5.268}$$

where $\rho(t)$ is the estimate of $\rho^* = b_p$ and

$$\xi(t) = N(z(t))\theta^T(t)\zeta(t) - \frac{1}{s + a_m}[N(z)\theta^T \omega](t), \tag{5.269}$$

$$\zeta(t) = \frac{1}{s + a_m}[\omega](t). \tag{5.270}$$

From (5.267)–(5.270), we have the error equation

$$\epsilon(t) = \rho^*(N(z(t))\theta(t) - \theta^*)^T \zeta(t) + \tilde{\rho}(t)\xi(t). \tag{5.271}$$

We choose the adaptive laws form $\theta(t)$ and $\rho(t)$ as

$$\dot{\theta}(t) = -\frac{\Gamma\epsilon(t)\zeta(t)}{m^2(t)}, \; \Gamma = \Gamma^T > 0, \tag{5.272}$$

$$\dot{\rho}(t) = -\frac{\gamma\epsilon(t)\xi(t)}{m^2(t)}, \; \gamma > 0, \tag{5.273}$$

where $m(t) = \sqrt{1 + \zeta^T(t)\zeta(t) + \xi^2(t)}$.

The time derivative of $V(\tilde{\theta}, \tilde{\rho}) = |b_p|\tilde{\theta}^T\Gamma^{-1}\tilde{\theta} + \tilde{\rho}^2\gamma^{-1}$, with $\tilde{\theta} = \theta - \text{sign}[b_p]\theta^*$, $\tilde{\rho} = \rho - \rho^*$, and $\rho^* = b_p$, as in (5.264), along (5.272)–(5.273), is

$$\begin{aligned} \dot{V} &= -2\left(\theta(t) - \text{sign}[b_p]\theta^*\right)^T \zeta(t)|b_p|\frac{\epsilon(t)}{m^2(t)} - 2\tilde{\rho}(t)\xi(t)\frac{\epsilon(t)}{m^2(t)} \\ &= -2\frac{\epsilon^2(t)}{m^2(t)} - 2|b_p|\frac{\theta^T(t)\zeta(t)\epsilon(t)}{m^2(t)} + 2b_p N(z(t))\frac{\theta^T(t)\zeta(t)\epsilon(t)}{m^2(t)}. \end{aligned} \tag{5.274}$$

Now we define the signal $z(t)$ from

$$\dot{z}(t) = \frac{\theta^T(t)\zeta(t)\epsilon(t)}{m^2(t)}, \; z(0) = z_0 \in R. \tag{5.275}$$

Integrating (5.724) gives

$$\begin{aligned} & V(\tilde{\theta}(t), \tilde{\rho}(t)) + 2\int_0^t \frac{\epsilon^2(\tau)}{m^2(\tau)}\, d\tau \\ & = V(\tilde{\theta}(0), \tilde{\rho}(0)) + 2b_p \int_{z_0}^{z} \xi^2 \sin\xi \, d\xi - 2|b_p| \int_{z_0}^{z} d\xi \triangleq \pi(z(t)), \end{aligned} \quad (5.276)$$

where

$$\begin{aligned} \pi(z(t)) = & V(\tilde{\theta}(0), \tilde{\rho}(0)) + z(t)(|b_p| + 2b_p \sin z(t)) + 2b_p \cos z(t) - b_p z^2(t) \cos z(t) \\ & - z_0(|b_p| + 2b_p \sin z_0) - 2b_p \cos z_0 + b_p z_0^2 \cos z_0 \end{aligned} \quad (5.277)$$

with $\pi(z(0)) = \pi(z_0) = V(\tilde{\theta}(0), \tilde{\rho}(0)) \geq 0$. Comparing (5.276) with (5.246), based on a similar argument to that in the Lyapunov design case, we conclude that $z(t)$ is bounded, and so is $\pi(z(t))$. Hence, $V(\tilde{\theta}(t), \tilde{\rho}(t))$ is bounded, and $\frac{\epsilon(t)}{m(t)} \in L^2$. Then, from (5.271)–(5.173), it follows that $\dot{\theta}(t) \in L^\infty \cap L^2$, $\dot{\rho}(t) \in L^\infty \cap L^2$, and $\frac{\epsilon(t)}{m(t)} \in L^\infty$. Based on these properties, all closed-loop signals are bounded and $\lim_{t\to\infty} e(t) = 0$ (see Section 5.3.5).

Designs for General Cases

The following generalizations of the above design examples can be made:

- A general version of the adaptive control system consisting of (5.227), (5.228), (5.236), and (5.238) for an n-dimensional state variable plant for state tracking (see Section 4.1).
- A general version of the adaptive control system consisting of (5.227), (5.228), (5.236), and (5.238) for an nth-order plant with relative degree $n^* = 1$ for output tracking (see Sections 5.3.2 and 5.3.3).
- A general version of the adaptive control system consisting of (5.227), (5.228), (5.236), and (5.238) for an nth-order plant with arbitrary relative degree n^* using a Lyapunov design (see Section 5.3.4).
- A general version of the adaptive control scheme consisting of (5.240)–(5.243) for an n-dimensional state variable plant in Section 4.1.
- A general version of the adaptive control scheme consisting of (5.240)–(5.243) for an nth-order plant with relative degree $n^* = 1$.
- A general version of the adaptive control scheme (5.240)–(5.243) for an nth-order plant with $n^* \geq 1$ using a Lyapunov design.

- A general version of the adaptive control scheme consisting of (5.256)–(5.263) for an nth-order plant with arbitrary relative degree n^* (see the gradient design in Section 5.3.5).

- A general version of the adaptive control scheme consisting of (5.265)–(5.273) for an nth-order plant with arbitrary relative degree n^*.

For more adaptive control and parameter estimation designs using Nussbaum gains, see [151], [267], [282]. It would be also interesting to develop a stable adaptive scheme using a Nussbaum gain for $\theta(t)$ and $\rho(t)$ based on the error model $\epsilon(t) = \rho^* \tilde{\theta}^T(t)\zeta(t) + \tilde{\rho}(t)\xi(t)$, where the sign of ρ^* is unknown.

5.6.2 An Adaptive Control System

In this subsection, we further study the use of the Nussbaum gain for adaptive control using an output feedback controller structure which is not based on an observer parametrization in Section 4.4.2 (i.e., it is different from that in (5.5)) and whose parameter vector size only depends on the number of sinusoidal signal components contained in the reference output signal $y_m(t)$ and not on the order the controlled plant (which is n for the plant (5.1)).

Plant Model and Reference Output

Consider the linear time-invariant plant

$$P(s)[y](t) = k_p Z(s)[u](t), \tag{5.278}$$

where $y(t) \in R$ is the measured plant output, $u(t) \in R$ is the applied input, and $P(s)$ and $Z(s)$ are monic polynomials of degrees n and $m = n-1$, respectively, that is, the plant transfer function $G_0(s) = k_p \frac{Z(s)}{P(s)}$ has relative degree $n^* = 1$. For model reference control, we assume that $Z(s)$ is stable (see Assumption (A5.1)), and in adaptive control, all plant parameters including k_p and those in $P(s)$ and $Z(s)$ are unknown. Unlike the adaptive control schemes in the previous sections, the adaptive design to be presented next does not need the knowledge of sign$[k_p]$, with the help of a Nussbaum gain.

As shown in [440], the plant (5.278) with $n^* = 1$ can be expressed by the minimal state representation

$$\begin{aligned} \dot{x}_1(t) &= A_1 x_1(t) + b_{12} y(t), \\ \dot{y}(t) &= c_{12} x_1(t) + \alpha_2 y(t) + k_p u(t) \end{aligned} \tag{5.279}$$

for some unknown constant $A_1 \in R^{(n-1)\times(n-1)}$, $b_{12} \in R^{n-1}$, $c_{12}^T \in R^{n-1}$, $\alpha_2 \in R$, where $x_1 \in R^{n-1}$ and the eigenvalues of A_1 are the zeros of $Z(s)$.

The control objective is to design an output feedback control law for $u(t)$ such that all closed-loop signals are bounded and the plant output $y(t)$ tracks a reference output $y_m(t)$ of the form

$$y_m(t) = y_0 + \sum_{i=1}^{l}(a_i \sin \omega_i t + b_i \cos \omega_i t), \tag{5.280}$$

where l, ω_i, $i = 1, \ldots, l$, are known constants and y_0, a_i, b_i, $i = 1, \ldots, l$, are constants whose knowledge is not needed in the design of the adaptive controller (i.e., only the reference signal $y_m(t)$ itself is needed).

Control Scheme

We propose the following adaptive controller for the plant (5.279):

$$u(t) = N(z(t))\theta^T(t)\omega(t), \tag{5.281}$$

$$\dot{z}(t) = \theta^T(t)\omega(t)e(t), \ z(0) = z_0 \in R, \tag{5.282}$$

$$N(z(t)) = z^2(t)\sin z(t), \tag{5.283}$$

$$\dot{\theta}(t) = \Gamma\omega(t)e(t), \ \Gamma = \Gamma^T > 0, \tag{5.284}$$

where $e(t) = y(t) - y_m(t)$ is the tracking error,

$$\theta(t) = [k_0(t), k_1(t), k_{11}(t), \ldots, k_{1l}(t), k_{21}(t), \ldots, k_{2l}(t)]^T \in R^{2+2l}, \tag{5.285}$$

$$\omega(t) = [e(t), 1, \sin\omega_1 t, \ldots, \sin\omega_l t, \cos\omega_1 t, \ldots, \cos\omega_l t]^T \in R^{2+2l}, \tag{5.286}$$

and $N(z(t))$ is a Nussbaum gain [293] (see Section 5.6.1). For the regulation problem when $y_m(t) = 0$, for all $t \geq 0$, $\theta(t) = k_0$ and $\omega(t) = e(t) = y(t)$.

Stability and Tracking Properties

This adaptive control system has the following desired properties.

Theorem 5.11 *All signals in the closed-loop system consisting of (5.281)–(5.286) are bounded and* $\lim_{t\to\infty}(y(t) - y_m(t)) = 0$.

Proof: For the purpose of analysis we define the system

$$\dot{x}_{1m}(t) = A_1 x_{1m}(t) + b_{12} y_m(t), \tag{5.287}$$

where $x_{1m} \in R^{n-1}$ represents the desired trajectory for $x_1(t)$. Defining $e_1(t) = x_1(t) - x_{1m}(t)$, from (5.279), (5.281), and (5.287), we obtain the error system

$$\begin{aligned}\dot{e}_1(t) &= A_1 e_1(t) + b_{12} e(t),\\ \dot{e}(t) &= c_{12} e_1(t) + \alpha_2 e(t) + (\alpha_2 y_m(t) - \dot{y}_m(t) + c_{12} x_{1m}(t))\\ &\quad + k_p N(z(t)) \theta^T(t) \omega(t).\end{aligned} \tag{5.288}$$

From (5.280) and the stability of A_1, the term $\alpha_2 y_m(t) - \dot{y}_m(t) + c_{12} x_{1m}(t)$ in (5.288), denoted as $f(t)$, can be expressed as

$$f(t) = f_0 + \sum_{i=1}^{l} \bar{a}_i \sin \omega_i t + \bar{b}_i \cos \omega_i t + \epsilon_m(t), \tag{5.289}$$

where f_0, $\bar{a}_i$, $\bar{b}_i$, $i = a, \ldots, l$, are some unknown constants and $\epsilon_m(t)$ is an exponentially decaying term due to system transient responses to $y_m(t)$.

Defining $\bar{\theta} = [0, f_0, \bar{a}_1, \ldots, \bar{a}_l, \bar{b}_1, \ldots, \bar{b}_l]^T$, we write (5.288) as

$$\begin{aligned}\dot{e}_1(t) &= A_1 e_1(t) + b_{12} e(t),\\ \dot{e}(t) &= c_{12} e_1(t) + \alpha_2 e(t) + \bar{\theta}^T \omega(t) + k_p N(z(t)) \theta^T(t) \omega(t) + \epsilon_m(t).\end{aligned} \tag{5.290}$$

Let us choose the positive definite function

$$V(e, e_1, \tilde{\theta}) = \frac{1}{2}\left(e^2 + e_1^T P_1 e_1 + |k_p| \tilde{\theta}^T \Gamma^{-1} \tilde{\theta}\right), \tag{5.291}$$

where $\tilde{\theta} = \theta - \bar{\theta}^*$, with

$$\bar{\theta}^* = \frac{\theta^*}{|k_p|}, \; \theta^* = [\bar{\theta}_1, 0, \ldots, 0]^T + \bar{\theta}, \tag{5.292}$$

for $\bar{\theta}_1 > 0$ sufficiently large, to be defined, and $P_1 = P_1^T > 0$ satisfying the Lyapunov equation $P_1 A_1 + A_1^T P_1 = -2I$.

The time derivative of $V(e, e_1, \tilde{\theta})$, along (5.284) and (5.290), is

$$\begin{aligned}\dot{V} &= -e_1^T(t) e_1(t) + e_1^T(t) P_1 b_{12} e(t) + e(t) c_{12} e_1(t) + \alpha_2 e^2(t) + \bar{\theta}^T \omega(t) e(t)\\ &\quad + \epsilon_m(t) e(t) + k_p N(z(t)) \theta^T(t) \omega(t) e(t) + |k_p| (\theta(t) - \bar{\theta}^*)^T \omega(t) e(t).\end{aligned} \tag{5.293}$$

Using (5.292) and (5.293), we have

$$\begin{aligned}\dot{V} &\le -(\bar{\theta}_1 - |\alpha_2|) e^2(t) - \|e_1(t)\|_2^2 + \|e_1(t)\|_2 \, |e(t)| \, \|P_1 b_{12} + c_{12}^T\|_2\\ &\quad + |k_p| \theta^T(t) \omega(t) e(t) + k_p N(z(t)) \theta^T(t) \omega(t) e(t) + \epsilon_m(t) e(t).\end{aligned} \tag{5.294}$$

For $\bar{\theta}_1 > |\alpha_2| + \frac{1}{2}\|P_1 b_{12} + c_{12}^T\|_2$, the 2×2 symmetric matrix

$$S = \begin{bmatrix} 1 & -\frac{1}{2}\|P_1 b_{12} + c_{12}^T\| \\ -\frac{1}{2}\|P_1 b_{12} + c_{12}^T\| & \frac{1}{2}(\bar{\theta}_1 - |\alpha_2|) \end{bmatrix} \tag{5.295}$$

is positive definite. Denoting λ_1 as the minimum eigenvalue of S ($\lambda_1 > 0$ as S is positive definite), we have from (5.294) that

$$\begin{aligned} \dot{V} \leq & -[\|e_1(t)\|_2, |e(t)|]\, S\, [\|e_1(t)\|_2, |e(t)|]^T + k_p N(z(t))\theta^T(t)\omega(t)e(t) \\ & + |k_p|\theta^T(t)\omega(t)e(t) - \frac{\bar{\theta}_1 - |\alpha_2|}{2}\left(|e(t)| - \frac{\epsilon_m(t)}{\bar{\theta}_1 - |\alpha_2|}\right)^2 + \frac{\epsilon_m^2(t)}{2(\bar{\theta}_1 - |\alpha_2|)} \\ \leq & -\lambda_1(\|e_1(t)\|^2 + e^2(t)) + k_p N(z(t))\theta^T(t)\omega(t)e(t) \\ & + |k_p|\theta^T(t)\omega(t) + \bar{\epsilon}^2(t), \end{aligned} \tag{5.296}$$

where $\bar{\epsilon}^2(t) = \frac{\epsilon_m^2(t)}{2(\bar{\theta}_1 - |\alpha_2|)}$ goes to zero exponentially fast.

Integrating on both sides of (5.296), we obtain

$$\begin{aligned} & V(e(t), e_1(t), \tilde{\theta}(t)) + \lambda_1 \int_0^t (\|e_1(\tau)\|^2 + e^2(\tau))\, d\tau \\ & \leq V_0 + \int_0^t \bar{\epsilon}^2(\tau)\, d\tau + |k_p| \int_{z_0}^{z} d\xi + k_p \int_{z_0}^{z} \xi^2 \sin\xi\, d\xi, \end{aligned} \tag{5.297}$$

where $V_0 = V(e(0), e_1(0), \tilde{\theta}(0))$. Letting $\int_0^t \bar{\epsilon}^2(\tau)\, d\tau \leq \epsilon_0$ and defining

$$\begin{aligned} \pi(z(t)) = & V_0 + \epsilon_0 + z(t)(|k_p| + 2k_p \sin z(t)) + 2k_p \cos z(t) - k_p z^2(t) \cos z(t) \\ & - z_0(|k_p| + 2k_p \sin z_0) - 2k_p \cos z_0 + k_p z_0^2 \cos z_0, \end{aligned} \tag{5.298}$$

where $\pi(z_0) = V_0 + \epsilon_0 \geq 0$, we have from (5.297) and (5.298) that

$$V(e(t), e_1(t), \tilde{\theta}(t)) + \lambda_1 \int_0^t (\|e_1(\tau)\|^2 + e^2(\tau))\, d\tau \leq \pi(z(t)). \tag{5.299}$$

Examination of $\pi(z(t))$ shows that there exists a closed, bounded interval $[z^-, z^+]$ containing $\pi(z_0)$ for which both $\pi(z^-)$ and $\pi(z^+)$ are negative. However, (5.299) implies that $z(t)$ cannot pass through either z^- or z^+; therefore $z(t)$ is bounded. From (5.298), $\pi(z(t))$ is bounded, and from (5.299), $V(e(t), e_1(t), \tilde{\theta}(t))$ is bounded and $e_1(t) \in L^2$, $e(t) \in L^2$. Hence $e(t)$, $e_1(t)$, and $\theta(t)$ are bounded. Since $\omega(t)$ is bounded, from (5.290), $\dot{e}_1(t)$ and $\dot{e}(t)$ are bounded. Hence $e_1(t)$, $e(t)$ go to zero as $t \to \infty$. Since $x_{1m}(t)$, $y_m(t)$ are bounded, it follows that $x_1(t)$ and $y(t)$ are also bounded. From (5.281), $u(t)$ is bounded and therefore all closed-loop signals are bounded. ∇

Remark 5.3 If the plant parameters are known, we can calculate $\bar{\theta}$ and set $\theta = \frac{\theta^*}{|g|}$ and $N(x) = -\text{sign}[g]$ in (5.281). From (5.290), we have

$$\begin{aligned} \dot{e}_1(t) &= A_1 e_1(t) + b_{12} e(t), \\ \dot{e}(t) &= c_{12} e_1(t) + (\alpha_2 - \bar{\theta}_1) e(t) + \epsilon_m(t). \end{aligned} \tag{5.300}$$

Since A_1 is stable, there exists a $\theta_1^* > 0$ such that the system (5.300) is exponentially stable for all $\bar{\theta}_1 \geq \theta_1^*$, ignoring the exponentially decaying $\epsilon_m(t)$. In this case the desired controller parameter θ^* is not unique. □

Remark 5.4 If the plant parameters are unknown but the sign of the high frequency gain g is known, the Nussbaum gain $N(z(t))$ can be replaced with $N = -\text{sign}[g]$ and Theorem 2.1 still holds. In this case,

$$\dot{V} \leq -\lambda_1(\|e_1(t)\|^2 + e^2(t)) + \bar{\epsilon}^2(t), \tag{5.301}$$

where $\bar{\epsilon}(t)$ goes to zero exponentially fast. Hence it follows from (5.301) that

$$V(e(t), e_1(t), \tilde{\theta}(t)) + \lambda_1 \int_0^t (\|e_1(\tau)\|^2 + e^2(\tau))\, d\tau \leq \epsilon_0 + V_0. \tag{5.302}$$

Hence $V(e(t), e_1(t), \tilde{\theta}(t)) \in L^\infty$ and $e_1(t) \in L^2$, $e(t) \in L^2$. □

Remark 5.5 The number of the adjustable controller parameters is independent of the order of the plant but depends on the number of the frequencies in the reference signal $y_m(t)$. For each additional frequency in $y_m(t)$, two additional parameters should be adjusted in the control law. This makes the number of adjustable parameters small for a wide class of $y_m(t)$, and therefore it is simpler, as compared with the observer based adaptive algorithms which require $2n$ adjustable parameters for any signal $y_m(t)$. □

Design and Analysis for Modeling Errors

As presented in Section 5.5, model reference adaptive control designs can be made robust with respect to bounded disturbances and stable unmodeled dynamics, by modifying adaptive laws for controller parameter updating. However, such robustness properties have not been established for the case when a Nussbaum gain is used to replace the sign of the high frequency gain. Modeling errors can introduce uncertainties in the expression of $\dot{V}$ in (5.245) or (5.274), which may destroy the integral form of (5.246)–(5.247) or (5.276)–(5.277) desirable for signal boundedness and L^2 error properties.

In this subsection we show that by adding a discontinuous component in the control signal an adaptive control design using a Nussbaum gain still has certain ideal properties in the presence of modeling errors such as disturbances or unmodeled dynamics in a controlled plant.

We first address the bounded disturbance case. Such disturbances may be present in the plant or may be introduced by inadequate modeling of the reference signal $y_m(t)$ as a sum of sinusoid signals.

Consider the linear time-invariant plant (5.279) with *disturbances*:

$$\begin{aligned}\dot{x}_1(t) &= A_1 x_1(t) + b_{12} y(t) + d_1(t),\\ \dot{y}(t) &= c_{12} x_1(t) + \alpha_2 y(t) + k_p u(t) + d_2(t),\end{aligned} \tag{5.303}$$

where $d_1(t)$, $d_2(t)$ are bounded disturbances. For $d_1(t) = 0$, $d_2(t) = 0$, (5.303) satisfies the same assumptions as (5.279). The reference signal is given by

$$y_m(t) = y_0 + \sum_{i=1}^{l} (a_i \sin \omega_i t + b_i \cos \omega_i t) + d_m(t), \tag{5.304}$$

where $d_m(t)$ is bounded, with bounded $\dot{d}_m(t)$. The function $d_m(t)$ may represent terms which cannot be modeled as a finite sum of sinusoidal functions. The parameters l, ω_i, a_i, b_i, $i = 1, \ldots, l$, are defined in (5.280).

The plant (5.303) is equivalent to

$$\begin{aligned}\dot{x}_1(t) &= A_1 x_1(t) + b_{12} y(t),\\ \dot{y}(t) &= c_{12} x_1(t) + \alpha_2 y(t) + k_p u(t) + d(t),\end{aligned} \tag{5.305}$$

where $d(t) = d_2(t) + c_{12}(sI - A_1)^{-1}[d_1](t)$ is bounded.[10]

To cancel the disturbance effect, we use the adaptive control scheme

$$u(t) = N(z(t))(\theta^T(t)\omega(t) + \mathrm{sgn}[e(t)]D_0), \tag{5.306}$$

$$N(z(t)) = z^2(t) \sin z(t), \tag{5.307}$$

$$\dot{z}(t) = \theta^T(t)\omega(t)e(t) + D_0|e(t)|,\ z(0) = z_0 \in R, \tag{5.308}$$

$$\dot{\theta}(t) = \Gamma\omega(t)e(t),\ \Gamma = \Gamma^T > 0, \tag{5.309}$$

[10] To show the effect of disturbances, we note that, with the adaptive control design (5.281)–(5.284) and for $d_m(t) = d_1(t) = 0$, the time derivative of the positive definite function V in (5.191) now, instead of (5.296), satisfies

$$\begin{aligned}\dot{V} \le& -\lambda_1(\|e_1(t)\|^2 + e^2(t)) + k_p N(z(t))\theta^T(t)\omega(t)e(t)\\ &+|k_p|\theta^T(t)\omega(t) + e(t)d_2(t) + \bar{\epsilon}^2(t),\end{aligned}$$

where $\lim_{t\to\infty} \bar{\epsilon}^2(t) = 0$ exponentially. For an unknown disturbance $d_2(t)$, the desired result in (5.297) and (5.298) does not hold anymore (however, it still holds if $d_2(t) \in L^2$).

where $e(t) = y(t) - y_m(t)$, $D_0 > 0$ is a design parameter to be specified, and $\text{sgn}[e(t)] = \text{sign}[e(t)]$ for $e(t) \neq 0$, $\text{sgn}[e(t)] = 0$ for $e(t) = 0$.

To analyze this scheme, it follows from (5.287) and (5.305)–(5.308) that

$$\begin{aligned}\dot{e}_1(t) &= A_1 e_1(t) + b_{12} e(t),\\ \dot{e}(t) &= c_{12} e_1(t) + \alpha_2 e(t) + \bar{\theta}^T \omega(t) + \epsilon(t)\\ &\quad + k_p N(z(t))(\theta^T(t)\omega(t) + \text{sgn}[e(t)] D_0) + d(t) + \bar{d}_m(t), \end{aligned} \tag{5.310}$$

where $\bar{d}_m(t) = \alpha_2 d_m(t) - \dot{d}_m(t)$ is bounded.

Consider the positive definite function

$$V(e, e_1, \tilde{\theta}) = \frac{1}{2}\left(e^2 + e_1^T P_1 e_1 + \tilde{\theta}^T \Gamma^{-1} \tilde{\theta}\right), \; \tilde{\theta} = \theta - \theta^*, \tag{5.311}$$

for θ^* in (5.292). We have from (5.309) and (5.310) that

$$\begin{aligned}\dot{V} &= -e_1^T(t) e_1(t) + e_1^T(t) P_1 b_{12} e(t) + e(t) c_{12} e_1(t) + \alpha_2 e^2(t) + \bar{\theta}^T \omega(t) e(t)\\ &\quad + \epsilon(t) e(t) + k_p N(z(t))(\theta^T(t)\omega(t) e(t) + |e(t)| D_0) + (\theta(t) - \theta^*)^T \omega(t) e(t)\\ &\quad + e(t)(d(t) + \bar{d}_m(t)). \end{aligned} \tag{5.312}$$

Following (5.295) and (5.296), we have

$$\begin{aligned}\dot{V} &\leq -\lambda_1(\|e_1(t)\|^2 + e^2(t)) + \bar{\epsilon}^2(t) + k_p N(z(t))(\theta^T(t)\omega(t) e(t) + D_0|e(t)|)\\ &\quad + \theta^T(t)\omega(t) e(t) + |e(t)|\,|d(t) + \bar{d}_m(t)|. \end{aligned} \tag{5.313}$$

Defining a design parameter bound D_0^* as

$$D_0^* = \sup_{t \geq 0} |d(t) + \bar{d}_m(t)|, \tag{5.314}$$

we have, from (5.313), that for all $D_0 \geq D_0^*$,

$$\begin{aligned}\dot{V} &\leq -\lambda_1(\|e_1(t)\|^2 + e^2(t)) + \bar{\epsilon}^2(t) + k_p N(z(t))(\theta^T(t)\omega(t) e(t) + D_0|e(t)|)\\ &\quad + \theta^T(t)\omega(t) e(t) + D_0|e(t)|. \end{aligned} \tag{5.315}$$

Using (5.307) and (5.308) and integrating on both sides of (5.315), we have

$$V(e(t), e_1(t), \tilde{\theta}(t)) + \lambda_1 \int_0^t (\|e_1(\tau)\|^2 + e^2(\tau))\, d\tau \leq \pi(z(t)), \tag{5.316}$$

where

$$\begin{aligned}\pi(z(t)) &= V(e(0), e_1(0), \tilde{\theta}(0)) + \epsilon_0 + z(t)(1 + 2k_p \sin z(t))\\ &\quad + 2k_p \cos z(t) - k_p z^2(t) \cos z(t) - z_0(1 + 2k_p \sin z_0)\\ &\quad - 2k_p \cos z_0 + k_p z_0^2 \cos z_0, \end{aligned} \tag{5.317}$$

with $\pi(z_0) = V(e(0), e_1(0), \tilde{\theta}(0)) + \epsilon_0 \geq 0$, and $\int_0^t \bar{\epsilon}_t^2(\tau)\, d\tau \leq \epsilon_0$ bounded.

Similar to that after (5.299) for Theorem 5.11, from the desired expressions (5.316) and (5.317), we may conclude that $z(t)$ is bounded, and so is $\pi(z(t))$, that is, $V(e(t), e_1(t), \tilde{\theta}(t))$ is bounded and $e_1(t) \in L^2$, $e(t) \in L^2$. We may also have closed-loop signal boundedness and asymptotic tracking.

Remark 5.6 The control law (5.305) uses a discontinuous signal component $\mathrm{sgn}[e(t)]D_0$. Existence of solutions for control systems of this type is characterized in an average sense [102], [424]. Continuous-time systems with discontinuities can be treated by using "sliding mode" methods which are well developed in the theory of variable structure systems [3], [29], [58], [69], [102], [131], [132], [133], [135], [352], [423], [424], [450], [451], [452], [460]. □

Remark 5.7 If we replace the Nussbaum gain $N(z(t))$ by $-\mathrm{sign}[g]$ in the control law (5.306), for $D_0 \geq \bar{D}_0$, where

$$\bar{D}_0 = \frac{1}{\bar{k}_0} \sup_{t \geq 0} |d(t) + \bar{d}_m(t)| \tag{5.318}$$

with $0 < \bar{k}_0 \leq |k_p|$, then Theorem 5.12 also holds, which can be verified by using the positive definite function $V(e(t), e_1(t), \tilde{\theta}(t))$ in (5.291). □

Now we consider the plant (5.279) with *unmodeled dynamics*:

$$y(t) = G(s)[u](t), \ \ G(s) = G_0(s)(1 + \mu\Delta(s)), \tag{5.319}$$

where $G(s)$ is a strictly proper rational function, $G_0(s) = k_p \frac{Z(s)}{P(s)}$, with relative degree 1 and $Z(s)$ being stable, is the modeled part of $G(s)$, and $\mu\Delta(s)$ is the multiplicative rational unmodeled dynamics with $\mu > 0$ and satisfies the following condition: there exists a known constant $q > 0$ such that

$$\|\Delta(s - q)s\|_\infty < k_m \tag{5.320}$$

for some constant $k_m > 0$, independent of μ, where $\Delta(s - q)$ is strictly proper and stable and $\|X(s)\|_\infty \stackrel{\triangle}{=} \sup_{\omega \in R} \|X(j\omega)\|$.

In this case, we propose the adaptive control scheme

$$u(t) = N(z(t)) \left(\theta^T(t)\omega(t) + g_0 \,\mathrm{sgn}[e(t)]m(t)\right), \tag{5.321}$$

$$N(z(t)) = z^2(t)\sin(z(t)), \tag{5.322}$$

$$\dot{z}(t) = \theta^T(t)\omega(t)e(t) + g_0|e(t)|m(t), \ \ z(0) = z_0 \in R, \tag{5.323}$$

$$\dot{m}(t) = -\delta_0 m(t) + \delta_1(|u(t)| + |y(t)|) + \delta_2, \ m(0) > 0, \tag{5.324}$$

$$\dot{\theta}(t) = \Gamma\omega(t)e(t), \ \Gamma = \Gamma^T > 0, \tag{5.325}$$

where $\delta_1 > 0$, $\delta_2 > 0$, $\delta_0 \in (0, q)$, and $g_0 > 0$ is a design parameter.

For this control scheme, using the positive definite function $V(e, e_1, \tilde{\theta})$ in (5.291), based on the property that $|\Delta(s)[u](t)| \leq c_1 m(t)$, $t \geq 0$, for some constant $c_1 > 0$, as ensured by the condition (5.320), we may show that there exists a $g_0^* > 0$ such that for any $g_0 \in (0, g_0^*)$ the existence of $\mu^* > 0$ is guaranteed such that for all $\mu \in [0, \mu^*)$ all closed-loop signals are bounded and the tracking error $e(t)$ goes to zero asymptotically. Such asymptotic tracking is only an ideal property for variable structure systems with discontinuous control signals, and actual system responses may experience certain chatterings around ideal trajectories, caused by signal discontinuities and system noises.

5.7 Discussion

Literature Overview

Model reference adaptive control (MRAC) is a popular, well-developed, and powerful adaptive control approach. The initial work of [306] refined the Lyapunov adaptive design and analysis for MRAC. The remarkable work of Monopoli [263] introduced the augmented error approach to the design of stable MRAC systems, which led to a considerable amount of research in developing globally stable adaptive algorithms for both the continuous- and discrete-time systems and in unifying the developed adaptive control algorithm [89], [91], [101], [265], [285], [289] (for discrete-time designs, see [125]).

The issue of robustness of adaptive control schemes with respect to modeling errors such as unmodeled dynamics, parameter variations, and bounded disturbances became a hot topic in the early 1980s. Many results were reported in the area of robust adaptive control for single-input, single-output (SISO) plants in the presence of unmodeled dynamics and bounded disturbances. Results in [19], [54], [90], [147], [148], [149], [194], [196], [331], [332], [334] showed that early adaptive control schemes may become unstable in the presence of unmodeled dynamics and/or bounded disturbances. The nonrobust behavior of the schemes of the 1970s attracted the interest of many researchers who concentrated on the development of robust adaptive controllers which perform in a stable and robust manner in the presence of modeling errors and bounded disturbances. Numerous robust adaptive laws have been proposed which guarantee global stability (that is, closed-loop signal boundedness) in the presence of modeling errors and bounded disturbances [7], [90], [145], [147], [148], [149], [150], [154], [155] (for a discrete-time design), [156], [193], [198], [200], [201], [277], [279], [281], [282], [307], [320], [321], [337], [339], [340], [414]. The dead-zone, σ, switching-σ, ϵ, and projection adaptive laws presented in this chapter and Chapter 3 are some commonly used adaptive schemes.

Instead of modifying the adaptive law for robustness, local robust stability results can be guaranteed by imposing persistent excitation conditions on certain signals within the adaptive loop as the means for achieving stability and robustness. It is shown that if a certain signal vector within the adaptive loop is persistently exciting, then the closed-loop adaptive control system is guaranteed to bc locally stable in the presence of small unmodeled dynamics and bounded disturbances [6], [7], [19], [54], [109], [196]. Such stability results are obtained using linearization [54], [280], averaging [6], [19], [109], small-gain theorem [194], integral manifold theory [6], [332], and other techniques [6], [33], [247], [249]. Both global and local results have contributed a great deal to development of robust adaptive control theory.

Adaptive control of time-varying systems has also been an active area of adaptive control [7], [11], [199], [253], [256], [295], [418]. Several adaptive control schemes have been developed to ensure closed-loop stability and bounded tracking errors which are small in a mean sense under the assumption that the plant parameter variations are small in a mean sense (this assumption can be relaxed if some parameter variation knowledge is available).

An important topic of adaptive control is system performance. As we have seen from this chapter, when there are no modeling errors such as unmodeled dynamics and disturbances in the controlled plant, a model reference adaptive controller is able to ensure closed-loop signal boundedness and asymptotic tracking of a given reference output $y_m(t)$ by the plant output $y(t)$ [121], [123], [265], [285]. In the presence of modeling errors, robust adaptive control schemes only ensure closed-loop signal boundedness and bounded tracking errors $y(t) - y_m(t)$ which are small in a mean sense (see Theorems 5.8, 5.9, and 5.10) when the modeling errors are small in a mean sense. Smallness in the mean does not imply smallness in absolute value or at steady state. It was shown in [5], [134] that the tracking error, although bounded, may be large at some time instant, even after a long time at steady state, which may affect the performance of the adaptive system and result in the so-called bursting phenomenon of robust adaptive control. If, however, the adaptive control system has strong persistency of excitation (that is, the level of persistency of excitation is larger than the magnitude of modeling errors), then the bursting phenomenon can be eliminated [5], [153], [192].

It is also important to improve the transient behavior of an adaptive control system. Due to initial parameter errors which may be large, the transient tracking error and control input can be larger than desired. There have been encouraging results in developing adaptive control schemes which ensure guaranteed desirable performance [67], [67], [258], [305], [360].

Recently, an adaptive backstepping design has been developed in [205], [206] for adaptive output tracking control of linear systems. This is a nonlinear feedback control design which is capable of improving system transient performance and leading to certain desirable closed-loop properties.

Proof of Theorem 4.2

Let us first assume the plant (4.37), that is, $\dot{x}(t) = Ax(t)+Bu(t)$, $y(t) = Cx(t)$, $x \in R^n$, $y(t) \in R$, is observable. From the design of a reduced-order observer (see Section 4.4.2) for this plant, we have

$$\lim_{t\to\infty} (x(t) - P^{-1}\hat{x}(t)) = 0 \text{ exponentially} \tag{5.326}$$

for some constant and nonsingular matrix $P \in R^{n\times n}$, where $\hat{x}(t) = [y(t), w(t)+L_r y(t)]^T$ for some vector $L_r \in R^{n-1}$, and

$$w(t) = \frac{G_u(s)}{\Lambda(s)}[u](t) + \frac{G_y(s)}{\Lambda(s)}[y](t) + \epsilon_0(t), \tag{5.327}$$

where $G_u(s) = [g_{u\,1}(s), \ldots, g_{u\,n-1}(s)]^T$ and $G_y(s) = [g_{y\,1}(s), \ldots, g_{y\,n-1}(s)]^T$ for some polynomials $g_{u\,i}(s)$ and $g_{u\,i}(s)$, $i = 1, 2, \ldots, n-1$, $\Lambda(s)$ is a stable and monic polynomial of degree $n-1$, and $\lim_{t\to\infty} \epsilon_w(t) = 0$ exponentially. If we define $a(s) = [1, s, \ldots, s^{n-2}]^T$ and

$$\omega_1(t) = \frac{a(s)}{\Lambda(s)}[u](t),\ \omega_2(t) = \frac{a(s)}{\Lambda(s)}[y](t), \tag{5.328}$$

it can be verified that

$$P^{-1}\hat{x}(t) = M^*[\omega_1^T(t), \omega_2^T(t), y(t)]^T + \hat{\epsilon}_x(t) \tag{5.329}$$

for some constant matrix $M^* \in R^{n\times(2n-1)}$ and $\lim_{t\to\infty} \hat{\epsilon}_x(t) = 0$ exponentially. Hence, for the control law (4.41), we have

$$k_1(t)x(t) = k_1(t)M^*[\omega_1^T(t), \omega_2^T(t), y(t)]^T + k_1(t)\epsilon_x(t) \tag{5.330}$$

for some $\epsilon_x(t)$ such that $\lim_{t\to\infty} \epsilon_x(t) = 0$ exponentially.

Introducing the parameter vector $\bar{\theta}(t) = [k_1(t)M^*, k_2(t)]^T$, we express the control law (4.41) as $u(t) = \bar{\theta}^T(t)\omega(t)$ for $\omega(t)$ in (5.38). From Lemma 4.1, the parameter vector $\bar{\theta}(t)$ has the same desired properties as those in Lemma 5.3 for $\theta(t)$ in (5.36): $\bar{\theta}(t) \in L^\infty$ and $\dot{\bar{\theta}}(t) \in L^2 \cap L^\infty$. Since $u(t) = \bar{\theta}^T(t)\omega(t)$

has the same structure as that in (5.34), the same analysis in the proof of Theorem 5.4 applies to the closed-loop system with the controller (4.41) so that closed-loop signals are bounded, and both (4.61) and (4.62) hold.

When the plant (4.37), that is, $\dot{x}(t) = Ax(t) + Bu(t)$, $y(t) = Cx(t)$, is only detectable, that is, there are unobservable but stable eigenvalues of A, this analysis is still valid, considering that some zeros of the filter polynomial $\Lambda(s)$ in (5.328) are equal to those unobservable but stable eigenvalues of A.

Swapping Lemma

The expression (5.83) or (5.138), generalized from its original version [285], is similar to the result known as the swapping lemma [151], [265], [340]: Let a stable and proper rational function $h(s)$ have a minimal realization $h(s) = c(sI - A)^{-1}b + d$ and $\theta(t)$ and $\omega(t)$ be two vector signals, and denote $h_c(s) = c(sI - A)^{-1}$ and $h_b(s) = (sI - A)^{-1}b$. Then one has the following equality:

$$\theta^T(t)h(s)[\omega](t) - h(s)[\theta^T\omega](t) = h_c(s)[h_b(s)[\omega^T]\dot{\theta}](t). \tag{5.331}$$

Controller Parameter Convergence

The ideal matching parameter vector θ^* is unique if and only if $\bar{n} = n$ and $P(s)$ and $Z(s)$ are coprime. With the controller (5.5), for the parameter estimate $\theta(t) \in R^{2n}$ to converge to its ideal value θ^* exponentially, it is necessary and sufficient for the reference input $r(t)$ to have $2n$ frequencies [36], [37].

When $q = \bar{n} - n > 0$ or/and $P(s)$ and $Z(s)$ have a common factor of degree l, the ideal matching parameter vector θ^* belongs to a linear variety $\mathcal{L}$ of dimension $q + l$. If the reference input $r(t)$ has $2\bar{n} - q - l$ frequencies, then the parameter estimate $\theta(t)$ converges to $\mathcal{L}$ asymptotically [36].

Stability Proof for Systems with Bounded Disturbance $d(t)$

The stability proof for robust (gradient) MRAC schemes applied to the plant (5.177) with $d(t) \neq 0$ can be found in [89], [151], [150], [282], [395].

Alternative Controller Structure

Based on the full-order observer parametrization scheme (4.140) of a state feedback control law used in Section 4.2 for adaptive control using system state variables, we can also use the following model reference controller structure:

$$u(t) = \theta_1^T\omega_1(t) + \theta_2^T\omega_2(t) + \theta_3 r(t), \tag{5.332}$$

where $\theta_1 \in R^{\bar{n}}$, $\theta_2 \in R^{\bar{n}}$, $\theta_3 \in R$, and

$$\omega_1(t) = \frac{a(s)}{\Lambda(s)}[u](t), \; \omega_2(t) = \frac{a(s)}{\Lambda(s)}[y](t) \tag{5.333}$$

with $a(s) = [1, s, \ldots, s^{\bar{n}-1}]^T$ and $\Lambda(s)$ a monic stable polynomial of degree $\bar{n}$.

This controller structure has one more parameter than that in the controller (5.5), leading to an increase of the adaptive system order. A model reference adaptive control scheme can be developed in the same way described in this chapter, under the same assumptions on the plant (5.177) with or without modeling errors, to ensure the same desired properties of those presented adaptive designs, in particular, closed-loop signal boundedness and asymptotic or small-in-the-mean output tracking properties. More precisely, in the presence of modeling errors $\mu\Delta_m(s)$, $\mu\Delta_a(s)$, and $d(t)$ in the plant (5.177), the adaptive system tracking performance is characterized by $e(t) \in B^{2,k_0\bar{d}_0^2}$ or $e(t) \in B^{2,k_0\mu^2}$, where $\bar{d}_0$ or μ is the size of the modeling errors. By Definition 5.1, $e(t) \in B^{2,\nu}$ if, for any $t_2 > t_1 \geq 0$ and some $\gamma_0 > 0$, $\int_{t_1}^{t_2} e^2(t)\, dt \leq \nu(t_2 - t_1) + \gamma_0$. Therefore, if ν is small, then $e(t)$ is also small in the mean. In fact, from the first inequality of (5.207), we have $\int_{t_1}^{t_2} e^2(t)\, dt \leq \gamma_0 + \int_{t_1}^{t_2} \frac{\bar{d}^2(t)}{a_m}\, dt$ for any $t_2 \geq t_1 \geq 0$, which indicates that the error $e^2(t)$ is bounded by the equivalent disturbance $\bar{d}^2(t)$ in the mean.

One advantage of the controller structure (5.332), however, is that with the gradient adaptive design (5.201)–(5.202) this controller admits a class of unmodeled dynamics $\mu\Delta_m(s)$, $\mu\Delta_a(s)$, larger than those characterized by Assumptions (A5.5a), (A5.6b) (see Section 5.5.1), which can be specified in deriving Corollary 5.2 and Lemma 5.5 with the controller (5.332) as follows.

With this controller structure, the gradient adaptive law presented in Section 5.3.5 can be used for the case when the controlled plant (5.177) is free of disturbances and unmodeled dynamics, that is, when $d(t) = 0$ and $\mu = 0$ so that (5.177) becomes $y(t) = G_0(s)[u](t)$ under the assumptions (A5.1)–(A5.4). The modified *gradient* adaptive law (5.201)–(5.202) can be used to ensure robust stability and tracking (see Theorem 5.10) for the case when $d(t) \neq 0$ and $\mu \neq 0$ in (5.177) under the assumptions (A5.5b), (A5.7b) and

(A5.6d): For $d_m = \lim_{s\to\infty} \frac{1}{P_m(s)}\Delta_m(s)$, $d_a = \lim_{s\to\infty} \Delta_a(s)$, there exists a known constant $q_\delta > 0$, independent of μ, such that the impulse response functions $h_m(t)$, $h_a(t)$ of $(\frac{1}{P_m(s-q_\delta)}\Delta_m(s - q_\delta) - d_m)s$, $(\Delta_a(s - q_\delta) - d_a)s$ satisfy

$$\|h_m(\cdot)\|_1 = \int_0^\infty |h_m(t)|\, dt \leq c, \; \|h_a(\cdot)\|_1 = \int_0^\infty |h_a(t)| dt \leq c \tag{5.334}$$

and also $|d_m| \leq c$, $|d_a| \leq c$ for a constant $c > 0$ independent of μ.

In this case, with $\theta_{20}(t) = \theta_{20}^* = 0$, we still have (5.193) for $\Delta(s)$ except for the order of $F_1(s)$ and $F_2(s)$ being increased by 1. We also have the expression (5.223) for the estimation error $\epsilon(t)$. However, unlike what was concluded in Corollary 5.1, the boundedness of $\frac{\Delta(s)[u](t)}{m(t)}$ is not ensured due to the new assumptions (A5.5b) and (A5.6d), which are *weaker* than (A5.5a) and (A5.6b). To prove the same results as that in Theorem 5.10, similar to that for (5.216), we use (5.193), (A5.5b), and (A5.6d) to express

$$\eta(t) = \eta_0(t) + \eta_1 u(t),\ \eta_0(t) = \Delta_0(s)\frac{1}{s+q_\delta}[u](t), \tag{5.335}$$

where $|\eta_1| \leq b_1$ and the impulse response function $h_0(t)$ of $\Delta_0(s-q_\delta)$ satisfies $\|h_0(\cdot)\|_1 = \int_0^\infty |h_0(t)|\,dt \leq b_2$ for some constant $b_1 > 0$, $b_2 > 0$ independent of μ. It then follows from Lemma 2.6 that $\frac{|\eta_0(t)|}{m(t)} \leq b_3$, $\frac{\omega(t)}{m(t)} \leq b_3$ for $\omega(t) = [\omega_1^T(t), \omega_2^T(t)]^T$ in (5.333) (in this case, the analysis does not involve the expression (5.211), due to the absence of $\theta_{20}(t)y(t)$ in the controller structure (5.332), which is crucial for relaxing the assumptions (A5.6b)–(A5.6d)), and $\frac{|u(t)|}{m(t)} \leq b_3\|\theta(t)\|_2$ for some constant $b_3 > 0$.

Then, from (5.224), we have

$$\begin{aligned}\dot V \leq & -\frac{\epsilon^2(t)}{m^2(t)} + 2\mu^2 b_3^2 + 2\mu^2 b_1^2 b_3^2 \|\theta(t)\|_2^2 \\ & - 2\sigma_1(t)|\rho^*|\tilde\theta^T(t)\theta(t) - 2\sigma_2(t)\tilde\rho(t)\rho(t).\end{aligned} \tag{5.336}$$

Similar to the proof of Lemma 5.4 (see (5.217)), from (5.336), we see that for any $\mu \in (0, \sqrt{\frac{|k_p|\sigma_{01}}{b_3^2 b_1^2}})$ the existence of constant $\theta_0 > 0$, $\rho_0 > 0$ is guaranteed so that $\dot V \leq 0$ whenever $\|\theta(t)\| > \theta_0$, $|\rho(t)| > \rho_0$. Therefore, we have that $\theta(t) \in L^\infty$, $\rho(t) \in L^\infty$, and $\epsilon(t) \in L^\infty$ and from (5.201) and (5.336) that $\frac{\epsilon(t)}{m(t)}$, $\dot\theta(t)$, and $\dot\rho(t)$ all belong to $B^{2,k_0\mu^2}$ for some constant $k_0 > 0$. Based on these desired properties of the adaptive law (5.201)–(5.202), we have the results of Theorem 5.10 (see its proof at the end of Section 5.5.4).

Adaptive Design (5.5) with (5.201)–(5.202) versus Proper $\Delta_a(s)$

The gradient adaptive law (5.201)–(5.202) ensures robust closed-loop stability in the presence of unmodeled dynamics $\mu\Delta_m(s)$ and $\mu\Delta_a(s)$ under the assumptions (A5.5a) and (A5.6b) when applied to the controller structure (5.5) (see Lemma 5.5) or under the assumptions (A5.5b) and (A5.6d) above when applied to the controller structure (5.332).

How about the controller (5.5) for $\mu\Delta_m(s)$ and $\mu\Delta_a(s)$ under the assumptions (A5.5b) and (A5.6d)? In view of (5.193), we consider $\Delta_a(s)[u](t)$. To apply the idea of (5.216) in proving Lemma 5.4, we express

$$\begin{aligned}\Delta_a(s)[u](t) &= (\Delta_a(s) - d_a)[u](t) + d_a u(t)\\ &= (\Delta_a(s) - d_a)(s + q_\delta)\frac{1}{s + q_\delta}[u](t) + d_a u(t).\end{aligned} \tag{5.337}$$

Under (A5.6d), the first term is bounded by $c_1 \frac{1}{s+q_\delta}[|u|](t)$ and thus by $c_2 m(t)$. The second term $d_a u(t)$, however, cannot be bounded by $c_3 d_a \|\theta(t)\| m(t)$ (i.e., $\omega(t)$ cannot be bounded by $c_4 m(t)$, different from the result of Corollary 5.2), as $y(t)$ in (5.211) has the term $\Delta(s)[u](t)$ which cannot be bounded by $c_5 m(t)$. This causes a difficulty in analysis, as the boundedness of $\frac{\omega(t)}{m(t)}$ is needed to lead to a term in $\dot{V}$, which is similar to $\mu^2 \frac{2b_3^2}{\kappa}\bar{\eta}_1^2 \|\theta(t)\|_2^2$ in (5.217), for establishing robust adaptation properties stated in Lemma 5.5.

Use of Input Filters for Larger Class of Unmodeled Dynamics

Although the controller structure (5.332) admits some proper unmodeled dynamics $\Delta_a(s)$ and $\frac{1}{P_m(s)}\Delta_m(s)$ (see (A5.5b) and (A5.6d) above, e.g., $\Delta_a(s) = \frac{s+2}{s+1}$), there are unmodeled dynamics which satisfy Assumption (A5.6c) (see Section 5.5.1) but not (A5.6d), for example, $\Delta_a(s) = d_a + \frac{1}{\mu s+1}$, $|d_a| \geq 0$, $(\Delta_a(s - q_\delta) - d_a)s = \frac{s}{\mu s + 1 - \mu q_\delta} = \frac{1}{\mu}(1 - \frac{1-\mu q_\delta}{\mu s+1-\mu q_\delta})$ $(0 < \mu < \frac{1}{q_\delta}$, $q_\delta > 0)$ has an impulse response function $h_a(t)$ such that $\|h_a(\cdot)\|_1 = \frac{2}{\mu}$.

To enlarge the class of unmodeled dynamics admissible to a robust MRAC scheme, we can employ an input compensator $H_c(s)$ [156] for the plant (5.177):

$$y(t) = G_c(s)[u_c](t),\ u(t) = H_c(s)[u_c](t),\ H_c(s) = \frac{1}{(s + a_c)^{n_c}},\ a_c > 0, \tag{5.338}$$

where $G_c(s) = G_{0c}(s)[1 + \mu\Delta_m(s)] + \mu\Delta_{ac}(s)$, $G_{0c}(s) = G_0(s)H_c(s)$, $\Delta_{ac}(s) = H_c(s)\Delta_a(s)$, and $n_c \geq 1$ to be chosen for different MRAC schemes applied to the plant (5.177) with different classes of $\mu\Delta_m(s)$ and $\mu\Delta_a(s)$.

The reference model transfer function now is $W_{mc}(s) = W_m(s)H_c(s)$, which has relative degree $n^* + n_c$ equal to the relative degree of $G_{0c}(s)$. With a proper choice of $n_c = 1, 2, 3$, the basic assumption on the unmodeled dynamics $\mu\Delta_m(s)$ and $\mu\Delta_a(s)$ admissible to robust MRAC, similar to (A5.6c), is

(A5.6e): $\|h_m(\cdot)\|_1 \leq c$, $\|h_a(\cdot)\|_1 \leq c$, for some constant $c > 0$ independent of μ, where $h_m(t)$ and $h_a(t)$ are the impulse response functions of $W_m(s - q_\delta)\Delta_m(s - q_\delta)$ and $\Delta_a(s - q_\delta)$, respectively, for some known constant $q_\delta > 0$.

Then, $\Delta_{ac}(s)$ and $W_{mc}(s)\Delta_m(s)$ satisfy those assumptions for a robust MRAC design with either controller (5.5) or (5.332). As an example, for $\Delta_a(s) = d_a + \frac{1}{\mu s+1}$, the choice of $H_c(s) = \frac{1}{s+a_c}$, $a_c > 0$, makes the controller structure (5.332), with order $n + n_c = n + 1$, applicable to this $\Delta_a(s)$.

Existence of Plant–Model Matching Parameters

The existence of the plant–model matching parameters θ_1^*, θ_2^*, θ_{20}^*, θ_3^* for the controller structure (5.5) was shown based on the solution to the polynomial equation (5.11) in a traditional way [151], [282], [340]. In fact, under the assumption that the state variable form of the plant (5.1),

$$\dot{x}(t) = Ax(t) + Bu(t),\ y(t) = Cx(t), \tag{5.339}$$

with $x(t) \in R^n$, $u(t) \in R$, $y(t) \in R$, is observable, the reduced-order observer theory presented in Section 4.4.2, applied to the state feedback controller structure (4.45), also verifies the existence of such plant–model matching parameters. Similarly, the full-order observer theory given in Section 4.4.1 verifies the existence of the plant–model matching parameters θ_1^*, θ_2^*, θ_3^* for the controller structure (5.332). The matching parameters θ_1^*, θ_2^* (and θ_{20}^*) can be explicitly specified using the derivations presented in Sections 4.4.1 and 4.4.2.

When the plant (5.334) is only detectable, that is, there are unobservable but stable eigenvalues of A, one cannot parametrize a state feedback $k_1^{*T}x(t)$ by $\theta_1^{*T}\omega_1(t) + \theta_2^{*T}\omega_2(t) + \theta_{20}^*y(t)$ with an *arbitrary* filter polynomial $\Lambda(s)$ for generating $\omega_1(t)$ and $\omega_2(t)$ in (5.229) or (5.333). However, since plant–model matching is from input to output, with output feedback, the unobservable modes of A are not changed and they are cancelled in the open loop. In this case, a partial state feedback using the observable states of $x(t)$ can be similarly parametrized with a reduced-order filter polynomial $\Lambda_r(s)$ and a set of reduced-dimension θ_1^* and θ_2^*. Then, the parametrization (5.5) or (5.332) can be restored by adding a common factor $\Lambda(s)/\Lambda_r(s)$ in the filters and parameter vectors. From the input–output point of view, the plant–model matching parameters still exist but, however, are not unique.

Other Robust Adaptive Schemes

Besides the switching σ-modification defined in (5.196)–(5.197) which needs the knowledge of upper bounds on $\|\theta^*\|_2$ and $|\rho^*|$, we can also use a fixed σ, or dead-zone, or ϵ modification presented in Section 3.8.1, with $m(t)$ given in (5.195), for robust adaptive control. In this case, since σ or d_1 or λ is present in (3.165) and (3.167) (or (3.175)–(3.176)) (which means that in the adaptive control case it results in $\frac{\epsilon(t)}{m(t)}, \dot{\theta}(t) \in B^{2,k_0(\mu^2+\nu)}$ ($\nu = \sigma$ or d_1 or λ) for some $k_0 > 0$), for closed-loop stability, there is an additional condition that $\nu \in (0, \nu^*)$, for some $\nu^* = \sigma^*$(or d_1^* or λ^*) > 0 which depends on the unknown $G_0(s)$ and cannot be large. On the other hand, the parameter projection algorithm (3.182) can be used to ensure the system properties as that from a switching σ-modification when the parameter bounds are available.

Feedback Unmodeled Dynamics

The results in Section 5.5 can be generalized to the following plant with more general unmodeled dynamics and disturbances:

$$\begin{aligned} y(t) = {} & G_0(s)[u](t) + \mu(G_0(s)\Delta_{mu}(s) + \Delta_{au}(s))[u](t) \\ & + \mu(G_0(s)\Delta_{my}(s) + \Delta_{ay}(s))[y](t) + G_0(s)H_d(s)[d_1](t) + d_2(t), \end{aligned} \tag{5.340}$$

where $\mu(G_0(s)\Delta_{my}(s) + \Delta_{ay}(s))[y](t))$ represents feedback unmodeled dynamics, and $H_d(s)[d_1](t)$ represents input disturbances [370]. This plant model captures some dynamics which cannot be modeled in the form of (5.177).

Robustness to $\lim_{t\to\infty} d(t) = 0$

A special class of disturbances are those $d(t) \in L^\infty$ such that $\lim_{t\to\infty} d(t) = 0$ but $d(t) \notin L^\alpha$ for any $\alpha \in [1, \infty)$ (e.g., $d(t) = \frac{1}{\ln(t+2)}$, $t \geq 0$). For such disturbances, the adaptive control schemes in Sections 5.3.2, 5.3.3, 5.3.4, 5.3.5, and 5.4 may not ensure closed-loop signal boundedness.

For robustness with respect to such disturbances, we should use the modified adaptive schemes in Section 5.5. Consider the plant (5.157), that is,

$$P(s)[y - d_y](t) = k_p Z(s)[u + d_u](t), \tag{5.341}$$

where $d_u(t) \in L^\infty$, $d_y(t) \in L^\infty$, and $\lim_{t\to\infty} d_u(t) = 0$, $\lim_{t\to\infty} d_y(t) = 0$. With the controller (5.34), the closed-loop error equation is (5.159) in which the equivalent output disturbance is

$$\begin{aligned} d(t) = {} & \left(1 + \frac{k_p}{P_m(s)}\left(\theta_2^{*T}\frac{a(s)}{\Lambda(s)} + \theta_{20}^*\right)\right)[d_y](t) \\ & + \frac{k_p}{P_m(s)}\left(1 - \theta_1^{*T}\frac{a(s)}{\Lambda(s)}\right)[d_u](t). \end{aligned} \tag{5.342}$$

Using the adaptive laws with a switching σ-modification (see Section 5.5), the controller (5.34) ensures closed-loop signal boundedness (see, e.g., Theorem

5.8 for $\dot{d}(t) \in L^\infty$). Consider the gradient algorithm (5.201)–(5.202). The time derivative of $V(\tilde{\theta}, \tilde{\rho})$ in (5.95), along (5.201)–(5.202), satisfies

$$\dot{V} \leq -\frac{\epsilon^2(t)}{m^2(t)} + \frac{d^2(t)}{m^2(t)} - 2\sigma_1(t)|\rho^*|\tilde{\theta}^T(t)\theta(t) - 2\sigma_2(t)\tilde{\rho}(t)\rho(t), \tag{5.343}$$

where $\sigma_1(t)|\rho^*|\tilde{\theta}^T(t)\theta(t) \geq 0$ and $\sigma_2(t)\tilde{\rho}(t)\rho(t) \geq 0$.

The question is whether the adaptive control scheme ensures $\lim_{t\to\infty}(y(t) - y_m(t)) = 0$ for $d_u(t) \in L^\infty$, $d_y(t) \in L^\infty$, $\lim_{t\to\infty} d_u(t) = 0$, $\lim_{t\to\infty} d_y(t) = 0$.

Clearly, $d(t) \in L^\infty$ in (5.337). How about $\lim_{t\to\infty} d(t) = 0$? The answer is "yes." To see this, it is sufficient to prove the following result.

Proposition 5.1 *If $x(t) \in L^\infty$ and $\lim_{t\to\infty} x(t) = 0$, and $a > 0$, then $w(t) = \frac{1}{s+a}[x](t) \in L^\infty$ and $\lim_{t\to\infty} w(t) = 0$.*

Proof: From $w(t) = \int_0^t e^{-a(t-\tau)}x(\tau)\,d\tau$, we have $|w(t)| \leq \frac{x_0}{a}$, $t \geq 0$, where $x_0 \geq |x(t)|$, $t \geq 0$. For any $\delta > 0$ there exists a $T_1 > 0$ such that $|x(t)| < \frac{\delta a}{2}$, $\forall t \geq T_1$. Let $T_2 > T_1 - \frac{1}{a}\ln\left(\frac{a\delta}{2x_0}\right)$. Then, for $t \geq T = \max\{T_1, T_2\}$, we have

$$\begin{aligned} |w(t)| &\leq \int_0^{T_1} e^{-a(t-\tau)}|x(\tau)|\,d\tau + \int_{T_1}^t e^{-a(t-\tau)}|x(\tau)|\,d\tau \\ &\leq \frac{x_0}{a}e^{-at}\left(e^{aT_1} - 1\right) + \frac{\delta}{2}\left(1 - e^{-a(t-T_1)}\right) < \frac{\delta}{2} + \frac{\delta}{2} = \delta. \end{aligned} \tag{5.344}$$

Hence we conclude that $\lim_{t\to\infty} w(t) = 0$. ∇

Using this result, it is not difficult to verify that the controller (5.34) with matching parameters $\theta_1 = \theta_1^*$, $\theta_2 = \theta_2^*$, $\theta_{20} = \theta_{20}^*$, and $\theta_3 = \theta_3^*$, for θ_1^*, θ_2^*, θ_{20}^*, θ_3^* satisfying (5.11), ensures closed-loop signal boundedness for $d_u(t) \in L^\infty$, $d_y(t) \in L^\infty$, $\lim_{t\to\infty} d_u(t) = 0$, $\lim_{t\to\infty} d_y(t) = 0$. However, with (5.342), it has not been concluded that $\lim_{t\to\infty}(y(t) - y_m(t)) = 0$ for the same class of $d_u(t)$ and $d_y(t)$ without any additional L^α ($\alpha \in [1, \infty)$) property.

Systems with Jumping Parameters

A typical system with jumping parameters can be modeled as

$$\dot{x}(t) = A(t)x(t) + B(t)u(t),\ y(t) = C(t)x(t), \tag{5.345}$$

where the system matrices $A(t) \in R^{n\times n}$, $B(t) \in R^n$, and $C(t) \in R^{1\times n}$ may take values from a set $\{A_i, B_i, C_i, i = 1, 2, \ldots, l\}$, that is,

$$A(t) = A_1\chi_1(t) + A_2\chi_2(t) + \cdots + A_l\chi_l(t),$$

$$
\begin{aligned}
B(t) &= B_1\chi_1(t) + B_2\chi_2(t) + \cdots + B_l\chi_l(t),\\
C(t) &= C_1\chi_1(t) + C_2\chi_2(t) + \cdots + C_l\chi_l(t),\\
\chi_i(t) &= \begin{cases} 1 & \text{for } (A(t), B(t), C(t)) = (A_i, B_i, C_i) \\ 0 & \text{otherwise,} \end{cases}
\end{aligned}
\tag{5.346}
$$

for $\chi_i(t)\chi_j(t) = 0$, $i \neq j$, and $\sum_{i=1}^{l} \chi_i(t) = 1$. There are two types of uncertainties: that of A_i, B_i, C_i and that of $\chi_i(t)$, $i = 1, 2 \ldots, l$.

If the knowledge of $\chi_i(t)$, $i = 1, 2 \ldots, l$, is available, adaptive control schemes can be designed to handle jumping parameters $A(t)$, $B(t)$, and $C(t)$. For example, for a reference model $\dot{x}_m(t) = A_m x_m(t) + B_m r(t)$, $x_m(t) \in R^n$, $r(t) \in R$, under the matching condition that there exist constant vectors $k_{1i}^* \in R^n$ and scalars k_{2i}^* such that $A_i + B_i k_{1i}^{*T} = A_m$, $B_i k_{2i}^* = B_m$, a stable adaptive control scheme, similar to that in Section 4.2, can be derived using the knowledge of $\text{sign}[k_{2i}^*]$, $i = 1, 2, \ldots, l$.

Problems

5.1 Work out the parametrization (4.150) for the $n = 3$ case.

5.2 Study an $n = 3$ and $m = 1$ case of (5.21).

5.3 Consider the plant

$$(s^2 + p_1 s + p_0)[y](t) = k_p(s + z_0)[u](t), \ \dot{y}(0) = 0, \ y(0) = 1,$$

where $z_0 > 0$, p_1, p_0, k_p are unknown but $\text{sign}[k_p]$ is known, and $y(t)$ is measured but not $\dot{y}(t)$, and choose the reference model

$$(s + 1)[y_m](t) = r(t), \ y_m(0) = 0.$$

(i) Study the plant–model matching equation (5.11).

(ii) Design an adaptive control scheme and prove that the tracking error $y(t) - y_m(t)$ converges to zero as $t \to \infty$.

(iii) Simulate the adaptive control system with $p_1 = -2$, $p_0 = 2$, $k_p = -1$, $z_0 = 1$ for $r(t) = 10$, $r(t) = 10\sin(t)$, $r(t) = 10\sin(t) + 13\sin(3.3t)$, respectively, and plot the tracking error for each case.

(iv) Comment on the simulation results.

5.4. Redo Problem 5.3 for the adaptive scheme in Section 5.3.3.

5.5 For the adaptive law (5.52), verify $\dot{\theta}(t) \in L^{\frac{1+\alpha}{\alpha}}$, that is, $\int_0^\infty \|\dot{\theta}(t)\|^{\frac{1+\alpha}{\alpha}} dt < \infty$, $\dot{\theta}(t) \in L^\infty$, and $\lim_{t\to\infty} \dot{\theta}(t) = 0$.

5.6 Consider the plant

$$(s^2 + p_1 s + p_0)[y](t) = k_p[u](t),\ \dot{y}(0) = 0,\ y(0) = 1,$$

where $y(t)$ is measured but not $\dot{y}(t)$, and choose the reference model

$$(s^2 + 3s + 2)[y_m](t) = r(t),\ \dot{y}_m(0) = y_m(0) = 0.$$

(i) Study the plant–model matching equation (5.11).
(ii) Design an adaptive controller for p_1, p_0, k_p unknown but sign$[k_p]$ known.
(iii) Design an adaptive controller for p_1, p_0 unknown but $k_p = 1$ known.
(iv) Simulate the adaptive system of part (i) with $p_1 = -2$, $p_0 = 3$, $k_p = 1$ for $r(t) = 10$, $r(t) = 10\sin(t)$, $r(t) = 10\sin(t) + 13\sin(3.3t)$, respectively, and plot the tracking errors and controller parameters.
(v) Comment on the simulation results and compare the controller parameters with the matching parameters calculated from part (i).

5.7 Verify (5.74), (5.79), and (5.80).

5.8 In (5.83), for the case of $n^* = 4$, from the expression

$$\theta^T(t)\zeta(t) = \theta^T(t)\frac{1}{L(s)}[\omega](t) = \frac{1}{L(s)}\left[L(s)\left[\theta^T\frac{1}{L(s)}[\omega]\right]\right](t),$$

first work out $L(s)[\theta^T\frac{1}{L(s)}[\omega]](t)$ and then verify (5.83). Using this experience, verify (5.83) for the general case. *Hint*: For $L(s) = s^{n^*-1} + l_{n^*-2}s^{n^*-2} + \cdots + l_1 s + l_0$,

$$\begin{aligned}
&L(s)\left[\theta^T\frac{1}{L(s)}[\omega]\right](t)\\
&= \left(s^{n^*-2} + l_{n^*-2}s^{n^*-3} + \cdots + l_1\right) s\left[\theta^T\frac{1}{L(s)}[\omega]\right](t) + \theta^T(t)\frac{l_0}{L(s)}[\omega](t)\\
&= \left(s^{n^*-2} + l_{n^*-2}s^{n^*-3} + \cdots + l_1\right)\left[\dot{\theta}^T\frac{1}{L(s)}[\omega]\right](t)\\
&\quad + \left(s^{n^*-2} + l_{n^*-2}s^{n^*-3} + \cdots + l_1\right)\left[\theta^T\frac{s}{L(s)}[\omega]\right](t) + \theta^T(t)\frac{l_0}{L(s)}[\omega](t).
\end{aligned}$$

5.9 Work out the closed-loop stability proof of the gradient adaptive design in Section 5.3.5 for the $n = 2$ case.

5.10 Consider the example in Section 5.4.1. Simulate the control system with a switching design for the adaptive law $\dot{k}(t) = \gamma|y(t)|^{1+\bar{\alpha}(t)}$ with $\gamma = 0.25$,

$y(0) = 1.25$, $k(0) = 1$, $\epsilon_0 = 1$, $\alpha_0 = 1$, and $\alpha_1 = 9$. The switching $\bar{\alpha}(t)$ is $\bar{\alpha}(t) = \alpha_0$ if $|y(t)| > 1$ and $\bar{\alpha}(t) = \alpha_1$ if $|y(t)| \leq 1$. Also simulate the control system with the adaptive law $\dot{k}(t) = 0.25y^{10}(t)$. As a comparison, simulate the control system with the adaptive law $\dot{k}(t) = 0.25y^2(t)$.

5.11 Verify the robustness properties of the adaptive law (5.91)–(5.92) when applied to the plant (5.157) with $d_u(t) \in L^\infty \cap L^2$ and $d_y(t) \in L^\infty \cap L^2$, that is, verify that $\frac{\epsilon(t)}{m(t)} \in L^\infty$ and $\dot{\theta}(t), \dot{\rho}(t) \in L^\infty \cap L^2$.

5.12 Consider the plant

$$(s^2 + p_1 s + p_0)[y](t) = k_p(1 - \mu s)[u](t),\ \mu > 0,\ \dot{y}(0) = u(0) = 0,\ y(0) = 1.$$

(i) Design a robust adaptive controller and state its robustness properties.

(ii) Simulate the adaptive control systems with $p_1 = -2$, $p_0 = 3$, $k_p = 1$, $r(t) = 10\sin(t)$ for $\mu = 0.01$, and plot the tracking error $e(t)$.

(iii) Simulate the adaptive controller of part (ii) of Problem 5.6 when applied to this plant and compare the results of (ii) and (iii).

5.13 Consider the plant

$$\dot{x}(t) = Ax(t) + bu(t) + d(t),$$

where $A \in R^{n \times n}$, $b \in R^n$ contain constant but unknown parameters and $d(t) \in R^n$ is a disturbance vector. Choose the reference model system

$$\dot{x}_m(t) = A_m x_m(t) + b_m r(t)$$

with $A_m \in R^{n \times n}$, $b_m \in R^n$ characterizing a desired reference system. The control objective is to design a *state* feedback control $u(t)$ to ensure

(a) the closed-loop signal boundedness in the presence of $d(t)$, and

(b) the asymptotic tracking of $x_m(t)$ by $x(t)$ in the absence of $d(t)$.

Design such a control scheme, specify the needed assumptions on A, b, $d(t)$, A_m, and b_m, and show that the proposed design has the desired properties.

5.14 Verify the error equation (5.159).

5.15 Prove the stability properties of the adaptive scheme (5.165)–(5.166).

5.16 Derive (5.192) and (5.193).

5.17 Consider an armature-control motor system with the electrical equation

$$L_a \frac{di_a}{dt} + R_a i_a = u - K_e \frac{d\theta_m}{dt}$$

and the mechanical equation

$$J_m \frac{d^2\theta_m}{dt^2} + b_m \frac{d\theta_m}{dt} = K_t i_a,$$

where L_a, R_a, and i_a are the armature inductance, resistance, and current; K_e and K_t are the back electromotive force (emf) constant and torque constant; J_m and b_m are the motor-load inertia and friction constant; and θ_m and u are the angular position and voltage (system input).

(i) For the output $\dot{\theta}_m$, design an adaptive controller for this system.

(ii) With $\mu = L_a$, formulate the robust adaptive control problem and develop or comment on a possible solution (for simplicity of analysis, set all parameters but L_a equal to 1).

(iii) Repeat (ii) with the new output θ_m.

(*Hint*: Use $\max_\omega |H(j\omega)| \leq \|h(\cdot)\|_1 \leq (2n+1)\max_\omega |H(j\omega)|$ for a stable and proper rational function $H(s)$ with n poles (see [78] for the first inequality and [156] for the second inequality) and $\|h(\cdot)\|_1 < \infty$ to verify Assumptions (A5.6a) and (A5.6b) for robust adaptive control.)

5.18 Develop adaptive laws for adaptive control and robust adaptive control based on the controller structure (5.332).

5.19 Develop a dead-zone, ϵ, or σ modification for robust adaptive control.

5.20 Develop a parameter projection algorithm for robust adaptive control.

5.21 In [334], Rohrs et al. presented the well-known Rohrs examples which stimulated the research on robustness of adaptive control systems. One example (also see [340]) is for the plant (5.177) with

$$y(t) = G(s)[u](t) + d(t), \; G(s) = G_0(s)(1 + \mu\Delta_m(s)),$$

$$G_0(s) = \frac{2}{s+1}, \; 1 + \mu\Delta_m(s) = \frac{229}{s^2 + 30s + 229}.$$

The control law is $u(t) = \theta_{20}(t)y(t) + \theta_3(t)r(t)$, the reference model is $y_m(t) = \frac{3}{s+3}[r](t)$, and the adaptive laws are $\dot{\theta}_{20}(t) = -\gamma_1 e(t)y(t)$, $\dot{\theta}_3(t) = -\gamma_2 e(t)r(t)$ for $\gamma_1 > 0$, $\gamma_2 > 0$, $e(t) = y(t) - y_m(t)$. The matching parameters based on the modeled plant $G_0(s)$ are $\theta_{20}^* = -1$ and $\theta_3^* = 1.5$.

As shown in [334], [340], the closed-loop system is unstable (that is, $y(t)$, $\theta_{20}(t)$, and $\theta_3(t)$ all diverge), for (i) $r(t) = 4.3$, $d(t) = 0$, (ii) $r(t) = 0.3 + 1.85\sin 16.1t$, $d(t) = 0$, or (iii) $r(t) = 2$, $d(t) = 0.5\sin 16.1t$, with $\gamma_1 = \gamma_2 = 1$, $\theta_{20}(0) = -0.65$, $\theta_3(0) = 1.14$, and other initial conditions are zero.

Design a simple σ modification (see Section 3.8.1) and a switching-σ modification for this adaptive control scheme, simulate the resulting adaptive systems (under the same conditions as above, with $\sigma = 0.3$ or $\sigma_0 = 0.3$), and compare the system tracking performance for the case when $G(s) = G_0(s)$.

5.22 Analyze and simulate the adaptive control system consisting of (5.227), (5.228), (5.236), and (5.238) and generalize it to an nth-order plant with relative degree $n^* = 1$ (see Sections 5.3.2 and 5.3.3) and to an n-dimensional state variable plant (see Sections 4.1).

5.23 Generalize the adaptive control system consisting of (5.227), (5.228), (5.236), and (5.238) to an nth-order plant with arbitrary relative degree n^*, using a Lyapunov design (see Section 5.3.4).

5.24 Generalize the adaptive control scheme consisting of (5.240)–(5.243) to an nth-order plant with relative degree $n^* = 1$ (see Sections 5.3.2 and 5.3.3) and to an n-dimensional state variable plant (see Sections 4.1).

5.25 Analyze the adaptive control scheme consisting of (5.240)–(5.243) using the positive definite function $V_0(e) = e^2$. (*Hint:* $\dot{z}(t) = -\frac{d}{dt}\left(\theta^T(t)\Gamma^{-1}\theta(t)\right)$.)

5.26 Generalize the adaptive control scheme (5.240)–(5.243) to an nth-order plant with $n^* \geq 1$ using a Lyapunov design (see Section 5.3.4).

5.27 Generalize the adaptive control scheme consisting of (5.256)–(5.263) to an nth-order plant with arbitrary relative degree n^* (see Section 5.3.5).

5.28 Generalize the adaptive control scheme consisting of (5.265)–(5.273) to an nth-order plant with arbitrary relative degree n^* (see Section 5.3.5).

5.29 Verify (5.279).

5.30 Analyze the adaptive control scheme (5.321)–(5.325) for the plant (5.319) and formulate a solution for the case with both $d(t)$ and $\mu\Delta(s)$.

5.31 Study the adaptive control schemes in Section 5.6.2 for

$$G_0(s) = \frac{k_p(s^3 + 6s^2 + 11s + 6)}{s^4 - s^3 - 2s^2 - 2s + 3},$$

which is unstable but has stable zeros.

The minimal state-space realization of this plant in the presence of bounded disturbances $d_{11}(t)$, $d_{12}(t)$, $d_{13}(t)$, $d_2(t)$ and unmodeled dynamics $\mu\Delta(s)$ can be described as (see (5.302), (5.318) and (5.328))

$$\dot{x}_1(t) = \begin{bmatrix} 0 & 1 & 0 \\ 0 & 0 & 1 \\ -6 & -11 & -6 \end{bmatrix} x_1(t) + \begin{bmatrix} 0 \\ 0 \\ 1 \end{bmatrix} y(t) + \begin{bmatrix} d_{11}(t) \\ d_{12}(t) \\ d_{13}(t) \end{bmatrix},$$
$$\dot{y}(t) = \begin{bmatrix} -45 & -69 & -29 \end{bmatrix} x_1(t) + 7y(t) + k_p(1 + \mu\Delta(s))u(t) + d_2(t).$$

Choose the reference signal as $y_m(t) = 3 + 5\sin(2t) + d_m(t)$, where $d_m(t) = d_m^0 \frac{\cos(t)}{1+\sin^2(0.5t)}$, the disturbances as $d_{11}(t) = d_{11}^0(\cos(3.5t) + \sin(4.4t))$, $d_{12}(t) = d_{12}^0(1+\sin(4.5t))$, $d_{13}(t) = d_{13}^0 \sin(2.5t)$, $d_2(t) = d_2^0 \sin(5.5t)$, and the unmodeled dynamics as $\mu\Delta(s) = \frac{\mu}{s+5}$.

When $d_m^0 = 0$, $y_m(t) = 3 + 5\sin(2t)$, that is, in (5.280), $y_0 = 3$, $l = 1$, $a_1 = 5$, $\omega_1 = 2$, $b_1 = 0$. Therefore, there are four controller parameters to be updated although the plant modeled part is of order $n = 4$.

Simulate different adaptive control designs for different cases with parameters $\theta(0) = [1, 1, 1, 1]^T$, $m(0) = 2$, $g = -5$, $\gamma = 10$, $z(0) = 0$, $\delta_0 = \delta_1 = 1$, $d_{11}^0 = d_{12}^0 = d_{13}^0 = d_m^0 = 2$, $d_2^0 = 3$, $D_0 = 20$, $\mu = 0.05$, $g_0 = 0.02$.

5.32 Redo Problem 5.3 with input (and output) disturbance $d_u(t)$ (and $d_y(t)$) $= \frac{1}{\ln(t+2)}$ and $\frac{1}{\ln\ln(t+5)}$, respectively, using the Lyapunov adaptive law (5.198).

5.33 Redo Problem 5.6 with $d_u(t)$ (and $d_y(t)$) $= \frac{1}{\ln(t+2)}$ and $\frac{1}{\ln\ln(t+5)}$, respectively, using the gradient adaptive law (5.201)–(5.202).

Advanced Topics

Actuator failure compensation. The actuator failure compensation problems formulated at the end of Chapter 4 are also important in the output feedback control case with the following linear time-invariant plant:

$$y(t) = G_1(s)[u_1](t) + \cdots + G_m(s)[u_m](t),$$

where $G_i(s)$, $i = 1, 2, \ldots, m$, are transfer functions with the same denominator $\det(sI - A)$ for some constant matrix $A \in R^{n\times n}$ and $u(t) = [u_1, \ldots, u_m]^T \in R^m$ is the input vector whose components may fail during system operation.

One type of actuator failure is

$$u_j(t) = \bar{u}_j,\ t \geq t_j,\ j \in \{1, 2, \ldots, m\},$$

where the constant value $\bar{u}_j$ and the failure time instant t_j are unknown.

The challenging task is to design an adaptive control scheme to ensure closed-loop signal boundedness and asymptotic output tracking in the presence of uncertainties in the failure values $\bar{u}_j$, failure time instants t_j, and failure actuators j as well as in plant parameters.

Relative degree condition. One of the assumptions needed for model reference adaptive control is that the relative degree n^* of the plant (5.1) is known (see (A5.4) in Section 5.1) for choosing the reference model (5.4) such that the controller (5.5) can achieve plant–model output matching, which is crucial for adaptive control. If n^* is unknown but $n_u^* \geq n^*$ is known, we may choose a reference model $y_m(t) = \frac{1}{P_m(s)}[r](t)$, where $P_m(s)$ is a stable polynomial of degree n_u^*, and the controller structure

$$u(t) = \theta_1^T \omega_1(t) + \theta_2^T \omega_2(t) + \theta_{20} y(t) + \theta_3^T b(s) y_m(t), \tag{5.347}$$

where $\theta_3 \in R^{n_u^*+1}$ and $b(s) = [1, s, \ldots, s^{n_u^*}]^T$.

The main issues such as plant–model matching, error equation, adaptive law and its stability analysis can be solved [385]. The key issue with this scheme is the stability of $\frac{1}{\theta_3^T b(s)}$, as an operator, for $\theta_3(t)$ from an adaptive law. Parameter projection of $\theta_3(t)$ onto a convex set which contains θ_3^* such that $\theta_3^{*T} b(s)$ is a stable polynomial of degree n^* for any possible $n^* > 0$ may do it. How does one specify such a convex set for $n_u^* > 2$ (or for $\tilde{n}^* = n_u^* - n_l^* > 2$, where $n_l^* > 0$ is known such that $n^* \geq n_l^*$, in this case a known stable polynomial of degree $n_l^* > 0$ is included as a part of $\theta_3^T b(s)$ [385])?

Time-varying systems. Adaptive control of systems with time-varying parameters is still an open area of adaptive control. Existing results [7], [199], [253], [256], [276], [295], [418], [433] have provided solutions to some general cases of small unknown parameter variations. For certain classes of systems with arbitrary finite unknown parameter variations, closed-loop signal boundedness can be guaranteed by an adaptive control scheme. Tracking performance has yet to be improved. For a general linear system with nonsmall parameter variations including unknown jumping parameters at unknown time instants, new adaptive control designs are yet to be developed.

Robust adaptive control with Nussbaum gains. As discussed in Section 5.6.2 (between Remark 5.5 and (5.302)), the unique analysis method (see (5.246)–(5.247), (5.276)–(5.277), or (5.297)–(5.298)) for adaptive systems with Nussbaum gains may not be able to handle modeling errors. New design and analysis techniques are needed for adaptive control using Nussbaum gains for plants with disturbances and unmodeled dynamics.

Chapter 6

Discrete-Time Model Reference Adaptive Control

In this chapter, we present the fundamentals of discrete-time model reference adaptive control theory, focusing on plant modeling, controller structure, error model, adaptive law, and stability and robustness analysis. While a discrete-time system is described by a difference equation (as compared with a differential equation in continuous time), the main difference between continuous-time and discrete-time control schemes is the system stability characterization, which leads to the unique choices of filters and adaptation gains as well as the unique stability and robust analysis, which are different from their continuous-time counterparts. The controller structure and matching conditions are similar for both continuous-time and discrete-time cases.

6.1 Control System Structure

Consider the discrete-time linear time-invariant plant

$$P(z)[y-d](t) = k_p Z(z)[u](t), \; t \in \{0, 1, 2, \ldots\}, \tag{6.1}$$

where $y(t)$ is the measured output, $u(t)$ is the applied control input, $d(t)$ is an unknown disturbance,[1] and

$$P(z) = z^n + p_{n-1}z^{n-1} + \cdots + p_1 z + p_0, \tag{6.2}$$

$$Z(z) = z^m + z_{m-1}z^{m-1} + \cdots + z_1 z + z_0 \tag{6.3}$$

with k_p, z_i, p_j, $i = 0, 1, \ldots, m-1$, $j = 0, 1, \ldots, n-1$, being constant parameters. The symbol z is used to denote, as the case may be, the z-transform

[1] An input disturbance $d_u(t)$ can be added by replacing $u(t)$ with $u(t) + d_u(t)$ in (6.1).

variable z or the time advance operator $z[x](t) = x(t+1)$ (i.e., z^{-1} is the time delay operator $z^{-1}[x](t) = x(t-1)$). The plant model (6.1) can also be expressed in the familiar transfer function form

$$y(t) = G(z)[u](t) + d(t),\ G(z) = \frac{k_p Z(z)}{P(z)}. \tag{6.4}$$

The objective of model reference adaptive control (MRAC) is to find an output feedback control signal $u(t)$ for the plant (6.1) with $G(z)$ unknown such that all closed-loop signals are bounded and the plant output $y(t)$ tracks a given reference output $y_m(t)$ as close as possible. The reference signal $y_m(t)$ may be generated from a reference model system

$$y_m(t) = W_m(z)[r](t), \tag{6.5}$$

where $W_m(z)$ is a stable transfer function and $r(t)$ is a bounded signal.

To design and analyze a model reference adaptive control scheme for the plant (6.1), we make the following assumptions on $G(z)$ and $d(t)$:

(A6.1): $Z(z)$ is a stable polynomial.

(A6.2): The degree n of $P(z)$ is known.

(A6.3): The sign of k_p, $\text{sign}[k_p]$, is known, and $|k_p| \leq k_p^0$ for some known constant $k_p^0 > 0$.

(A6.4): The relative degree $n^* = n - m > 0$ is known.

(A6.5): The disturbance $d(t)$ is bounded.

Assumption (A6.1) means that all the zeros of $Z(z)$ are inside the unit circle of the complex z-plane. Assumption (A6.2) can be relaxed as: (A6.2a) an upper bound $\bar{n}$ on n, $\bar{n} \geq n$, is known. Assumption (A6.3) indicates that the knowledge of an upper bound k_p^0 on $\text{sign}[k_p]k_p$ is needed in the discrete-time case, in addition to the sign of k_p. With (A6.4), we can take $W_m(z) = z^{-n^*}$ in the reference model (6.5) so that $y_m(t+n^*) = r(t)$. We will develop different control schemes whose stability and tracking performance are robust with respect to disturbances $d(t)$ satisfying (A6.5). We will also study robustness of standard MRAC schemes with respect to those $d(t)$ which belong to $L^{1+\alpha}$, $0 < \alpha < \infty$ (as we did in the continuous-time case).

The model reference controller structure is chosen as

$$u(t) = \theta_1^T \omega_1(t) + \theta_2^T \omega_2(t) + \theta_3 y_m(t+n^*), \tag{6.6}$$

where $\theta_1 \in R^{n-1}$, $\theta_2 \in R^n$, $\theta_3 \in R$ are parameters and

$$\omega_1(t) = a_\lambda(z)[u](t),\ a_\lambda(z) = [z^{-n+1}, \ldots, z^{-1}]^T, \tag{6.7}$$

$$\omega_2(t) = b_\lambda(z)[y](t),\ b_\lambda(z) = [z^{-n+1}, \ldots, z^{-1}, 1]^T. \tag{6.8}$$

If only an upper bound $\bar{n} \geq n$ is known, then we design the controller (6.6) with n replaced by $\bar{n}$ so that $\theta_1 \in R^{\bar{n}-1}$, $\theta_2 \in R^{\bar{n}}$.[2]

6.2 Model Reference Control

Before solving the adaptive control problem for the case when $G(z)$ is unknown, we first solve the nonadaptive control problem for $G(z)$ known.

Lemma 6.1 *There exist constant parameters $\theta_1^* \in R^{n-1}$, $\theta_2^* \in R^n$ such that*

$$\theta_1^{*T} a_\lambda(z) P(z) + \theta_2^{*T} b_\lambda(z) k_p Z(z) = P(z) - Z(z) z^{n^*}. \tag{6.9}$$

Proof: To solve (6.9), it is equivalent to solving the matching equation

$$\theta_1^{*T} z^{n-1} a_\lambda(z) P(z) + \theta_2^{*T} z^{n-1} b_\lambda(z) k_p Z(z) = z^{n-1}(P(z) - Z(z) z^{n^*}). \tag{6.10}$$

If $Z(z)$ and $P(z)$ are coprime, we can express (6.10) as

$$v(z) S^* \bar{\theta}^* = v(z) q^*,\ \bar{\theta}^* = [\theta_1^{*T}, \theta_2^{*T}]^T \in R^{2n-1}, \tag{6.11}$$

where, as in Section 5.1 for a continuous-time model reference controller, $S^* \in R^{(2n-1)\times(2n-1)}$ depends on $P(z)$ and $k_p Z(z)$, $q^* \in R^{2n-1}$ depends on the coefficients of $z^{n-1}(P(z) - Z(z) z^{n^*})$, and $v(z) = [z^{2n-2}, z^{2n-3}, \ldots, z, 1]$. The matrix S^* is nonsingular due to the coprimeness of $P(z)$ and $Z(z)$, so there is a unique $\bar{\theta}^*$ to satisfy (6.11) or (6.10).

---

[2] A more general model reference control structure, similar to that in (5.5) for the continuous-time case, is

$$u(t) = \theta_1^T \omega_1(t) + \theta_2^T \omega_2(t) + \theta_{20} y(t) + \theta_3 r(t),$$

where

$$\omega_1(t) = \frac{a(z)}{\Lambda(z)}[u](t),\ \omega_2(t) = \frac{a(z)}{\Lambda(z)}[y](t),$$

$$a(z) = [1, z, \ldots, z^{\bar{n}-2}]^T,$$

$\theta_1 \in R^{\bar{n}-1}$, $\theta_2 \in R^{\bar{n}-1}$, $\theta_{20} \in R$, $\theta_3 \in R$, and $\Lambda(z)$ is a monic stable polynomial of degree $\bar{n}-1$ (i.e., all zeros of $\Lambda(z)$ are inside the unit circle of the complex z-plane). The choice of $\Lambda(z) = z^{\bar{n}-1}$ with $\bar{n} = n$ results in the controller (6.6).

If $Z(z)$ and $P(z)$ have a common factor $C(z)$ of degree n_c, that is, $Z(z) = C(z)\bar{Z}(z)$, $P(z) = C(z)\bar{P}(z)$, where $\bar{Z}(z)$ and $\bar{P}(z)$ are coprime, then (6.10) with the common factor $C(z)$ cancelled out is expressed as

$$\bar{v}(z)\bar{S}^*\bar{\theta}^* = \bar{v}(z)\bar{q}^*, \tag{6.12}$$

where $S^* \in R^{(2n-n_c-1)\times(2n-1)}$ has rank $2n - n_c - 1$ due to the coprimeness of $\bar{Z}(z)$ and $\bar{P}(z)$, $\bar{q}^* \in R^{2n-n_c-1}$ depends on $z^{n-1}(\bar{P}(z) - \bar{Z}(z)z^{n^*})$, and $\bar{v}(z) = [z^{2n-n_c-2}, z^{2n-3}, \ldots, z, 1]$. Thus the solution $\bar{\theta}^*$ to (6.12) or (6.10) exists but is not unique (see the proof of Lemma 5.1). ∇

Similar to the continuous-time case in the proof of Lemma 5.1, the parameters θ_1^*, θ_2^* satisfying (6.9) are always nonunique if $\bar{n} > n$ is used in (6.6) (i.e., when $\theta_1^* \in R^{\bar{n}-1}$, $\theta_2^* \in R^{\bar{n}}$), because, in this case, (6.10) is expressed as (6.12) with $S^* \in R^{(2n-n_c-1)\times(2\bar{n}-1)}$ of rank $2n - n_c - 1$ and $\bar{q}^* \in R^{2n-n_c-1}$.

Example 6.1 Choose $k_p = 1$, $Z(z) = z + z_0$, and $P(z) = z^2 + p_1 z + p_0$.

For $P(z)$ and $Z(z)$ coprime, that is, $z_0^2 - p_1 z_0 + p_0 \neq 0$, we write (6.11) as

$$[z^2, z, 1]\begin{bmatrix} 1 & 0 & 1 \\ p_1 & 1 & z_0 \\ p_0 & z_0 & 0 \end{bmatrix}\begin{bmatrix} \theta_{11}^* \\ \theta_{21}^* \\ \theta_{22}^* \end{bmatrix} = [z^2, z, 1]\begin{bmatrix} p_1 - z_0 \\ p_0 \\ 0 \end{bmatrix}. \tag{6.13}$$

Since $z_0^2 - p_1 z_0 + p_0 \neq 0$, (6.13) has the unique solution as $\bar{\theta}^* = (S^*)^{-1}q^*$.

If $P(z) = (z + z_1)(z + z_0)$ for some constant z_1, then (6.12) has the form

$$[z, 1]\begin{bmatrix} 1 & 0 & 1 \\ z_1 & 1 & 0 \end{bmatrix}\begin{bmatrix} \theta_{11}^* \\ \theta_{21}^* \\ \theta_{22}^* \end{bmatrix} = [z, 1]\begin{bmatrix} z_1 \\ 0 \end{bmatrix}. \tag{6.14}$$

The solution to (6.14) is not unique:

$$\bar{\theta}^* = \begin{bmatrix} \theta_{11}^* \\ \theta_{21}^* \\ \theta_{22}^* \end{bmatrix} = \begin{bmatrix} 0 \\ 0 \\ z_1 \end{bmatrix} + \alpha\begin{bmatrix} 1 \\ -z_1 \\ -1 \end{bmatrix} \tag{6.15}$$

for any $\alpha \in R$, that is, $\bar{\theta}^*$ lies in a one-dimensional linear variety. □

The following theorem states the properties of a model reference controller.

Theorem 6.1 *There exist unique θ_1^* and θ_2^* which satisfy (6.9) as well as ensure that with $\theta_3^* = k_p^{-1}$ the controller*

$$u(t) = \theta_1^{*T}\omega_1(t) + \theta_2^{*T}\omega_2(t) + \theta_3^* y_m(t + n^*) \tag{6.16}$$

leads to closed-loop signal boundedness and the tracking error

$$y(t+n^*) - y_m(t+n^*) = d_0(t),\ t \in \{0, 1, 2, \ldots\}, \tag{6.17}$$

where $d_0(t) = (z^{n^*} + \theta_2^{*T} b_\lambda(z) k_p)[d](t)$ *is bounded.*

Proof: Let us first consider the case when $Z(z)$ and $P(z)$ are coprime. From (6.9), it follows that there exists some polynomial

$$F(z) = -z^{-m} + f_{n-m-2} z^{-m-1} + \cdots + f_0 z^{-n+1} \tag{6.18}$$

which satisfies the equality

$$\theta_1^{*T} a_\lambda(z) - 1 = F(z) Z(z). \tag{6.19}$$

Using this result, we can rewrite (6.9) with unique θ_1^* and θ_2^* as

$$F(z)P(z) + \theta_2^{*T} b_\lambda(z) k_p = -z^{n^*}. \tag{6.20}$$

Operating both sides of (6.20) on $y(t)$ and using (6.1), we obtain

$$F(z)(k_p Z(z)[u](t) + P(z)[d](t)) + \theta_2^{*T} b_\lambda(z) k_p [y](t) = -y(t+n^*). \tag{6.21}$$

Substituting (6.19) into (6.21) and using (6.7) and (6.8), we obtain

$$k_p \theta_1^{*T} \omega_1(t) + F(z)P(z)[d](t) + k_p \theta_2^{*T} \omega_2(t) = k_p u(t) - y(t+n^*). \tag{6.22}$$

Combining (6.16) and (6.22), we obtain

$$y(t+n^*) = y_m(t+n^*) - F(z)P(z)[d](t). \tag{6.23}$$

Defining $d_0(t) = -F(z)P(z)[d](t)$ and using (6.9) and (6.19), we have

$$d_0(t) = (z^{n^*} + \theta_2^{*T} b_\lambda(z) k_p)[d](t). \tag{6.24}$$

The boundedness of $y(t)$ follows from (6.23) and Assumption (A6.5). The control $u(t)$ is bounded because the plant zeros are stable.

Now we consider the case when $Z(z)$ and $P(z)$ are not coprime. In this case, we write $Z(z) = Z_1(z) Z_2(z)$ such that $Z_2(z)$ and $P(z)$ are coprime and $Z_1(z)$ has degree n_1. Then, because of the coprimeness of $P(z)$ and $Z_2(z)$, there exist unique parameters $\bar{\theta}_1^* \in R^{n-n_1-1}$, $\theta_2^* \in R^n$ such that

$$\begin{aligned} &z^{n-n_1-1} \bar{\theta}_1^{*T} \bar{a}_\lambda(z) P(z) + \theta_2^{*T} z^{n-1} b_\lambda(z) k_p Z_2(z) \\ &= z^{n-n_1-1} P(z) - z^{n-1} Z_2(z) z^{n^*} \end{aligned} \tag{6.25}$$

for $\bar{a}_\lambda(z) = [z^{-n+n_1+1}, \ldots, z^{-1}]^T$. Equivalently, we can write (6.25) as

$$(\bar{\theta}_1^{*T}\bar{a}_\lambda(z) - 1)z^{-n_1}P(z) + \theta_2^{*T}b_\lambda(z)k_pZ_2(z) = -Z_2(z)z^{n^*}. \tag{6.26}$$

Then, from (6.26), we have

$$\bar{\theta}_1^{*T}\bar{a}_\lambda(z) - 1 = \bar{F}(z)Z_2(z) \tag{6.27}$$

for some $\bar{F}(z) = -z^{-m+n_1} + \bar{f}_{n-m+n_1-2}z^{-m+n_1-1} + \cdots + \bar{f}_0 z^{-n+n_1+1}$. Using (6.26) and (6.27) and denoting $F(z) = \bar{F}(z)z^{-n_1}$, we also obtain (6.20), based on which the theorem's results follow. In this case the unique parameter vector θ_1^* is determined from

$$\theta_1^{*T}a_\lambda(z) - 1 = (\bar{\theta}_1^{*T}a_\lambda(z) - 1)Z_1(z)z^{-n_1} = F(z)Z(z). \tag{6.28}$$

Multiplying (6.26) by $Z_1(z)$ and using (6.28), we also have (6.10) or (6.9). ∇

Remark 6.1 When $P(z) = \bar{P}(z)C(z)$ and $Z(z) = \bar{Z}(z)C(z)$ have a common factor $C(z)$ of degree $n_c > 0$, the controller (6.16) with $\theta_3^* = k_p^{-1}$ and an arbitrary pair of parameters θ_1^*, θ_2^* satisfying (6.9) (whose solutions are nonunique when $P(z)$ and $Z(z)$have a common factor) may only lead to closed signal boundedness and the tracking error equation

$$C(z)[y - y_m](t + n^*) = C(z)[d_0](t),\ t \in \{0, 1, 2, \ldots\}, \tag{6.29}$$

which adds a transient response to $y(t) - y_m(t)$ due to the dynamics of $C(z)$, as compared with (6.17). To see (6.29), from (6.9), we obtain

$$\theta_1^{*T}a_\lambda(z) - 1 = \hat{F}(z)\bar{Z}(z) \tag{6.30}$$

for some polynomial

$$\hat{F}(z) = -z^{-m+n_c} + \bar{f}_{n-m+n_c-2}z^{-m+n_c-1} + \cdots + \hat{f}_0 z^{-n+n_c+1}. \tag{6.31}$$

With (6.30), from (6.9), we have

$$\hat{F}(z)\bar{P}(z) + \theta_2^{*T}b_\lambda(z)k_p = -z^{n^*}, \tag{6.32}$$

in which $\hat{F}(z)$ may not be expressible as $\hat{F}(z) = F(z)C(z)$ for $F(z)$ in (6.20). Instead of (6.22), which leads to (6.17), now from (6.30) and (6.32), we have

$$\begin{aligned} &k_pC(z)[\theta_1^{*T}\omega_1](t) + F(z)P(z)[d](t) + k_pC(z)[\theta_2^{*T}\omega_2](t) \\ &= k_pC(z)[u](t) - C(z)[y](t + n^*). \end{aligned} \tag{6.33}$$

Using this equality and (6.16) operated by $C(z)$, we have (6.29). Since $C(z)$ is stable by Assumption (A6.1), (6.29) also implies the closed-loop signal boundedness. In particular, when the disturbance $d(t)$ disappears, (6.17) becomes $y(t+n^*) - y_m(t+n^*) = 0$, which implies exact tracking, while (6.29) becomes $C(z)[y - y_m](t+n^*) = 0$, which only ensures (exponentially) asymptotic tracking when not all zeros of $C(z)$ are at the origin of the z-plane.

If, instead of n, $\bar{n} > n$ is used in the controller (6.16), then θ_1^* and θ_2^* satisfying the conditions of Theorem 6.1 exist but are not unique. □

6.3 Adaptive Control Systems

When $G(z)$ is unknown, an adaptive version of the controller (6.6) is implemented with $\theta_1 = \theta_1(t)$, $\theta_2 = \theta_2(t)$, and $\theta_3 = \theta_3(t)$, the estimates of θ_1^*, θ_2^*, and θ_3^*, to be updated from an adaptive law, that is,

$$u(t) = \theta_1^T(t)\omega_1(t) + \theta_2^T(t)\omega_2(t) + \theta_3(t)y_m(t+n^*), \tag{6.34}$$

where $\theta_1(t) \in R^{n-1}$, $\theta_2(t) \in R^n$, $\theta_3(t) \in R$, $\omega_1(t) = a_\lambda(z)[u](t)$, and $\omega_2(t) = b_\lambda(z)[y](t)$ with $a_\lambda(z) = [z^{-n+1}, \ldots, z^{-1}]^T$ and $b_\lambda(z) = [z^{-n+1}, \ldots, z^{-1}, 1]^T$.

To design a desired adaptive law, we first define an estimation error and derive an error equation. Introducing $\rho^* = k_p$ and

$$\omega(t) = [\omega_1^T(t), \omega_2^T(t), y_m(t+n^*)]^T, \tag{6.35}$$

$$\theta(t) = [\theta_1^T(t), \theta_2^T(t), \theta_3(t)]^T, \; \theta^* = [\theta_1^{*T}, \theta_2^{*T}, \theta_3^*]^T \in R^{2n}, \tag{6.36}$$

$$e(t) = y(t) - y_m(t), \; \tilde{\theta}(t) = \theta(t) - \theta^* \tag{6.37}$$

and combining (6.34) and (6.22), we have

$$e(t) = -\rho^*(\theta^{*T}\omega(t-n^*) - \theta^T(t-n^*)\omega(t-n^*)) + \bar{d}(t), \tag{6.38}$$

where $\bar{d}(t) = z^{-n^*}[d_0](t)$. In view of (6.38), we define the estimation error

$$\epsilon(t) = e(t) + \rho(t)\xi(t), \tag{6.39}$$

where $\rho(t)$ is the estimate of ρ^* and

$$\xi(t) = \theta^T(t)\omega(t-n^*) - \theta^T(t-n^*)\omega(t-n^*). \tag{6.40}$$

Substituting (6.38) and (6.40) into (6.39), we obtain the error equation

$$\epsilon(t) = \rho^*\tilde{\theta}^T(t)\omega(t-n^*) + \tilde{\rho}(t)\xi(t) + \bar{d}(t), \tag{6.41}$$

where $\tilde{\rho}(t) = \rho(t) - \rho^*$. This error equation is linear in the parameter errors $\tilde{\theta}(t)$ and $\tilde{\rho}(t)$, with a bounded disturbance term $\bar{d}(t)$.

6.3.1 Adaptive Control for Disturbance $d(t) = 0$

We now present the adaptive design for the case when $G(z)$ is unknown and the disturbance $d(t)$ is absent: $d(t) = 0$. In this case, the plant (6.1) is

$$P(z)[y](t) = k_p Z(z)[u](t) \tag{6.42}$$

with $P(z)$ and $Z(z)$ in (6.2) and (6.3), the controller is (6.34), the estimation error is (6.39), and the error equation (6.41) becomes

$$\epsilon(t) = \rho^* \tilde{\theta}^T(t)\omega(t-n^*) + \tilde{\rho}(t)\xi(t). \tag{6.43}$$

We choose the gradient adaptive law for $\theta(t)$:

$$\theta(t+1) = \theta(t) - \frac{\text{sign}[k_p]\Gamma\epsilon(t)\omega(t-n^*)}{m^2(t)}, \; 0 < \Gamma = \Gamma^T < \frac{2}{k_p^0} I_{2n}, \tag{6.44}$$

$$\rho(t+1) = \rho(t) - \frac{\gamma\epsilon(t)\xi(t)}{m^2(t)}, \; 0 < \gamma < 2, \tag{6.45}$$

where $\text{sign}[k_p]$ and $k_p^0 \geq |k_p|$ are defined in Assumption (A6.3) and

$$m(t) = \sqrt{1 + \omega^T(t-n^*)\omega(t-n^*) + \xi^2(t)}. \tag{6.46}$$

This adaptive design has the following desired properties.

Lemma 6.2 *The adaptive law (6.44)–(6.45) ensures that $\theta(t) \in L^\infty$, $\rho(t) \in L^\infty$, $\frac{\epsilon(t)}{m(t)} \in L^2 \cap L^\infty$, and $\theta(t+i_0) - \theta(t) \in L^2$ for any finite integer $i_0 > 0$.*

Proof: In view of (6.50), the time increment of the positive definite function

$$V(\tilde{\theta}, \tilde{\rho}) = |\rho^*|\tilde{\theta}^T\Gamma^{-1}\tilde{\theta} + \gamma^{-1}\tilde{\rho}^2 \tag{6.47}$$

along the trajectories of (6.44)–(6.45) is

$$\begin{aligned} &V(\tilde{\theta}(t+1), \tilde{\rho}(t+1)) - V(\tilde{\theta}(t), \tilde{\rho}(t)) \\ &= -\left(2 - \frac{|k_p|\omega^T(t-n^*)\Gamma\omega(t-n^*) + \gamma\xi^2(t)}{m^2(t)}\right)\frac{\epsilon^2(t)}{m^2(t)} \leq -\alpha_1 \frac{\epsilon^2(t)}{m^2(t)} \end{aligned} \tag{6.48}$$

for some constant $\alpha_1 > 0$. This implies that $\theta(t) \in L^\infty$, $\rho(t) \in L^\infty$, and $\frac{\epsilon(t)}{m(t)} \in L^2$. From (6.43), we have $\frac{\epsilon(t)}{m(t)} \in L^\infty$, and from (6.44), we have $\theta(t+1) - \theta(t) \in L^2$. Finally, using the inequality

$$\|\theta(t+i_0) - \theta(t)\|_2 \leq \sum_{i=0}^{i_0-1} \|\theta(t+i+1) - \theta(t+i)\|_2, \tag{6.49}$$

we have that $\theta(t+i_0) - \theta(t) \in L^2$ for any finite integer i_0. ∇

Theorem 6.2 *All signals in the closed-loop system with the plant (6.1) with $d(t) = 0$, reference model (6.5) with $W_m(z) = z^{n^*}$, controller (6.34), and adaptive law (6.44)–(6.45) are bounded, and $\lim_{t\to\infty}(y(t) - y_m(t)) = 0$.*

Proof: For $\omega_1(t)$ and $\omega_2(t)$ in (6.7) and (6.8), we introduce

$$\bar{\omega}(t) = [\omega_1^T(t), \omega_2^T(t)]^T. \tag{6.50}$$

From $\theta_3^* = k_p^{-1}$ and (6.22) with $d(t) = 0$, we have

$$\theta_1^{*T}\omega_1(t) + \theta_2^{*T}\omega_2(t) = u(t) - \theta_3^* y(t + n^*). \tag{6.51}$$

Using (6.50), (6.51), and (6.1) with $d(t) = 0$, we express

$$\bar{\omega}(t+1) = A^*\bar{\omega}(t) + b^* y(t + n^*), \tag{6.52}$$

where

$$A^* = \begin{bmatrix} E_{n-2} & 0_{(n-2)\times n} \\ \theta_1^{*T} & \theta_2^{*T} \\ 0_{(n-1)\times(n-1)} & E_{n-1} \\ A_1^{*T} & A_2^{*T} \end{bmatrix} \in R^{(2n-1)\times(2n-1)}, b^* = \begin{bmatrix} b_1^* \\ b_2^* \end{bmatrix} \in R^{2n-1}, \tag{6.53}$$

$$E_i = \begin{bmatrix} 0 & 1 & 0 & 0 & \cdots & 0 \\ 0 & 0 & 1 & 0 & \cdots & 0 \\ \vdots & \vdots & \vdots & \vdots & \vdots & \vdots \\ 0 & 0 & \cdots & 0 & 0 & 1 \end{bmatrix} \in R^{i\times(i+1)}, \; b_1^* = \begin{bmatrix} 0 \\ \vdots \\ 0 \\ \theta_3^* \end{bmatrix} \in R^{n-1}, \tag{6.54}$$

$$\begin{aligned} &\{A_1^{*T} \in R^{n-1}, A_2^{*T} \in R^n, b_2^* \in R^n\} \\ &= \begin{cases} \{[z^{*T}, k_p, 0, \ldots, 0], -p^{*T}, [0, \ldots, 0]\} & \text{if } n - m > 1, \\ \{z^{*T} + k_p\theta_1^{*T}, -p^{*T} + k_p\theta_2^{*T}, [0, \ldots, 0, 1]\} & \text{if } n - m = 1, \end{cases} \end{aligned} \tag{6.55}$$

$$p^* = [p_0, p_1, \ldots, p_{n-1}]^T \in R^n, \; z^* = [z_0, z_1, \ldots, z_{m-1}]^T \in R^m. \tag{6.56}$$

Letting $c^* = [1, 0, \ldots, 0] \in R^{1\times(2n-1)}$, from (6.52) and the plant (6.1) with $d(t) = 0$: $P(z)[y](t) = k_p Z(z)[u](t)$, we have

$$\begin{aligned} P(z)z^{-n+1}[u](t) &= c^*(zI_{2n-1} - A^*)^{-1}b^* z^{n^*} P(z)[y](t) \\ &= c^*(zI_{2n-1} - A^*)^{-1}b^* z^{n^*} Z(z)[u](t). \end{aligned} \tag{6.57}$$

This expression implies that $\det(zI_{2n-1} - A^*) = z^{n+n^*-1}Z(z)$. By Assumption (A6.1), the eigenvalues of A^* are inside the unit circle, so that there exists a nonsingular $Q^* \in C^{(2n-1)\times(2n-1)}$ such that $\|Q^*A^*Q^{*-1}\|_2 < 1$, where $\|\cdot\|_2$ is the induced Euclidean matrix norm (see Section 2.7.3). With this matrix Q^*, we then define the vector norm $\|\cdot\|$ in R^{2n-1} as $\|x\| = \|Q^* x\|_2$.

To proceed, from (6.39) and (6.52), we obtain

$$\begin{aligned}\bar{\omega}(t+1) = A^*\bar{\omega}(t) + b^*(y_m(t+n^*)\\ + \epsilon(t+n^*) - \rho(t+n^*)(\theta(t+n^*) - \theta(t))^T\omega(t)).\end{aligned} \tag{6.58}$$

Introducing the auxiliary signal

$$x_s(t) = |\frac{\epsilon(t+n^*)}{m(t+n^*)}| + \|\theta(t+n^*) - \theta(t)\|_2, \tag{6.59}$$

we see from Lemma 6.2 that $x_s(t) \in L^2$. Using (6.58) and (6.59), we obtain

$$\|\bar{\omega}(t+1)\| \leq (a_0 + c_1 x_s(t))\,\|\bar{\omega}(t)\| + c_2 \tag{6.60}$$

for some constants $a_0 \in (0,1)$, $c_1 > 0$, and $c_2 > 0$. Since $x_s(t) \in L^2$, we have

$$\sum_{t=t_0}^{t_0+j} x_s(t) \leq \sqrt{j+1}\sqrt{\sum_{t=t_0}^{t_0+j} x_s^2(t)} \leq c_3\sqrt{j+1} \tag{6.61}$$

for any $j \geq 1$ and some constant $c_3 > 0$. Using this property, we have

$$\begin{aligned}\prod_{t=t_0}^{t_0+j}(a_0 + c_1 x_s(t)) &\leq \left(a_0 + \frac{c_1}{j+1}\sum_{t=t_0}^{t_0+j} x_s(t)\right)^{j+1} \leq \left(a_0 + \frac{c_1 c_3}{\sqrt{j+1}}\right)^{j+1}\\ &\leq a_0^{j+1}\left(1 + \frac{c_1 c_3}{a_0\sqrt{j+1}}\right)^{j+1} \leq a_0^{j+1} e^{\frac{c_1 c_3}{a_0}\sqrt{j+1}},\end{aligned} \tag{6.62}$$

from which it follows that

$$\lim_{k\to\infty}\sum_{j=1}^{k}\prod_{t=t_0}^{t_0+j}(a_0 + c_5 x_s(t)) < \infty. \tag{6.63}$$

Using (6.60) and (6.63), we conclude that $\|\bar{\omega}(t)\|$ is bounded; that is, all closed-loop signals are bounded. Then we have $\epsilon(t) \in L^2$, $\theta(t+1) - \theta(t) \in L^2$ so that $\lim_{t\to\infty}\epsilon(t) = 0$, $\lim_{t\to\infty}(\theta(t+1) - \theta(t)) = 0$, and $\lim_{t\to\infty}\xi(t) = 0$. Finally, from (6.39), we have $\lim_{t\to\infty} e(t) = 0$. ∇

6.3.2 Robustness of MRAC with $d(t) \in L^2$

For the case when a disturbance $d(t) \in L^2$ is present in the plant (6.1), we also use the controller (6.34) and define the estimation error $\epsilon(t)$ from (6.39), and the adaptive law is (6.44)–(6.45). In this case, the time increment of $V(\tilde{\theta}, \tilde{\rho})$ in (6.46), along the trajectories of (6.44)–(6.45), is

$$\begin{aligned} & V(\tilde{\theta}(t+1), \tilde{\rho}(t+1)) - V(\tilde{\theta}(t), \tilde{\rho}(t)) \\ &= -\left(2 - \frac{|k_p|\omega^T(t-n^*)\Gamma\omega(t-n^*) + \gamma\xi^2(t)}{m^2(t)}\right)\frac{\epsilon^2(t)}{m^2(t)} + \frac{2\epsilon(t)\bar{d}(t)}{m^2(t)} \\ &\leq -\frac{\alpha_1}{2}\frac{\epsilon^2(t)}{m^2(t)} + \frac{2}{\alpha_1}\frac{\bar{d}^2(t)}{m^2(t)} \end{aligned} \tag{6.64}$$

for some constant $\alpha_1 > 0$. Since $d(t) \in L^2$, from (6.24) and (6.38), $\bar{d}(t) = (1 + z^{-n^*}\theta_2^{*T}b_\lambda(z)k_p)[d](t) \in L^2$, and so does $\frac{\bar{d}(t)}{m(t)}$. Hence, from (6.64), we have

$$V(\tilde{\theta}(t+1), \tilde{\rho}(t+1)) + \frac{\alpha_1}{2}\sum_{i=0}^{t}\frac{\epsilon^2(i)}{m^2(i)} \leq V(\tilde{\theta}(0), \tilde{\rho}(0)) + \frac{2}{\alpha_1}\sum_{i=0}^{t}\frac{\bar{d}^2(i)}{m^2(i)}, \tag{6.65}$$

which implies that $V(\tilde{\theta}(t), \tilde{\rho}(t)) \in L^\infty$ and $\frac{\epsilon(t)}{m^2(t)} \in L^2$, so Lemma 6.2 also holds.

To show the closed-loop stability and tracking performance in the presence of $d(t) \in L^2$, we note that, in this case, (6.51) becomes

$$\theta_1^{*T}\omega_1(t) + \theta_2^{*T}\omega_2(t) = u(t) - \theta_3^*(y(t+n^*) - d_0(t)) \tag{6.66}$$

for $d_0(t) = (z^{n^*} + \theta_2^{*T}b_\lambda(z)k_p)[d](t)$ in (6.24), and (6.58) becomes

$$\begin{aligned} \bar{\omega}(t+1) = A^*\bar{\omega}(t) + b^*(y_m(t+n^*) - d_0(t) + \epsilon(t+n^*) \\ - \rho(t+n^*)(\theta(t+n^*) - \theta(t))^T\omega(t)) + d_1(t), \end{aligned} \tag{6.67}$$

where $d_1(t) = [0, \ldots, 0, z^{-n+1}P(z)[d](t)]^T \in R^{2n-1}$. Therefore, we have (6.60) for a different c_2 due to $d_0(t)$ in (6.67). From (6.60), we also have the boundedness of $\bar{\omega}(t)$ and then the desired closed-loop properties stated in Theorem 6.2 even in the presence of an L^2 disturbance $d(t)$.

This result characterizes an important inherent property of the adaptive control scheme developed in Section 6.3.1: robustness of MRAC with respect to L^2 disturbances. In the presence of the more general class of disturbances satisfying Assumption (A6.5), however, the developed MRAC scheme needs to be modified with robust adaptive update laws.

6.3.3 Robust Adaptation for Bounded $d(t)$

When the disturbance $d(t)$ is only bounded (not necessarily in L^2), a robust MRAC scheme is constructed with the use of the controller (6.34), estimation error $\epsilon(t)$ in (6.39), and the modified adaptive law for $\theta(t)$ and $\rho(t)$:

$$\theta(t+1) = \theta(t) - \frac{\text{sign}[k_p]\Gamma\epsilon(t)\omega(t-n^*)}{m^2(t)} + f_\theta(t), \tag{6.68}$$

$$\rho(t+1) = \rho(t) - \frac{\gamma\epsilon(t)\xi(t)}{m^2(t)} + f_\rho(t), \tag{6.69}$$

where $f_\theta(t)$ and $f_\rho(t)$ are modification terms for robustness of the adaptive law with respect to $d(t)$ in (6.1) or $\bar{d}(t)$ in (6.41), which is only assumed to be bounded. These modification terms can be designed using a dead-zone [125], [200] or σ-modification [155] (see Section 3.8.2). Here we present a parameter projection design for which the step size parameters are chosen as

$$\Gamma = \text{diag}\{\gamma_1, \ldots, \gamma_{n_\theta}\},\ 0 < \gamma_j < \frac{2}{k_p^0},\ j = 1, \ldots, n_\theta,\ 0 < \gamma < 2, \tag{6.70}$$

where n_θ is the dimension of $\theta(t)$. For parameter projection, we need the knowledge of a convex set containing the unknown parameters $\theta^* \in R^{n_\theta}$ and ρ^*. Such a convex set can be characterized by the region defined in the assumption:

> (A6.6): The lower bound θ_j^a and upper bound θ_j^b of the jth component θ_j^* of θ^*, $j = 1, \ldots, n_\theta$, and those of ρ^*, ρ^a, and ρ^b, are known: $\theta_j^* \in [\theta_j^a, \theta_j^b]$, $\rho^* \in [\rho^a, \rho^b]$.

Then, denoting the jth component of $f_\theta(t)$ as $f_{\theta j}(t)$, $j = 1, 2, \ldots, n_\theta$, we choose $\theta_j(0) \in [\theta_j^a, \theta_j^b]$, $\rho(0) \in [\rho^a, \rho^b]$ and let

$$f_{\theta j}(t) = \begin{cases} 0 & \text{if } \theta_j(t) + g_{\theta j}(t) \in [\theta_j^a, \theta_j^b], \\ \theta_j^b - \theta_j(t) - g_{\theta j}(t) & \text{if } \theta_j(t) + g_{\theta j}(t) > \theta_j^b, \\ \theta_j^a - \theta_j(t) - g_{\theta j}(t) & \text{if } \theta_j(t) + g_{\theta j}(t) < \theta_j^a, \end{cases} \tag{6.71}$$

$$f_\rho(t) = \begin{cases} 0 & \text{if } \rho(t) + g_\rho(t) \in [\rho^a, \rho^b], \\ \rho^b - \rho(t) - g_\rho(t) & \text{if } \rho(t) + g_\rho(t) > \rho^b, \\ \rho^a - \rho(t) - g_\rho(t) & \text{if } \rho(t) + g_\rho(t) < \rho^a, \end{cases} \tag{6.72}$$

where $g_{\theta j}(t)$ is the jth component of $g_\theta(t)$ defined by

$$g_\theta(t) = -\frac{\text{sign}[k_p]\Gamma\omega(t-n^*)\epsilon(t)}{m^2(t)} \tag{6.73}$$

for $j = 1, \ldots, n_\theta$ and

$$g_\rho(t) = -\frac{\gamma\xi(t)\epsilon(t)}{m^2(t)}. \tag{6.74}$$

This modified adaptive law has the following properties.

Lemma 6.3 *The adaptive law (6.68)–(6.74) ensures that $\theta(t) \in L^\infty$, $\rho(t) \in L^\infty$, $\frac{\epsilon(t)}{m(t)} \in L^\infty$, and*

$$\sum_{t=t_1}^{t_2} \frac{\epsilon^2(t)}{m^2(t)} \le a_1 + b_1 \sum_{t=t_1}^{t_2} \frac{\bar{d}^2(t)}{m^2(t)}, \tag{6.75}$$

$$\sum_{t=t_1}^{t_2} \|\theta(t+i_0) - \theta(t)\|_2^2 \leq a_2 + b_2 \sum_{t=t_1}^{t_2} \frac{\bar{d}^2(t)}{m^2(t)} \tag{6.76}$$

for any finite i_0, some constants $a_i > 0$, $b_i > 0$, $i = 1, 2$, and all $t_2 > t_1 \geq 0$.

Proof: Letting $\gamma_m = \max\{\gamma_1 k_p^0, \ldots, \gamma_{n_\theta} k_p^0, \gamma\} < 2$ and using (6.41), we express the time increment of $V(\tilde{\theta}, \tilde{\rho})$ in (6.47), along (6.68)–(6.69), as

$$\begin{aligned}
& V(\tilde{\theta}(t+1), \tilde{\rho}(t+1)) - V(\tilde{\theta}(t), \tilde{\rho}(t)) \\
&= -\left(2 - \frac{|k_p|\omega^T(t-n^*)\Gamma\omega(t-n^*) + \gamma\xi^2(t)}{m^2(t)}\right) \frac{\epsilon^2(t)}{m^2(t)} \\
&\quad + 2|k_p| f_\theta^T(t)\Gamma^{-1}(\tilde{\theta}(t) + g_\theta(t) + f_\theta(t)) - |k_p| f_\theta^T(t)\Gamma^{-1} f_\theta(t) \\
&\quad + 2f_\rho(t)\gamma^{-1}(\tilde{\rho}(t) + g_\rho(t) + f_\rho(t)) - f_\rho(t)\gamma^{-1} f_\rho(t) + \frac{2\epsilon(t)\bar{d}(t)}{m^2(t)} \\
&\leq -\frac{2-\gamma_m}{2} \frac{\epsilon^2(t)}{m^2(t)} + \frac{4}{2-\gamma_m} \frac{\bar{d}^2(t)}{m^2(t)} \\
&\quad + 2|k_p| f_\theta^T(t)\Gamma^{-1}(\tilde{\theta}(t) + g_\theta(t) + f_\theta(t)) - |k_p| f_\theta^T(t)\Gamma^{-1} f_\theta(t) \\
&\quad + 2f_\rho(t)\gamma^{-1}(\tilde{\rho}(t) + g_\rho(t) + f_\rho(t)) - f_\rho(t)\gamma^{-1} f_\rho(t).
\end{aligned} \tag{6.77}$$

Since Γ in (6.70) is a diagonal matrix, we can rewrite

$$\begin{aligned}
& 2|k_p| f_\theta^T(t)\Gamma^{-1}(\tilde{\theta}(t) + g_\theta(t) + f_\theta(t)) \\
&= 2|k_p| \sum_{j=1}^{n_\theta} \gamma_j^{-1} f_{\theta j}(t)(\theta_j(t) - \theta_j^* + g_{\theta j}(t) + f_{\theta j}(t)).
\end{aligned} \tag{6.78}$$

Using (6.72), we have

$$f_{\theta j}(t)(\theta_j(t) - \theta_j^* + g_{\theta j}(t) + f_{\theta j}(t)) \leq 0 \tag{6.79}$$

for all $j = 1, \ldots, n_\theta$. Similarly, using (6.72), we have

$$f_\rho(t)(\rho(t) - \rho^* + g_\rho(t) + f_\rho(t)) \leq 0. \tag{6.80}$$

Hence, with (6.78)–(6.80), (6.85) becomes

$$\begin{aligned}
& V(\tilde{\theta}(t+1), \tilde{\rho}(t+1)) - V(\tilde{\theta}(t), \tilde{\rho}(t)) \\
&\leq -\frac{2-\gamma_m}{2} \frac{\epsilon^2(t)}{m^2(t)} + \frac{4}{2-\gamma_m} \frac{\bar{d}^2(t)}{m^2(t)} \\
&\quad - |k_p| f_\theta^T(t)\Gamma^{-1} f_\theta(t) - f_\rho(t)\gamma^{-1} f_\rho(t).
\end{aligned} \tag{6.81}$$

Since the projection algorithm (6.72)–(6.74) ensures that $\theta_j(t) \in [\theta_j^a, \theta_j^b]$, $j = 1, \ldots, n_\theta$, and $\rho \in [\rho^a, \rho^b]$, it follows that $\theta(t)$ and $\rho(t)$ are bounded. Then

(6.62) implies that $\frac{\epsilon(t)}{m(t)}$ is bounded. Using (6.81), we have (6.75). Finally, from (6.68), there exist constants $k_i > 0$, $i = 1, 2$, such that

$$\|\theta(t+1) - \theta(t)\|_2^2 \leq k_1 \frac{\epsilon^2(t)}{m^2(t)} + k_2 f_\theta^T(t) f_\theta(t). \tag{6.82}$$

Using (6.81), (6.82), (6.49) and the boundedness of $\bar{d}(t)$, we have (6.76). ∇

Theorem 6.3 *All signals in the closed-loop system with the plant (6.1) with $d(t)$ bounded, reference model (6.5) with $W_m(z) = z^{n^*}$, controller (6.34), and adaptive law (6.68)–(6.69) are bounded, and*

$$\sum_{t=t_1}^{t_2} (y(t) - y_m(t))^2 \leq c_1 + c_2 \sum_{t=t_1}^{t_2} d^2(t) \tag{6.83}$$

for some constants $c_1 > 0$, $c_2 > 0$.

Proof: For $x_s(t)$ defined in (6.59), from Lemma 6.3 and $\bar{d}(t) \in L^\infty$, we have

$$\sum_{t=t_1}^{t_1+t_2} x_s^2(t) \leq c_3 + \sum_{t=t_1-n^*}^{t_1+t_2} \frac{c_4}{1 + \|\bar{\omega}(t)\|^2} \tag{6.84}$$

for some constants $c_3 > 0$, $c_4 > 0$. In view of (6.67), we see that there exist constants $a_0 \in (0,1)$, $c_5 > 0$, $c_6 > 0$ such that

$$\|\bar{\omega}(t+1)\| \leq (a_0 + c_5 x_s(t)) \|\bar{\omega}(t)\| + c_6. \tag{6.85}$$

Substituting (6.38) and (6.39) into (6.67), recalling that $k_p \theta_3^* = 1$ and $\bar{\theta}^* = [\theta_1^{*T}, \theta_2^{*T}]^T$, and defining $\bar{\theta}(t) = [\theta_1^T(t), \theta_2^T(t)]^T$, we obtain

$$\bar{\omega}(t+1) = \left(A^* + b^* k_p (\bar{\theta}(t) - \bar{\theta}^*)^T\right) \bar{\omega}(t) + b^* k_p \theta_3(t) y_m(t+n^*) + d_1(t). \tag{6.86}$$

Since $\theta(t)$, $y_m(t)$, and $d_1(t)$ are all bounded, (6.86) implies that $\bar{\omega}(t)$ grows at most exponentially; that is, there exist constants $c_7 > 0$, $c_8 > 0$ such that

$$\|\bar{\omega}(t+1)\| \leq c_7 \|\bar{\omega}(t)\| + c_8, \ \forall t \geq 0. \tag{6.87}$$

Now we show the boundedness of $\bar{\omega}(t)$ by contradiction. Assume that $\bar{\omega}(t)$ grows unboundedly. Then, in view of (6.87), given any $\delta_0 > 0$ and $t_2 > 0$, we can find $\delta \in (0, \delta_0]$ and $t_1 > 0$ such that

$$\|\bar{\omega}(t)\| \geq \frac{1}{\delta_0}, \ t \in \{t_1 - n^*, \ldots, t_1 - 1\}, \tag{6.88}$$

$$\|\bar{\omega}(t)\| = \frac{1}{\delta}, \; t = t_1, \tag{6.89}$$

$$\|\bar{\omega}(t)\| \geq \frac{1}{\delta}, \; t \in \{t_1+1, \ldots, t_1+t_2+1\}. \tag{6.90}$$

Therefore, in view of (6.84), (6.88)–(6.90), for $j \in \{0, \ldots, t_2\}$, the state transition function $\phi(t_1, t_1+j)$ of the difference equation

$$\|\bar{\omega}(t+1)\| = (a_0 + c_5 x_s(t)) \, \|\bar{\omega}(t)\| \tag{6.91}$$

has the stability property

$$\begin{aligned} \phi(t_1, t_1+j) &= \prod_{t=t_1}^{t_1+j} (a_0 + c_5 x_s(t)) \leq \left(a_0 + \frac{c_5}{j+1} \sum_{t=t_1}^{t_1+j} x_s(t) \right)^{j+1} \\ &\leq \left(a_0 + c_5 \delta_0 \sqrt{c_3(n^*+1)} + c_5 \frac{\sqrt{c_4}}{\sqrt{j+1}} \right)^{j+1} \\ &\leq \left(a_0 + c_5 \delta_0 \sqrt{c_3(n^*+1)} \right)^{j+1} e^{\frac{c_5 \sqrt{c_4}}{a_0 + c_5 \delta_0 \sqrt{c_3(n^*+1)}} \sqrt{j+1}}. \end{aligned} \tag{6.92}$$

For any δ_0 such that

$$a_0 + c_5 \delta_0 \sqrt{c_3(n^*+1)} < 1 \tag{6.93}$$

the inequalities (6.85) and (6.92) imply that

$$\|\bar{\omega}(t_1+j+1)\| \leq \prod_{t=t_1}^{t_1+j} (a_0 + c_5 x_s(t)) \|\bar{\omega}(t_1)\| + c_0 \tag{6.94}$$

for some constant $c_0 > 0$ and that

$$\prod_{t=t_1}^{t_1+j} (a_0 + c_5 x_s(t)) < \frac{1}{2} \tag{6.95}$$

for any $j \geq j_1$ and some $j_1 \geq 0$. Hence, for $t_2 \geq j_1$ and

$$\delta_0 \in \left(0, \min \left\{ \frac{1 - a_0}{c_5 \sqrt{c_3(n^*+1)}}, \frac{1}{2c_0} \right\} \right), \tag{6.96}$$

(6.94) and (6.95) imply that

$$\|\bar{\omega}(t_1+j+1)\| < \frac{1}{\delta}, \quad j \in \{j_1, \ldots, t_2\}, \tag{6.97}$$

which contradicts (6.88)–(6.90). Hence $\bar{\omega}(t)$ is bounded, and so is $\omega(t)$. Thus, all closed-loop signals are bounded. Finally, (6.83) follows from signal boundedness, (6.39), (6.40), (6.75), (6.76), and the boundedness of $d(t)$. ∇

Thus far, we have developed stable adaptive control schemes for the cases when the disturbance $d(t) = 0$, $d(t) \in L^2$, and $d(t) \in L^\infty$. Next, we study the $L^{1+\alpha}$ $(1 < \alpha < \infty)$ robustness properties of a MRAC scheme.

6.4 Robustness of MRAC with $L^{1+\alpha}$ Errors

In this section, a new adaptive parameter update law is studied for discrete-time model reference adaptive control which ensures robustness of an adaptive control system with respect to $L^{1+\alpha}$ $(1 \leq \alpha < \infty)$ parameter variations and disturbances in the sense that all closed-loop signals are bounded and the tracking error belongs to $L^{1+\bar{\alpha}}$ $(\alpha \leq \bar{\alpha} < \infty)$ and converges to zero asymptotically. Some basic theory of time-varying operations on signals is presented.

6.4.1 Plant with Modeling Errors

Consider a discrete-time linear time-varying plant with parameter variations and disturbances described by the difference equation

$$\bar{P}(z,t)[y-d](t) = \bar{Z}(z,t)[u](t), \tag{6.98}$$

where $y(t)$ is the measured output, $d(t)$ is an output disturbance which belongs to $L^{1+\alpha}$ with $1 \leq \alpha < \infty$ (i.e., $\sum_{t=0}^{\infty} |d(t)|^{1+\alpha} \leq \bar{d}$ for some constant $\bar{d}$), $u(t)$ is the control input, and the corresponding operators are

$$\begin{aligned}
\bar{P}(z,t) &= P(z) + \tilde{P}(z,t),\\
P(z) &= z^n + p_{n-1}z^{n-1} + \cdots + p_1 z + p_0,\\
\tilde{P}(z,t) &= \tilde{p}_{n-1}(t)z^{n-1} + \tilde{p}_{n-2}(t)z^{n-2} + \cdots + \tilde{p}_1(t)z + \tilde{p}_0(t)
\end{aligned} \tag{6.99}$$

$$\begin{aligned}
\bar{Z}(z,t) &= k_p Z(z) + \tilde{Z}(z,t),\\
Z(z) &= z^m + z_{m-1}z^{m-1} + \cdots + z_1 z + z_0,\ m < n,\\
\tilde{Z}(z,t) &= \tilde{z}_m(t)z^m + \tilde{z}_{m-1}(t)z^{m-1} + \cdots + \tilde{z}_1(t)z + \tilde{z}_0(t).
\end{aligned} \tag{6.100}$$

In this plant model, the symbol z is the time advance operator: $z[x](t) = x(t+1)$, k_p, p_i, $i = 0, 1, \ldots, n-1$, and z_j, $j = 0, 1, \ldots, m-1$, are unknown constant parameters, and $\tilde{p}_i(t)$, $i = 0, 1, \ldots, n-1$, and $\tilde{z}_j(t)$, $j = 0, 1, \ldots, m$, are unknown time-varying parameters which represent plant parameter variations and also belong to $L^{1+\alpha}$ with $1 \leq \alpha < \infty$.

The control objective is to design an output feedback control $u(t)$ to ensure the closed-loop signal boundedness and tracking of a reference output $y_m(t)$ by the plant output $y(t)$, despite the presence of the parameter uncertainties and the unknown $L^{1+\alpha}$ parameter variations and disturbances. This is an adaptive output tracking plus $L^{1+\alpha}$ modeling error attenuation problem.

We make the standard MRAC assumptions (A6.1)–(A6.4) (see Section 6.1) for the modeled parts $P(z)$ and $Z(z)$ of the unknown plant (6.98). For the unmodeled parts $\tilde{P}(z,t)$, $\tilde{Z}(z,t)$, and $d(t)$, as already stated, we assume that

(A6.5a): $d(t) \in L^{1+\alpha}$, $\tilde{p}_i(t) \in L^{1+\alpha}$, $i = 0, 1, \ldots, n-1$, and $\tilde{z}_j(t) \in L^{1+\alpha}$, $j = 0, 1, \ldots, m$, $1 \le \alpha < \infty$.[3]

6.4.2 Adaptive Control Scheme

The model reference adaptive controller structure is the same as (6.34):

$$u(t) = \theta_1^T(t)\omega_1(t) + \theta_2^T(t)\omega_2(t) + \theta_3(t)y_m(t+n^*), \tag{6.101}$$

where $\theta_1 \in R^{n-1}$, $\theta_2 \in R^n$, and $\theta_3 \in R$ are parameters and

$$\omega_1(t) = a_\lambda(z)[u](t),\ a_\lambda(z) = [z^{-n+1}, \ldots, z^{-1}]^T, \tag{6.102}$$

$$\omega_2(t) = b_\lambda(z)[y](t),\ b_\lambda(z) = [z^{-n+1}, \ldots, z^{-1}, 1]^T. \tag{6.103}$$

Introducing the parameter vector

$$\theta(t) = [\theta_1^T(t), \theta_2^T(t), \theta_3(t)]^T \in R^{2n} \tag{6.104}$$

and the corresponding regressor vector

$$\omega(t) = [\omega_1^T(t), \omega_2^T(t), y_m(t+n^*)]^T \in R^{2n} \tag{6.105}$$

and defining the estimation error

$$\epsilon(t) = e(t) + \rho(t)\xi(t), \tag{6.106}$$

where $e(t) = y(t) - y_m(t)$, $\rho(t)$ is the estimate of $\rho^* = k_p$, and

$$\xi(t) = \theta^T(t)\omega(t-n^*) - \theta^T(t-n^*)\omega(t-n^*), \tag{6.107}$$

we choose the adaptive update laws for $\theta(t)$ and $\rho(t)$:

$$\theta(t+1) = \theta(t) - \frac{\text{sign}[k_p]\Gamma\epsilon(t)\omega(t-n^*)}{m^2(t)}f(t),\ 0 < \Gamma = \Gamma^T \le \frac{2}{k_p^0}I_{2n}, \tag{6.108}$$

$$\rho(t+1) = \rho(t) - \frac{\gamma\epsilon(t)\xi(t)}{m^2(t)}f(t),\ 0 < \gamma < 2, \tag{6.109}$$

where

$$m(t) = \sqrt{1 + \omega^T(t-n^*)\omega(t-n^*) + \xi^2(t) + u^2(t-n^*) + \epsilon^2(t)} \tag{6.110}$$

$$f(t) = \left(\frac{|\epsilon(t)|}{m(t)}\right)^{\bar{\alpha}-1},\ \alpha \le \bar{\alpha} < \infty. \tag{6.111}$$

The new features of this adaptive law are the use of $u^2(t-n^*)$ and $\epsilon^2(t)$ in the normalizing signal $m(t)$ and the use of the time-varying adaptation gain $f(t)$ such that $0 \le f(t) < 1$, $\forall t \ge 0$. We should note that for (6.108)–(6.109), unlike that for (5.165)–(5.166), the design parameter $\bar{\alpha} \ge 1$ is needed even if the modeling error $d(t)$ belongs to $L^{1+\alpha}$ with $0 < \alpha < 1$.

[3]In discrete time, an L^α signal is also an L^β signal for $0 < \alpha < \beta \le \infty$.

6.4.3 Robustness Analysis

There exist unique parameters $\theta_1^* \in R^{n-1}$, $\theta_2^* \in R^n$ and polynomial

$$F(z) = -z^{-m} + f_{n-m-2}z^{-m-1} + \cdots + f_0 z^{-n+1} \tag{6.112}$$

(see the proof of Theorem 6.1) such that the following equations hold:

$$\theta_1^{*T} a_\lambda(z) - 1 = F(z)Z(z), \tag{6.113}$$

$$F(z)P(z) + \theta_2^{*T} b_\lambda(z) k_p = -z^{n^*}. \tag{6.114}$$

Operating both sides of (6.114) on $y(t)$ and using (6.113) and the plant equation (6.98), we have

$$\begin{aligned} &k_p\theta_1^{*T}\omega_1(t) + F(z)[\tilde{Z}(z,\cdot)[u] + \bar{P}(z,\cdot)[d](t) - \tilde{P}(z,\cdot)[y]](t) + k_p\theta_2^{*T}\omega_2(t) \\ &= k_p u(t) - y(t+n^*). \end{aligned} \tag{6.115}$$

Substituting the controller equation (6.101) into (6.114), we obtain

$$e(t) = -\rho^*(\theta^{*T}\omega(t-n^*) - \theta^T(t-n^*)\omega(t-n^*)) + \delta(t), \tag{6.116}$$

where

$$\theta^* = [\theta_1^{*T}, \theta_2^{*T}, \theta_3^*]^T \in R^{2n}, \tag{6.117}$$

$$\delta(t) = z^{-n^*}F(z)[\tilde{P}(z,\cdot)[y] - \bar{P}(z,\cdot)[d] - \tilde{Z}(z,\cdot)[u]](t). \tag{6.118}$$

The modeling error $\delta(t)$ has the following property.

Lemma 6.4 *The modeling error $\delta(t)$ in (6.118) can be expressed as*

$$\delta(t) = \delta_\phi(t) + \delta_d(t), \tag{6.119}$$

where $\delta_\phi(t)$ and $\delta_d(t)$ are such that

$$|\delta_\phi(t)| \le \tilde{\delta}(t)\|\phi(t)\|_2,\ \tilde{\delta}(t) \in L^{1+\alpha},\ \delta_d(t) \in L^{1+\alpha}, \tag{6.120}$$

$$\begin{aligned} \phi(t) = (&u(t-n^*), \ldots, u(t-n^*-n+1), \\ &y(t-n^*), \ldots, y(t-n^*-n+1))^T. \end{aligned} \tag{6.121}$$

Proof: For $F(z)$ in (6.113) and $\tilde{Z}(z,t)$ in (6.100), we have

$$\begin{aligned} &z^{-n^*}F(z)[\tilde{Z}(z,\cdot)[u]](t) \\ &= -\tilde{z}_m(t-n^*-m)u(t-n^*) + \cdots \\ &\quad + f_0\tilde{z}_0(t-n^*-n+1)u(t-n^*-n+1), \end{aligned} \tag{6.122}$$

whose coefficients in the front of $u(\cdot)$ all belong to $L^{1+\alpha}$ because the coefficients of $\tilde{Z}(z,t)$ belong to $L^{1+\alpha}$. Similarly, for $\tilde{P}(z,t)$ in (6.99), we have

$$\begin{aligned} & z^{-n^*}F(z)[\tilde{P}(z,\cdot)[y]](t) \\ &= -\tilde{p}_{n-1}(t-n^*-m)y(t-1)+\cdots \\ &\quad + f_0\tilde{p}_0(t-n^*-n+1)y(t-n^*-n+1), \end{aligned} \tag{6.123}$$

whose coefficients in the front of $y(\cdot)$ all belong to $L^{1+\alpha}$ because the coefficients of $\tilde{P}(z,t)$ belong to $L^{1+\alpha}$. However, the expression of $z^{-n^*}F(z)[\tilde{P}(z,\cdot)[y]](t)$ contains the terms $y(t-i)$, $i=1,\ldots,n^*-1$, which are not contained by $\phi(t)$ in (6.121), in addition to the terms $y(t-j)$, $j=n^*,n^*+1,\ldots,n^*+n-1$, which are contained in $\phi(t)$. To express $y(t-i)$, $i=1,\ldots,n^*-1$, in terms of those in $\phi(t)$, we use the plant equation (6.98) to express

$$\begin{aligned} y(t-n^*+1) &= -\bar{P}_{n-1}(z,t-n^*-n+1)[y](t-n^*-n+1) \\ &\quad + \bar{P}(z,t-n^*-n+1)[d](t-n^*-n+1) \\ &\quad + \bar{Z}(z,t-n^*-n+1)[u](t-n^*-n+1), \end{aligned} \tag{6.124}$$

where

$$\begin{aligned} \bar{P}_{n-1}(z,t) &= -(p_{n-1}+\tilde{p}_{n-1}(t))z^{n-1} - \cdots \\ &\quad -(p_1+\tilde{p}_1(t))z - (p_0+\tilde{p}_0(t)). \end{aligned} \tag{6.125}$$

Using (6.98) and (6.124), we can express $y(t-n^*+2)$ in terms of $y(t-n^*-i)$, $i=0,1,\ldots,n-1$, $u(t-n^*-j)$, $j=1,\ldots,n-1$, and $d(\cdot)$. Similarly, using (6.98) and the obtained expressions of $y(t-n^*+k)$, $k=1,2,\ldots,q$, for $2\le q\le n^*-2$, we can express $y(t-n^*+k+1)$ in terms of $y(t-n^*-i)$, $i=0,1,\ldots,n-1$, $u(t-n^*-j)$, $j=1,\ldots,n-1$, and $d(\cdot)$ such that all the coefficients in those expressions are bounded. Following this argument and letting $\delta_d(t)$ include all terms dependent on $d(\cdot)$, we have the lemma's result. ∇

To analyze the robustness properties of the adaptive law (6.108)–(6.109), we use (6.106) and (6.116) to express

$$\epsilon(t) = k_p\tilde{\theta}^T(t)\omega(t-n^*) + \tilde{\rho}(t)\xi(t) + \delta(t), \tag{6.126}$$

where

$$\tilde{\theta}(t) = \theta(t)-\theta^*,\ \tilde{\rho}(t) = \rho(t)-\rho^*. \tag{6.127}$$

In view of (6.1260), the time increment of the positive definite function

$$V(\tilde{\theta},\tilde{\rho}) = |\rho^*|\tilde{\theta}^T\Gamma^{-1}\tilde{\theta} + \gamma^{-1}\tilde{\rho}^2 \tag{6.128}$$

along the trajectories of (6.108)–(6.109) is

$$
\begin{aligned}
& V(\tilde{\theta}(t+1), \tilde{\rho}(t+1)) - V(\tilde{\theta}(t), \tilde{\rho}(t)) \\
&= -\left(2 - \frac{|k_p|\omega^T(t-n^*)\Gamma\omega(t-n^*) + \gamma\xi^2(t)}{m^2(t)} f(t)\right) \frac{\epsilon^2(t)}{m^2(t)} f(t) \\
&+ \frac{2\epsilon(t)\delta(t)}{m^2(t)} f(t).
\end{aligned} \tag{6.129}
$$

From (6.110), (6.111), and Γ and γ, we have that $0 \leq f(t) < 1$ and

$$
0 \leq \frac{|k_p|\omega^T(t-n^*)\Gamma\omega(t-n^*) + \gamma\xi^2(t)}{m^2(t)} f(t) < 2, \ \forall t \geq 0. \tag{6.130}
$$

Hence, using (6.111) again, from (6.129) and (6.130), we have

$$
\begin{aligned}
& V(\tilde{\theta}(t+1), \tilde{\rho}(t+1)) - V(\tilde{\theta}(t), \tilde{\rho}(t)) \\
& \leq -c_0 \left(\frac{|\epsilon(t)|}{m(t)}\right)^{1+\bar{\alpha}} + 2\left(\frac{|\epsilon(t)|}{m(t)}\right)^{\bar{\alpha}} \frac{|\delta(t)|}{m(t)}
\end{aligned} \tag{6.131}
$$

for some constant $c_0 > 0$. From (6.131), it follows that

$$
\begin{aligned}
& c_0 \sum_{\tau=0}^{t} \frac{|\epsilon(\tau)|^{1+\bar{\alpha}}}{m^{1+\bar{\alpha}}(\tau)} + V(\tilde{\theta}(t+1), \tilde{\rho}(t+1)) \\
& \leq 2\sum_{\tau=0}^{t} \frac{|\epsilon(\tau)|^{\bar{\alpha}}|\delta(\tau)|}{m^{1+\bar{\alpha}}(\tau)} + V(\tilde{\theta}(0), \tilde{\rho}(0)), \ \forall t \geq 0.
\end{aligned} \tag{6.132}
$$

Using the Hölder inequality (see Section 2.7), we have

$$
\sum_{\tau=0}^{t} \frac{|\epsilon(\tau)|^{\bar{\alpha}}|\delta(\tau)|}{m^{1+\bar{\alpha}}(\tau)} \leq \left(\sum_{\tau=0}^{t} \left(\frac{|\epsilon(\tau)|^{\bar{\alpha}}}{m^{\bar{\alpha}}(\tau)}\right)^{\frac{1+\bar{\alpha}}{\bar{\alpha}}}\right)^{\frac{\bar{\alpha}}{1+\bar{\alpha}}} \left(\sum_{\tau=0}^{t} \left(\frac{|\delta(\tau)|}{m(\tau)}\right)^{1+\bar{\alpha}}\right)^{\frac{1}{1+\bar{\alpha}}}. \tag{6.133}
$$

From (6.110), (6.119), and (6.120), we have that $\frac{|\delta(t)|}{m(t)} \in L^{1+\alpha}$, which also implies that $\frac{|\delta(t)|}{m(t)} \in L^{\infty}$. Since $\bar{\alpha} \geq \alpha$, it follows that

$$
\left(\sum_{\tau=0}^{t} \left(\frac{|\delta(\tau)|}{m(\tau)}\right)^{1+\bar{\alpha}}\right)^{\frac{1}{1+\bar{\alpha}}} = \left(\sum_{\tau=0}^{t} \left(\frac{|\delta(\tau)|}{m(\tau)}\right)^{1+\alpha} \left(\frac{|\delta(\tau)|}{m(\tau)}\right)^{\bar{\alpha}-\alpha}\right)^{\frac{1}{1+\bar{\alpha}}} \leq k_\delta \tag{6.134}
$$

for some constant $k_\delta > 0$. From (6.132)–(6.134), we obtain

$$
\begin{aligned}
& c_0 \sum_{\tau=0}^{t} \frac{|\epsilon(\tau)|^{1+\bar{\alpha}}}{m^{1+\bar{\alpha}}(\tau)} + V(\tilde{\theta}(t+1), \tilde{\rho}(t+1)) \\
& \leq 2k_\delta \left(\sum_{\tau=0}^{t} \frac{|\epsilon(\tau)|^{1+\bar{\alpha}}}{m^{1+\bar{\alpha}}(\tau)}\right)^{\frac{\bar{\alpha}}{1+\bar{\alpha}}} + V(\tilde{\theta}(0), \tilde{\rho}(0)), \ \forall t \geq 0.
\end{aligned} \tag{6.135}
$$

Define $h(t) = \sum_{\tau=0}^{t} \frac{|\epsilon(\tau)|^{1+\bar{\alpha}}}{m^{1+\bar{\alpha}}(\tau)}$. Then, using (6.135), we get

$$c_0 h(t) \leq 2k_\delta (h(t))^{\frac{\bar{\alpha}}{1+\bar{\alpha}}} + V(\tilde{\theta}(0), \tilde{\rho}(0)), \tag{6.136}$$

which means that $h(t) \in L^\infty$, that is, $\frac{\epsilon(t)}{m(t)} \in L^{1+\bar{\alpha}}$. With this result, using (6.135) again, we have that $V(\tilde{\theta}(t), \tilde{\rho}(t)) \in L^\infty$, so that $\theta(t), \rho(t) \in L^\infty$. The fact that $\frac{\epsilon(t)}{m(t)} \in L^{1+\bar{\alpha}}$ implies that $\frac{\epsilon(t)}{m(t)} \in L^\infty$. In view of (6.108) and (6.109), we have that $\theta(t+1) - \theta(t), \rho(t+1) - \rho(t) \in L^{\frac{1+\bar{\alpha}}{\bar{\alpha}}}$. Thus we have:

Lemma 6.5 *The adaptive law (6.108)–(6.109) ensures that $\theta(t) \in L^\infty$, $\rho(t) \in L^\infty$, $\frac{\epsilon(t)}{m(t)} \in L^{1+\bar{\alpha}} \cap L^\infty$, $\theta(t+1) - \theta(t) \in L^{\frac{1+\bar{\alpha}}{\bar{\alpha}}}$, and $\rho(t+1) - \rho(t) \in L^{\frac{1+\bar{\alpha}}{\bar{\alpha}}}$.*

Then we can establish the following robust stability and tracking result.

Theorem 6.4 *All signals in the adaptive control system consisting of the plant (6.98) and the controller (6.101) updated by the adaptive law (6.108)–(6.109) are bounded and the tracking error $e(t) = y(t) - y_m(t)$ belongs to $L^{1+\bar{\alpha}}$ and converges to zero asymptotically.*

Proof: For $\omega_1(t)$ and $\omega_2(t)$ in (6.102) and (6.103), we introduce

$$\bar{\omega}(t) = [\omega_1^T(t), \omega_2^T(t)]^T. \tag{6.137}$$

Using (6.137) and (6.115), we can express

$$\begin{aligned} \bar{\omega}(t+1) = A^* \bar{\omega}(t) + b^* (y_m(t+n^*) + \delta_0(t) + \epsilon(t+n^*) \\ - \rho(t+n^*)(\theta(t+n^*) - \theta(t))^T \omega(t)) + \delta_1(t) \end{aligned} \tag{6.138}$$

for some $A^* \in R^{(2n-1)\times(2n-1)}$ which depends on θ^* and the coefficients of $P(z)$ and $Z(z)$ and has all its eigenvalues inside the complex unit circle (see the proof of Theorem 6.2 in Section 6.3) (this implies that there exists a nonsingular constant (possibly complex) $Q^* \in C^{(2n-1)\times(2n-1)}$ such that $\|Q^* A^* Q^{*-1}\|_2 < 1$ for the induced Euclidean matrix norm $\|\cdot\|_2$, see Section 2.7.3), some $b^* \in R^{2n-1}$ dependent on k_p, and

$$\delta_0(t) = F(z)[\tilde{Z}(z,\cdot)[u] - \tilde{P}(z,\cdot)[y] + \bar{P}(z,\cdot)[d]](t) \tag{6.139}$$

$$\begin{aligned} \delta_1(t) = [0, \ldots, 0, z^{-n+1}[\tilde{Z}(z,\cdot)[u] + \bar{P}(z,\cdot)[d] \\ - \tilde{P}(z,\cdot)[y]](t)]^T \in R^{2n-1}. \end{aligned} \tag{6.140}$$

With the above matrix Q^*, we define the vector norm $\|\cdot\|$ in R^{2n-1} as $\|x\| = \|Q^* x\|_2$. Using Lemma 6.4, from (6.139), we see that $\frac{\delta_0(t)}{m_0(t+n^*)} \in L^{1+\alpha}$

(and $\frac{\delta_0(t)}{m_0(t+n^*)} \in L^\infty$), where

$$m_0(t) = \sqrt{1 + \omega^T(t-n^*)\omega(t-n^*) + u^2(t-n^*)}$$
$$= \sqrt{1 + \phi^T(t)\phi(t)} \tag{6.141}$$

for $\phi(t)$ defined in (6.121). From (6.140), we have that $\frac{\delta_1(t)}{m_0(t+n^*)} \in L^{1+\alpha}$ (and $\frac{\delta_1(t)}{m_0(t+n^*)} \in L^\infty$). From (6.101) and the boundedness of $\theta(t)$ and $y_m(t)$, we have $|u(t)| \leq k_1\|\bar{\omega}(t)\| + k_2$, $|\xi(t+n^*)| \leq k_1\|\bar{\omega}(t)\| + k_2$, and $|m_0(t+n^*)| \leq k_1\|\bar{\omega}(t)\| + k_2$ for some constants k_1 and k_2. Because $\frac{1}{\bar{\alpha}} \leq \bar{\alpha}$ for $\bar{\alpha} \geq 1$, that is, $L^{\frac{1+\bar{\alpha}}{\bar{\alpha}}} \subset L^{1+\bar{\alpha}}$, from Lemma 6.5, we also have $\theta(t+1) - \theta(t) \in L^{1+\bar{\alpha}}$.

Introducing the auxiliary signal

$$x_s(t) = \left|\frac{\epsilon(t+n^*)}{m(t+n^*)}\right| + \|\theta(t+n^*) - \theta(t)\|_2$$
$$+ \frac{|\delta_0(t)|}{m_0(t+n^*)} + \frac{\|\delta_1(t)\|}{m_0(t+n^*)}, \tag{6.142}$$

we see from Lemma 6.5 and the above analysis that $x_s(t) \in L^{1+\bar{\alpha}}$. From (6.126), Lemma 6.4, and the boundedness of $\theta(t)$ and $\rho(t)$, it follows that $\frac{m(t)}{m_0(t)} \in L^\infty$. From (6.138) and (6.142), we obtain

$$\|\bar{\omega}(t+1)\| \leq (a_0 + c_1 x_s(t))\|\bar{\omega}(t)\| + c_2 \tag{6.143}$$

for some constants $a_0 \in (0,1)$, $c_1 > 0$, and $c_2 > 0$, where $\|\bar{\omega}(t)\| = \|Q^*\bar{\omega}(t)\|_2$ as defined above. Since $x_s(t) \in L^{1+\bar{\alpha}}$, using the Hölder inequality, we have

$$\sum_{t=t_0}^{t_0+j} x_s(t) \leq (j+1)^{\frac{\bar{\alpha}}{1+\bar{\alpha}}} \left(\sum_{t=t_0}^{t_0+j} (x_s(t))^{1+\bar{\alpha}}\right)^{\frac{1}{1+\bar{\alpha}}} \leq c_3 (j+1)^{\frac{\bar{\alpha}}{1+\bar{\alpha}}} \tag{6.144}$$

for any $j \geq 1$ and some constant $c_3 > 0$. Using this property, we have

$$\prod_{t=t_0}^{t_0+j} (a_0 + c_1 x_s(t)) \leq \left(a_0 + \frac{c_1}{j+1}\sum_{t=t_0}^{t_0+j} x_s(t)\right)^{j+1} \leq \left(a_0 + \frac{c_1 c_3}{(j+1)^{\frac{1}{1+\bar{\alpha}}}}\right)^{j+1}$$
$$\leq a_0^{j+1}\left(1 + \frac{c_1 c_3}{a_0 (j+1)^{\frac{1}{1+\bar{\alpha}}}}\right)^{j+1} \leq a_0^{j+1} \exp\left(\frac{c_1 c_3}{a_0}(j+1)^{\frac{\bar{\alpha}}{1+\bar{\alpha}}}\right). \tag{6.145}$$

Defining $\beta = \frac{\bar{\alpha}}{1+\bar{\alpha}} < 1$ and $\lambda = \exp\left(\frac{c_1 c_3}{a_0}\right) > 1$, we have

$$a_0^{j+1} \exp\left(\frac{c_1 c_3}{a_0}(j+1)^{\frac{\bar{\alpha}}{1+\bar{\alpha}}}\right) = a_0^{j+1}\lambda^{(j+1)^\beta} = a_0^{j+1}\lambda^{(j+1)^{\beta-1}(j+1)}$$
$$= a_0^{j+1}(\lambda^{(j+1)^{\beta-1}})^{j+1} = a_0^{j+1}\left(\lambda^{\frac{1}{(j+1)^{1-\beta}}}\right)^{j+1}. \tag{6.146}$$

Since $1-\beta > 0$, there exists a finite $j_0 > 0$ such that $\lambda^{\frac{1}{(j+1)^{1-\beta}}} < \frac{1}{a_0}$ for any $j \geq j_0$, that is, $a_0\lambda^{\frac{1}{(j+1)^{1-\beta}}} < 1$ for any $j \geq j_0$. Hence, from (6.145), we have

$$\lim_{k\to\infty} \sum_{j=1}^{k} \prod_{t=t_0}^{t_0+j} (a_0 + c_5 x_s(t)) < \infty. \tag{6.147}$$

Using (6.143) and (6.147), we conclude that $\|\bar{\omega}(t)\|$ is bounded; that is, all closed-loop signals are bounded. Then we have $\epsilon(t) \in L^{1+\bar{\alpha}}$, $\theta(t+1)-\theta(t) \in L^{1+\bar{\alpha}}$ so that $\lim_{t\to\infty} \epsilon(t) = 0$, $\lim_{t\to\infty}(\theta(t+1)-\theta(t)) = 0$, and $\lim_{t\to\infty} \xi(t) = 0$. Finally, from (6.106), we have $\lim_{t\to\infty} e(t) = 0$. ∇

The proof of Theorem 6.4 is based on the L^∞ property of $\theta(t)$ and $\rho(t)$ and the $L^{1+\bar{\alpha}}$ property of $\frac{\epsilon(t)}{m(t)}$, $\theta(t+1)-\theta(t)$ in Lemma 6.5, as well as on the $L^{1+\alpha}$ property of $\delta_d(t)$, $\tilde{\delta}(t)$ in Lemma 6.4. In brief, the parameter boundedness property ensures a well-defined feedback control system, while the $L^{1+\bar{\alpha}}$ property leads to a small gain of the feedback loop, sufficient for closed-loop signal boundedness with which an $L^{1+\bar{\alpha}}$ tracking error is ensured.

6.4.4 Discussion

Let us examine the selection of the gain $f(t)$ in the adaptive law (6.108)–(6.109). From (6.110) and (6.111), $0 \leq f(t) < 1$, and from Lemma 6.5, $\frac{\epsilon(t)}{m(t)} \in L^{1+\bar{\alpha}}$ for $\bar{\alpha} \geq \alpha \geq 1$ and $\lim_{t\to\infty} f(t) = 0$. This means that the adaptive law (6.108)–(6.109) with such an $f(t)$ leads to a slow adaptation.

Why do we use a slow adaptation? If $f(t) = 1$, that is, $\bar{\alpha} = 1$, then $\theta(t+1)-\theta(t)$ in (6.108) would be directly proportional to $\epsilon(t)$ and in turn to the modeling error $\delta(t)$ (see (6.126)). Since an $L^{1+\alpha}$ ($\alpha > 1$) modeling error can be more "violent" than an L^2 modeling error (which can be handled with $f(t) = 1$), we need a slow adaptation to reduce the effect of the modeling error $\delta(t)$ on $\theta(t+1)-\theta(t)$. How slow should the adaptation be (how large should $\bar{\alpha}$ in (6.111) be)? The proof for Lemma 6.5 (especially, Equation (6.134)) suggests that $\alpha \leq \bar{\alpha}$ is needed. Finally, to define $f(t)$ in (6.111), $\bar{\alpha} < \infty$ is also needed, which means that the case when $\alpha = \infty$ should be excluded.

We note that when $\bar{\alpha} = \alpha = 1$, the design signal $f(t) \equiv 1$ and the normalizing signal $m(t)$ for (6.108)–(6.109) can be chosen as

$$m(t) = \sqrt{1 + \omega^T(t-n^*)\omega(t-n^*) + \xi^2(t) + u^2(t-n^*)}. \tag{6.148}$$

Furthermore, if there is no variation to the gain k_p, that is, $\tilde{z}_m(t) \equiv 0$ in (6.100) and (6.122), then we can choose

$$m(t) = \sqrt{1 + \omega^T(t-n^*)\omega(t-n^*) + \xi^2(t)} \tag{6.149}$$

for the adaptive law (6.108)–(6.109). In this case, we have a standard gradient adaptive law for the ideal case when there is no plant parameter variation or disturbance: $P(z)[y](t) = k_p Z(z)[u](t)$. Therefore, our result has also shown that the standard discrete-time model reference adaptive control system is robust with respect to L^2 parameter variations and L^2 output disturbances in the sense that all closed-loop signals are bounded and the tracking error belongs to L^2 and converges to zero asymptotically.

Problems

6.1 Work out the parametrization (4.258) for the $n = 3$ case.

6.2 Consider the plant

$$(z^2 + p_1 z + p_0)[y](t) = k_p(z + z_0)[u](t)$$

with zero initial conditions, where $|z_0| < 1$, p_1, p_0, k_p are unknown.

(i) Study the plant–model matching conditions in Section 6.2.

(ii) Design and analyze an adaptive control scheme for this plant.

(iii) Simulate the adaptive control system with $p_1 = -2$, $p_0 = 2$, $k_p = 1$, $z_0 = 0.5$ for $r(t) = 10$, $r(t) = 10\sin(0.1t)$, $r(t) = 10\sin(t) + 13\sin(1.33t)$, and plot the tracking error and parameter errors for each case.

(iv) Comment on the simulation results.

6.3 Redo Problem 6.2 with $\Lambda(z) = (z + 0.2)^2$ and $a(z) = [1, z]^T$ in the controller structure given in footnote 2 in Section 6.2.

6.4 Derive (6.48) and show that $\theta(t+1) - \theta(t) \in L^2$.

6.5 Verify (6.77) and (6.79).

6.6 Verify (6.130) and (6.133).

6.7 Design and analyze a switching σ-modification for $f_\theta(t)$ and $f_\rho(t)$ in (6.68) and (6.69). (*Hint*: Choose $\Gamma = \gamma_\theta I_{n_\theta}$, $0 < \gamma_\theta < 1/k_p^0$, $0 < \gamma < 1$, $0 < \sigma_0 < (1 - \gamma_m)/2$ for $\gamma_m = \max\{\gamma_\theta k_p^0, \gamma\} < 1$; see (A6.3) in Section 6.1 and the switching σ-modification in Section 3.8.2.)

6.8 Assume that the parameter variations $\tilde{p}_i(t)$, $i = 0, 1, \ldots, n-1$, $\tilde{z}_j(t)$, $j = 0, 1, \ldots, m$, and the disturbance $d(t)$ in the plant (6.98) are bounded only (not in L^β for any $0 < \beta < \infty$). Verify (6.118) and (6.119) with $\bar{\delta}(t) \in L^\infty$ and $\delta_d(t) \in L^\infty$. Design and analyze a robust adaptive controller using a parameter projection or a switching σ-modification for the adaptive law.

Chapter 7

Indirect Adaptive Control

The adaptive control schemes in Chapters 4, 5, and 6 are called *direct adaptive control* designs, as the parameters of a controller are updated *directly* from an adaptive law. In this chapter, we address *indirect adaptive control*, where the parameters of a controller are calculated from the adaptive estimates of the parameters of a plant to be controlled, that is, the controller parameters are updated *indirectly*. A typical indirect adaptive control design has two steps: estimation of the plant (i.e., the controlled system) parameters by an adaptive law and calculation of the controller parameters using some design equations. An indirect adaptive control scheme can be developed to solve the model reference adaptive control problem formulated in the previous chapters. Furthermore, an indirect adaptive control scheme can be designed using a pole placement method to adaptively control plants whose zeros may be unstable (for which a model reference controller cannot be applied). To introduce the basic issues and solutions of indirect adaptive control, we first address indirect model reference adaptive control and then present indirect pole placement control designs in both continuous time and discrete time.

7.1 Model Reference Designs

The procedure of indirect model reference adaptive control consists of two main steps: estimation of the parameters of a controlled plant and calculation of the controller parameters from the plant parameter estimates using a matching equation, while the controller structure and reference model are the same as that for a direct adaptive design (see Chapters 5 and 6), with the same plant assumptions except that the estimate of the high frequency gain k_p needs to be ensured to be away from zero for implementing a nonsingular control law. To develop such an adaptive approach, we start with two illustrative examples.

7.1.1 Simple Adaptive Control Systems

We first complete the closed-loop stability (signal boundedness) analysis of the adaptive system in Example 1.3 and then present the design and analysis of an indirect adaptive control system with a two-parameter plant $\dot{y}(t) = a_p y(t) + b_p u(t)$ for which a parameter projection algorithm is employed to ensure that the estimate of the plant high frequency gain $k_p = b_p$ is away from zero, as needed for implementing a stable adaptive control system.

Stability Analysis of Example 1.3

As in Example 1.3, to control the first-order plant

$$\dot{y}(t) = a_p y(t) + u(t), \ t \geq 0, \tag{7.1}$$

with an unknown parameter a_p, for the plant output $y(t)$ to asymptotically track the output $y_m(t)$ of the reference model system

$$\dot{y}_m(t) = -a_m y_m(t) + r(t), \ t \geq 0, \tag{7.2}$$

we use the controller structure

$$u(t) = k(t)y(t) + r(t), \tag{7.3}$$

where the controller parameter $k(t)$ is *calculated* from

$$k(t) = -\hat{a}_p(t) - a_m \tag{7.4}$$

with $\hat{a}_p(t)$ being an adaptive estimate of a_p.

Introducing a stable filter $\frac{1}{s+a_f}$, $a_f > 0$, we express the plant (7.1) as

$$\begin{aligned} y(t) &= (a_f + a_p)\frac{1}{s+a_f}[y](t) + \frac{1}{s+a_f}[u](t) \\ &= \theta_p^* \phi(t) + \frac{a_f}{s+a_f}[y](t) + \frac{1}{s+a_f}[u](t), \end{aligned} \tag{7.5}$$

where

$$\theta_p^* = a_p, \ \phi(t) = \frac{1}{s+a_f}[y](t). \tag{7.6}$$

We then define the estimation error

$$\epsilon(t) = \theta_p(t)\phi(t) - y(t) + \frac{a_f}{s+a_f}[y](t) + \frac{1}{s+a_f}[u](t), \tag{7.7}$$

where $\theta_p(t) = \hat{a}_p(t)$ is the estimate of $\theta_p^* = a_p$.

As in Section 3.3 and Example 1.3, we use the adaptive law for $\theta_p(t)$:

$$\dot{\theta}_p(t) = -\frac{\gamma\epsilon(t)\phi(t)}{m^2(t)},\ m(t) = \sqrt{1+\phi^2(t)},\ \gamma > 0,\ t \geq 0, \tag{7.8}$$

which ensures that $\theta_p(t) \in L^\infty$, and $\frac{\epsilon(t)}{m(t)} \in L^2 \cap L^\infty$ and $\dot{\theta}_p(t) \in L^2 \cap L^\infty$.

With the controller (7.3), the closed-loop system becomes

$$\dot{y}(t) = -a_m y(t) + (a_p - \hat{a}_p(t))y(t) + r(t). \tag{7.9}$$

From (7.5) and (7.7), we have

$$\epsilon(t) = (\hat{a}_p(t) - a_p)y_f(t),\ y_f(t) \triangleq \frac{1}{s+a_f}[y](t). \tag{7.10}$$

Using the equality

$$y(t) = \dot{y}_f(t) + a_f y_f(t), \tag{7.11}$$

we derive

$$\dot{\epsilon}(t) + a_f\epsilon(t) = \dot{\hat{a}}_p(t)y_f(t) + (\hat{a}_p(t) - a_p)y(t). \tag{7.12}$$

Using (7.9)–(7.12), we have

$$\begin{aligned} &\ddot{y}_f(t) + a_f\dot{y}_f(t) + a_m\dot{y}_f(t) + a_f a_m y_f(t) \\ &= -(\dot{\epsilon}(t) + a_f\epsilon(t)) + \dot{\hat{a}}_p(t)y_f(t) + r(t), \end{aligned} \tag{7.13}$$

from which we express

$$y_f(t) = -\frac{1}{s+a_m}[\epsilon](t) + \frac{1}{(s+a_m)(s+a_f)}[\dot{\hat{a}}_p y_f](t) + \frac{1}{s+a_f}[y_m](t). \tag{7.14}$$

From the properties of the adaptive law (7.7), we have

$$m(t) \leq 1 + |y_f(t)|,\ |\dot{\hat{a}}_p(t)| \leq a_3\frac{|\epsilon(t)|}{m(t)} \tag{7.15}$$

for some constant $a_3 > 0$, and from (7.14), we have

$$|y_f(t)| \leq a_4 + a_4\int_0^t e^{-a_5(t-\tau)}\frac{|\epsilon(t)|}{m(\tau)}|y_f(\tau)|\,d\tau \tag{7.16}$$

for some constants $a_4 > 0$ and $a_5 > 0$. This inequality has the form (2.102) with $z_1(t) = |y_f(t)|$ and $x_1(t) = \frac{|\epsilon(t)|}{m(t)}$. Since $\frac{|\epsilon(t)|}{m(t)} \in L^2$, from Lemma 2.3, we conclude that $y_f(t)$ is bounded.

Using the equality

$$(s+a_f)[ky_f](t) - \dot{k}(t)y_f(t) = k(t)y(t) \tag{7.17}$$

and (7.3), we have

$$\begin{aligned} u_f(t) \triangleq \frac{1}{s+a_f}[u](t) &= \frac{1}{s+a_f}[ky](t) + \frac{1}{s+a_f}[r](t) \\ &= k(t)y_f(t) - \frac{1}{s+a_f}[\dot{k}y_f](t) + \frac{1}{s+a_f}[r](t), \end{aligned} \tag{7.18}$$

which is bounded as $y_f(t)$, $k(t)$, and $\dot{k}(t)$ are. Then, from (7.5), $y(t)$ is bounded, and from (7.3), $u(t)$ is bounded. Hence, $\epsilon(t) \in L^2$, and from (7.12), $\dot{\epsilon}(t) \in L^\infty$, so that $\lim_{t\to\infty} \epsilon(t) = 0$ as from Lemma 2.14

From (7.13), with $(s+a_f)[y_f](t) = y(t)$ and $s^2 + a_f s + a_m s + a_f a_m = (s+a_m)(s+a_f)$, we have

$$(s+a_m)[y](t) = -(s+a_f)[\epsilon](t) + \dot{\hat{a}}_p(t)y_f(t) + r(t). \tag{7.19}$$

Using (7.2) and Lemma 2.18, from (7.19), we have $\lim_{t\to\infty}(y(t) - y_m(t)) = 0$.

Design and Analysis for a Two-Parameter System

We now consider the first-order plant

$$\dot{y}(t) = a_p y(t) + b_p u(t) \tag{7.20}$$

with two constant parameters a_p and b_p and the reference model

$$\dot{y}_m(t) = -a_m y_m(t) + b_m r(t) \tag{7.21}$$

with $a_m > 0$. Let $\hat{a}_p(t)$ and $\hat{b}_p(t)$ be some estimates (from an adaptive law) of the unknown plant parameters a_p and b_p, respectively, with $|\hat{b}_p(t)| \geq b_0 > 0$, $\forall t \geq 0$, for some constant $b_0 > 0$. Then, an indirect adaptive controller is

$$u(t) = k_1(t)y(t) + k_2(t)r(t), \tag{7.22}$$

where

$$k_1(t) = \frac{-\hat{a}_p(t) - a_m}{\hat{b}_p(t)}, \quad k_2(t) = \frac{b_m}{\hat{b}_p(t)}. \tag{7.23}$$

To generate desired parameter estimates $\hat{a}_p(t)$ and $\hat{b}_p(t)$, we choose a stable filter $\frac{1}{s+a_f}$ with $a_f > 0$ and express the plant (7.20) as

$$\begin{aligned} y(t) &= (a_f + a_p)\frac{1}{s+a_f}[y](t) + b_p \frac{1}{s+a_f}[u](t) \\ &= \theta_p^{*T}\phi(t) + \frac{a_f}{s+a_f}[y](t), \end{aligned} \tag{7.24}$$

where

$$\begin{aligned}\theta_p^* &= [a_p, b_p]^T,\\ \phi(t) &= [\phi_1(t), \phi_2(t)]^T\\ &= \left[\frac{1}{s+a_f}[y](t), \frac{1}{s+a_f}[u](t)\right]^T \triangleq [y_f(t), u_f(t)]^T.\end{aligned} \tag{7.25}$$

Letting $\theta_p(t)$ be the estimate of the parameter vector θ_p^* (i.e., $\theta_p(t) = [\hat{a}_p, \hat{b}_p]^T$), we introduce the estimation error

$$\epsilon(t) = \theta_p^T(t)\phi(t) - y(t) + \frac{a_f}{s+a_f}[y](t). \tag{7.26}$$

It follows from (7.24) and (7.26) that

$$\epsilon(t) = \tilde{\theta}_p^T(t)\phi(t),\ \tilde{\theta}_p(t) = \theta_p(t) - \theta_p^*. \tag{7.27}$$

Then, we choose the gradient adaptive law for $\theta_p(t)$:

$$\dot{\theta}_p(t) = -\frac{\Gamma\epsilon(t)\phi(t)}{m^2(t)},\ \Gamma = \Gamma^T > 0, \tag{7.28}$$

where

$$m(t) = \sqrt{1 + \phi^T(t)\phi(t)}. \tag{7.29}$$

However, this adaptive law does not ensure that the estimate $|\hat{b}_p| \geq b_0$ for some constant $b_0 > 0$, which is needed to ensure that $k_1(t)$ and $k_2(t)$, solved from the design equation (7.23), are bounded. We need some knowledge of the unknown parameter b_p: $\mathrm{sign}[b_p]$ is known and $b_0 > 0$ is known such that $\mathrm{sign}[b_p]b_p \geq b_0$. Using this knowledge, we modify the adaptive law (7.28) as

$$\dot{\theta}_p(t) = -\frac{\Gamma\epsilon(t)\phi(t)}{m^2(t)} + f(t), \tag{7.30}$$

where $\Gamma = \mathrm{diag}\{\gamma_1, \gamma_2\}$ with $\gamma_1 > 0$ and $\gamma_2 > 0$ and

$$f(t) = [0, f_2(t)], \tag{7.31}$$

$$f_2(t) = \begin{cases} 0 & \text{if } \mathrm{sign}[b_p]\hat{b}_p(t) > b_0 \text{ or} \\ & \text{if } \mathrm{sign}[b_p]\hat{b}_p(t) = b_0,\ -\frac{\mathrm{sign}[b_p]\gamma_2\epsilon(t)\phi_2(t)}{m^2(t)} \geq 0, \\ \frac{\gamma_2\epsilon(t)\phi_2(t)}{m^2(t)} & \text{otherwise} \end{cases} \tag{7.32}$$

for $\phi_2(t) = \frac{1}{s+a_f}[u](t)$ in (7.25), with $\hat{b}_p(0)\mathrm{sign}[b_p] \geq b_0$.

Consider the measure of the parameter error $\tilde{\theta}_p$,

$$V(\tilde{\theta}_p) = \tilde{\theta}_p^T \Gamma^{-1} \tilde{\theta}_p, \tag{7.33}$$

and its time derivative

$$\dot{V} = \frac{-2\epsilon^2(t)}{m^2(t)} + 2(1 - \chi(t))(\hat{b}_p(t) - b_p)\frac{\epsilon(t)\phi_2(t)}{m^2(t)}, \tag{7.34}$$

where

$$\chi(t) = \begin{cases} 1 & \text{if } \mathrm{sign}[b_p]\hat{b}_p(t) > b_0 \text{ or} \\ & \text{if } \mathrm{sign}[b_p]\hat{b}_p(t) = b_0 \text{ and } -\frac{\mathrm{sign}[b_p]\gamma_2\epsilon(t)\phi_2(t)}{m^2(t)} \geq 0, \\ 0 & \text{otherwise} \end{cases} \tag{7.35}$$

such that $\chi(t) = 0$ if and only if $\mathrm{sign}[b_p]\hat{b}_p(t) = b_0$ and $-\frac{\mathrm{sign}[b_p]\gamma_2\epsilon(t)\phi_2(t)}{m^2(t)} < 0$ (the "only if" property holds, because $\theta_p(t)$ is continuous and $\hat{b}_p(0)\mathrm{sign}[b_p] \geq b_0$ so that the projection (7.32) ensures that the situation when $\mathrm{sign}[b_p]\hat{b}_p(t) < b_0$ does not occur). It then follows that

$$(1 - \chi(t))(\hat{b}_p(t) - b_p)\frac{\epsilon(t)\phi_2(t)}{m^2(t)} \leq 0, \tag{7.36}$$

$$\mathrm{sign}[b_p]\hat{b}_p(t) \geq b_0. \tag{7.37}$$

Then, from (7.34), we have $\dot{V} \leq \frac{-2\epsilon^2(t)}{m^2(t)}$, from which we conclude that $\theta_p(t)$ is bounded, and so are $\frac{\epsilon(t)}{m(t)}$ and $\dot{\theta}_p(t)$. Moreover, from (7.30) and (7.34), we have that $\frac{\epsilon(t)}{m(t)} \in L^2$ and $\dot{\theta}_p(t) \in L^2$.

To prove closed-loop signal boundedness, we substitute the control law (7.22)–(7.23) into the plant (7.20) to obtain the closed-loop control system

$$\begin{aligned} \dot{y}(t) &= a_p y(t) + \hat{b}_p(t)u(t) + (b_p - \hat{b}_p(t))u(t) \\ &= -a_m y(t) + (a_p - \hat{a}_p(t))y(t) + (b_p - \hat{b}_p(t))u(t) + b_m r(t). \end{aligned} \tag{7.38}$$

Using $y_f(t)$ and $u_f(t)$ in (7.25), we can write (7.24) as

$$y(t) = (a_f + a_p)y_f(t) + b_p u_f(t). \tag{7.39}$$

Using $\theta_p(t) = [\hat{a}_p(t), \hat{b}_p(t)]^T$, (7.26), and (7.39), we have

$$\begin{aligned} \epsilon(t) &= (a_f + \hat{a}_p(t))y_f(t) + \hat{b}_p(t)u_f(t) - y(t) \\ &= (\hat{a}_p(t) - a_p)y_f(t) + (\hat{b}_p(t) - b_p)u_f(t). \end{aligned} \tag{7.40}$$

From the definitions of $y_f(t)$ and $u_f(t)$ in (7.25) again, we have

$$y(t) = \dot{y}_f(t) + a_f y_f(t), \; u(t) = \dot{u}_f(t) + a_f u_f(t). \tag{7.41}$$

Then it follows from (7.40) and (7.41) that

$$\begin{aligned}
&\dot{\epsilon}(t) + a_f \epsilon(t) \\
&= \dot{\hat{a}}_p(t) y_f(t) + \dot{\hat{b}}_p(t) u_f(t) + (\hat{a}_p(t) - a_p)(\dot{y}_f(t) + a_f y_f(t)) \\
&\quad + (\hat{b}_p(t) - b_p)(\dot{u}_f(t) + a_f u_f(t)) \\
&= \dot{\hat{a}}_p(t) y_f(t) + \dot{\hat{b}}_p(t) u_f(t) + (\hat{a}_p(t) - a_p) y(t) + (\hat{b}_p(t) - b_p) u(t).
\end{aligned} \tag{7.42}$$

Using (7.38), the first equation of (7.41), and (7.42), we obtain

$$\begin{aligned}
&\ddot{y}_f(t) + a_f \dot{y}_f(t) \\
&= -a_m y(t) - (\dot{\epsilon}(t) + a_f \epsilon(t)) + \dot{\hat{a}}_p(t) y_f(t) + \dot{\hat{b}}_p(t) u_f(t) + b_m r(t).
\end{aligned} \tag{7.43}$$

Substituting the first equation of (7.41) into (7.43), we have

$$\begin{aligned}
&\ddot{y}_f(t) + a_f \dot{y}_f(t) + a_m \dot{y}_f(t) + a_f a_m y_f(t) \\
&= -(\dot{\epsilon}(t) + a_f \epsilon(t)) + \dot{\hat{a}}_p(t) y_f(t) + \dot{\hat{b}}_p(t) u_f(t) + b_m r(t).
\end{aligned} \tag{7.44}$$

Recall that the adaptive law (7.30) with the parameter projection $f(t)$ ensures that $\theta_p(t) = [\hat{a}_p(t), \hat{b}_p(t)]^T$ is bounded with $|\hat{b}_p(t)| \geq b_0 > 0, \forall t \geq 0$, and so are $\frac{\epsilon(t)}{m(t)}$ and $\dot{\theta}_p(t)$ (i.e., $\dot{\hat{a}}_p(t)$ and $\dot{\hat{b}}_p(t)$ are bounded), and that $\frac{\epsilon(t)}{m(t)} \in L^2$ and $\dot{\theta}_p(t) \in L^2$ (i.e., $\dot{\hat{a}}_p(t) \in L^2$, $\dot{\hat{b}}_p(t) \in L^2$). From (7.23), that is, $k_1(t) = \frac{-\hat{a}_p(t) - a_m}{\hat{b}_p(t)}$, we have $k_1(t) \in L^\infty$ and $\dot{k}_1(t) = \frac{-\dot{\hat{a}}_p(t)\hat{b}_p(t) + \dot{\hat{b}}_p(t)(\hat{a}_p(t) - a_m)}{\hat{b}_p^2(t)} \in L^\infty \cap L^2$.

From the control law (7.22), the equality

$$(s + a_f)[k_1 y_f](t) - \dot{k}_1(t) y_f(t) = k_1(t) y(t), \tag{7.45}$$

and the definition of $u_f(t)$ in (7.25), we have

$$\begin{aligned}
u_f(t) &= \frac{1}{s + a_f}[k_1 y](t) + \frac{1}{s + a_f}[k_2 r](t) \\
&= k_1(t) y_f(t) - \frac{1}{s + a_f}[\dot{k}_1 y_f](t) + \frac{1}{s + a_f}[k_2 r](t).
\end{aligned} \tag{7.46}$$

From (7.46), there exist constants $a_1 > 0$ and $a_2 > 0$ such that

$$|u_f(t)| \leq a_1 + a_1 |y_f(t)| + a_1 \int_0^t e^{-a_2(t-\tau)} |y_f(\tau)| \, d\tau. \tag{7.47}$$

From (7.44), with $a_0(t) = \frac{1}{s+a_f}[y_m](t)$, we can express

$$y_f(t) = -\frac{1}{s+a_m}[\epsilon](t) + \frac{1}{(s+a_m)(s+a_f)}[\dot{\hat{a}}_p y_f + \dot{\hat{b}}_p u_f](t) + a_0(t). \tag{7.48}$$

From (7.29), (7.30), and $\dot{\theta}_p(t) = [\dot{\hat{a}}_p(t), \dot{\hat{b}}_p(t)]^T$, we have

$$|\epsilon(t)| = \frac{|\epsilon(t)|}{m(t)} m(t) \leq \frac{|\epsilon(t)|}{m(t)}(1 + |y_f(t)| + |u_f(t)|) \tag{7.49}$$

$$|\dot{\hat{a}}_p(t)| \leq a_3 \frac{|\epsilon(t)|}{m(t)}, \quad |\dot{\hat{b}}_p(t)| \leq a_3 \frac{|\epsilon(t)|}{m(t)} \tag{7.50}$$

for some constant $a_3 > 0$. Then, from (7.48)–(7.50), we obtain

$$\begin{aligned} |y_f(t)| \leq\; & a_4 + a_4 \int_0^t e^{-a_5(t-\tau)} \frac{|\epsilon(t)|}{m(\tau)} |y_f(\tau)|\, d\tau \\ & + a_4 \int_0^t e^{-a_5(t-\tau)} \frac{|\epsilon(t)|}{m(\tau)} \int_0^\tau e^{-a_2(\tau-\sigma)} |y_f(\sigma)|\, d\sigma\, d\tau \end{aligned} \tag{7.51}$$

for some constants $a_4 > 0$ and $a_5 > 0$. Defining

$$z_1(t) = |y_f(t)| + \int_0^t e^{-a_2(t-\sigma)} |y_f(\sigma)|\, d\sigma, \tag{7.52}$$

from (7.51), we have

$$z_1(t) \leq a_6 + a_6 \int_0^t e^{-a_7(t-\tau)} \frac{|\epsilon(t)|}{m(\tau)} z_1(\tau)\, d\tau \tag{7.53}$$

for some constants $a_6 > 0$ and $a_7 > 0$.

This inequality has the form (2.102) with $x_1(t) = \frac{|\epsilon(t)|}{m(t)}$. Since $\frac{|\epsilon(t)|}{m(t)} \in L^2$, from Lemma 2.3, we conclude that $z_1(t)$ is bounded, and from (7.51), that is,

$$|y_f(t)| \leq a_4 + a_4 \int_0^t e^{-a_5(t-\tau)} \frac{|\epsilon(t)|}{m(\tau)} z_1(\tau)\, d\tau \tag{7.54}$$

we have that $y_f(t)$ is bounded. Furthermore, from (7.47), $u_f(t)$ is bounded (so that $m(t)$ in (7.29) is bounded and $\epsilon(t) \in L^2$), from (7.39), $y(t)$ is bounded, and from (7.22), $u(t)$ is bounded. Then, from (7.41), $\dot{y}_f(t)$ and $\dot{u}_f(t)$ are bounded, and from (7.42), $\dot{\epsilon}(t)$ is bounded. From Lemma 2.14, a bounded $\dot{\epsilon}(t)$ and a finite $\int_0^\infty \epsilon^2(t)\, dt$ imply that $\lim_{t\to\infty} \epsilon(t) = 0$ so that $\lim_{t\to\infty} \dot{\hat{a}}_p(t) = \lim_{t\to\infty} \dot{\hat{b}}_p(t) = 0$. Finally, from (7.44), we have

$$\dot{y}(t) + a_m y(t) = -(\dot{\epsilon}(t) + a_f \epsilon(t)) + \dot{\hat{a}}_p(t) y_f(t) + \dot{\hat{b}}_p(t) u_f(t) + b_m r(t), \tag{7.55}$$

and from (7.21), that is, $\dot{y}_m(t) + a_m y_m(t) = b_m r(t)$, we obtain

$$\dot{y}(t)-\dot{y}_m(t)+a_m(y(t)-y_m(t)) = -(\dot{\epsilon}(t)+a_f\epsilon(t))+\dot{\hat{a}}_p(t)y_f(t)+\dot{\hat{b}}_p(t)u_f(t). \quad (7.56)$$

Using Lemma 2.7 and Lemma 2.18, we conclude that

$$\lim_{t\to\infty}(y(t) - y_m(t)) = 0, \int_0^\infty (y(t) - y_m(t))^2\, dt < \infty. \quad (7.57)$$

7.1.2 General Design Procedure

An indirect model reference adaptive control design procedure consists of

- plant parameter estimation,
- reference model selection,
- controller structure construction, and
- controller parameter caluculation (also see [151], [340]).

Plant Parameter Estimation

Consider a linear time-invariant plant described by the differential equation

$$P(s)[y](t) = k_p Z(s)[u](t), \quad (7.58)$$

where $y(t) \in R$ and $u(t) \in R$ are the measured plant input and output,

$$P(s) = s^n + p_{n-1}s^{n-1} + \cdots + p_1 s + p_0, \quad (7.59)$$

$$Z(s) = s^m + z_{m-1}s^{m-1} + \cdots + z_1 s + z_0 \quad (7.60)$$

are polynomials in s with s being the time differentiation operator $s[x](t) = \dot{x}(t)$, and p_i, $i = 0, 1, \ldots, n-1$, k_p, and z_i, $i = 0, 1, \ldots, m-1$, with $n > m$, are the unknown but constant parameters to be estimated.

Choosing a stable polynomial $\Lambda_e(s) = s^n + \lambda_{n-1}^e s^{n-1} + \cdots + \lambda_1^e s + \lambda_0^e$ and defining the parameter and regressor vectors

$$\theta_p^* = [k_p z_0, k_p z_1, \ldots, k_p z_{m-1}, k_p, \\ -p_0, -p_1, \ldots, -p_{n-2}, -p_{n-1}]^T \in R^{n+m+1}, \quad (7.61)$$

$$\phi(t) = \left[\frac{1}{\Lambda(s)}[u](t), \frac{s}{\Lambda_e(s)}[u](t), \ldots, \frac{s^{m-1}}{\Lambda_e(s)}[u](t), \frac{s^m}{\Lambda_e(s)}[u](t), \right. \\ \left. \frac{1}{\Lambda_e(s)}[y](t), \frac{s}{\Lambda_e(s)}[y](t), \ldots, \frac{s^{n-2}}{\Lambda_e(s)}[y](t), \frac{s^{n-1}}{\Lambda_e(s)}[y](t)\right]^T, \quad (7.62)$$

as in Section 3.1, we express the plant (7.58) as

$$y(t) - \frac{\Lambda_{n-1}(s)}{\Lambda_e(s)}[y](t) = \theta_p^{*T}\phi(t), \tag{7.63}$$

where $\Lambda_{n-1}(s) = \lambda_{n-1}^e s^{n-1} + \cdots + \lambda_1^e s + \lambda_0^e$.

To generate the vector signal $\phi(t)$, we construct two dynamic systems

$$\dot{\omega}_1^e(t) = A_e\omega_1^e(t) + b_e u(t), \tag{7.64}$$

$$\dot{\omega}_2^e(t) = A_e\omega_2^e(t) + b_e y(t), \tag{7.65}$$

where $\omega_1^e(t) \in R^n$, $\omega_1^e(t) \in R^n$, and

$$A_e = \begin{bmatrix} 0 & 1 & 0 & \cdots & 0 & 0 \\ 0 & 0 & 1 & 0 & \cdots & 0 \\ \vdots & \vdots & \vdots & \vdots & \vdots & \vdots \\ 0 & 0 & \cdots & \cdots & 0 & 1 \\ -\lambda_0^e & -\lambda_1^e & \cdots & \cdots & -\lambda_{n-2}^e & -\lambda_{n-1}^e \end{bmatrix}, \; b = \begin{bmatrix} 0 \\ \vdots \\ 0 \\ 1 \end{bmatrix} \in R^n. \tag{7.66}$$

Then, we generate the regressor vector $\phi(t)$ from

$$\phi(t) = \left[(\bar{\omega}_1^e(t))^T, (\omega_2^e(t))^T\right]^T, \tag{7.67}$$

where $\bar{\omega}_1^e(t) \in R^{m+1}$ is the first $m+1$ components subvector of $\omega_1^e(t)$.

Denoting $\theta_p(t)$ as the estimate of θ_p^*, we define the estimation error

$$\epsilon(t) = \theta_p^T(t)\phi(t) - y(t) + \frac{\Lambda_{n-1}(s)}{\Lambda_e(s)}[y](t), \; t \geq t_0, \tag{7.68}$$

where $\frac{\Lambda_{n-1}(s)}{\Lambda_e(s)}[y](t)$ is implemented as

$$\frac{\Lambda_{n-1}(s)}{\Lambda_e(s)}[y](t) = \lambda_e^T\omega_2^e(t), \; \lambda_e = [\lambda_0^e, \lambda_1^e, \ldots, \lambda_{n-2}^e, \lambda_{n-1}^e]^T \in R^n. \tag{7.69}$$

Then, to update the parameter estimate $\theta_p(t)$, we use the following gradient algorithm with parameter projection:

$$\dot{\theta}_p(t) = -\frac{\Gamma\phi(t)\epsilon(t)}{m^2(t)} + f(t), \; \theta_p(t_0) = \theta_0, \; t \geq t_0, \tag{7.70}$$

where $\Gamma = \text{diag}\{\Gamma_1, \gamma_{m+1}, \Gamma_2\}$ with $\Gamma_1 \in R^{m\times m}$, $\Gamma_1 = \Gamma_1^T > 0$, $\gamma_{m+1} > 0$ and $\Gamma_2 \in R^{n\times n}$, $\Gamma_2 = \Gamma_2^T > 0$, θ_0 is an initial estimate of $\theta_p^* \in R^{n+m+1}$, and

$$m(t) = \sqrt{1 + \kappa\phi^T(t)\phi(t)} \tag{7.71}$$

with $\kappa > 0$. The parameter projection term $f(t)$ has the form

$$f(t) = [0_{1\times m}, f_{m+1}(t), 0_{1\times n}]^T \in R^{n+m+1}. \tag{7.72}$$

To design $f_{m+1}(t)$ for projecting $\theta_{p\,m+1}(t) = \hat{k}_p(t)$, the $(m+1)$th component of $\theta_p(t)$, to ensure that $\hat{k}_p(t)$ is away from zero, we assume that

(A7.1C): The lower bound k_{p0} of $|k_p|$, such that $0 < k_{p0} \leq |k_p|$, is known, and so is $\text{sign}[k_p]$, the sign of k_p.

Then, we define $\theta^a_{p\,m+1} = k_{p0}$ and $\theta^b_{p\,m+1} = \infty$ for $k_p > 0$ and $\theta^a_{p\,m+1} = -\infty$ and $\theta^b_{p\,m+1} = -k_{p0}$ for $k_p < 0$, choose the initial estimate $\theta_{p\,m+1}(0)$ to satisfy

$$\theta_{p\,m+1}(0) \in (\theta^a_{p\,m+1}, \theta^b_{p\,m+1}), \tag{7.73}$$

denote $\phi(t) = [\phi_1(t), \ldots, \phi_m(t), \phi_{m+1}(t), \phi_{m+2}(t), \ldots, \phi_{n+m+1}(t)]^T$, $\phi_{m+1}(t) = \frac{s^m}{\Lambda_e(s)}[u](t)$ as the $(m+1)$th component of $\omega^e_1(t)$ defined in (7.64), introduce

$$g_{m+1}(t) = -\frac{\gamma_{m+1}\phi_{m+1}(t)\epsilon(t)}{m^2(t)}, \tag{7.74}$$

and set the projection function component $f_{m+1}(t)$ as

$$f_{m+1}(t) = \begin{cases} 0 & \text{if } \theta_{p\,m+1}(t) \in (\theta^a_{p\,m+1}, \theta^b_{p\,m+1}) \text{ or} \\ & \text{if } \theta_{p\,m+1}(t) = \theta^a_{p\,m+1},\ g_{m+1}(t) \geq 0 \text{ or} \\ & \text{if } \theta_{p\,m+1}(t) = \theta^b_{p\,m+1},\ g_{m+1}(t) \leq 0, \\ -g_{m+1}(t) & \text{otherwise.} \end{cases} \tag{7.75}$$

This $f(t)$ ensures that $\theta_{p\,m+1}(t) \in [\theta^a_{p\,m+1}, \theta^b_{p\,m+1}]$ and $|\theta_{p\,m+1}(t)| < \infty$ (i.e., $k_{p0} \leq |\hat{k}_p(t)| < \infty$) and that $(\theta_{p\,m+1}(t) - \theta^*_{p\,m+1})f_{m+1}(t) \leq 0$. The time derivative of the positive definite function $V(\tilde{\theta}_p) = \tilde{\theta}_p^T\Gamma^{-1}\tilde{\theta}_p$ with $\tilde{\theta}_p = \theta_p - \theta^*$ is

$$\begin{aligned} \dot{V} &= -\frac{2\epsilon^2(t)}{m^2(t)} + 2\tilde{\theta}_p^T(t)\Gamma^{-1}f(t) \\ &= -\frac{2\epsilon^2(t)}{m^2(t)} + 2(\theta_{p\,m+1}(t) - \theta^*_{p\,m+1})\gamma^{-1}_{m+1}f_{m+1}(t) \leq -\frac{2\epsilon^2(t)}{m^2(t)}, \end{aligned} \tag{7.76}$$

from which we can verify that

(i) $\theta_p(t)$, $\dot{\theta}_p(t)$, $\frac{\epsilon(t)}{m(t)}$ are bounded; and
(ii) $\frac{\epsilon(t)}{m(t)}$ and $\dot{\theta}_p(t)$ belong to L^2.

Based on the parameter vector $\theta_p(t)$ obtained from (7.70), we identify

$$\begin{aligned} &[\widehat{k_p z_0}(t), \widehat{k_p z_1}(t), \ldots, \widehat{k_p z_{m-1}}(t), \hat{k}_p(t), -\hat{p}_0(t), -\hat{p}_1(t), \ldots, -\hat{p}_{n-2}(t), -\hat{p}_{n-1}(t)] \\ &= \theta_p^T(t) \in R^{1\times(n+m+1)} \end{aligned} \tag{7.77}$$

and define the parameter estimates of z_i as

$$\hat{z}_i(t) = \frac{\widehat{k_p z_i}(t)}{\hat{k}_p(t)}, \ i = 0, 1, \ldots, m-1. \tag{7.78}$$

Then, we introduce the estimates of $P(s)$ and $Z(s)$ in (7.59)–(7.60) as

$$\hat{P}(s, \hat{p}) = s^n + \hat{p}_{n-1}s^{n-1} + \cdots + \hat{p}_1 s + \hat{p}_0, \tag{7.79}$$

$$\hat{Z}(s, \hat{z}) = s^m + \hat{z}_{m-1}s^{m-1} + \cdots + \hat{z}_1 s + \hat{z}_0, \tag{7.80}$$

where $\hat{z} = [\hat{z}_0, \hat{z}_1, \ldots, \hat{z}_{m-1}]^T$ and $\hat{p} = [\hat{p}_0, \hat{p}_1, \ldots, \hat{p}_{n-1}]^T$ are the estimates of $z^* = [z_0, z_1, \ldots, z_{m-1}]^T \in R^m$ and $p^* = [p_0, p_1, \ldots, p_{n-1}]^T \in R^n$.

Reference Model System

The reference model system is the same as that in (5.4), that is,

$$P_m(s)[y_m](t) = r(t), \tag{7.81}$$

where $P_m(s)$ is a monic stable polynomial of degree $n-m$ and $r(t)$ is a bounded and piecewise continuous reference input signal.

Controller Structure

The same model reference controller structure (5.5) is used:

$$u(t) = \theta_1^T \omega_1(t) + \theta_2^T \omega_2(t) + \theta_{20} y(t) + \theta_3 r(t), \tag{7.82}$$

where $\theta_1 \in R^{n-1}$, $\theta_2 \in R^{n-1}$, $\theta_{20} \in R$, and $\theta_3 \in R$ are parameters, and

$$\omega_1(t) = \frac{a(s)}{\Lambda_c(s)}[u](t), \ \omega_2(t) = \frac{a(s)}{\Lambda_c(s)}[y](t) \tag{7.83}$$

for $a(s) = [1, s, \ldots, s^{n-2}]^T$, $\Lambda_c(s)$ a monic stable polynomial of degree $n-1$.

For $\Lambda_c(s) = s^{n-1} + \lambda_{n-2}^c s^{n-2} + \cdots + \lambda_1^c s + \lambda_0^c$ stable, the signals $\omega_1^c(t)$ and $\omega_2^c(t)$ are obtained as the states of the auxiliary systems

$$\dot{\omega}_1^c(t) = A_c \omega_1^c(t) + b_c u(t), \tag{7.84}$$

$$\dot{\omega}_2^c(t) = A_c \omega_2^c(t) + b_c y(t), \tag{7.85}$$

where $\omega_1(t) \in R^{n-1}$, $\omega_2(t) \in R^{n-1}$, and

$$A_c = \begin{bmatrix} 0 & 1 & 0 & \cdots & 0 & 0 \\ 0 & 0 & 1 & 0 & \cdots & 0 \\ \vdots & \vdots & \vdots & \vdots & \vdots & \vdots \\ 0 & 0 & \cdots & \cdots & 0 & 1 \\ -\lambda_0^c & -\lambda_1^c & \cdots & \cdots & -\lambda_{n-3}^c & -\lambda_{n-2}^c \end{bmatrix}, \ b_c = \begin{bmatrix} 0 \\ \vdots \\ 0 \\ 1 \end{bmatrix} \in R^{n-1}. \tag{7.86}$$

Controller Parameter Calculation

The controller parameters $\theta_1 \in R^{n-1}$, $\theta_2 \in R^{n-1}$, $\theta_{20} \in R$, and $\theta_3 \in R$ are obtained from the design equation

$$\begin{aligned} &\theta_1^T a(\lambda)\hat{P}(\lambda,\hat{p}) + (\theta_2^T a(\lambda) + \theta_{20}\Lambda_c(\lambda))\hat{k}_p\hat{Z}(\lambda,\hat{z}) \\ &= \Lambda_c(\lambda)(\hat{P}(\lambda,\hat{p}) - \hat{k}_p\theta_3\hat{Z}(\lambda,\hat{z})P_m(\lambda)). \end{aligned} \tag{7.87}$$

As from Lemma 5.1, this equation has a solution $\{\theta_1, \theta_2, \theta_{20}, \theta_3\}$ for any $\hat{P}(\lambda,\hat{p})$ and $\hat{Z}(\lambda,\hat{z})$ in (7.79)–(7.80) with s replaced by λ. While the solution to (7.87) is not unique when $\hat{P}(\lambda,\hat{p})$ and $\hat{Z}(\lambda,\hat{z})$ are not coprime, to implement the adaptive controller (7.82), one can divide (7.87) by $\hat{P}(\lambda,\hat{p})$ to obtain

$$\begin{aligned} &\theta_1^T a(\lambda) + (\theta_2^T a(\lambda) + \theta_{20}\Lambda_c(\lambda))\hat{k}_p\frac{\hat{Z}(\lambda,\hat{z})}{\hat{P}(\lambda,\hat{p})} \\ &= \Lambda_c(\lambda) - \hat{k}_p\theta_3\hat{Z}(\lambda,\hat{z})\left(\hat{Q}(\lambda,\hat{q}) + \frac{\hat{R}(\lambda,\hat{r})}{\hat{P}(\lambda,\hat{p})}\right) \end{aligned} \tag{7.88}$$

for some polynomials $\hat{Q}(\lambda,\hat{q})$, monic, of degree n^*-1 and parametrized by some parameters in $\hat{q} \in R^{n^*-1}$, and $\hat{R}(\lambda,\hat{r})$ of degree $n-1$ and parametrized by some $\hat{r} \in R^n$, such that $\frac{\Lambda_c(\lambda)P_m(\lambda)}{\hat{P}(\lambda,\hat{p})} = \hat{Q}(\lambda,\hat{q}) + \frac{\hat{R}(\lambda,\hat{r})}{\hat{P}(\lambda,\hat{p})}$. With this parametrization, one can choose the solution $\{\theta_1, \theta_2, \theta_{20}, \theta_3\}$ to (7.87) from

$$\theta_1^T a(\lambda) = \Lambda_c(\lambda) - \hat{Q}(\lambda,\hat{q})\hat{Z}(\lambda,\hat{z}), \tag{7.89}$$

$$\theta_2^T a(\lambda) + \theta_{20}\Lambda_c(\lambda) = -\theta_3\hat{R}(\lambda,\hat{r}), \tag{7.90}$$

and $\theta_3 = \frac{1}{\hat{k}_p}$. Note that $\hat{R}(\lambda,\hat{r}) = \Lambda_c(\lambda)P_m(\lambda) - \hat{Q}(\lambda,\hat{q})\hat{P}(\lambda,\hat{p})$ [151], [340].

Stability and Tracking Properties

This adaptive control scheme has the following desired properties.

Theorem 7.1 *Under the condition that all zeros of $Z(s)$ are stable, all signals in the adaptive control system consisting of (7.58), (7.70), (7.81), (7.82), and (7.87) are bounded and $y(t) - y_m(t) \in L^2$, $\lim_{t\to\infty}(y(t) - y_m(t)) = 0$.*

Proof: In view of (7.82), we define

$$\theta(t) = [\theta_1^T(t), \theta_2^T(t), \theta_{20}(t), \theta_3(t)]^T \in R^{2n}, \tag{7.91}$$

$$\omega(t) = [\omega_1^T(t), \omega_2^T(t), y(t), r(t)]^T \in R^{2n}, \tag{7.92}$$

$$\theta^* = [\theta_1^{*T}, \theta_2^{*T}, \theta_{20}^*, \theta_3^*]^T \in R^{2n}, \tag{7.93}$$

$$\tilde{\theta}(t) = \theta(t) - \theta^*,\ e(t) = y(t) - y_m(t), \tag{7.94}$$

where θ_i^*, $i = 1, 2, 3$, and θ_{20}^* are the nominal plant–model matching parameters satisfying (5.11) with $\bar{n} = n$. As in Section 5.3.1, for the controller structure (7.82), we also have the tracking error equation

$$e(t) = y(t) - y_m(t) = \frac{k_p}{P_m(s)}[\tilde{\theta}^T \omega](t). \tag{7.95}$$

From the adaptive law (7.70) with parameter projection (7.75), we had that $\theta_p(t) \in L^\infty$, $0 < k_{p0} \leq |\hat{k}_p(t)| \leq k^0 < \infty$, for some constants k_{p0} and k^0, and $\dot{\theta}_p(t) \in L^2 \cap L^\infty$. From the design equations (7.87)–(7.90), it follows that $\theta(t) \in L^\infty$, $\dot{\theta}(t) \in L^2 \cap L^\infty$, and $\rho(t) = \hat{k}_p(t) \in L^\infty$.

For $W_m(s) = \frac{1}{P_m(s)}$ with $P_m(s)$ being a stable polynomial of degree n^*, a constant $\alpha > 0$, and the three polynomials with time-varying coefficients,

$$\hat{P}(s, \hat{p}(t)) = s^n + \hat{p}_{n-1}(t)s^{n-1} + \cdots + \hat{p}_1(t)s + \hat{p}_0(t), \tag{7.96}$$

$$\hat{Z}(s, \hat{z}(t)) = s^m + \hat{z}_{m-1}(t)s^{m-1} + \cdots + \hat{z}_1(t)s + \hat{z}_0(t), \tag{7.97}$$

$$\hat{Q}(s, \hat{q}(t)) = s^{n^*-1} + \hat{q}_{n^*-2}(t)s^{n^*-2} + \cdots + \hat{q}_1(t)s + \hat{q}_0(t), \tag{7.98}$$

where $\hat{q} = [\hat{q}_0, \hat{q}_1, \ldots, \hat{q}_{n^*-2}]^T$ is the estimate of a plant–model matching parameter vector q^* containing the coefficients of the polynomial part $Q(s)$ (of degree $n^* - 1$) of $\frac{\Lambda_c(s)P_m(s)}{P(s)}$, we define the operator $\frac{1}{\hat{k}_p(t)}\hat{Q}(s, \hat{q}(t)) \circ W_m(s)(s + \alpha) \circ \hat{k}_p(t)\hat{Z}(s, \hat{z}(t))$ (similarly, $\frac{1}{\hat{k}_p(t)}\hat{Q}(s, \hat{q}(t)) \circ W_m(s)(s + \alpha) \circ \hat{P}(s, \hat{p}(t))$) as

$$\begin{aligned}
&\frac{1}{\hat{k}_p(t)}\hat{Q}(s, \hat{q}(t)) \circ W_m(s)(s + \alpha) \circ \hat{k}_p(t)\hat{Z}(s, \hat{z}(t))[x](t) \\
&\triangleq \frac{1}{\hat{k}_p(t)}\hat{Q}(s, \hat{q}(t))[W_m(s)(s + \alpha)[\hat{k}_p\hat{Z}(s, \hat{z})[x]]](t) \\
&= \sum_{i=0}^{n^*-1} \frac{1}{\hat{k}_p(t)}\hat{q}_i(t)s^i \left[W_m(s)(s + \alpha)\left[\sum_{j=0}^{m} \hat{k}_p\hat{z}_j s^j[x]\right]\right](t)
\end{aligned} \tag{7.99}$$

with $\hat{q}_{n^*-1}(t) = \hat{z}_m(t) = 1$ ($\hat{p}_n(t) = 1$) for (7.96)–(7.98).

With the definition (7.99), we use the design equations (7.89) and (7.90) to construct the corresponding operator equations [151]. Letting $\alpha > 0$ be constant and operating both sides of (7.89) on $\frac{W_m(s)}{\Lambda_c(s)}[u](t)$, we have

$$\begin{aligned}
\theta_1^T(t)W_m(s)[\omega_1](t) &= W_m(s)[u](t) + \frac{1}{\hat{k}_p(t)}\epsilon_u(t) \\
&- \frac{1}{\hat{k}_p(t)}\hat{Q}(s, \hat{q}(t))\left[W_m(s)(s + \alpha)\left[\hat{k}_p\hat{Z}(s, \hat{z})\left[\frac{1}{\Lambda_c(s)(s + \alpha)}[u]\right]\right]\right](t),
\end{aligned} \tag{7.100}$$

where, for $\hat{\rho}_j(t) = \hat{k}_p(t)\hat{z}_j(t)$, $j = 0, 1, 2, \ldots, m-1$, and $\hat{\rho}_m(t) = \hat{k}_p(t)$,

$$\epsilon_u(t) \triangleq \hat{Q}(s, \hat{q}(t)) \left[W_m(s)(s+\alpha)\left[\hat{k}_p\hat{Z}(s,\hat{z})\left[\frac{1}{\Lambda_c(s)(s+\alpha)}[u]\right]\right]\right](t)$$
$$- \hat{Q}(s, \hat{q}(t))\hat{k}_p\hat{Z}(s,\hat{z})\left[W_m(s)(s+\alpha)\frac{1}{\Lambda_c(s)(s+\alpha)}[u]\right](t) \quad (7.101)$$

Using the explicit swapping lemma,[1] for $W_m(s) = \frac{1}{P_m(s)}$ with a stable monic polynomial $P_m(s)$ of degree n^* and $i = 0, 1, \ldots, n^*-1$, we express

$$s^i \left[W_m(s)(s+\alpha)\left[\sum_{j=0}^{m} \hat{\rho}_j s^j \left[\frac{1}{\Lambda_c(s)(s+\alpha)}[u]\right]\right]\right](t)$$
$$= \sum_{j=0}^{m} \hat{\rho}_j(t) s^i \left[W_m(s)(s+\alpha)\left[\sum_{j=0}^{m} s^j \left[\frac{1}{\Lambda_c(s)(s+\alpha)}[u]\right]\right]\right](t)$$

[1] Denoting $P_m(s) = s^{n^*} + a_{n^*-1}s^{n^*-1} + \cdots + a_1 s + a_0$, for two general vector signals $\theta(t)$ and $\omega(t)$, from (5.138) and with $a_{n^*} = 1$, we have

$$\theta^T(t)\frac{1}{P_m(s)}[\omega](t) - \frac{1}{P_m(s)}[\theta^T\omega](t) = \sum_{i=1}^{n^*}\left(\frac{\sum_{j=0}^{n^*-i} a_{n^*-j}s^{n^*-i-j}}{P_m(s)}\right)\left[\dot{\theta}^T\frac{s^{i-1}}{P_m(s)}[\omega]\right](t).$$

Introducing $F(s) = f_{n^*-1}s^{n^*-1} + \cdots + f_1 s + f_0$, we express

$$F(s)\left[\theta^T(t)\frac{1}{P_m(s)}[\omega]\right](t)$$
$$= \sum_{i=1}^{n^*-1}\left(\sum_{j=1}^{n^*-i} f_{n^*-j}s^{n^*-i-j}\right)\left[\dot{\theta}^T\frac{s^{i-1}}{P_m(s)}[\omega]\right](t) + \theta^T(t)\frac{F(s)}{P_m(s)}[\omega](t).$$

Hence, we obtain the following explicit swapping lemma:

$$\frac{F(s)}{P_m(s)}[\theta^T\omega](t) - \theta^T(t)\frac{F(s)}{P_m(s)}[\omega](t)$$
$$= \sum_{i=1}^{n^*-1}\frac{\alpha_i(s)}{P_m(s)}\left[\dot{\theta}^T\frac{s^{i-1}}{P_m(s)}[\omega]\right](t) - \frac{F(s)}{P_m(s)}\left[\dot{\theta}^T\frac{s^{n^*-1}}{P_m(s)}[\omega]\right](t) \triangleq \sum_{i=1}^{n^*}\frac{\alpha_i(s)}{P_m(s)}\left[\dot{\theta}^T\frac{s^{i-1}}{P_m(s)}[\omega]\right](t),$$

where $\alpha_i(s) = P_m(s)\sum_{j=1}^{n^*-i} f_{n^*-j}s^{n^*-i-j} - F(s)\sum_{j=0}^{n^*-i} a_{n^*-j}s^{n^*-i-j}$, $i = 1, 2, \ldots, n^*-1$, are polynominals of degrees n^*-1 or less, and $\alpha_{n^*}(s) \triangleq -F(s)$.

For a proper transfer function $h(s) = f_{n^*} + \frac{F(s)}{P_m(s)}$, it follows that

$$h(s)[\theta^T\omega](t) - \theta^T(t)h(s)[\omega](t) = \sum_{i=1}^{n^*}\frac{\alpha_i(s)}{P_m(s)}\left[\dot{\theta}^T\frac{s^{i-1}}{P_m(s)}[\omega]\right](t).$$

For the swapping lemma [151], [265], [340], see (5.331).

$$+\sum_{j=0}^{m}\sum_{k=1}^{n^*}\frac{\alpha_{ik}(s)}{P_m(s)}\left[\dot{\hat{\rho}}_j\frac{s^{k-1}}{P_m(s)}s^j\left[\frac{1}{\Lambda_c(s)(s+\alpha)}[u]\right]\right](t), \tag{7.102}$$

for some polynomials $\alpha_{ik}(s)$ of degrees n^*-1 or less, so that, in view of (7.99),

$$\epsilon_u(t)=\sum_{i=0}^{n^*-1}\hat{q}_i(t)\sum_{j=0}^{m}\sum_{k=1}^{n^*}\frac{\alpha_{ik}(s)}{P_m(s)}\left[\dot{\hat{\rho}}_j\frac{s^{k-1}}{P_m(s)}s^j\left[\frac{1}{\Lambda_c(s)(s+\alpha)}[u]\right]\right](t). \tag{7.103}$$

Similarly, for $\alpha>0$, operating both sides of (7.101) on $\frac{W_m(s)}{\Lambda_c(s)}[y](t)$, we have

$$\begin{aligned}\theta_2^T(t)W_m(s)[\omega_2](t)+\theta_{20}(t)W_m(s)[y](t)&=-\frac{1}{\hat{k}_p(t)}y(t)+\frac{1}{\hat{k}_p(t)}\epsilon_y(t)\\&+\frac{1}{\hat{k}_p(t)}\hat{Q}(s,\hat{q}(t))\left[W_m(s)(s+\alpha)\left[\hat{P}(s,\hat{p})\left[\frac{1}{\Lambda_c(s)(s+\alpha)}[y]\right]\right]\right](t),\end{aligned} \tag{7.104}$$

where

$$\begin{aligned}\epsilon_y(t)&\triangleq\hat{Q}(s,\hat{q}(t))\left[W_m(s)(s+\alpha)\left[\hat{P}(s,\hat{p})\left[\frac{1}{\Lambda_c(s)(s+\alpha)}[y]\right]\right]\right](t)\\&\quad-\hat{Q}(s,\hat{q}(t))\hat{P}(s,\hat{p})\left[W_m(s)(s+\alpha)\frac{1}{\Lambda_c(s)(s+\alpha)}[y]\right](t)\\&=\sum_{i=0}^{n^*-1}\hat{q}_i(t)\sum_{j=0}^{n}\sum_{k=1}^{n^*}\frac{\alpha_{ik}(s)}{P_m(s)}\left[\dot{\hat{p}}_j\frac{s^{k-1}}{P_m(s)}s^j\left[\frac{1}{\Lambda_c(s)(s+\alpha)}[y]\right]\right](t).\end{aligned} \tag{7.105}$$

In (7.100) and (7.104), replacing all parameter estimates by their nominal values for plant–model matching, we obtain

$$\begin{aligned}\theta_1^{*T}W_m(s)[\omega_1](t)&=W_m(s)[u](t)\\&\quad-\frac{1}{k_p}Q(s)W_m(s)(s+\alpha)k_pZ(s)\left[\frac{1}{\Lambda_c(s)(s+\alpha)}[u]\right](t)\end{aligned} \tag{7.106}$$

$$\begin{aligned}\theta_2^{*T}W_m(s)[\omega_2](t)+\theta_{20}^*W_m(s)[y](t)&=-\frac{1}{k_p}y(t)\\&\quad+\frac{1}{k_p}Q(s)W_m(s)(s+\alpha)P(s)\left[\frac{1}{\Lambda_c(s)(s+\alpha)}[y]\right](t).\end{aligned} \tag{7.107}$$

Introducing $\rho(t)=\hat{k}_p(t)$ and

$$\xi(t)=\theta^T(t)\zeta(t)-W_m(s)[\theta^T\omega](t),\ \zeta(t)=W_m(s)[\omega](t), \tag{7.108}$$

$$\begin{aligned}\bar{\epsilon}(t)&=\hat{Q}(s,\hat{q}(t))\left[W_m(s)(s+\alpha)\left[\hat{P}(s,\hat{p})\left[\frac{1}{\Lambda_c(s)(s+\alpha)}[y]\right]\right]\right](t)\\&\quad-\hat{Q}(s,\hat{q}(t))\left[W_m(s)(s+\alpha)\left[\hat{k}_p\hat{Z}(s,\hat{z})\left[\frac{1}{\Lambda_c(s)(s+\alpha)}[u]\right]\right]\right](t),\end{aligned} \tag{7.109}$$

from (7.100), (7.104), and (7.106)–(7.109), we can derive

$$\tilde{\theta}^T(t)\zeta(t) + (\theta_3(t) - \theta_3^*)(y(t) - y_m(t)) = \frac{1}{\hat{k}_p(t)}(\bar{\epsilon}(t) + \epsilon_u(t) + \epsilon_y(t)), \quad (7.110)$$

where $\theta_3^* k_p = 1$. Using (7.95) and (7.110), we obtain

$$y(t) - y_m(t) + \rho(t)\xi(t) = \bar{\epsilon}(t) + \epsilon_u(t) + \epsilon_y(t). \quad (7.111)$$

From (7.62), (7.68), and (7.78)–(7.80), we have

$$\epsilon(t) = \hat{k}_p(t)\hat{Z}(s, \hat{z}(t))\left[\frac{1}{\Lambda_e(s)}[y]\right](t) - \hat{P}(s, \hat{p}(t))\left[\frac{1}{\Lambda_e(s)}[y]\right](t), \quad (7.112)$$

where $\frac{\epsilon(t)}{m(t)} \in L^2 \cap L^\infty$, for $m(t)$ in (7.71), and $\dot{\hat{z}}(t) \in L^2 \cap L^\infty$, $\dot{\hat{p}}(t) \in L^2 \cap L^\infty$, that is, $\dot{\hat{\rho}}_i(t) \in L^2 \cap L^\infty$, $i = 0, 1, \ldots, m$. For simplicity of analysis, we take $\Lambda_e(s) = \Lambda_c(s)(s + \alpha)$ for $\alpha > 0$. Then, we express $\bar{\epsilon}(t)$ in (7.109) as

$$\begin{aligned}\bar{\epsilon}(t) &= \hat{Q}(s, \hat{q}(t))\left[W_m(s)(s+\alpha)\left[\hat{P}(s,\hat{p})\left[\frac{1}{\Lambda_e(s)}[y]\right]\right]\right](t) \\ &\quad - \hat{Q}(s, \hat{q}(t))\left[W_m(s)(s+\alpha)\left[\hat{k}_p\hat{Z}(s,\hat{z})\left[\frac{1}{\Lambda_e(s)(s+\alpha)}[u]\right]\right]\right](t) \\ &= -\hat{Q}(s, \hat{q}(t))\left[W_m(s)(s+\alpha)[\epsilon]\right](t). \end{aligned} \quad (7.113)$$

Since $W_m(s)(s+\alpha)$ is stable and has relative degree $n^* - 1$ and $\hat{Q}(s, \hat{q})$ in (7.98) is monic and of degree $n^* - 1$, with $\hat{q}(t) = [\hat{q}_0(t), \hat{q}_1(t), \ldots, \hat{q}_{n^*-2}(t)]^T \in L^\infty$, it follows that $\hat{Q}(s, \hat{q}(t)) \circ W_m(s)(s + \alpha)$, defined as

$$\begin{aligned}&\hat{Q}(s, \hat{q}(t)) \circ W_m(s)(s+\alpha)[x](t) \\ &\triangleq \hat{Q}(s, \hat{q}(t))[W_m(s)(s+\alpha)[x]](t) = \sum_{i=0}^{n^*-1} \hat{q}_i(t)s^i[W_m(s)(s+\alpha)[x]](t), \end{aligned} \quad (7.114)$$

is a stable and proper operator (see Definition 2.23).

Comparing (7.111) with (5.88) and defining $z(t) = \frac{1}{s+a_0}[y](t)$ for $a_0 > 0$, we can also obtain (5.139) with $\epsilon(t)$ replaced by $\bar{\epsilon}(t)$ in (7.112):

$$z(t) = \frac{1}{s+a_0}[y_m](t) + \frac{1}{s+a_0}[\bar{\epsilon} + \epsilon_u + \epsilon_y - \rho\xi](t), \quad (7.115)$$

from which the results of Theorem 7.1 can be derived, in a similar way to that for proving Theorem 5.4 in Section 5.3.5. ∇

Remark 7.1 When $n^* = n - m = 1$, the design equation (7.87) (or (7.89)–(7.90)) has a simple solution, as $\hat{Q}(s, \hat{p}) = 1$, such that

$$\theta_1 = \lambda^c - \hat{z}, \tag{7.116}$$

$$\begin{bmatrix} \theta_2 + \theta_{20}\lambda^c \\ \theta_{20} \end{bmatrix} = \frac{1}{\hat{k}_p}(\hat{p} - a), \tag{7.117}$$

where $\lambda^c = [\lambda_0^c, \lambda_1^c, \ldots, \lambda_{n-2}^c]^T$, $\hat{z} = [\hat{z}_0, \hat{z}_1, \ldots, \hat{z}_{n-2}]^T$, $\hat{p} = [\hat{p}_0, \hat{p}_1, \ldots, \hat{p}_{n-1}]^T$, and $a = [a_0, a_1, \ldots, a_{n-2}]^T$, for $\Lambda_c(s)P_m(s) = s^n + a_{n-1}s^{n-1} + \cdots + a_0$. □

Remark 7.2 If we choose $\Lambda_e(s) = (s + \alpha)\Lambda_c(s)$, $\alpha > 0$, we can reduce the order of the adaptive control system using

$$\dot{y}_f(t) = -\alpha y_f(t) + y(t), \tag{7.118}$$

$$\dot{u}_f(t) = -\alpha u_f(t) + u(t), \tag{7.119}$$

$$\dot{\omega}_1(t) = A_c\omega_1(t) + b_c u_f(t), \tag{7.120}$$

$$\dot{\omega}_2(t) = A_c\omega_2(t) + b_c y_f(t), \tag{7.121}$$

where A_c and b_c are defined in (7.86). Then, we can construct

$$\omega_1^e(t) = [\omega_1^T(t), -[\lambda_0^c, \lambda_1^c, \ldots, \lambda_{n-2}^c]\omega_1(t) + u_f(t)], \tag{7.122}$$

$$\omega_2^e(t) = [\omega_2^T(t), -[\lambda_0^c, \lambda_1^c, \ldots, \lambda_{n-2}^c]\omega_2(t) + y_f(t)], \tag{7.123}$$

$$\omega_1^c(t) = M_\alpha \omega_1^e(t), \tag{7.124}$$

$$\omega_2^c(t) = M_\alpha \omega_2^e(t), \tag{7.125}$$

$$M_\alpha = \{m_{ij}\} \in R^{(n-1)\times n} : m_{ii} = \alpha,\ m_{i\,i+1} = 1. \tag{7.126}$$

The order of (7.91)–(7.94) is $2n$ ($4n - 2$ for (7.64)–(7.65), (7.84)–(7.85)). □

Remark 7.3 The model reference controller structure (5.332) can also be used for indirect model reference adaptive control, that is, we can choose

$$u(t) = \theta_1^T \omega_1^c(t) + \theta_2^T \omega_2^c(t) + \theta_3 r(t), \tag{7.127}$$

where $\theta_1 \in R^n$, $\theta_2 \in R^n$, $\theta_3 \in R$, and

$$\omega_1(t) = \frac{a(s)}{\Lambda_c(s)}[u](t),\ \omega_2(t) = \frac{a(s)}{\Lambda_c(s)}[y](t) \tag{7.128}$$

with $a(s) = [1, s, \ldots, s^{n-1}]^T$ and $\Lambda_c(s)$ a monic stable polynomial of degree n. We can choose $\Lambda_e(s) = \Lambda_c(s)$ for both the parameter estimator and controller filters so that $\omega_1^c(t) = \omega_1^e(t)$ and $\omega_2^c(t) = \omega_2^e(t)$ in (7.84) and (7.85). □

7.2 Pole Placement Designs

A crucial assumption for a model reference controller is that all zeros of the controlled plant are stable (that is, the plant is minimum phase). This condition is necessary, because a model reference controller is to make the plant output track the output of a *reference system* so that the plant zeros are to be cancelled by a feedback control law, and such a cancellation must be stable.

By solving a different design equation—a closed-loop pole placement equation, a control scheme can be developed for a nonminimum phase plant, a plant whose zeros may be unstable (i.e., they may be in $\mathrm{Re}[s] \geq 0$), to ensure closed-loop signal boundedness and output tracking of a class of reference outputs. Such a control design, called a *pole placement* design, can be made adaptive by using adaptive estimates of the plant parameters for the case when the plant parameters are unknown (see, e.g., [8], [23], [92], [151], [222], [237]).

7.2.1 Control System Structure

Plant Parametrization and Reference Output

Consider the linear time-invariant plant

$$P(s)[y](t) = Z(s)[u](t), \tag{7.129}$$

where $y(t) \in R$, $u(t) \in R$ are the measured plant output and input,

$$P(s) = s^n + p_{n-1}s^{n-1} + \cdots + p_1 s + p_0, \tag{7.130}$$

$$Z(s) = z_{n-1}s^{n-1} + z_{n-2}s^{n-2} + \cdots + z_1 s + z_0 \tag{7.131}$$

are polynomials in s,[2] the differentiation operator or Laplace transform operator as the case may be, and p_i, z_i, $i = 0, 1, \ldots, n-1$, are the unknown but constant parameters. We do not assume that $Z(s)$ is a stable polynomial.

The control objective is: given a reference output signal $y_m(t)$ satisfying

$$Q_m(s)[y_m](t) = 0, \tag{7.132}$$

where $Q_m(s)$ is a monic polynomial of degree n_q with either nonrepeated zeros on the $j\omega$-axis or zeros in the open left-half of the complex plane, so that $y_m(t)$ is bounded, find an output feedback control $u(t)$ such that all closed-loop signals are bounded and $y(t)$ tracks $y_m(t)$ asymptotically as $t \to \infty$.

[2]This form of $Z(s)$, as compared with that in (7.58), is more convenient for presenting a pole placement design to be developed in this section, which leads to the definition of z^* in (7.133), different from that in the presentation of (7.79)–(7.80).

Example 7.1 As examples of the polynomial $Q_m(s)$ and the corresponding reference signal $y_m(t)$, we have

$Q_m(s) = 1$ for $y_m(t) = 0$;

$Q_m(s) = s$ for $y_m(t) = c \neq 0$, a constant;

$Q_m(s) = s + a$ with $a > 0$ for $y_m(t) = c e^{-at}$, $c \neq 0$;

$Q_m(s) = s^2 + \sigma^2$ for $y_m(t) = a\sin(\sigma t) + b\cos(\sigma t)$ with $\sigma \neq 0$ and not both a, b being zero;

$Q_m(s) = (s^2 + \sigma_1^2)(s^2 + \sigma_2^2)$ for $y_m(t) = a_1\sin(\sigma_1 t) + a_2\sin(\sigma_2 t) + b_1\cos(\sigma_1 t) + b_2\cos(\sigma_2 t)$ with $\sigma_1 \neq 0$, $\sigma_2 \neq 0$, $\sigma_1 \neq \sigma_2$, not both a_1, b_1 being zero, and not both a_2, b_2 being zero; and

$Q_m(s) = s(s^2 + \sigma^2)$ for $y_m(t) = a\sin(\sigma t) + b\cos(\sigma t) + c$ with $c \neq 0$, $\sigma \neq 0$ and not both a, b being zero. □

Without the need that the plant zeros are stable as in the model reference adaptive control case (see Chapters 5 and 6), the following basic assumptions are needed for solving the pole placement control problem:

(A7.2C): $Q_m(s)P(s)$ and $Z(s)$ are coprime.

(A7.3C): The order n of $P(s)$ is known.

To meet the control objective, we use an indirect approach to design the adaptive controller: first estimate the plant parameters in

$$p^* = [p_0, p_1, \ldots, p_{n-1}]^T \in R^n, \; z^* = [z_0, z_1, \ldots, z_{n-1}]^T \in R^n \tag{7.133}$$

and then map their estimates to controller parameters using a design equation which places the closed-loop system poles at any desired locations.

Design Equation

To present the pole placement equation, let the estimates of p^* and z^* be

$$\hat{p} = [\hat{p}_0, \hat{p}_1, \ldots, \hat{p}_{n-1}]^T \in R^n, \; \hat{z} = [\hat{z}_0, \hat{z}_1, \ldots, \hat{z}_{n-1}]^T \in R^n \tag{7.134}$$

and the estimates of $P(s)$ and $Z(s)$, with $s = \lambda$, be

$$\hat{P}(\lambda, \hat{p}) = \lambda^n + \hat{p}_{n-1}\lambda^{n-1} + \cdots + \hat{p}_1\lambda + \hat{p}_0, \tag{7.135}$$

$$\hat{Z}(\lambda, \hat{z}) = \hat{z}_{n-1}\lambda^{n-1} + \hat{z}_{n-2}\lambda^{n-2} + \cdots + \hat{z}_1\lambda + \hat{z}_0, \tag{7.136}$$

and assume that $Q_m(\lambda)\hat{P}(\lambda, \hat{p})$ and $\hat{Z}(\lambda, \hat{z})$ are coprime polynomials in the indeterminate λ. Note that $Z(s) = \hat{Z}(s, z^*)$ and $P(s) = \hat{P}(s, p^*)$.

Choose the desired closed-loop characteristic polynomial as a monic polynomial $A^*(s)$ of degree $2n + n_q - 1$, which has all its zeros in $\mathrm{Re}[s] < 0$, based on the knowledge of the prespecified (desired) closed-loop system poles, where n_q is the degree of $Q_m(s)$ in (7.132).

Solve the Diophantine equation [3]

$$\hat{C}(\lambda, \psi_c)Q_m(\lambda)\hat{P}(\lambda, \hat{p}) + \hat{D}(\lambda, \psi_d)\hat{Z}(\lambda, \hat{z}) = A^*(\lambda) \tag{7.137}$$

to find the polynomials $\hat{C}(\lambda, \psi_c)$, $\hat{D}(\lambda, \psi_d)$ in the form

$$\hat{C}(\lambda, \psi_c) = \lambda^{n-1} + c_{n-2}\lambda^{n-2} + \cdots + c_1\lambda + c_0, \tag{7.138}$$

$$\hat{D}(\lambda, \psi_d) = d_{n_q+n-1}\lambda^{n_q+n-1} + \cdots + d_1\lambda + d_0, \tag{7.139}$$

where the parameter vectors ψ_c and ψ_d are

$$\psi_c = [c_0, c_1, \ldots, c_{n-2}]^T \in R^{n-1}, \ \psi_d = [d_0, d_1, \ldots, d_{n+n_q-1}]^T \in R^{n+n_q}. \tag{7.140}$$

Obtain $\hat{C}(s, \psi_c)$ and $\hat{D}(s, \psi_d)$ by making the substitution

$$\hat{C}(s, \psi_c) = \hat{C}(\lambda, \psi_c)|_{\lambda=s} = s^{n-1} + c_{n-2}s^{n-2} + \cdots + c_1 s + c_0, \tag{7.141}$$

$$\hat{D}(s, \psi_d) = \hat{D}(\lambda, \psi_d)|_{\lambda=s} = d_{n_q+n-1}s^{n_q+n-1} + \cdots + d_1 s + d_0. \tag{7.142}$$

Note that, as an operator, $c_i s^i$, $i = 1, \ldots, n-1$, is different from $s^i c_i$ when $c_i = c_i(t)$ is a time-varying coefficient, and so is $d_j s^j$, $j = 1, \ldots, n_q+n-1$, from $s^j d_j$ when $d_j = d_j(t)$ is a time-varying coefficient. For example, $c_1 s[x](t) = c_1(t)\dot{x}(t)$, while $sc_1[x](t) = s[c_1 x](t) = \dot{c}_1(t)x(t) + c_1(t)\dot{x}(t)$.

Controller Structure

We then choose the controller structure as

$$u(t) = (\Lambda_c(s) - \hat{C}(s, \psi_c)Q_m(s))\frac{1}{\Lambda_c(s)}[u](t) + \hat{D}(s, \psi_d)\frac{1}{\Lambda_c(s)}[y_m - y](t), \tag{7.143}$$

[3] If $Q_m(\lambda)\hat{P}(\lambda, \hat{p})$ and $\hat{Z}(\lambda, \hat{z})$ are coprime, then for any $A^*(\lambda)$ of degree $n_q + n - 1$ this equation has a unique solution $\{\hat{C}(\lambda, \phi_c), \hat{D}(\lambda, \psi_d)\}$ of the form (7.138)–(7.139) (see the proof of Lemma 5.1). In this case, the solution parameters ψ_c and ψ_d in (7.140) are obtained from an algebraic equation (see (7.170)) similar to that in (5.21).

On the other hand, if $Q_m(\lambda)\hat{P}(\lambda, \hat{p})$ and $\hat{Z}(\lambda, \hat{z})$ are not coprime, then for an arbitrary $A^*(s)$ the solution $\{\hat{C}(\lambda, \phi_c), \hat{D}(\lambda, \psi_d)\}$ may not exist.

where $\Lambda_c(s)$ is a monic stable polynomial of degree $n + n_q - 1$. Note that because $A^*(s)$ is monic and with degree $2n + n_q - 1$ and $\hat{P}(s, \hat{p})$ is monic and with degree n, $\hat{C}(s, \psi_c)Q_m(s)$ is always monic and of degree $n + n_q - 1$ so that $(\Lambda_c(s) - \hat{C}(s, \psi_c)Q_m(s))\frac{1}{\Lambda_c(s)}$ is stable and strictly proper (the polynomial $\hat{C}(\lambda, \psi_c)$ is monic because $Q_m(\lambda)$ and $A^*(\lambda)$ are both monic in (7.137)).

To write (7.143) in a parametrized form, denoting

$$Q_m(s) = s^{n_q} + q_{n_q-1}s^{n_q-1} + \cdots + q_1 s + q_0, \tag{7.144}$$

$$\Lambda_c(s) = s^{n_q+n-1} + \lambda^c_{n_q+n-2}s^{n_q+n-2} + \cdots + \lambda^c_1 s + \lambda^c_0, \tag{7.145}$$

and $\bar{\psi}_c = [\psi_c^T, 1]^T$, we introduce the parameters

$$\psi_1 = \begin{bmatrix} \psi_c & 0 & \cdots & 0 & 0 \\ 1 & \psi_c & \cdots & \vdots & 0 \\ 0 & 1 & \cdots & 0 & \vdots \\ \vdots & 0 & \cdots & 0 & 0 \\ 0 & \vdots & \cdots & 0 & 0 \\ 0 & 0 & \cdots & \psi_c & 0 \\ 0 & 0 & \cdots & 1 & \psi_c \end{bmatrix} \begin{bmatrix} q_0 \\ q_1 \\ \vdots \\ q_{n_q-1} \\ 1 \end{bmatrix} - \begin{bmatrix} \lambda^c_0 \\ \lambda^c_1 \\ \vdots \\ \lambda^c_{n_q+n-2} \end{bmatrix}, \tag{7.146}$$

$$\psi_2 = -\begin{bmatrix} d_0 \\ d_1 \\ \vdots \\ d_{n_q+n-2} \end{bmatrix} + d_{n_q+n-1} \begin{bmatrix} \lambda_0 \\ \lambda_1 \\ \vdots \\ \lambda_{n_q+n-3} \\ \lambda_{n_q+n-2} \end{bmatrix}, \tag{7.147}$$

$$\psi_3 = -d_{n_q+n-1} \tag{7.148}$$

and vector function $a_1(s) = [1, s, \ldots, s^{n_q+n-2}]^T$ and express (7.143) as

$$u(t) = \psi_1^T \frac{a_1(s)}{\Lambda_c(s)}[u](t) + \psi_2^T \frac{a_1(s)}{\Lambda_c(s)}[y - y_m](t) + \psi_3(y(t) - y_m(t)), \tag{7.149}$$

which is similar to the model reference controller structure (7.82).

7.2.2 Pole Placement Control

We first analyze the pole placement control scheme for the case when the parameters p^* and z^* of $P(s)$ and $Z(s)$ in (7.129) are known. We solve

$$C(\lambda)Q_m(\lambda)P(\lambda) + D(\lambda)Z(\lambda) = A^*(\lambda) \tag{7.150}$$

(which is the ideal (nonadaptive) version of (7.137)) for $C(\lambda) = \hat{C}(\lambda, \psi_c^*)$ and $D(\lambda) = \hat{D}(\lambda, \psi_d^*)$ of the forms (7.138) and (7.139) for some parameter vectors ψ_c^* and ψ_d^* of the form (7.140), where $Z(\lambda) = \hat{Z}(\lambda, z^*)$ and $P(\lambda) = \hat{P}(\lambda, p^*)$, and generate the control signal $u(t)$ for the plant (7.129) from

$$u(t) = (\Lambda_c(s) - C(s)Q_m(s))\frac{1}{\Lambda_c(s)}[u](t) + D(s)\frac{1}{\Lambda_c(s)}[y_m - y](t). \tag{7.151}$$

This control scheme has the following properties.

Theorem 7.2 *The pole placement controller (7.151) ensures that all signals in the closed-loop system are bounded and the tracking error $y(t) - y_m(t)$ converges to zero exponentially as t goes to infinity.*

Proof: From (7.151), we have

$$C(s)Q_m(s)[u](t) = D(s)[y_m - y](t). \tag{7.152}$$

Replacing λ by s in (7.150) leads to

$$C(s)Q_m(s)P(s) + D(s)Z(s) = A^*(s) \tag{7.153}$$

and operating both sides of (7.153) on $y(t)$ results in

$$C(s)Q_m(s)P(s)[y](t) + D(s)Z(s)[y](t) = A^*(s)[y](t). \tag{7.154}$$

Hence, with (7.129), (7.152), and (7.154), we have

$$\begin{aligned} A^*(s)[y](t) &= C(s)Q_m(s)P(s)[y](t) + Z(s)[D(s)[y_m] - C(s)Q_m(s)[u]](t) \\ &= Z(s)D(s)[y_m](t), \end{aligned} \tag{7.155}$$

which implies that $y(t) \in L^\infty$ because $y_m(t) \in L^\infty$ and $A^*(s)$ is stable.

Operating both sides of (7.153) on $u(t)$ results in

$$C(s)Q_m(s)P(s)[u](t) + D(s)Z(s)[u](t) = A^*(s)[u](t) \tag{7.156}$$

and substituting (7.129) and (7.152) into (7.156) leads to

$$A^*(s)[u](t) = P(s)D(s)[y_m](t), \tag{7.157}$$

which implies that $u(t) \in L^\infty$.

Using (7.132), (7.153), and (7.155), we have

$$\begin{aligned} A^*(s)[y - y_m](t) &= Z(s)D(s)[y_m](t) - A^*(s)[y_m](t) \\ &= -C(s)Q_m(s)P(s)[y_m](t) \\ &= -C(s)P(s)Q_m(s)[y_m](t) = 0, \end{aligned} \tag{7.158}$$

which implies that $\lim_{t\to\infty}(y(t) - y_m(t)) = 0$ exponentially. ∇

When the parameter vectors $p^* \in R^n$ and $z^* \in R^n$ are unknown, we cannot solve (7.150) to implement the pole placement controller (7.151), and we need an adaptive scheme to update the parameters of (7.143) for implementing an adaptive pole placement controller to achieve the control objective.

7.2.3 Controller Parameter Adaptation

Adaptive laws for updating the estimates $\hat{p}$ and $\hat{z}$ of the unknown plant parameter vectors $p^* \in R^n$ and $z^* \in R^n$ are based on the standard gradient or least-squares algorithm developed in Chapter 3 (the gradient algorithm has been used in Section 7.1.2 for the model reference adaptive control case). The key requirement, however, in the adaptive pole placement control case is that the two polynomials $Q_m(\lambda)\hat{P}(\lambda,\hat{p})$ and $\hat{Z}(\lambda,\hat{z})$ must be coprime for each pair of estimates $\hat{p}$ and $\hat{z}$ as functions of the time t. To ensure this, special treatments to the standard adaptive algorithms are needed [151]. With such estimates $\hat{p}$ and $\hat{z}$, that is, with $\hat{P}(\lambda,\hat{p})$ and $\hat{Z}(\lambda,\hat{z})$ in (7.135) and (7.136), we solve (7.137) for $\hat{C}(\lambda,\psi_c)$ and $\hat{D}(\lambda,\psi_d)$ of the form (7.138)–(7.139) to construct $\hat{C}(s,\psi_c)$, $\hat{D}(s,\psi_d)$ for implementing the controller (7.143).

Standard Adaptive Laws

Choosing a stable polynomial $\Lambda_e(s) = s^n + \lambda_{n-1}^e s^{n-1} + \cdots + \lambda_1^e s + \lambda_0^e$ and defining the parameter and regressor vectors

$$\begin{aligned}\theta_p^* &= [z^{*T}, -p^{*T}]^T \\ &= [z_0, z_1, \ldots, z_{n-1}, -p_0, -p_1, \ldots, -p_{n-1}]^T \in R^{2n},\end{aligned} \tag{7.159}$$

$$\begin{aligned}\phi(t) = \Bigg[&\frac{1}{\Lambda_e(s)}[u](t), \frac{s}{\Lambda_e(s)}[u](t), \ldots, \frac{s^{n-1}}{\Lambda_e(s)}[u](t), \\ &\frac{1}{\Lambda_e(s)}[y](t), \frac{s}{\Lambda_e(s)}[y](t), \ldots, \frac{s^{n-1}}{\Lambda_e(s)}[y](t)\Bigg]^T \in R^{2n},\end{aligned} \tag{7.160}$$

as in Section 3.1, we express the plant (7.129) as

$$y(t) - \frac{\Lambda_{n-1}(s)}{\Lambda_e(s)}[y](t) = \theta_p^{*T}\phi(t), \tag{7.161}$$

where $\Lambda_{n-1}(s) = \lambda_{n-1}^e s^{n-1} + \cdots + \lambda_1^e s + \lambda_0^e$ and $\phi(t) = [(\omega_1^e(t))^T, (\omega_2^e(t))^T]^T$ for $\omega_1^e(t)$ and $\omega_2^e(t)$ defined from (7.64) and (7.65).

Denoting $\theta_p(t)$ as the estimate of θ_p^*, we define the estimation error

$$\epsilon(t) = \theta_p^T(t)\phi(t) - y(t) + \frac{\Lambda_{n-1}(s)}{\Lambda_e(s)}[y](t),\ t \geq t_0, \tag{7.162}$$

where $\frac{\Lambda_{n-1}(s)}{\Lambda_e(s)}[y](t)$ is implemented as

$$\frac{\Lambda_{n-1}(s)}{\Lambda_e(s)}[y](t) = \lambda_e^T \omega_2^e(t),\ \lambda_e = [\lambda_0^e, \lambda_1^e, \ldots, \lambda_{n-2}^e, \lambda_{n-1}^e]^T \in R^n. \tag{7.163}$$

Then, to update the parameter estimate $\theta_p(t)$, we use either a gradient algorithm (see Section 3.3) or a least-squares algorithm (see Section 3.4).

Gradient algorithm:

$$\dot{\theta}_p(t) = -\frac{\Gamma\phi(t)\epsilon(t)}{m^2(t)},\ \theta_p(0) = \theta_0,\ t \geq 0, \tag{7.164}$$

where the adaptation gain matrix $\Gamma \in R^{2n\times 2n}$ is such that $\Gamma = \Gamma > 0$, θ_0 is an initial estimate of $\theta_p^* \in R^{n+m+1}$, and

$$m(t) = \sqrt{1 + \kappa\phi^T(t)\phi(t)},\ \kappa > 0. \tag{7.165}$$

As in Lemma 3.1, this gradient algorithm guarantees that $\theta_p(t)$, $\dot{\theta}_p(t)$, $\frac{\epsilon(t)}{m(t)}$ are bounded and $\dot{\theta}_p(t)$, $\frac{\epsilon(t)}{m(t)}$ belong to L^2.

Least-squares algorithm:

$$\dot{\theta}_p(t) = -\frac{P(t)\phi(t)\epsilon(t)}{m^2(t)},\ \theta(0) = \theta_0, \tag{7.166}$$

$$\dot{P}(t) = -\frac{P(t)\phi(t)\phi^T(t)P(t)}{m^2(t)},\ P(0) = P_0 = P_0^T > 0, \tag{7.167}$$

$$m(t) = \sqrt{1 + \kappa\phi^T(t)P(t)\phi(t)},\ \kappa > 0. \tag{7.168}$$

As in Lemma 3.2, this least-squares algorithm guarantees that $P(t)$, $\theta_p(t)$, $\dot{\theta}_p(t)$, $\frac{\epsilon(t)}{m(t)}$ are bounded and $\dot{\theta}_p(t)$, $\frac{\epsilon(t)}{m(t)}$ belong to L^2.

The desired estimates $\hat{p}(t)$ and $\hat{z}(t)$ are obtained from

$$\hat{p}(t) = -[0_{n\times n}, I_n]\theta_p(t),\ \hat{z}(t) = [I_n, 0_{n\times n}]\theta_p(t) \tag{7.169}$$

such that $\hat{p}(t)$, $\dot{\hat{p}}(t)$, $\hat{z}(t)$, $\dot{\hat{z}}(t)$ are bounded and $\dot{\hat{p}}(t)$ and $\dot{\hat{z}}(t)$ belong to L^2.

Singularity Problem

To implement the control law (7.143), we need to solve (7.137) on-line for the polynomials $\hat{C}(\lambda, \psi_c)$, $\hat{D}(\lambda, \psi_d)$ in the form (7.138)–(7.139). Given

$$A^*(\lambda) = \lambda^{2n+n_q-1} + a^*_{2n+n_q-2}\lambda^{2n+n_q-2} + \cdots + a^*_1\lambda + a^*_0,$$

$$Q_m(\lambda)\hat{P}(\lambda, \hat{p}(t)) \triangleq \lambda^{n+n_q} + \hat{\rho}_{n+n_q-1}(t)\lambda^{n+n_q-1} + \cdots + \hat{\rho}_1(t)\lambda + \hat{\rho}_0, \quad (7.170)$$

we introduce the parameter vectors

$$a^* = [1, a^*_{2n+n_q-2}, \ldots, a^*_1, a^*_0]^T \in R^{2n+n_q}, \quad (7.171)$$

$$\hat{\theta}_{qp}(t) = [1, \hat{\rho}_{n+n_q-1}, \hat{\rho}_{n+n_q-2}, \ldots, \hat{\rho}_1, \hat{\rho}_0]^T \in R^{n+n_q}, \quad (7.172)$$

$$\hat{\theta}_z(t) = [\hat{z}_{n-1}, \hat{z}_{n-2}, \ldots, \hat{z}_1, \hat{z}_0]^T \in R^n, \quad (7.173)$$

$$\hat{\theta}_c(t) = [1, \hat{c}_{n-1}, \hat{c}_{n-2}, \ldots, \hat{c}_1, \hat{c}_0]^T \in R^n, \quad (7.174)$$

$$\hat{\theta}_d(t) = [1, \hat{d}_{n+n_q-1}, \hat{d}_{n+n_q-2}, \ldots, \hat{d}_1, \hat{d}_0]^T \in R^{n+n_q}, \quad (7.175)$$

$$\hat{\theta}(t) = [\hat{\theta}_c^T(t), \hat{\theta}_d^T(t)]^T \in R^{2n+n_q} \quad (7.176)$$

and, similar to (5.21), express (7.137) as

$$S(\hat{p}(t), \hat{z}(t))\hat{\theta}(t) = a^*, \quad (7.177)$$

where the Sylvester matrix of $Q_m(\lambda)\hat{P}(\lambda, \hat{p}(t))$ and $\hat{Z}(\lambda, \hat{z}(t))$ is

$$\begin{aligned} &S(\hat{p}(t), \hat{z}(t)) \\ &= \left[\, S_1(\hat{p}(t)) \in R^{(2n+n_q)\times n} \quad \middle| \; S_2(\hat{z}(t)) \in R^{(2n+n_q)\times(n+n_q)} \right] \\ &= \left[\begin{array}{ccccc|ccccc} \hat{\theta}_{qp} & 0 & \cdots & 0 & 0 & \hat{\theta}_z & 0 & \cdots & 0 & 0 \\ 0 & \hat{\theta}_{qp} & \cdots & \vdots & 0 & 0 & \hat{\theta}_z & \cdots & \vdots & 0 \\ \vdots & 0 & \cdots & 0 & \vdots & \vdots & 0 & \cdots & 0 & \vdots \\ 0 & \vdots & \cdots & 0 & 0 & 0 & \vdots & \cdots & 0 & 0 \\ 0 & 0 & \cdots & \hat{\theta}_{qp} & 0 & 0 & 0 & \cdots & \hat{\theta}_z & 0 \\ 0 & 0 & \cdots & 0 & \hat{\theta}_{qp} & 0 & 0 & \cdots & 0 & \hat{\theta}_z \end{array}\right]. \end{aligned} \quad (7.178)$$

To ensure the boundedness of $\hat{\theta}(t)$, $Q_m(\lambda)\hat{P}(\lambda, \hat{p}(t))$ and $\hat{Z}(\lambda, \hat{z}(t))$ are required to be coprime uniformly in t, that is, $|\det[S(\hat{p}(t), \hat{z}(t))]| \geq s_0$ for some constant $s_0 > 0$ and all $t \geq 0$, for $\hat{p}(t)$ and $\hat{z}(t)$ in (7.169), obtained from the adaptive update law (7.164) or (7.166). The singularity of the indirect adaptive pole placement control law occurs when $\det[S(\hat{p}(t), \hat{z}(t))] = 0$, and it is near singularity when $\det[S(\hat{p}(t), \hat{z}(t))]$ is close to 0, and either case should be avoided for implementation of adaptive control.

It has been an important issue in indirect adaptive pole placement control to develop adaptive control schemes which are singularity free, as in general neither the gradient algorithm (7.164) nor the least-squares algorithm (7.166) can ensure a singularity-free adaptive pole placement controller, and certain modification of the standard adaptive parameter update algorithms is needed (see [27], [92], [118], [197], [198], [202], [257], [272], [276]).

There arc two cases in which such a uniform coprimeness condition is ensured, with the use of additional knowledge of the plant parameters z^* and p^* (see (7.159)) in an adaptive law for updating their estimates $\hat{z}(t)$ and $\hat{p}(t)$.

Case I: Since $Q_m(\lambda)\hat{P}(\lambda, p^*) = Q_m(\lambda)P(\lambda)$ and $\hat{Z}(\lambda, z^*)$ are coprime (by assumption), there is a ball in the parameter space, centered at θ_p^*,

$$B_h = \{\theta_p : \|\theta_p - \theta_p^*\|_2 \leq h, h > 0\}, \tag{7.179}$$

such that, for any $\theta_p = [\hat{z}^T, -\hat{p}^T]^T \in B_h$, $Q_m(\lambda)\hat{P}(\lambda, \hat{p})$ and $\hat{Z}(\lambda, \hat{z})$ are coprime. Then the uniform coprimeness condition on $Q_m(\lambda)\hat{P}(\lambda, \hat{p})$ and $\hat{Z}(\lambda, \hat{z})$ is ensured if the parameter estimates are initiated within this ball: $\theta_p(0) \in B_h$ (with $\Gamma = cI$ for the gradient algorithm (7.164), and $P(0) = cI$ for the least-squares algorithm (7.166), for $c > 0$ being a scalar), as verified below.

(i) For the gradient algorithm (7.164), it follows that $\frac{d}{dt}((\theta_p(t)-\theta_p^*)^T(\theta_p(t)-\theta_p^*)) = -\frac{2c\epsilon^2(t)}{m^2(t)} \leq 0$; hence, $\theta_p(0) \in B_h$ implies that $\theta_p(t) \in B_h$, $\forall t \geq 0$.

(ii) For the least-squares algorithm (7.166), $\frac{d}{dt}((\theta_p(t) - \theta_p^*)^T P^{-1}(t)(\theta_p(t) - \theta_p^*)) = -\frac{\epsilon^2(t)}{m^2(t)} \leq 0$, which implies $(\theta_p(t) - \theta_p^*)^T P^{-1}(t)(\theta_p(t) - \theta_p^*) \leq (\theta_p(0) - \theta_p^*)^T P^{-1}(0)(\theta_p(0) - \theta_p^*)$. Since $P^{-1}(t) \geq P^{-1}(0) = cI$, $\forall t \geq 0$ (see Section 3.4), it follows that $(\theta_p(t) - \theta_p^*)^T(\theta_p(t) - \theta_p^*) \leq (\theta_p(0) - \theta_p^*)^T(\theta_p(0) - \theta_p^*)$, $\forall t \geq 0$, that is, if $\theta_p(0) \in B_h$, then $\theta_p(t) \in B_h$, $\forall t \geq 0$.

Case II: Another method is to use projection for parameter adaptation using *a priori* knowledge about the true plant parameters. Let θ_{pi}, θ_{pi}^* be the ith component of θ_p, θ_p^*, $i = 1, \ldots, 2n$. Since $Q_m(\lambda)\hat{P}(\lambda, p^*)$ and $\hat{Z}(\lambda, z^*)$ are coprime, there are constants θ_{pi}^a, θ_{pi}^b, $i = 1, \ldots, 2n$, such that $\theta_{pi}^a \leq \theta_{pi}^* \leq \theta_{pi}^b$, and for each $\theta_{pi} \in [\theta_{pi}^a, \theta_{pi}^b]$, $i = 1, \ldots, 2n$, $Q_m(\lambda)\hat{P}(\lambda, \hat{p})$ and $\hat{Z}(\lambda, \hat{z})$ are coprime. With the knowledge of θ_{pi}^a, θ_{pi}^b, we can modify the gradient algorithm (7.164) with parameter projection to ensure $\theta_{pi} \in [\theta_{pi}^a, \theta_{pi}^b]$ for $i = 1, \ldots, 2n$.

Singularity-free designs. From the above analysis, we see that *a priori* knowledge about the plant parameters for implementing a singularity-free adaptive pole placement controller is characterized by the assumptions:

(A7.4a): B_h defined in (7.179) is known.

(A7.4b): θ_i^a, θ_i^b, $i = 1, \ldots, 2n$, are known.

Design I: With (A7.4a), we choose $\theta_p(0) \in B_h$ to initiate the gradient algorithm (7.164) with $\Gamma = cI$ or the least-squares algorithm (7.166) with $P(0) = cI$ with $c > 0$, which leads to $\theta_p(t) \in B_h$, $\forall t \geq 0$.

Design II: With (A7.4b), we choose $\theta_{pi}(0) \in [\theta_{pi}^a, \theta_{pi}^b]$, $i = 1, \ldots, 2n$ and modify the gradient algorithm (7.164) by parameter projection:

Let $f(t) = [f_1(t), \ldots, f_{2n}(t)]^T = -\frac{\Gamma\phi(t)\epsilon(t)}{m^2(t)}$ with $\Gamma = \text{diag}\{\gamma_1, \ldots, \gamma_{2n}\} > 0$ and choose the adaptive law for $\theta(t)$ as

$$\dot{\theta}(t) = g(t),\ g(t) = [g_1(t), \ldots, g_{2n}(t)]^T \in R^{2n}, \tag{7.180}$$

$$g_i(t) = \begin{cases} f_i(t) & \text{if } \theta_i(t) \in (\theta_i^a, \theta_i^b) \text{ or} \\ & \text{if } \theta_i(t) = \theta_i^a \text{ and } f_i(t) \geq 0 \text{ or} \\ & \text{if } \theta_i(t) = \theta_i^b \text{ and } f_i(t) \leq 0, \\ 0 & \text{otherwise.} \end{cases} \tag{7.181}$$

This modified adaptive law ensures that $\theta_{pi}(t) \in [\theta_{pi}^a, \theta_{pi}^b]$, $\forall t \geq 0$, $i = 1, \ldots, 2n$, in addition to the properties of the standard gradient algorithm (7.164). For the least-squares algorithm, such a modification can also be done [151], which is more complicated than (7.181). Note that the projection algorithm (7.181) also ensures the desired property that $(\theta_{pi} - \theta_{pi}^*)(g_i - f_i) \leq 0$ for $i = 1, \ldots, 2n$.

With these designs, $Q_m(\lambda)\hat{P}(\lambda, \hat{p})$ and $\hat{Z}(\lambda, \hat{z})$ are ensured to be coprime uniformly in t for $\hat{p} = \hat{p}(t)$ and $\hat{z} = \hat{z}(t)$ from $\theta_p(t) = [\hat{z}^T(t), -\hat{p}^T(t)]^T \in R^{2n}$.

Stability Properties

The developed indirect adaptive pole placement control scheme has the following desired stability and tracking properties.

Theorem 7.3 *Under the singularity-free condition, all signals in the closed-loop system are bounded and* $\lim_{t\to\infty}(y(t) - y_m(t)) = 0$.

Proof: For simplicity of analysis, we let $(s + a_0)\Lambda_c(s) = \Lambda_e(s)\Lambda_0(s)$ for $\Lambda_c(s)$ in (7.143), $\Lambda_e(s)$ in (7.160), some monic polynomial $\Lambda_0(s)$ of degree n_q, and some $a_0 > 0$. For $a_1(s)$ in (7.149), similar to that in [151], we define

$$\omega(t) = \left[\frac{a_1^T(s)}{(s + a_0)\Lambda_c(s)}[u](t), \frac{a_1^T(s)}{(s + a_0)\Lambda_c(s)}[y](t), \frac{1}{s + a_0}[y](t)\right]^T. \tag{7.182}$$

From (7.135), (7.136), and (7.162), we have

$$\frac{s^n}{\Lambda_e(s)}[y](t) = \theta_p^T(t)\phi(t) - \epsilon(t), \tag{7.183}$$

that is, $\hat{P}(s,\hat{p})\frac{1}{\Lambda_e(s)}[y](t) = \hat{Z}(s,\hat{z})\frac{1}{\Lambda_e(s)}[u](t) - \epsilon(t)$, or

$$\hat{P}(s,\hat{p})\frac{\Lambda_0(s)}{(s+a_0)\Lambda_c(s)}[y](t) = \hat{Z}(s,\hat{z})\frac{\Lambda_0(s)}{(s+a_0)\Lambda_c(s)}[u](t) - \epsilon(t) \tag{7.184}$$

for $(s+a_0)\Lambda_c(s) = \Lambda_e(s)\Lambda_0(s)$, which can be rewritten as

$$\frac{s^{n+n_q}}{(s+a_0)\Lambda_c(s)}[y](t) = \bar{\theta}_p^T(t)\omega(t) - \epsilon(t) \tag{7.185}$$

for some $\bar{\theta}_p(t) \in R^{2(n+n_q)-1}$. From (7.149), we express the control input as

$$u(t) = \psi^T(t)\bar{\omega}(t) + \bar{y}_m(t), \tag{7.186}$$

where $\bar{y}_m(t) = \hat{D}(s,\psi_d)\frac{1}{\Lambda_c(s)}[y_m](t)$ is bounded, and

$$\psi(t) = [\psi_1^T(t), \psi_2^T(t), \psi_3(t)]^T \in R^{2(n+n_q)-1}, \tag{7.187}$$

$$\bar{\omega}(t) = \left[\frac{a_1^T(s)}{\Lambda_c(s)}[u](t), \frac{a_1^T(s)}{\Lambda_c(s)}[y](t), y(t)\right]^T \in R^{2(n+n_q)-1} \tag{7.188}$$

such that $\omega(t) = \frac{1}{s+a_0}[\bar{\omega}](t) \in R^{2n+2n_q-1}$. It can be verified (see (5.83)) that

$$\frac{1}{s+a_0}[\psi^T\bar{\omega}](t) = \psi^T(t)\omega(t) - \frac{1}{s+a_0}[\dot{\psi}^T\omega](t). \tag{7.189}$$

Introducing $\lambda_c = [\lambda_0^c, \lambda_1^c, \ldots, \lambda_{n+n_q-2}^c]^T \in R^{n+n_q-1}$ for $\Lambda_c(s) = s^{n+n_q-1} + \lambda_{n+n_q-2}^c s^{n+n_q-2} + \cdots + \lambda_1^c s + \lambda_0^c$, we express

$$\frac{s^{n+n_q-1}}{(s+a_0)\Lambda_c(s)}[x](t) = \frac{1}{s+a_0}[x](t) - \frac{\lambda_c^T a_1(s)}{(s+a_0)\Lambda_c(s)}[x](t) \tag{7.190}$$

for $x(t) = u(t)$ or $y(t)$. Similarly, we have

$$\begin{aligned}\frac{s^{n+n_q}}{(s+a_0)\Lambda_c(s)}[y](t) &= \frac{s}{s+a_0}[y](t) - \frac{\lambda_c^T s a_1(s)}{(s+a_0)\Lambda_c(s)}[y](t) \\ &= \frac{s}{s+a_0}[y](t) - \frac{\bar{\lambda}_c^T a_1(s)}{(s+a_0)\Lambda_c(s)}[y](t) - \frac{\bar{\lambda}_0}{s+a_0}[y](t) \quad (7.191)\end{aligned}$$

for some constant $\bar{\lambda}_c \in R^{n+n_q-1}$ and $\bar{\lambda}_0 \in R$.

Using (7.185) and (7.188)–(7.191) with $x(t) = u(t)$ or $y(t)$, we have

$$\frac{s^{n+n_q-1}}{(s+a_0)\Lambda_c(s)}[u](t) = \left(\psi^T(t) - [\lambda_c, 0_{1\times(n+n_q)}]\right)\omega(t) + \frac{1}{s+a_0}[\bar{y}_m - \dot{\psi}^T\omega](t),$$

$$\frac{s^{n+n_q-1}}{(s+a_0)\Lambda_c(s)}[y](t) = [-\lambda_c^T, 1]\left[\frac{a_1^T(s)}{(s+a_0)\Lambda_c(s)}[y](t), \frac{1}{s+a_0}[y](t)\right]^T,$$

$$\frac{s}{s+a_0}[y](t) = \left(\hat{\theta}_p^T(t) + [0_{n+n_q-1}, \bar{\lambda}_c^T, \bar{\lambda}_0]\right)\omega(t) - \epsilon(t). \tag{7.192}$$

Introducing $\omega_0(t) = \frac{1}{s+a_0}[\dot{\psi}^T\omega](t)$, we have

$$\dot{\omega}_0(t) = -a_0\omega_0(t) + \dot{\psi}^T(t)\omega(t). \tag{7.193}$$

Letting $a_2(s) = [1, s, s^2, \ldots, s^{n-1}]^T$, from (7.160), we have

$$\begin{aligned}\phi(t) &= \left[\frac{a_2^T(s)}{\Lambda_e(s)}[u](t), \frac{a_2^T(s)}{\Lambda_e(s)}[y](t)\right]^T \\ &= \left[\frac{\Lambda_0(s)a_2^T(s)}{(s+a_0)\Lambda_c(s)}[u](t), \frac{\Lambda_0(s)a_2^T(s)}{(s+a_0)\Lambda_c(s)}[y](t)\right]^T.\end{aligned} \tag{7.194}$$

Denoting $\Lambda_0(s) = s^{n_q} + \lambda_{n_q-1}^0 s^{n_q-1} + \cdots + \lambda_1^0 s + \lambda_0^0$, we have

$$\frac{\Lambda_0(s)a_2(s)}{(s+a_0)\Lambda_c(s)} = M_0\frac{a_1(s)}{(s+a_0)\Lambda_c(s)} + B_0\frac{s^{n+n_q-1}}{(s+a_0)\Lambda_c(s)}, \tag{7.195}$$

where $a_1(s) = [1, s, \ldots, s^{n_q+n-2}]^T$ as before,

$$M_0 = \begin{bmatrix} \lambda_0^0 & \lambda_1^0 & \cdots & \lambda_{n_q-1}^0 & 0 & 0 & \cdots & 0 \\ 0 & \lambda_0^0 & \lambda_1^0 & \cdots & \lambda_{n_q-1}^0 & 0 & \cdots & 0 \\ \vdots & \vdots & \vdots & \vdots & \vdots & \vdots & \vdots & \vdots \\ 0 & \cdots & 0 & \lambda_0^0 & \lambda_1^0 & \cdots & \lambda_{n_q-1}^0 & 0 \\ 0 & 0 & 0 & \cdots & 0 & \lambda_0^0 & \cdots & \lambda_{n_q-2}^0 \end{bmatrix}, \tag{7.196}$$

and $B_0 = [0, 0, \ldots, 0, \lambda_{n_q-1}^0]^T \in R^n$. From (7.192), (7.194), and (7.195), we can express $\phi(t)$ in terms of $\omega(t)$ in (7.182) and $\omega_0(t)$ in (7.193):

$$\begin{aligned}\phi(t) &= A_\phi\omega(t) + [B_0^T, 0, \ldots, 0]^T\frac{1}{s+a_0}[\bar{y}_m - \dot{\psi}^T\omega](t) \\ &= A_\phi\omega(t) + [B_0^T, 0, \ldots, 0]^T\left(\frac{1}{s+a_0}[\bar{y}_m](t) - \omega_0(t)\right),\end{aligned} \tag{7.197}$$

where

$$A_\phi = \begin{bmatrix} [M_0, 0_{n\times(n+n_q)}] + B_0\left(\psi^T(t) - [\lambda_c, 0_{1\times(n+n_q)}]\right) \\ [0_{n\times(n+n_q-1)}, M_0, 0_{n\times 1}] + B_0[0_{1\times(n+n_q-1)}, -\lambda_c^T, 1] \end{bmatrix}. \tag{7.198}$$

From (7.182), (7.192), and (7.193), we express the closed-loop system as

$$\dot{\omega}(t) = A(t)\omega(t) + B_1\omega_0(t) - B_1\frac{1}{s+a_0}[\bar{y}_m](t) + B_2\epsilon(t),$$
$$\dot{\omega}_0(t) = -a_0\omega_0(t) + \dot{\psi}^T(t)\omega(t), \tag{7.199}$$

where $B_1 = [0_{1\times(n+n_q-2)}, -1, 0_{1\times(n+n_q)}]^T$, $B_2 = [0_{1\times(2(n+n_q)-2)}, -1]^T$, and

$$A(t) = \begin{bmatrix} \begin{matrix}[0_{(n+n_q-2)\times 1}, I_{n+n_q-2}] \\ \psi_1^T(t) - \lambda_c^T\end{matrix} & \begin{matrix}0_{(n+n_q-2)\times(n+n_q)} \\ [\psi_2^T(t), \psi_3(t)]\end{matrix} \\ \begin{matrix}0_{(n+n_q-2)\times(n+n_q-1)} \\ 0_{1\times(n+n_q-1)}\end{matrix} & \begin{matrix}[0_{(n+n_q-2)\times 1}, I_{n+n_q-2}] \\ [-\lambda_c^T, 1]\end{matrix} \\ \multicolumn{2}{c}{\hat{\theta}_p^T(t) + [0_{n+n_q-1}, \bar{\lambda}_c^T, \bar{\lambda}_0]} \end{bmatrix}. \tag{7.200}$$

From the adaptive law (7.164) or (7.166), it follows that $\theta_p(t) \in L^\infty$, $\frac{\epsilon(t)}{\sqrt{1+\phi^T(t)\phi(t)}} \in L^\infty \cap L^2$, and $\dot{\theta}_p(t) \in L^\infty \cap L^2$. Under the singularity-free condition for (7.177) (that is, $|\det[S(\hat{p}(t), \hat{z}(t))]| \geq s_0$ for some constant $s_0 > 0$ and all $t \geq 0$), $\hat{\theta}(t)$ in (7.177) and $\psi(t)$ in (7.149) and (7.187) have the desired properties that $\hat{\theta}(t) \in L^\infty$, $\psi(t) \in L^\infty$ and $\dot{\hat{\theta}}(t) \in L^\infty \cap L^2$, $\dot{\psi}(t) \in L^\infty \cap L^2$, as $\hat{\theta}(t) = S^{-1}(\hat{p}(t), \hat{z}(t))a^*$ and $\dot{\hat{\theta}}(t) = -S^{-1}(\hat{p}(t), \hat{z}(t))\dot{S}(\hat{p}(t), \hat{z}(t))S^{-1}(\hat{p}(t), \hat{z}(t))a^*$. So, we have $A(t) \in L^\infty$, $\dot{A}(t) \in L^\infty \cap L^2$, and $A_\phi(t) \in L^\infty$ in (7.197).

It can be verified that $\det[\lambda I - A(t)] = \hat{C}(\lambda, \psi_c)Q_m(\lambda)\hat{P}(\lambda, \hat{p})\Lambda_0(\lambda) + \hat{D}(\lambda, \psi_d)\hat{Z}(\lambda, \hat{z})\Lambda_0(\lambda) = A^*(\lambda)\Lambda_0(\lambda)$ at each time t (see (7.137)), that is, all eigenvalues of $A(t)$ have real parts less than $-\sigma_0$ for some constant $\sigma_0 > 0$ and all $t \geq 0$ [151]. This condition plus the properties that $A(t) \in L^\infty$ and $\dot{A}(t) \in L^\infty \cap L^2$ ensure that $\dot{\omega}(t) = A(t)\omega(t)$ is exponentailly stable as a special case of the general result established in the following lemma.

Lemma 7.1 *For a linear system $\dot{z}(t) = A(t)z(t)$, $z \in R^n$, where $A(t) \in R^{n\times n}$ is differentiable and bounded, and all eigenvalues of $A(t)$ have real parts less than $-\sigma_0$ for some constant $\sigma_0 > 0$ and all $t \geq 0$, its equilibrium state $z_e = 0$ is exponentially stable if*

$$\int_t^{t+T} \|\dot{A}(\tau)\|^p \, d\tau \leq \nu T + \gamma_0 \tag{7.201}$$

for any $t \geq 0$, any $T > 0$, some $\nu > 0$ small, and some $\gamma_0 > 0$, $p \in [1, \infty)$.

Proof: The $p = 1$ case is proved in [151], [426], while the $p = 2$ case is proved in [151]. The proof here for the $p \in (1, \infty)$ case consists of two steps: (i) the existence of a bounded and uniformly positive definite matrix $P(t) = P^T(t) \in R^{n\times n}$ (as defined in Section 2.4.4) satisfying

$$P(t)A(t) + A^T(t)P(t) = -I \tag{7.202}$$

and having its eigenvalues all in some interval $[\lambda^a, \lambda^b]$ for some constants $\lambda^b > \lambda^a > 0$ (this follows from a known linear algebra result [44] that there is a unique $P(t) > 0$ to satisfy (7.202) for any $A(t) \in R^{n\times n}$ whose eigenvalues all have negative real parts at each time t), and (ii) the derivation of

$$\int_0^t \|\dot{P}(\tau)\|_2 \, d\tau \leq \frac{\alpha}{2\beta} t + \frac{\gamma_0 a_0^p (2\beta)^{p-1}}{\alpha^{p-1}}, \ \forall t > 0, \tag{7.203}$$

using $\|\dot{P}(t)\|_2 \leq a_0 \|\dot{A}(t)\|$ for some $a_0 > 0$ (this is from the result that

$$\dot{P}(t)A(t) + A^T(t)\dot{P}(t) = -P(t)\dot{A}(t) - \dot{A}^T(t)P(t) \tag{7.204}$$

has a unique solution $\dot{P}(t)$ if all eigenvalues of $A(t)$ have negative real parts, that is, elements of $\dot{P}(t)$ are linear in elements of $\dot{A}(t)$, the Hölder inequality (see Section 2.3.2), and the analysis (2.119)–(2.120) applied to the version of (7.201) for $\|\dot{P}(t)\|_2$ with $\nu \in (0, \left(\frac{\alpha}{2a_0\beta}\right)^p]$, where $\alpha = \frac{1}{\lambda^b}$ and $\beta = \frac{1}{\lambda^a}$). Then, for the positive definite function $V(z,t) = z^T P(t) z$, we have

$$\dot{V} = -z^T z + z^T \dot{P}(t) z \leq -\left(\alpha - \beta \|\dot{P}(t)\|_2\right) V, \tag{7.205}$$

from which we obtain

$$V(z(t),t) \leq V(z(0),0) e^{-\alpha t + \beta \int_0^t \|\dot{P}(\tau)\|_2 \, d\tau}. \tag{7.206}$$

From (7.203) and (7.206), we have

$$V(z(t),t) \leq V(z(0),0) e^{-\frac{\alpha}{2}t} e^{\frac{\beta\gamma_0 a_0^p (2\beta)^{p-1}}{\alpha^{p-1}}}, \tag{7.207}$$

which implies exponential stability of the equilibrium state $z_e = 0$.

For $p = 1$, (7.203) can also be used, with $\alpha^0 = 1$. In fact, exponential stability is ensured for a larger range of ν: $\nu \in (0, \left(\frac{\alpha}{a_0\beta}\right)^p)$, for which we have $V(z(t),t) \leq V(z(0),0) e^{-(\alpha - \beta a_0 \nu^{\frac{1}{p}})t} e^{\frac{\beta\gamma_0 a_0^p \beta^{p-1}}{\alpha^{p-1}}}$, where $\alpha - \beta a_0 \nu^{\frac{1}{p}} > 0$. ∇

To proceed, we note that in our case $\nu = 0$ and $p = 2$, and $\dot{\omega}(t) = A(t)\omega(t)$ is indeed exponentailly stable. Moreover, we see that the system

$$\begin{aligned} &\dot{\Omega}(t) = A_\Omega(t)\Omega(t), \ \Omega(t) = [\omega^T(t), \omega_0(t)]^T, \\ &A_\Omega(t) = \begin{bmatrix} A(t) & B_1 \\ 0_{1\times(2(n+n_q)-1)} & -a_0 \end{bmatrix} \end{aligned} \tag{7.208}$$

is exponentially stable, and so is the system

$$\dot{\Omega}(t) = A_\Omega(t)\Omega(t) + \left([\dot{\psi}^T(t), 0] + \bar{\epsilon}(t)[A_\phi(t), -[B_0^T, 0, \ldots, 0]]\right)\Omega(t) \quad (7.209)$$

for $[B_0^T, 0, \ldots, 0]$ in (7.197) and $\bar{\epsilon}(t) = \frac{\epsilon(t)}{\sqrt{1+\phi^T(t)\phi(t)}}$, as $\bar{\epsilon}(t) \in L^\infty \cap L^2$ and $\dot{\psi}(t) \in L^\infty \cap L^2$ (see Problem 2.14). Finally, in view of the boundedness of $\bar{y}_m(t)$, the exponential stability of (7.204), and the inequality

$$|\epsilon(t)| \leq \frac{|\epsilon(t)|}{\sqrt{1+\phi^T(t)\phi(t)}}(1 + \|\phi(t)\|_2) \quad (7.210)$$

for $\|\phi(t)\|_2 = \sqrt{\phi^T(t)\phi(t)}$, we conclude that $\omega(t)$ and $\omega_0(t)$ are bounded. Then, we have from (7.197) that $\phi(t) \in L^\infty$, from (7.161) and (7.162) that $\epsilon(t) = (\theta_p(t) - \theta_p^*)^T\phi(t) \in L^\infty$, from (7.162) that $y(t) \in L^\infty$, from (7.199) that $\dot{\omega}(t) \in L^\infty$ so that $\bar{\omega}(t) = \dot{\omega}(t) + a_0\omega(t) \in L^\infty$, and from (7.186) that $u(t) \in L^\infty$, that is, all closed-loop signals are bounded. From $\dot{\epsilon}(t) = \dot{\theta}_p^T(t)\phi(t) + (\theta_p(t) - \theta_p^*)^T\dot{\phi}(t)$ with $\dot{\phi}(t) \in L^\infty$ (as from (7.160) and the boundedness of $u(t)$ and $y(t)$), we have $\dot{\epsilon}(t) \in L^\infty$. Since $\epsilon(t) \in L^\infty \cap L^2$ (as $\bar{\epsilon}(t) = \frac{\epsilon(t)}{\sqrt{1+\phi^T(t)\phi(t)}}$ does and $\phi(t) \in L^\infty$), it follows from Lemma 2.14 that $\lim_{t\to\infty} \epsilon(t) = 0$, which implies that $\lim_{t\to\infty} \dot{\theta}_p(t) = 0$ and, similarly, that $\lim_{t\to\infty} \dot{\psi}(t) = 0$.

To prove that $\lim_{t\to\infty}(y(t) - y_m(t)) = 0$, we define the composite polynomial operators $\hat{A}\hat{B}$ and $\hat{A} \circ \hat{B}$ of two time-varying polynomial operators $\hat{A}(s,t) = \sum_{i=0}^n \hat{a}_i(t)s^i$ and $\hat{B}(s,t) = \sum_{j=0}^m \hat{b}_j(t)s^j$, characterized by the operations

$$\hat{A}\hat{B}[x](t) \triangleq \sum_{i=0}^{n} \hat{a}_i(t) \sum_{j=0}^{m} \hat{b}_j(t)s^{i+j}[x](t), \quad (7.211)$$

$$\hat{A} \circ \hat{B}[x](t) \triangleq \hat{A}[\hat{B}[x]](t) = \sum_{i=0}^{n} \hat{a}_i(t) \sum_{j=0}^{m} s^i \left[\hat{b}_j s^j[x]\right](t), \quad (7.212)$$

where $s^k[x](t) = \frac{d^k}{dt^k}x(t)$, $k = 1, 2, \ldots$, and $s[xy](t) = s[x](t)y(t) + x(t)s[y](t)$. By definition, $\hat{A}(s,t)\hat{B}(s,t) = \hat{B}(s,t)\hat{A}(s,t)$ and, in general, $\hat{A}(s,t) \circ \hat{B}(s,t) \neq \hat{B}(s,t) \circ \hat{A}(s,t)$ [151]. However, if $\hat{B}(s,t) = B(s) = \sum_{j=0}^m b_j s^j$ is a polynomial with constant coeffficients b_j, then $\hat{A}(s,t) \circ B(s) = \hat{A}(s,t)B(s)$, and $B(s)\hat{A}(s,t) = \hat{A}(s,t)B(s) = \hat{A}(s,t) \circ B(s)$, that is,

$$\hat{A} \circ B[x](t) \triangleq \hat{A}[B[x]](t) = \hat{A}B[x](t) = B\hat{A}[x](t). \quad (7.213)$$

Then, using $\hat{C}(s,\psi_c)Q_m(s)\hat{P}(s,\hat{p}) + \hat{D}(s,\psi_d)\hat{Z}(s,\hat{z}) = A^*(s)$ (the operator version of the Diophantine equation (7.137)), we have

$$A^*[\Lambda_c^{-1}[y - y_m]](t) = (\hat{C}Q_m\hat{P} + \hat{D}\hat{Z})[\Lambda_c^{-1}[y]](t) - A^*[\Lambda_c^{-1}[y_m]](t), \quad (7.214)$$

$$A^*[\Lambda_c^{-1}[y_m]](t) = (\hat{C}Q_m\hat{P} + \hat{D}\hat{Z})[\Lambda_c^{-1}[y_m]](t). \tag{7.215}$$

From the control law (7.143), we have

$$\hat{C}Q_m[\Lambda_c^{-1}[u]](t) + \hat{D}[\Lambda_c^{-1}[y]](t) = \hat{D}[\Lambda_c^{-1}[y_m]](t). \tag{7.216}$$

Operating both sides of (7.216) by $\hat{Z} = \hat{Z}(s, \hat{z})$, we obtain

$$\hat{Z} \circ \hat{C}Q_m[\Lambda_c^{-1}[u]](t) + \hat{Z} \circ \hat{D}[\Lambda_c^{-1}[y]](t) = \hat{Z} \circ \hat{D}[\Lambda_c^{-1}[y_m]](t). \tag{7.217}$$

From (7.132), $Q_m(s)[y_m](t) = 0$, we have $Q_m(s)[\Lambda_c^{-1}(s)[y_m]](t) = 0$, and

$$\hat{C}Q_m\hat{P}[\Lambda_c^{-1}[y_m]](t) = \hat{C}\hat{P}Q_m[\Lambda_c^{-1}[y_m]](t) = \hat{C}\hat{P}[Q_m[\Lambda_c^{-1}[y_m]]](t) = 0, \tag{7.218}$$

as from (7.213). Using (7.215), (7.217), and (7.218) in (7.214), we obtain

$$\begin{aligned}
&A^*[\Lambda_c^{-1}[y - y_m]](t)\\
&= \hat{C}Q_m\hat{P}[\Lambda_c^{-1}[y]](t) + \hat{D}\hat{Z}[\Lambda_c^{-1}[y]](t) - \hat{D}\hat{Z}[\Lambda_c^{-1}[y_m]](t)\\
&= \hat{C}Q_m\hat{P}[\Lambda_c^{-1}[y]](t) + \hat{D}\hat{Z}[\Lambda_c^{-1}[y]](t) - \hat{D}\hat{Z}[\Lambda_c^{-1}[y_m]](t)\\
&\quad + \hat{Z} \circ \hat{D}[\Lambda_c^{-1}[y_m]](t) - \hat{Z} \circ \hat{C}Q_m[\Lambda_c^{-1}[u]](t) - \hat{Z} \circ \hat{D}[\Lambda_c^{-1}[y]](t)\\
&= \left(\hat{C}Q_m\hat{P}[\Lambda_c^{-1}[y]](t) - \hat{C}Q_m\hat{Z}[\Lambda_c^{-1}[u]](t)\right)\\
&\quad + \left(\hat{D}\hat{Z}[\Lambda_c^{-1}[y]](t) - \hat{Z} \circ \hat{D}[\Lambda_c^{-1}[y]](t)\right)\\
&\quad + \left(\hat{Z} \circ \hat{D}[\Lambda_c^{-1}[y_m]](t) - \hat{D}\hat{Z}[\Lambda_c^{-1}[y_m]](t)\right)\\
&\quad + \left(\hat{C}Q_m\hat{Z}[\Lambda_c^{-1}[u]](t) - \hat{Z} \circ \hat{C}Q_m[\Lambda_c^{-1}[u]](t)\right).
\end{aligned} \tag{7.219}$$

From the properties that $u(t) \in L^\infty$, $y(t) \in L^\infty$, $y_m(t) \in L^\infty$, $\psi(t) \in L^\infty$, $\theta_p(t) \in L^\infty$, $\dot{\psi}(t) \in L^2 \cap L^\infty$, $\dot{\theta}_p(t) \in L^2 \cap L^\infty$, $\epsilon(t) \in L^2 \cap L^\infty$, $\lim_{t\to\infty} \epsilon(t) = 0$, $\lim_{t\to\infty} \dot{\theta}_p(t) = 0$, and $\lim_{t\to\infty} \dot{\psi}(t) = 0$, it can be shown that

$$\begin{aligned}
&\lim_{t\to\infty} A^{*-1}[\hat{C}Q_m\hat{P}[\Lambda_c^{-1}[y]] - \hat{C}Q_m\hat{Z}[\Lambda_c^{-1}[u]]](t)\\
&= \lim_{t\to\infty} A^{*-1}[\hat{D}\hat{Z}[\Lambda_c^{-1}[y]] - \hat{Z} \circ \hat{D}[\Lambda_c^{-1}[y]]](t)\\
&= \lim_{t\to\infty} A^{*-1}[\hat{Z} \circ \hat{D}[\Lambda_c^{-1}[y_m]] - \hat{D}\hat{Z}[\Lambda_c^{-1}[y_m]]](t)\\
&= \lim_{t\to\infty} A^{*-1}[\hat{C}Q_m\hat{Z}[\Lambda_c^{-1}[u]] - \hat{Z} \circ \hat{C}Q_m[\Lambda_c^{-1}[u]]](t) = 0,
\end{aligned} \tag{7.220}$$

that is, $\lim_{t\to\infty} \Lambda_c^{-1}(s)[y - y_m](t) = 0$. In addition, from (7.161), we have $\dot{y}(t) \in L^\infty$, and from (7.132), it follows that $\dot{y}_m(t) \in L^\infty$.

Defining the fictitious filters $K_1(s)$ and $K(s)$ from

$$sK_1(s) = 1 - K(s),\ K(s) = \frac{a^{n+n_q-1}}{(s+a)^{n+n_q-1}}, \tag{7.221}$$

where $a > 0$ is a constant, for $\partial\Lambda_c(s) = n + n_q - 1$, we have the identity

$$y(t) - y_m(t) = K_1(s)s[y - y_m](t) + K(s)\Lambda_c(s)[\Lambda_c^{-1}(s)[y - y_m]](t). \quad (7.222)$$

Recall from Proposition 2.10 that $K_1(s)$ is stable and strictly proper and its impulse response $k_1(t)$ satisfies $\|k_1(\cdot)\|_1 = \int_0^\infty |k_1(t)|\,dt = \frac{n+n_q-1}{a}$. It follows from (7.221), (7.222), and Proposition 5.1 (see Section 5.7) that

$$|y(t) - y_m(t)| \leq \frac{c_0}{a} + a^{n+n_q-1}c_1(t),\ t \geq 0, \quad (7.223)$$

for some constant $c_0 > 0$ (as $\dot{y}(t) - \dot{y}_m(t) \in L^\infty$) and some function $c_1(t) \in L_\infty$ such that $\lim_{t\to\infty} c_1(t) = 0$ (as $\lim_{t\to\infty} \Lambda_c^{-1}(s)[y - y_m](t) = 0$). Since $a > 0$ is a fictitious parameter, we can make it large to show that $y(t) - y_m(t)$ is small for large t. More precisely, for any given $\delta > 0$, we let $a = \frac{2c_0}{\delta}$. Then, there is a $T > 0$ such that $c_1(t) < \frac{\delta}{2a^{n+n_q-1}} = \frac{\delta^{n+n_q}}{2(2c_0)^{n+n_q-1}}$ for any $t > T$, that is, $|y(t) - y_m(t)| < \frac{\delta}{2} + \frac{\delta}{2} = \delta$ for any $t > T$. Hence, we can conclude that $\lim_{t\to\infty}(y(t) - y_m(t)) = 0$. This completes the proof of Theorem 7.3. ∇

Such an adaptive or nonadaptive pole placement control design uses the knowledge of $Q_m(s)$ which defines the reference output $y_m(t)$ (see (7.132)), and it is also called an internal model principle design [125], [151].

7.3 Discrete-Time Adaptive Control Systems

The discrete-time counterparts of the indirect model reference adaptive control scheme developed in Section 7.1 and the indirect adaptive pole placement control scheme developed in Section 7.2 can be readily derived by combining a similar controller structure and a discrete-time adaptive parameter estimator for a linear time-invariant plant modeled in discrete time.

7.3.1 Model Reference Designs

Design Example

Consider a first-order discrete-time time-invariant plant

$$y(t+1) = a_p y(t) + b_p u(t) \quad (7.224)$$

with unknown parameters a_p and b_p. The control objective is to find a feedback control input $u(t)$ which ensures that all closed-loop signals are bounded, and the plant output $y(t)$ tracks the output $y_m(t)$ of a reference model system

$$y_m(t+1) = a_m y_m(t) + b_m r(t), \quad (7.225)$$

where $|a_m| < 1$ and $r(t)$ is a bounded reference input.

Let $\hat{a}_p(t)$ and $\hat{b}_p(t)$ be the adaptive estimates of the unknown plant parameters a_p and b_p, respectively, with $|\hat{b}_p(t)| \geq b_0 > 0$, $\forall t \geq 0$, for some constant $b_0 > 0$. Then, an indirect adaptive controller is

$$u(t) = k_1(t)y(t) + k_2(t)r(t), \tag{7.226}$$

where

$$k_1(t) = \frac{-\hat{a}_p(t) - a_m}{\hat{b}_p(t)}, \; k_2(t) = \frac{b_m}{\hat{b}_p(t)}. \tag{7.227}$$

To generate such desired parameter estimates, we choose a stable filter $\frac{1}{z+a_f}$ with $|a_f| < 1$ and express the plant (7.224) as

$$y(t+1) = -a_f y(t) + (a_f + a_p)y(t) + b_p u(t), \tag{7.228}$$

which can be further written as

$$y(t) = (a_f + a_p)\frac{1}{z+a_f}[y](t) + b_p \frac{1}{z+a_f}[u](t) = \theta_p^{*T}\phi(t), \tag{7.229}$$

where

$$\theta_p^* = [a_f + a_p, b_p]^T, \tag{7.230}$$

$$\begin{aligned} \phi(t) &= [\phi_1(t), \phi_2(t)]^T = [y_f(t), u_f(t)]^T \\ &= \left[\frac{1}{z+a_f}[y](t), \frac{1}{z+a_f}[u](t)\right]^T. \end{aligned} \tag{7.231}$$

Letting $\theta_p(t)$ be the estimate of the parameter vector θ_p^* in (7.230), that is, $\theta_p(t) = [\theta_{p1}(t), \theta_{p2}(t)]^T = [a_f + \hat{a}_p, \hat{b}_p]^T$, we introduce the estimation error

$$\epsilon(t) = \theta_p^T(t)\phi(t) - y(t). \tag{7.232}$$

It follows from (7.230) and (7.232) that

$$\epsilon(t) = \tilde{\theta}_p^T(t)\phi(t), \; \tilde{\theta}_p(t) = \theta_p(t) - \theta_p^*. \tag{7.233}$$

Based on this error equation, we choose the adaptive law for $\theta_p(t)$ as

$$\theta_p(t+1) = \theta_p(t) - \frac{\Gamma\epsilon(t)\phi(t)}{m^2(t)} + f(t), \tag{7.234}$$

where $\Gamma = \text{diag}\{\gamma_1, \gamma_2\}$ with $0 < \gamma_1 < 2$ and $0 < \gamma_2 < 2$ and

$$m(t) = \sqrt{1 + \phi^T(t)\phi(t)}, \tag{7.235}$$

$$f(t) = [0, f_2(t)]^T. \tag{7.236}$$

To choose $f_2(t)$, we initialize $\theta_{p2}(0)$ to meet the condition $\text{sign}[b_p]\theta_{p2}(0) \geq b_0$, where $\text{sign}[b_p]$ and $b_0 > 0$ are known such that $\text{sign}[b_p]b_p \geq b_0$, and let

$$\bar{\theta}_{p2}(t) = \theta_{p2}(t) + g_2(t),\ g_2(t) = -\frac{\gamma_2 \epsilon(t)\phi_2(t)}{m^2(t)} \tag{7.237}$$

and set

$$f_2(t) = \begin{cases} 0 & \text{if } \text{sign}[b_p]\bar{\theta}_{p2}(t) \geq b_0, \\ b_0\, \text{sign}[b_p] - \bar{\theta}_{p2}(t) & \text{if } \text{sign}[b_p]\bar{\theta}_{p2}(t) < b_0. \end{cases} \tag{7.238}$$

The time increment of the parameter error measure $V(\tilde{\theta}_p) = \tilde{\theta}_p^T \Gamma^{-1} \tilde{\theta}_p$, along the trajectories of (7.234), is

$$\begin{aligned} & V(\tilde{\theta}_p(t+1)) - V(\tilde{\theta}_p(t)) \\ = & -\left(2 - \frac{\phi^T(t)\Gamma\phi(t)}{m^2(t)}\right)\frac{\epsilon^2(t)}{m^2(t)} \\ & + 2f_2(t)\gamma_2^{-1}(\tilde{\theta}_{p2}(t) + g_2(t) + f_2(t)) - f_2^2(t)\gamma_2^{-1} \\ \leq & -\left(2 - \frac{\phi^T(t)\Gamma\phi(t)}{m^2(t)}\right)\frac{\epsilon^2(t)}{m^2(t)} \leq -\alpha_1 \frac{\epsilon^2(t)}{m^2(t)} \end{aligned} \tag{7.239}$$

for some constant $\alpha_1 > 0$. The last inequality is from $0 < \gamma_1 < 2$, $0 < \gamma_2 < 2$, and the parameter projection property that

$$f_2(t)(\theta_{p2}(t) - b_p + g_2(t) + f_2(t)) \leq 0. \tag{7.240}$$

From (7.239), we can derive the desired properties of (7.234): $\theta_p(t) \in L^\infty$, $\frac{\epsilon(t)}{m(t)} \in L^2 \cap L^\infty$ and $\theta_p(t+i_0) - \theta_p(t) \in L^2$ for any integer $i_0 > 0$. In addition, the projection term $f_2(t)$ in (7.238) ensures that $\text{sign}[b_p]\hat{b}_p(t) \geq b_0$, $t \geq 0$.

For $a_m = 0$ and $b_m = 1$ in (7.225) and $a_f = 0$ in (7.228), we have

$$y_m(t+1) = r(t),\ \theta_p^* = [a_p, b_p]^T,\ \phi(t) = [y(t-1), u(t-1)]^T \tag{7.241}$$

and, with (7.232), the controller (7.226) can be expressed as

$$\begin{aligned} r(t) &= \theta_p^T(t)\phi(t+1) = \theta_p^T(t+1)\phi(t+1) - (\theta_p(t+1) - \theta_p(t))^T\phi(t+1) \\ &= y(t+1) + \epsilon(t+1) - (\theta_p(t+1) - \theta_p(t))^T\phi(t+1). \end{aligned} \tag{7.242}$$

For $m(t)$ in (7.235), we have

$$m(t+1) \leq 1 + |y(t)| + |u(t)|, \tag{7.243}$$

and from (7.226), it follows that

$$|u(t)| \leq |k_1(t)|\,|y(t)| + |k_2(t)r(t)|. \tag{7.244}$$

From (7.242)–(7.244), we obtain

$$\begin{aligned} |y(t+1)| &\leq \frac{|\epsilon(t+1)|}{m(t+1)} m(t+1) + \|\theta_p(t+1) - \theta_p(t)\|_2 (|y(t)| + |u(t)|) + |r(t)| \\ &\leq \left(\frac{|\epsilon(t+1)|}{m(t+1)} + \|\theta_p(t+1) - \theta_p(t)\|_2 \right) (1 + |k_1(t)|)|y(t)| + |r(t)| \\ &+ \frac{|\epsilon(t+1)|}{m(t+1)} (1 + |k_2(t)r(t)|) + \|\theta_p(t+1) - \theta_p(t)\|_2 |k_2(t)r(t)|. \end{aligned} \tag{7.245}$$

From (7.239) and (7.245), we can derive the desired properties of (1.139).

Since $\frac{\epsilon(t)}{m(t)} \in L^2 \cap L^\infty$ and $\theta_p(t+1) - \theta_p(t) \in L^2$, we have that $\lim_{t\to\infty} \frac{\epsilon(t)}{m(t)} = 0$ and $\lim_{t\to\infty} \|\theta_p(t+1) - \theta_p(t)\|_2 = 0$, which, together with the boundedness of $k_1(t)$, $k_2(t)$, and $r(t)$, imply that $y(t) \in L^\infty$, and and so is $u(t)$ as from (7.244). Then we have $\lim_{t\to\infty} \epsilon(t) = 0$ and $\lim_{t\to\infty} (\theta_p(t+1) - \theta_p(t))^T \phi(t+1) = 0$, so that $\lim_{t\to\infty} e(t) = 0$ for $e(t) = y(t) - y_m(t)$, as from (7.242).

General Design Procedure

Consider the discrete-time linear time-invariant plant

$$P(z)[y](t) = k_p Z(z)[u](t), \tag{7.246}$$

where $y(t) \in R$ and $u(t) \in R$ are the measured plant input and output,

$$P(z) = z^n + p_{n-1} z^{n-1} + \cdots + p_1 z + p_0, \tag{7.247}$$

$$Z(z) = z^m + z_{m-1} z^{m-1} + \cdots + z_1 z + z_0 \tag{7.248}$$

are polynomials in z with z being the time advance operator $z[x](t) = x(t+1)$ (i.e., z^{-1} is the time delay operator $z^{-1}[x](t) = x(t-1)$), and p_i, $i = 0, 1, \ldots, n-1$, k_p, z_j, $j = 0, 1, \ldots, m-1$, with $n > m$, are the unknown but constant parameters to be estimated.

The control objective is to develop an indirect adaptive control scheme by combining a feedback control structure with an adaptive estimator for the unknown plant parameters p_i and z_j, for the plant (7.246), such that all closed-loop signals are bounded and $\lim_{t\to\infty}(y(t) - y_m(t)) = 0$ for some given bounded reference output $y_m(t)$ under the assumptions:

(A6.1): $Z(z)$ is stable, that is, all its zeros are in $|z| < 1$.

(A7.1D): The lower bound k_{p0} of $|k_p|$, such that $0 < k_{p0} \leq |k_p|$, is known, and so is $\mathrm{sign}[k_p]$, the sign of k_p.

Plant parameter estimation. The plant (7.246) can be expressed as

$$y(t) = \theta_p^{*T} \phi(t), \ t \in \{0, 1, \ldots\}, \tag{7.249}$$

similar to that in Section 3.6.1, where

$$\begin{aligned}\theta_p^* = [&k_p z_0, k_p z_1, \ldots, k_p z_{m-1}, k_p, \\ &-p_0, -p_1, \ldots, -p_{n-2}, -p_{n-1}]^T \in R^{n+m+1}\end{aligned} \tag{7.250}$$

$$\begin{aligned}\phi(t) = [&u(t-n), u(t-n+1), \ldots, u(t-n+m-1), u(t-n+m), \\ &y(t-n), y(t-n+1), \ldots, y(t-2), y(t-1)]^T \in R^{n+m+1}.\end{aligned} \tag{7.251}$$

Letting $\theta_p(t)$ be the estimate of θ_p^*, we define the estimation error

$$\epsilon(t) = \theta_p^T(t)\phi(t) - y(t). \tag{7.252}$$

The adaptive law for updating $\theta_p(t)$ is

$$\theta_p(t+1) = \theta_p(t) - \frac{\Gamma\phi(t)\epsilon(t)}{m^2(t)} + f(t), \ \theta(0) = \theta_0, \ t \in \{0, 1, \ldots\}, \tag{7.253}$$

where the adaptation gain matrix $\Gamma = \text{diag}\{\Gamma_1, \gamma_{m+1}, \Gamma_2\}$ is with $\Gamma_1 \in R^{m\times m}$, $0 < \Gamma_1 = \Gamma_1^T < 2I_m$, $0 < \gamma_{m+1} < 2$, and $\Gamma_2 \in R^{n\times n}$, $0 < \Gamma_2 = \Gamma_2^T < 2I_n$, θ_0 is an initial estimate of $\theta_p^* \in R^{n+m+1}$, and

$$m(t) = \sqrt{\kappa + \phi^T(t)\phi(t)} \tag{7.254}$$

with $\kappa > 0$. The parameter projection term $f(t)$ has the form

$$f(t) = [0_{1\times m}, f_{m+1}(t), 0_{1\times n}]^T \in R^{n+m+1}, \tag{7.255}$$

where $f_{m+1}(t)$ is for projecting $\theta_{p\,m+1}(t) = \hat{k}_p(t)$, the $(m+1)$th component of $\theta_p(t)$, to ensure that $\hat{k}_p(t)$ is away from zero. Using Assumption (A7.1D), we define $\theta_{p\,m+1}^a = k_{p0}$ and $\theta_{p\,m+1}^b = \infty$ for $k_p > 0$ and $\theta_{p\,m+1}^a = -\infty$ and $\theta_{p\,m+1}^b = -k_{p0}$ for $k_p < 0$, choose the initial estimate $\theta_{p\,m+1}(0)$ to satisfy

$$\theta_{p\,m+1}(0) \in (\theta_{p\,m+1}^a, \theta_{p\,m+1}^b), \tag{7.256}$$

denote $\phi_{m+1}(t) = u(t-m)$, introduce

$$g_{m+1}(t) = \theta_{p\,m+1}(t) - \frac{\gamma_{m+1}\phi_{m+1}(t), \epsilon(t)}{m^2(t)} \tag{7.257}$$

and set the projection function component $f_{m+1}(t)$ as

$$f_{m+1}(t) = \begin{cases} 0 & \text{if } g_{m+1}(t) \in [\theta^a_{p\,m+1}, \theta^b_{p\,m+1}], \\ \theta^b_{p\,m+1} - g_{m+1}(t) & \text{if } g_{m+1}(t) > \theta^b_{p\,m+1}, \\ \theta^a_{p\,m+1} - g_{m+1}(t) & \text{if } g_{m+1}(t) < \theta^a_{p\,m+1}. \end{cases} \tag{7.258}$$

This scheme has the properties that $\theta_{p\,m+1}(t) \in [\theta^a_{p\,m+1}, \theta^b_{p\,m+1}]$ and $|\theta_{p\,m+1}(t)| < \infty$, that is, $k_{p0} \le |\hat{k}_p(t)| < \infty$, and that $f_{m+1}(t)(g_{m+1}(t) - \theta^*_{p\,m+1} + f_{m+1}(t)) \le 0$. The time increment of $V(\tilde{\theta}_p) = \tilde{\theta}_p^T \Gamma^{-1} \tilde{\theta}_p$ with $\tilde{\theta}_p(t) = \theta_p(t) - \theta_p^*$ is

$$\begin{aligned} &V(\tilde{\theta}(t+1)) - V(\tilde{\theta}(t)) \\ &\le -\alpha_1 \frac{\epsilon^2(t)}{m^2(t)} + 2 f_{m+1}(t)(g_{m+1}(t) - \theta^*_{p\,m+1} + f_{m+1}(t)) \end{aligned} \tag{7.259}$$

for some constant $\alpha_1 > 0$. From (7.259), it follows that $\theta_p(t)$ and $\frac{\epsilon(t)}{m(t)}$ are bounded and $\frac{\epsilon(t)}{m(t)}$ and $\theta_p(t+1) - \theta_p(t)$ belong to L^2.

From the components $\theta_{pj}(t)$, $j = 1, 2, \ldots, n+m+1$, of $\theta_p(t)$, we obtain the estimates of k_p, $z_i, i = 0, 1, 2, \ldots, m-1$, and p_j, $j = 0, 1, 2, \ldots, n-1$, as

$$\begin{aligned} &\hat{k}_p(t) = \theta_{p\,m+1}(t), \; \hat{z}_i(t) = \frac{\theta_{p\,i+1}(t)}{\theta_{p\,m+1}(t)}, \; i = 0, 1, \ldots, m-1, \\ &\hat{p}_j(t) = -\theta_{p\,j+m+2}(t), \; j = 0, 1, \ldots, n-1, \end{aligned} \tag{7.260}$$

and the estimates of $P(z)$ and $Z(z)$ as

$$\hat{P}(z, \hat{p}) = z^n + \hat{p}_{n-1} z^{n-1} + \cdots + \hat{p}_1 z + \hat{p}_0, \tag{7.261}$$

$$\hat{Z}(z, \hat{z}) = z^m + \hat{z}_{m-1} z^{m-1} + \cdots + \hat{z}_1 z + \hat{z}_0, \tag{7.262}$$

where $\hat{z} = [\hat{z}_0, \hat{z}_1, \ldots, \hat{z}_{m-1}]^T$ and $\hat{p} = [\hat{p}_0, \hat{p}_1, \ldots, \hat{p}_{n-1}]^T$ are the estimates of $z^* = [z_0, z_1, \ldots, z_{m-1}]^T \in R^m$ and $p^* = [p_0, p_1, \ldots, p_{n-1}]^T \in R^n$.

Feedback controller. The adaptive controller structure is

$$u(t) = \theta_1^T \omega_1(t) + \theta_2^T \omega_2(t) + \theta_3 y_m(t + n^*), \tag{7.263}$$

where $n^* = n - m$, $\theta_1 \in R^{n-1}$, $\theta_2 \in R^n$, $\theta_3 \in R$ are parameters and

$$\omega_1(t) = a_c(z)[u](t), \; a_c(z) = [z^{-n+1}, \ldots, z^{-1}]^T, \tag{7.264}$$

$$\omega_2(t) = b_c(z)[y](t), \; b_c(z) = [z^{-n+1}, \ldots, z^{-1}, 1]^T. \tag{7.265}$$

The controller parameters θ_1, θ_2, and θ_3 satisfy the design equation

$$\theta_1^T a_c(z)\hat{P}(z,\hat{p}) + \theta_2^T b_c(z)\hat{k}_p\hat{Z}(z,\hat{z}) = \hat{P}(z,\hat{p}) - \hat{k}_p\theta_3\hat{Z}(z,\hat{z})z^{n^*}. \quad (7.266)$$

With $|\hat{k}_p(t)| \geq k_{p0} > 0$, the solution $\{\theta_1, \theta_2, \theta_3\}$ always exists for any $\hat{p}$ and $\hat{z}$ and is not unique when $\hat{P}(z,\hat{p})$ and $\hat{Z}(z,\hat{z})$ are not coprime. Similar to their continuous-time counterparts (7.89)–(7.90), the following design equations can be used to calculate a solution $\{\theta_1, \theta_2, \theta_3\}$ to implement (7.263):

$$\theta_3 = \frac{1}{\hat{k}_p}, \quad \theta_1^T a_c(\lambda)\lambda^{n-1} = \lambda^{n-1} - \hat{Q}(\lambda,\hat{q})\hat{Z}(\lambda,\hat{z}), \quad (7.267)$$

$$\theta_2^T b_c(\lambda)\lambda^{n-1} = -\theta_3\hat{R}(\lambda,\hat{r}), \quad (7.268)$$

where $\hat{Q}(\lambda,\hat{q})$ (monic) and $\hat{R}(\lambda,\hat{r})$ are polynomials of degrees n^*-1 and $n-1$, parametrized by $\hat{q} \in R^{n^*-1}$ and $\hat{r} \in R^n$, respectively, such that

$$\frac{\lambda^{n+n^*-1}}{\hat{P}(\lambda,\hat{p})} = \hat{Q}(\lambda,\hat{q}) + \frac{\hat{R}(\lambda,\hat{r})}{\hat{P}(\lambda,\hat{p})}. \quad (7.269)$$

Note that the projection algorithm (7.258) ensures that $|\hat{k}_p(t)| \geq k_{p0} > 0$ so that (7.267)–(7.268) is singularity free, that is, no division by zero.

Remark 7.4 The solution $\{\theta_1, \theta_2, \theta_3\}$ to (7.267)–(7.269) can also be obtained by the following recursive scheme [125]. Defining the λ^{-1} polynomials

$$\hat{A}(\lambda^{-1}) = \lambda^{-n}\hat{P}(\lambda,\hat{p}) \triangleq 1 + a_1\lambda^{-1} + \cdots + a_n\lambda^{-n}, \quad (7.270)$$

$$\hat{B}'(\lambda^{-1}) = \lambda^{-m}\hat{Z}(\lambda,\hat{z}) \triangleq b_0 + b_1\lambda^{-1} + \cdots + b_m\lambda^{-m}, \quad (7.271)$$

one solves the polynomial equation

$$1 = \hat{F}(\lambda^{-1})\hat{A}(\lambda^{-1}) + \lambda^{-n^*}\hat{G}(\lambda^{-1}) \quad (7.272)$$

for the unique polynomial solutions

$$\hat{F}(\lambda^{-1}) = f_0 + f_1\lambda^{-1} + \cdots + f_{n^*-1}\lambda^{-n^*+1}, \quad (7.273)$$

$$\hat{G}(\lambda^{-1}) = g_0 + g_1\lambda^{-1} + \cdots + g_{n-1}\lambda^{-n+1}, \quad (7.274)$$

where $f_0 = 1$ and

$$f_i = -\sum_{j=0}^{i-1} f_j a_{i-j}, \; i = 1, 2, \ldots, n^* - 1, \quad (7.275)$$

$$g_i = -\sum_{j=0}^{n^*-1} f_j a_{i+n^*-j}, \; i = 0, 1, 2, \ldots, n-1, \quad (7.276)$$

with $a_{n+1} = a_{n+2} = \cdots = a_{n+n^*-1} = 0$, and sets

$$\hat{\alpha}(\lambda^{-1}) = \hat{G}(\lambda^{-1}) \triangleq \alpha_0 + \alpha_1\lambda^{-1} + \cdots + \alpha_{n-1}\lambda^{-n+1}, \tag{7.277}$$

$$\hat{\beta}(\lambda^{-1}) = F(\lambda^{-1})B'(\lambda^{-1}) \triangleq \beta_0 + \beta_1\lambda^{-1} + \cdots + \beta_{m+n^*-1}\lambda^{-m-n^*+1}, \tag{7.278}$$

where $\beta_0 = \hat{k}_p$. Then, it can be verified that

$$\theta_1^T a_c(\lambda) = -\frac{1}{\beta_0}\hat{\alpha}(\lambda^{-1}), \tag{7.279}$$

$$\theta_2^T b_c(\lambda) = \frac{1}{\beta_0}(\beta_0 - \hat{\beta}(\lambda^{-1})), \tag{7.280}$$

that is, the controller structure can be expressed as

$$\hat{\beta}(z^{-1})[u](t) + \hat{\alpha}(z^{-1})[y](t) = y_m(t+n^*). \tag{7.281}$$

With this formulation, for $\hat{p} = p^*$ and $\hat{z} = z^*$, the model reference control scheme of Section 6.2 can be expressed as

$$\beta(z^{-1})[u](t) + \alpha(z^{-1})[y](t) = y_m(t+n^*), \tag{7.282}$$

where $\beta(z^{-1})$ and $\alpha(z^{-1})$ are the nominal versions of $\hat{\beta}(z^{-1})$ and $\hat{\alpha}(z^{-1})$ with $\hat{p} = p^*$ and $\hat{z} = z^*$. The plant (7.246) can be parametrized as

$$y(t+n^*) = \beta(z^{-1})[u](t) + \alpha(z^{-1})[y](t) \tag{7.283}$$

(see [125]), and the controller (7.263) leads to $y(t) = y_m(t)$ for $t \geq n^*$. When $\alpha(z^{-1})$ and $\beta(z^{-1})$ are unknown, an adaptive version (7.281) is used. □

Adaptive system properties. This model reference adaptive controller ensures closed-loop signal boundedness and $\lim_{t\to\infty}(y(t) - y_m(t)) = 0$.

The proof of these properties can be derived in a similar way to that for proving Theorem 6.2 using the desired properties that $\theta_i(t)$ is bounded and $\theta_i(t+1) - \theta_i(t) \in L^2$, $i = 1, 2, 3$, and $\frac{\epsilon(t)}{m(t)} \in L^\infty \cap L^2$ and the discrete-time version of the estimation error equation (7.111) with

$$\bar{\epsilon}(t) = -\hat{Q}(z, \hat{q}(t))[z^{-n^*+1}[\epsilon]](t), \tag{7.284}$$

that is, $y(t) - y_m(t) + \rho(t)\xi(t) = \bar{\epsilon}(t) + \epsilon_u(t) + \epsilon_y(t)$ (with $\hat{Q}(z, \hat{q}(t))z^{-n^*+1}$ being a stable and proper operator), which is similar to (6.39): $y(t) - y_m(t) + \rho(t)\xi(t) = \epsilon(t)$, where $\bar{\epsilon}(t)$ and $\epsilon(t)$ have similar properties (see Lemma 6.2 and that following (7.259)), while the controller structures are the same for both designs. The extra terms $\epsilon_u(t)$ and $\epsilon_y(t)$ are similar to that in (7.103) and (7.105) with $\dot{\hat{p}}_j(t)$ and $\dot{\hat{\rho}}_i(t)$ replaced by $\hat{p}_j(t+1) - \hat{p}_j(t)$ and $\hat{\rho}_i(t+1) - \hat{\rho}_i(t)$, and the associated $\frac{s^{k-1}}{P_m(s)}$ replaced by $\frac{z^k}{P_m(z)}$ (with all other s-functions replaced by z-functions), as from the discrete-time swapping lemma (also see page 411).

7.3.2 Pole Placement Designs

Consider the discrete-time linear time-invariant plant

$$P(z)[y](t) = Z(z)[u](t), \tag{7.285}$$

where $y(t) \in R$, $u(t) \in R$ are the measured plant output and input and

$$P(z) = z^n + p_{n-1}z^{n-1} + \cdots + p_1 z + p_0, \tag{7.286}$$

$$Z(z) = z_{n-1}z^{n-1} + z_{n-2}z^{n-2} + \cdots + z_1 z + z_0, \tag{7.287}$$

with p_i, z_i, $i = 0, 1, \ldots, n-1$, being the unknown but constant parameters.

The control objective is: given a reference output signal $y_m(t)$ satisfying

$$Q_m(z)[y_m](t) = 0, \tag{7.288}$$

where $Q_m(z)$ is a monic polynomial of degree n_q with either nonrepeated zeros on the unit circle, $|z| = 0$, or zeros inside the unit circle, $|z| < 1$, so that $y_m(t)$ is bounded, find an output feedback control $u(t)$ such that all closed-loop signals are bounded and $y(t)$ tracks $y_m(t)$ asymptotically with time.

Example 7.2 As examples of the polynomial $Q_m(z)$ and the corresponding reference signal $y_m(t)$, we have

$Q_m(z) = 1$ for $y_m(t) = 0$;

$Q_m(z) = z - 1$ for $y_m(t) = c \neq 0$, a constant;

$Q_m(z) = z - e^{-a}$ with $a > 0$ for $y_m(t) = c\,e^{-at}$, $c \neq 0$;

$Q_m(z) = z^2 - 2z\cos\sigma + 1$ for $y_m(t) = a\sin(\sigma t) + b\cos(\sigma t)$ with $\sigma \neq 0$ and not both a, b being zero;

$Q_m(z) = (z^2 - 2z\cos\sigma_1 + 1)(z^2 - 2z\cos\sigma_2 + 1)$ for $y_m(t) = a_1\sin(\sigma_1 t) + a_2\sin(\sigma_2 t) + b_1\cos(\sigma_1 t) + b_2\cos(\sigma_2 t)$ with $\sigma_1 \neq 0$, $\sigma_2 \neq 0$, $\sigma_1 \neq \sigma_2$, not both a_1, b_1 being zero, and not both a_2, b_2 being zero;

$Q_m(z) = (z-1)(z^2 - 2z\cos\sigma + 1)$ for $y_m(t) = a\sin(\sigma t) + b\cos(\sigma t) + c$ with $c \neq 0$, $\sigma \neq 0$ and not both a, b being zero; and

$Q_m(z) = (z^2 - 2ze^{-a}\cos\sigma + e^{-2a})$ for $y_m(t) = e^{-at}(b\sin(\sigma t) + c\cos(\sigma t))$ with $a > 0$, $\sigma \neq 0$ and not both a, b being zero. □

Similar to the continuous-time case in Section 7.2.1, we assume:

(A7.2D): $Q_m(z)P(z)$ and $Z(z)$ are coprime.

(A7.3D): The order n of $P(z)$ is known.

To implement an indirect adaptive pole placement control design for the plant (7.285) with possible unstable zeros, we first estimate

$$p^* = [p_0, p_1, \ldots, p_{n-1}]^T \in R^n, \;\; z^* = [z_0, z_1, \ldots, z_{n-1}]^T \in R^n \tag{7.289}$$

and then use their estimates to solve a pole placement design equation to obtain the parameters of a feedback controller for meeting the control objective.

Parameter Estimation

Defining the parameter and regressor vectors

$$\begin{aligned}\theta_p^* &= [z^{*T}, -p^{*T}]^T \\ &= [z_0, z_1, \ldots, z_{n-2}, z_{n-1}, \; -p_0, -p_1, \ldots, -p_{n-2}, -p_{n-1}]^T \in R^{2n},\end{aligned} \tag{7.290}$$

$$\begin{aligned}\phi(t) = [&u(t-n), u(t-n+1), \ldots, u(t-2), u(t-1), \\ &y(t-n), y(t-n+1), \ldots, y(t-2), y(t-1)]^T \in R^{2n},\end{aligned} \tag{7.291}$$

we express the plant (7.285) as

$$y(t) = \theta_p^{*T}\phi(t), \; t \in \{0, 1, 2, \ldots\}. \tag{7.292}$$

Letting $\theta_p(t)$ be the estimate of θ_p^*, we define the estimation error

$$\epsilon(t) = \theta_p^T(t)\phi(t) - y(t). \tag{7.293}$$

The standard gradient algorithm for updating $\theta_p(t)$ is

$$\theta_p(t+1) = \theta_p(t) - \frac{\Gamma\phi(t)\epsilon(t)}{m^2(t)}, \; \theta(0) = \theta_0, \; t \in \{0, 1, \ldots\}, \tag{7.294}$$

where the adaptation gain matrix $\Gamma \in R^{2n\times 2n}$ is such that $0 < \Gamma = \Gamma^T < 2I_{2n}$, θ_0 is an initial estimate of $\theta_p^* \in R^{2n}$, and

$$m(t) = \sqrt{\kappa + \phi^T(t)\phi(t)}, \; \kappa > 0. \tag{7.295}$$

The standard normalized least-squares algorithm for updating $\theta(t)$ is

$$\theta_p(t+1) = \theta_p(t) - \frac{P(t-1)\phi(t)\epsilon(t)}{m^2(t)}, \; \theta_p(0) = \theta_0, \; t \in \{0, 1, 2, \ldots\}, \tag{7.296}$$

$$\begin{aligned}P(t) = P(t-1) - \frac{P(t-1)\phi(t)\phi^T(t)P(t-1)}{m^2(t)}, \\ P(-1) = P_0 = P_0^T > 0, \; t \in \{0, 1, 2, \ldots\},\end{aligned} \tag{7.297}$$

$$m(t) = \sqrt{\kappa + \phi^T(t)P(t-1)\phi(t)}, \; \kappa > 0, \tag{7.298}$$

where θ_0 is the initial estimate of $\theta_p^* \in R^{2n}$, P_0 is the initial value of the gain matrix $P(t) \in R^{2n \times 2n}$ at $t = -1$, and κ is a design parameter.

These adaptive laws ensure that $\theta_p(t)$ and $\frac{\epsilon(t)}{m_0(t)}$ are bounded and $\frac{\epsilon(t)}{m_0(t)}$ and $\theta_p(t+1) - \theta_p(t)$ belong to L^2, where $m_0(t) = \sqrt{1 + \phi^T(t)\phi(t)}$.

From $\theta_p(t) = [\hat{z}^T(t), -\hat{p}^T(t)]^T$, we obtain the estimates of p^* and z^* as

$$\hat{p} = [\hat{p}_0, \hat{p}_1, \ldots, \hat{p}_{n-1}]^T \in R^n, \; \hat{z} = [\hat{z}_0, \hat{z}_1, \ldots, \hat{z}_{n-1}]^T \in R^n \tag{7.299}$$

and the estimates of $P(z)$ and $Z(z)$, with $z = \lambda$, as

$$\hat{P}(\lambda, \hat{p}) = \lambda^n + \hat{p}_{n-1}\lambda^{n-1} + \cdots + \hat{p}_1\lambda + \hat{p}_0, \tag{7.300}$$

$$\hat{Z}(\lambda, \hat{z}) = \hat{z}_{n-1}\lambda^{n-1} + \hat{z}_{n-2}\lambda^{n-2} + \cdots + \hat{z}_1\lambda + \hat{z}_0. \tag{7.301}$$

Design Equation

Choose the desired closed-loop characteristic polynomial as a monic polynomial $A^*(z)$ of degree $2n + n_q - 1$, which has all its zeros in $|z| < 1$ (a common choice for a discrete-time control design is $A^*(z) = z^{2n+n_q-1}$), where n_q is the degree of the reference output characteristic polynomial $Q_m(z)$ in (7.288).

Assume that $Q_m(\lambda)\hat{P}(\lambda, \hat{p})$ and $\hat{Z}(\lambda, \hat{z})$ are coprime polynomials in the indeterminate λ, and solve the Diophantine equation

$$\hat{C}(\lambda, \psi_c)Q_m(\lambda)\hat{P}(\lambda, \hat{p}) + \hat{D}(\lambda, \psi_d)\hat{Z}(\lambda, \hat{z}) = A^*(\lambda) \tag{7.302}$$

to find the polynomials $\hat{C}(\lambda, \psi_c)$, $\hat{D}(\lambda, \psi_d)$ in the form

$$\hat{C}(\lambda, \psi_c) = \lambda^{n-1} + c_{n-2}\lambda^{n-2} + \cdots + c_1\lambda + c_0, \tag{7.303}$$

$$\hat{D}(\lambda, \psi_d) = d_{n_q+n-1}\lambda^{n_q+n-1} + \cdots + d_1\lambda + d_0, \tag{7.304}$$

where the parameter vectors ψ_c and ψ_d are

$$\psi_c = [c_0, c_1, \ldots, c_{n-2}]^T \in R^{n-1}, \; \psi_d = [d_0, d_1, \ldots, d_{n+n_q-1}]^T \in R^{n+n_q}. \tag{7.305}$$

Obtain $\hat{C}(z, \psi_c)$ and $\hat{D}(z, \psi_d)$ by making the substitutions:

$$\hat{C}(z, \psi_c) = \hat{C}(\lambda, \psi_c)|_{\lambda=z} = z^{n-1} + c_{n-2}z^{n-2} + \cdots + c_1 z + c_0, \tag{7.306}$$

$$\hat{D}(z, \psi_d) = \hat{D}(\lambda, \psi_d)|_{\lambda=z} = d_{n_q+n-1}z^{n_q+n-1} + \cdots + d_1 z + d_0. \tag{7.307}$$

Note that, as an operator, $c_i z^i$, $i = 1, \ldots, n-1$, is different from $z^i c_i$ when $c_i = c_i(t)$ is a time-varying coefficient, and so is $d_j z^j$, $j = 1, \ldots, n_q+n-1$, from $z^j d_j$ when $d_j = d_j(t)$ is a time-varying coefficient. For example, $c_1 z[x](t) = c_1(t)x(t+1)$, while $zc_1[x](t) = z[c_1 x](t) = c_1(t+1)x(t+1)$.

Controller Structure

We then choose the controller structure as

$$u(t) = (\Lambda_c(z) - \hat{C}(z,\psi_c)Q_m(z))\frac{1}{\Lambda_c(z)}[u](t) + \hat{D}(z,\psi_d)\frac{1}{\Lambda_c(z)}[y_m - y](t), \quad (7.308)$$

where $\Lambda_c(z)$ is a monic stable polynomial of degree $n + n_q - 1$. For a discrete-time design, a common choice is $\Lambda_c(z) = z^{n+n_q-1}$. With this choice, the controller (7.308) can be simplified as

$$u(t) = \psi_1^T(t)\omega_1(t) + \psi_2^T(t)\omega_{2e}(t) + \psi_3(t)(y(t) - y_m(t)), \quad (7.309)$$

where $\psi_1 = [\psi_{10}, \psi_{11}, \ldots, \psi_{1\,n+n_q-2}]^T \in R^{n+n_q-1}$, for

$$\hat{C}(\lambda,\psi_c)Q_m(\lambda) \triangleq \lambda^{n+n_q-1} - \psi_{1\,n+n_q-2}\lambda^{n+n_q-2} - \cdots - \psi_{11}\lambda - \psi_{10}, \quad (7.310)$$

$\psi_2 = -[d_0, d_1, \ldots, d_{n+n_q-2}]^T \in R^{n+n_q-1}$, $\psi_3 = -d_{n+n_q-1}$, for d_i in (7.307), and

$$\omega_1(t) = [u(t - n - n_q + 1), \ldots, u(t-1)]^T \in R^{n+n_q-1}, \quad (7.311)$$

$$\omega_2(t) = [e(t - n - n_q + 1), \ldots, e(t-1)]^T \in R^{n+n_q-1} \quad (7.312)$$

with $e(t) = y(t) - y_m(t)$ being the tracking error.

Stability Properties

Pole placement control. When the parameters p^* and z^* of $P(z)$ and $Z(z)$ in (7.285) are known, the design equation (7.302) becomes

$$C(\lambda)Q_m(\lambda)P(\lambda) + D(\lambda)Z(\lambda) = A^*(\lambda), \quad (7.313)$$

and the controller (7.308) becomes

$$u(t) = (\Lambda_c(z) - C(z)Q_m(z))\frac{1}{\Lambda_c(z)}[u](t) + D(z)\frac{1}{\Lambda_c(z)}[y_m - y](t). \quad (7.314)$$

This pole placement control scheme ensures that all signals in the closed-loop system are bounded and $\lim_{t\to\infty}(y(t) - y_m(t)) = 0$ exponentially.

Adaptive pole placement control. When the parameters p^* and z^* in $P(s)$ and $Z(s)$ are unknown, the adaptive design equation (7.302) is used, with $\hat{P}(\lambda,\hat{p})$ and $\hat{Z}(\lambda,\hat{z})$ in (7.300) and (7.301) being estimates of $P(s)$ and $Z(s)$, respectively, updated from an adaptive law. To solve (7.302), we need the singularity-free condition that $Q_m(\lambda)\hat{P}(\lambda,\hat{p})$ and $\hat{Z}(\lambda,\hat{z})$ are coprime, uniformly in time t. This condition is not automatically ensured by the standard adaptive algorithms (7.294) and (7.296). In two special cases, as presented in (7.179)–(7.181) in Section 7.2.3 for the continuous-time case, such a condition can be met. For a general case, either some modifications can be added to (7.294) and (7.296) or some special conditions can be imposed on the adaptive system, so that such a key condition can be satisfied [8], [60], [117], [125], [233], [235], [237], [247], [310]. Under the above singularity-free condition, all signals in the closed-loop system are bounded and $\lim_{t\to\infty}(y(t) - y_m(t)) = 0$.

To prove this desired adaptive control system property, we use the notation of (7.309)–(7.312) to express the controller (7.308) with $\Lambda_c(z) = z^{n+n_q-1}$ as

$$u(t) = \psi^T(t)\omega(t) + \bar{y}_m(t), \tag{7.315}$$

where $\bar{y}_m(t) = \hat{D}(z, \psi_d)\frac{1}{\Lambda_c(z)}[y_m](t)$ is bounded, and

$$\psi(t) = [\psi_1^T(t), \psi_2^T(t), \psi_3(t)]^T \in R^{2(n+n_q)-1}, \tag{7.316}$$

$$\omega(t) = [\omega_1^T(t), y(t-n-n_q+1), \ldots, y(t-1), y(t)]^T \in R^{2(n+n_q)-1}. \tag{7.317}$$

From (7.291) for $\phi(t)$ and (7.317) for $\omega(t)$, we have

$$\phi(t+1) = A_\phi(t)\omega(t) + B_0\bar{y}_m(t), \tag{7.318}$$

where $B_0 = [0_{1\times(n-1)}, 1, 0_{1\times n}]^T \in R^{2n}$ and

$$A_\phi(t) = \begin{bmatrix} [0_{(n-1)\times n_q}, I_{n-1}] & 0_{(n-1)\times(n+n_q)} \\ \psi_1^T(t) & [\psi_2^T(t), \psi_3(t)] \\ 0_{n\times(n+n_q-1)} & [0_{n\times n_q}, I_n] \end{bmatrix}. \tag{7.319}$$

Using (7.318), the estimation error expression (7.293), that is, $\epsilon(t) = \theta_p^T(t)\phi(t) - y(t)$, and the control law expression (7.315), we express

$$\begin{aligned} \omega(t+1) = {} & A(t)\omega(t) + \Delta(t)\omega(t) + B_1\bar{y}_m(t) \\ & + B_2\left(\epsilon(t+1) - \theta_p^T(t+1)B_0\bar{y}_m(t)\right), \end{aligned} \tag{7.320}$$

where $B_1 = [0_{1\times(n+n_q-2)}, 1, 0_{1\times(n+n_q)}]^T$, $B_2 = -[0_{1\times(2(n+n_q)-2)}, 1]^T$, and

$$A(t) = \begin{bmatrix} [0_{(n+n_q-2)\times 1}, I_{n+n_q-2}] & 0_{(n+n_q-2)\times(n+n_q)} \\ \psi_1^T(t) & [\psi_2^T(t), \psi_3(t)] \\ 0_{(n+n_q-1)\times(n+n_q-1)} & [0_{(n+n_q-1)\times 1}, I_{n+n_q-1}] \\ \multicolumn{2}{c}{\theta_p^T(t)A_\phi(t)} \end{bmatrix}, \tag{7.321}$$

$$\Delta(t) = \begin{bmatrix} 0_{(2(n+n_q)-2)\times(2(n+n_q)-1)} \\ (\theta_p(t+1) - \theta_p(t))^T A_\phi(t) \end{bmatrix}. \tag{7.322}$$

Under the singularity-free condition, from (7.302), it follows that $\psi(t) \in L^\infty$ and $\psi(t+1) - \psi(t) \in L^2$ (so that $\lim_{t\to\infty}(\psi(t+1) - \psi(t)) = 0$). From the adaptive law (7.294) or (7.296), we have $\theta_p(t) \in L^\infty$ and $\theta_p(t+1) - \theta_p(t) \in L^2$ (so that $\lim_{t\to\infty}(\theta_p(t+1) - \theta_p(t)) = 0$). Therefore, $A(t) \in L^\infty$ and $A(t+1) - A(t) \in L^2$ (so that $\lim_{t\to\infty}(A(t+1) - A(t)) = 0$).

With (7.302) for $A^*(\lambda) = \lambda^{2n+n_q-1}$, it can be shown that at each fixed t, $\det(\lambda I - A(t)) = \lambda^{2(n+n_q)-1}$. It follows that the system $\omega(t+1) = A(t)\omega(t)$ is exponentially stable [77].[4] Since $A_\phi(t) \in L^\infty$, it follows that $\lim_{t\to\infty}\Delta(t) = 0$, so that the system $\omega(t+1) = A(t)\omega(t) + \Delta(t)\omega(t)$ is exponentially stable.

Given that $\frac{\epsilon(t)}{m_0(t)} \in L^\infty \cap L^2$, where $m_0(t) = \sqrt{1 + \phi^T(t)\phi(t)}$, we have

$$\begin{aligned} |\epsilon(t+1)| &\le \frac{|\epsilon(t)|}{m_0(t)}(1 + \|\phi(t+1)\|_2) \\ &\le \frac{|\epsilon(t)|}{m_0(t)}(1 + \|A_\phi(t)\|_2\|\omega(t)\|_2 + |\bar{y}_m(t)|) \end{aligned} \tag{7.323}$$

as from (7.318). Since $\frac{|\epsilon(t)|}{m_0(t)}\|A_\phi(t)\|_2 \in L^2$ (so that $\lim_{t\to\infty}\frac{|\epsilon(t)|}{m_0(t)}\|A_\phi(t)\|_2 = 0$), the system $\omega(t+1) = A(t)\omega(t) + \Delta(t)\omega(t) + \frac{\epsilon(t)}{m_0(t)}A_\phi(t)\omega(t)$ is exponetially stable. Hence, the solution $\omega(t)$ of system (7.320) is bounded, and so are $u(t)$ and $y(t)$, that is, all closed-loop signals are bounded.

From closed-loop signal boundedness, with $\frac{\epsilon(t)}{m_0(t)} \in L^2$, it follows that $\epsilon(t) \in L^2$ and $\lim_{t\to\infty}\epsilon(t) = 0$, and from (7.295) and (7.299)–(7.301), that $\epsilon(t) = -\hat{P}(z,\hat{p})[z^{-n}[y]](t) + \hat{Z}(z,\hat{z})[z^{-n}[u]](t)$, so that

$$\begin{aligned} &\lim_{t\to\infty}(\hat{P}(z,\hat{p})[z^{-n}[y]](t) - \hat{Z}(z,\hat{z})[z^{-n}[u]](t)) \\ &= \lim_{t\to\infty}(\hat{P}(z,\hat{p})z^{-k}[y](t) - \hat{Z}(z,\hat{z})z^{-k}[u](t)) = 0 \end{aligned} \tag{7.324}$$

for any finite k; in particular, for $k = n+n_q-1$, $z^k = z^{n+n_q-1} = \Lambda_c(z)$ (the first equality of (7.324) follows from $\lim_{t\to\infty}\epsilon(t+n-k) = 0$, $\lim_{t\to\infty}(\hat{P}(z,\hat{p}(t)) - \hat{P}(z,\hat{p}(t+n-k)))[z^{-k}[y]](t) = 0$, and $\lim_{t\to\infty}(\hat{Z}(z,\hat{z}(t)) - \hat{Z}(z,\hat{z}(t+n-k)))[z^{-k}[u]](t) = 0$ for $\hat{P}(z,\hat{p}(t)) = \hat{P}(z,\hat{p})$ and $\hat{Z}(z,\hat{z}(t)) = \hat{Z}(z,\hat{z})$).

To prove asymptotic output tracking of $y_m(t)$ by $y(t)$, we first introduce the following composite polynomial operators [125] of two time-varying polynomial

---

[4] A linear system $x(t+1) = A(t)x(t)$ is exponentially stable if $A(t)$ is bounded, $\max_i |\lambda_i[A(t)]| \le 1-\delta$ for some $\delta > 0$, all i, and any $t \in \{0,1,2,\ldots\}$, and $\|A(t+1) - A(t)\|$ is sufficiently small for any $t \in \{0,1,2,\ldots\}$ [77] (or for any $t \ge t_1$ and some $t_1 > 0$).

This is a special case of the discrete-time counterpart of Lemma 7.1, with $\sum_{\tau=t}^{t+T-1}\|A(\tau+1) - A(\tau)\|^p \le \nu T + \gamma_0$ for any $t \ge 0$, $T > 0$, some $\gamma_0 > 0$, $\nu > 0$ small, and $p \in [1,\infty)$.

operators $\hat{A}(z,t) = \sum_{i=0}^{n} \hat{a}_i(t) z^i$ and $\hat{B}(z,t) = \sum_{j=0}^{m} \hat{b}_j(t) z^j$:

$$\hat{A}(z,t)\hat{B}(z,t) = \sum_{i=0}^{n} \hat{a}_i(t) \sum_{j=0}^{m} \hat{b}_j(t) z^{i+j}, \tag{7.325}$$

$$\hat{A}(z,t) \circ \hat{B}(z,t) = \sum_{i=0}^{n} \sum_{j=0}^{m} \hat{a}_i(t) \hat{b}_j(t+i) z^{i+j}. \tag{7.326}$$

The corresponding two operations are defined as

$$\hat{A}\hat{B}[x](t) = \sum_{i=0}^{n} \sum_{j=0}^{m} \hat{a}_i(t)\hat{b}_j(t)x(t+i+j), \tag{7.327}$$

$$\hat{A} \circ \hat{B}[x](t) = \hat{A}[\hat{B}[x]](t) = \sum_{i=0}^{n} \sum_{j=0}^{m} \hat{a}_i(t)\hat{b}_j(t+i)x(t+i+j). \tag{7.328}$$

It follows that, as operators, in general, $\hat{A}(z,t)\hat{B}(z,t) = \hat{B}(z,t)\hat{A}(z,t)$ and $\hat{A}(z,t) \circ \hat{B}(z,t) \neq \hat{B}(z,t) \circ \hat{A}(z,t)$. Furthermore, if $\hat{B}(z,t) = B(z) = \sum_{j=0}^{m} b_j z^j$ is a polynomial with constant coeffficients b_j, then $\hat{A}(z,t) \circ B(z) = \hat{A}(z,t)B(z)$ and $B(z)\hat{A}(z,t) = \hat{A}(z,t) \circ B(z)$, that is, $B\hat{A}[x](t) = \hat{A}B[x](t) = \hat{A}[B[x]](t)$.

Then, using $\hat{C}(z,\psi_c)Q_m(z)\hat{P}(z,\hat{p}) + \hat{D}(z,\psi_d)\hat{Z}(z,\hat{z}) = A^*(z)$ (the operator version of the Diophantine equation (7.302)), we have

$$A^*[\Lambda_c^{-1}[y - y_m]](t) = (\hat{C}Q_m\hat{P} + \hat{D}\hat{Z})[\Lambda_c^{-1}[y]](t) - A^*[\Lambda_c^{-1}[y_m]](t), \tag{7.329}$$

$$A^*[\Lambda_c^{-1}[y_m]](t) = (\hat{C}Q_m\hat{P} + \hat{D}\hat{Z})[\Lambda_c^{-1}[y_m]](t). \tag{7.330}$$

From the control law (7.314), we have

$$\hat{C}Q_m[\Lambda_c^{-1}[u]](t) + \hat{D}[\Lambda_c^{-1}[y]](t) = \hat{D}[\Lambda_c^{-1}[y_m]](t). \tag{7.331}$$

Operating both sides of this equation by $\hat{Z} = \hat{Z}(z,\hat{z})$, we obtain

$$\hat{Z} \circ \hat{C}Q_m[\Lambda_c^{-1}[u]](t) + \hat{Z} \circ \hat{D}[\Lambda_c^{-1}[y]](t) = \hat{Z} \circ \hat{D}[\Lambda_c^{-1}[y_m]](t). \tag{7.332}$$

From (7.288), $Q_m(z)[y_m](t) = 0$, we have $Q_m(z)[\Lambda_c^{-1}(z)[y_m]](t) = 0$, and

$$\hat{C}Q_m\hat{P}[\Lambda_c^{-1}[y_m]](t) = \hat{C}\hat{P}Q_m[\Lambda_c^{-1}[y_m]](t) = \hat{C}\hat{P}[Q_m[\Lambda_c^{-1}[y_m]]](t) = 0. \tag{7.333}$$

Using (7.330), (7.332), and (7.333) in (7.329), we obtain

$$\begin{aligned} &A^*[\Lambda_c^{-1}[y - y_m]](t) \\ &= \hat{C}Q_m\hat{P}[\Lambda_c^{-1}[y]](t) + \hat{D}\hat{Z}[\Lambda_c^{-1}[y]](t) - \hat{D}\hat{Z}[\Lambda_c^{-1}[y_m]](t) \\ &= \hat{C}Q_m\hat{P}[\Lambda_c^{-1}[y]](t) + \hat{D}\hat{Z}[\Lambda_c^{-1}[y]](t) - \hat{D}\hat{Z}[\Lambda_c^{-1}[y_m]](t) \end{aligned}$$

$$
\begin{aligned}
&+ \hat{Z} \circ \hat{D}[\Lambda_c^{-1}[y_m]](t) - \hat{Z} \circ \hat{C}Q_m[\Lambda_c^{-1}[u]](t) - \hat{Z} \circ \hat{D}[\Lambda_c^{-1}[y]](t) \\
&= \Big(\hat{C}Q_m\hat{P}[\Lambda_c^{-1}[y]](t) - \hat{C}Q_m\hat{Z}[\Lambda_c^{-1}[u]](t)\Big) \\
&+ \Big(\hat{D}\hat{Z}[\Lambda_c^{-1}[y]](t) - \hat{Z} \circ \hat{D}[\Lambda_c^{-1}[y]](t)\Big) \\
&+ \Big(\hat{Z} \circ \hat{D}[\Lambda_c^{-1}[y_m]](t) - \hat{D}\hat{Z}[\Lambda_c^{-1}[y_m]](t)\Big) \\
&+ \Big(\hat{C}Q_m\hat{Z}[\Lambda_c^{-1}[u]](t) - \hat{Z} \circ \hat{C}Q_m[\Lambda_c^{-1}[u]](t)\Big). \qquad (7.334)
\end{aligned}
$$

From (7.324) and the properties that $u(t) \in L^\infty$, $y(t) \in L^\infty$, $y_m(t) \in L^\infty$, $\psi(t) \in L^\infty$, $\theta_p(t) \in L^\infty$, $\psi(t+1) - \psi(t) \in L^2$, and $\theta_p(t+1) - \theta_p(t) \in L^2$ (that is, $\lim_{t\to\infty}(\psi(t+1) - \psi(t)) = 0$ and $\lim_{t\to\infty}(\theta_p(t+1) - \theta_p(t)) = 0$, which means the two operations defined in (7.327) and (7.328) are asymptotically equivalent for all related operators in (7.334)), it follows that in (7.334) the four terms in the parentheses all converge to zero as $t \to \infty$. Therefore, from (7.335) with $A^*(z)$ stable, we have that $\lim_{t\to\infty}(y(t) - y_m(t)) = 0$.

7.4 Discussion

Pole Placement Design with Reference Input

The pole placement control designs in Sections 7.2 and 7.3.2 are developed for asymptotic tracking of a reference output $y_m(t)$ by the plant output $y(t)$. The polynomial $Q_m(D)$ such that $Q_m(D)[y_m](t) = 0$ in (7.132) or (7.288) is used in the control design equations (7.137) and (7.143) or (7.302) and (7.308).

An alternative pole placement controller structure [23], [125] uses a reference input signal $r(t)$, as in a model reference controller, that is,

$$
u(t) = (\Lambda_c(D) - \hat{C}(D, \psi_c))\frac{1}{\Lambda_c(D)}[u](t) - \hat{D}(D, \psi_d)\frac{1}{\Lambda_c(D)}[y](t) + r(t), \quad (7.335)
$$

where $\Lambda_c(D)$ is a monic stable polynomial of degree $n-1$ and

$$
\hat{C}(\lambda, \psi_c) = \lambda^{n-1} + c_{n-2}\lambda^{n-2} + \cdots + c_1\lambda + c_0, \quad (7.336)
$$

$$
\hat{D}(\lambda, \psi_d) = d_{n-1}\lambda^{n-1} + \cdots + d_1\lambda + d_0 \quad (7.337)
$$

are the solutions of the Diophantine equation

$$
\hat{C}(\lambda, \psi_c)\hat{P}(\lambda, \hat{p}) + \hat{D}(\lambda, \psi_d)\hat{Z}(\lambda, \hat{z}) = A^*(\lambda) \quad (7.338)
$$

for a chosen stable monic polynomial $A^*(D)$ of degree $2n-1$ and the estimates $\hat{P}(D, \hat{p})$ and $\hat{Z}(D, \hat{z})$ of $P(D)$ and $Z(D)$ of degrees n and $n-1$ with $P(D)$ monic, in the controlled plant $P(D)[y](t) = Z(D)[u](t)$. This controller, when implemented with $\hat{C}(\lambda, \psi_c) = C(D)$ and $\hat{D}(\lambda, \psi_d) = D(D)$, which satisfy $C(\lambda)P(\lambda) + D(\lambda)Z(\lambda) = A^*(\lambda)$, leads to the closed-loop system

$$y(t) = \frac{Z(D)\Lambda_c(D)}{A^*(D)}[r](t), \; u(t) = \frac{P(D)\Lambda_c(D)}{A^*(D)}[r](t). \tag{7.339}$$

In this case, the plant output $y(t)$ depends on the plant zeros in $Z(D)$. The controller (7.335), with a standard gradient or least-squares algorithm for updating the estimates $\hat{P}(D, \hat{p})$ and $\hat{Z}(D, \hat{z})$ and under a singularity-free condition for solving the Diophantine equation (7.338), ensures that all signals in the closed-loop system are bounded, similar to (7.143) or (7.308).

Direct Pole Placement Control

As a comparison, we now present a direct adaptive pole placement control scheme: direct adaptation of parameters of a pole placement controller, developed by Elliott et al. [92], which avoids the singularity problem of an indirect design and, however, still has open issues for stability and tracking.

Consider the linear time-invariant plant

$$P(D)[y](t) = Z(D)[u](t), \tag{7.340}$$

where $u(t)$ is the input, $y(t)$ is the output, and $Z(D)$, $P(D)$ are coprime polynomials in D (with $D[x](t) = \dot{x}(t)$, $t \in [0, \infty)$ for the continuous-time case or $D[x](t) = x(t+1)$, $t \in \{0, 1, 2, 3, \dots\}$, for the discrete-time case):

$$P(D) = D^n + p_{n-1}D^{n-1} + \cdots + p_1 D + p_0, \tag{7.341}$$

$$Z(D) = z_{n-1}D^{n-1} + \cdots + z_1 D + z_0, \tag{7.342}$$

with p_i, z_i, $i = 0, 1, \dots, n-1$, being constant parameters.

We use the following familiar controller structure:

$$u(t) = k(D)\frac{1}{\Lambda(D)}[u](t) + h(D)\frac{1}{\Lambda(D)}[y](t) + v(t), \tag{7.343}$$

where $\Lambda(D)$ is a stable and monic polynomial in D of degree n,

$$k(D) = k_{n-1}D^{n-1} + \cdots + k_1 D + k_0, \tag{7.344}$$

$$h(D) = h_{n-1}D^{n-1} + \cdots + h_1 D + h_0, \tag{7.345}$$

and $v(t)$ is a bounded external input signal.

Pole placement. Letting $P_d(D)$ be a stable and monic polynomial in D of degree n and with the desired closed-loop poles, solving the following equation for polynomials $k^*(D)$ and $h^*(D)$ of degree $n-1$:

$$k^*(D)P(D) + h^*(D)Z(D) = \Lambda(D)(P(D) - p_d(D)), \tag{7.346}$$

replacing $k(D)$ and $h(D)$ in (7.344) by $k^*(D)$ and $h^*(D)$, respectively, and using (7.341) and (7.344), we obtain the closed-loop system

$$\Lambda(D)P_d(D)[y](t) = \Lambda(D)Z(D)[v](t), \tag{7.347}$$

whose poles are the zeros of $p_d(D)$ and $\Lambda(D)$. Therefore, there exists a parameter vector θ_1^* containing the coefficients of $k^*(D)$ and $h^*(D)$ such that for $\theta_1 = \theta_1^*$ in (7.346) all closed-loop signals are bounded and the closed-loop poles are placed at the desired zeros of $p_d(D)$.

Adaptive design. When the parameters of the plant (7.341) are unknown, an adaptive scheme may be developed to update the estimate θ_1 of θ_1^*, based on an augmented parametrization. Since $Z(D)$ and $P(D)$ are coprime, there exist polynomials $b^*(D) = b_{n-1}D^{n-1} + \cdots + b_1 D + b_0$ and $c^*(D) = c_{n-1}D^{n-1} + \cdots + c_1 D + c_0$ [92] such that

$$b^*(D)Z(D) + c^*(D)P(D) = 1. \tag{7.348}$$

Then, choosing a stable and monic polynomial $f(D)$ of degree n and using (7.346) and (7.348), we obtain

$$\begin{aligned} k^*(D)\frac{1}{f(D)p_d(D)}\frac{1}{\Lambda(D)}P(D) + h^*(D)\frac{1}{f(D)p_d(D)}\frac{1}{\Lambda(D)}Z(D) \\ + c^*(D)\frac{1}{f(D)}P(D) + b^*(D)\frac{1}{f(D)}Z(D) = \frac{1}{f(D)p_d(D)}P(D). \end{aligned} \tag{7.349}$$

Introducing $w(t)$ from $P(D)[w](t) = u(t)$ so that $y(t) = Z(D)[w](t)$, and operating both sides of (7.349) on $w(t)$, and using (7.340), one can derive

$$\frac{1}{f(D)p_d(D)}[u](t) = \theta_1^{*T}\zeta_1(t) + \theta_2^{*T}\zeta_2(t) \tag{7.350}$$

for some parameter vectors θ_1^*, θ_2^*, and measured vector signals $\zeta_1(t)$, $\zeta_2(t)$. Based on this parametrization, either a gradient or least-squares algorithm of Chapter 3 can be used to update the parameter estimate $\theta(t) = [\theta_1^T(t), \theta_2^T(t)]^T$, either in continuous time or in discrete time.

It is shown in [92] that for the discrete-time case under certain block-data adaptation (i.e., the controller parameters are calculated once over each block of parameter estimates), if the external input signal $v(t)$ has $2n$ frequencies and a normalized least-squares algorithm is used, then all closed-loop signals are bounded and the closed-loop system converges to that characterized by (7.347). The similar result is also given in [92] for the continuous-time case with the use of a hybrid control scheme (i.e., controller implementation and controller parameter adaptation are done based on a sampled-data model of the original continuous-time plant). Some stability results without richness of $v(t)$ are given in [18], [161], [202], [203], [319].

Other Indirect Adaptive Control Designs

Using the similar idea of first estimating the plant parameters and then calculating the controller parameters from a design equation, one can also develop adaptive versions of linear quadratic control [151], [362], predictive control [23], [125], [275], PID control [20], [364], backstepping [206], [251], and other control designs [33], [115], [213], [286], [311].

Adaptive versions of robust control designs, [53], [68], [447], [448], can also be developed, for which some controller gain parameters are adaptively tuned (so that it is more in a sense of direct adaptive control).

Problems

7.1 Simulate the indirect adaptive controller for the plant (7.20), with $a_p = 1.5$ and $b_p = -2$, for $r(t) = 1, \sin t, \sin t + \sin 1.7t$, respectively. Also try to simulate the system without parameter projection (7.32) for $\theta_p(0) = [1, 1]$

7.2 Verify the properties of the parameter projection (7.75).

7.3 Verify (7.89)–(7.90) as a solution of (7.87) and test an $n = 3$ case.

7.4 For $\theta(t)$ defined in (7.91), prove that $\theta(t) \in L^\infty$, $\dot{\theta}(t) \in L^2 \cap L^\infty$, and $\rho(t) = \hat{k}_p(t) \in L^\infty$ using the properties of the adaptive law (7.70).

7.5 Verify (7.100) and (7.104).

7.6 Derive (7.110)–(7.113).

7.7 Derive (7.116) and (7.117).

7.8 Derive (7.123)–(7.126).

7.9 Develop an indirect model reference adaptive control scheme based on the controller structure (7.128) and analyze its stability properties.

7.10 Consider the plant

$$(s^3 + p_2 s^2 + p_1 s + p_0)[y](t) = u(t),$$

where p_2, p_1, p_0 are the unknown parameters.

(i) Design an indirect model reference adaptive control scheme, with

$$(s^3 + 3s^2 + 3s + 1)[y_m](t) = r(t).$$

(ii) Simulate your adaptive control system (for $r(t) = 10$ and $r(t) = 6.37\sin(1.34t)$) and plot the tracking errors and parameter errors for an unstable plant $s^3 + p_2 s^2 + p_1 s + p_0 = (s-1)(s+2)(s+3)$).

7.11 Consider the plant

$$(s^2 + p_1 s + p_0)[y](t) = k_p(s + z_0)[u](t),$$

where z_0, p_1, p_0, k_p are unknown and $y(t)$ is measured but not $\dot{y}(t)$.

(i) Simulate the model reference adaptive control scheme designed for Problem 5.3 and applied to this plant with $z_0 = 0.01, 0, -0.01, \bar{z}_0 < 0$ which makes the closed-loop system unstable, while $p_1 = -2$, $p_0 = 2$, $k_p = -1$, $y_m(t) = 10\sin(t)$ (note that $r(t) = (s+1)[y_m](t) = 10(\sin(t) + \cos(t))$).

(ii) For $y_m(t) = 10\sin(t)$, design an indirect adaptive pole placement control scheme including the controller structure and adaptive law.

(iii) Simulate the adaptive control system of (ii) with $p_1 = -2$, $p_0 = 2$, $k_p = -1$, $z_0 = 1, 0.01, 0, -0.01, \bar{z}_0$, and $y_m(t) = 10\sin(t)$ and plot the tracking error and parameter errors.

(iv) Compare the simulation results of (i), (iii) and that of Problem 5.1.

7.12 Assuming that $Q_m(\lambda)\hat{P}(\lambda, \hat{p})$ and $\hat{Z}(\lambda, \hat{z})$ are coprime, verify the solution formula (7.177) for $\{\ddot{C}(\lambda, \phi_c), \ddot{D}(\lambda, \psi_d)\}$ in the design equation (7.137).

7.13 Simulate the indirect adaptive controller for the plant (7.224), with $a_p = -1.5$ and $b_p = -0.2$, for $r(t) = 1, \sin 0.1t, \sin 0.1t + \sin 0.17t$, respectively. Also try to simulate the adaptive system without parameter projection (7.238) for $\theta_p(0) = [1, 1]$, and compare the results with that in Problem 7.1.

7.14 Analyze the properties of the pole placement controller (7.314).

7.15 Verify that in (7.335) the four terms in the parentheses converge to zero as $t \to \infty$ under the signal boundedness condition.

7.16 Complete the adaptive parameter estimation design for (7.350).

Chapter 8

A Comparison Study

When adaptive control designs are derived for systems with arbitrary parametric uncertainties, they may lead to high-order complex control systems. Very often the use of system structural, state, or physical information can help in reducing parametric uncertainties and controller complexity. In this chapter, five adaptive controllers for a two-body system with joint flexibility and damping are studied, including direct (state feedback), direct (output feedback), indirect, direct–indirect, backstepping designs, with detailed design and analysis for the last two designs. With different complexity, they all ensure closed-loop signal boundedness and asymptotic output tracking.

8.1 Benchmark Example

Consider the two-body plant described in Section 4.2.1 (see Figure 8.1):

$$\begin{aligned} J_m\ddot{\theta}_m(t) + b_m\dot{\theta}_m(t) &= u(t) - u_k(t), \\ J_l\ddot{\theta}_l(t) + b_l\dot{\theta}_l(t) &= u_k(t) \end{aligned} \tag{8.1}$$

with flexibility and damping force

$$u_k(t) = k(\theta_m(t) - \theta_l(t)) + b_k(\dot{\theta}_m(t) - \dot{\theta}_l(t)) \tag{8.2}$$

and unknown but constant parameters $J_m > 0$ (motor inertia), $J_l > 0$ (load inertia), $b_m \geq 0$ (motor viscous friction coefficient), $b_l \geq 0$ (load viscous friction coefficient), $k > 0$ (stiffness coefficient of the two flexible bodies), and $b_k > 0$ (damping coefficient of the two moving bodies). Assume that all position and velocity signals $\theta_l(t)$, $\theta_m(t)$, $\dot{\theta}_l(t)$, and $\dot{\theta}_l(t)$ are measured.

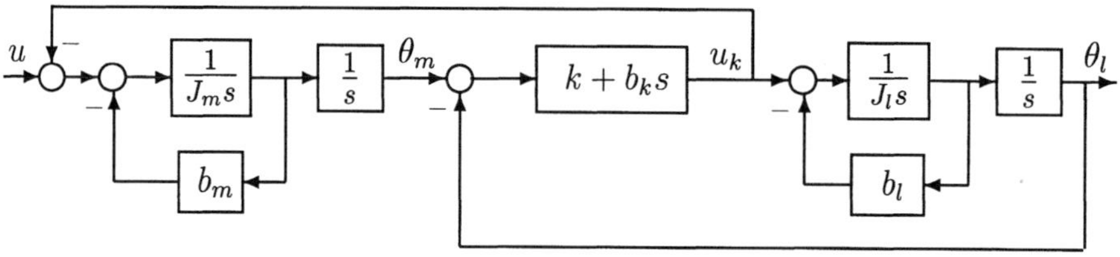

Figure 8.1: A two-body dynamic system.

The plant (8.1) has the state-space form

$$\dot{x}(t) = Ax(t) + Bu(t),$$
$$y(t) = Cx(t), \tag{8.3}$$

where $x(t) = [\theta_m(t), \dot{\theta}_m(t), \theta_l(t), \dot{\theta}_l(t)]^T$, $C = [0, 0, 1, 0]$, and

$$A = \begin{bmatrix} 0 & 1 & 0 & 0 \\ -\frac{k}{J_m} & -\frac{b_m+b_k}{J_m} & \frac{k}{J_m} & \frac{b_k}{J_m} \\ 0 & 0 & 0 & 1 \\ \frac{k}{J_l} & \frac{b_k}{J_l} & -\frac{k}{J_l} & -\frac{b_l+b_k}{J_l} \end{bmatrix}, \quad B = \begin{bmatrix} 0 \\ \frac{1}{J_m} \\ 0 \\ 0 \end{bmatrix}. \tag{8.4}$$

The plant transfer function is

$$G(s) = \frac{y(s)}{u(s)} = \frac{\frac{b_k}{J_m J_l}(s + \frac{k}{b_k})}{s^4 + \frac{J_0 b_k + J_l b_m + J_m b_l}{J_m J_l} s^3 + \frac{b_0 b_k + k J_0 + b_m b_l}{J_m J_l} s^2 + \frac{b_0 k}{J_m J_l} s}, \tag{8.5}$$

where $J_0 = J_m + J_l$ and $b_0 = b_m + b_l$. It can be verified that the controllability of (8.3) is lost when $k = k_c = \frac{b_k b_l}{J_l}$, that is, the uncontrollable mode is $s_c = -\frac{k_c}{b_k} = -\frac{b_l}{J_l} < 0$, so that system (8.3) is stabilizable. Similarly, system (8.3) becomes unobservable when $k = k_o = \frac{b_k b_m}{J_m}$. The unobservable mode is $s_o = -\frac{k_o}{b_k} = -\frac{b_m}{J_m} < 0$ so that system (8.3) is detectable.

The control objective is to find a feedback control signal $u(t)$ such that all the closed-loop signals are bounded and the plant output $y(t) = \theta_l(t)$ tracks a reference signal $y_m(t) = \theta_{ld}(t)$ generated from the reference model

$$\theta_{ld}(t) = \frac{1}{(s + \alpha_1)(s + \alpha_2)(s + \alpha_3)}[r](t) \triangleq W_m(s)[r](t) \tag{8.6}$$

for some chosen design parameters $\alpha_i > 0$, $i = 1, 2, 3$, and a desired bounded reference input signal $r(t)$.

Given that the state vector $x(t) = [\theta_m(t), \dot{\theta}_m(t), \theta_l(t), \dot{\theta}_l(t)]^T$ is available for measurement, one would like to employ a state feedback adaptive controller to reduce the control system order. However, the adaptive control scheme developed in Section 4.1 is not applicable to (8.3), because for any feedback gains $k_1^* = [k_{11}^*, k_{12}^*, k_{13}^*, k_{14}^*]^T$ and k_2^*, with the feedback controller $u(t) = k_2^{*T} y(t) + k_2^* r(t)$, the resulting closed-loop system matrix $A + Bk_1^{*T}$ cannot be made equal to a stable matrix A_m independent of the parameters in A so that the matching condition $A + Bk^{*T} = A_m$ (see (4.16)) cannot be satisfied.

Next, we will compare five other adaptive control schemes applied to the two-body system (8.1), including the direct state feedback design of Section 4.2, direct output feedback design of Section 5.3.5, indirect output feedback design of Section 7.1, direct–indirect design to be developed in Section 8.4, and a backstepping design to be developed in Section 8.5.

8.2 Direct Adaptive Control Designs

A direct adaptive controller is designed with direct adaptation of controller parameters based on performance error (tracking error), which employs either state feedback (see (4.41)) or output feedback (see (5.5)).

8.2.1 State Feedback Design

A state feedback model reference adaptive controller structure is

$$u(t) = k_1^T(t)x(t) + k_2(t)r(t) = \theta^T(t)\omega(t), \tag{8.7}$$

where $k_1(t) \in R^n, k_2 \in R$ ($n = 4$ for (8.3)), and

$$\theta(t) = [k_1^T(t), k_2(t)]^T,\ \omega(t) = [x^T(t), r(t)]^T. \tag{8.8}$$

Introducing the estimation error

$$\epsilon(t) = e(t) + \rho(t)\xi(t), \tag{8.9}$$

where $e(t) = y(t) - y_m(t)$, $\rho(t)$ is the estimate of $\rho^* = \frac{b_k}{J_m J_l}$, and

$$\xi(t) = \theta^T(t)\zeta(t) - W_m(s)[\theta^T\omega](t),\ \zeta(t) = W_m(s)[\omega](t), \tag{8.10}$$

we use the adaptive laws

$$\dot\theta(t) = -\frac{\text{sign}[\rho^*]\Gamma\zeta(t)\epsilon(t)}{m^2(t)},\ \Gamma = \Gamma^T > 0, \tag{8.11}$$

$$\dot\rho(t) = -\frac{\gamma\xi(t)\epsilon(t)}{m^2(t)},\ \gamma > 0, \tag{8.12}$$

where $m(t) = \sqrt{1 + \zeta^T(t)\zeta(t) + \xi^2(t)}$.

In Section 4.2 we showed that this adaptive control scheme ensures closed-loop signal boundedness and asymptotic tracking of $y_m(t)$ by $y(t)$. The order of such an adaptive control scheme is $n + 6n^* + 2$: n^* for $W_m(s)[r](t)$ in (8.6), n^* for $W_m(s)[\theta^T\omega](t)$ in (8.10), $4n^*$ for $\zeta(t)$ in (8.10), and $n+2$ for $\dot{\theta}(t)$ and $\dot{\rho}(t)$ in (8.11) and (8.12). For $n = 4$, $n^* = 3$ in (8.5), the order is 24.

8.2.2 Output Feedback Design

An output feedback model reference adaptive controller structure is

$$u(t) = \theta_1^T\omega_1(t) + \theta_2^T\omega_2(t) + \theta_{20}y(t) + \theta_3 r(t), \tag{8.13}$$

where $\theta_1 \in R^{n-1}$, $\theta_2 \in R^{n-1}$ ($n = 4$ for (8.3)), $\theta_{20} \in R$, and $\theta_3 \in R$, and

$$\omega_1(t) = \frac{a(s)}{\Lambda(s)}[u](t),\ \omega_2(t) = \frac{a(s)}{\Lambda(s)}[y](t) \tag{8.14}$$

with $a(s) = [1, s, \ldots, s^{n-2}]^T$ and $\Lambda(s)$ a monic stable polynomial of degree 3. The adaptive parameter update scheme is

$$\dot{\theta}(t) = -\frac{\text{sign}[\rho^*]\Gamma\epsilon(t)\zeta(t)}{m^2(t)},\ \Gamma = \Gamma^T > 0, \tag{8.15}$$

$$\dot{\rho}(t) = -\frac{\gamma\epsilon(t)\xi(t)}{m^2(t)},\ \gamma > 0, \tag{8.16}$$

where $m(t) = \sqrt{1 + \zeta^T(t)\zeta(t) + \xi^2(t)}$ and

$$\epsilon(t) = e(t) + \rho(t)\xi(t), \tag{8.17}$$

$$\xi(t) = \theta^T(t)\zeta(t) - W_m(s)[\theta^T\omega](t),\ \zeta(t) = W_m(s)[\omega](t). \tag{8.18}$$

In Section 5.3.5, we showed that this adaptive control scheme has the desired stability and tracking properties. In general, the order of such an adaptive control scheme is $4n+5n^*-1$: n^* for $W_m(s)[\theta^T\omega](t)$, n^* for $W_m(s)[r](t)$, $2n+1$ for $\dot{\theta}(t)$ and $\dot{\rho}(t)$, $2n-2$ for $\omega_1(t)$ and $\omega_2(t)$, and $3n^*$ or more for $\zeta(t)$ ($3n^*$ when $n^* = n-1$). For $n = 4$, $n^* = 3$, the order is 30.

8.3 Indirect Adaptive Control Design

An indirect model reference adaptive controller structure is the same as that in (8.13), with $\Lambda(s) = \Lambda_c(s)$ of degree $n-1$ and $\theta_1 \in R^{n-1}$, $\theta_2 \in R^{n-1}$,

$\theta_{20} \in R$, and $\theta_3 \in R$, solved from a design equation (see Section 7.1.2) using the estimates $\theta_p(t)$ of the plant parameters θ_p^*, updated from the gradient adaptive algorithm (see Section 3.3 and Section 7.1.2)

$$\dot{\theta}_p(t) = -\frac{\Gamma\phi(t)\epsilon(t)}{m^2(t)}, \ \Gamma = \Gamma^T > 0, \tag{8.19}$$

where $m(t) = \sqrt{1 + \kappa\phi^T(t)\phi(t)}$, $\kappa > 0$, and

$$\epsilon(t) = \theta_p^T(t)\phi(t) - y(t) + \frac{\Lambda_e(s) - s^n}{\Lambda_e(s)}[y](t) \tag{8.20}$$

for the parametrized plant model of an nth-order plant (3.1):

$$y(t) - \frac{\Lambda_e(s) - s^n}{\Lambda_e(s)}[y](t) = \theta_p^{*T}\phi(t) \tag{8.21}$$

with $\theta_p^* \in R^{n+m+1}$ and $\Lambda_e(s)$ stable and of degree n.

The order of such an adaptive control scheme is $4n - n^* + 1$: $2n - n^* + 1$ for $\dot{\theta}_p(t)$, $2n$ for $\omega_1^e(t)$, $\omega_2^e(t)$ and $\omega_1^c(t)$, $\omega_2^c(t)$ (when the controller filter $\Lambda_c(s)$ is a part of the estimator filter $\Lambda_e(s)$, see Section 7.1.2). For $n = 4$, $n^* = 3$ for the plant (8.3), the order is 14.

8.4 Direct–Indirect Adaptive Control Design

In this section, a reduced-order adaptive control scheme using a direct state feedback control law and an indirect feedforward compensation design is presented for the two-body system with joint flexibility and damping. A detailed stability analysis is given for such a two-step adaptive control design which ensures closed-loop signal boundedness and asymptotic output tracking. With this design, the adaptive control system order is 12.

8.4.1 Direct Adaptive Control for Motor Dynamics

We first develop a direct adaptive control scheme for the motor dynamics of the two-body system (8.1), which is the second subsystem of (8.3):

$$\dot{x}_2(t) = -A_2 x(t) + B_2 u(t), \tag{8.22}$$

where

$$A_2 = \begin{bmatrix} \frac{k}{J_m} & \frac{b_m + b_k}{J_m} & -\frac{k}{J_m} & -\frac{b_k}{J_m} \end{bmatrix}, \ B_2 = \frac{1}{J_m}. \tag{8.23}$$

Controller Structure

For the subsystem (8.22), we propose the state feedback controller

$$u(t) = K_2(t)x(t) + k_r(t)(-\alpha_1 x_2(t) - \alpha_2(x_2(t) + \alpha_1 x_1(t)) + \hat{r}(t)), \tag{8.24}$$

where $\hat{r}(t)$ is a bounded design signal to be specified, $\alpha_1 > 0$, $\alpha_2 > 0$ are design parameters in (8.6), and $K_2(t) \in R^{1\times 4}$ and $k_r(t) \in R$ are estimates of

$$K_2^* = \frac{A_2}{B_2},\ k_r^* = \frac{1}{B_2}. \tag{8.25}$$

The controller (8.24) leads the subsystem (8.22) to

$$\begin{aligned}\dot{x}_2(t) = &-\alpha_1 x_2(t) - \alpha_2(x_2(t) + \alpha_1 x_1(t)) + \hat{r}(t) + B_2(K_2(t) - K_2^*)x(t)\\ &+ B_2(k_r(t) - k_r^*)(-\alpha_1 x_2(t) - \alpha_2(x_2(t) + \alpha_1 x_1(t)) + \hat{r}(t)).\end{aligned} \tag{8.26}$$

Defining the auxiliary signals

$$\bar{x}_2(t) = x_2(t) + \alpha_1 x_1(t),\ \bar{r}(t) = -\alpha_1 x_2(t) - \alpha_2(x_2(t) + \alpha_1 x_1(t)) + \hat{r}(t) \tag{8.27}$$

and using the definition $x_2(t) = \dot{x}_1(t)$, we rewrite (8.26) as

$$\dot{\bar{x}}_2(t) = -\alpha_2 \bar{x}_2(t) + \hat{r}(t) + B_2(K_2(t) - K_2^*)x(t) + B_2(k_r(t) - k_r^*)\bar{r}(t). \tag{8.28}$$

Introduce the reference signal

$$\theta_{md}(t) = x_{1m}(t) = \frac{1}{(s+\alpha_1)(s+\alpha_2)}[\hat{r}](t) \tag{8.29}$$

and use it to define

$$x_{2m}(t) = \dot{x}_{1m}(t),\ \bar{x}_{2m}(t) = x_{2m}(t) + \alpha_1 x_{1m}(t). \tag{8.30}$$

It follows from (8.29) and (8.30) that

$$\dot{\bar{x}}_{2m}(t) = -\alpha_2 \bar{x}_{2m}(t) + \hat{r}(t). \tag{8.31}$$

Defining the error signal

$$\bar{e}_2(t) \triangleq \bar{x}_2(t) - \bar{x}_{2m}(t) = e_2(t) + \alpha_1 e_1(t), \tag{8.32}$$

where $e_1(t) = x_1(t) - x_{1m}(t)$, $e_2(t) = x_2(t) - x_{2m}(t) = \dot{e}_1(t)$, from (8.28) and (8.32), we obtain the error equation

$$\dot{\bar{e}}_2(t) = -\alpha_2 \bar{e}_2(t) + B_2(K_2(t) - K_2^*)x(t) + B_2(k_r(t) - k_r^*)\bar{r}(t). \tag{8.33}$$

Controller Parameter Adaptation

We choose the adaptive laws for $K_2(t)$ and $k_r(t)$ as

$$\dot{K}_2^T(t) = -\Gamma \bar{e}_2(t) x(t), \; \Gamma = \Gamma^T > 0, \tag{8.34}$$

$$\dot{k}_r(t) = -\gamma \bar{e}_2(t) \bar{r}(t), \; \gamma > 0. \tag{8.35}$$

For the positive definite function

$$V(\bar{e}_2, \tilde{K}_2, \tilde{k}_r) = \frac{1}{2}(\bar{e}_2^2 + B_2 \tilde{K}_2 \Gamma^{-1} \tilde{K}_2^T + B_2 \gamma^{-1} \tilde{k}_r^2), \tag{8.36}$$

where $\tilde{K}_2(t) = K_2(t) - K_2^*$, $\tilde{k}_r(t) = k_r(t) - k_r^*$, from (8.33)–(8.35), we have

$$\begin{aligned} &\frac{d}{dt} V(\bar{e}_2(t), \tilde{K}_2(t), \tilde{k}_r(t)) \\ &= -\alpha_2 \bar{e}_2^2(t) + \bar{e}_2(t)(B_2(K_2(t) - K_2^*)x(t) + B_2(k_r(t) - k_r^*)\bar{r}(t)) \\ &\quad + B_2 \tilde{K}_2(t) \Gamma^{-1} \dot{\tilde{K}}_2^T(t) + B_2 \tilde{k}_r(t) \gamma^{-1} \dot{\tilde{k}}_r(t) = -\alpha_2 \bar{e}_2^2(t), \end{aligned} \tag{8.37}$$

from which we conclude that $\bar{e}_2(t) \in L^\infty \cap L^2$, $K_2(t) \in L^\infty$, $k_r(t) \in L^\infty$. Then, from the boundedness of $x_{1m}(t)$ and $\dot{x}_{1m}(t)$ in (8.29) and definitions of $e_1(t)$ and $e_2(t)$ in (8.32), we have that $e_1(t) \in L^\infty \cap L^2$, $e_2(t) \in L^\infty \cap L^2$, $x_1(t) \in L^\infty$, $x_2(t) \in L^\infty$, and from (8.1) (i.e., $(J_l s^2 + (b_l + b_k)s + k)[\theta_l](t) = (b_k s + k)[x_1](t)$ is stable with input $x_1 = \theta_m(t)$, output θ_l), we have $x_3(t) = \theta_l(t) \in L^\infty$, $x_4(t) = \dot{\theta}_l(t) \in L^\infty$, that is, $x(t) \in L^\infty$. Finally, from (8.24), we have $u(t) \in L^\infty$, which, with (8.22) and the boundedness of $x(t)$, implies that $\dot{x}_2(t) \in L^\infty$. Since $e_1(t) = x_1(t) - x_{1m}(t) = \theta_m(t) - \theta_{md}(t) \in L^2$, $\dot{e}_1(t) = e_2(t) \in L^\infty$, and $e_2(t) = x_2(t) - x_{2m}(t) = \dot{\theta}_m(t) - \dot{\theta}_{md}(t) \in L^2$, $\dot{e}_2(t) = \dot{x}_2(t) - \ddot{\theta}_{md}(t) \in L^\infty$, we have that

$$\lim_{t \to \infty} (\theta_m(t) - \theta_{md}(t)) = 0, \; \lim_{t \to \infty} (\dot{\theta}_m(t) - \dot{\theta}_{md}(t)) = 0. \tag{8.38}$$

In summary, we have the following results.

Lemma 8.1 *For the plant (8.1) and the reference signal $\theta_{md}(t)$ in (8.29), the adaptive controller (8.24) updated with (8.34)–(8.35) ensures that $x(t) \in L^\infty$ and $\lim_{t\to\infty}(\theta_m(t) - \theta_{md}(t)) = \lim_{t\to\infty}(\dot{\theta}_m(t) - \dot{\theta}_{md}(t)) = 0$.*

8.4.2 Indirect Adaptive Control for Load Dynamics

To ensure asymptotic tracking for the load output $y(t) = \theta_l(t)$, we need to design the signal $\hat{r}(t)$ in (8.24). From the fourth subsystem of (8.3), that is,

$$\dot{x}_4(t) = \ddot{\theta}_l(t) = \begin{bmatrix} \frac{k}{J_l} & \frac{b_k}{J_l} & -\frac{k}{J_l} & -\frac{b_l + b_k}{J_l} \end{bmatrix} x(t), \tag{8.39}$$

we obtain the relationship between $\theta_l(t)$ and $\theta_m(t)$:

$$(J_l s^2 + (b_l + b_k)s + k)[\theta_l](t) = (b_k s + k)[\theta_m](t), \tag{8.40}$$

which can be expressed as

$$(s^2 + p_3^* s + p_2^*)[\theta_l](t) = (p_1^* s + p_2^*)[\theta_m](t), \tag{8.41}$$

where $p_1^* = \frac{b_k}{J_l} > 0$, $p_2^* = \frac{k}{J_l} > 0$, $p_3^* = \frac{b_l + b_k}{J_l} > 0$, so that both $p_1^* s + p_2^*$ and $s^2 + p_3^* s + p_2^*$ are stable. For the given reference signal $\theta_{ld}(t)$ in (8.6), if we were able to make $\theta_m(t)$ to satisfy

$$(p_1^* s + p_2^*)[\theta_m](t) = (s^2 + p_3^* s + p_2^*)[\theta_{ld}](t), \tag{8.42}$$

then the desired tracking $(s^2 + p_3^* s + p_2^*)[\theta_l - \theta_{ld}](t) = 0$ (which means that $\lim_{t\to\infty}(\theta_l(t) - \theta_{ld}(t)) = 0$) would be achieved. In view of (8.38), the condition (8.42) is asymptotically equivalent to

$$(p_1^* s + p_2^*)[\theta_{md}](t) = (s^2 + p_3^* s + p_2^*)[\theta_{ld}](t), \tag{8.43}$$

which would suggest that an ideal $\hat{r}(t)$ in (8.24) and (8.29) should be

$$\hat{r}(t) = \frac{s^2 + p_3^* s + p_2^*}{(p_1^* s + p_2^*)(s + \alpha_3)}[r](t). \tag{8.44}$$

However, such an ideal design cannot be implemented because the parameters $p_i^* > 0$, $i = 1, 2, 3$, are unknown. The adaptive control objective here is to use some desirable adaptive estimates p_1, p_2, and p_3 of the unknown parameters p_1^*, p_2^*, and p_3^* to generate an implementable input signal $\hat{r}(t)$ in (8.24) and (8.29) to achieve asymptotic tracking of $\theta_{ld}(t)$ by $\theta_l(t)$.

Reference Input

To develop such an adaptive control scheme, we choose the design signal $\hat{r}(t)$ in the controller (8.24) and the reference system (8.29) as

$$\hat{r}(t) = (s + \alpha_2)\left[\frac{1}{p_1 s + p_2}\left[(s^2 + p_3 s + p_2)\left[\frac{1}{(s + \alpha_2)(s + \alpha_3)}[r]\right]\right]\right](t), \tag{8.45}$$

where p_1, p_2, and p_3 are time-varying estimates of p_1^*, p_2^*, and p_3^*, generated from an adaptive update law to satisfy the following key properties:

(P1) $p_i(t) \in L^\infty$, $p_i(t) \geq p^0$, $i = 1, 2, 3$, for some constant $p^0 > 0$;

(P2) $\dot{p}_i(t) \in L^2$, $i = 2, 3$;

(P3) for some $\delta_1(t)$ such that $\lim_{t\to\infty}\delta_1(t)=0$,

$$(p_1 s + p_2)[\theta_{md}](t) = (s^2 + p_3 s + p_2)[\theta_{ld}](t) + \delta_1(t); \tag{8.46}$$

(P4) for some $\delta_2(t)$ such that $\lim_{t\to\infty}\delta_2(t)=0$,

$$(p_1 s + p_2)[\theta_{md}](t) = (s^2 + p_3 s + p_2)[\theta_l](t) + \delta_2(t). \tag{8.47}$$

Remark 8.1 (i) The term $s+\alpha_2$ in (8.45) is not cancelled out because the parameter estimates $p_1(t)$, $p_2(t)$, and $p_3(t)$ are time varying.

(ii) The property (P1) ensures that $p_1(t)s+p_2(t)$ is a stable polynomial in the sense that $\frac{1}{p_1(t)s+p_2(t)}$ is a stable operator, while the properties (P1) and (P2) ensure that $s^2+p_3(t)s+p_2(t)$ is a stable polynomial in the sense that $\frac{1}{s^2+p_3(t)s+p_2(t)}$ is a stable operator (see Section 2.5.3 and Lemma 8.2).

(iii) The properties (P3) and (P4) imply

$$(s^2 + p_3 s + p_2)[\theta_{ld}](t) = (s^2 + p_3 s + p_2)[\theta_l](t) \tag{8.48}$$

asymptotically, which, together with the properties (P1) and (P2) (i.e., $s^2 + p_3 s + p_2$ is stable), implies that $\lim_{t\to\infty}(\theta_l(t) - \theta_{ld}(t)) = 0$. □

Proposition 8.1 *The property (P1) is ensured by a parameter projection which results in $p_i(t) \in [p_i^a, p_i^b]$ for some constants $p_i^b > p_i^a > 0$.*

Proposition 8.2 *The property (P1) ensures that $\tilde{r}(t) \in L^\infty$.*

Proof: In view of (8.45), for a bounded $r(t)$ and with the property (P1),

$$\tilde{r}(t) \triangleq (s^2 + p_3 s + p_2)\left[\frac{1}{(s+\alpha_2)(s+\alpha_3)}[r]\right](t) \tag{8.49}$$

is also bounded because

$$\begin{aligned}\tilde{r}(t) &= r(t) + ((p_3 - \alpha_2 - \alpha_3)s + p_2 - \alpha_2\alpha_3)\left[\frac{1}{(s+\alpha_2)(s+\alpha_3)}[r]\right](t)\\ &= r(t) + \begin{bmatrix} p_3 - \alpha_2 - \alpha_3 & p_2 - \alpha_2\alpha_3 \end{bmatrix}\begin{bmatrix} x_{r1}(t) \\ x_{r2}(t)\end{bmatrix},\end{aligned} \tag{8.50}$$

where the state variables $x_{r1}(t)$ and $x_{r2}(t)$ are defined from

$$\begin{bmatrix} \dot{x}_{r1}(t) \\ \dot{x}_{r2}(t)\end{bmatrix} = \begin{bmatrix} 0 & 1 \\ -\alpha_2\alpha_3 & -\alpha_2 - \alpha_3\end{bmatrix}\begin{bmatrix} x_{r1}(t) \\ x_{r2}(t)\end{bmatrix} + \begin{bmatrix} 0 \\ 1 \end{bmatrix} r(t). \tag{8.51}$$

The boundedness of the signal

$$w(t) \triangleq \frac{1}{p_1 s + p_2}[\tilde{r}](t) \tag{8.52}$$

follows from its differential equation form

$$\dot{w}(t) = -\frac{p_2}{p_1} w(t) + \frac{1}{p_1}\tilde{r}(t) \tag{8.53}$$

with $p_i \in [p_i^a, p_i^b]$, $i = 1, 2$, for some constants $p_i^b > p_i^a > 0$, that is, $\frac{p_2}{p_1} \in [\frac{p_2^a}{p_1^b}, \frac{p_2^b}{p_1^a}]$. Finally, from (8.45), (8.49), (8.52), and (8.53), we have the boundedness of

$$\hat{r}(t) = (s + \alpha_2) w(t) = \dot{w}(t) + \alpha_2 w(t) = \left(\alpha_2 - \frac{p_2}{p_1}\right) w(t) + \frac{1}{p_1}\tilde{r}(t). \tag{8.54}$$

This completes the proof. ∇

Remark 8.2 Propositions 8.1 and 8.2 illustrate the basic realization of the reference input signal $\hat{r}(t)$ in (8.45): with $p_i(t) \in [p_i^a, p_i^b]$ for some constants $p_i^b > p_i^a > 0$, $\hat{r}(t)$ is realized through (8.54) and (8.50), with a third-order system consisting of (8.51) and (8.53). To implement a parameter projection in an adaptive law for $p_i(t)$, $i = 1, 2, 3$, the bounds p_i^a and p_i^b are assumed to be known such that $p_i^* \in [p_i^a, p_i^b]$, $i = 1, 2, 3$. It is important to note that the desired property that $p_i(t) \in [p_i^a, p_i^b]$ can be obtained independently of the boundedness of other signals in the closed-loop control system. $\Box$

Lemma 8.2 *If $p_i(t) \in L^\infty$, $p_i(t) \geq p^0$ for some constant $p^0 > 0$, $\dot{p}_i(t) \in L^2$, $i = 2, 3$, and $p_j(t) \in L^\infty$, $j = 0, 1$, then $(p_1(t)s + p_0(t)) \frac{1}{s^2 + p_3(t)s + p_2(t)}$ is a stable and strictly proper operator.*[1]

Proof: Introducing the state variable equation

$$\begin{bmatrix} \dot{z}_1(t) \\ \dot{z}_2(t) \end{bmatrix} = \begin{bmatrix} 0 & 1 \\ -p_2(t) & -p_3(t) \end{bmatrix} \begin{bmatrix} z_1(t) \\ z_2(t) \end{bmatrix} + \begin{bmatrix} 0 \\ 1 \end{bmatrix} u(t),$$
$$y(t) = \begin{bmatrix} p_1(t) & p_0(t) \end{bmatrix} \begin{bmatrix} z_1(t) \\ z_2(t) \end{bmatrix}, \tag{8.55}$$

we have $y(t) = (p_1(t)s + p_0(t)) [\frac{1}{s^2 + p_3(\cdot)s + p_2(\cdot)}[u]](t)$, which defines the operator $(p_1(t)s + p_0(t)) \frac{1}{s^2 + p_3(t)s + p_2(t)}$. The stability property of the system

$$\dot{z}(t) = A(t) z(t), \ A(t) = \begin{bmatrix} 0 & 1 \\ -p_2(t) & -p_3(t) \end{bmatrix} \tag{8.56}$$

[1]See Definition 2.23 in Section 2.5.3.

is crucial for Lemma 8.2 to hold. Consider the Lyapunov equation

$$P(t)A(t) + A^T(t)P(t) = -Q(t) \tag{8.57}$$

for a given $Q(t) = Q^T(t)$ and denote a possible solution as

$$P(t) = \begin{bmatrix} p_{11}(t) & p(t) \\ p(t) & p_{22}(t) \end{bmatrix}. \tag{8.58}$$

From (8.57) and (8.58), we obtain

$$M_a(t) \begin{bmatrix} p_{11}(t) \\ p(t) \\ p_{22}(t) \end{bmatrix} = - \begin{bmatrix} q_{11}(t) \\ q(t) \\ q_{22}(t) \end{bmatrix}, \tag{8.59}$$

for $Q(t) = \{q_{ij}(t)\} \in R^{2\times 2}$ with $q_{12} = q_{21} = q$, where

$$M_a(t) = \begin{bmatrix} 0 & -2p_2(t) & 0 \\ 1 & -p_3(t) & -p_2(t) \\ 0 & 2 & -2p_3(t) \end{bmatrix} \tag{8.60}$$

with $\det[M_a(t)] = -4p_2(t)p_3(t) \leq -4(p^0)^2 \neq 0$. Hence, for any $Q(t)$, there is a unique solution $P(t) \in L^\infty$. As a well-known result [151], [426] for the general case of dimension n, if $Q(t) > 0$, then there is a unique $P(t) > 0$ to satisfy (8.57) for any $A(t) \in R^{n\times n}$ whose eigenvalues all have negative real parts.

Let $Q(t) = I_2$, the 2×2 identity matrix, and $P(t) = P^T(t) > 0$ satisfy (8.57). Then, from (8.57), we have that

$$\dot{P}(t)A(t) + A^T(t)\dot{P}(t) = -P(t)\dot{A}(t) - \dot{A}^T(t)P(t), \tag{8.61}$$

which can written as

$$M_a(t) \begin{bmatrix} \dot{p}_{11}(t) \\ \dot{p}(t) \\ \dot{p}_{22}(t) \end{bmatrix} = \begin{bmatrix} 2p(t)\dot{p}_2(t) \\ p(t)\dot{p}_3(t) + p_{22}(t)\dot{p}_2(t) \\ 2p_{22}(t)\dot{p}_3(t) \end{bmatrix}, \tag{8.62}$$

from which we conclude that $\dot{P}(t) \in L^2$, as $p_i(t) \in L^2$, $i = 2, 3$, and $P(t) \in L^\infty$ and $\det[M_a(t)] \leq -4(p^0)^2 \neq 0$.

Since $p_i(t) \geq p^0 > 0$, $i = 2, 3$, for some constant p^0, that is, the eigenvalues of $A(t)$ in (8.56) are in $\mathrm{Re}[s] \leq -\sigma_0$ for some constant $\sigma_0 > 0$, we have that all eigenvalues of $P(t) \in L^\infty$ from (5.57) with $Q(t) = I_2$ are in some interval $[\lambda^a, \lambda^b]$ for some constants $\lambda^b > \lambda^a > 0$. Hence, $\lambda^a z^T z \leq z^T P(t) z \leq \lambda^b z^T z$.

Introduce the positive definite function $V(z,t) = z^T P(t) z$ and derive

$$\dot{V} = -z^T z + z^T \dot{P}(t) z \leq -\left(\alpha - \beta\|\dot{P}(t)\|_2\right) V, \tag{8.63}$$

where $\alpha = \frac{1}{\lambda^b}$ and $\beta = \frac{1}{\lambda^a}$. Then, from (8.63), we have

$$V(z(t),t) \leq V(z(0),0)e^{-\alpha t+\beta\int_0^t \|\dot{P}(\tau)\|_2\,d\tau}. \tag{8.64}$$

Since $\dot{P}(t) \in L^2$, there is a constant $\gamma_0 > 0$ such that $\int_0^\infty \|\dot{P}(\tau)\|_2^2\,d\tau \leq \gamma_0$. Using the Schwarz inequality (see Section 2.3.2), for $\delta = (\frac{\alpha}{2\beta})^2$, we have

$$\int_0^t \|\dot{P}(\tau)\|_2\,d\tau \leq \sqrt{t}\sqrt{\gamma_0} \leq \sqrt{t+\frac{\gamma_0}{\delta}}\sqrt{\delta t+\gamma_0} = \frac{\alpha}{2\beta}t + \frac{2\gamma_0\beta}{\alpha},\ \forall t > 0. \tag{8.65}$$

From (8.64) and (8.65), we have $V(z(t),t) \leq V(z(0),0)e^{-\frac{\alpha}{2}t}e^{\frac{2\gamma_0\beta^2}{\alpha}}$, which implies that the equilibrium state $z_e = 0$ of system (8.56) is exponentially stable, so that the state transition matrix $\Phi(t,\tau)$ of (8.56) satisfies $\|\Phi(t,\tau)\| \leq \beta_0 e^{-\alpha_0(t-\tau)}$ for some constants $\beta_0 > 0$, $\alpha_0 > 0$ and any $t \geq \tau \geq 0$.

From here on, Definition 2.23 for a stable and strictly proper operator in Section 2.5.3 can be verified, using (2.11), assuming zero initial conditions. ∇

Remark 8.3 The above proof is applicable to the case when $A(t) \in R^{n\times n}$ is differentiable and bounded, all eigenvalues of $A(t)$ have real parts less than $-\sigma_0$ for some constant $\sigma_0 > 0$ and all $t \geq 0$, and $\dot{A}(t) \in L^2$, to conclude that the state transition matrix $\Phi(t,\tau)$ of $\dot{z}(t) = A(t)z(t)$ satisfies $\|\Phi(t,\tau)\| \leq \beta_0 e^{-\alpha_0(t-\tau)}$ for some constants $\beta_0 > 0$, $\alpha_0 > 0$, and $\forall t \geq \tau \geq 0$.

In general, for a linear system $\dot{z}(t) = A(t)z(t)$, $z \in R^n$, where $A(t) \in R^{n\times n}$ is differentiable and bounded and all eigenvalues of $A(t)$ have real parts less than $-\sigma_0$ for some constant $\sigma_0 > 0$ and all $t \geq 0$, its equilibrium state $z_e = 0$ is exponentially stable if

$$\int_t^{t+T} \|\dot{A}(\tau)\|^p\,d\tau \leq \nu T + \gamma_0 \tag{8.66}$$

for any $t \geq 0$ and $T > 0$, some $\gamma_0 > 0$, some $\nu > 0$ small, and $p \in [1,\infty)$.

The $p = 1$ case is proved in [151], [426], A proof for the $p = 2$ case is given in [151], and the general case for $p \in [1,\infty)$ is proved in Lemma 7.1 in two steps: the existence of a bounded and positive definite matrix $P(t) = P^T(t) \in R^{n\times n}$ satisfying $P(t)A(t) + A^T(t)P(t) = -I$ and with its eigenvalues between two constants $\lambda^b > \lambda^a > 0$, and the derivation of

$$\int_0^t \|\dot{P}(\tau)\|_2\,d\tau \leq \frac{\alpha}{2\beta}t + \frac{\gamma_0 a_0^p(2\beta)^{p-1}}{\alpha^{p-1}},\ \forall t > 0, \tag{8.67}$$

using $\|\dot{P}(t)\|_2 \leq a_0\|\dot{A}(t)\|$ for some $a_0 > 0$ and from (8.66) with $\nu \in (0, \left(\frac{\alpha}{2a_0\beta}\right)^p]$, where $\alpha = \frac{1}{\lambda^b}$ and $\beta = \frac{1}{\lambda^a}$. Then, from the n-dimensional version of (8.64), we have that $V(z(t),t) \leq V(z(0),0)e^{-\frac{\alpha}{2}t}e^{\frac{\beta\gamma_0 a_0^p(2\beta)^{p-1}}{\alpha^{p-1}}}$. □

Parameter Estimation

To develop an adaptive law for the parameter estimates p_1, p_2, and p_3 with the desired properties (P1)–(P4), we rewrite (8.41) as

$$\ddot{\theta}_l(t) + p_3^* \dot{\theta}_l(t) + p_2^* \theta_l(t) = p_1^* \dot{\theta}_m(t) + p_2^* \theta_m(t). \tag{8.68}$$

In addition to $\dot{\theta}_l(t)$, $\theta_l(t)$, $\dot{\theta}_m(t)$, and $\theta_m(t)$, we assume that $\ddot{\theta}_l(t)$ is also available. This assumption is needed for obtaining parameter estimates p_1, p_2, and p_3, based on (8.68), to meet (8.47). If we just want to estimate p_1^*, p_2^*, and p_3^* of the subsystem (8.68), without the need to meet (8.47), we do not need the knowledge of $\ddot{\theta}_l(t)$; in fact, the knowledge of $\theta_l(t)$ and $\theta_m(t)$ is sufficient for a gradient algorithm of Section 3.3 for estimating p_i^*, $i = 1, 2, 3$.

To proceed, from (8.68), we obtain the linear parametrization model

$$\ddot{\theta}_l(t) = p^* \omega_p(t), \tag{8.69}$$

where

$$p^* = [p_1^*, p_2^*, p_3^*], \tag{8.70}$$

$$\omega_p(t) = [\dot{\theta}_m(t), \theta_m(t) - \theta_l(t), -\dot{\theta}_l(t)]^T. \tag{8.71}$$

Here the signals $\ddot{\theta}_l(t)$ and $\omega_p(t)$ are bounded because $x(t)$ is, and so is $\dot{\omega}_p(t)$. Such boundedness depends on the boundedness of $\hat{r}(t)$, which is ensured by letting $p_i(t) \in [p_i^a, p_i^b]$ for some constants $p_i^b > p_i^a > 0$ independent of other signals in the adaptive system (see Remark 8.2).

Following the gradient adaptive law design in Section 3.3, we let $p(t) = [p_1(t), p_2(t), p_3(t)]$ be an estimate of p^* and define the estimation error as

$$\epsilon(t) = \ddot{\theta}_l(t) - p(t)\omega_p(t). \tag{8.72}$$

Using the parametrized model (8.69), we obtain the error equation

$$\epsilon(t) = -\tilde{p}(t)\omega_p(t), \tag{8.73}$$

where $\tilde{p}(t) = p(t) - p^*$. Based on the error equation (8.72), we use the gradient algorithm for updating the parameter estimate $p(t)$:

$$\dot{p}(t) = \gamma_p \epsilon(t) \omega_p^T(t) + f(t), \tag{8.74}$$

where $\gamma_p > 0$ is an adaptation gain and $f(t) = [f_1(t), f_2(t), f_3(t)]$ is a projection modification to ensure that $p_i(t) \in [p_i^a, p_i^b]$ for some known positive bounds $p_i^b > p_i^a > 0$ such that $p_i^* \in [p_i^a, p_i^b]$, $i = 1, 2, 3$. To construct such a projection term, as in Section 3.8, we define

$$g(t) = \gamma_p \omega_p(t) \epsilon(t) = [g_1(t), g_2(t), g_3(t)] \tag{8.75}$$

and choose $p_i(0) \in [p_i^a, p_i^b]$, and set

$$f_i(t) = \begin{cases} 0 & \text{if } p_i(t) \in (p_i^a, p_i^b) \text{ or} \\ & \text{if } p_i(t) = p_i^a,\ g_i(t) \geq 0 \text{ or} \\ & \text{if } p_i(t) = p_i^b,\ g_i(t) \leq 0, \\ -g_i(t) & \text{otherwise.} \end{cases} \tag{8.76}$$

For the positive definite function

$$V(\tilde{p}) = \frac{1}{2\gamma_p}\tilde{p}\tilde{p}^T, \tag{8.77}$$

in view of (8.73) and (8.74), we derive its time derivative as

$$\begin{aligned} \frac{d}{dt}V(\tilde{p}(t)) &= \tilde{p}(t)(\epsilon(t)\omega_p(t) + \gamma_p^{-1}f^T(t)) \\ &= -\epsilon^2(t) + \gamma_p^{-1}\tilde{p}f^T(t). \end{aligned} \tag{8.78}$$

From the projection algorithm (8.76), $p_i(t) \in [p_i^a, p_i^b]$, $i = 1, 2, 3$, and

$$\tilde{p}f^T(t) = \tilde{p}_1(t)f_1(t) + \tilde{p}_2(t)f_2(t) + \tilde{p}_3(t)f_3(t), \tag{8.79}$$

$$\tilde{p}_i(t)f_i(t) = (p_i(t) - p_i^*)f_i(t) \leq 0,\ i = 1, 2, 3, \tag{8.80}$$

that is, $\gamma_p^{-1}\tilde{p}f^T(t) \leq 0$. This, together with (8.78), leads to

$$\frac{d}{dt}V(\tilde{p}(t)) \leq -\epsilon^2(t), \tag{8.81}$$

from which we have $\epsilon(t) \in L^2$. This result, together with the boundedness of $p(t)$, (8.73), and (8.74), implies that $\epsilon(t) \in L^\infty$, $\dot{p}(t) \in L^2 \cap L^\infty$. Since $\dot{\omega}_p(t) \in L^\infty$ (see (8.71) with the boundedness of $x(t)$), $\dot{\epsilon}(t) \in L^\infty$, and

$$\lim_{t\to\infty} \epsilon(t) = \lim_{t\to\infty} (\ddot{\theta}_l(t) - p(t)\omega_p(t)) = 0, \tag{8.82}$$

$$\lim_{t\to\infty} \dot{p}(t) = 0. \tag{8.83}$$

In summary, we have the following results.

Lemma 8.3 *The adaptive algorithm (8.74)–(8.76) ensures that*
(i) $p_i(t) \in [p_i^a, p_i^b]$ *for some constants* $p_i^b > p_i^a > 0$ *and* $\dot{p}(t) \in L^2 \cap L^\infty$*;*
(ii) $\epsilon(t) \in L^2 \cap \in L^\infty$ *and* $\dot{\epsilon}(t) \in L^\infty$*; and*
(iii) $\lim_{t\to\infty} \epsilon(t) = 0$ *and* $\lim_{t\to\infty} \dot{p}(t) = 0$.

To show that the properties (P3) and (P4) are satisfied by the adaptive algorithm (8.74)–(8.76), we first prove the following proposition.

Proposition 8.3 *The stable operator* $\frac{1}{p_1 s + p_2}$ *has the property*

$$(s+\alpha_1)\left[\frac{1}{p_1 s + p_2}[z]\right](t) = \frac{1}{p_1 s + p_2}\left[(s+\alpha_1)[z] - \dot{p}_1 s\left[\frac{1}{p_1 s + p_2}[z]\right] - \dot{p}_2 \frac{1}{p_1 s + p_2}[z]\right](t). \quad (8.84)$$

Proof: This result follows from operating $\frac{1}{p_1 s + p_2}$ on both sides of the identity

$$\begin{aligned}(s+\alpha_1)[z](t) &= (s+\alpha_1)\left[(p_1 s + p_2)\left[\frac{1}{p_1 s + p_2}[z]\right]\right](t) \\ &= (p_1 s + p_2)\left[(s+\alpha_1)\left[\frac{1}{p_1 s + p_2}[z]\right]\right](t) \\ &\quad + \dot{p}_1 s\left[\frac{1}{p_1 s + p_2}[z]\right](t) + \dot{p}_2 \frac{1}{p_1 s + p_2}[z](t), \end{aligned} \quad (8.85)$$

with the identity $(p_1 s + p_2)\left[\frac{1}{p_1 s + p_2}[z]\right](t) = z(t)$ for a stable $\frac{1}{p_1 s + p_2}$. ∇

Lemma 8.4 *The adaptive algorithm (8.74)–(8.76) ensures the properties (P3) and (P4) characterized by (8.46)–(8.47).*

Proof: With $\hat{r}(t)$ chosen from (8.45), the reference signal in (8.29) now is

$$\begin{aligned}\theta_{md}(t) &= \frac{1}{s+\alpha_1}\left[\frac{1}{p_1 s + p_2}\left[(s^2 + p_3 s + p_2)\left[\frac{1}{(s+\alpha_2)(s+\alpha_3)}[r]\right]\right]\right](t) \\ &= \frac{1}{s+\alpha_1}\left[\frac{1}{p_1 s + p_2}\left[(s^2 + p_3 s + p_2)[(s+\alpha_1)[\theta_{ld}]]\right]\right](t), \end{aligned} \quad (8.86)$$

where $\theta_{ld}(t)$ is defined in (8.6). Then, letting

$$z(t) = (s^2 + p_3 s + p_2)[\theta_{ld}](t), \quad (8.87)$$

we have that $z(t) \in L^\infty$ and

$$\begin{aligned}(s+\alpha_1)\left[\frac{1}{p_1 s + p_2}[z]\right](t) &= \frac{1}{p_1 s + p_2}\left[(s^2 + p_3 s + p_2)[(s+\alpha_1)[\theta_{ld}]]\right](t) \\ &+ \frac{1}{p_1 s + p_2}\left[\dot{p}_3\dot{\theta}_{ld} + \dot{p}_2\theta_{ld} - \dot{p}_1 s\left[\frac{1}{p_1 s + p_2}[z]\right] - \dot{p}_2\frac{1}{p_1 s + p_2}[z]\right](t), \end{aligned} \quad (8.88)$$

which implies that

$$\begin{aligned}\frac{1}{p_1 s + p_2}[z](t) &= \frac{1}{s+\alpha_1}\left[\frac{1}{p_1 s + p_2}\left[(s^2 + p_3 s + p_2)[(s+\alpha_1)[\theta_{ld}]]\right]\right](t) \\ &+ \frac{1}{s+\alpha_1}\left[\frac{1}{p_1 s + p_2}\left[\dot{p}_3\dot{\theta}_{ld} + \dot{p}_2\theta_{ld} - \dot{p}_1 s\left[\frac{1}{p_1 s + p_2}[z]\right] - \dot{p}_2\frac{1}{p_1 s + p_2}[z]\right]\right](t) \\ &= \theta_{md}(t) + \epsilon_p(t), \end{aligned} \quad (8.89)$$

where $\epsilon_p(t)$ is the sum of all terms containing $\dot{p}_1(t)$, $\dot{p}_2(t)$, or $\dot{p}_3(t)$.

Since $\dot{p}(t) \in L^2 \cap L^\infty$ and $\lim_{t\to\infty} \dot{p}(t) = 0$, from closed-loop signal boundedness, we also have that $\epsilon_p(t) \in L^2 \cap L^\infty$, $\lim_{t\to\infty} \epsilon_p(t) = 0$, and $(p_1 s + p_2)[\epsilon_p](t) \in L^2 \cap L^\infty$, $\lim_{t\to\infty}(p_1 s + p_2)[\epsilon_p](t) = 0$.

From Lemma 8.1, we have the motor subsystem performance:

$$\theta_m(t) - \theta_{md}(t) \in L^2 \cap L^\infty,\ \dot{\theta}_m(t) - \dot{\theta}_{md}(t) \in L^2 \cap L^\infty, \tag{8.90}$$

$$\lim_{t\to\infty} (\theta_m(t) - \theta_{md}(t)) = \lim_{t\to\infty} (\dot{\theta}_m(t) - \dot{\theta}_{md}(t)) = 0. \tag{8.91}$$

In view of (8.87) and (8.89), the reference input signal $\hat{r}(t)$ in (8.45) results in

$$(p_1 s + p_2)[\theta_{md}](t) = (s^2 + p_3 s + p_2)[\theta_{ld}](t) - (p_1 s + p_2)[\epsilon_p](t). \tag{8.92}$$

The adaptive algorithm (8.74)–(8.76) ensures that

$$(s^2 + p_3(t)s + p_2(t))[\theta_l](t) = (p_1(t)s + p_2(t))[\theta_{md}](t) + \epsilon_2(t) + \epsilon(t), \tag{8.93}$$

where

$$\epsilon_2(t) = (p_1(t)s + p_2(t))[\theta_m - \theta_{md}](t) \tag{8.94}$$

has the desired properties that $\epsilon_2(t) \in L^2 \cap L^\infty$ and $\lim_{t\to\infty} \epsilon_2(t) = 0$, and so does $\epsilon(t)$: $\epsilon(t) \in L^2 \cap L^\infty$ and $\lim_{t\to\infty} \epsilon(t) = 0$, see Lemma 8.3. Therefore, the properties (P3) and (P4) are satisfied with

$$\delta_1(t) = -(p_1 s + p_2)[\epsilon_p](t),\ \delta_2(t) = -\epsilon_2(t) - \epsilon(t) \tag{8.95}$$

such that $\lim_{t\to\infty} \delta_1(t) = \lim_{t\to\infty} \delta_2(t) = 0$. ∇

To analyze system tracking performance, from (8.92) and (8.93), we have

$$(s^2 + p_3(t)s + p_2(t))[\theta_l - \theta_{ld}](t) = \epsilon_2(t) + \epsilon(t) - (p_1 s + p_2)[\epsilon_p](t). \tag{8.96}$$

The properties of the parameter estimates p_i that $p_i(t) \in [p_i^a, p_i^b]$, $0 < p_i^a < p_i^b$, and $\dot{p}_i(t) \in L^2$, $i = 2, 3$, ensure that $\frac{1}{s^2+p_3 s+p_2}$ is a stable operator (see Lemma 8.2), which, together with (8.96), implies that

$$\theta_l(t) - \theta_{ld}(t) \in L^2 \cap L^\infty,\ \lim_{t\to\infty} (\theta_l(t) - \theta_{ld}(t)) = 0. \tag{8.97}$$

In summary, we have the following main result.

Theorem 8.1 *The adaptive control scheme which consists of (8.24), (8.34), (8.35), (8.45), and (8.74), when applied to the plant (8.1), ensures that all closed-loop signals are bounded and the tracking error $\theta_l(t) - \theta_{ld}(t)$ satisfies $\theta_l(t) - \theta_{ld}(t) \in L^2 \cap L^\infty$ and $\lim_{t\to\infty}(\theta_l(t) - \theta_{ld}(t)) = 0$.*

Remark 8.4 To implement the adaptive law (8.34)–(8.35) (whose order is 5), we need $\bar{e}_2(t) = \bar{x}_2(t) - \bar{x}_{2m}(t)$ as in (8.32), where $\bar{x}_2(t)$ is from (8.27) and $\bar{x}_{2m}(t)$ is from (8.31) (whose order is 1): $\bar{x}_{2m}(t) = \frac{1}{s+\alpha_2}[\hat{r}](t)$, with $\hat{r}(t)$ in (8.45) (whose order is 3) in which the parameters p_1, p_2, and p_3 are updated from (8.74) (whose order is 3). Therefore, the order of the adaptive control scheme consisting of (8.24), (8.34), (8.35), (8.45), and (8.72) is $5+1+3+3=12$, as compared with 24 for the direct adaptive state feedback control scheme (8.7)–(8.12), 30 for the direct adaptive output feedback control scheme (8.13)–(8.18), and 14 for the indirect adaptive control scheme in Section 8.3. □

8.4.3 Simulation Results

We now present some simulation results to verify the desired performance of this direct–indirect adaptive control system. The values of the parameters J_m, J_l, b_m, b_l, k, b_k of a real-life gun turret system [408] are

$$J_m = 0.01,\ J_l = 0.2,\ b_m = 0.5,\ b_l = 1.7,\ k = 246.2,\ b_k = 0.6 \tag{8.98}$$

(with their units omitted). The initial parameters were taken as

$$K_2(0) = \kappa_0 K_2^*,\ k_r(0) = \kappa_0 k_r^*,\ p(0) = 1.2p^*,\ p^a = 0.5p^*,\ p^b = 2.5p^*, \tag{8.99}$$

where κ_0 is 1.2 or 0.5. For $\alpha_1 = 1$, $\alpha_2 = 1.5$, $\alpha_3 = 2$, $\Gamma = \gamma I$ with $\gamma = 100$ and $\gamma_p = 10$, the tracking error $e(t) = \theta_l(t) - \theta_{ld}(t)$, for $r(t) = 10.0\sin(1.11t)$ with $t \in [0,5]$ seconds is shown in Figure 8.2(a) for $K_2(0) = 1.2K_2^*$, $k_r(0) = 1.2k_r^*$ and in Figure 8.2(b) for $K_2(0) = 0.5K_2^*$, $k_r(0) = 0.5k_r^*$. In both cases, the tracking error $e(t)$ converges to small values.

8.5 Adaptive Backstepping Design

Consider the plant (8.1). For a representation convenient for a backstepping design, we introduce $x(t) = [\theta_l(t), \dot{\theta}_l(t), \theta_m(t), \dot{\theta}_m(t)]^T$ and express (8.1) as

$$\dot{x}(t) = Ax(t) + Bu(t),\ y(t) = x_1(t),$$
$$A = \begin{bmatrix} 0 & 1 & 0 & 0 \\ -\frac{k}{J_l} & -\frac{b_l+b_k}{J_l} & \frac{k}{J_l} & \frac{b_k}{J_l} \\ 0 & 0 & 0 & 1 \\ \frac{k}{J_m} & \frac{b_k}{J_m} & -\frac{k}{J_m} & -\frac{b_m+b_k}{J_m} \end{bmatrix},\ B = \begin{bmatrix} 0 \\ 0 \\ 0 \\ \frac{1}{J_m} \end{bmatrix}. \tag{8.100}$$

Introducing the parameters

$$p_1 = -\frac{k}{J_l},\ p_2 = -\frac{b_l+b_k}{J_l},\ p_3 = \frac{b_k}{J_l},\ p_4 = \frac{k}{J_m},$$

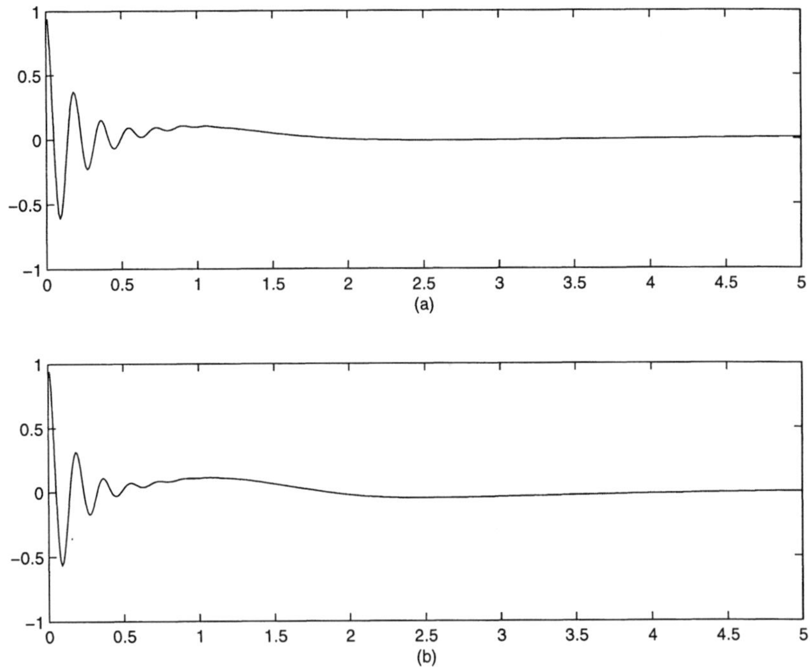

Figure 8.2: Tracking error $e(t) = \theta_l(t) - \theta_{ld}(t)$ for different initial conditions: (a) $[K_2(0), k_r(0)] = 1.2[K_2^*, k_r^*]$; (b) $[K_2(0), k_r(0)] = 0.5[K_2^*, k_r^*]$ (t in sec).

$$p_5 = \frac{b_k}{J_m}, p_6 = -\frac{b_m + b_k}{J_m}, p_0 = \frac{1}{J_m}, \tag{8.101}$$

we reparametrize the plant (8.100) as

$$\begin{aligned}
\dot{x}_1 &= x_2, \\
\dot{x}_2 &= x_3 + p_1 x_1 + p_2 x_2 + (p_1 + 1)(-x_3) + p_3 x_4, \\
\dot{x}_3 &= x_4, \\
\dot{x}_4 &= p_4 x_1 + p_5 x_2 + p_4(-x_3) + p_6 x_4 + p_0 u.
\end{aligned} \tag{8.102}$$

The control objective is to generate a feedback control input $u(t)$ for the plant (8.102) with unknown parameters p_i, $i = 0, 1, 2, \ldots, 6$, such that all closed-loop signals are bounded, and the plant output $y(t) = x_1(t)$ tracks a given bounded reference output $y_m(t)$ with bounded derivatives $\dot{y}_m(t)$, $\ddot{y}_m(t)$, $\dddot{y}_m(t)$.

For the plant (8.102) with relative degree 3, an adaptive backstepping control design procedure consists of three steps.

Step 1: Let the tracking error be $z_1 = x_1 - y_m$ and introduce $z_2 = x_2 - \alpha_1$, where α_1 is a function to be designed. Then, from (8.102), we get

$$\dot{z}_1 = z_2 + \alpha_1 - \dot{y}_m. \tag{8.103}$$

Choosing the design function α_1 as

$$\alpha_1 = -c_1 z_1 + \dot{y}_m \tag{8.104}$$

and considering the first partial positive definite function

$$V_1 = \frac{z_1^2}{2}, \tag{8.105}$$

we derive the time derivative of V_1 as

$$\dot{V}_1 = -c_1 z_1^2 + z_1 z_2. \tag{8.106}$$

Step 2: Let $z_3 = x_3 - \alpha_2$, where

$$\alpha_2 = -z_1 - c_2 z_2 - \hat{\beta}_1 - c_1 x_2 + c_1 \dot{y}_m + \ddot{y}_m, c_2 > 0. \tag{8.107}$$

Then, using the second equation of (8.102), we express

$$\dot{z}_2 = -z_1 - c_2 z_2 + z_3 + \beta_1 - \hat{\beta}_1, \tag{8.108}$$

where $\hat{\beta}$ is the estimate of β:

$$\beta_1 = p_1 x_1 + p_2 x_2 + (p_1 + 1)(-x_3) + p_3 x_4, \tag{8.109}$$

$$\hat{\beta}_1 = \hat{p}_1 x_1 + \hat{p}_2 x_2 + (\hat{p}_1 + 1)(-x_3) + \hat{p}_3 x_4. \tag{8.110}$$

With the second positive definite function

$$V_2 = V_1 + \frac{z_2^2}{2} + \sum_{i=1}^{3} \frac{(\hat{p}_i - p_i)^2}{2\gamma_i}, \gamma_i > 0, \tag{8.111}$$

from (8.106), (8.108), (8.111), we have

$$\dot{V}_2 = -\sum_{i=1}^{2} c_i z_i^2 + z_2 z_3 + \sum_{i=1}^{3} (\hat{p}_i - p_i)\gamma_i^{-1}(\dot{\hat{p}}_i - \gamma_i \tau_{2i}), \tag{8.112}$$

where $\tau_{21} = z_2(x_1 - x_3)$, $\tau_{22} = z_2 x_2$, $\tau_{23} = z_2 x_4$.

Step 3: Introducing

$$\beta_2 = p_4 x_1 + p_5 x_2 + p_4(-x_3) + p_6 x_4, \tag{8.113}$$

$$\hat{\beta}_2 = \hat{p}_4 x_1 + \hat{p}_5 x_2 + \hat{p}_4(-x_3) + \hat{p}_6 x_4, \tag{8.114}$$

from (8.102), we express $\dot{x}_4 = p_0 u + \beta_2$ and, using (8.109), $\dot{x}_2 = x_3 + \beta_1$. From $z_3 = x_3 - \alpha_2$ and (8.107), we obtain

$$\begin{aligned}
\dot{z}_3 &= x_4 + \dot{z}_1 + c_2\dot{z}_2 + \frac{d}{dt}\hat{\beta}_1 + c_1\dot{x}_2 - c_1\ddot{y}_m - \dddot{y}_m \\
&= x_4 + \dot{z}_1 + c_2(-z_1 - c_2 z_2 + z_3 + \beta_1 - \hat{\beta}_1) + \hat{p}_1 x_2 \\
&\quad + \hat{p}_2(x_3 + (\beta_1 - \hat{\beta}_1) + \hat{\beta}_1) + (\hat{p}_1 + 1)(-x_4) + \hat{p}_3(p_0 u + (\beta_2 - \hat{\beta}_2) + \hat{\beta}_2) \\
&\quad + \dot{\hat{p}}_1 x_1 + \dot{\hat{p}}_2 x_2 + \dot{\hat{p}}_1(-x_3) + \dot{\hat{p}}_3 x_4 + c_1(x_3 + (\beta_1 - \hat{\beta}_1) + \hat{\beta}_1) - c_1\ddot{y}_m - \dddot{y}_m \\
&= x_4 + \dot{z}_1 + c_2(-z_1 - c_2 z_2 + z_3) + c_1(\beta_1 - \hat{\beta}_1) + \hat{p}_1 x_2 + \hat{p}_2(\beta_1 - \hat{\beta}_1) \\
&\quad + \hat{p}_2(x_3 + \hat{\beta}_1) + (\hat{p}_1 + 1)(-x_4) + \hat{p}_3 p_0 u + \hat{p}_3(\beta_2 - \hat{\beta}_2) + \hat{p}_3\hat{\beta}_2 \\
&\quad + \dot{\hat{p}}_1(x_1 - x_3) + \dot{\hat{p}}_2 x_2 + \dot{\hat{p}}_3 x_4 + c_1(\beta_1 - \hat{\beta}_1) + c_1(x_3 + \hat{\beta}_1) - c_1\ddot{y}_m - \dddot{y}_m \\
&= \beta_3 + \beta_4 - z_2 - c_3 z_3 + (c_1 + c_2 + \hat{p}_2)(\beta_1 - \hat{\beta}_1) + \hat{p}_3(\beta_2 - \hat{\beta}_2) \\
&\quad + (\dot{\hat{p}}_1 - \gamma_1\tau_{31})(x_1 - x_3) + (\dot{\hat{p}}_2 - \gamma_2\tau_{32})x_2 - c_5 z_3 x_4^4 \\
&\quad + \hat{p}_3 p_0 u - c_4 z_3 x_4^4(c_1 + c_2 + \hat{p}_2)^2 + (\dot{\hat{p}}_3 - \gamma_3\tau_{33})x_4,
\end{aligned} \tag{8.115}$$

where

$$\begin{aligned}
\beta_3 &= x_4 + \dot{z}_1 + c_2(-z_1 - c_2 z_2 + z_3) + \hat{p}_1 x_2 + (\hat{p}_1 + 1)(-x_4) + \hat{p}_3\hat{\beta}_2 \\
&\quad + (c_1 + \hat{p}_2)(x_3 + \hat{\beta}_1) - c_1\ddot{y}_m - \dddot{y}_m,
\end{aligned} \tag{8.116}$$

$$\begin{aligned}
\beta_4 &= z_2 + c_3 z_3 + (x_1 - x_3)\gamma_1\tau_{31} + x_2\gamma_2\tau_{32} + x_4\gamma_3\tau_{33} \\
&\quad + c_4 z_3 x_4^4(c_1 + c_2 + \hat{p}_2)^2 + c_5 z_3 x_4^4
\end{aligned} \tag{8.117}$$

with $c_3 > 0, c_4 > 0, c_5 > 0$. Introducing $\beta = \beta_3 + \beta_4$ and

$$\begin{aligned}
\tau_{31} &= (c_1 + c_2 + \hat{p}_2)z_3(x_1 - x_3) + \tau_{21}, \\
\tau_{32} &= (c_1 + c_2 + \hat{p}_2)z_3 x_2 + \tau_{22}, \\
\tau_{33} &= (c_1 + c_2 + \hat{p}_2)z_3 x_4 + \tau_{23}, \\
\tau_{34} &= z_3(x_1 - x_3)\hat{p}_3, \\
\tau_{35} &= z_3 x_2 \hat{p}_3, \\
\tau_{36} &= z_3 x_4 \hat{p}_3,
\end{aligned} \tag{8.118}$$

we choose the control law

$$u = -\frac{\hat{\theta}_0\beta}{\hat{p}_3} \tag{8.119}$$

and the adaptive laws

$$\dot{\hat{\theta}}_0 = -\gamma_0 \operatorname{sign}[p_0]\beta z_3, \tag{8.120}$$

$$\dot{\hat{p}}_i = \gamma_i\tau_{3i}, \quad i = 1, 2, 4, 5, 6, \tag{8.121}$$

$$\dot{\hat{p}}_3 = \gamma_3\tau_{33} + f_3, \tag{8.122}$$

where $\hat{\theta}_0$ is the estimate of $\theta_0 = \frac{1}{p_0}$, $\gamma_i > 0$, $i = 0, 1, 2, \ldots, 6$, and

$$f_3 = \begin{cases} -\gamma_3 \tau_{33} & \text{if } \hat{p}_3 = p_3^a > 0,\ \gamma_3 \tau_{33} < 0, \\ 0 & \text{otherwise} \end{cases} \tag{8.123}$$

and p_3^a is a lower bound of $p_3 = \frac{b_k}{J_l}$: $0 < p_3^a \leq p_3$. The projection f_3 ensures that $\hat{p}_3 \geq p_3^a$ (see Section 3.8), so that u is well defined.

Consider the third positive definite function

$$V_3 = V_2 + \frac{z_3^2}{2} + \sum_{i=4}^{6} \frac{(\hat{p}_i - p_i)^2}{2\gamma_i} + \frac{(\hat{\theta}_0 - \theta_0)^2}{2\gamma_0} |p_0| \tag{8.124}$$

and its time derivative

$$\begin{aligned} \dot{V}_3(t) = &-c_1 z_1^2 - c_2 z_2^2 - c_3 z_3^2 + \frac{(\hat{p}_3 - p_3)}{\gamma_3} f_3 + z_3 x_4 f_3 \\ &- c_4 z_3^2 x_4^4 (c_1 + c_2 + \hat{p}_2)^2 - c_5 z_3 x_4^4. \end{aligned} \tag{8.125}$$

Since the projection (8.123) also ensures that $(\hat{p}_3 - p_3) f_3 \leq 0$ (see Section 3.8), and $|f_3| \leq \gamma_3 |\tau_{33}|$, from (8.125), we have

$$\dot{V}_3 \leq -c_1 z_1^2 - \left(c_2 - \frac{\gamma_3^2}{4c_5}\right) z_2^2 - \left(c_3 - \frac{\gamma_3^2}{4c_4}\right) z_3^2. \tag{8.126}$$

For some properly chosen $c_2, c_3, c_4, c_5, \gamma_3$, we have

$$\dot{V}_3 \leq -c_1 z_1^2 - \bar{c}_2 z_2^2 - \bar{c}_3 z_3^2 \leq 0 \tag{8.127}$$

for some $\bar{c}_i > 0$, which implies that V_3 in (8.124) is bounded, that is, $\hat{p}_i(t) \in L^\infty$, $i = 0, 1, 2, \ldots, 6$, and $z_j(t) \in L^\infty$, $j = 1, 2, 3$, and that $z_j \in L^2$, $j = 1, 2, 3$. Hence, $x_1 = z_1 + y_m$ is bounded, α_1 in (8.104) is bounded, $x_2 = z_2 + \alpha_1$ is bounded, and x_3 is bounded from the third equation of (8.102): $p_3 \dot{x}_3 - p_1 x_3 = \dot{x}_2 - p_2 x_2 - p_1 x_1$ for $p_3 > 0$ and $p_1 < 0$, that is, $\alpha_2 = x_3 - z_3$ is bounded, which implies that $\hat{\beta}_1$ in (8.107) is bounded and x_4 in (8.110) is bounded as $\hat{p}_3 \geq p_3^a > 0$. Thus we have proved that $x(t)$ is bounded so that $u(t)$ in (8.119) is bounded. Finally, from (8.103), we have $\dot{z}_1$ is bounded so that $\lim_{t\to\infty} z_1(t) = 0$—the same result as that in Theorem 8.1.

The dynamic order of this adaptive control scheme is basically the order of the parameter estimator (8.120)–(8.122), which is 7, plus the order of a dynamic system generating $y_m(t)$ with bounded derivatives $\dot{y}_m(t)$, $\ddot{y}_m(t)$, $\dddot{y}_m(t)$, which is 3; therefore, the total control scheme order is 10, as compared with the orders 30, 24, 14, and 12, with the previous control schemes (8.13)–(8.18), (8.7)–(8.12), that in Section 8.3, and that in Section 8.4, respectively.

Problems

8.1 Simulate the five studied adaptive control schemes for the plant (8.1) with the parameter values given in Section 8.4.3, with same initial conditions, and comment on your simulation results.

8.2 Prove the linear time-varying system stability results stated in Remark 8.3 using a procedure similar to that for the proof of Lemma 8.2.

8.3 Verify (8.122) and (8.125).

8.4 Develop the five adaptive control schemes for the plant (8.1) with $b_k = 0$.

8.5 Formulate the robust adaptive control problem with a nominal plant which has $b_k > 0$ and an unmodeled dynamics which includes the $b_k = 0$ case.

8.6 Formulate the robust adaptive control problem with a nominal plant which has $b_k = 0$ and an unmodeled dynamics which includes the $b_k > 0$ case.

Advanced Topics

Friction compensation. Consider the two-body system

$$J_m\ddot{\theta}_m(t) + b_m\dot{\theta}_m(t) = u(t) - u_k(t),$$
$$J_l\ddot{\theta}_l(t) + F(\dot{\theta}_l(t)) = u_k(t),$$

where the load friction [16] is $F(\dot{\theta}_l(t)) = F_a(u_k, \dot{\theta}_l) + b_l\dot{\theta}_l$,

$$F_a(u_k, \dot{\theta}_l) = \begin{cases} \left(f_c + (f_s - f_c)\exp(-\frac{\dot{\theta}_l^2}{\beta_s^2})\right)\text{sign}[\dot{\theta}_l] & \text{if } \dot{\theta}_l \neq 0, \\ F_0(u_k) & \text{if } \dot{\theta}_l = 0. \end{cases}$$

Here, f_c is the Coulomb (kinetic) friction level, f_s is the static friction level, $(f_s - f_c)\exp(-\frac{\dot{\theta}_l^2}{\beta_s^2})$ is the Stribeck friction component, and the static friction is

$$F_0(u_k) = \begin{cases} -f_s & \text{if } u_k < -f_s, \\ u_k & \text{if } -f_s \leq u_k \leq f_s, \\ f_s & \text{if } u_k > f_s. \end{cases}$$

Develop control schemes for compensating the effects of load friction to achieve closed-loop stability and output tracking.

Due to joint flexibility k and damping b_k, the main difficulty in this control problem is that the control input effort needs to pass through a dynamic block before effectively compensating the friction effects, which is essentially different from that without joint flexibility and damping, that is, when $k = \infty$ and $b_k = 0$, $\theta_m(t) = \theta_l(t) = \theta(t)$, and $(J_m + J_l)\ddot{\theta}(t) + b_m\dot{\theta}(t) + F(\dot{\theta}(t)) = u(t)$, where the input $u(t)$ has direct access to $F(\dot{\theta}(t))$.

Chapter 9

Multivariable Adaptive Control

Adaptive control of multi-input, multi-output (MIMO) systems is an important research area with both theoretical challenges and practical significance (e.g., aircraft control systems and process control systems usually have multiple inputs and multiple outputs). The key technical issue in multivariable adaptive control is to deal with dynamic interactions between system inputs and outputs, in addition to other issues in dealing with parametric, structural, and environmental uncertainties as in the single-input, single-output cases. In this chapter, we start with presenting an adaptive state feedback control design for multi-input, multi-state systems. We then develop a unified theory for model reference adaptive control (MARC) of linear time-invariant systems with M inputs and M outputs to show some key design and analysis techniques for multivariable adaptive control in a framework suitable for both continuous-time and discrete-time cases. We also develop advanced adaptive control designs, based on LDU parametrization of system high frequency gain matrices, for systems with input and output delays, using a variable system structure design, or using a backstepping method. In addition, we will present a robust adaptive control theory for control of robot manipulator systems and discuss some issues in multivariable pole placement control.

9.1 Adaptive State Feedback Control

Consider an M-input linear time-invariant plant

$$\dot{x}(t) = Ax(t) + Bu(t),\ x(t) \in R^n,\ u(t) \in R^M,\ M > 1,\ t \geq 0, \tag{9.1}$$

with $x(0) = x_0$, where $A \in R^{n\times n}$, $B \in R^{n\times M}$ are unknown constant parameter matrices and the state vector $x(t)$ is available for measurement (recall that the case when $M = 1$ was addressed in Section 4.1).

9.1.1 Design for State Tracking

The control objective is to design a feedback control law for the plant input $u(t)$ to ensure that the closed-loop system is stable (all system signals are bounded) and that the system state $x(t)$ asymptotically tracks a given reference state $x_m(t)$ generated from the reference model system

$$\dot{x}_m(t) = A_m x_m(t) + B_m r(t),\ x_m(t) \in R^n,\ r(t) \in R^M,\ t \geq 0, \tag{9.2}$$

with $x_m(0) = x_{m0}$, where $A_m \in R^{n\times n}$ with eigenvalues all in the open left-half complex plane, $B_m \in R^{n\times M}$, and $r(t)$ is bounded and piecewise continuous.

To achieve such a control objective, similar to the single-input case in Section 4.1, we make the following assumptions:

(A9s.1): There exist a constant matrix $K_1^* \in R^{n\times M}$ and a nonsingular constant matrix $K_2^* \in R^{M\times M}$ such that

$$A + BK_1^{*T} = A_m,\ BK_2^* = B_m. \tag{9.3}$$

(A9s.2): There is a known matrix $S \in R^{M\times M}$ such that $K_2^* S$ is symmetric and positive definite: $M_s = K_2^* S = (K_2^* S)^T = S^T K_2^{*T} > 0$ (as a comparison, when $M = 1$, $S = \mathrm{sign}[k_2^*]$ as in Assumption (A4.2) of Section 4.1, so that $Sk_2^* = |k_2^*| > 0$).

Remark 9.1 If the parameters A and B were known and the matching equations in (9.3) are satisfied, then the control law

$$u(t) = K_1^{*T} x(t) + K_2^* r(t) \tag{9.4}$$

would result in the closed-loop system

$$\dot{x}(t) = Ax(t) + B\left(K_1^{*T} x(t) + K_2^* r(t)\right) = A_m x(t) + B_m r(t),\ t \geq 0, \tag{9.5}$$

which has a (uniformly) bounded solution $x(t)$. The tracking error $e(t) = x(t) - x_m(t)$ under this control law is described by

$$\dot{e}(t) = A_m e(t),\ e(0) = x_0 - x_{m0} \tag{9.6}$$

so that $\lim_{t\to\infty} e(t) = 0$ exponentially. □

In the adaptive control problem, the parameters of A, B are unknown, and we need to use the adaptive version of (9.4):

$$u(t) = K_1^T(t)x(t) + K_2(t)r(t), \tag{9.7}$$

where $K_1(t)$ and $K_2(t)$ are the estimates of K_1^*, K_2^*, respectively.

To design adaptive update laws for $K_1(t)$ and $K_2(t)$ which ensure the stability of the closed-loop system, we define the parameter errors as

$$\tilde{K}_1(t) = K_1(t) - K_1^*,\ \tilde{K}_2(t) = K_2(t) - K_2^* \tag{9.8}$$

and use (9.1) and (9.7) to obtain

$$\begin{aligned}\dot{x}(t) &= Ax(t) + B\left(K_1^T(t)x(t) + K_2(t)r(t)\right)\\ &= A_m x(t) + B_m r(t) + B\left(\tilde{K}_1^T(t)x(t) + \tilde{K}_2(t)r(t)\right)\\ &= A_m x(t) + B_m r(t) + B_m\left(K_2^{*-1}\tilde{K}_1^T(t)x(t) + K_2^{*-1}\tilde{K}_2(t)r(t)\right).\end{aligned} \tag{9.9}$$

Substituting (9.2) into (9.9), we have the tracking error equation

$$\dot{e}(t) = A_m e(t) + B_m\left(K_2^{*-1}\tilde{K}_1^T(t)x(t) + K_2^{*-1}\tilde{K}_2(t)r(t)\right),\ t \geq 0. \tag{9.10}$$

For matrices $K_1(t)$ and $K_2(t)$, the state vector of the closed-loop system is

$$e_c(t) = \left[e^T(t), \tilde{k}_{11}^T(t), \ldots, \tilde{k}_{1n}^T(t), \tilde{k}_{21}^T(t), \ldots, \tilde{k}_{2M}^T(t)\right]^T \in R^{(M+1)n+M^2}, \tag{9.11}$$

where $\tilde{k}_{1i}(t) \in R^M$ is the ith column of $\tilde{K}_1^T(t)$, $i = 1, 2, \ldots, n$, and $\tilde{k}_{2j}(t) \in R^M$ is the ith column of $\tilde{K}_2(t)$, $j = 1, 2, \ldots, M$, that is,

$$\begin{aligned}\tilde{K}_1^T(t) &= \left[\tilde{k}_{11}(t), \ldots, \tilde{k}_{1n}(t)\right] \in R^{M\times n},\\ \tilde{K}_2(t) &= \left[\tilde{k}_{21}(t), \ldots, \tilde{k}_{2M}(t)\right] \in R^{M\times M}.\end{aligned} \tag{9.12}$$

Consider the positive definite function

$$V(e_c) = e^T P e + \sum_{i=1}^{n} \tilde{k}_{1i}^T M_s^{-1} \tilde{k}_{1i} + \sum_{j=1}^{M} \tilde{k}_{2j}^T M_s^{-1} \tilde{k}_{2j} \tag{9.13}$$

as a measure of these tracking errors and parameter errors, where $P \in R^{n\times n}$ is a constant matrix such that $P = P^T > 0$ and satisfies

$$PA_m + A_m^T P = -Q \tag{9.14}$$

for some constant matrix $Q \in R^{n\times n}$ such that $Q = Q^T > 0$, and $M_s = M_s^T > 0$ satisfies Assumption (A9s.2).

With tr$[X]$ denoting the trace of a square matrix X, we express $V(e_c)$ as

$$V(e_c) = e^T P e + \mathrm{tr}[\tilde{K}_1 M_s^{-1} \tilde{K}_1^T] + \mathrm{tr}[\tilde{K}_2^T M_s^{-1} \tilde{K}_2]. \tag{9.15}$$

The time derivative of $V(e_c)$ is

$$\begin{aligned}
\dot{V} &= \frac{d}{dt}V(e_c) \\
&= \left(\frac{\partial V(e_c)}{\partial e}\right)^T \dot{e}(t) + \sum_{i=1}^{n}\left(\frac{\partial V(e_c)}{\partial \tilde{k}_{1i}}\right)^T \dot{\tilde{k}}_{1i}(t) + \sum_{j=1}^{M}\left(\frac{\partial V(e_c)}{\partial \tilde{k}_{2j}}\right)^T \dot{\tilde{k}}_{2j}(t) \\
&= 2e^T(t)P\dot{e}(t) + 2\sum_{i=1}^{n}\tilde{k}_{1i}^T(t)M_s^{-1}\dot{\tilde{k}}_{1i}(t) + 2\sum_{j=1}^{M}\tilde{k}_{2j}^T(t)M_s^{-1}\dot{\tilde{k}}_{2j}(t) \\
&= 2e^T(t)P\dot{e}(t) + 2\,\mathrm{tr}[\tilde{K}_1(t)M_s^{-1}\dot{\tilde{K}}_1^T(t)] + 2\,\mathrm{tr}[\tilde{K}_2^T(t)M_s^{-1}\dot{\tilde{K}}_2(t)]. \qquad (9.16)
\end{aligned}$$

Substituting (9.10) and (9.14) into (9.16), we have

$$\begin{aligned}
\dot{V} = &-e^T(t)Qe(t) + 2e^T(t)PB_mK_2^{*-1}\tilde{K}_1^T(t)x(t) + 2e^T(t)PB_mK_2^{*-1}\tilde{K}_2(t)r(t) \\
&+ 2\,\mathrm{tr}[\tilde{K}_1(t)M_s^{-1}\dot{\tilde{K}}_1^T(t)] + 2\,\mathrm{tr}[\tilde{K}_2^T(t)M_s^{-1}\dot{\tilde{K}}_2(t)]. \qquad (9.17)
\end{aligned}$$

Using the condition that $M_s = K_2^*S = M_s^T > 0$ and the properties that $\mathrm{tr}[M_1M_2] = \mathrm{tr}[M_2M_1]$, $\mathrm{tr}[M_3] = \mathrm{tr}[M_3^T]$ for any matrices M_1, M_2, and M_3 of appropriate dimensions, we obtain

$$\begin{aligned}
&e^T(t)PB_mK_2^{*-1}\tilde{K}_1(t)x(t) \\
&= \mathrm{tr}[e^T(t)PB_mK_2^{*-1}\tilde{K}_1^T(t)x(t)] = \mathrm{tr}[e^T(t)PB_mSM_s^{-1}\tilde{K}_1^T(t)x(t)] \\
&= \mathrm{tr}[x(t)e^T(t)PB_mSM_s^{-1}\tilde{K}_1^T(t)] = \mathrm{tr}[\tilde{K}_1(t)M_s^{-1}S^TB_m^TPe(t)x^T(t)], \qquad (9.18)
\end{aligned}$$

$$\begin{aligned}
&e^T(t)PB_mK_2^{*-1}\tilde{K}_2(t)r(t) \\
&= \mathrm{tr}[e^T(t)PB_mK_2^{*-1}\tilde{K}_2(t)r(t)] = \mathrm{tr}[e^T(t)PB_mSM_s^{-1}\tilde{K}_2(t)r(t)] \\
&= \mathrm{tr}[r(t)e^T(t)PB_mSM_s^{-1}\tilde{K}_2(t)] = \mathrm{tr}[\tilde{K}_2^T(t)M_s^{-1}S^TB_m^TPe(t)r^T(t)]. \qquad (9.19)
\end{aligned}$$

To make $\dot{V} \leq 0$, (9.17)–(9.19) suggest that we choose the following adaptive laws for the controller parameters $K_1(t)$ and $K_2(t)$:

$$\dot{K}_1^T(t) = -S^TB_m^TPe(t)x^T(t),\ t \geq 0, \qquad (9.20)$$

$$\dot{K}_2(t) = -S^TB_m^TPe(t)r^T(t),\ t \geq 0 \qquad (9.21)$$

with S satisfying Assumption (A9s.2) and $K_1(0)$, $K_2(0)$ being arbitrary.

With this choice, from (9.17)–(9.19), we have

$$\dot{V} = -e^T(t)Qe(t) \leq -\alpha_q\|e(t)\|_2^2 \leq 0 \qquad (9.22)$$

for some $\alpha_q > 0$. Hence the equilibrium state $e_c = 0$ of the closed-loop system consisting of (9.10), (9.20), and (9.21) is uniformly stable and its solution $e_c(t)$ is uniformly bounded. That is, $y(t)$, $K_1(t)$, $K_2(t)$, and $\dot{e}(t)$ all are bounded. Furthermore (9.22) implies $e(t) \in L^2$ and so $\lim_{t\to\infty} e(t) = 0$.

9.1.2 Design Based on LDU Parametrization

The adaptive design of Section 9.1.1 for the M-input plant (9.1) needs Assumption (A9s.2): $M_s = K_2^*S = (K_2^*S)^T = S^TK_2^{*T} > 0$ for some known matrix $S \in R^{M\times M}$. This S matrix is analogous to the sign of k_2^* for the case of $M = 1$; however, the knowledge of such an S matrix is more difficult to obtain than that of the sign of k_2^*. Relaxation of such knowledge for multivariable adaptive control is important. In this section, we present an adaptive control scheme for the plant (9.1), using different knowledge for the gain matrix K_2^*. This adaptive design employs a modified controller parametrization based on an LDU decomposition [126], [137], [270] of the gain matrix K_2^{*-1}.

From the LDU decomposition of Section 2.2 (see Proposition 2.4), we first express the gain matrix K_2^{*-1} (which satisfies Assumption (A9s.1)) as

$$K_2^{*-1} = LD^*U \tag{9.23}$$

for some $M \times M$ unit (i.e., with all diagonal elements being 1) lower triangular matrix L and unit upper triangular matrix U and

$$D^* = \mathrm{diag}\{d_1^*, d_2^*, \ldots, d_M^*\} = \mathrm{diag}\left\{\Delta_1, \frac{\Delta_2}{\Delta_1}, \ldots, \frac{\Delta_M}{\Delta_{M-1}}\right\} \tag{9.24}$$

with Δ_i, $i = 1, 2, \ldots, M$, as the leading principal minors [1] of K_2^{*-1} in (9.3). To use this decomposition, we assume

(A9s.2a): All leading principal minors Δ_i of the matrix K_2^{*-1} are nonzero, and their signs, $\mathrm{sign}[d_i^*]$, $i = 1, 2, \ldots, M$, are known.

(A9s.2b): The matrix L in (9.23) is known.

Using (9.3) and (9.23), we express the plant (9.2) as

$$\begin{aligned}
\dot{x}(t) &= A_m x(t) + B_m r(t) + B\left(u(t) - K_1^{*T}x(t) - K_2^* r(t)\right)\\
&= A_m x(t) + B_m r(t) + B_m K_2^{*-1}\left(u(t) - K_1^{*T}x(t) - K_2^* r(t)\right)\\
&= A_m x(t) + B_m r(t) + B_m LD^*\left(Uu(t) - UK_1^{*T}x(t) - UK_2^* r(t)\right)\\
&= A_m x(t) + B_m r(t)\\
&\quad + B_m LD^*\left(u(t) - (I-U)u(t) - UK_1^{*T}x(t) - UK_2^* r(t)\right),
\end{aligned} \tag{9.25}$$

[1]The kth leading principal minor of a square matrix $A = \{a_{ij}\} \in R^{M\times M}$ is

$$\det\begin{bmatrix} a_{11} & \cdots & a_{1k}\\ \vdots & & \vdots\\ a_{k1} & \cdots & a_{kk}\end{bmatrix}, \quad k = 1, 2, \ldots, M.$$

where $I - U$ is an upper triangular matrix with zero diagonal elements as U is a unit upper triangular matrix (whose diagonal elements are 1). From (9.2) and (9.25), we have the tracking error equation

$$\dot{e}(t) = A_m e(t) + B_m L D^* \left(u(t) - \Phi_0^* u(t) - \Phi_1^{*T} x(t) - \Phi_2^* r(t)\right), \; t \geq 0, \quad (9.26)$$

where

$$\Phi_0^* = I - U, \; \Phi_1^{*T} = U K_1^{*T}, \; \Phi_2^* = U K_2^*. \quad (9.27)$$

This error equation motivates the controller structure

$$u(t) = \Phi_0(t) u(t) + \Phi_1^T(t) x(t) + \Phi_2(t) r(t), \quad (9.28)$$

where $\Phi_0(t)$, $\Phi_1(t)$, and $\Phi_2(t)$ are the estimates of Φ_0^*, Φ_1^*, and Φ_2^*, respectively; in particular, the parameter matrix Φ_0 has the same special upper triangular form as that of $\Phi_0^* = I - U$, that is,

$$\Phi_0 = \begin{bmatrix} 0 & \phi_{12} & \phi_{13} & \cdots & \phi_{1M} \\ 0 & 0 & \phi_{23} & \cdots & \phi_{2M} \\ \vdots & \vdots & \vdots & \vdots & \vdots \\ 0 & 0 & \cdots & 0 & \phi_{M-1\,M} \\ 0 & \cdots & \cdots & 0 & 0 \end{bmatrix}. \quad (9.29)$$

This special form ensures that the control signal $u(t)$ is implementable from (9.28) without singularity, that is,

$$\begin{aligned} u_M(t) &= [\Phi_1^T(t) x(t) + \Phi_2(t) r(t)]_M, \\ u_{M-1}(t) &= \phi_{M-1\,M} u_M(t) + [\Phi_1^T(t) x(t) + \Phi_2(t) r(t)]_{M-1}, \\ u_{M-2}(t) &= \phi_{M-2\,M-1} u_{M-1}(t) + \phi_{M-2\,M} u_M(t) + [\Phi_1^T(t) x(t) + \Phi_2(t) r(t)]_{M-2}, \\ &\vdots \\ u_2(t) &= \sum_{i=3}^{M} \phi_{2\,i} u_i(t) + [\Phi_1^T(t) x(t) + \Phi_2(t) r(t)]_2, \\ u_1(t) &= \sum_{i=2}^{M} \phi_{1\,i} u_i(t) + [\Phi_1^T(t) x(t) + \Phi_2(t) r(t)]_1, \end{aligned} \quad (9.30)$$

where $[v]_i$ denotes the ith row of the vector v.

To arrive at a compact and exact expression of the control law (9.28) with the special parameter structure (9.29), we let $\Phi_{1i}^T(t)$ be the ith row of $\Phi_1^T(t)$ and $\Phi_{2i}^T(t)$ be the ith row of $\Phi_2^T(t)$, $i = 1, 2, \ldots, M$, and define

$$\theta_1(t) = [\phi_{12}(t), \phi_{13}(t), \ldots, \phi_{1M}(t), \Phi_{11}^T(t), \Phi_{21}^T(t)]^T \in R^{n+2M-1},$$

$$\theta_2(t) = [\phi_{23}(t), \phi_{24}(t), \ldots, \phi_{2M}(t), \Phi_{12}^T(t), \Phi_{22}^T(t)]^T \in R^{n+2M-2},$$

$$\vdots$$

$$\begin{aligned}
\theta_{M-2}(t) &= [\phi_{M-2\,M-1}(t), \phi_{M-2\,M}(t), \Phi_{1\,M-2}^T(t), \Phi_{2\,M-2}^T(t)]^T \in R^{n+M+2},\\
\theta_{M-1}(t) &= [\phi_{M-1\,M}(t), \Phi_{1\,M-1}^T(t), \Phi_{2\,M-1}^T(t)]^T \in R^{n+M+1},\\
\theta_M(t) &= [\Phi_{1\,M}^T(t), \Phi_{2\,M}^T(t)]^T \in R^{n+M},
\end{aligned} \tag{9.31}$$

which are the estimates of the corresponding θ_i^* from the rows of Φ_0^*, Φ_1^{*T}, and Φ_2^* in (9.27). It follows from (9.26), (9.28), (9.31) that

$$\dot{e}(t) = A_m e(t) + B_m L D^* \begin{bmatrix} \tilde{\theta}_1^T(t)\omega_1(t) \\ \tilde{\theta}_2^T(t)\omega_2(t) \\ \vdots \\ \tilde{\theta}_{M-1}^T(t)\omega_{M-1}(t) \\ \tilde{\theta}_M^T(t)\omega_M(t) \end{bmatrix}, \tag{9.32}$$

where $\tilde{\theta}_i(t) = \theta_i(t) - \theta_i^*$, $i = 1, 2, \ldots, M$, and

$$\begin{aligned}
\omega_1(t) &= [u_2(t), u_3(t), \ldots, u_M(t), x^T(t), r^T(t)]^T \in R^{n+2M-1},\\
\omega_2(t) &= [u_3(t), u_4(t), \ldots, u_M(t), x^T(t), r^T(t)]^T \in R^{n+2M-2},\\
&\vdots\\
\omega_{M-2}(t) &= [u_{M-1}(t), u_M(t), x^T(t), r^T(t)]^T \in R^{n+M+2},\\
\omega_{M-1}(t) &= [u_M(t), x^T(t), r^T(t)]^T \in R^{n+M+1},\\
\omega_M(t) &= [x^T(t), r^T(t)]^T \in R^{n+M}.
\end{aligned} \tag{9.33}$$

We now choose the following adaptive laws for $\theta_i(t)$, $i = 1, 2, \ldots, M$:

$$\dot{\theta}_i(t) = -\text{sign}[d_i^*]\Gamma_i \bar{e}_i(t)\omega_i(t), \ t \geq 0, \tag{9.34}$$

where $\bar{e}_i(t)$ is the ith component of $e^T(t)PB_mL$ with $P = P^T > 0$, satisfying (9.14), $\Gamma_i = \Gamma_i^T > 0$, and $\text{sign}[d_i^*]$ is from Assumption (A9s.2a).

To analyze the stability properties of the adaptive scheme (9.34), we consider the positive definite function

$$V(e, \tilde{\theta}_i, i = 1, 2, \ldots, M) = e^T P e + \sum_{i=1}^{M} |d_i^*| \tilde{\theta}_i^T \Gamma_i^{-1} \tilde{\theta}_i. \tag{9.35}$$

The time derivative of $V(e, \tilde{\theta}_i)$, along the trajectory of (9.34), is

$$\begin{aligned}
\dot{V} &= 2e^T(t)P\dot{e}(t) + 2\sum_{i=1}^{M} |d_i^*| \tilde{\theta}_i^T(t)\Gamma_i^{-1}\dot{\theta}_i(t)\\
&= 2e^T(t)Pe(t) + 2\sum_{i=1}^{M} \bar{e}_i(t) d_i^* \tilde{\theta}_i^T(t)\omega_i(t) + 2\sum_{i=1}^{M} |d_i^*| \tilde{\theta}_i^T(t)\Gamma_i^{-1}\dot{\theta}_i(t)\\
&= -e^T(t)Qe(t).
\end{aligned} \tag{9.36}$$

Since $Q = Q^T > 0$, (9.36) implies that the equilibrium state $e_c = 0$, with $e_c = [e^T, \tilde{\theta}_1^T, \ldots, \tilde{\theta}_M^T]^T$, of the closed-loop system consisting of (9.32) and (9.34) is uniformly stable and its solution $e_c(t)$ is uniformly bounded. That is, $y(t)$, $\theta_i(t)$, $i = 1, 2, \ldots, M$, and $\dot{e}(t)$ all are bounded. From (9.36), we also have $e(t) = x(t) - x_m(t) \in L^2$ and so $\lim_{t\to\infty} e(t) = 0$ as from Lemma 2.14.

Remark 9.2 To derive the above adaptive control scheme, we assumed that the matrix L in (9.23) is known (see Assumption (A9s.2b)). The issue with an unknown L for the state tracking problem is yet to be solved (we will present a solution for the output tracking problem in Chapter 9). On the other hand, this assumption is satisfied if K_2^{*-1} is already in an upper triangular form (i.e., $L = I$). Furthermore, the matrix L in (9.23) is allowed to be a nontriangular and known matrix for the adaptive design to be applicable. □

9.1.3 System Identification

A related issue in adaptive control is system identification: design of an adaptive system which generates estimates of the unknown plant parameter matrices A, B, based on the plant input and state information. A solution to this problem can be derived from the previous adaptive control scheme.

Considering system (9.1) with multiple inputs and $x(t)$ available for measurement, we now develop an estimation scheme for identifying (estimating) the system matrices A and B. In view of (9.3), if the estimates $K_2(t)$ of K_2^* is nonsingular, then the estimates $\hat{A}(t)$, $\hat{B}(t)$ of A, B are

$$\hat{A}(t) = A_m - B_m K_2^{-1}(t) K_1^T(t), \;\; \hat{B}(t) = B_m K_2^{-1}(t), \tag{9.37}$$

where $K_1(t)$ and $K_2(t)$ are adaptively updated from (9.20) and (9.21).

Two problems of using (9.37) as an identification scheme are:

(i) The estimate $K_2(t)$ has to be nonsingular.

(ii) The matching condition (9.3) which was previously used for the adaptive control scheme consisting of (9.7), (9.20), and (9.21) may not be satisfied.

We need to modify this adaptive scheme to resolve these problems to derive an algorithm for system identification. To this end, we start with the reference model (9.2) by replacing $r(t)$ with

$$r(t) = -K_2^{-1}(t) K_1^T(t) x(t) + K_2^{-1}(t) u(t), \tag{9.38}$$

and using (9.37), we have

$$\dot{x}_m(t) = A_m x_m(t) + (\hat{A}(t) - A_m) x(t) + \hat{B}(t) u(t). \tag{9.39}$$

In this equation, the estimates $K_1(t)$ and $K_2(t)$ disappear. A further examination of (9.39) indicates that if $\hat{A} = A$, $\hat{B} = B$, then (9.1) and (9.39) imply that $\frac{d}{dt}(x_m(t) - x(t)) = A_m(x_m(t) - x(t))$ so that $\lim_{t\to\infty}(x(t) - x_m(t)) = 0$ if A_m is chosen to have all its eigenvalues in $\mathrm{Re}[s] < 0$. Since $x(t)$ and $x_m(t)$ are the available signals, adaptive laws to be developed for $\hat{A}(t)$ and $\hat{B}(t)$ are based on the error $x(t) - x_m(t)$ and are to make $\lim_{t\to\infty}(x(t) - x_m(t)) = 0$. In (9.39), $x_m(t)$ depends on $\hat{A}(t)$ and $\hat{B}(t)$ and is no longer a given reference signal, and its boundedness remains to be established.

Letting $\hat{A}(t)$, $\hat{B}(t)$ be the estimates of A and B, respectively (independent of (9.37)), introducing the error signals

$$e(t) = x_m(t) - x(t),\ \tilde{A}(t) = \hat{A}(t) - A,\ \tilde{B}(t) = \hat{B}(t) - B \tag{9.40}$$

and using (9.1) and (9.39), we have the error equation

$$\dot{e}(t) = A_m e(t) + \tilde{A}(t)x(t) + \tilde{B}(t)u(t),\ t \geq 0. \tag{9.41}$$

To design adaptive laws for $\hat{A}(t)$ and $\hat{B}(t)$ in (9.41), we assume that A_m has all its eigenvalues in $\mathrm{Re}[s] < 0$ and the system input $u(t)$ and state $x(t)$ are bounded. Then we choose the adaptive laws for $\hat{A}(t)$ and $\hat{B}(t)$ as

$$\dot{\hat{A}}(t) = -\Gamma_1 P e(t) x^T(t),\ t \geq 0, \tag{9.42}$$

$$\dot{\hat{B}}(t) = -\Gamma_2 P e(t) u^T(t),\ t \geq 0, \tag{9.43}$$

with $\hat{A}(0)$ and $\hat{B}(0)$ being arbitrary, where $\Gamma_1 \in R^{n\times n}$, $\Gamma_2 \in R^{n\times n}$ are arbitrary constant matrices such that $\Gamma_1 = \Gamma_1^T > 0$, $\Gamma_2 = \Gamma_2^T > 0$ and $P \in R^{n\times n}$ is a constant matrix such that $P = P^T > 0$ and satisfies (9.14) for some constant matrix $Q \in R^{n\times n}$ such that $Q = Q^T > 0$.

The state vector of the closed-loop system consisting of (9.41)–(9.43) is

$$e_c(t) = \left[e^T(t), \tilde{a}_1^T(t), \ldots, \tilde{a}_n^T(t), \tilde{b}_1^T(t), \ldots, \tilde{b}_M^T(t)\right]^T \in R^{(M+1)n+n^2}, \tag{9.44}$$

where $\tilde{a}_i(t)$ is the ith column of $\tilde{A}(t)$, $i = 1, 2, \ldots, n$, and $\tilde{b}_j(t)$ is the jth column of $\tilde{B}(t)$, $j = 1, 2, \ldots, M$, that is,

$$\tilde{A}(t) = [\tilde{a}_1(t), \ldots, \tilde{a}_n(t)] \in R^{n\times n},\ \tilde{B}(t) = \left[\tilde{b}_1(t), \ldots, \tilde{b}_M(t)\right] \in R^{n\times M}. \tag{9.45}$$

To establish the stability properties of the adaptive scheme (9.42)–(9.43) applied to the error system (9.41), we use the positive definite function

$$\begin{aligned} V(e_c) &= e^T P e + \sum_{i=1}^{n} \tilde{a}_i^T \Gamma_1^{-1} \tilde{a}_i + \sum_{j=1}^{M} \tilde{b}_j^T \Gamma_2^{-1} \tilde{b}_j \\ &= e^T P e + \mathrm{tr}[\tilde{A}^T \Gamma_1^{-1} \tilde{A}] + \mathrm{tr}[\tilde{B}^T \Gamma_2^{-1} \tilde{B}]. \end{aligned} \tag{9.46}$$

The time derivative of $V(e_c)$ is

$$\begin{aligned}\dot{V} &= \frac{d}{dt}V(e_c)\\ &= \left(\frac{\partial V(e_c)}{\partial e}\right)^T \dot{e}(t) + \sum_{i=1}^{n}\left(\frac{\partial V(e_c)}{\partial \tilde{a}_i}\right)^T \dot{\tilde{a}}_i(t) + \sum_{j=1}^{M}\left(\frac{\partial V(e_c)}{\partial \tilde{b}_j}\right)^T \dot{\tilde{b}}_j(t)\\ &= 2e^T(t)P\dot{e}(t) + 2\sum_{i=1}^{n}\tilde{a}_i^T(t)\Gamma_1^{-1}\dot{\tilde{a}}_i(t) + 2\sum_{j=1}^{M}\tilde{b}_j^T(t)\Gamma_2^{-1}\dot{\tilde{b}}_j(t)\\ &= 2e^T(t)P\dot{e}(t) + 2\mathrm{tr}[\tilde{A}^T(t)\Gamma_1^{-1}\dot{\tilde{A}}(t)] + 2\mathrm{tr}[\tilde{B}^T(t)\Gamma_2^{-1}\dot{\tilde{B}}(t)]. \end{aligned} \tag{9.47}$$

Substituting (9.41)–(9.43) into (9.47) and using

$$\begin{aligned}& e^T(t)P\tilde{A}(t)x(t)\\ &= \mathrm{tr}[e^T(t)P\tilde{A}(t)x(t)] = \mathrm{tr}[e^T(t)P\Gamma_1\Gamma_1^{-1}\tilde{A}(t)x(t)]\\ &= \mathrm{tr}[x(t)e^T(t)P\Gamma_1\Gamma_1^{-1}\tilde{A}(t)] = \mathrm{tr}[\tilde{A}^T(t)\Gamma_1^{-1}\Gamma_1 Pe(t)x^T(t)], \end{aligned} \tag{9.48}$$

$$\begin{aligned}& e^T(t)P\tilde{B}(t)u(t)\\ &= \mathrm{tr}[e^T(t)P\tilde{B}(t)u(t)] = \mathrm{tr}[e^T(t)P\Gamma_2\Gamma_2^{-1}\tilde{B}(t)u(t)]\\ &= \mathrm{tr}[u(t)e^T(t)P\Gamma_2\Gamma_2^{-1}\tilde{B}(t)] = \mathrm{tr}[\tilde{B}^T(t)\Gamma_2^{-1}\Gamma_2 Pe(t)u^T(t)], \end{aligned} \tag{9.49}$$

we have

$$\begin{aligned}\dot{V} &= -e^T(t)Qe(t) + 2e^T(t)P\tilde{A}(t)x(t) + 2e^T(t)P\tilde{B}(t)r(t)\\ &\quad + 2\,\mathrm{tr}[\tilde{A}^T(t)\Gamma_1^{-1}\dot{\tilde{A}}(t)] + 2\,\mathrm{tr}[\tilde{B}^T(t)\Gamma_2^{-1}\dot{\tilde{B}}(t)]\\ &= -e^T(t)Qe(t) \le -\alpha_q\|e(t)\|_2^2 \le 0 \end{aligned} \tag{9.50}$$

for some $\alpha_q > 0$. Hence the equilibrium state $e_c = 0$ of the closed-loop system consisting of (9.41)–(9.43) is uniformly stable and its solution $e_c(t)$ is uniformly bounded, that is, $y_m(t)$, $\hat{A}(t)$, $\hat{B}(t)$, and $\dot{e}(t)$ all are bounded. Furthermore, (9.50) implies $e(t) = x(t) - x_m(t) \in L^2$ and so $\lim_{t\to\infty} e(t) = 0$.

Remark 9.3 To ensure parameter convergence, that is, $\lim_{t\to\infty}\tilde{A}(t) = 0$ and $\lim_{t\to\infty}\tilde{B}(t) = 0$, certain persistent excitation conditions are needed for the adaptive system (9.41). Such conditions are related to the richness of the input signal $u(t)$, that is, the number of frequency components in $u(t)$. Given a stable plant (i.e., all eigenvalues of A are in $\mathrm{Re}[s] < 0$), it can be shown, when $m = 1$, (A, B) is controllable, and $u(t) = \sum_{i=1}^{m_1} a_i \sin(\omega_i t)$, with scalars $a_i \neq 0$, $\omega_i \neq \omega_j$ for $i \neq j$, $m_1 \geq \frac{n+1}{2}$ for $A \in R^{n\times n}$, then $\lim_{t\to\infty}\tilde{A}(t) = 0$ and $\lim_{t\to\infty}\tilde{B}(t) = 0$ exponentially fast [151]. □

9.2 Model Reference Adaptive Control

In this section, the design and analysis of model reference adaptive control schemes for both continuous- and discrete-time multivariable plants with additive and multiplicative unmodeled dynamics are presented in a unified framework. Issues such as plant and controller parametrizations, design of adaptive laws, stability, robustness, and performance are clarified. We also present designs for plants with input and output delays, designs using a plant right interactor matrix, designs using LDU parametrizations of the plant high frequency gain matrix, and continuous-time Lyapunov designs.

9.2.1 Description of Multivariable Systems

To give a unified presentation for both continuous-time and discrete-time multivariable adaptive control systems, we first give the following description of multivariable linear time-invariant systems (plants) and time-varying operators, applicable in both continuous time and discrete time.

- The symbol D is used, in the continuous-time (CT) case, as the Laplace transform variable or the time differentiation operator $D[x](t) = \dot{x}(t)$, $t \in [0, \infty)$, as the case may be or, in the discrete-time (DT) case, as the z-transform variable or the time advance operator $D[x](t) = x(t+1)$, $t \in \{0, 1, 2, 3, \ldots\}$, as the case may be.

- A continuous-time linear time-invariant plant can be described as $\dot{x}(t) = Ax(t) + Bu(t)$, $y(t) = Cx(t)$, with $x(t) \in R^n$, $u(t) \in R^M$, and $y(t) \in R^p$, where $A \in R^{n\times n}$, $B \in R^{n\times M}$, and $C \in R^{p\times n}$ are constant matrices. The system transfer matrix is $G(s) = C(sI - A)^{-1}B$. Similarly, a discrete-time linear time-invariant system can be described as $x(t+1) = Ax(t) + Bu(t)$, $y(t) = Cx(t)$, and its transfer matrix is $G(z) = C(zI - A)^{-1}B$. In a unified notation, $G(D) = C(DI - A)^{-1}B$.

- A transfer matrix (or a polynomial or a square matrix) is called stable if all its poles (or zeros or eigenvalues) are in the open left-half of the complex s-plane for the CT case and on the open unit disk of the complex z-plane for the DT case.

- For a vector signal $x(t)$, $\|x(t)\|$ denotes a vector norm of $x(t)$ (typically, a Euclidean (l^2) norm), $\|x(\cdot)\|_p$ denotes a signal norm of $x(t)$, and L^p denotes the associated signal space (see Sections 2.3 and 2.7).

- The $M \times M$ ($M > 1$) identity matrix is denoted as I_M.

- The symbol ∂ is used either in $\frac{\partial f(x)}{\partial x}$, the partial derivative of $f(x)$ with respect to x, or in $\partial[P(s)]$, the degree of a polynomial (matrix) $P(s)$.

- A vector signal $x(t)$ is bounded if $x(t) \in L^\infty$ (i.e., $\|x(t)\| \leq c$, $\forall t \geq 0$, for some constant $c > 0$) for both the CT case and DT case.

- The L^1 signal norm of a vector signal $x(t)$ is $\|x(\cdot)\|_1 \triangleq \int_0^\infty \|x(t)\|\, dt$ in the CT case and $\|x(\cdot)\|_1 \triangleq \sum_{t=0}^\infty \|x(t)\|$ in the DT case.

- An operator $T(D,t)$ of certain dimensions is linear if

$$T(D,\cdot)[\alpha_1 u_1 + \alpha_2 u_2](t) = \alpha_1 T(D,\cdot)[u_1](t) + \alpha_2 T(D,\cdot)[u_2](t) \quad (9.51)$$

  for any constants α_1, α_2, and any (vector) signals $u_1(t)$, $u_2(t)$.

- A linear operator $T(D,t)$ is stable and proper if for $y(t) \triangleq T(D,\cdot)[u](t)$, some constants $\beta \geq 0$, $\alpha > 0$, and $\gamma > 0$, all $t \geq 0$, and any $u(t)$,

$$\|y(t)\| = \|T(D,\cdot)[u](t)\| \leq \beta \int_0^t e^{-\alpha(t-\tau)} \|u(\tau)\|\, d\tau + \gamma \|u(t)\| \quad (9.52)$$

  in the CT case or

$$\|y(t)\| = \|T(D,\cdot)[u](t)\| \leq \beta \sum_{\tau=0}^{t-1} e^{-\alpha(t-1-\tau)} \|u(\tau)\| + \gamma \|u(t)\| \quad (9.53)$$

  in the DT case, where $y(t) = T(D,\cdot)[u](t)$ denotes the convolution of the impulse response of $T(D,\cdot)$ with $u(\cdot)$ at t. A linear operator $T(D,t)$ is stable and strictly proper if it is stable with $\gamma = 0$.

- A linear operator $T(D,t)$ has an impulse response matrix $\hat{T}(t,\tau)$ in the sense that $y(t) = T(D,\cdot)[u](t) = \int_{t_0}^t \hat{T}(t,\tau) u(\tau)\, d\tau$ in the CT case or $y(t) = T(D,\cdot)[u](t) = \sum_{\tau=t_0}^t \hat{T}(t,\tau) u(\tau)$ in the DT case. For a linear time-invariant system, its impulse response matrix is $\hat{T}(t,\tau) = \hat{T}(t-\tau)$.

- In this section, equations whose numbers have the letter "C" (or "D") are used only for the CT (or the DT) case; otherwise equations are used for both the CT and DT cases (as a reminder, "CT" stands for continuous time and "DT" stands for discrete time).

- Consider a rational transfer matrix $G_0(D) \in R^{M\times M}(D)$, with a right coprime polynomial matrix decomposition, $G_0(D) = Z_r(D) P_r^{-1}(D)$. As in [171], [442], the column degrees of $P_r(D) = \{p_{ij}(D)\}$ are

$$\mu_j = \partial_{cj}[P_r(D)] = \max_{1\leq i\leq M} \{\partial[p_{ij}(D)]\},\ j = 1, 2, \ldots, M. \quad (9.54)$$

Introduce $S_c(D) = \mathrm{diag}\{D^{-\mu_1}, D^{-\mu_2}, \ldots, D^{-\mu_M}\}$ and define

$$\Gamma_c[P_r(D)] = \lim_{D\to\infty} P_r(D)S_c(D). \tag{9.55}$$

Then, $P_r(D)$ is called column reduced [171], [340] (or column proper [282]), if Γ_c is nonsingular.

- Similar to that in Section 2.2 for the case of rational functions, a rational matrix $G_0(D)$ is proper if $G_\infty = \lim_{D\to\infty} G_0(D)$ is nonzero and finite and $G_0(D)$ is strictly proper if $G_\infty = 0$ and is improper if $G_\infty = \infty$ (i.e., some components of G_∞ are ∞). As shown in [335], given $\Gamma_c[P_r(D)]$ nonsingular, $G_0(D) = Z_r(D)P_r^{-1}(D)$ is proper if and only if $\partial_{cj}[Z_r(D)] \leq \partial_{cj}[P_r(D)]$, $j = 1, 2, \ldots, M$, and $G_0(D)$ is strictly proper if and only if $\partial_{cj}[Z_r(D)] < \partial_{cj}[P_r(D)]$, $j = 1, 2, \ldots, M$.

- Consider a left coprime polynomial matrix decomposition of $G_0(D) = P_l^{-1}(D)Z_l(D)$, define the row degrees of $P_l(D) = \{\hat{p}_{ij}(D)\}$ as

$$\nu_i = \partial_{ri}[P_l(D)] = \max_{1\leq j\leq M}\{\partial[\hat{p}_{ij}(D)]\},\ i = 1, 2, \ldots, M, \tag{9.56}$$

and, with $S_r(D) = \mathrm{diag}\{D^{-\nu_1}, D^{-\nu_2}, \ldots, D^{-\nu_M}\}$, introduce

$$\Gamma_r[P_l(D)] = \lim_{D\to\infty} S_r(D)P_l(D). \tag{9.57}$$

Then, $P_l(D)$ is called row reduced (proper) if Γ_r is nonsingular. It can be verified that given $\Gamma_r[P_l(D)]$ nonsingular, $G_0(D) = P_l^{-1}(D)Z_l(D)$ is proper if and only if $\partial_{ri}[Z_l(D)] \leq \partial_{ri}[P_l(D)]$, $i = 1, 2, \ldots, M$, and $G_0(D)$ is strictly proper if and only if $\partial_{ri}[Z_r(D)] < \partial_{ri}[P_l(D)]$, $i = 1, 2, \ldots, M$.

- It can be verified [171] that for a minimal (controllable and observable) realization ($A \in R^{n\times n}, B \in R^{n\times M}, C \in R^{M\times n}$) of $G_0(D)$, it follows that

$$n = \sum_{j=1}^{M} \mu_j = \sum_{i=1}^{M} \nu_i, \tag{9.58}$$

where μ_j, $j = 1, 2, \ldots, M$, are called the controllability indices of $G_0(D)$ with $\max_{j=1,2,\ldots,M} \mu_j$ called the controllability index of $G_0(D)$, while ν_i, $i = 1, 2, \ldots, M$, are called the observability indices of $G_0(D)$, which can be equivalently defined from (A, B) or (A, C) [171]. In particular, $\nu = \max_{1\leq i\leq M} \nu_i$ is called the observability index of $G_0(D)$, which is equal to that defined in (2.30), and $\partial_{ci}[P_l(D)] \leq \nu$.

9.2.2 Plant and Controller Parametrizations

Consider the linear time-invariant plant with M inputs and M outputs:

$$y(t) = G(D)[u](t), \tag{9.59}$$

where $y(t) \in R^M$ is the plant output, $u(t) \in R^M$ is the plant input, with $t \in [0, \infty)$ for the CT case and $t \in \{0, 1, 2, 3, \ldots\}$ for the DT case,

$$G(D) = G_0(D)[I + \mu\Delta_m(D)] + \mu\Delta_a(D) \tag{9.60}$$

is the plant transfer matrix, $G_0(D) = Z_0(D)P_0^{-1}(D)$ is the modeled or nominal part of $G(D)$, $Z_0(D)$ and $P_0(D)$ are $M \times M$ right coprime polynomial matrices with $P_0(D)$ being column proper and of degree equal to the controllability index of $G_0(D)$, and $\mu\Delta_m(D)$, $\mu\Delta_a(D)$ are the multiplicative and additive plant unmodeled dynamics, respectively, with $\mu \geq 0$.

The objective of model reference adaptive control (MRAC) can be stated as follows: given a reference model system of the form

$$y_m(t) = W_m(D)[r](t), \tag{9.61}$$

where $W_m(D)$ is an $M \times M$ rational transfer matrix, $y_m(t) \in R^M$ is the reference output, and $r(t) \in R^M$ is an external input signal, find a feedback control signal $u(t)$ for the plant (9.51) with unknown $G(D)$ such that $y(t)$ tracks $y_m(t)$ as close as possible and the closed-loop plant is globally stable in the sense that all signals in the system are bounded for any bounded initial conditions, in spite of the presence of the unmodeled dynamics $\mu\Delta_m(D)$, $\mu\Delta_a(D)$.

Plant Zero Structure at Infinity

An important concept used in designing multivariable model reference control (MRC) schemes is the zero structure at infinity of the nominal plant $G_0(D)$. This structure is characterized by the Hermite normal form [266], [349] or the interactor matrix [93], [443]. Next we present the concept of a modified interactor matrix (see Proposition 2.3), which has a stable inverse and can be used in the design of MRC schemes in both the CT and DT cases.

Lemma 9.1 [377], [379] *For any $M \times M$ strictly proper full rank rational matrix $G_0(D)$,*[2] *there exists a lower triangular polynomial matrix $\xi_m(D)$, defined*

[2]An $M \times M$ rational matrix $G_0(D)$ is full rank if $\det[G_0(D)] \not\equiv 0$ for all complex D.

as the modified left interactor (MLI) matrix of $G_0(D)$, *of the form*

$$\xi_m(D) = \begin{bmatrix} d_1(D) & 0 & \cdots & \cdots & 0 \\ h_{21}^m(D) & d_2(D) & 0 & \cdots & 0 \\ \vdots & \vdots & \vdots & \vdots & \vdots \\ h_{M1}^m(D) & \cdots & \cdots & h_{M\,M-1}^m(D) & d_M(D) \end{bmatrix}, \tag{9.62}$$

where $h_{ij}^m(D)$, $j = 1, \ldots, M-1$, $i = 2, \ldots, M$, *are polynomials, and* $d_i(D)$, $i = 1, \ldots, M$, *are monic stable polynomials of degrees* $l_i > 0$, *such that the high frequency gain matrix of* $G_0(D)$, *defined as*

$$K_p = \lim_{D \to \infty} \xi_m(D) G_0(D), \tag{9.63}$$

is finite and nonsingular.

Proof: The proof is the extension of that in [443] for the interactor matrix.

Let $g_i(D)$ denote the ith row of $G_0(D)$. Then there exist unique integers $\bar{l}_i$, $i = 1, \ldots, M$, such that $\lim_{D\to\infty} p_i(D) g_i(D) = z_i$, where $p_i(D)$, $i = 1, \ldots, M$, are arbitrary stable monic polynomials of degree $\bar{l}_i > 0$ and $z_i \in R^{1\times M}$, $i = 1, \ldots, M$, are finite, nonzero row vectors.

Let $x_i(D)$ be the ith row of $P(D) = \mathrm{diag}[p_1(D), \ldots, p_M(D)]$ and define $\xi_1(D) = x_1(D)$. If z_2 and z_1 are linearly independent, then define $\xi_2(D) = x_2(D)$, otherwise let $x_2^1(D) = p_2^1(D)(x_2(D) + \alpha_1^1 x_1^1(D))$, where $\alpha_1^1 \neq 0$ is such that $z_2 + \alpha_1^1 z_1 = 0$ since z_1 and z_2 are linearly dependent, $x_1^1(D) = [p_1^1(D), 0, \ldots, 0]$ with $p_1^1(D)$ an arbitrary monic polynomial of degree $\bar{l}_1$, and $p_2^1(D)$ is an arbitrary stable monic polynomial of degree $l_2^1 > 0$ such that $\lim_{D\to\infty} x_2^1(D) G_0(D) = z_2^1 \in R^{1\times M}$ is a finite, nonzero row vector. If z_2^1 and z_1 are linearly independent, then define $\xi_2(D) = x_2^1(D)$, otherwise let $x_2^2(D) = p_2^2(D)(x_2^1(D) + \alpha_1^2 x_1^2(D))$, where $\alpha_1^2 \neq 0$ is such that $z_2^1 + \alpha_1^2 z_1 = 0$ since z_1 and z_2^1 are linearly dependent, $x_1^2(D) = [p_1^2(D), 0, \ldots, 0]$, with $p_1^2(D)$ being an arbitrary monic polynomial of degree $\bar{l}_1$, and $p_2^2(D)$ is an arbitrary stable monic polynomial of degree $l_2^2 > 0$ such that $\lim_{D\to\infty} x_2^2(D) G_0(D) = z_2^2$ is a finite, nonzero row vector. If z_2^2 and z_1 are linearly independent, then define $\xi_2(D) = x_2^2(D)$, otherwise we repeat the procedure until linear independence is obtained. This procedure must terminate since $G_0(D)$ has full rank and $\bar{l}_i > 0$ for $i = 1, \ldots, M$. Similarly, we obtain $\xi_j(D)$ for $2 < j \leq M$ such that the vectors $\lim_{D\to\infty} \xi_i(D) G_0(D)$, $i = 1, \ldots, M$, are linearly independent. Then we define $\xi_m(D)$ to be the polynomial matrix whose ith row is $\xi_i(D)$.
∇

For a DT adaptive control design, the polynomials $d_i(D)$ in $\xi_m(D)$ can be chosen as $d_i(D) = D^{l_i}$ for some integer l_i, $i = 1, 2, \ldots, M$. In this case, the modified interactor matrix $\xi_m(D)$ becomes the interactor matrix [125], [443]:

$$\xi(D) = \begin{bmatrix} D^{l_1} & 0 & \cdots & \cdots & 0 \\ D^{l_1}h_{21}(D) & D^{l_2} & 0 & \cdots & 0 \\ \vdots & \vdots & \vdots & \vdots & \vdots \\ D^{l_1}h_{M1}(D) & D^{l_2}h_{M2}(D) & \cdots & D^{l_{M-1}}h_{M\,M-1}(D) & D^{l_M} \end{bmatrix}, \quad (9.64)$$

where $h_{ij}(D)$, $j = 1, \ldots, M-1$, $i = 2, \ldots, M$, are polynomials divisible by s and $l_i \geq 0$, $i = 1, \ldots, M$, are integers such that

$$\lim_{s\to\infty} \xi(D)G(D) = K_p^0 \quad (9.65)$$

is finite and nonsingular (see Proposition 2.2).

To obtain a modified interactor matrix $\xi_m(D)$ from the interactor matrix $\xi(D)$, we can choose arbitrary stable polynomials

$$d_i(D) = D^{l_i} + a_1^i D^{l_i-1} + \cdots + a_{l_i}^i \quad (9.66)$$

for $i = 1, \ldots, M$ and set

$$\begin{aligned} h_{ij}^m(D) = {} & \text{polynomial part of} \\ & D^{l_j}(h_{ij}(D) + a_1^i D^{-1}h_{ij}(D) + \cdots + a_{l_i}^i D^{-l_i}h_{ij}(D)) \end{aligned} \quad (9.67)$$

for $j = 1, \ldots, M-1$, $i = 2, \ldots, M$. For this choice of $h_{ij}^m(D)$, we have

$$\lim_{s\to\infty} \xi_m(D)G_0(D) = K_p^0, \quad (9.68)$$

which can proved by direct verification (see Proposition 2.3).

Example 9.1 Consider the continuous-time transfer matrix

$$G_0(s) = \begin{bmatrix} \frac{1}{(s+a_1)(s+a_2)} & \frac{1}{(s+b)} \\ \frac{k}{(s+c_1)(s+c_2)} & \frac{h}{(s+d)} \end{bmatrix}, \quad h \neq k, \quad (9.69)$$

where $a_i, c_i, i = 1, 2, b, d, k, h$ are constants. Its modified interactor matrix is

$$\xi_m(s) = \begin{bmatrix} s + p_0 & 0 \\ -h(s^2 + q_1 s) & s^2 + q_1 s + q_0 \end{bmatrix}, \quad (9.70)$$

where p_0, q_1, q_0 are arbitrary positive constants. Hence $\xi_m(s)$ is not completely specified unless h is known. The high frequency gain matrix

$$K_p = \begin{bmatrix} 0 & 1 \\ k-h & h(b-d) \end{bmatrix} \quad (9.71)$$

depends on b, d in addition to the high frequency gains k, h. □

Lemma 9.2 *The maximum degree d of $\xi_m(D)$ satisfies*

$$d \leq n^* - M + 1 \leq n_0 - M + 1 \leq n - M + 1, \tag{9.72}$$

where n^* is the relative degree of $\det(G_0(D))$, *n_0 is the order of any minimal realization of $G_0(D)$, and n is the order of the physical realization of $G_0(D)$. The equalities in (9.72) can be achieved by some $G_0(D)$.*

Proof: (i) $d \leq n^* - M + 1$: Applying the result of [80] for the interactor matrix to the modified interactor matrix (see Lemma 9.1), we have that

$$\text{degree of } h_{ij}^m(D) \leq \sum_{k=j}^{i} l_k - (i-j) \tag{9.73}$$

for $i > j$, $i, j = 1, \ldots, M$. Since $l_k \geq 1$, $k = 1, \ldots, M$, it follows from (9.73) that $d \leq \sum_{k=1}^{m} l_k - M + 1$. From (9.62), we have $n^* = \sum_{k=1}^{M} l_k$.

(ii) $n^* - M + 1 \leq n_0 - M + 1$: $n^* \leq n_0$ is from linear system theory [44].

(iii) $n_0 - M + 1 \leq n - M + 1$: it is easy to see that $n_0 \leq n$.

(iv) Consider the following example similar to that presented in [80]:

$$G_0(D) = \begin{bmatrix} D^{-3} & 0 & \cdots & \cdots & 0 \\ -D^{-1} & D^{-5} & 0 & \cdots & 0 \\ 0 & -D^{-1} & D^{-7} & 0 & \cdots \\ \vdots & \vdots & \vdots & \vdots & \vdots \\ 0 & \cdots & 0 & -D^{-1} & D^{-2M-1} \end{bmatrix}, \tag{9.74}$$

where the (i,i)th element is D^{-2i-1}, the $(j+1,j)$th element is $-D^{-1}$, $i = 1, \ldots, M$, $j = 1, \ldots, M-1$, and other elements in $G_0(D)$ are all zero.

It can be shown that for such a $G_0(D)$ the inverse of $G_0(D)$ uniquely defines the interactor matrix of $G_0(D)$. More precisely, the interactor matrix $\xi(D)$ is an $M \times M$ lower triangular polynomial matrix with its (i,i)th element being D^{2i+1} and the (i,j)th $(i > j)$ element being $D^{i(i+1)-j(j-1)+1}$. From the procedure presented in the proof of Lemma 9.1, we can construct the modified interactor matrix $\xi_m(D)$. It can be verified that the modified interactor matrix for this $G_0(D)$ satisfies the equalities in (9.72).

We only give the details for the case when $M = 3$. The left interactor matrix $\xi(D)$ and the modified interactor matrix $\xi_m(D)$ are

$$\xi(D) = \begin{bmatrix} D^3 & 0 & 0 \\ D^7 & D^5 & 0 \\ D^{13} & D^{11} & D^7 \end{bmatrix}, \tag{9.75}$$

$$\xi_m(D) = \begin{bmatrix} d_1(D) & 0 & 0 \\ D^2 d_2(D) & d_2(D) & 0 \\ D^6 d_3(D) & D^4 d_3(D) & d_3(D) \end{bmatrix}, \tag{9.76}$$

where $d_i(D)$ is an arbitrary monic stable polynomial of degree $2i+1$, $i = 1, 2, 3$. For this example, $d = 13$, $n^* - M + 1 = 15 - 3 + 1 = 13$, $n_0 - M + 1 = 15 - 3 + 1 = 13$ ($n_0 = 15$ as from linear system theory [44]). The last equality is achieved if the physical realization of $G_0(D)$ is minimal. ∇

Remark 9.4 The observability index ν of $G_0(D)$ satisfies

$$\nu \leq n_0 - M + 1 \leq n - M + 1. \tag{9.77}$$

This can be verified by considering a minimal realization (A, B, C) of $G_0(D) = C(sI - A)^{-1}B$, whose subobservability matrix $[C^T, A^T C^T, \ldots, (A^T)^{n-k} C^T]$ does not increase its rank when $k \leq M$. Thus we also have an estimate for the upper bound $\bar{\nu}$ on ν. However, a relation between d and ν cannot be specified, as shown in Example 9.2, where a plant transfer matrix can have a modified interactor matrix with its maximum degree $d = 1, 2, 3$ for different parameters but with $\nu = 2$ for all three cases. □

Plant Assumptions

To design MRAC schemes to achieve the control objective, we make the following assumptions about the plant (9.51) and the reference model (9.53):

(A9.1): All zeros of $G_0(D)$ (i.e., zeros of $\det[Z_0(D)]$) are stable.

(A9.2): The observability index ν of $G_0(D)$ is known.[3]

(A9.3): An $M \times M$ matrix S_p is known such that $K_p S_p = (K_p S_p)^T > 0$ in the CT case or $2I > K_p S_p = (K_p S_p)^T > 0$ in the DT case.

(A9.4): $G_0(D)$ is strictly proper and full rank and has a known modified interactor matrix $\xi_m(D)$, and all poles of $W_m(D)$ are stable and the zero structure at infinity of $W_m(D)$ is the same as that of $G_0(D)$, that is, $\lim_{D\to\infty} \xi_m(D) W_m(D)$ is finite and nonsingular.

(A9.5): $G_0(D)\Delta_m(D)$ and $\Delta_a(D)$ are proper rational matrices.

(A9.6): The matrices

$$D_m = \lim_{D\to\infty} W_m(D) K_p \Delta_m(D), \; D_a = \lim_{D\to\infty} \Delta_a(D) \tag{9.78}$$

[3] For our MIMO case here, an upper bound $\bar{\nu}$ on the observability index ν of $G_0(D)$ (i.e., $\bar{\nu} \geq \nu$) is also sufficient for constructing a suitable controller structure (9.79) or (9.102) with ν replaced by $\bar{\nu}$, similar to the SISO case where only an upper bound $\bar{n}$ on the plant order n is needed for a controller structure (see Chapters 5 and 6).

have finite norm bounds independent of μ, and the transfer matrices $(W_m(s-q)K_p\Delta_m(s-q)-D_m)s$ and $(\Delta_a(s-q)-D_a)s$ are stable and proper with finite L^1 operator norm (upper) bounds independent of μ for some known $q > 0$ in the CT case, or the transfer matrices $(W_m(qz)K_p\Delta_m(qz)-D_m)(z+1)$ and $(\Delta_a(qz)-D_a)(z+1)$ are stable and proper with finite L^1 operator norm (upper) bounds independent of μ, for some known $q \in (0,1)$ in the DT case.[4]

From (A9.4), without loss of generality, we choose $W_m(D) = \xi_m^{-1}(D)$ for the reference model (9.53), with $\xi_m^{-1}(s-q_m)$ stable for some $q_m \geq q > 0$ in the CT case, or with $\xi_m^{-1}(q_m z)$ stable for some $q_m \in (0,q) \subset (0,1)$ in the DT case.

Controller Structure

With (A9.2), we choose the adaptive controller structure as

$$u(t) = \Theta_1^T\omega_1(t) + \Theta_2^T\omega_2(t) + \Theta_3 r(t), \tag{9.79}$$

[4]The L^1 operator norm of a continuous-time $M \times M$ rational transfer matrix $H(s)$ is

$$\|H\|_1 = \max_{1\leq j\leq M} \int_0^\infty \sum_{i=1}^M |h_{ij}(\tau)|\, d\tau,$$

where $h_{ij}(t)$ is the impulse response of $H_{ij}(s)$, the (i,j)th element of $H(s)$. The L^1 operator norm of a discrete-time $M \times M$ rational transfer matrix $H(z)$ is

$$\|H\|_1 = \max_{1\leq j\leq M} \sum_{\tau=0}^\infty \sum_{i=1}^M |h_{ij}(\tau)|,$$

where $h_{ij}(t)$ is the impulse response of $H_{ij}(z)$, the (i,j)th element of $H(z)$ (see Section 2.8).

Recall from Section 2.8 that for a stable and proper rational transfer matrix $H(D)$ its L^1 operator norm (upper) bound is independent of μ if and only if its H^∞ norm (that is, its L^2 operator norm) bound is independent of μ. The H^∞ norm (L^2 operator norm) of a continuous-time $M \times M$ stable rational transfer matrix $H(s)$ is

$$\|H(s)\|_\infty = \sqrt{\sup_{\omega\geq 0} \max_{1\leq i\leq M} \lambda_i[H^T(-j\omega)H(j\omega)]},$$

where $\lambda_i[H^T(-j\omega)H(j\omega)]$ is the ith eigenvalue of $H^T(-j\omega)H(j\omega)$. The H^∞ norm (L^2 operator norm) of a discrete-time $M \times M$ stable rational transfer matrix $H(z)$ is

$$\|H(z)\|_\infty = \sqrt{\sup_{\Omega\in[0,2\pi]} \max_{1\leq i\leq M} \lambda_i[H^T(e^{-j\Omega})H(e^{j\Omega})]}.$$

By the norm definition, Assumption (A9.6) is satisfied if and only if the individual components of the corresponding transfer matrices satisfy the norm bound conditions.

Assumption (A9.6) implies that the rational transfer matrices $W_m(D)K_p\Delta_m(D)$ and $\Delta_a(D)$ are stable (in particular, all finite poles of $\Delta_m(D)$ which may be improper are stable) and have known stability margins with certain bounded operator norms.

where $\Theta_1 = [\Theta_{11}, \ldots, \Theta_{1\nu}]^T \in R^{\nu M \times M}$, $\Theta_2 = [\Theta_{21}, \ldots, \Theta_{2\nu}]^T \in R^{\nu M \times M}$, with $\Theta_{ij} \in R^{M \times M}$, $i = 1, 2$, $j = 1, \ldots, \nu$, $\Theta_3 \in R^{M \times M}$, and

$$\omega_1(t) = F(D)[u](t),\ \omega_2(t) = F(D)[y](t), \tag{9.80}$$

$$F(D) = \frac{A(D)}{\Lambda(D)},\ A(D) = [I_M, DI_M, \ldots, D^{\nu-1} I_M]^T, \tag{9.81}$$

for a monic polynomial $\Lambda(D)$ of degree ν such that $\Lambda(s - q_0)$ is stable for some $q_0 > 0$ in the CT case or $\Lambda(q_0 z)$ is stable for some $q_0 \in (0, 1)$ in the DT case.

With the above specification of $\Lambda(D)$, $\xi_m(D)$, $P_0(D)$, $Z_0(D)$, we have the following polynomial matrix matching result.

Lemma 9.3 *There exist constant parameter matrices* $\Theta_1^* \in R^{\nu M \times M}$, $\Theta_2^* \in R^{\nu M \times M}$, $\Theta_3^* = K_p^{-1} \in R^{M \times M}$ *such that*

$$\Theta_1^{*T} A(D) P_0(D) + \Theta_2^{*T} A(D) Z_0(D) = \Lambda(D)(P_0(D) - \Theta_3^* \xi_m(D) Z_0(D)). \tag{9.82}$$

Proof: As in [340], we let $G_0(D) = P_l^{-1}(D) Z_l(D)$, where $Z_l(D)$ and $P_l(D)$ are left coprime (see Section 2.2), with $P_l(D)$ being row reduced and $\partial_{ri}[P_l(D)] = \nu_i \le \nu$, $i = 1, 2, \ldots, M$ (see Section 9.2.1). Divide $\Lambda(D) K_p^{-1} \xi_m(D)$ on the right by $P_l(D)$, that is, express

$$\Lambda(D) K_p^{-1} \xi_m(D) = Q_l(D) P_l(D) + R_l(D) \tag{9.83}$$

for some $M \times M$ polynomial matrices $Q_l(D)$ and $R_l(D)$ such that $\partial_{ci}[R_l(D)] < \partial_{ci}[P_l(D)] \le \nu$. Then, we define Θ_1^* and Θ_2^* from

$$\Theta_2^{*T} A(D) = -R_l(D) = Q_l(D) P_l(D) - \Lambda(D) K_p^{-1} \xi_m(D), \tag{9.84}$$

$$\Theta_1^{*T} A(D) = \Lambda(D) I_M - Q_l(D) Z_l(D). \tag{9.85}$$

With this choice, (9.82) is satisfied as $Z_l(D) P_0(D) = P_l(D) Z_0(D)$, from which we obtain the plant–model transfer matrix matching equation

$$I_M - \Theta_1^{*T} F(D) - \Theta_2^{*T} F(D) G_0(D) = \Theta_3^* W_m^{-1}(D) G_0(D), \tag{9.86}$$

where $W_m(D) = \xi_m^{-1}(D)$. From (9.86) with $\lim_{D \to \infty} \Theta_3^* W_m^{-1}(D) G_0(D) = I_M$ as from the definition of $\xi_m(D)$, we have

$$\lim_{D \to \infty} \Theta_1^{*T} F(D) = 0, \tag{9.87}$$

which implies that $\partial[\Theta_1^{*T} A(D)] \le \nu - 1$. ∇

Example 9.2 Consider the continuous-time transfer matrix

$$G_0(s) = \begin{bmatrix} \frac{1}{s-1} & \frac{1}{s+a} \\ \frac{1}{s+3} & \frac{c}{s+b} \end{bmatrix}, \ c \geq 1, \tag{9.88}$$

whose parameters $a, b, c \in R$ are constant.

Plant zeros: The plant zeros are defined as the zeros of

$$\det[G_0(s)] = \frac{(c-1)s^2 + ((3+a)c + 1 - b)s + 3ac + b}{(s-1)(s+3)(s+a)(s+b)}. \tag{9.89}$$

If $c > 1$, then for a minimum phase plant with a full rank $G_0(s)$, the parameters a, b, and c need to satisfy $(3+a)c + 1 - b > 0$, $3ac + b > 0$. That is, in the parameter space R^3, the parameter vector $(a, b, c)^T$ needs to be in the region: $c > 1$, $(3+a)c + 1 - b > 0$, $3ac + b > 0$. If $c = 1$, then for a minimum phase plant with a full rank $G_0(s)$, the parameters a and b need to satisfy $(3a+b)(a-b+4) > 0$ or $3a + b \neq 0$ when $a - b + 4 = 0$. In the parameter space R^2, the parameter vector $(a, b)^T$ needs to be in one of the two disjoint regions: (a) $3a + b > 0$, $a - b + 4 \geq 0$; (b) $3a + b < 0$, $a - b + 4 \leq 0$. To design a model reference adaptive controller, we need to assume that the plant has minimum phase and with a full rank $G_0(s)$; see Assumption (A9.1).

Interactor matrices: The interactor matrices of the plant transfer matrix $G_0(s)$ depend on the parameters a and b.

Case 1: $c > 1$, $(3+a)c + 1 - b > 0$, $3ac + b > 0$. The modified interactor matrix of $G_0(s)$ is

$$\xi_{m1}(s) = \begin{bmatrix} s + c_1 & 0 \\ 0 & s + c_1 \end{bmatrix}, \tag{9.90}$$

where $c_1 > 0$, and the associated high frequency gain matrix is

$$K_{p1} = \lim_{s\to\infty} \xi_{m1}(s) G_0(s) = \begin{bmatrix} 1 & 1 \\ 1 & c \end{bmatrix} \tag{9.91}$$

such that $\det[K_{p1}] = c - 1 \neq 0$.

Case 2: $c = 1$, $a - b + 4 \neq 0$. The modified interactor matrix of $G_0(s)$ is

$$\xi_{m2}(s) = \begin{bmatrix} s + c_1 & 0 \\ -s(s+c_1) & (s+c_1)(s+c_2) \end{bmatrix}, \tag{9.92}$$

where $c_1 > 0$, $c_2 > 0$, and the associated high frequency gain matrix is

$$K_{p2} = \lim_{s\to\infty} \xi_{m2}(s) G_0(s) = \begin{bmatrix} 1 & 1 \\ -4 + c_2 & a - b + c_2 \end{bmatrix} \tag{9.93}$$

such that $\det[K_{p2}] = a - b + 4 \neq 0$.

Case 3: $c = 1$, $a - b + 4 = 0$. The modified interactor matrix of $G_0(s)$ is

$$\xi_{m3}(s) = \begin{bmatrix} s + c_1 & 0 \\ -s(s + c_2 + c_3 - 4)(s + c_1) & (s + c_1)(s + c_2)(s + c_3) \end{bmatrix}, \quad (9.94)$$

where $c_1 > 0$, $c_2 > 0$, and $c_3 > 0$ (or $c_2 + c_3 > 0$, $c_2c_3 > 0$, $c_2 = c_3^*$, with "$*$" denoting "complex conjugate" for the case of a complex pair c_2 and c_2), and the associated high frequency gain matrix is

$$K_{p3} = \lim_{s\to\infty} \xi_{m2}(s)G_0(s)$$
$$= \begin{bmatrix} 1 & 1 \\ 12 + c_2c_3 - 4(c_2 + c_3) & 4b + c_2c_3 - 4(c_2 + c_3) \end{bmatrix}, \quad (9.95)$$

where $\det[K_{p3}] = -4(3 - b) \neq 0$ because $3a + b \neq 0$ ($3a + b > 0$ or $3a + b < 0$ for a minimum phase $G_0(s)$) and $a - b + 4 = 0$, that is, $b \neq 3$.

Note that the maximum degree of $\xi_{m1}(s)$ is 1, the maximum degree of $\xi_{m2}(s)$ is 2, and the maximum degree of $\xi_{m3}(s)$ is 3.

Controller structure: The largest class of $G_0(s)$ admissible by model reference adaptive control are those with minimum phase and full rank: $(3 + a)c + 1 - b > 0$, $3ac + b > 0$, when $c > 1$; $3a + b > 0$ and $a - b + 4 \geq 0$, or $3a + b < 0$ and $a - b + 4 \leq 0$, when $c = 1$.

The plant transfer matrix can be expressed as $G_0(s) = Z_0(s)P_0^{-1}(s)$, where

$$Z_0(s) = \begin{bmatrix} s + 3 & s + b \\ s - 1 & c(s + a) \end{bmatrix}, \quad (9.96)$$

$$P_0(s) = \begin{bmatrix} (s - 1)(s + 3) & 0 \\ 0 & (s + a)(s + b) \end{bmatrix}, \quad (9.97)$$

whose observability index is $\nu = 2$. For simplicity, we take $\bar{\nu} = \nu = 2$ in (A9.2). The controller structure (9.79) has the form

$$u(t) = \Theta_1^T(t)\omega_1(t) + \Theta_2^T(t)\omega_2(t) + \Theta_3(t)r(t), \quad (9.98)$$

where

$$\omega_1(t) = \frac{A(s)}{\Lambda(s)}[u](t), \ \omega_2(t) = \frac{A(s)}{\Lambda(s)}[y](t), \quad (9.99)$$

$A(s) = [I_2, sI_2]^T$, $\Lambda(s) = s^2 + \lambda_1 s + \lambda_0$ with $\lambda_1 > 0$ and $\lambda_0 > 0$, and

$$\Theta_1(t) \in R^{4\times 2}, \ \Theta_2(t) \in R^{4\times 2}, \ \Theta_3(t) \in R^{2\times 2}. \quad (9.100)$$

For the controller (9.98), the matching equation

$$\Theta_1^{*T} A(s)P_0(s) + \Theta_2^{*T} A(s)Z_0(s) = \Lambda(s)(P_0(s) - \Theta_3^* \xi_m(s)Z_0(s)) \tag{9.101}$$

defines the matching parameters Θ_1^*, Θ_2^*, and Θ_3^*. □

Remark 9.5 As the multivariable version of the adaptive controller structure (5.5) for the single-input, single-output (SISO) case, an alternative adaptive controller structure for the multivariable plant is

$$u(t) = \Theta_1^T \omega_1(t) + \Theta_2^T \omega_2(t) + \Theta_{20} y(t) + \Theta_3 r(t), \tag{9.102}$$

where $\Theta_1 = [\Theta_{11}, \ldots, \Theta_{1\,\nu-1}]^T$, $\Theta_2 = [\Theta_{21}, \ldots, \Theta_{2\,\nu-1}]^T$, with $\Theta_{ij} \in R^{M\times M}$, $i = 1, 2$, $j = 1, \ldots, \nu - 1$, $\Theta_{20} \in R^{M\times M}$, and $\Theta_3 \in R^{M\times M}$, and

$$\omega_1(t) = F(D)[u](t),\ \omega_2(t) = F(D)[y](t), \tag{9.103}$$

$$F(D) = \frac{A(D)}{\Lambda(D)},\ A(D) = [I_M, DI_M, \ldots, D^{\nu-2} I_M]^T, \tag{9.104}$$

for a monic polynomial $\Lambda(D)$ of degree $\nu - 1$ such that $\Lambda(s - q_0)$ is stable for a $q_0 > 0$ in the CT case or $\Lambda(q_0 z)$ is stable for a $q_0 \in (0, 1)$ in the DT case.

This controller structure is commonly used in the literature [93], [282], [340]. With the specification of $\Lambda(D)$, $\xi_m(D)$, $P_0(D)$, $Z_0(D)$, it follows that there exist Θ_1^*, Θ_2^*, Θ_{20}^*, $\Theta_3^* = K_p^{-1}$ [93] such that

$$\begin{aligned} &\Theta_1^{*T} A(D)P_0(D) + (\Theta_2^{*T} A(D) + \Theta_{20}^* \Lambda(D))Z_0(D) \\ &= \Lambda(D)(P_0(D) - \Theta_3^* \xi_m(D)Z_0(D)). \end{aligned} \tag{9.105}$$

The proof of this result can be given based on a matrix version of the Bezout identity [282] or a direct calculation [340] used for the proof of (9.82).

From (9.102), we have the plant–model transfer matrix matching equation

$$I - \Theta_1^{*T} F(D) - (\Theta_2^{*T} F(D) + \Theta_{20}^*)G_0(D) = \Theta_3^* W_m^{-1}(D)G_0(D). \tag{9.106}$$

This controller parametrization is suitable for achieving the model reference adaptive control objective when $\mu = 0$ in (9.60) (i.e., closed-loop signal boundedness and asymptotic output tracking) and when $\mu \neq 0$ (i.e., closed-loop signal boundedness and small tracking error in the mean; see Theorem 9.2). However, for $\mu \neq 0$, the controller (9.102) admits a different (smaller) class of plant unmodeled dynamics $\mu\Delta_m(D)$ and $\mu\Delta_a(D)$ for robust adaptive control than (9.79) does, when a gradient design is used, as presented in the rest of this section. This can be verified in a similar way to that (see the subsection

"An Alternative Controller Structure" in Section 5.7) used to compare the controller structure (5.5) in Section 5.1 (which is the single-input, single-output (SISO) version of (9.102)) with the controller structure (5.333) in Section 5.7 (which is the SISO version of (9.79)).

For the plant in Example 9.2, given $\bar{\nu} = \nu = 2$, the controller (9.102) is

$$u(t) = \Theta_1^T(t)\omega_1(t) + \Theta_2^T(t)\omega_2(t) + \Theta_{20}(t)y(t) + \Theta_3(t)r(t), \tag{9.107}$$

where

$$\omega_1(t) = \frac{1}{\Lambda(s)}[u](t),\ \omega_2(t) = \frac{1}{\Lambda(s)}[y](t), \tag{9.108}$$

$\Lambda(s) = s + \lambda$ with $\lambda > 0$, and

$$\Theta_1(t) \in R^{2\times 2},\ \Theta_2(t) \in R^{2\times 2},\ \Theta_{20}(t) \in R^{2\times 2},\ \Theta_{30}(t) \in R^{2\times 2}. \tag{9.109}$$

For the controller (9.107), the matching equation

$$\Theta_1^{*T} P_0(s) + (\Theta_2^{*T} + \Theta_{20}^*\Lambda(s))Z_0(s) = \Lambda(s)(P_0(s) - \Theta_3^*\xi_m(s)Z_0(s)) \tag{9.110}$$

defines the matching parameters Θ_1^*, Θ_2^*, Θ_{20}^*, and Θ_{30}^*. □

9.2.3 Robust Model Reference Control

When $G_0(D)$ in the plant (9.59) is known, we can calculate the parameter matrices Θ_i^*, $i = 1, 2, 3$, by solving the matching equation (9.82), using the division scheme (9.84)–(9.85), and use them in the controller (9.79), that is, set $\Theta_i = \Theta_i^*$, $i = 1, 2, 3$, to implement a nominal model reference controller.

Using the matching equation (9.86), $I - \Theta_1^{*T}F(D) - \Theta_2^{*T}F(D)G_0(D) = \Theta_3^* W_m^{-1}(D)G_0(D)$, and denoting I_M simply by I, from (9.59), (9.60), (9.61) with $W_m(D) = \xi_m^{-1}(D)$, and (9.79) with $\Theta_i = \Theta_i^*$, $i = 1, 2, 3$, we obtain

$$\begin{aligned}
& y(t) - y_m(t) \\
&= G(D)\left(I - \Theta_1^{*T}F(D) - \Theta_2^{*T}F(D)G(D)\right)^{-1}\Theta_3^*[r](t) \\
&\quad - G_0(D)\left(I - \Theta_1^{*T}F(D) - \Theta_2^{*T}F(D)G_0(D)\right)^{-1}\Theta_3^*[r](t) \\
&= \mu\left(W_m(D)\Theta_3^{*-1}\Theta_2^{*T}F(D) + I\right)(G_0(D)\Delta_m(D) + \Delta_a(D)) \\
&\quad \left(I - \mu G_0^{-1}(D)W_m(D)\Theta_3^{*-1}\Theta_2^{*T}F(D)\left(G_0(D)\Delta_m(D) + \Delta_a(D)\right)\right)^{-1} \\
&\quad G_0^{-1}(D)W_m(D)[r](t),
\end{aligned} \tag{9.111}$$

where $G(D) = G_0(D) + \mu\left(G_0(D)\Delta_m(D) + \Delta_a(D)\right)$ as in (9.60).

Using (9.86), we have

$$\begin{aligned}&\left(W_m(D)\Theta_3^{*-1}\Theta_2^{*T}F(D)+I\right)G_0(D)\Delta_m(D)\\&=W_m(D)\Theta_3^{*-1}\left(I-\Theta_1^{*T}F(D)\right)\Delta_m(D),\end{aligned}\tag{9.112}$$

which is proper (see Assumption (A9.5)) and stable (if all finite poles of $\Delta_m(D)$ are stable; see Assumption (A9.6)), and

$$\begin{aligned}&G_0^{-1}(D)W_m(D)\Theta_3^{*-1}\Theta_2^{*T}F(D)G_0(D)\Delta_m(D)\\&=\left(G_0^{-1}(D)W_m(D)\Theta_3^{*-1}\left(I-\Theta_1^{*T}F(D)\right)-I\right)\Delta_m(D),\end{aligned}\tag{9.113}$$

which is also proper and stable (if all finite poles of $\Delta_m(D)$ are stable).

From (9.59) and (9.79) with $\Theta_i=\Theta_i^*$, $i=1,2,3$, we also obtain

$$\begin{aligned}u(t)=&\left(I-\mu G_0^{-1}(D)W_m(D)\Theta_3^{*-1}\Theta_2^{*T}F(D)\left(G_0(D)\Delta_m(D)+\Delta_a(D)\right)\right)^{-1}\\&G_0^{-1}(D)W_m(D)[r](t).\end{aligned}\tag{9.114}$$

From (9.111) and (9.114), we may state the following robustness result.

Theorem 9.1 *The controller (9.79), if implemented with $\Theta_i=\Theta_i^*$, $i=1,2,3$, which satisfy (9.82), and applied to the plant (9.59) with $\mu\in[0,\mu_0)$, where*

$$\mu_0=\frac{1}{\|G_0^{-1}(D)W_m(D)\Theta_3^{*-1}\Theta_2^{*T}F(D)(G_0(D)\Delta_m(D)+\Delta_a(D))\|_\infty},\tag{9.115}$$

ensures the closed-loop signal boundedness and exponential convergence of the tracking error $y(t)-y_m(t)$ to the set

$$S_0=\{e:\ \|e\|\le b_1\mu\,\|r(\cdot)\|_\infty\}\tag{9.116}$$

for some constant $b_1>0$.

This result is the multivariable counterpart of Theorem 5.7. It follows that $\left(I-\mu G_0^{-1}(D)W_m(D)\Theta_3^{*-1}\Theta_2^{*T}F(D)\left(G_0(D)\Delta_m(D)+\Delta_a(D)\right)\right)^{-1}$ is stable for any $\mu\in[0,\mu_0)$, while other transfer matrices are all stable and proper.

9.2.4 Error Model

In adaptive control, the modeled part $G_0(D)$ of the plant (9.59) is assumed to be unknown, in addition to the uncertainties of $\mu\Delta_m(D)$ and $\mu\Delta_a(D)$. In this case, we need to design an adaptive law to update the parameters $\Theta_i(t)$, $i=1,2,3$, of the controller (9.79) so that the desired closed-loop signal boundedness and tracking performance can still be achieved.

To develop an adaptive law for the controller (9.79), we need to express the closed-loop system in terms of the tracking error $e(t) = y(t) - y_m(t)$ and the parameter error $\tilde{\Theta}(t) = \Theta(t) - \Theta^*$, where

$$\Theta(t) = [\Theta_1^T(t), \Theta_2^T(t), \Theta_3(t)]^T \tag{9.117}$$

is the estimate of the unknown parameter matrix

$$\Theta^* = [\Theta_1^{*T}, \Theta_2^{*T}, \Theta_3^*]^T. \tag{9.118}$$

Operating both side of (9.86) on $u(t)$ and using (9.59), we obtain

$$\begin{aligned} & u(t) - \Theta_1^{*T} F(D)[u](t) - \Theta_2^{*T} F(D)[y](t) \\ &= \Theta_3^* W_m^{-1}(D)[y](t) - \Theta_3^* W_m^{-1}(D)(G(D) - G_0(D))[u](t) \\ &\quad - \Theta_2^{*T} F(D)(G(D) - G_0(D))[u](t) \\ &= \Theta_3^* W_m^{-1}(D)[y](t) \\ &\quad - \mu(I - \Theta_1^{*T} F(D)) \left(\Delta_m(D) + G_0^{-1}(D)\Delta_a(D)\right)[u](t) \end{aligned} \tag{9.119}$$

(for the second equality, we used (9.86) again). It follows from (9.119) that

$$\begin{aligned} & \Theta_3^* W_m^{-1}(D)[y - y_m](t) \\ &= u(t) - \Theta^{*T}\omega(t) + \mu(I - \Theta_1^{*T} F(D)) \left(\Delta_m(D) + G_0^{-1}(D)\Delta_a(D)\right)[u](t), \end{aligned} \tag{9.120}$$

where Θ^* is given in (9.118) and

$$\omega(t) = [\omega_1^T(t), \omega_2^T(t), r^T(t)]^T. \tag{9.121}$$

Hence, from (9.120) we derive the tracking error equation

$$\begin{aligned} y(t) - y_m(t) &= W_m(D)\Theta_3^{*-1}[u - \Theta^{*T}\omega](t) + \mu\eta(t) \\ &= W_m(D)\Theta_3^{*-1}[(\Theta - \Theta^*)^T\omega](t) + \mu\eta(t), \end{aligned} \tag{9.122}$$

where $u(t) = \Theta^T(t)\omega(t)$ as in (9.79) and

$$\eta(t) = \Delta(D)\, u(t), \tag{9.123}$$

$$\Delta(D) = W_m(D)\Theta_3^{*-1}(I - \Theta_1^{*T} F(D)) \left(\Delta_m(D) + G_0^{-1}(D)\Delta_a(D)\right). \tag{9.124}$$

To proceed, with $K_p = \Theta_3^{*-1}$, we rewrite (9.120) as

$$\begin{aligned} & \xi_m(D)[y - y_m](t) \\ &= K_p(\Theta - \Theta^*)^T\omega(t) \\ &\quad + \mu K_p(I - \Theta_1^{*T} F(D)) \left(\Delta_m(D) + G_0^{-1}(D)\Delta_a(D)\right)[u](t)). \end{aligned} \tag{9.125}$$

Let d_m be the maximum degree of $\xi_m(D)$, $\Psi^* = \Theta_3^{*-1} = K_p$, $f(D)$ be any polynomial of $D = s$ or z with degree d_m such that $f(s - q_0)$ is stable in the CT case for some $q_0 > 0$ or $f(q_0 z)$ is stable in the DT case for some $0 < q_0 < 1$, and $h(D) = \frac{1}{f(D)}$. Filtering both sides of (9.125) by $h(D)$, we have

$$\begin{aligned} &\xi_m(D)h(D)[y - y_m](t) \\ &= \Psi^*(h(D)[u](t) - \Theta^{*T}h(D)[\omega](t)) + \mu\eta_1(t), \end{aligned} \tag{9.126}$$

where

$$\eta_1(t) = \Delta_1(D)[u](t), \tag{9.127}$$

$$\Delta_1(D) = \Theta_3^{*-1}h(D)(I - \Theta_1^{*T}F(D))(\Delta_m(D) + G_0^{-1}(D)\Delta_a(D)). \tag{9.128}$$

Let $\Psi(t)$ be the estimate of Ψ^*, define the normalized estimation error

$$\epsilon(t) = \frac{\xi_m(D)h(D)[y - y_m](t) + \Psi(t)\xi(t)}{1 + \beta(\zeta^T(t)\zeta(t) + \xi^T(t)\xi(t)) + \alpha m^2(t)}, \tag{9.129}$$

where $\beta > 0$ for the CT case and $\beta \geq 1$ for the DT case, $\alpha > 0$,

$$\xi(t) = \Theta^T(t)\zeta(t) - h(D)[\Theta^T\omega](t),\ \zeta(t) = h(D)[\omega](t),\ h(D) = \frac{1}{f(D)}, \tag{9.130}$$

and introduce the dynamic normalizing signal $m(t)$ generated from

$$(D - \sigma)[m](t) = -\delta_0 m(t) + \delta_1(\|u(t)\| + \|y(t)\|) + \delta_2,\ m(0) > 0, \tag{9.131}$$

where $\sigma = 0$, $0 < \delta_0 < \min\{q, q_0\}$ for the CT case, or $\sigma = 1$, $\max\{q, q_0\} < 1 - \delta_0 < 1$ for the DT case, $\delta_1 > 0$, and $\delta_2 > 0$ (recall that q is defined in Assumption (A9.6) and q_0 is defined in the controller and estimation error filter polynomials $\Lambda(D)$ and $f(D)$; see (9.81) and (9.130)).

9.2.5 Adaptive Laws

The adaptive laws for updating the controller parameters are chosen as

$$\dot{\Theta}^T(t) = -S_p\epsilon(t)\zeta^T(t) - G_1^T(t), \tag{9.132C}$$

$$\dot{\Psi}(t) = -\Gamma\epsilon(t)\xi^T(t) - G_2(t) \tag{9.133C}$$

for the CT case, and

$$\Theta^T(t+1) = \Theta^T(t) - S_p\epsilon(t)\zeta^T(t) - G_1^T(t), \tag{9.132D}$$

$$\Psi(t+1) = \Psi(t) - \Gamma\epsilon(t)\xi^T(t) - G_2(t) \tag{9.133D}$$

for the DT case, where $G_1(t)$, $G_2(t)$ are the modifying terms for robustness with respect to the plant unmodeled dynamics $\mu\Delta_m(D)$ and $\mu\Delta_a(D)$ and Γ and S_P are $M \times M$ constant adaptation gain matrices such that S_p satisfies Assumption (A9.3) and $\Gamma = \Gamma^T > 0$ ($2\beta I > \Gamma > 0$ for the DT case).

We introduce the matrix version of the switching σ-modification (5.196)–(5.197) for the adaptive laws (9.132) and (9.133):

$$G_1^T(t) = \sigma_1(t)\Theta^T(t),\ G_2(t) = \sigma_2(t)\Gamma\Psi(t), \tag{9.134}$$

where, with the matrix norm $\|\cdot\|$ being the Frobenius norm[5] (which is con-

[5]The Frobenius norm of an $m \times n$ matrix $A = \{a_{ij}\}$ is defined as

$$\|A\| = \|A\|_F = \sqrt{\text{trace}[A^TA]} = \sqrt{\sum_{i=1}^{m}\sum_{j=1}^{n}|a_{ij}|^2}.$$

The Frobenius norm has the following properties:

(i) $|\text{tr}[A^T\Gamma B]| \leq \lambda_{\max}[\Gamma]\|A\|\,\|B\|$, and

(ii) $\text{tr}[A^T\Gamma A] \geq \lambda_{\min}[\Gamma]\text{tr}[A^TA]$,
where $\Gamma = \Gamma^T \in R^{m\times m}$ is positive definite and $\lambda_{\max}[\Gamma] \geq \lambda_{\min}[\Gamma] > 0$ are the maximum and minimum eigenvalues of Γ, respectively.

Proof: (i) Denoting $A = [a_1, a_2, \ldots, a_n]$, $B = [b_1, b_2, \ldots, b_n] \in R^{m\times n}$, we have

$$\begin{aligned}|\text{tr}[A^T\Gamma B]| &= \left|\sum_{i=1}^{n} a_i^T\Gamma b_i\right| \leq \sum_{i=1}^{n}\|a_i\|_2\|\Gamma\|_2\|b_i\|_2 = \sum_{i=1}^{n}\|a_i\|_2\lambda_{\max}[\Gamma]\|b_i\|_2 \\ &\leq \lambda_{\max}[\Gamma]\sqrt{\sum_{i=1}^{n}\|a_i\|_2^2}\sqrt{\sum_{i=1}^{n}\|b_i\|_2^2} = \lambda_{\max}[\Gamma]\|A\|\,\|B\|.\end{aligned}$$

Here, $\|\Gamma\|_2 = \lambda_{\max}[\Gamma]$ as $\Gamma = \Gamma^T > 0$, and the last inequality is from the Schwarz inequality.

(ii) Similarly, for $A = [a_1, a_2, \ldots, a_n]$, we have

$$\text{tr}[A^T\Gamma A] = \sum_{i=1}^{n} a_i^T\Gamma a_i \geq \sum_{i=1}^{n}\|a_i\|_2^2\lambda_{\min}[\Gamma] = \lambda_{\min}[\Gamma]\|A\|^2 = \lambda_{\min}[\Gamma]\text{tr}[A^TA].$$

Note that for an $m \times m$ matrix $\Gamma = \Gamma^T > 0$ there is a matrix $P \in R^{m\times m}$ [298] such that $P^TP = I$ and $\Gamma = P^TD_\lambda P$ with $D_\lambda = \text{diag}\{\lambda_1, \lambda_2, \ldots, \lambda_m\}$, the eigenvalues of Γ, so that $a_i^T\Gamma a_i = (Pa_i)^TD_\lambda Pa_i \geq \lambda_{\min}[\Gamma](Pa_i)^TPa_i = \lambda_{\min}[\Gamma]a_i^Ta_i = \lambda_{\min}[\Gamma]\|a_i\|_2^2$. ∇

For the Frobenius norm and l^2-induced norm $\|\cdot\|_2$, with $A \in R^{m\times n}$, we have [126]

$$\|A\|_2 \leq \|A\|_F \leq \sqrt{n}\,\|A\|_2.$$

venient for implementation of the matrix σ-modification), for the CT case,

$$\sigma_1(t) = \begin{cases} 0 & \text{if } \|\Theta(t)\| < M_1, \\ \sigma_{10}(\frac{\|\Theta(t)\|}{M_1} - 1) & \text{if } M_1 \leq \|\Theta\| < 2M_1, \\ \sigma_{10} & \text{if } \|\Theta(t)\| \geq 2M_1, \end{cases} \tag{9.135C}$$

$$\sigma_2(t) = \begin{cases} 0 & \text{if } \|\Psi(t)\| < M_2, \\ \sigma_{20}(\frac{\|\Psi(t)\|}{M_2} - 1) & \text{if } M_2 \leq \|\Psi(t)\| < 2M_2, \\ \sigma_{20} & \text{if } \|\Psi(t)\| \geq 2M_2 \end{cases} \tag{9.136C}$$

for $M_1 > 0$ and $M_2 > 0$ such that

$$\frac{\|\Theta^*\| \, \|\Gamma_p\|_2}{\lambda_1} < M_1, \ \|\Psi^*\| < M_2 \tag{9.137C}$$

$$\Gamma_p = K_p^T S_p^{-1} = (S_p^{-1})^T (K_p S_p)^T S_p^{-1}, \tag{9.138C}$$

with λ_1 being the minimum eigenvalue of Γ_p, $\sigma_{10} > 0$, $\sigma_{20} > 0$ are arbitrary, and for the DT case,

$$\sigma_1(t) = \begin{cases} 0 & \text{if } \|\Theta(t)\| < M_1, \\ \sigma_{10} & \text{if } \|\Theta(t)\| \geq M_1, \end{cases} \tag{9.135D}$$

$$\sigma_2(t) = \begin{cases} 0 & \text{if } \|\Psi(t)\| < M_2, \\ \sigma_{20} & \text{if } \|\Psi(t)\| \geq M_2 \end{cases} \tag{9.136D}$$

for $M_1 > 0$, $M_2 > 0$, $\sigma_{10} > 0$, $\sigma_{20} > 0$ such that

$$\frac{3\|\Theta^*\| \, \|\Gamma_p\|_2}{2\lambda_1} < M_1, \ \frac{3\|\Psi^*\|}{2} < M_2, \tag{9.137D}$$

$$\sigma_{10} \leq \frac{\lambda_1 \lambda_2}{2(\lambda_2 \|\Gamma_p\|_2 + \|S_p\|_2^2)}, \ \sigma_{20} \leq \frac{\lambda_3}{2(\lambda_3 + 1)}, \tag{9.138D}$$

with λ_1, λ_2, λ_3 being the minimum eigenvalues of Γ_p, $2\beta I - S_p^T K_p^T > 0$, $2\beta I - \Gamma > 0$, respectively, for Γ_p given in (9.138C) (note that $\|\Gamma_p\|_2 = \lambda_{\max}[\Gamma_p]$).

Defining $\tilde{\Psi}(t) = \Psi(t) - \Psi^*$, from (9.126) and (9.129), we obtain

$$\epsilon(t) = \frac{K_p \tilde{\Theta}^T(t)\zeta(t) + \tilde{\Psi}(t)\xi(t) + \mu\eta_1(t)}{1 + \beta(\zeta^T(t)\zeta(t) + \xi^T(t)\xi(t)) + \alpha m^2(t)}. \tag{9.139}$$

From (9.86), (9.127), and (9.128), it follows that

$$\begin{aligned} \Delta_1(D) = {} & \Theta_3^{*-1}(I - \Theta_1^{*T} F(D)) K_p^{-1} h(D) \xi_m(D) W_m(D) K_p \Delta_m(D) \\ & + h(D)\xi_m(D) \left(I + W_m(D)\Theta_3^{*-1}\Theta_2^{*T} F(D)\right) \Delta_a(D). \end{aligned} \tag{9.140}$$

Using (A9.5), (A9.6), and (9.140), we can express $\Delta_1(D)$ in (9.128) as

$$\begin{aligned}\Delta_1(D) &= (\Delta_1(D) - D_1)(D+q_1)\frac{1}{D+q_1} + D_1\\ &\triangleq \bar{\Delta}_1(D)\frac{1}{D+q_1} + D_1\end{aligned} \tag{9.141}$$

for $D_1 = \lim_{D\to\infty}\Delta_1(D)$ with a finite norm bound independent of μ, and

- in the CT case, for some $q_1 \in (\delta_0, \min\{q_0, q\})$ (for $q > 0$ in Assumption (A9.6), $\delta_0 > 0$ in (9.131) for the CT case, and $q_0 > 0$ for the controller filter polynomial $\Lambda(s)$ and reference model polynomial matrix $\xi_m(s)$ in the CT case) such that $\bar{\Delta}_1(s - q_1)$ is stable and proper with a finite operator norm (upper) bound independent of μ; or

- in the DT case, for some $q_1 \in (\max\{q, q_0\}, 1-\delta_0)$ (for $q > 0$ in Assumption (A9.6), $\delta_0 > 0$ in (9.131) for the DT case such that $1 - \delta_0 < 1$, and $q_0 > 0$ for the controller filter polynomial $\Lambda(z)$ and reference model polynomial matrix $\xi_m(z)$ in the DT case) such that $\bar{\Delta}_1(q_1 z)$ is stable and proper with a finite operator norm (upper) bound independent of μ.

Using the multivariable version of Lemma 2.6,[6] from Assumptions (A9.5) and (A9.6), (9.141), and (9.131), for both the CT and DT cases, there exist

[6]**Multivariable version of Lemma 2.6**: If $x_1(t) = H(s)\frac{1}{s+\delta}[x_2](t)$, with $\delta > 0$, $x_1(t) \in R^M, x_2(t) \in R^M$, and the impulse response matrix $h_\delta(t)$ of $H(s-\delta)$ satisfies $\|h_\delta(\cdot)\|_1 \triangleq \max_{1\le j\le M}\int_0^\infty \sum_{i=1}^M |h_{\delta ij}(\tau)|\,d\tau < \infty$, where $h_{\delta ij}(t)$ is the (i,j)th element of $h_\delta(t)$, then

$$\|x_1(t)\| \le \|h_\delta(\cdot)\|_1 \frac{1}{s+\delta}[\|x_2\|](t).$$

Proof: Introducing $x_3(t) = e^{\delta t}x_2(t)$, we have

$$\begin{aligned}\|x_1(t)\| &\le \frac{1}{s+\delta}[\|H(s)[x_2]\|](t) = \int_0^t e^{-\delta(t-\tau)}\|H(s)[x_2](\tau)\|\,d\tau = e^{-\delta t}\int_0^t \|e^{\delta\tau}H(s)[x_2](\tau)\|\,d\tau\\ &= e^{-\delta t}\int_0^t \|H(s-\delta)[x_3](\tau)\|\,d\tau \le e^{-\delta t}\|h_\delta(\cdot)\|_1\int_0^t e^{\delta t}\|x_2(\tau)\|\,d\tau\\ &= \|h_\delta(\cdot)\|_1 \int_0^t e^{-\delta(t-\tau)}\|x_2(\tau)\|\,d\tau = \|h_\delta(\cdot)\|_1 \frac{1}{s+\delta}[\|x_2\|](t).\end{aligned}$$

Note that $\mathcal{L}[e^{\delta t}h(t)] = H(s-\delta)$ and $H(s-\delta)[x_3](\tau) = \int_0^\tau e^{\delta(\tau-\sigma)}h(\tau-\sigma)x_3(\sigma)\,d\sigma = e^{\delta\tau}\int_0^\tau h(\tau-\sigma)x_2(\sigma)\,d\sigma = e^{\delta\tau}H(s)[x_2](\tau)$ and that (2.251) holds with $p = 1$. ∇

Discrete-time version of Lemma 2.6: If $x_1(t) = H(z)\frac{1}{z-\delta}[x_2](t)$, with $|\delta| < 1$, $x_1(t) \in R, x_2(t) \in R$, and the impulse response function $h_\delta(t)$ of $H(\delta z)$ satisfies $\|h_\delta(\cdot)\|_1 = \sum_{\tau=0}^\infty |h_\delta(\tau)| < \infty$, then,

$$|x_1(t)| \le \|h_\delta(\cdot)\|_1 \frac{1}{z-\delta}[|x_2|](t).$$

constants $k_1 > 0$, $k_2 > 0$ such that

$$\frac{\|\eta_1(t)\|}{m(t)} \le k_1 + k_2\|\Theta(t)\|, \tag{9.142}$$

where $\eta_1(t) = \Delta_1(D)[u](t)$ with $\Delta_1(D)$ in (9.141). Here we used the control signal expression $u(t) = \Theta^T(t)\omega(t)$ and the fact that $\frac{\|\omega(t)\|}{m(t)} \le k_0$ for some $k_0 > 0$. We note that $\Delta_m(D)$ and $\Delta_a(D)$ satisfying Assumptions (A9.5) and (A9.6) with q also satisfy the assumptions of the same form but with q replaced by any $q_1 \in (0, q)$ (see the footnote for (5.195) in Section 5.5.3).

Remark 9.6 As similar to that in Section 5.3.4, for the CT case, we can define an alternate estimation error $\epsilon(t)$ from the dynamic equation

$$\epsilon(t) = \xi_m(s)h(s)[y - y_m](t) + h(s)f(s)[\Psi\xi - \alpha\,\epsilon\, m^2](t), \tag{9.143C}$$

where $\alpha > 0$ is arbitrary, $h(s) = \frac{1}{f(s)(s+a_m)}$ for $a_m > q_0 > 0$, $f(s)$ is a polynomial of degree $d_m - 1$ such that $f(s - q_0)$ is stable, and

$$\xi(t) = \Theta^T(t)\zeta(t) - \frac{1}{f(s)}[\Theta^T\omega](t),\ \zeta(t) = \frac{1}{f(s)}[\omega](t). \tag{9.144C}$$

In this case, $h(s)f(s) = \frac{1}{s+a_m}$, and the error model is of the form

$$\dot\epsilon(t) = -a_m\epsilon(t) + K_p\tilde\Theta^T(t)\zeta(t) + \tilde\Psi(t)\xi(t) - \alpha\,\epsilon(t)\,m^2(t) + \mu\eta_2(t), \tag{9.145C}$$

where

$$\eta_2(t) = \Delta_2(s)[u](t), \tag{9.146C}$$

$$\begin{aligned}\Delta_2(s) = \frac{s+a_m}{s+q}\Big(&h(s)W_m^{-1}(s)\left(I + W_m(s)\Theta_3^{*-1}\Theta_2^{*T}F(s)\right)(s+q)\Delta_a(s)\\ &+ K_p(I - \Theta_1^{*T}F(s))h(s)K_p^{-1}W_m^{-1}(s)(s+q)W_m(s)K_p\Delta_m(s)\Big).\end{aligned} \tag{9.147C}$$

Proof: Introducing $x_3(t) = \delta^{-t}x_2(t)$, we have

$$\begin{aligned}|x_1(t)| &\le \frac{1}{z-\delta}[|H(z)[x_2]|](t) = \sum_{\tau=0}^{t-1}\delta^{t-\tau-1}|H(z)[x_2](\tau)| = \delta^{t-1}\sum_{\tau=0}^{t-1}|\delta^{-\tau}H(z)[x_2](\tau)|\\ &= \delta^{t-1}\sum_{\tau=0}^{t-1}|H(\delta z)[x_3](\tau)| \le \delta^{t-1}\|h_\delta(\cdot)\|_1\sum_{\tau=0}^{t-1}\delta^{-\tau}|x_2(\tau)|\\ &= \|h_\delta(\cdot)\|_1\sum_{\tau=0}^{t-1}\delta^{t-\tau-1}|x_2(\tau)| = \|h_\delta(\cdot)\|_1\frac{1}{z-\delta}[|x_2|](t).\end{aligned}$$

(Here we used the properties that $\mathcal{Z}[\delta^{-t}h(t)] = H(\delta z)$ and $H(\delta z)[x_3](\tau) = \sum_{\sigma=0}^{\tau-1}\delta^{-\tau+\sigma}h(\tau-\sigma)x_3(\sigma) = \delta^{-\tau}\sum_{\sigma=0}^{\tau-1}h(\tau-\sigma)x_2(\sigma) = \delta^{-\tau}H(z)[x_2](\tau)$ and (2.252) for $p = 1$.) ∇

The multivariable version of this discrete-time result can be derived in a similar way. The SISO version of (9.142) in the CT case is (5.216), as from Lemma 2.6.

For a design based on the estimation error $\epsilon(t)$ in (9.143C), we assume

(A9.5a): $\Delta_m(s)$, $\Delta_a(s)$ are strictly proper rational matrices.

(A9.6a): The matrices

$$D_m = \lim_{s\to\infty} sW_m(s)K_p\Delta_m(s), \; D_a = \lim_{s\to\infty} s\Delta_a(s) \tag{9.148C}$$

are finite, and the transfer matrices $(sW_m(s-q)K_p\Delta_m(s-q) - D_m)s$, $(s\Delta_a(s-q) - D_a)s$ are stable and proper with finite operator norms for some known $q > 0$, independent of μ.

Then, the bounding property (9.142) also holds, as from (A9.5a), (A9.6a), (9.146C), (9.147C), (9.131). We should note that the class of unmodeled dynamics satisfying (A9.5a) and (A9.6a) is a subset of those satisfying (A9.5) and (A9.6). However, if the estimation error $\epsilon(t)$ in (9.143C) is used, then the reduction of the order of the adaptive control system is up to $3M$, where M is the dimension of the plant output $y(t)$.

9.2.6 Stability and Robustness Analysis

In this section we analyze the stability, robustness, and tracking performance of the proposed adaptive control schemes.

We first derive the stability properties of the adaptive laws (9.132)–(9.138).

Lemma 9.4 *(i) The adaptive law (9.132C)–(9.138C) ensures that $\Theta(t)$, $\Psi(t)$, $\epsilon(t) \in L^\infty$, $\dot\Theta(t)$, $\dot\Psi(t)$, $\epsilon(t)m(t) \in L^\infty$, and*

$$\int_{t_1}^{t_2} x^2(t)\, dt \le \alpha_c\mu^2(t_2 - t_1) + a_c \tag{9.149C}$$

for some $\alpha_c > 0$, $a_c > 0$, any $t_2 > t_1 \ge 0$, $x(t) = \|\dot\Theta(t)\| + \|\epsilon(t)\|m(t)$.

(ii) The adaptive law (9.132D)–(9.138D) ensures that $\Theta(t)$, $\Psi(t)$, $\epsilon(t) \in L^\infty$, $\epsilon(t)m(t) \in L^\infty$, and

$$\sum_{t=t_1}^{t_2-1} x^2(t) \le \alpha_d\mu^2(t_2 - t_1) + a_d \tag{9.149D}$$

for some $\alpha_d > 0$, $a_d > 0$, any $t_2 > t_1 \ge 0$, $x(t) = \|\Theta(t+1) - \Theta(t)\| + \|\epsilon(t)\|m(t)$.

Proof: (i) For the CT case, we consider the positive definite function

$$V = \mathrm{tr}[\tilde\Theta(t)\Gamma_p\tilde\Theta^T(t)] + \mathrm{tr}[\tilde\Psi^T(t)\Gamma^{-1}\tilde\Psi(t)], \; t \in [0, \infty). \tag{9.150C}$$

Using (9.132C)–(9.138C), (9.139), (9.142), and (9.150C), we derive

$$\begin{aligned}\dot{V} &= -2\epsilon^T(t)\epsilon(t) - 2\alpha\epsilon^T(t)\epsilon(t)m^2(t) + 2\mu\epsilon^T(t)\eta_1(t) \\ &\quad - 2\beta\epsilon^T(t)\epsilon(t)(\zeta^T(t)\zeta(t) + \xi^T(t)\xi(t)) \\ &\quad - 2\sigma_1(t)\mathrm{tr}[\tilde{\Theta}(t)\Gamma_p\Theta^T(t)] - 2\sigma_2(t)\mathrm{tr}[\tilde{\Psi}^T(t)\Psi(t)] \\ &\leq -2\epsilon^T(t)\epsilon(t) - \alpha\epsilon^T(t)\epsilon(t)m^2(t) + k_3\mu^2 + k_4\mu^2\|\Theta(t)\|^2 \\ &\quad - 2\sigma_1(t)\mathrm{tr}[\tilde{\Theta}(t)\Gamma_p\Theta^T(t)] - 2\sigma_2(t)\mathrm{tr}[\tilde{\Psi}^T(t)\Psi(t)] \end{aligned} \tag{9.151C}$$

for some k_3, $k_4 > 0$. For λ_1, σ_1, $\sigma_2(t)$ in (9.135C)–(9.138C), we have

$$\begin{aligned}&\sigma_1(t)\mathrm{tr}[\tilde{\Theta}(t)\Gamma_p\Theta^T(t)] + \sigma_2(t)\mathrm{tr}[\tilde{\Psi}^T(t)\Psi(t)] \\ &\geq \sigma_1(t)(\lambda_1\|\Theta(t)\|^2 - \|\Gamma_p\|_2\,\|\Theta(t)\|\,\|\Theta^*\|) \\ &\quad + \sigma_2(t)(\|\Psi(t)\|^2 - \|\Psi(t)\|\,\|\Psi^*\|) \geq 0. \end{aligned} \tag{9.152C}$$

It follows from (9.151C) and (9.152C) that there exists $\mu_1^* > 0$ such that for any $\mu \in [0, \mu_1^*)$ the existence of $k_5 > 0$ is guaranteed such that $\dot{V} \leq 0$ whenever $\|\Theta(t)\| + \|\Psi(t)\| \geq k_5$, so that $\Theta(t) \in L^\infty$, $\Psi(t) \in L^\infty$. Then, from (9.131), (9.132C), (9.133C), (9.139), and (9.142), we have that $\epsilon(t) \in L^\infty$, $\dot{\Theta}(t)$, $\dot{\Psi}(t)$, $\epsilon(t)m(t) \in L^\infty$.

From (9.135C) and (9.136C), there exists k_6, $k_7 > 0$ such that

$$\|G_1(t)\| \leq k_6\sigma_1(t)\,\mathrm{tr}[\tilde{\Theta}(t)\Gamma_p\Theta^T(t)], \tag{9.153C}$$

$$\|G_2(t)\| \leq k_7\sigma_2(t)\,\mathrm{tr}[\tilde{\Psi}^T(t)\Psi(t)]. \tag{9.154C}$$

From (9.151C), it follows that there exists $k_8 > 0$ such that

$$2\sigma_1(t)\,\mathrm{tr}[\tilde{\Theta}(t)\Gamma_p\Theta^T(t)] + 2\sigma_2(t)\,\mathrm{tr}[\tilde{\Psi}^T(t)\Psi(t)] \leq -\dot{V} + k_8\mu^2. \tag{9.155C}$$

Therefore, using (9.132C), (9.151C)–(9.154C), we have (9.149C).

(ii) For the DT case, we consider the positive definite function

$$V = \mathrm{tr}[\tilde{\Theta}(t)\Gamma_p\tilde{\Theta}^T(t)] + \mathrm{tr}[\tilde{\Psi}^T(t)\Gamma^{-1}\tilde{\Psi}(t)],\ t \in \{0, 1, 2, \ldots\}. \tag{9.150D}$$

Using (9.132D)–(9.138D), (9.139), (9.142), and (9.150D) and the definitions of Γ_p in (9.138C) and λ_i in (9.138D), $i = 1, 2, 3$, we have

$$\begin{aligned}&V(\tilde{\Theta}(t+1), \tilde{\Psi}(t+1)) - V(\tilde{\Theta}(t), \tilde{\Psi}(t)) \\ &= -2\epsilon^T(t)\epsilon(t) - 2\alpha\epsilon^T(t)\epsilon(t)m^2(t) + 2\mu\epsilon^T(t)\eta_1(t) \\ &\quad - \zeta^T(t)\zeta(t)\epsilon^T(t)(2\beta I - S_p^T K_p^T)\epsilon(t)\end{aligned}$$

$$
\begin{aligned}
&+ 2\sigma_1(t)\epsilon^T(t)S_p^T\Theta(t)\zeta(t) + \sigma_1^2(t)\,\mathrm{tr}[\Theta(t)\Gamma_p\Theta^T(t)] \\
&- \frac{1}{2}\sigma_1(t)\,\mathrm{tr}[\Theta(t)\Gamma_p\Theta^T(t)] - \frac{1}{2}\sigma_1(t)\,\mathrm{tr}[\tilde{\Theta}(t)\Gamma_p\Theta^T(t)] \\
&- \frac{1}{2}\sigma_1(t)\,\mathrm{tr}[(2\Theta(t) - 3\Theta^*)\Gamma_p\Theta^T(t)] \\
&- \xi^T(t)\xi(t)\epsilon^T(t)(2\beta I - \Gamma)\epsilon(t) + 2\sigma_2(t)\epsilon^T(t)\Gamma\Psi(t)\xi(t) \\
&+ \sigma_2^2(t)\,\mathrm{tr}[\Psi^T(t)\Gamma\Psi(t)] - \frac{1}{2}\sigma_2(t)\,\mathrm{tr}[\Psi^T(t)\Psi(t)] \\
&- \frac{1}{2}\sigma_2(t)\,\mathrm{tr}[\tilde{\Psi}^T(t)\Psi(t)] - \frac{1}{2}\sigma_2(t)\,\mathrm{tr}[(2\Psi(t) - 3\Psi^*)^T\Psi(t)] \\
\leq{}& -2\epsilon^T(t)\epsilon(t) - \alpha\epsilon^T(t)\epsilon(t)m^2(t) + k_3\mu^2 + k_4\mu^2\|\Theta(t)\|^2 \\
&- \frac{1}{2}\sigma_1(t)\,\mathrm{tr}[\tilde{\Theta}(t)\Gamma_p\Theta^T(t)] - \frac{1}{2}\sigma_2(t)\,\mathrm{tr}[\tilde{\Psi}^T(t)\Psi(t)] \qquad (9.151\mathrm{D})
\end{aligned}
$$

for some k_3, $k_4 > 0$.[7] Similar to (9.152C), we have

$$
\begin{aligned}
&\sigma_1(t)\,\mathrm{tr}[\tilde{\Theta}(t)\Gamma_p\Theta^T(t)] + \sigma_2(t)\,\mathrm{tr}[\tilde{\Psi}^T(t)\Psi(t)] \\
\geq{}& \frac{\sigma_1(t)}{2}\left(2\lambda_1\|\Theta(t)\|^2 - 3\|\Gamma_p\|_2\,\|\Theta(t)\|\,\|\Theta^*\|\right) \\
&+ \frac{\sigma_2(t)}{2}\left(2\|\Psi(t)\|^2 - 3\|\Psi(t)\|\,\|\Psi^*\|\right) \geq 0. \qquad (9.152\mathrm{D})
\end{aligned}
$$

It follows from (9.151D) and (9.152D) that there exists $\mu_1^* > 0$ such that for any $\mu \in [0, \mu_1^*)$ the existence of $k_5 > 0$ is guaranteed such that $V(\tilde{\Theta}(t+1), \tilde{\Psi}(t+1)) - V(\tilde{\Theta}(t), \tilde{\Psi}(t)) \leq 0$ whenever $\|\Theta(t)\| + \|\Psi(t)\| \geq k_5$. Hence, we conclude that $\Theta(t) \in L^\infty$, $\Psi(t) \in L^\infty$. From (9.131), (9.132D), (9.133D), (9.139), and (9.142), we have that $\epsilon(t) \in L^\infty$ and $\epsilon(t)m(t) \in L^\infty$.

Based on (9.135D) and (9.136D), there exists k_6, $k_7 > 0$ such that

$$\|G_1(t)\| \leq k_6\sigma_1(t)\,\mathrm{tr}[\tilde{\Theta}(t)\Gamma_p\Theta^T(t)], \qquad (9.153\mathrm{D})$$

$$\|G_2(t)\| \leq k_7\sigma_2(t)\,\mathrm{tr}[\tilde{\Psi}^T(t)\Psi(t)]. \qquad (9.154\mathrm{D})$$

From (9.151D), it follows that there exists $k_8 > 0$ such that

$$
\begin{aligned}
&\frac{\sigma_1(t)}{2}\,\mathrm{tr}[\tilde{\Theta}(t)\Gamma_p\Theta^T(t)] + \frac{\sigma_2(t)}{2}\,\mathrm{tr}[\tilde{\Psi}^T(t)\Psi(t)] \\
\leq{}& -V(\tilde{\Theta}(t+1), \tilde{\Psi}(t+1)) + V(\tilde{\Theta}(t), \tilde{\Psi}(t)) + k_8\mu^2. \qquad (9.155\mathrm{D})
\end{aligned}
$$

Therefore, from (9.132D), (9.151D)–(9.154D), we have (9.149D). ∇

---

[7]From (9.135D)–(9.138D), it follows that

$$\sigma_1(t)\,\mathrm{tr}[(2\Theta(t) - 3\Theta^*)\Gamma_p\Theta^T(t)] \geq \sigma_1(t)\left(2\lambda_1\|\Theta(t)\|^2 - 3\|\Gamma_p\|_2\,\|\Theta(t)\|\,\|\Theta^*\|\right) \geq 0,$$

and similarly, $\sigma_2(t)\,\mathrm{tr}[(2\Psi(t) - 3\Psi^*)^T\Psi(t)] \geq 0$.

Remark 9.7 For the alternate choice of the estimation error $\epsilon(t)$ in (9.143C) for the CT case, we consider the positive definite function

$$V(\epsilon, \tilde{\Theta}, \tilde{\Psi}) = \epsilon^T \epsilon + \mathrm{tr}[\tilde{\Theta} \Gamma_p \tilde{\Theta}^T] + \mathrm{tr}[\tilde{\Psi}^T \Gamma^{-1} \tilde{\Psi}] \tag{9.156C}$$

and use (9.132C), (9.133C), (9.142), (9.145C) to derive

$$\begin{aligned} \dot{V} &= -2a_m \epsilon^T(t)\epsilon(t) - 2\alpha\epsilon^T(t)\epsilon(t)m^2(t) + 2\mu\epsilon^T(t)\eta_2(t) \\ &\quad - 2\sigma_1(t)\,\mathrm{tr}[\tilde{\Theta}(t)\Gamma_p\Theta^T(t)] - 2\sigma_2(t)\,\mathrm{tr}[\tilde{\Psi}^T(t)\Psi(t)] \\ &\leq -(2a_m + \alpha m^2(t))\epsilon^T(t)\epsilon(t) + k_3\mu^2 + k_4\mu^2\|\Theta(t)\|^2 \\ &\quad - 2\sigma_1(t)\,\mathrm{tr}[\tilde{\Theta}(t)\Gamma_p\Theta^T(t)] - 2\sigma_2(t)\,\mathrm{tr}[\tilde{\Psi}^T(t)\Psi(t)] \end{aligned} \tag{9.157C}$$

for some k_3, $k_4 > 0$. Hence for some $\mu_1^* > 0$ and any $\mu \in [0, \mu_1^*)$, there exists $k_5 > 0$ such that $\dot{V} \leq 0$ whenever $\|\Theta(t)\| + \|\Psi(t)\| \geq k_5$, so that $\Theta(t) \in L^\infty$, $\Psi(t) \in L^\infty$, and $\epsilon(t) \in L^\infty$. Similarly, we also have (9.149C). □

The stability and robustness properties of the adaptive control schemes (9.79) and (9.132)–(9.138) are given in the following theorem.

Theorem 9.2 *There exists $\mu^* > 0$ such that for any $\mu \in [0, \mu^*)$ all signals in the closed-loop plant (9.59), (9.61), (9.79), (9.132)–(9.133) are bounded and the tracking error $e(t) = y(t) - y_m(t)$ satisfies*

$$\int_{t_1}^{t_2} \bar{e}^T(t)\bar{e}(t)\,dt \leq \beta_c\mu^2(t_2 - t_1) + b_c \tag{9.158C}$$

for some $\beta_c > 0$, $b_c > 0$ and any $t_2 > t_1 \geq 0$, in the CT case, and

$$\sum_{t=t_1}^{t_2-1} \bar{e}^T(t)\bar{e}(t) \leq \beta_d\mu^2(t_2 - t_1) + b_d \tag{9.158D}$$

for some $\beta_d > 0$, $b_d > 0$ and any $t_2 > t_1 \geq 0$ in the DT case, where $\bar{e}(t) = h(D)\xi_m(D)[e](t)$. Moreover, $\lim_{t\to\infty} e(t) = 0$ if $\mu = 0$ in the plant (9.59).

Proof: Letting d_m be the maximum degree of the modified interactor matrix $\xi_m(D)$ of $G_0(D)$, we introduce fictitious filters $H_i(D)$, $K_i(D)$ defined from

$$DH_i(D) = 1 - K_i(D),\ K_i(D) = \frac{a_i^{d_m}}{(D + a_i)^{d_m}},\quad i = 1, 2, 3, \tag{9.159C}$$

for the CT case (with $D = s$), or from

$$(D-1)H_1(D) = 1 - K_1(D),\ K_1(D) = \frac{(1 - \frac{1}{a_1})^{d_m} D^{d_m}}{(D - \frac{1}{a_1})^{d_m}}, \tag{9.159D}$$

for the DT case (with $D = z$), where $a_i > 0$ is chosen to be sufficiently large but finite, for $i = 1, 2, 3$. Denoting $h_i(t)$ as the impulse response functions of the transfer functions $H_i(D)$, $i = 1, 2, 3$, we have the L^1 operator norms [8]

$$\|h_i(\cdot)\|_1 = \begin{cases} \frac{d_m}{a_i} & \text{for the CT case } (a_i > 0,\ i = 1, 2, 3), \\ \frac{d_m}{a_i - 1} & \text{for the DT case } (i = 1,\ a_1 > 1). \end{cases} \tag{9.160}$$

From (9.59), $\omega_1(t)$ in (9.80), $H_1(D)$ and $K_1(D)$ in (9.159), we obtain

$$\begin{aligned} F(D)G_0^{-1}(D)[y](t) = K_1^{-1}(D)[\omega_1 - H_1(D)(D - \sigma)[\omega_1]](t) \\ + \mu F(D)\left(\Delta_m(D) + G_0^{-1}(D)\Delta_a(D)\right)[u](t), \end{aligned} \tag{9.161}$$

where $\sigma = 0$ for the CT case or $\sigma = 1$ for the DT case. Let $\omega_1(t) = F(D)[u](t)$ (where $F(D) = \frac{A(D)}{\Lambda(D)}$) have a controllable realization (A, B), that is,

$$D[\omega_1](t) = (A + \sigma I)\omega_1(t) + Bu(t), \tag{9.162}$$

where $A + \sigma I$ is a stable matrix. From (9.79), (9.161), and (9.162), we have

$$\begin{aligned} \omega_1(t) = K_1(D)F(D)G_0^{-1}(D)[y](t) + H_1(D)[A\omega_1](t) \\ + H_1(D)B[\Theta_1^T\omega_1 + \Theta_2^T(\cdot)F(D)[y] + \Theta_3^T r](t) \\ - \mu K_1(D)F(D)\left(\Delta_m(D) + G_0^{-1}(D)\Delta_a(D)\right)[u](t). \end{aligned} \tag{9.163}$$

---

[8] The CT result of (9.160) is from Proposition 2.10. For the DT case, we have

$$\frac{1}{a_1}\frac{1}{D - \frac{1}{a_1}}\left(1 + \frac{(1 - \frac{1}{a_1})D}{D - \frac{1}{a_1}} + \cdots + \frac{(1 - \frac{1}{a_1})^{d_m - 1}D^{d_m - 1}}{(D - \frac{1}{a_1})^{d_m - 1}}\right)$$

$$= \frac{1}{a_1}\frac{1}{D - \frac{1}{a_1}}\frac{1 - \frac{(1-\frac{1}{a_1})^{d_m}D^{d_m}}{(D-\frac{1}{a_1})^{d_m}}}{1 - \frac{(1-\frac{1}{a_1})D}{D-\frac{1}{a_1}}} = \frac{1 - \frac{(1-\frac{1}{a_1})^{d_m}D^{d_m}}{(D-\frac{1}{a_1})^{d_m}}}{D - 1} = H_1(D).$$

With $D = z$ as the z-transform variable, we consider $f_j(t) = \mathcal{Z}[\frac{1}{D-\frac{1}{a_1}}\frac{(1-\frac{1}{a_1})^j D^j}{(D-\frac{1}{a_1})^j}]$, $j = 0, 1, \ldots, d_m - 1$, and its sum $g_j(t) = \sum_{\tau=0}^{t} f_j(\tau)$. We see that $f_j(t) \geq 0$, $\forall t \geq 0$, as $f_j(t)$ is the convolution of $f_{j-1}(t)$ and $\mathcal{Z}^{-1}[\frac{(1-\frac{1}{a_1})D}{D-\frac{1}{a_1}}]$, which are nonnegative, $j = 1, \ldots, d_m - 1$. We also see that $\lim_{t\to\infty} g_j(t)$ exists and is finite, as $a_1 > 1$. Hence, with $G_j(D) = \mathcal{Z}[g_j(t)] = \frac{D}{D-1}F_j(D)$ (as $g_j(t)$ is the convolution of $f_j(t)$ and the unit step function $u_s(t)$ whose z-transform is $\frac{D}{D-1}$) and $F_j(D) = \mathcal{Z}[f_j(t)] = \frac{1}{D-\frac{1}{a_1}}\frac{(1-\frac{1}{a_1})^j D^j}{(D-\frac{1}{a_1})^j}$, we have

$$\sum_{\tau=0}^{\infty} f_j(\tau) = \lim_{t\to\infty} g_j(t) = \lim_{D\to 1}\left(1 - \frac{1}{D}\right)G_j(D) = \lim_{D\to 1} F_j(D) = \frac{1}{1 - \frac{1}{a_1}} = \frac{a_1}{a_1 - 1}.$$

Then, we have that $\|h_1(\cdot)\|_1 = \sum_{j=1}^{d_m}\frac{1}{a_1}\sum_{\tau=0}^{\infty} f_j(\tau) = \frac{d_m}{a_1 - 1}$ for $a_1 > 1$.

For the *CT case*, since $H_1(D)$ satisfies (9.160) and $\Theta_1(t)$ is bounded, there exists $a_1^0 > 0$ such that $(I - H_1(D)(A + B\Theta_1^T(t)))^{-1}$ is a stable and proper operator with a finite gain[9] for any finite $a_1 > a_1^0$ (which, in view of (9.52), can be proved in a similar way to that for proving Lemma 2.4), where $a_1 > 0$ is in (9.159C). For $0 < a_1^0 < a_1$ it follows from (9.163) that

$$\omega_1(t) = G_1(D,\cdot)[y](t) + G_2(D,\cdot)[r](t) + \mu G_{\delta 1}(D,\cdot)[u](t), \tag{9.164C}$$

where, for $T_1(D,t) \triangleq (I - H_1(D)(A + B\Theta_1^T(t)))^{-1}$,

$$G_1(D,t) = T_1(D,t)\left(K_1(D)F(D)G_0^{-1}(D) + H_1(D)B\Theta_2^T(t)F(D)\right), \tag{9.165C}$$

$$G_2(D,t) = T_1(D,t)H_1(D)B\Theta_3(t), \tag{9.166C}$$

$$G_{\delta 1}(D,t) = -T_1(D,t)K_1(D)F(D)\left(\Delta_m(D) + G_0^{-1}(D)\Delta_a(D)\right) \tag{9.167C}$$

are stable and proper operators with finite gains. Hence it follows from (9.164C), (9.80), and (9.121) that

$$\omega(t) = G_3(D,\cdot)[y](t) + G_4(D,\cdot)[r](t) + \mu G_{\delta 2}(D,\cdot)[u](t), \tag{9.168C}$$

where

$$G_3(D,t) = \begin{bmatrix} G_1(D,t) \\ F(D) \\ 0 \end{bmatrix}, \tag{9.169C}$$

$$G_4(D,t) = \begin{bmatrix} G_2(D,t) \\ 0 \\ I \end{bmatrix}, \tag{9.170C}$$

$$G_{\delta 2}(D,t) = \begin{bmatrix} G_{\delta 1}(D,t) \\ 0 \\ 0 \end{bmatrix}. \tag{9.171C}$$

Then, from (9.120), we express

$$\begin{aligned} &D[y](t) - D[y_m](t) \\ &= DW_m(D)\Theta_3^{*-1}[\tilde{\Theta}^T\omega \\ &\quad + \mu\left(I - \Theta_1^{*T}F(D)\right)\left(\Delta_m(D) + G_0^{-1}(D)\Delta_a(D)\right)[u]](t). \end{aligned} \tag{9.172C}$$

[9] As defined in (9.52), a linear operator $T(D,t)$ is stable and proper if

$$\|T(D,\cdot)[x](t)\| \le \beta \int_0^t e^{-\alpha(t-\tau)}\|x(\tau)\|\,d\tau + \gamma\|x(t)\|$$

for some constants $\beta \ge 0$, $\alpha > 0$, and $\gamma > 0$, all $t \ge 0$, and any $x(t)$. The operator has a finite gain as $\frac{\beta}{\alpha}+\gamma$ is finite. (Note: In view of (9.52), the L^1 operator norm of $h(t) = \beta e^{-\alpha t} + \gamma\delta(t)$, where $\delta(t)$ is the Dirac δ-function in (2.15), is $\frac{\beta}{\alpha} + \gamma$.)

Defining $\bar{y}(t) = h(D)[y](t)$ and using (9.86), (9.168C), (9.172C), and $H_2(D)$, $K_2(D)$ in (9.159C), we have that

$$\begin{aligned} y(t) = {} & K_2(D)h^{-1}(D)[\bar{y}](t) + H_2(D)DW_m(D)\Theta_3^{*-1}\tilde{\Theta}^T(\cdot)G_3(D,\cdot)[y](t) \\ & + H_2(D)DW_m(D)\Theta_3^{*-1}\tilde{\Theta}^T(\cdot)G_4(D,\cdot)[r](t) \\ & + \mu H_2(D)D\left(W_m(D)\Theta_3^{*-1}\tilde{\Theta}^T(\cdot)G_{\delta 2}(D,\cdot) + \Delta_0(D)\right)[u](t), \quad (9.173C) \end{aligned}$$

where

$$\begin{aligned} \Delta_0(D) = {} & W_m(D)\Theta_3^{*-1}\left(I - \Theta_1^{*T}F(D)\right)\Delta_m(D) \\ & + \left(W_m(D)\Theta_3^{*-1}\Theta_2^{*T}F(D) + I\right)\Delta_a(D) \end{aligned} \quad (9.174C)$$

with $\Theta_3^{*-1} = K_p$. From (A9.5), (A9.6), (9.86), and (9.174C), we have that $\Delta_0(D)$ is stable and proper with a finite gain.

Similarly, $(1 - H_2(D)DW_m(D)\Theta_3^{*-1}\tilde{\Theta}^T(t)G_3(D,t))^{-1}$ is a stable and proper operator with a finite gain for any finite $a_2 > a_2^0$ and some $a_2^0 > 0$. For $0 < a_2^0 < a_2$, it follows from (9.173C) that

$$y(t) = G_5(D,\cdot)[\bar{y}](t) + G_6(D,\cdot)[r](t) + \mu G_{\delta 3}(D,\cdot)[u](t), \quad (9.175C)$$

where, for $T_2(D,t) \triangleq (I - H_2(D)DW_m(D)\Theta_3^{*-1}\tilde{\Theta}^T(t)G_3(D,t))^{-1}$,

$$G_5(D,t) = T_2(D,t)K_2(D)h^{-1}(D), \quad (9.176C)$$

$$G_6(D,t) = T_2(D,t)H_2(D)DW_m(D)\left(\Theta_3^{*-1}\tilde{\Theta}^T(t)G_4(D,t) + I\right), \quad (9.177C)$$

$$G_{\delta 3}(D,t) = T_2(D,t)H_2(D)D\left(W_m(D)\Theta_3^{*-1}\tilde{\Theta}^T(t)G_{\delta 2}(D,t) + \Delta_0(D)\right) \quad (9.178C)$$

are stable and proper operators with finite gains. Therefore, it follows from (9.168C) and (9.175C) that

$$\begin{aligned} \omega(t) - {} & G_3(D,\cdot)G_5(D,\cdot)[\bar{y}](t) + (G_3(D,\cdot)G_6(D,\cdot) + G_4(D,\cdot))[r](t) \\ & + \mu(G_3(D,\cdot)G_{\delta 3}(D,\cdot) + G_{\delta 2}(D,\cdot))[u](t). \end{aligned} \quad (9.179C)$$

From (9.129), we express

$$\bar{y}(t) = \bar{y}_m(t) + W_m(D)[\epsilon - \Psi\xi + \beta(\zeta^T\zeta + \xi^T\xi) + \alpha\epsilon m^2](t), \quad (9.180C)$$

where $\bar{y}_m(D) = h(D)[y_m](t)$. From (9.79) and (9.179C), we obtain

$$\begin{aligned} u(t) &= \Theta^T(t)\omega(t) \\ &= \Theta^T(t)G_3(D,\cdot)G_5(D,\cdot)[\bar{y}](t) \\ &\quad + \Theta^T(t)(G_3(D,\cdot)G_6(D,\cdot) + G_4(D,\cdot))[r](t) \\ &\quad + \mu\Theta^T(t)(G_3(D,\cdot)G_{\delta 3}(D,\cdot) + G_{\delta 2}(D,\cdot))[u](t). \end{aligned} \quad (9.181C)$$

From (9.181C), it follows that

$$u(t) = G_7(D,\cdot)\Theta^T(\cdot)G_3(D,\cdot)G_6(D,\cdot)[\bar{y}](t) + G_7(D,\cdot)\Theta^T(\cdot)\left(G_3(D,\cdot)G_6(D,\cdot) + G_4(D,\cdot)\right)[r](t), \quad (9.182C)$$

where $G_7(D,t) = \left(I - \mu\Theta^T(t)(G_3(D,t)G_{\delta 3}(D,t) + G_{\delta 2}(D,t))\right)^{-1}$ is stable and proper with finite gain for any $\mu \in [0, \mu_2^*)$ and some $\mu_2^* > 0$ because $\Theta(t) \in L^\infty$ and $G_3(D,t)G_{\delta 3}(D,t) + G_{\delta 2}(D,t)$ is stable and proper with finite gain.

For the *DT case*, we define $\bar{\omega}(t) = [\omega_1^T(t), \omega_2^T(t)]^T$, $\bar{\Theta}^* = [\Theta_1^{*T}, \Theta_2^{*T}]^T$, $\bar{\Theta}(t) = [\Theta_1^T(t), \Theta_2^T(t)]^T$, $\tilde{\bar{\Theta}}(t) = \bar{\Theta}(t) - \bar{\Theta}^*$, $\tilde{\Theta}_3(t) = \Theta_3(t) - \Theta_3^*$. It follows from (9.120) and the definition of $\bar{\omega}(t)$ that

$$D[\bar{\omega}](t) = P_1(D,\cdot)[\bar{\omega}](t) + b_1(t), \quad (9.164D)$$

where

$$P_1(D,t) = \begin{bmatrix} DF(D)\bar{\Theta}^T(t) \\ DF(D)\left(W_m(D)\Theta_3^{*-1}\tilde{\bar{\Theta}}^T(t) + \mu\Delta_0(D)\bar{\Theta}(t)\right) \end{bmatrix} \quad (9.165D)$$

is stable and proper with a finite gain and

$$b_1(t) = \begin{bmatrix} DF(D)[\Theta_3 r](t) \\ DF(D)\left(W_m(D)(\Theta_3^{*-1}\tilde{\Theta}_3(\cdot) + I) + \mu\Delta_0(D)\Theta_3(\cdot)\right)[r](t) \end{bmatrix} \quad (9.166D)$$

is bounded, with $\Delta_0(D)$ in the same form as that in (9.174C) for the CT case.

Using the fact that the signals $r(t+k)$, $\Theta(t+k)$ are bounded for any finite $k > 0$, we have that

$$D^2[\bar{\omega}](t) = P_2(D,\cdot)[\bar{\omega}](t) + b_2(t), \quad (9.167D)$$

where $P_2(D,t) = P_1(D,t+1)P_1(D,t)$ is stable and proper with finite gain and $b_2(t) = \bar{P}_1(D,\cdot)[b_1](t) + D[b_1](t)$ is bounded with $\bar{P}_1(D,t) = P_1(D,t+1)$. By deduction, we have that

$$D^{d_m}[\bar{\omega}](t) = P_0(D,\cdot)[\bar{\omega}](t) + b_0(t), \quad (9.168D)$$

where $P_0(D,t)$ is stable and proper with finite gain and $b_0(t)$ is bounded.

Expressing $P_0(D,\cdot)[\bar{\omega}](t) = P_{01}(D,\cdot)[\omega_1](t) + P_{02}(D,\cdot)[\omega_2](t)$ and using (9.79) and (9.168D), we have

$$D^{d_m}[u](t) = \bar{\Theta}^T(t+d_m)P_{01}(D,\cdot)[\omega_1](t) + \bar{\Theta}^T(t+d_m)P_{02}(D,\cdot)[\omega_2](t) + \bar{\Theta}^T(t+d_m)b_0(t) + D^{d_m}[\Theta_3 r](t). \quad (9.169D)$$

Since the filter $H_1(D)$ satisfies (9.160) and $\Theta_1(t)$ is bounded, there exists a constant $a_1^0 > 0$ such that

$$T_1(D,t) = \Big(I - H_1(D)(A + B\Theta_1^T(t)) + \mu K_1(D)F(D)D^{-d_m} \times(\Delta_m(D) + G_0^{-1}(D)\Delta_a(D))\bar{\Theta}(t+d_m)P_{01}(D,t)\Big)^{-1} \quad (9.170D)$$

is a stable and proper operator with a finite gain for any fixed $a_1 > a_1^0$, any $\mu \in [0, \mu_2^*)$, and some $\mu_2^* = \mu_2^*(a_1) > 0$. For a finite and fixed $a_1 > a_1^0$ and a new signal $\hat{y}(t) \triangleq \frac{\xi_m(D)}{D+a_0}[y](t)$ where $0 < a_0 < 1$ is arbitrary, it follows from (9.163) and (9.169D) that

$$\omega_1(t) = G_1(D,\cdot)[\hat{y}](t) + b_3(t), \quad (9.171D)$$

where

$$G_1(D,t) = T_1(D,t)\Big(K_1(D)F(D)(D+a_0)G_0^{-1}(D)\xi_m^{-1}(D) + \Big(H_1(D)B\Theta_2^T(t) - \mu K_1(D)F(D)D^{-d_m}(\Delta_m(D) + G_0^{-1}(D)\Delta_a(D)) \bar{\Theta}(t+d_m)P_{02}(D,t)\Big)(D+a_0)F(D)\xi_m^{-1}(D)\Big) \quad (9.172D)$$

is stable and proper with a finite gain and

$$b_3(t) = T_1(D,\cdot)[H_1(D)B[\Theta_3 r] - \mu K_1(D)F(D)D^{-d_m} \times(\Delta_m(D) + G_0^{-1}(D)\Delta_a(D))[D^{d_m}[\Theta_3 r] + D^{d_m}[\bar{\Theta}^T]b_0]](t) \quad (9.173D)$$

is bounded. From (9.129), we have

$$\hat{y}(t) = \frac{1}{D+a_0}[r](t) + \frac{1}{D+a_0}[\epsilon - \Psi\xi + \beta(\zeta^T\zeta + \xi^T\xi) + \alpha\epsilon m^2](t). \quad (9.174D)$$

From (9.130), we denote $\xi(t) = [\xi_1(t), \ldots, \xi_M(t)]^T$, $\Theta(t) = [\theta_1(t), \ldots, \theta_M(t)]$ with $\theta_i(t) \in R^{M\times(2\nu+1)}$, $i = 1, \ldots, M$, and $f(D) = D^{d_m} + a_{d_m-1}D^{d_m-1} + \cdots + a_1 D + a_0$. Then, $\xi_i(t) = \theta_i^T(t)\zeta(t) - \frac{1}{f(D)}[\theta_i^T\omega](t)$, $i = 1, \ldots, M$.

For the CT case, similar to (5.138), we express $\xi_i(t)$ as

$$\begin{aligned}\xi_i(t) = {} & \frac{D^{d_m-1} + a_{d_m-1}D^{d_m-2} + \cdots + a_2 D + a_1}{f(D)}\left[D[\theta_i^T]\frac{1}{f(D)}[\omega]\right](t) \\ & + \frac{D^{d_m-2} + a_{d_m-1}D^{d_m-3} + \cdots + a_2}{f(D)}\left[D[\theta_i^T]\frac{D}{f(D)}[\omega]\right](t) + \cdots \\ & + \frac{D + a_{d_m-1}}{f(D)}\left[D[\theta_i^T]\frac{D^{d_m-2}}{f(D)}[\omega]\right](t) \\ & + \frac{1}{f(D)}\left[D[\theta_i^T]\frac{D^{d_m-1}}{f(D)}[\omega]\right](t),\end{aligned} \quad (9.183C)$$

where $D[\theta_i^T](t) = \dot{\theta}_i^T(t)$, $i = 1, \ldots, M$.

For the DT case, as an analogous version of (9.183C), we express $\xi_i(t)$ as

$$\begin{aligned}\xi_i(t) = & \frac{D^{d_m-1} + a_{d_m-1}D^{d_m-2} + \cdots + a_2 D + a_1}{f(D)}\left[(D-1)[\theta_i^T]\frac{D}{f(D)}[\omega]\right](t) \\ & + \frac{D^{d_m-2} + a_{d_m-1}D^{d_m-3} + \cdots + a_2}{f(D)}\left[(D-1)[\theta_i^T]\frac{D^2}{f(D)}[\omega]\right](t) + \cdots \\ & + \frac{D + a_{d_m-1}}{f(D)}\left[(D-1)[\theta_i^T]\frac{D^{d_m-1}}{f(D)}[\omega]\right](t) \\ & + \frac{1}{f(D)}\left[(D-1)[\theta_i^T]\frac{D^{d_m}}{f(D)}[\omega]\right](t), \end{aligned} \tag{9.183D}$$

where $(D-1)[\theta_i^T](t) = \theta_i^T(t+1) - \theta_i^T(t)$, $i = 1, \ldots, M$.[10]

Hence, for the CT case, from (9.131), Lemma 9.4, (9.175C), (9.182C), and (9.183C), we have that

$$\|\bar{y}(t)\| \leq x_0(t) + \beta_1 \int_0^t e^{-\alpha_1(t-\tau)} x(\tau) \left(\int_0^\tau e^{-\alpha_2(\tau-w)}\|\bar{y}(w)\|\, dw\right) d\tau \tag{9.184C}$$

with $x(t) = \|\dot{\Theta}(t)\| + \|\epsilon(t)\|m(t)$; and for the DT case, from (9.131), Lemma 9.4, (9.171D), (9.174D), and (9.183D), we have

$$\|\hat{y}(t)\| \leq x_0(t) + \beta_1 \sum_{\tau=0}^{t-1} e^{-\alpha_1(t-\tau-1)} x(\tau) \left(\sum_{w=0}^{\tau-1} e^{-\alpha_2(\tau-w-1)}\|\hat{y}(w)\|\right), \tag{9.184D}$$

where $x(t) = \|\Theta(t+1) - \Theta(t)\| + \|\epsilon(t)\|m(t)$, for some β_1, α_1, $\alpha_2 > 0$, $x_0(t)$ is bounded. Here we used the fact that $\frac{\|\zeta(t)\|}{m(t)}$, $\frac{\|\xi(t)\|}{m(t)}$ are bounded.

For the CT case, applying Lemma 2.3 on (9.184C), we conclude that there exists μ^*, $0 < \mu^* < \min\{\mu_1^*, \mu_2^*\}$, such that $\bar{y}(t)$ or $\hat{y}(t)$ is bounded for any $\mu \in [0, \mu^*)$, so are $u(t)$, $y(t)$, and all signals in the closed-loop system. For $\bar{e}(t) = \xi_m(D)h(D)[e](t)$, from (9.129), (9.149C), (9.183C), we have (9.158C).

For the DT case, using the discrete-time version (see Section 2.7) of the Bellman-Gronwall lemma (Lemma 2.1), we can establish the discrete-time

---

[10] **Discrete-Time Swapping Lemma**: The expression (9.183C) is the same as (5.138), generalized from its original version [285], whereas (9.183D) is the counterpart of (9.183C) in discrete time. The discrete-time version of the swapping lemma [151], [265], [340] (see (5.332)) can be stated as: Let a stable and proper rational function $h(z)$ have a minimal realization $h(z) = c(zI - A)^{-1}b + d$ and $\theta(t)$ and $\omega(t)$ be two vector signals, and denote $h_c(z) = c(zI - A)^{-1}$ and $h_b(z) = (zI - A)^{-1}b$. Then,

$$\theta^T(t)h(z)[\omega](t) - h(z)[\theta^T\omega](t) = h_c(z)[(h_b(z)z)[\omega^T](z-1)[\theta]](t).$$

Note that the "d" term in $h(z)$ is swapped away.

version of Lemma 2.3. Then, we can show that there exists μ^*, $0 < \mu^* < \min\{\mu_1^*, \mu_2^*\}$, such that $\bar{y}(t)$ or $\hat{y}(t)$ is bounded for any $\mu \in [0, \mu^*)$, so are $u(t)$, $y(t)$, and all signals in the closed-loop system. For $\bar{e}(t) = \xi_m(D)h(D)[e](t)$, from (9.129), (9.149D), (9.183D), we also have (9.158D).

When $\mu = 0$, from (9.149), we have that $\|x^2(\cdot)\|_1$ (the L^1 signal norm of $x^2(t)$) is finite, and from (9.129) and (9.183), $\|\bar{e}^T(\cdot)\bar{e}(\cdot)\|_1$ is finite.

For the CT case, from (9.120) with $\mu = 0$, using $H_3(D)$ and $K_3(D)$ defined in (9.159), for $e(t) = y(t) - y_m(t)$, we have

$$\begin{aligned} e(t) &= W_m(D)\Theta_3^{*-1}[\tilde{\Theta}^T\omega](t) \\ &= H_3(D)DW_m(D)\Theta_3^{*-1}[\tilde{\Theta}^T\omega](t) + W_m(D)K_3(D)h^{-1}(D)[\bar{e}](t), \quad (9.185C) \end{aligned}$$

where

$$\lim_{t\to\infty} W_m(D)K_3(D)h^{-1}(D)[\bar{e}](t) = 0 \quad (9.186C)$$

for any finite $a_3 > 0$ in $K_3(D)$, and $DW_m(D)\Theta_3^{*-1}[\tilde{\Theta}^T\omega](t)$ is bounded. Hence from (9.185C) and (9.186C) it follows that

$$\|e(t)\| \le c_3\|h_3(t)\|_1 + z_1(t) \le \frac{c_4}{a_3} + z_1(t), \quad (9.187C)$$

where c_3, $c_4 > 0$, $\lim_{t\to\infty} z_1(t) = 0$. Since $a_3 > 0$ in $H_3(D)$ can be set arbitrarily large, it follows from (9.187C) that $\lim_{t\to\infty} e(t) = 0$.

For the DT case, that $\|\bar{e}^T(\cdot)\bar{e}(\cdot)\|_1$ is finite implies that $\lim_{t\to\infty} \bar{e}(t) = 0$, where $\bar{e}(t) = h(D)\xi_m(D)[e](t)$. Since $\xi_m(D)$ has a stable inverse, it follows that $\lim_{t\to\infty} e(t) = 0$, where $e(t) = y(t) - y_m(t)$. ∇

Thus far we have developed a unified framework for controller design and stability, robustness, and tracking performance analysis for multivariable model reference adaptive control systems. The adaptive law (9.132)–(9.133) guarantees that the estimation error $\epsilon(t)$ and the rate of the parameter adaptation, $\dot{\Theta}(t)$ in the CT case or $\Theta(t+1) - \Theta(t)$ in the DT case, are of the order of μ in a mean sense (see (9.149)), in addition to the boundedness of the estimation error and the parameter estimates. This leads to the closed-loop system (9.184) whose loop gain is small as from (9.149) if $\mu \ge 0$ is small, so that the closed-loop signal boundedness is ensured.

Recall that in generating the estimation error $\epsilon(t)$ in (9.129), we used the signal $\bar{e}(t) = h(D)\xi_m(D)[e](t)$, which is crucial for the derivation of a suitable error model for the design and analysis of multivariable adaptive controllers. The adaptive law (9.132)–(9.133) ensures that $\epsilon(t)$ satisfies (9.149), so we have (9.158) for the filtered tracking error $\bar{e}(t)$. It can be shown that in the

CT case, if $D^{d_m}W_m(D)$ is proper, then $\int_{t_1}^{t_2} e^T(t)e(t)\,dt \leq \bar{\beta}_c\mu^2(t_2 - t_1) + \bar{b}_c$ for some $\bar{\beta}_c$, $\bar{b}_c > 0$, which is the same tracking error property as that in the SISO case when $M = 1$, and that in the DT case (9.158D) implies that $\sum_{t=t_1}^{t_2-1} e^T(t)e(t) \leq \bar{\beta}_d\mu^2(t_2 - t_1) + \bar{b}_d$ for some $\bar{\beta}_d$, $\bar{b}_d > 0$.

Remark 9.8 In the presence of plant unmodeled dynamics, the developed adaptive control schemes guarantee that the tracking error $e(t) = y(t) - y_m(t)$ is small (that is, of the order of the unmodeled dynamics, characterized by μ) in the mean sense (see (9.158). As in the SISO case (see Section 5.7), smallness in the mean does not imply smallness in absolute value. The tracking error $e(t)$ may be large over short time intervals even in the steady state [5], [134]. However, if the reference input $r(t)$ belongs a class of dominantly rich signals [153], [379] so that the vector signal $\zeta(t)$ has a high level of persistent excitation, then the tracking error $e(t)$ and the parameter error $\tilde{\Theta}(t)$ can be guaranteed to be small in the steady state so that the performance of adaptive control systems can be considerably improved. □

Remark 9.9 We should note that in the absence of the plant unmodeled dynamics, that is, $\mu = 0$ in the plant (9.59), we can design an adaptive controller by setting $G_1(t) = 0$, $G_2(t) = 0$ in (9.132)–(9.133). The above unified theory can be used to show that $\|x^2(\cdot)\|_1$ is finite where $x(t)$ is defined in (9.149) and $\Theta(t)$, $\Psi(t)$, $\epsilon(t)$ are bounded, which leads to (9.184), which has a small loop gain due to the L^2 property of $x(t)$, and, in turn, to the closed-loop stability and asymptotic output tracking. In this case, similar to the SISO case without modeling errors (in Section 5.3.5 where normalization is implemented in the adaptive law, instead of the estimation error $\epsilon(t)$), we can set $\alpha = 0$ in (9.129). Note that for the CT case the condition on S_p in (9.132C) and Γ in (9.133C) is that $K_pS_p = (K_pS_p)^T > 0$ and $\Gamma = \Gamma^T > 0$, and for the DT case the condition on S_p in (9.132D) and Γ in (9.133D) is that $K_pS_p = (K_pS_p)^T > 0$, $2\beta I - S_p^T K_p^T > 0$, $\Gamma = \Gamma^T > 0$ and $2\beta I - \Gamma > 0$. Similarly, we can use

$$m(t) = m_0(t) = \sqrt{\zeta^T(t)\zeta(t) + \xi^T(t)\xi(t)} \tag{9.188}$$

for the error model (9.143C) (whose SISO version is in Section 5.3.4). □

9.2.7 MRAC Using Right Interactor Matrices

The modified left interactor (MLI) matrix $\xi_m(D)$ plays an important role in the design and analysis of multivariable MRAC schemes. In fact, any $M \times M$ polynomial matrix $\xi_0(D)$ such that $\xi_0^{-1}(D)$ is stable and $\lim_{D\to\infty} \xi_0(D)G_0(D)$

is finite and nonsingular can take the role of $\xi_m(D)$. The triangular form of $\xi_m(D)$ makes it possible to design a stable adaptive controller for multivariable plants with only the information about the degrees of the diagonal elements and the maximum degree of $\xi_m(D)$, which has been reported in [85], [94], [125] for the discrete-time case where an augmented estimator is used, based on the interactor matrix $\xi(D)$. The estimation on the maximum degree of $\xi(D)$ was given in [63], [80]. In general, $\xi_m(D)$ depends on the parameters of $G_0(D)$, which are unknown in the adaptive control problem. However, $\xi_m(D)$ can be specified independent of the parameters of $G_0(D)$ if $\xi_m(D)$ is diagonal. The MLI matrix $\xi_m(D)$ can be constructed from the interactor matrix $\xi(D)$ [377], and $\xi_m(D)$ is diagonal if and only if $\xi(D)$ is diagonal. Designing a stable adaptive controller with less *a priori* plant knowledge has been an important topic in multivariable adaptive control. In [349], a method for designing an input compensator for a multivariable plant so that the compensated plant has a diagonal Hermite normal form (equivalently the interactor matrix) was presented using the knowledge of the relative degrees of the elements of $G_0(D)$. In [367] a stable adaptive control scheme was developed for plants with unknown interactor matrix and only the knowledge of an upper bound on the maximum degree of the interactor matrix is needed for implementation.

On the other hand, for many plants whose MLI matrix may be nondiagonal, the right interactor matrix [419] may be diagonal and thus can be specified without the knowledge of the plant parameters [377]. In this case a parametrization technique can be used to modify the plant zero structure at infinity so that the MLI matrix of the modified plant is diagonal and an adaptive controller can be designed with available plant knowledge.

A dual concept to the MLI matrix $\xi_m(D)$ defined in (9.62) is the modified right interactor (MRI) matrix $\xi_m^r(D)$ of an $M \times M$ nonsingular, strictly proper rational transfer matrix $G_0(D)$, which is defined by an upper triangular polynomial matrix of the form

$$\xi_m^r(D) = \begin{bmatrix} d_1^r(D) & h_{12}^r(D) & \cdots & \cdots & h_{1M}^r(D) \\ 0 & d_2^r(D) & h_{23}^r(D) & \cdots & h_{2M}^r(D) \\ \vdots & \vdots & \vdots & \vdots & \vdots \\ 0 & \cdots & 0 & d_{M-1}^r(D) & h_{M-1\,M}^r(s) \\ 0 & \cdots & \cdots & 0 & d_M^r(D) \end{bmatrix}, \tag{9.189}$$

where $h_{ij}^r(D)$, $i = 1, \ldots, M-1$, $j = 2, 3, \ldots, M$, are polynomials and $d_i^r(D)$, $i = 1, \ldots, M$, are any monic stable polynomials of certain degrees $r_i > 0$ such that $\lim_{D\to\infty} G_0(D)\xi_m^r(D) = K_p^r$ is nonsingular and finite. The existence of such a MRI matrix $\xi_m^r(D)$ is guaranteed by Lemma 9.1 applied to $G_0^T(D)$.

If the MLI matrix is unknown but the MRI matrix of $G_0(D)$ is known, we can reparametrize the plant (9.59) so that the MLI matrix of the new nominal plant is diagonal and can be specified based on only the knowledge of the maximum degree of the MRI matrix $\xi_m^r(D)$ of $G_0(D)$.

For such a parametrization, we express the plant (9.59) as

$$y(t) = G_r(D)[v](t),\ u(t) = g^{-1}(D)\xi_m^r(D)[v](t), \tag{9.190}$$

where

$$G_r(D) = G_{0r}(D)[I + \mu\Delta_{mr}(D)] + \mu\Delta_{ar}(D), \tag{9.191}$$

$$G_{0r}(D) = G_0(D)\xi_m^r(D)g^{-1}(D), \tag{9.192}$$

$$\Delta_{mr}(D) = (\xi_m^r(D))^{-1}\Delta_m(D)\xi_m^r(D), \tag{9.193}$$

$$\Delta_{ar}(D) = \Delta_a(D)g^{-1}(D)\xi_m^r(D), \tag{9.194}$$

and $g(D)$ is any monic stable polynomial of degree d_r which is the maximum degree of $\xi_m^r(D)$. The MLI matrix $\bar{\xi}_m(D)$ of $G_{0r}(D)$ is $g(D)I$ and the associated high frequency gain matrix equals K_p^r. We note that any known $M \times M$ polynomial matrix $\xi_0^r(D)$ such that $(\xi_0^r(D))^{-1}$ is stable and $\lim_{D\to\infty} G_0(D)\xi_0^r(D)$ is finite and nonsingular can be used to take the role of $\xi_m^r(D)$ in this plant reparametrization for the design of a model reference adaptive controller.

We then can design an adaptive controller to generate the signal $v(t)$ based on the remodeled plant (9.190), the controller structure (9.79) or (9.102), and the adaptive law (9.132)–(9.133). The control signal $u(t)$ is obtained through the input compensator $g^{-1}(D)\xi_m^r(D)$, that is,

$$u(t) = g^{-1}(D)\xi_m^r(D)[v](t). \tag{9.195}$$

The assumptions on the remodeled nominal plant $G_{0r}(D)$, the unmodeled dynamics $\mu\Delta_{mr}(D)$, $\mu\Delta_{ar}(D)$ can be made similar to that in Section 9.2.2 for the plant (9.59). The above unified analysis is applicable to this case to prove the desired closed-loop stability and asymptotic output tracking properties. Moreover, when defining an estimation error $\epsilon(t)$, we can choose $h(D) = g(D)$ so that $\bar{\xi}_m(D)h(D) = I$, leading to the estimation error

$$\epsilon(t) = \frac{y(t) - y_m(t) + \Psi(t)\xi(t)}{1 + \beta(\zeta^T(t)\zeta(t) + \xi^T(t)\xi(t)) + \alpha m^2(t)}. \tag{9.196}$$

It can be shown that the use of (9.196) ensures that in the CT case the tracking error $e(t) = y(t) - y_m(t)$ satisfies that $\int_{t_1}^{t_2} e^T(t)e(t)\,dt \leq \beta_c\mu^2(t_2 - t_1) + b_c$ for some β_c, $b_c > 0$ in terms of $e(t)$, which is different from that in (9.158C).

We also note that for the continuous-time case the adaptive scheme based on an alternative estimation error, as described in Remarks 9.3 and 9.4, can also be developed, using a modified right interactor matrix $\xi_m^r(s)$.

9.2.8 Continuous-Time Lyapunov Designs

In this subsection, we present model reference adaptive control schemes based on Lyapunov designs for multi-input, multi-output continuous-time plants whose modeled parts are of relative degree 1, unmodeled parts satisfy some assumptions similar to (A9.5) and (A9.6) in Section 9.1.3 (see (9.78)), and disturbances are bounded with bounded derivatives (see Section 5.5.3).

Consider the M-input and M-output linear time-invariant plant

$$y(t) = G_0(s)(I + \mu\Delta_m(s))[u](t) + d(t), \tag{9.197}$$

where $G_0(s) = Z_0(s)P_0^{-1}(s)$ with $Z_0(s)$, $P_0(s)$ being $M \times M$ right coprime polynomial matrices such that $G_0(s)$ is strictly proper, $y(t)$, $u(t)$, and $d(t) \in R^M$ are the plant output, input, and disturbance, respectively, and $\mu\Delta_m(s)$ is the plant (multiplicative) unmodeled dynamics. The modified left interactor (MLI) matrix $\xi_m(s)$ of $G_0(s)$ is an $M \times M$ lower triangular polynomial matrix defined in (9.62) which has a stable inverse and is such that $\lim_{s\to\infty} \xi_m(s)G_0(s) = K_p$, the high frequency gain matrix of $G_0(s)$, is nonsingular.

We note that the additive unmodeled dynamics $\mu\Delta_a(s)$ in the plant (9.60) can be also treated, similar to that in this section so far. In fact, the additive unmodeled dynamics $\mu\Delta_a(s)$ is equivalent to the multiplicative unmodeled dynamics in the sense that $\Delta_a(s) = G_0(s)\bar{\Delta}_m(s)$, where $\bar{\Delta}_m(s) = G_0^{-1}(s)\Delta_a(s)$, for $G_0^{-1}(s)$ stable and additionally with a known stability margin.

The control objective is to find $u(t)$ for the plant (9.197) with unknown parameters such that the closed-loop plant is globally stable and $y(t)$ tracks a reference signal $y_m(t)$ as close as possible, where

$$y_m(t) = W_m(s)[r](t) \tag{9.198}$$

with $W_m(s)$ being an $M \times M$ strictly proper and stable transfer matrix and $r(t) \in R^M$ being a piecewise continuous and bounded known signal.

Plant Assumptions and Controller Structure

We assume that the plant (9.197) satisfies the assumptions (A9.1)–(A9.6) state in Section 9.1.3, that is, $G_0(s)$ is of full rank and minimum phase, an upper bound $\bar{\nu}$ on the observability index ν of $G_0(s)$ is known, the high frequency gain matrix K_p of $G_0(s)$ is such that $SK_p = (SK_p)^T > 0$ for some known matrix S, and especially for a Lyapunov design,

> (A9.4a): The modified left interactor (MLI) matrix $\xi_m(s)$ of $G_0(s)$ is known and is of the form $\xi_m(s) = \xi_1 s + \xi_0$ such that $\xi_m^{-1}(s)$ is strictly positive real (SPR) (so that $\xi_1 = \xi_1^T > 0$ as ξ_1 is the associated residue matrix (at $s = \infty$) of $\xi_m(s)$, which is also SPR, and $\xi_m(s)$ is a MLI matrix of $G(s)$ to make K_p nonsingular, and $\xi_0 + \xi_0^T > 0$ as from the SPR property of $\xi_m(s)$).[11]

[11] A positive real (strictly positive real) matrix corresponds to the driving point impedance matrix of a circuit network with passive (dissipative) elements.

Definition 9.1 [10] *An $M \times M$ rational matrix $H(s)$ is positive real (PR) if*
(i) $H(s)$ does not have any poles in $\text{Re}[s] > 0$*; and*
(ii) $H^T(s^*) + H(s) \geq 0$ for $\text{Re}[s] \geq 0$.

Lemma 9.5 [10] *An $M \times M$ rational matrix $H(s)$ is positive real (PR) if*
(i) $H(s)$ does not have any poles in $\text{Re}[s] > 0$*; and*
(ii) poles of any element of $H(s)$ on the $j\omega$-axis are distinct, and the associated residue matrix of $H(s)$ is nonnegative definite Hermitian; and
(iii) $H(j\omega) + H^T(-j\omega) \geq 0$ $\forall \omega$ such that $j\omega$ are not a pole of any element of $H(j\omega)$.

A matrix $H = \{h_{ij}\} \in C^{M \times M}$ is Hermitian if $H = H^{*T} = \{h_{ji}^*\}$, the complex conjugate transpose of H, with $(\sigma + j\omega)^* = \sigma - j\omega$ for real σ, ω. All eigenvalues of a Hermitian matrix are real numbers. A Hermitian matrix H is nonnegative (positive) definite if $x^{*T}Hx \geq (>)\, 0$, $\forall x \in C^M$ (if and only if all its eigenvalues are nonnegative (positive)).

Definition 9.2 [288] *A rational matrix $H(s)$ is strictly positive real (SPR) if $H(s - \epsilon)$ is positive real (PR) for some $\epsilon > 0$.*

Theorem 9.3 [378] *An $M \times M$ rational matrix $H(s) = C(sI - A)^{-1}B + D$ such that $H(s) + H^T(-s)$ has full rank is strictly positive real if and only if*
(i) $H(s)$ does not have any poles in $\text{Re}[s] \geq 0$*;*
(ii) $H(j\omega) + H^T(-j\omega) > 0$, $\forall \omega \in R$; and
(iii) (a) $\lim_{\omega\to\infty} \omega^2[H(j\omega) + H^T(-j\omega)] > 0$, $D + D^T \geq 0$ for $\det[D + D^T] = 0$, or
(b) $\lim_{\omega\to\infty}[H(j\omega) + H^T(-j\omega)] > 0$, for $\det[D + D^T] \neq 0$.

Theorem 9.4 [378] *Given a scalar $\delta > 0$, a matrix A such that all zeros of $\det(sI - A)$ are in the open left-half plane, a real matrix B such that (A, B) is controllable, real matrices C and D, and an arbitrary real matrix $L = L^T > 0$. Then real matrices Q and K, a real matrix $P = P^T > 0$ exist such that*

$$PA + A^T P = -QQ^T - \delta L,\ PB = C^T - QK,\ K^T K = D^T + D \tag{9.199}$$

if and only if $H(s) = C(sI - A)^{-1}B + D$ is strictly positive real such that $H(s) + H^T(-s)$ has full rank and δ is sufficiently small.

(A9.5-6a): For some constants $k_m > 0$ and $q > 0$, independent of μ, the rational matrix $\Delta_m(s-q)s$ is stable and proper such that $\|\Delta_m(s-q)s\|_\infty \leq k_m$ (see Section 2.8 for the H^∞ norm).

(A9.7): $d(t)$ and $\dot{d}(t)$ are bounded.

Assumption (A9.5-6a) is equivalent to the condition that the transfer matrix $\Delta_m(s-q)s$ has a finite L^1 operator norm (upper) bound independent of μ, or the condition that individual components of $\Delta_m(s-q)s$ have L^1 or L^2 (H^∞) norm bounds independent of μ. Note that $\Delta_m(s-q)s$ has a finite L^1 or L^2 operator norm (upper) bound independent of μ implies that $\Delta_m(s-q_1)s$ has the same property for any $q_1 \in (0, q)$ (see the footnote for (5.195)).

In view of (A9.4a), we choose the reference model transfer matrix as

$$W_m(s) = \xi_m^{-1}(s). \tag{9.200}$$

We propose to use the controller structure

$$\begin{aligned} u(t) = {} & \Theta_1^T \omega_1(t) + \Theta_2^T \omega_2(t) + \Theta_{20} y(t) + \Theta_3 r(t) \\ & - k_0 h(e(t)) - g_0 f(e(t)) m(t), \end{aligned} \tag{9.201}$$

where $\Theta_1 = [\Theta_{11}, \ldots, \Theta_{1\bar{\nu}-1}]^T$, $\Theta_2 = [\Theta_{21}, \ldots, \Theta_{2\bar{\nu}-1}]^T$, with $\Theta_{ij} \in R^{M\times M}$, $i = 1, 2$, $j = 1, \ldots, \bar{\nu}-1$, $\Theta_{20} \in R^{M\times M}$ and $\Theta_3 \in R^{M\times M}$,

$$\omega_1(t) = F(s)[u](t),\ \omega_2(t) = F(s)[y](t), \tag{9.202}$$

$$F(s) = \frac{A(s)}{\Lambda(s)},\ A(s) = [I_M, sI_M, \ldots, s^{\bar{\nu}-2} I_M]^T, \tag{9.203}$$

for a monic polynomial $\Lambda(s)$ of degree $\bar{\nu}-1$, $m(t)$ is generated from

$$\dot{m}(t) = -\delta_0 m(t) + \delta_1 \|u(t)\| + \delta_2,\ m(0) > 0, \tag{9.204}$$

with $\delta_1 > 0$, $\delta_2 > 0$, $\delta_0 \in (0, \min\{q, q_0\})$ for $q_0 > 0$ such that $\xi_m(s-q_0)$ and $\Lambda(s-q_0)$ are stable, the design parameters $k_0 \geq 1$, $g_0 \in (0, g_0^*)$ for some constant $g_0^* > 0$, and $h(e(t)) \in R^M$ and $f(e(t)) \in R^M$ for $e(t) = y(t) - y_m(t)$ are some design functions, to be specified.

Remark 9.10 The term $-k_0 h(e(t))$ in the control law (9.201) is to deal with the bounded disturbance $d(t)$ in the plant model (9.197), while the term $-g_0 f(e(t)) m(t)$ is to deal with the unmodeled dynamics $\mu\Delta_m(s)$. For the class of plants considered and the Lyapunov analysis method used, the design parameters in these terms can be properly chosen to improve the closed-loop stability and tracking performance (see Theorem 9.5). □

Error Model

When $d(t) = 0$ and $\mu = 0$, constant matrices Θ_1^*, Θ_2^*, Θ_{20}^*, and $\Theta_3^* = K_p^{-1}$ exist and satisfy (9.105) such that for $\Theta_1 = \Theta_1^*$, $\Theta_2 = \Theta_2^*$, $\Theta_{20} = \Theta_{20}^*$, and $\Theta_3 = \Theta_3^*$, the controller (9.201), with $f(e(t)) = h(e(t)) = 0$, makes the closed-loop plant transfer matrix from $r(t)$ to $y(t)$ equal to $W_m(s)$ and all closed-loop poles stable (see Remark 9.5 and Section 9.2.3).

When $d(t) \neq 0$ and $\mu \neq 0$, using (9.105), we express the plant (9.197) as

$$\begin{aligned} u(t) &= \Theta_1^{*T}\omega_1(t) + \Theta_2^{*T}\omega_2(t) + \Theta_{20}^* y(t) + \Theta_3^*\xi_m(s)[y](t) \\ &\quad - [\Theta_2^{*T}, \Theta_{20}^*]d_1(t) - \Theta_3^*\xi_m(s)[d](t) - \mu\Delta_m(s)[u](t) + \mu\Theta_1^{*T}\eta_1(t), \end{aligned} \tag{9.205}$$

where

$$d_1(t) = \left[\left(\frac{A(s)}{\Lambda(s)}[d](t)\right)^T, d^T(t)\right]^T, \; \eta_1(t) = \frac{A(s)}{\Lambda(s)}\Delta_m(s)[u](t). \tag{9.206}$$

Define the tracking error $e(t) = y(t) - y_m(t)$ and the parameter matrix

$$\Theta(t) = [\Theta_1^T(t), \Theta_2^T(t), \Theta_2(t), \Theta_3^T(t)]^T, \tag{9.207}$$

which is the estimate of $\Theta^* = [\Theta_1^{*T}, \Theta_2^{*T}, \Theta_2^*, \Theta_3^{*T}]^T$ at time t, and let

$$\omega(t) = [\omega_1^T(t), \omega_2^T(t), y^T(t), r^T(t)]. \tag{9.208}$$

From (9.201) and (9.207), we have the tracking error equation

$$\begin{aligned} &\xi_1\dot{e}(t) + \xi_0 e(t) \\ &= K_p\tilde{\Theta}^T\omega(t) + \bar{d}(t) + \mu\eta(t) - K_p(k_0 h(e(t)) + g_0 f(e(t))m(t)), \end{aligned} \tag{9.209}$$

where $\tilde{\Theta}(t) = \Theta(t) - \Theta^*$ is the parameter error matrix and

$$\bar{d}(t) = K_p\left([\Theta_2^{*T}, \Theta_{20}^*]d_1(t) + \Theta_3^{*T}\xi_m(s)[d](t)\right), \tag{9.210}$$

$$\eta(t) = \Delta(s)[u](t), \; \Delta(s) = K_p\left(I - \Theta_1^{*T}\frac{A(s)}{\Lambda(s)}\right)\Delta_m(s). \tag{9.211}$$

It follows from Assumptions (A9.5-6a) and (A9.7) that $\bar{d}(t)$ is bounded and that, for $m(t)$ in (9.204), $\frac{\|\eta(t)\|}{m(t)} \leq g_1$ for some constant $g_1 > 0$, $\forall t \geq 0$ (see (9.141) and (9.142) with $D_1 = 0$ and $k_2 = 0$ and the multivariable version of Lemma 2.6 stated in the footnote for (9.142); note that $\eta(t) = \Delta(s)[u](t) = \Delta(s)(s+\delta_0)\frac{1}{s+\delta_0}[u](t)$ and $\Delta(s-\delta_0)s$ has a finite L^1 operator norm bound independent of μ as from Assumption (A9.5-6a)).

Adaptive Control Design

The adaptive law for updating $\Theta(t)$ is chosen as

$$\dot{\Theta}^T(t) = -(\xi_1^{-1}S)^{-T}e(t)\omega^T(t) - \sigma(t)\Theta^T(t), \tag{9.212}$$

where S and ξ_1 are defined in Assumptions (A9.3) and (A9.4a),

$$\sigma(t) = \begin{cases} 0 & \text{if } \|\Theta(t)\| < M_\theta, \\ \sigma_0(\frac{\|\Theta(t)\|}{M_\theta} - 1) & \text{if } M_\theta \le \|\Theta\| < 2M_\theta, \\ \sigma_0 & \text{if } \|\Theta(t)\| \ge 2M_\theta, \end{cases} \tag{9.213}$$

$$M_\theta > \frac{\|\Theta^*\| \, \|SK_p\|_2}{\lambda_{\min}[SK_p]}, \tag{9.214}$$

with $\lambda_{\min}[SK_p] > 0$ being the minimum eigenvalue of $SK_p = (SK_p)^T > 0$ and $\sigma_0 > 0$ being arbitrary, and the matrix norm $\|\cdot\|$ is the Frobenius norm (see Section 9.2.5) (while the l^2 induced matrix norm $\|\cdot\|_2$ is $\|SK_p\|_2 = \lambda_{\max}[SK_p]$).

The parameters k_0 and g_0 in (9.201) are chosen as

$$k_0 \ge 1, \; g_0 \in (0, g_0^*), \tag{9.215}$$

where $g_0^* > 0$ is to be specified. To specify the design functions $f(e(t))$ and $h(e(t))$ in (9.201), assuming an $M \times M$ matrix S_r is known such that $K_pS_r = \text{diag}\{s_1, \ldots, s_M\}$, for some nonzero $s_i \in R$, $i = 1, \ldots, M$, we choose

$$h(e(t)) = S_rD_0 \, \text{sat}[\xi_1 e(t)], \; f(e(t)) = S_rM_0 \, \text{sat}[\xi_1 e(t)m(t)], \tag{9.216}$$

where

$$D_0 = \text{diag}\left\{\frac{\text{sign}[s_1]}{s_{01}} d_{01}, \ldots, \frac{\text{sign}[s_M]}{s_{0M}} d_{0M}\right\}, \tag{9.217}$$

$$M_0 = \text{diag}\left\{\frac{\text{sign}[s_1]}{s_{01}} \lambda_{01}, \ldots, \frac{\text{sign}[s_m]}{s_{0M}} \lambda_{0M}\right\}, \tag{9.218}$$

$$\text{sat}[\xi_1 e(t)] \triangleq [\text{sat}[(\xi_1 e(t))_1; \epsilon_1], \ldots, \text{sat}[(\xi_1 e(t))_M; \epsilon_M]]^T, \tag{9.219}$$

$$\text{sat}[\xi_1 e(t)m(t)] \triangleq [\text{sat}[(\xi_1 e(t))_1 m(t); \epsilon_1], \ldots, \text{sat}[(\xi_1 e(t))_M m(t); \epsilon_M]]^T, \tag{9.220}$$

$s_{0i} \in (0, |s_i|]$, $\lambda_{0i} > 0$, $i = 1, \ldots, M$, are constants, d_{0i} is the upper bound for $|\bar{d}_i(t)|$ for $\bar{d}(t) = [\bar{d}_1(t), \ldots, \bar{d}_M(t)]^T$, and $(\xi_1 e(t))_i$ is the ith component of $\xi_1 e(t)$, $i = 1, \ldots, M$, and $\text{sat}[w; \epsilon_i]$ for some $\epsilon_i > 0$ is defined for $w \in R$ as

$$\text{sat}[w; \epsilon_i] = \begin{cases} 1 & \text{if } w \ge \epsilon_i, \\ w & \text{if } |w| < \epsilon_i, \\ -1 & \text{if } w \le -\epsilon_i. \end{cases} \tag{9.221}$$

Stability Analysis

The above adaptive control scheme has the following desired properties.

Theorem 9.5 *There exists a constant $g_0^* > 0$ such that for any $g_0 \in (0, g_0^*)$ the existence of $\mu^* > 0$ is ensured such that for any $\mu \in [0, \mu^*)$ all signals in the closed-loop system (9.197), (9.198), (9.201) with (9.216), and (9.212) are bounded, and the tracking error $e(t) = y(t) - y_m(t)$ satisfies*

$$\int_{t_1}^{t_2} \|e(t)\|_2^2 \, dt \le c_0 + (c_1 + c_2\mu)\epsilon_0(t_2 - t_1) \tag{9.222}$$

for $\epsilon_0 = \max_i \epsilon_i > 0$ in (9.221), and some constants $c_i > 0$, $i = 0, 1, 2$. When $d(t) = 0$ and $\mu\Delta_m(s) = 0$ in the plant (9.197), $\lim_{t\to\infty} e(t) = 0$.

Proof: Consider the positive definite function

$$V(e, \tilde{\Theta}) = e^T(t)\xi_1 e(t) + \mathrm{tr}[\tilde{\Theta}(t) S K_p \tilde{\Theta}^T(t)], \tag{9.223}$$

where $\xi_1 = \xi_1^T > 0$ due to the SPR property of $W_m(s) = \xi_m^{-1}(s)$ and $SK_p = (SK_p)^T > 0$ as in Assumption (A9.3). Using (9.209) and (9.212), we obtain

$$\begin{aligned} \dot{V} = {} & -e^T(t)(\xi_0 + \xi_0^T)e(t) + 2e^T(t)\xi_1(\bar{d}(t) - K_p k_0 h(e(t))) \\ & + 2e^T(t)\xi_1(\mu\eta(t) - g_0 K_p f(e(t))m(t)) - 2\sigma(t)\,\mathrm{tr}[\tilde{\Theta}(t) S K_p \Theta^T(t)]. \end{aligned} \tag{9.224}$$

For the design functions $f(e(t))$ and $h(e(t))$ in (9.216), it follows that

$$\begin{aligned} & 2e^T(t)\xi_1(\bar{d}(t) - K_p k_0 h(e(t))) \\ & = 2e^T(t)\xi_1\bar{d}(t) - 2k_0 e^T(t)\xi_1 K_p S_r D_0 \,\mathrm{sat}[\xi_1 e(t)] \\ & = 2\sum_{i=1}^{M}\left((\xi_1 e(t))_i \bar{d}_i(t) - k_0(\xi_1 e(t))_i \frac{|s_i|}{s_{0i}} d_{0i}\,\mathrm{sat}[(\xi_1 e(t))_i; \epsilon_i]\right) \\ & \le 2\sum_{i=1}^{M} \epsilon_i d_{0i}. \end{aligned} \tag{9.225}$$

The last inequality is from the definition of D_0 in (9.117), (9.220), and (9.221). Similarly, for $\eta(t) = [\eta_1(t), \ldots, \eta_M(t)]^T$, we have

$$\begin{aligned} & 2e^T(t)\xi_1(\mu\eta(t) - K_p g_0 f(e(t))m(t)) \\ & = 2e^T(t)\xi_1\mu\eta(t) - 2g_0 e^T(t)\xi_1 K_p S_r M_0 \,\mathrm{sat}[\xi_1 e(t) m(t)] m(t) \\ & = 2\sum_{i=1}^{M}\left(\mu(\xi_1 e(t))_i \eta_i(t) - g_0(\xi_1 e(t))_i \frac{|s_i|}{s_{0i}} \lambda_{0i}\,\mathrm{sat}[(\xi_1 e(t))_i m(t); \epsilon_i] m(t)\right) \\ & \le 2\sum_{i=1}^{M}\left(\mu|(\xi_1 e(t))_i| m(t) g_1 - g_0(\xi_1 e(t))_i \frac{|s_i|}{s_{0i}} \lambda_{0i}\,\mathrm{sat}[(\xi_1 e(t))_i m(t); \epsilon_i] m(t)\right) \\ & \le 2\mu\sum_{i=1}^{M} \epsilon_i g_1 \end{aligned} \tag{9.226}$$

for $\mu \in (0, \mu^*)$, where $\mu^* = \frac{g_0 \min_i\{\lambda_{0i}\}}{g_1}$.

From (9.224)–(9.226), we have

$$\dot{V} \leq -e^T(t)(\xi_0 + \xi_0^T)e(t) - 2\sigma(t)\,\mathrm{tr}[\tilde{\Theta}(t)SK_p\Theta^T(t)] + 2\sum_{i=1}^{M}\epsilon_i d_{0i} + 2\mu\sum_{i=1}^{M}\epsilon_i g_1 \tag{9.227}$$

for any $\mu \in (0, \mu^*)$. From (9.213) and (9.214), we have

$$\sigma(t)\,\mathrm{tr}[\tilde{\Theta}(t)SK_p\Theta^T(t)] \geq 0, \quad \lim_{\|\Theta\|\to\infty} \sigma(t)\,\mathrm{tr}[\tilde{\Theta}(t)SK_p\Theta^T(t)] = \infty. \tag{9.228}$$

Hence, we conclude that $\Theta(t) \in L^\infty$ and $e(t) \in L^\infty$ and satisfies (9.228).

From (9.204), it follows that

$$m(t) \leq a_1 + \delta_1 \int_0^t e^{-\delta_0(t-\tau)}\|u(\tau)\|\,d\tau \tag{9.229}$$

for some constant $a_1 > 0$. From (9.197), we obtain

$$u(t) = G_0^{-1}(s)[y - d](t) - \mu\Delta_m(s)[u](t). \tag{9.230}$$

From (9.202), we can express $\omega_1(t)$ as

$$\omega_1(t) = \frac{A(s)}{\Lambda(s)}G_0^{-1}(s)[y - d](t) - \mu\frac{A(s)}{\Lambda(s)}\Delta_m(s)[u](t). \tag{9.231}$$

From (A9.5-6a), it follows that $\|\frac{A(s)}{\Lambda(s)}\Delta_m(s)[u](t)\| \leq a_2 m(t)$ for some constant $a_2 > 0$. From (9.201), (9.229), and (9.231) with (A9.4a), we derive

$$\|u(t)\| \leq a_3\|\omega_1(t)\| + a_4 + g_0 a_5 m(t) \leq a_6 + (a_2a_3\mu + a_5g_0)\delta_1 \int_0^t e^{-\delta_0(t-\tau)}\|u(\tau)\|\,d\tau \tag{9.232}$$

for some constants $a_i > 0$, $i = 3, \ldots, 6$. Using the Bellman-Gronwall lemma (Lemma 2.1; also see (2.106) in the proof of Lemma 2.3), we obtain

$$\|u(t)\| \leq a_6 + a_6 \int_0^t e^{-\delta_0(t-\tau)+(a_2a_3\mu+a_5g_0)\delta_1(t-\tau)}\,d\tau. \tag{9.233}$$

Hence, $u(t)$ is bounded if

$$(a_2a_3\mu + a_5g_0)\delta_1 < (a_2a_3\mu^* + a_5g_0)\delta_1 < \delta_0, \tag{9.234}$$

where $\mu^* = \frac{g_0 \min_i\{\lambda_{0i}\}}{g_1}$, as defined form (9.226). The condition (9.234) is satisfied for $g_0 \in (0, g_0^*)$, where $g_0^* = \frac{\delta_0 g_1}{(a_2a_3\min_i\{\lambda_{0i}\}+a_5g_1)\delta_1}$.

When $d(t) = 0$ and $\mu\Delta_f(s) = 0$ in (9.197), we have

$$\begin{aligned} & 2e^T(t)\xi_1(\bar{d}(t) - K_p k_0 h(e(t))) \\ & = -2\sum_{i=1}^{M} k_0(\xi_1 e(t))_i \frac{|s_i|}{s_{0i}} d_{0i}\,\mathrm{sat}[(\xi_1 e(t))_i; \epsilon_i] \le 0, \end{aligned} \tag{9.235}$$

$$\begin{aligned} & 2e^T(t)\xi_1(\mu\eta(t) - K_p g_0 f(e(t))m(t)) \\ & = -2\sum_{i=1}^{M} g_0(\xi_1 e(t))_i \frac{|s_i|}{s_{0i}} \lambda_{0i}\,\mathrm{sat}[(\xi_1 e(t))_i m(t); \epsilon_i] m(t) \le 0 \end{aligned} \tag{9.236}$$

(see (9.225) and (9.226)), so that from (9.224) and (9.228), we obtain

$$\dot{V} = -e^T(t)(\xi_0 + \xi_0^T)e(t) \le 0, \tag{9.237}$$

that is, $e(t) \in L_2$, so that, with $\dot{e}(t) \in L^\infty$, $\lim_{t\to\infty} e(t) = 0$. ∇

When $d(t) = 0$ and $\mu\Delta_m(s) = 0$, the design parameter $k_0 > 0$ can be arbitrary, while the design parameter g_0 needs to satisfy $0 < g_0 < g_0^* = \frac{\delta_0}{a_5\delta_1}$.

In the SISO case (see Section 5.5), that is, when $M = 1$, we can choose $S_r = \mathrm{sign}[k_p]$, $D_0 = \frac{d_{01}}{s_p}$, $M_0 = \lambda_{01}$, where $s_p > 0$ is the lower bound for $|k_p|$. In this case the plant observability index is equal to the plant order, $\xi_m(s)$ is any stable polynomial of degree 1, and S in (A9.3) is $\mathrm{sign}[k_p]$.

Remark 9.11 The limit case of the design (9.216)–(9.221) with $\epsilon_i = 0$, $i = 1, \ldots, M$, is presented as: assuming an $M \times M$ matrix S_r is known such that $K_p S_r = \mathrm{diag}\{s_1, \ldots, s_M\}$, for some nonzero $s_i \in R$, $i = 1, \ldots, M$, we choose

$$h(e(t)) = S_r D_0\,\mathrm{sgn}[\xi_1 e(t)], \quad f(e(t)) = S_r M_0\,\mathrm{sgn}[\xi_1 e(t)], \tag{9.238}$$

where

$$D_0 = \mathrm{diag}\left\{\frac{\mathrm{sign}[s_1]d_{01}}{s_{01}}, \ldots, \frac{\mathrm{sign}[s_M]d_{0M}}{s_{0M}}\right\}, \tag{9.239}$$

$$M_0 = \mathrm{diag}\left\{\frac{\mathrm{sign}[s_1]}{s_{01}}\lambda_{01}, \ldots, \frac{\mathrm{sign}[s_m]}{s_{0M}}\lambda_{0M}\right\}, \tag{9.240}$$

$$\mathrm{sgn}[\xi_1 e(t)] \triangleq [\mathrm{sgn}[(\xi_1 e(t))_1], \ldots, \mathrm{sgn}[(\xi_1 e(t))_M]]^T, \tag{9.241}$$

$s_{0i} \in (0, |s_i|]$, $\lambda_{0i} > 0$, $i = 1, \ldots, m$, are arbitrary constants, d_{0i} is the upper bound for $|\bar{d}_i(t)|$ for $\bar{d}(t) = [\bar{d}_1(t), \ldots, \bar{d}_M(t)]^T$, and $(\xi_1 e(t))_i$ is the ith component of $\xi_1 e(t)$, $i = 1, \ldots, M$, and $\mathrm{sgn}[w]$ is defined for $w \in R$ as

$$\mathrm{sgn}[w] = \begin{cases} 1 & \text{if } w > 0, \\ 0 & \text{if } w = 0, \\ -1 & \text{if } w < 0. \end{cases} \tag{9.242}$$

The discontinuous nature of the sgn function (9.242) makes this design of $f(e(t))$ and $h(e(t))$ for the controller (9.201) to have a feature of variable structure systems [3], [29], [58], [69], [102], [131], [132], [133], [135], [352], [423], [424], [450], [451], [452], [460]. Such a feature essentially is that the system solution exists in an average sense [102], [423], that chattering in system response signals may happen due to the discontinuity of the control signal, and that a so-called sliding mode motion may be reached to make the closed-loop system insensitive to modeling errors, as indicated by the ideal properties

$$\begin{aligned} & 2e^T(t)\xi_1(\bar{d}(t) - K_p k_0 h(e(t))) \\ & = 2e^T(t)\xi_1\bar{d}(t) - 2k_0 e^T(t)\xi_1 K_p S_r D_0 \operatorname{sgn}[\xi_1 e(t)] \\ & = 2\sum_{i=1}^{M}\left((\xi_1 e(t))_i \bar{d}_i(t) - k_0|(\xi_1 e(t))_i|\frac{|s_i|}{s_{0i}} d_{0i}\right) \leq 0, \end{aligned} \tag{9.243}$$

$$\begin{aligned} & 2e^T(t)\xi_1(\mu\eta(t) - K_p g_0 f(e(t))m(t)) \\ & = 2e^T(t)\xi_1\mu\eta(t) - 2g_0 e^T(t)\xi_1 K_p S_r M_0 \operatorname{sgn}[\xi_1 e(t)]m(t) \\ & = 2\sum_{i=1}^{M}\left(\mu(\xi_1 e(t))_i\eta_i(t) - g_0(\xi_1 e(t))_i\frac{|s_i|}{s_{0i}}\lambda_{0i}\operatorname{sgn}[(\xi_1 e(t))_i]m(t)\right) \\ & \leq 2\sum_{i=1}^{M}\left(\mu|(\xi_1 e(t))_i|m(t)g_1 - g_0|(\xi_1 e(t))_i|\frac{|s_i|}{s_{0i}}\lambda_{0i}m(t)\right) \leq 0 \end{aligned} \tag{9.244}$$

for $\mu \in (0, \mu_1^*)$, where $\mu_1^* = \frac{g_0 \min_i\{\lambda_{0i}\}}{g_1}$. □

Remark 9.12 The parameter adaptation for the control law (9.201) can be eliminated if a fixed estimate Θ_0 of Θ^* is used and an upper bound for $\|K_p(\Theta_0 - \Theta^*)^T\|$ is known. Then the control law (9.201) can be modified as

$$u(t) = \Theta_0^T\omega(t) - f_0(e(t)) - k_0 h(e(t)) - g_0 f(e(t))m(t), \tag{9.245}$$

where Θ_0 is a fixed parameter matrix and $f_0(e(t))$ is a design function to be specified. With (9.205) and (9.245), the tracking error equation now is

$$\begin{aligned} \dot{e}(t) + \xi_1^{-1}\xi_0 e(t) = {} & K_p\tilde{\Theta}_0^T\omega(t) + \bar{d}(t) + \eta(t) \\ & - K_p(f_0(e(t)) + k_0 h(e(t)) + g_0 f(e(t))m(t)), \end{aligned} \tag{9.246}$$

where $\tilde{\Theta}_0 = \Theta_0 - \Theta^*$. The time derivative of $V_0(t) = e^T(t)\xi_1 e(t)$ is

$$\begin{aligned} \dot{V}_0 = {} & -e^T(t)(\xi_0 + \xi_0^T)e(t) + 2e^T(t)\xi_1(\bar{d}(t) - K_p k_0 h(e(t))) \\ & + 2e^T(t)\xi_1(\eta(t) - g_0 K_p f(e(t))m(t)) \\ & + 2e^T(t)\xi_1(K_p\tilde{\Theta}_0^T\omega(t) - K_p f_0(e(t))). \end{aligned} \tag{9.247}$$

In view of S_r in (9.216), we choose the design function $f_0(e(t))$ as

$$f_0(e(t)) = \gamma_0 S_r S_0 \,\mathrm{sat}[\xi_1 e(t)\|\omega(t)\|_2]\, \|\omega(t)\|_2, \tag{9.248}$$

where $\gamma_0 \geq 1$ and

$$S_0 = \mathrm{diag}\left\{\frac{\mathrm{sign}[s_1]}{s_{01}}\,\|(K_p\tilde{\Theta}_0^T)_1\|_2, \ldots, \frac{\mathrm{sign}[s_M]}{s_{0M}}\,\|(K_p\tilde{\Theta}_0^T)_M\|_2\right\}, \tag{9.249}$$

$$\begin{aligned}&\mathrm{sat}[\xi_1 e(t)\|\omega(t)\|_2]\\ &\triangleq [\mathrm{sat}[(\xi_1 e(t))_1\|\omega(t)\|_2; \epsilon_1], \ldots, \mathrm{sat}[(\xi_1 e(t))_M\|\omega(t)\|_2; \epsilon_M]]^T,\end{aligned} \tag{9.250}$$

with $(K_p\tilde{\Theta}_0^T)_i$ being the ith row of $K_p\tilde{\Theta}_0^T$, $i = 1, \ldots, M$, and the sat function $\mathrm{sat}[\cdot; \epsilon_i]$ defined in (9.221) for some design parameter $\epsilon_i > 0$.

With this choice of $f_0(e(t))$, we have

$$\begin{aligned}&2e^T(t)\xi_1(K_p\tilde{\Theta}_0^T\omega(t) - K_p f_0(e(t)))\\ &= 2\sum_{i=1}^{M}\Big((\xi_1 e(t))_i(K_p\tilde{\Theta}_0^T)_i\omega(t)\\ &\quad - \gamma_0(\xi_1 e(t))_i\frac{|s_i|}{s_{0i}}\,\|(K_p\tilde{\Theta}_0^T)_i\|_2\,\mathrm{sat}[(\xi_1 e(t))_i\|\omega(t)\|_2; \epsilon_i]\,\|\omega(t)\|_2\Big)\\ &\leq 2\sum_{i=1}^{M}\Big(|(\xi_1 e(t))_i|\,\|(K_p\tilde{\Theta}_0^T)_i\|_2\,\|\omega(t)\|_2\\ &\quad - \gamma_0(\xi_1 e(t))_i\frac{|s_i|}{s_{0i}}\,\|(K_p\tilde{\Theta}_0^T)_i\|_2\,\mathrm{sat}[(\xi_1 e(t))_i\|\omega(t)\|_2; \epsilon_i]\,\|\omega(t)\|_2\Big)\\ &\leq 2\sum_{i=1}^{M}\epsilon_i\|(K_p\tilde{\Theta}_0^T)_i\|_2.\end{aligned} \tag{9.251}$$

From (9.225), (9.226), (9.247), and (9.251), we obtain

$$\dot{V}_0 \leq -e^T(t)(\xi_0+\xi_0^T)e(t) + 2\sum_{i=1}^{M}\epsilon_i d_{0i} + 2\mu\sum_{i=1}^{M}\epsilon_i g_1 + 2\sum_{i=1}^{M}\epsilon_i\|(K_p\tilde{\Theta}_0^T)_i\|_2 \tag{9.252}$$

for $\mu \in (0, \mu^*)$, where $\mu^* = \frac{g_0 \min_i\{\lambda_{0i}\}}{g_1}$. From (9.252), we can conclude that $e(t) = y(t) - y_m(t) \in L^\infty$ and $e(t)$ exponentially converges to the set

$$S_0 = \{e : \ \|e\| \leq b_1\epsilon_0\} \tag{9.253}$$

for some constant $b_1 > 0$, where $\epsilon_0 = \max_i \epsilon_i > 0$.

It then can be shown that there exists a $g_0^* > 0$ such that for any $g_0 \in (0, g_0^*)$ and all $\mu \in [0, \mu^*)$ the signal $u(t)$ is bounded. □

Design for Feedback Unmodeled Dynamics

Let us now consider the M-input and M-output linear time-invariant plant

$$y(t) = G_0(s)[u](t) + \mu\Delta_f(s)[y](t) + d(t), \tag{9.254}$$

where $\mu\Delta_f(s)$ is the plant (feedback) unmodeled dynamics, under the assumptions (A9.1)–(A9.3), (A9.4a), (A9.7), and

> (A9.5-6b): For some constants $k_f > 0$, $a > q > 0$, independent of μ, the rational matrix $(s+a-q)\Delta_f(s-q)s$ is stable and proper such that $\|(s+a-q)\Delta_f(s-q)s\|_\infty \le k_f$ (or the L^1 operator norm of $(s+a-q)\Delta_f(s-q)s$ is not larger than a constant $\bar{k}_f$).[12]

Reference model and control law. For adaptive control, we use the reference model (9.198) with transfer matrix given in (9.200), that is,

$$y_m(t) = \xi_m^{-1}(s)[r](t), \tag{9.255}$$

for $\xi_m(s) = \xi_1 s + \xi_0$ being SPR, and the controller structure (9.201),

$$\begin{aligned} u(t) = {} & \Theta_1^T\omega_1(t) + \Theta_2^T\omega_2(t) + \Theta_{20}y(t) + \Theta_3 r(t) \\ & - k_0 h(e(t)) - g_0 f(e(t))m(t), \end{aligned} \tag{9.256}$$

with the normalizing signal $m(t)$ generated from

$$\dot{m}(t) = -\delta_0 m(t) + \delta_1\|y(t)\| + \delta_2,\ m(0) > 0 \tag{9.257}$$

for $\delta_1 > 0$, $\delta_2 > 0$, $\delta_0 \in (0, \min\{q, q_0\})$ such that $\xi_m(s-q_0)$ and $\Lambda(s-q_0)$ are stable, and with design parameters $k_0 \ge 0$, $g_0 > 0$ and design functions

$$h(e(t)) = S^{-T}\xi_1 e(t),\ f(e(t)) = S^{-T}\xi_1 e(t)m(t) \tag{9.258}$$

for $e(t) = y(t) - y_m(t)$ and S, ξ_1 in Assumptions (A9.3), (A9.4a).

Error model. Similar to (9.205), using (9.105), we express (9.254) as

$$\begin{aligned} u(t) = {} & \Theta_1^{*T}\omega_1(t) + \Theta_2^{*T}\omega_2(t) + \Theta_{20}^*y(t) + \Theta_3^*\xi_m(s)[y](t) \\ & - \left(\Theta_2^{*T}\frac{A(s)}{\Lambda(s)} + \Theta_{20}^* + \Theta_3^*\xi_m(s)\right)[d](t) \\ & -\mu\left(I - \Theta_1^{*T}\frac{A(s)}{\Lambda(s)}\right)G_0^{-1}(s)\Delta_f(s)[y](t). \end{aligned} \tag{9.259}$$

[12]Under this assumption, for any $q_1 \in (0, q)$, the rational matrix $(s+a-q_1)\Delta_f(s-q_1)s$ is also stable and proper such that $\|(s+a-q_1)\Delta_f(s-q_1)s\|_\infty \le k_f$.

From (9.106), we have

$$\left(I - \Theta_1^{*T}\frac{A(s)}{\Lambda(s)}\right) G_0^{-1}(s) = \left(\Theta_2^{*T}\frac{A(s)}{\Lambda(s)} + \Theta_{20}^*\right) + \Theta_3^*\xi_m(s). \tag{9.260}$$

With $\Theta(t)$, Θ^* and $\omega(t)$ defined in (9.207) and (9.208) and $\tilde{\Theta}(t) = \Theta(t) - \Theta^*$, $e(t) = y(t) - y_m(t)$, we also have the tracking error equation (9.209):

$$\begin{aligned} &\dot{e}(t) + \xi_1^{-1}\xi_0 e(t) \\ &= K_p\tilde{\Theta}^T\omega(t) + \bar{d}(t) + \mu\eta(t) - K_p(k_0 h(e(t)) + g_0 f(e(t))m(t)), \end{aligned} \tag{9.261}$$

where

$$\bar{d}(t) = K_p\left(\Theta_2^{*T}\frac{A(s)}{\Lambda(s)} + \Theta_{20}^* + \Theta_3^*\xi_m(s)\right), \tag{9.262}$$

$$\eta(t) = \Delta(s)[y](t),\ \Delta(s) = K_p\left(\Theta_2^{*T}\frac{A(s)}{\Lambda(s)} + \Theta_{20}^* + \Theta_3^*\xi_m(s)\right)\Delta_f(s). \tag{9.263}$$

It follows from Assumptions (A9.5-6b) and (A9.7) that $\bar{d}(t)$ is bounded and that, for $m(t)$ in (9.257), $\frac{\|\eta(t)\|}{m(t)} \le g_1$ for some constant $g_1 > 0$, $\forall t \ge 0$.[13]

Adaptive law. The adaptive law for updating $\Theta(t)$ is the same as (9.212):

$$\dot{\Theta}^T(t) = -(\xi_1^{-1}S)^{-T}e(t)\omega^T(t) - \sigma(t)\Theta^T(t), \tag{9.264}$$

where S and ξ_1 are defined in Assumptions (A9.3) and (A9.4a), that is, $SK_p = (SK_p)^T > 0$ and $\xi_1 = \xi_1^T > 0$, and $\sigma(t)$ is defined in (9.213) and (9.214).

Stability analysis. For the positive definite function $V(e, \tilde{\Theta})$ defined in (9.223), using (9.260) and (9.263), we obtain

$$\begin{aligned} \dot{V} = &-e^T(t)(\xi_0 + \xi_0^T)e(t) + 2e^T(t)\xi_1(\bar{d}(t) - K_p k_0 h(e(t))) \\ &+ 2e^T(t)\xi_1(\mu\eta(t) - g_0 K_p f(e(t))m(t)) - \sigma(t)\,\mathrm{tr}[\tilde{\Theta}(t)SK_p\Theta^T(t)]. \end{aligned} \tag{9.265}$$

[13]This is from the multivariable version of Lemma 2.6 (see (9.142)) and the expression

$$\begin{aligned} \eta(t) = \Delta(s)[y](t) &= K_p\left(\Theta_2^{*T}\frac{A(s)}{\Lambda(s)} + \Theta_{20}^* + \Theta_3^*\xi_m(s)\right)\Delta_f(s)[y](t) \\ &= K_p\left(\Theta_2^{*T}\frac{A(s)}{\Lambda(s)} + \Theta_{20}^* + \Theta_3^*\xi_m(s)\right)\frac{1}{s+a}(s+a)\Delta_f(s)(s+\delta_0)\frac{1}{s+\delta_0}[y](t), \end{aligned}$$

where, for $G_1(s) = K_p\left(\Theta_2^{*T}\frac{A(s)}{\Lambda(s)} + \Theta_{20}^* + \Theta_3^*\xi_m(s)\right)\frac{1}{s+a}$ and $G_2(s) = (s+a)\Delta_f(s)(s+\delta_0)$, both $G_1(s-\delta_0)$ and $G_2(s-\delta_0)$ have finite L^1 operator norm bounds independent of μ.

For the design in (9.258), it follows that

$$\begin{aligned}
&2e^T(t)\xi_1(\bar{d}(t) - K_p k_0 h(e(t))) \\
&= 2e^T(t)\xi_1\bar{d}(t) - 2k_0 e^T(t)\xi_1 K_p S^{-T}\xi_1 e(t) \\
&\le -2k_0\lambda_{\min}[S_p]\,\|e(t)\|_2^2 + 2\|e(t)\|_2\,\|\xi_1\bar{d}(t)\|_2 \\
&= -k_0\lambda_{\min}[S_p]\,\|e(t)\|_2^2 + \frac{\|\xi_1\bar{d}(t)\|_2^2}{k_0\lambda_{\min}[S_p]},
\end{aligned} \tag{9.266}$$

where $\lambda_{\min}[S_p] > 0$ is the minimum eigenvalue of

$$S_p = \xi_1 K_p S^{-T}\xi_1 = \xi_1 S^{-1} S K_p S^{-T}\xi_1^T = S_p^T > 0 \tag{9.267}$$

as $SK_p = (SK_p)^T > 0$ and $\xi_1 = \xi_1^T > 0$. Similarly, we have

$$\begin{aligned}
&2e^T(t)\xi_1(\mu\eta(t) - K_p g_0 f(e(t))m(t)) \\
&= 2e^T(t)\xi_1\mu\eta(t) - 2g_0 e^T(t)\xi_1 K_p S^{-T}\xi_1 e(t) m^2(t) \\
&\le -2g_0\lambda_{\min}[S_p]\,\|e(t)\|_2^2\, m^2(t) + 2\mu\|e(t)\|_2\,\|\xi_1\|_2\,\|\eta(t)\|_2 \\
&\le -2g_0\lambda_{\min}[S_p]\,\|e(t)\|_2^2\, m^2(t) + 2\mu g_1\|e(t)\|_2\,\|\xi_1\|_2\, m(t) \\
&= -g_0\lambda_{\min}[S_p]\,\|e(t)\|_2^2\, m^2(t) + \mu^2\frac{g_1^2\|\xi_1\|_2^2}{g_0\lambda_{\min}[S_p]},
\end{aligned} \tag{9.268}$$

where $\|\eta(t)\|_2 \le g_1 m(t)$ for some constant $g_1 > 0$.

From (9.265)–(9.268), we have

$$\begin{aligned}
\dot{V} \le\; & -e^T(t)(\xi_0 + \xi_0^T)e(t) - 2\sigma(t)\,\mathrm{tr}[\tilde{\Theta}(t) S K_p \Theta^T(t)] \\
& + \frac{\|\xi_1\bar{d}(t)\|_2^2}{k_0\lambda_{\min}[S_p]} + \mu^2\frac{g_1^2\|\xi_1\|_2^2}{g_0\lambda_{\min}[S_p]}.
\end{aligned} \tag{9.269}$$

From (9.228) and (9.269), we conclude that $\Theta(t) \in L^\infty$, and $e(t) = y(t) - y_m(t) \in L^\infty$ and satisfies

$$\int_{t_1}^{t_2} \|e(t)\|_2^2\, dt \le c_0 + \left(\frac{c_1}{k_0} + c_2\mu^2\right)(t_2 - t_1) \tag{9.270}$$

for some constants $c_i > 0$, $i = 0, 1, 2$, and any $\mu > 0$.

From (9.257), it follows that $m(t) \in L^\infty$. From (9.254), we have

$$u(t) = G_0^{-1}(s)[y - d](t) - \mu\Delta_f(s)[y](t), \tag{9.271}$$

and from (9.202), we express $\omega_1(t)$ as

$$\omega_1(t) = \frac{A(s)}{\Lambda(s)}G_0^{-1}(s)[y - d](t) - \mu\frac{A(s)}{\Lambda(s)}\Delta_f(s)[y](t), \tag{9.272}$$

which, in view of Assumption (A9.4a) and the boundedness of $y(t)$ and $d(t)$, implies that $\omega_1(t)$ is bounded. Finally, from (9.256), $u(t)$ is bounded.

When $d(t) = 0$ and $\mu\Delta_f(s) = 0$ in (9.254), from (9.269), we have that $\dot{V} = -e^T(t)(\xi_0 + \xi_0^T)e(t) \leq 0$, that is, $e(t) \in L_2$, which, together with the closed-loop signal boundedness, implies that $\lim_{t\to\infty} e(t) = 0$.

This adaptive control scheme with (9.256) for the plant (2.354) with feedback unmodeled dynamics *does not* need either the plant uncertainty level μ or the design parameter $g_0 > 0$ to be small, essentially different from the case of multiplicative unmodeled dynamics in the plant (9.197).

9.2.9 MRAC Designs for Input and Output Delays

In this section, we present model reference adaptive control designs for multivariable plants with input or output delays. A controller structure which employs a distributed function for stable plant–model matching in the presence of delays and an adaptive parameter update law which ensures stable parameter adaptation using a bounding normalization signal are developed. Closed-loop stability and asymptotic output tracking are achieved.

MRAC of Multivariable Plants with Output Delays

Consider the M-input and M-output linear time-invariant plant

$$y(t) = G(s)[u](t),\ G(s) = D(s)G_0(s), \tag{9.273}$$

where $G_0(s)$ is an $M \times M$ unknown and strictly proper rational transfer matrix satisfying Assumptions (A9.1)–(A9.4) stated in Section 9.2.2 (that is, $G_0(s)$ is of full rank and with all its zeros stable; an upper bound $\bar{\nu}$ on the observability index ν of $G_0(s)$ is known; for the high frequency gain matrix K_p of $G_0(s)$, SK_p is a symmetric and positive definite matrix for some known matrix S; and the modified left interactor matrix $\xi_m(s)$ of $G_0(s)$ is known) and $D(s)$ is an $M \times M$ diagonal matrix characterizing the plant output delays,

$$D(s) = \text{diag}\{e^{-\tau_1 s}, \ldots, e^{-\tau_M s}\}, \tag{9.274}$$

where $\tau_i \geq 0$, $i = 1, \ldots, M$, are delay times at outputs.

For a chosen reference model system

$$y_m(t) = W_m(s)[r](t), \tag{9.275}$$

where $W_m(s)$ is an $M \times M$ stable transfer matrix and $r(t) \in R^M$ is a piecewise continuous and bounded reference input signal, our objective is to find a feedback control signal $u(t)$ to ensure closed-loop stability and asymptotic output tracking: $\lim_{t\to\infty} y(t) - y_m(t) = 0$.

To solve this model reference adaptive control problem, similar to the SISO case [139], [300], the following assumptions are used:

(A9.8): τ_i, $i = 1, \ldots, M$, in $D(s)$ are known.

(A9.9): The elements of $G_0(s)$ do not have multiple poles.

We choose the reference model transfer matrix as

$$W_m(s) = D(s)\xi_m^{-1}(s). \tag{9.276}$$

Letting $\tau = \max_{1\le i\le M}\{\tau_i\}$ and $d(s) = e^{-\tau s}$, and introducing

$$y_d(t) \triangleq d(s)D^{-1}(s)[y](t) = d(s)G_0(s)[u](t), \tag{9.277}$$

which has a simple delay structure $d(s)I_M = e^{-\tau s}I_M$, we define

$$y_{md}(t) = d(s)D^{-1}(s)[y_m](t) = d(s)\xi_m^{-1}(s)[r](t). \tag{9.278}$$

We then design an adaptive controller based on the signals $u(t)$ and $y_d(t)$. If all signals in the closed-loop system are bounded and $y_d(t)$ tracks $y_{md}(t)$ asymptotically with time, then the control objective is achieved.

Controller structure. We choose the controller structure

$$\begin{aligned} u(t) &= \Theta_1^T\omega_{1d}(t) + \Theta_2^T\omega_{2d}(t) + \Theta_{20}y_d(t) + \Theta_3 r(t) \\ &\quad + \int_{-\tau}^{0} \Lambda(\sigma)u(t+\sigma)\,d\sigma, \end{aligned} \tag{9.279}$$

where $\Theta_1^T = [\Theta_{11}, \ldots, \Theta_{1\bar{\nu}-1}]$, $\Theta_2^T = [\Theta_{21}, \ldots, \Theta_{2\bar{\nu}-1}]$, with Θ_3, $\Theta_{ij} \in R^{M\times M}$,

$$\omega_{1d}(t) = \frac{A(s)}{\Lambda(s)}d(s)[u](t),\ \omega_{2d}(t) = \frac{A(s)}{\Lambda(s)}[y_d](t),\ d(s) = e^{-\tau s}, \tag{9.280}$$

with $A(s) = [I_M, sI_M, \ldots, s^{\bar{\nu}-2}I_M]^T$ and $\Lambda(s)$ being any monic stable polynomial of degree $\bar{\nu} - 1$, and $\Lambda(\sigma)$, $\sigma \in [-\tau, 0]$, is a distributed function to be specified. The new feature of this controller, as compared with all previous ones, is the term $\int_{-\tau}^{0} \Lambda(\sigma)u(t+\sigma)\,d\sigma$ in (9.279) for handling output delays.

The controller (9.279), when applied to the plant (9.273), has the following desired properties suitable for MRAC.

Lemma 9.6 *(i) Constant parameter matrices* Θ_i^*, $i = 1,2,3$, Θ_{20}^*, *and a bounded distributed function* $\Lambda^*(\sigma)$, $\sigma \in [-\tau, 0]$, *exist such that for* $\Theta_i = \Theta_i^*$, $i = 1,2,3$, $\Theta_{20} = \Theta_{20}^*$, *and* $\Lambda(\sigma) = \Lambda^*(\sigma)$, *the controller (9.279) makes the closed-loop system transfer function from* $r(t)$ *to* $y(t)$ *equal to* $W_m(s)$ *and all of the closed-loop system poles stable.*

(ii) For any estimates $\Theta(t)$ *and* $\Lambda(\sigma,t)$ *of* $\Theta^* = [\Theta_1^{*T}, \Theta_2^{*T}, \Theta_{20}^*, \Theta_3^*]^T$ *and* $\Lambda^*(\sigma)$, *respectively, with parameter errors* $\tilde{\Theta}(t) = \Theta(t) - \Theta^*$ *and* $\tilde{\Lambda}(\sigma,t) = \Lambda(\sigma,t) - \Lambda^*(\sigma)$, *and the regressor signal*

$$\omega_d(t) = [\omega_{1d}^T(t), \omega_{2d}^T(t), y_d^T(t), r^T(t)]^T, \tag{9.281}$$

the tracking error equation is

$$K_p^{-1} d^{-1}(s)\xi_m(s)[y_d(t) - y_{md}](t) = \tilde{\Theta}^T \omega_d(t) + \int_{-\tau}^{0} \tilde{\Lambda}(\sigma,t)u(t+\sigma)\, d\sigma. \tag{9.282}$$

Proof: (i) We first define the polynomial matrix $F_0(s)$ from

$$F_0(s)P_0^{-1}(s) \triangleq \sum_{i=1}^{n} \frac{A_i}{s-\lambda_i} = I - K_p^{-1}\xi_m(s)G_0(s) \tag{9.283}$$

for some constant $M \times M$ matrices A_i, $i = 1, \ldots, n$, where $n > 0$ is the degree of the common denominator polynomial of $G_0(s)$ (note that the knowledge of n is not needed in implementing the MRAC scheme), $F_0(s)$ is such that $F_0(s)P_0^{-1}(s)$ is strictly proper, for $G_0(s) = Z_0(s)P_0^{-1}(s)$ for some right coprime $M \times M$ polynomial matrices $Z_0(s)$ and $P_0(s)$ with $P_0(s)$ being column proper (see Section 9.2.1). The last equality in (9.283) is due to the definition of K_p and $\xi_m(s)$, that is, $\lim_{s\to\infty}(I - K_p^{-1}\xi_m(s)G_0(s)) = 0$. We should note that $\Lambda^*(\sigma)$ is real even if some of λ_i, $i = 1, \ldots, n$, may be complex (without loss of generality, we assume that λ_i, $i = 1, \ldots, n$, are all real).

We then choose the matching distributed function as

$$\Lambda^*(\sigma) = \sum_{i=1}^{n} A_i e^{-\lambda_i \sigma}. \tag{9.284}$$

From (9.283) and (9.284), we have

$$\begin{aligned}\mathcal{L}\left\{\int_{-\tau}^{0} \Lambda^*(\sigma)u(t+\sigma)\, d\sigma\right\} &= \left(\sum_{i=1}^{n} A_i \int_{-\tau}^{0} e^{-\lambda_i\sigma}e^{\sigma s}\, d\sigma\right) u(s) \\ &= \left(F_0(s)P_0^{-1}(s) - d(s)F_d(s)P_0^{-1}(s)\right) u(s),\end{aligned} \tag{9.285}$$

where $F_d(s)$ is an $M \times M$ real polynomial matrix satisfying

$$F_d(s)P_0^{-1}(s) = \sum_{i=1}^{n} \frac{A_i e^{\lambda_i \tau}}{s-\lambda_i}. \tag{9.286}$$

Hence, in the frequency s-domain, it follows from (9.279) and (9.283) that

$$u(s) = \Theta_1^T \frac{A(s)}{\Lambda(s)} e^{-\tau s} u(s) + \left(\Theta_2^T \frac{A(s)}{\Lambda(s)} + \Theta_{20} \right) y_d(s) + \Theta_3 r(s)$$
$$+ \left(F_0(s) P_0^{-1}(s) - d(s) F_d(s) P_0^{-1}(s) \right) u(s). \tag{9.287}$$

From (9.277) and (9.287), we have the closed-loop system as

$$y_d(s) = d(s) Z_0(s) \Lambda(s) ((P_0(s) - F_0(s)) \Lambda(s)$$

$$+ (F_d(s)\Lambda(s) - \Theta_1^T A(s) P_0(s) - (\Theta_2^T A^T(s) + \Theta_{20}\Lambda(s)) Z_0(s)) d(s))^{-1} \Theta_3 r(s). \tag{9.288}$$

Since $Z_0(s)$ and $P_0(s)$ are right coprime with $P_0(s)$ being column proper and $F_d(s)P_0^{-1}(s)$ is strictly proper, there exist constant Θ_1^*, Θ_2^*, Θ_{20}^* such that

$$\Theta_1^{*T} A(s) P_0(s) + (\Theta_2^{*T} A(s) + \Theta_{20}^* \Lambda(s)) Z_0(s) = F_d(s) \Lambda(s). \tag{9.289}$$

For $\Theta_1 = \Theta_1^*$, $\Theta_2 = \Theta_2^*$, $\Theta_{20} = \Theta_{20}^*$, it follows from (9.288) and (9.289) that

$$y_d(s) = d(s) Z_0(s) (P_0(s) - F_0(s))^{-1} \Theta_3 r(s). \tag{9.290}$$

Setting $\Theta_3^* = K_p^{-1}$ and using (9.283), from (9.290), we have

$$y_d(s) = d(s) \xi_m^{-1}(s) r(s). \tag{9.291}$$

Since $y_d(s) = d(s) D^{-1}(s) y(s)$, it follows from (9.291) that

$$y(s) = D(s) \xi_m^{-1}(s) r(s), \tag{9.292}$$

which is the desired plant–model matching. Since $Z_0(s)$ is stable, the cancellation of $Z_0(s)$ and $Z_0^{-1}(s)$ in (9.290) does not cause internal instability.

(ii) From (9.289), we obtain

$$\frac{\Theta_1^{*T} A(s) d(s)}{\Lambda(s)} + (\frac{\Theta_2^{*T} A(s)}{\Lambda(s)} + \Theta_{20}^*) d(s) Z_0(s) P_0^{-1}(s) = F_d(s) P_0^{-1}(s) d(s). \tag{9.293}$$

Operating both sides of (9.293) on $u(t)$, in the frequency s-domain, we have

$$\left(\frac{\Theta_1^{*T} A(s) d(s)}{\Lambda(s)} + \left(\frac{\Theta_2^{*T} A(s)}{\Lambda(s)} + \Theta_{20}^* \right) d(s) Z_0(s) P_0^{-1}(s) \right) u(s)$$
$$= F_d(s) P_0^{-1}(s) d(s) u(s). \tag{9.294}$$

Using (9.285) and (9.294), we have

$$\left(\frac{\Theta_1^{*T} A(s)d(s)}{\Lambda(s)} + \left(\frac{\Theta_2^{*T} A(s)}{\Lambda(s)} + \Theta_{20}^*\right) d(s) Z_0(s) P_0^{-1}(s)\right) u(s)$$
$$= F_0(s) P_0^{-1}(s) u(s) - \mathcal{L}\left\{\int_{-\tau}^{0} \Lambda^*(\sigma) u(t+\sigma)\, d\sigma\right\}. \quad (9.295)$$

Using (9.295), (9.283), and (9.277), we obtain

$$\left(\frac{\Theta_1^{*T} A(s)d(s)}{\Lambda(s)} + \left(\frac{\Theta_2^{*T} A(s)}{\Lambda(s)} + \Theta_{20}^*\right) d(s) Z_0(s) P_0^{-1}(s)\right) u(s)$$
$$+ \mathcal{L}\left\{\int_{-\tau}^{0} \Lambda^*(\sigma) u(t+\sigma)\, d\sigma\right\} + \Theta_3^* d^{-1}(s) \xi_m(s) y_d(s) = u(s). \quad (9.296)$$

Expressing (9.296) in the time domain, we have

$$u(t) = \Theta_1^{*T} \omega_{1d}(t) + \Theta_2^{*T} \omega_{2d}(t) + \Theta_{20}^* y_d(t) + \Theta_3^* d^{-1}(s) \xi_m(s)[y_d](t)$$
$$+ \int_{-\tau}^{0} \Lambda^*(\sigma) u(t+\sigma)\, d\sigma. \quad (9.297)$$

Hence, from (9.279) and (9.297), we have (9.282). ∇

Error model. To develop an error equation suitable for deriving an adaptive law for controller parameter adaptation, we rewrite (9.282) as

$$\xi_m(s) h(s)[y_d - y_{md}](t)$$
$$= \Psi^* d(s) h(s) \left[\tilde{\Theta}^T \omega_d + \int_{-\tau}^{0} \tilde{\Lambda}(\sigma, \cdot) u(\cdot + \sigma)\, d\sigma\right](t), \quad (9.298)$$

where $h(s) = \frac{1}{f(s)}$ with $f(s)$ being any monic stable polynomial of degree d_m, the maximum degree of $\xi_m(s)$, and $\Psi^* = K_p$.

Let $\Psi(t)$ be the estimate of Ψ^* and define the estimation error as

$$\epsilon(t) = \xi_m(s) h(s)[y_d - y_{md}](t) - \Psi(t) \Big(d(s) h(s)[\Theta^T \omega_d](t) - \Theta^T(t) \zeta_d(t)$$
$$+ h(s) d(s) \left[\int_{-\tau}^{0} \Lambda(\sigma, \cdot) u(\cdot + \sigma)\, d\sigma\right](t) - \int_{-\tau}^{0} \Lambda(\sigma, t) u_d(t+\sigma)\, d\sigma\Big), \quad (9.299)$$

where $\zeta_d(t) = h(s) d(s) I[\omega_d](t)$ and $u_d(t) = h(s) d(s)[u](t)$. Then, from (9.298) and (9.299), we have the estimation error equation

$$\epsilon(t) = \tilde{\Psi}(t) \xi_d(t) + \Psi^* \left(\tilde{\Theta}^T(t) \zeta_d(t) + \int_{-\tau}^{0} \tilde{\Lambda}(\sigma, t) u_d(t+\sigma)\, d\sigma\right), \quad (9.300)$$

where the parameter errors are $\tilde{\Psi}(t) = \Psi(t) - \Psi^*$, $\tilde{\Theta}(t) = \Theta(t) - \Theta^*$, $\tilde{\Lambda}(\sigma, t) = \Lambda(\sigma, t) - \Lambda^*(\sigma)$, $\sigma \in [-\tau, 0]$ and the auxiliary signal $\xi_d(t)$ is

$$\xi_d(t) = \Theta^T(t) \zeta_d(t) - h(s) d(s)[\Theta^T \omega_d](t)$$
$$+ \int_{-\tau}^{0} \Lambda(\sigma, t) u_d(t+\sigma)\, d\sigma - h(s) d(s) \left[\int_{-\tau}^{0} \Lambda(\sigma, \cdot) u(\cdot + \sigma)\, d\sigma\right](t). \quad (9.301)$$

Adaptive law. For adaptive control with $G_0(s)$ unknown, we choose the following gradient adaptive law for updating the controller parameters:

$$\dot{\Theta}^T(t) = \frac{-S^{-T}\epsilon(t)\zeta_d^T(t)}{m^2(t)}, \tag{9.302}$$

$$\dot{\Psi}(t) = \frac{-\Gamma\epsilon(t)\xi_d^T(t)}{m^2(t)}, \tag{9.303}$$

$$\frac{\partial}{\partial t}\Lambda(\sigma, t) = \frac{-S^{-T}\epsilon(t)u_d^T(t+\sigma)}{m^2(t)}, \ -\tau \le \sigma \le 0, \tag{9.304}$$

where S satisfies (A9.3), that is, $SK_p = (SK_p)^T > 0$, $\Gamma = \Gamma^T > 0$, and

$$m(t) = \sqrt{1 + \zeta_d^T(t)\zeta_d(t) + \xi_d^T(t)\xi_d(t) + (\sup_{-\tau\le\sigma\le 0} \|u_d(t+\sigma)\|)^2}. \tag{9.305}$$

The adaptive scheme has the following desired stability properties.

Lemma 9.7 *The adaptive law (9.302)–(9.304) guarantees that* $\Theta(t)$, $\Psi(t)$, $\Lambda(\sigma,t) \in L_\infty$ *and* $\frac{\epsilon(t)}{m(t)}$, $\dot{\Theta}(t)$, $\dot{\Psi}(t)$, $\frac{\partial}{\partial t}\Lambda(\sigma,t) \in L_2 \cap L_\infty$, $\sigma \in [-\tau, 0]$.

Proof: For the positive definite function

$$V(\tilde{\Theta}, \tilde{\Psi}, \tilde{\Lambda}) = \mathrm{tr}[\tilde{\Theta}(t)SK_p\tilde{\Theta}^T(t)] + \mathrm{tr}[\tilde{\Psi}^T(t)\Gamma^{-1}\tilde{\Psi}(t)] + \mathrm{tr}\left[\int_{-\tau}^{0} \tilde{\Lambda}^T(\sigma,t)SK_p\tilde{\Lambda}(\sigma,t)\,d\sigma\right], \tag{9.306}$$

from (9.300), (9.302)–(9.304), we obtain

$$\dot{V} = -\frac{2\epsilon^T(t)\epsilon(t)}{m^2(t)} \le 0. \tag{9.307}$$

Hence $V(\tilde{\Theta}(t), \tilde{\Psi}(t), \tilde{\Lambda}(t))$ is bounded, that is, $\Theta(t)$, $\Psi(t)$, and $\Lambda(\sigma,t)$ are bounded, for $\sigma \in [-\tau, 0]$. Since $\tilde{\Lambda}(\sigma,t)$ is bounded, $\sigma \in [-\tau, 0]$, we have

$$\left\|\int_{-\tau}^{0} \tilde{\Lambda}(\sigma,t)u_d(t+\sigma)\,d\sigma\right\| \le k_\lambda \sup_{-\tau\le\sigma\le 0} \|u_d(t+\sigma)\| \tag{9.308}$$

for some constant $k_\lambda > 0$. Therefore from (9.300), (9.302)–(9.304), (9.307), and (9.308), the boundedness of $\Theta(t)$ and $\Psi(t)$, it follows that $\frac{\epsilon(t)}{m(t)}$, $\dot{\Theta}(t)$, $\dot{\Psi}(t)$, $\frac{\partial}{\partial t}\Lambda(\sigma,t) \in L_2 \cap L_\infty$ for $\sigma \in [-\tau, 0]$. ∇

We now summarize the stability and tracking properties of the developed MRAC scheme, whose proof can be found in [371].

Theorem 9.6 *In the MRAC system with the plant (9.273) under Assumptions (A9.1)–(A9.4), (A9.8), and (A9.9), the reference model (9.275), and the controller (9.279) with the adaptive law (9.302)–(9.304), all signals are bounded and* $\lim_{t\to\infty} y(t) - y_m(t) = 0$.

MRAC of Multivariable Plants with Input Delays

Similar results can be obtained for MRAC of multivariable plants with input delays. To show this, let us consider the linear time-invariant plant

$$y(t) = G(s)[u](t), \; G(s) = G_0(s)D(s), \tag{9.309}$$

where $G_0(s)$ is an $M \times M$ strictly proper rational transfer matrix satisfying Assumptions (A9.1)–(A9.5) and (A9.9) and $D(s)$ is an $M \times M$ diagonal matrix characterizing the input delays in the plant, of the form

$$D(s) = \text{diag}\{e^{-\tau_1 s}, \ldots, e^{-\tau_m s}\}, \tag{9.310}$$

with $\tau_i \geq 0$, $i = 1, \ldots, M$, which are assumed to be known.

We choose the reference model as

$$y_m(t) = W_m(s)[r](t), \; W_m(s) = \xi_m^{-1}(s)d(s), \tag{9.311}$$

where $d(s) = e^{-\tau s}$ with $\tau = \max\{\tau_i, i = 1, \ldots, m\}$.

Defining $v(t) = d^{-1}(s)D(s)[u](t)$, from (9.309), we have

$$y(t) = G_0(s)d(s)[v](t), \; u(t) = D^{-1}(s)d(s)[v](t). \tag{9.312}$$

Since the transformed plant (9.312) has a simple delay structure $d(s)I_M = e^{-\tau s}I_M$, using the above MRAC results for the output delay case, we can design a model reference adaptive controller for the plant (9.312).

Controller structure. We choose the controller structure as

$$\begin{aligned} v(t) &= \Theta_1^T \omega_{1d}(t) + \Theta_2^T \omega_2(t) + \Theta_{20} y(t) + \Theta_3 r(t) + \int_{-\tau}^{0} \Lambda(\sigma) v(t+\sigma)\, d\sigma, \\ u(t) &= d(s)D^{-1}(s)[v](t), \end{aligned} \tag{9.313}$$

where $\Theta_1^T = [\Theta_{11}, \ldots, \Theta_{1\bar{\nu}-1}]$, $\Theta_2^T = [\Theta_{21}, \ldots, \Theta_{2\bar{\nu}-1}]$, with Θ_3, $\Theta_{ij} \in R^{M \times M}$,

$$\omega_{1d}(t) = \frac{A(s)}{\Lambda(s)} d(s)[v](t), \; \omega_2(t) = \frac{A(s)}{\Lambda(s)}[y](t) \tag{9.314}$$

with $A(s) = [I_M, sI_M, \ldots, s^{\bar{\nu}-2}I_M]^T$ and $\Lambda(s)$ being any monic stable polynomial of degree $\bar{\nu} - 1$, and $\Lambda(\sigma)$, $\sigma \in [-\tau, 0]$, is a distributed function.

For this controller structure, we can show the following desired properties in a similar way to that for the output delay case (see Lemma 9.6).

Lemma 9.8 *(i) Constant parameter matrices Θ_i^*, $i = 1,2,3$, Θ_{20}^*, and a bounded distributed function $\Lambda^*(\sigma)$, $\sigma \in [-\tau, 0]$, exist such that for $\Theta_i = \Theta_i^*$, $i = 1,2,3$, $\Theta_{20} = \Theta_{20}^*$, and $\Lambda(\sigma) = \Lambda^*(\sigma)$, the controller (9.313) makes the closed-loop system transfer function from $r(t)$ to $y(t)$ equal to $W_m(s)$ and all of the closed-loop system poles stable.*

(ii) For any estimates $\Theta(t)$ and $\Lambda(\sigma,t)$ of $\Theta^* = [\Theta_1^{*T}, \Theta_2^{*T}, \Theta_{20}^*, \Theta_3^*]^T$ and $\Lambda^*(\sigma)$, respectively, with parameter errors $\tilde{\Theta}(t) = \Theta(t) - \Theta^*$ and $\tilde{\Lambda}(\sigma,t) = \Lambda(\sigma,t) - \Lambda^*(\sigma)$, and the regressor signal

$$\omega_d(t) = [\omega_{1d}^T(t), \omega_2^T(t), y^T(t), r^T(t)]^T, \tag{9.315}$$

the tracking error equation is

$$K_p^{-1} d^{-1}(s)\xi_m(s)[y - y_m](t) = \tilde{\Theta}^T(t)\omega_d(t) + \int_{-\tau}^{0} \tilde{\Lambda}(\sigma,t)v(t+\sigma)\,d\sigma. \tag{9.316}$$

Error model and adaptive law. Let $\Psi(t)$ be the estimate of $\Psi^* = K_P$, and, for a stable filter $h(s) = \frac{1}{f(s)}$ with $f(s)$ being any monic stable polynomial of degree d_m, the maximum degree of $\xi_m(s)$, define the estimation error

$$\begin{aligned}\epsilon(t) = {} & \xi_m(s)h(s)[y - y_m](t) - \Psi(t)\Big(d(s)h(s)[\Theta^T\omega_d](t) - \Theta^T(t)\zeta_d(t) \\ & + h(s)d(s)\left[\int_{-\tau}^{0} \Lambda(\sigma,\cdot)v(\cdot+\sigma)\,d\sigma\right](t) - \int_{-\tau}^{0} \Lambda(\sigma,t)v_d(t+\sigma)\,d\sigma\Big),\end{aligned} \tag{9.317}$$

where $\zeta_d(t) = h(s)d(s)I[\omega_d](t)$ and $v_d(t) = h(s)d(s)[v](t)$.

From (9.316) and (9.317), we have

$$\epsilon(t) = \tilde{\Psi}(t)\xi_d(t) + \Psi^*\left(\tilde{\Theta}^T(t)\zeta_d(t) + \int_{-\tau}^{0} \tilde{\Lambda}(\sigma,t)v_d(t+\sigma)\,d\sigma\right), \tag{9.318}$$

where $\tilde{\Psi}(t) = \Psi(t) - \Psi^*$ and

$$\begin{aligned}\xi_d(t) = {} & \Theta^T(t)\zeta_d(t) - h(s)d(s)[\Theta^T\omega_d](t) \\ & + \int_{-\tau}^{0} \Lambda(\sigma,t)v_d(t+\sigma)\,d\sigma - h(s)d(s)\left[\int_{-\tau}^{0} \Lambda(\sigma,\cdot)v(\cdot+\sigma)\,d\sigma\right](t).\end{aligned} \tag{9.319}$$

Based on the error equation (9.318), we choose the adaptive laws

$$\dot{\Theta}^T(t) = \frac{-S^{-T}\epsilon(t)\zeta_d^T(t)}{m^2(t)}, \tag{9.320}$$

$$\dot{\Psi}(t) = \frac{-\Gamma\epsilon(t)\xi_d^T(t)}{m^2(t)}, \tag{9.321}$$

$$\frac{\partial}{\partial t}\Lambda(\sigma,t) = \frac{-S^{-T}\epsilon(t)v_d^T(t+\sigma)}{m^2(t)}, \quad -\tau \le \sigma \le 0, \tag{9.322}$$

where S satisfies (A9.3), that is, $SK_p = (SK_p)^T > 0$, $\Gamma = \Gamma^T > 0$, and

$$m(t) = \sqrt{1 + \zeta_d^T(t)\zeta_d(t) + \xi_d^T(t)\xi_d(t) + (\sup_{-\tau\leq\sigma\leq 0} \|v_d(t+\sigma)\|)^2}. \qquad (9.323)$$

The following stability and tracking properties of this adaptive law and the closed-loop system can be derived, similar to the output delay case.

Lemma 9.9 *The adaptive law (9.320)–(9.322) guarantees that* $\Theta(t)$, $\Psi(t)$, $\Lambda(\sigma, t) \in L_\infty$ *and* $\frac{\epsilon(t)}{m(t)}$, $\dot{\Theta}(t)$, $\dot{\Psi}(t)$, $\frac{\partial}{\partial t}\Lambda(\sigma, t) \in L_2 \cap L_\infty$, $\sigma \in [-\tau, 0]$.

Theorem 9.7 *In the MRAC system with the plant (9.309) under Assumptions (A9.1)–(A9.4), (A9.8), and (A9.9), the reference model (9.311), and the controller (9.313) with the adaptive law (9.320)–(9.322), all signals are bounded and* $\lim_{t\to\infty} y(t) - y_m(t) = 0$.

Remark 9.13 When there are both input and output delays in the plant

$$y(t) = D_o(s)G_0(s)D_i(s)[u](t), \qquad (9.324)$$

where $D_o(s)$, $D_i(s)$ are the output and input delay matrices of the form (9.274) or (9.310), we assume that $G_0(s)$ satisfies Assumptions (A9.1)–(A9.4) and (A9.9) and that $d_o(s)D_o^{-1}(s)$ and $D_i^{-1}(s)d_i(s)$ are known and physically realizable, with $d_o(s) = e^{-\tau^o s}$, $d_i(s) = e^{-\tau^i s}$, for some known $\tau^o > 0$, $\tau^i > 0$.

We then use the transformations $y_d(t) = d_o(s)D_o^{-1}(s)[y](t)$ and $u(t) = D_i^{-1}(s)d_i(s)[v](t)$ to transform the plant (9.324) to

$$y_d(t) = d(s)G_0(s)[v](t), \qquad (9.325)$$

where $d(s) = d_o(s)d_i(s) = e^{-\tau s}$ with $\tau = \tau^o + \tau^i > 0$, to design a stable MRAC for this new plant. The reference model can be chosen as $y_m(t) = W_m(s)[r](t)$, where $r(t)$ is the external reference input signal and $W_m(s) = D_o(s)\xi_m^{-1}(s)d_i(s)$ with $\xi_m(s)$ being the left modified interactor matrix of $G_0(s)$. □

Illustrative Example

We now present an $M = 2$ example to highlight some main points in developing the MRAC scheme for the output delay case by considering the plant

$$y(t) = D(s)G_0(s)[u](t),\ y(t) \in R^2,\ u(t) \in R^2, \qquad (9.326)$$

where

$$D(s) = \begin{bmatrix} e^{-\tau_1 s} & 0 \\ 0 & e^{-\tau_2 s} \end{bmatrix},\ G_0(s) = \begin{bmatrix} \frac{1}{s-1} & \frac{1}{s-4} \\ \frac{1}{s-2} & \frac{1}{s-6} \end{bmatrix}. \qquad (9.327)$$

For clarity of presentation, we take $0 < \tau_1 \leq \tau_2 = \ln(2) = 0.693$ so that $\tau = \max\{\tau_1, \tau_2\} = \tau_2$. For $d(s) = e^{-\tau s}$, we define the delayed output signal $y_d(t) = d(s)D^{-1}(s)[y](t) = [y_1(t - \tau_2 + \tau_1), y_2(t)]^T$.

The transfer matrix $G_0(s)$ is expressed as $G_0(s) = Z_0(s)P_0^{-1}(s)$, where

$$Z_0(s) = \begin{bmatrix} s-2 & s-6 \\ s-1 & s-4 \end{bmatrix}, P_0(s) = \begin{bmatrix} (s-1)(s-2) & 0 \\ 0 & (s-4)(s-6) \end{bmatrix}. \quad (9.328)$$

This $G_0(s)$ is of minimum phase as $\det[Z_0(s)] = s + 2$ and has observability index $\nu = 2$. The (nonunique) modified left interactor matrix $\xi_m(s)$ of $G_0(s)$ and associated high frequency gain matrix K_p are

$$\xi_m(s) = \begin{bmatrix} s+1 & 0 \\ -s^2 & (s+1)(s+2) \end{bmatrix}, K_p = \lim_{s\to\infty} \xi_m(s)G_0(s) = \begin{bmatrix} 1 & 1 \\ 4 & 5 \end{bmatrix}. \quad (9.329)$$

The reference model is (9.275) with (9.276), and the controller structure is (9.279) with $\bar{\nu} = \nu = 2$ and $\Lambda(s) = s + 3$, so that

$$\omega_{1d}(t) = \frac{d(s)}{\Lambda(s)}[u](t) = \frac{1}{\Lambda(s)}[u](t-\tau),\ \omega_{2d}(t) = \frac{1}{\Lambda(s)}[y_d](t) \quad (9.330)$$

and $\Theta_i(t)$, $\Theta_{20}(t) \in R^{2\times 2}$, $i = 1, 2, 3$, $\Lambda(\sigma, t) \in R^{2\times 2}$, $\sigma \in [-\tau, 0]$.

Next we derive the plant–model matching controller parameters Θ_i^*, Θ_{20}^*, $i = 1, 2, 3$, and $\Lambda^*(\sigma)$, assuming that the parameters of $G_0(s)$ are known.

As in (9.283) and (9.284), expressing

$$I_2 - K_p^{-1}\xi_m(s)G_0(s) = \begin{bmatrix} s+10 & 15s+22 \\ -3s-6 & -20s+8 \end{bmatrix} P_0^{-1}(s) = \begin{bmatrix} -11 & 0 \\ 9 & 0 \end{bmatrix}\frac{1}{s-1}$$

$$+\begin{bmatrix} 12 & 0 \\ -12 & 0 \end{bmatrix}\frac{1}{s-2} + \begin{bmatrix} 0 & -44 \\ 0 & 36 \end{bmatrix}\frac{1}{s-4} + \begin{bmatrix} 0 & 56 \\ 0 & -56 \end{bmatrix}\frac{1}{s-6}, \quad (9.331)$$

we obtain

$$F_0(s) = \begin{bmatrix} s+10 & 15s+22 \\ -3s-6 & -20s+8 \end{bmatrix}, \quad (9.332)$$

$$\Lambda^*(\sigma) = \begin{bmatrix} -11 & 0 \\ 9 & 0 \end{bmatrix} e^{-\sigma} + \begin{bmatrix} 12 & 0 \\ -12 & 0 \end{bmatrix} e^{-2\sigma}$$

$$+ \begin{bmatrix} 0 & -44 \\ 0 & 36 \end{bmatrix} e^{-4\sigma} + \begin{bmatrix} 0 & 56 \\ 0 & -56 \end{bmatrix} e^{-6\sigma}. \quad (9.333)$$

From (9.286) with $e^\tau = e^{ln(2)} = 2$, we have that

$$F_d(s)P_0^{-1}(s) = \begin{bmatrix} -11 & 0 \\ 9 & 0 \end{bmatrix}\frac{e^\tau}{s-1} + \begin{bmatrix} 12 & 0 \\ -12 & 0 \end{bmatrix}\frac{e^{2\tau}}{s-2} + \begin{bmatrix} 0 & -44 \\ 0 & 36 \end{bmatrix}\frac{e^{4\tau}}{s-4}$$

$$+\begin{bmatrix} 0 & 56 \\ 0 & -56 \end{bmatrix} \frac{e^{6\tau}}{s-6} = \begin{bmatrix} 26s-4 & 2880s-10112 \\ -30s+12 & -3008s+11200 \end{bmatrix} P_0^{-1}(s). \tag{9.334}$$

Hence we find the polynomial matrix $F_d(s)$ as

$$F_d(s) = \begin{bmatrix} 26s-4 & 2880s-10112 \\ -30s+12 & -3008s+11200 \end{bmatrix}. \tag{9.335}$$

Finally we solve (9.289) with $A(s) = I$ for Θ_1^*, Θ_2^*, Θ_{20}^* to obtain, with $\Theta_3^* = K_p^{-1}$, the unique plant–model matching parameters

$$\Theta^* = [\Theta_1^{*T}, \Theta_2^{*T}, \Theta_{20}^*, \Theta_3^*]^T = \frac{1}{53}\left[\begin{matrix} -336118 & -184856 \\ 361682 & 203848 \end{matrix}\right.$$
$$\left.\begin{matrix} -1241776 & -513480 & -437648 & -100152 & 265 & -53 \\ 1327376 & 596760 & -480080 & 116808 & -212 & 53 \end{matrix}\right]. \tag{9.336}$$

For adaptive control, we use the adaptive law (9.302)–(9.304) to update the estimates $\Theta(t)$ and $\Lambda(\sigma, t)$ of Θ^* and $\Lambda^*(\sigma)$ on-line.

9.2.10 Adaptation and High Frequency Gain Matrix

To adaptively control a multivariable plant $y(t) = G_0(D)[u](t)$ with an $M \times M$ plant transfer matrix $G_0(D)$, the knowledge of the high frequency gain matrix $K_p = \lim_{D\to\infty} \xi_m(D)G_0(D)$ is crucial for constructing a stable adaptive controller parameter update law, where $\xi_m(D)$ is the modified left interactor matrix of $G_0(D)$ (see Lemma 9.1 in Section 9.2.2).

Different cases from the previous subsections are summarized here.

State Feedback Designs

As in Section 9.1.1, we had the error model (9.10), that is,

$$\dot{e}(t) = A_m e(t) + B_m \left(K_2^{*-1}\tilde{K}_1^T(t)x(t) + K_2^{*-1}\tilde{K}_2(t)r(t)\right), \tag{9.337}$$

where K_2^{*-1} is equivalent to K_p, and the adaptive laws

$$\dot{\tilde{K}}_1^T(t) = -S^T B_m^T P e(t) x^T(t), \tag{9.338}$$

$$\dot{\tilde{K}}_2(t) = -S^T B_m^T P e(t) r^T(t), \tag{9.339}$$

where $P = P^T > 0$ satisfying $PA_m + A_m^T P = -Q$, for $Q = Q^T > 0$, and $M_s = K_2^* S = (K_2^* S)^T = S^T K_2^{*T} > 0$. The time derivative of

$$V(e_c) = e^T P e + \text{tr}[\tilde{K}_1 M_s^{-1} \tilde{K}_1^T] + \text{tr}[\tilde{K}_2^T M_s^{-1} \tilde{K}_2] \tag{9.340}$$

is $\dot{V} = -e^T(t)Qe(t) \leq 0$, from which we have closed-loop signal boundedness and asymptotic tracking $\lim_{t\to\infty} e(t) = 0$.

Design based on LDU decomposition of K_2^{*-1}. If all leading principal minors Δ_i of the matrix K_2^{*-1} are nonzero, we have the following LDU decomposition (see (2.37) in Proposition 2.4):

$$K_2^{*-1} = LD^*U \tag{9.341}$$

for some $M \times M$ unit lower triangular matrix L and $M \times M$ unit upper triangular matrix U, and $M \times M$ diagonal matrix

$$D^* = \mathrm{diag}\{d_1^*, d_2^*, \ldots, d_M^*\} = \mathrm{diag}\left\{\Delta_1, \frac{\Delta_2}{\Delta_1}, \ldots, \frac{\Delta_M}{\Delta_{M-1}}\right\}. \tag{9.342}$$

As in Section 9.1.2, using (9.341), we can reparametrize (9.337) as

$$\dot{e}(t) = A_m e(t) + B_m L D^* \begin{bmatrix} \tilde{\theta}_1^T(t)\omega_1(t) \\ \vdots \\ \tilde{\theta}_M^T(t)\omega_M(t) \end{bmatrix} \tag{9.343}$$

for some *vector* parameter errors $\tilde{\theta}_i(t) = \theta_i(t) - \theta_i^*$, $i = 1, 2, \ldots, M$.

Then the adaptive laws

$$\dot{\theta}_i(t) = -\mathrm{sign}[d_i^*]\Gamma_i \bar{e}_i(t)\omega_i(t),\ t \geq 0, \tag{9.344}$$

where $\bar{e}_i(t)$ is the ith component of $e^T(t)PB_mL$ and $\Gamma_i = \Gamma_i^T > 0$, $i = 1, 2, \ldots, M$, make the time derivative of

$$V = e^T P e + \sum_{i=1}^{M} |d_i^*| \tilde{\theta}_i^T \Gamma_i^{-1} \tilde{\theta}_i \tag{9.345}$$

equal to $\dot{V} = -e^T(t)Qe(t)$, the desired result.

Output Feedback Designs

With an output feedback MRAC design, from (9.139) (with $\mu = 0$, $\beta = 1$, and $\alpha = 0$), we have the estimation error equation

$$\epsilon(t) = \frac{K_p \tilde{\Theta}^T(t)\zeta(t) + \tilde{\Psi}(t)\xi(t)}{1 + \zeta^T(t)\zeta(t) + \xi^T(t)\xi(t)}. \tag{9.346}$$

For the continuous-time case, the adaptive laws are

$$\dot{\Theta}^T(t) = -S_p \epsilon(t)\zeta^T(t), \tag{9.347}$$

$$\dot{\Psi}(t) = -\Gamma \epsilon(t)\xi^T(t) \tag{9.348}$$

for $K_pS_p = (K_pS_p)^T > 0$ and $\Gamma = \Gamma^T > 0$. The time derivative of

$$V = \mathrm{tr}[\tilde{\Theta}(t)\Gamma_p\tilde{\Theta}^T(t)] + \mathrm{tr}[\tilde{\Psi}^T(t)\Gamma^{-1}\tilde{\Psi}(t)],\ t \in [0, \infty), \tag{9.349}$$

with $\Gamma_p = K_p^T S_p^{-1} = (S_p^{-1})^T(K_pS_p)^T S_p^{-1} = \Gamma_p^T > 0$, is as desired: $\dot{V} = -2\epsilon^T(t)\epsilon(t) - 2\epsilon^T(t)\epsilon(t)(\zeta^T(t)\zeta(t) + \xi^T(t)\xi(t)) \leq 0$.

For the discrete-time case, the adaptive laws are

$$\Theta^T(t+1) = \Theta^T(t) - S_p\epsilon(t)\zeta^T(t), \tag{9.350}$$

$$\Psi(t+1) = \Psi(t) - \Gamma\epsilon(t)\xi^T(t) \tag{9.351}$$

for $2I > K_pS_p = (K_pS_p)^T > 0$ and $2I > \Gamma = \Gamma^T > 0$. The time increment of

$$V = \mathrm{tr}[\tilde{\Theta}(t)\Gamma_p\tilde{\Theta}^T(t)] + \mathrm{tr}[\tilde{\Psi}^T(t)\Gamma^{-1}\tilde{\Psi}(t)],\ t \in \{0, 1, 2, \ldots\} \tag{9.352}$$

is $V(\tilde{\Theta}(t+1), \tilde{\Psi}(t+1)) - V(\tilde{\Theta}(t), \tilde{\Psi}(t)) = -2\epsilon^T(t)\epsilon(t) - \zeta^T(t)\zeta(t)\epsilon^T(t)(2I - S_p^T K_p^T)\epsilon(t) - \xi^T(t)\xi(t)\epsilon^T(t)(2I - \Gamma)\epsilon(t)$, as desired too.

Discussion. In the SISO case when $M = 1$ (see Chapters 4, 5, and 6), the high frequency gain matrix K_p becomes a scalar whose sign information is sufficient for designing a stable adaptive law. In the MIMO case, *a priori* knowledge of the high frequency gain matrix K_p for a stable adaptive law is more difficult to determine. The condition that $K_pS_p = (K_pS_p)^T > 0$ for some known matrix S_p can be relaxed when some structural information of K_p is available (see the adaptive state feedback state tracking control scheme in Section 9.1.2 using the LDU decomposition of K_p, which leads to the adaptive system (9.243)–(9.345)). On the other hand, for a simpler error model, for example, $\epsilon(t) = K_p\tilde{\Theta}^T(t)\zeta(t)$ [120], [124], [282], the condition on K_p may be relaxed to $\Gamma K_p + K_p^T\Gamma^T > 0$ for some known matrix Γ or $\Gamma K_p + K_p^T\Gamma^T > 0$ for some matrix $\Gamma = \Gamma^T > 0$ in the continuous-time case (see Problem 9.34).

In the next subsection, we present three output feedback MRAC schemes based on the LDU (SDU, LDS) decompositions of K_p with relaxed conditions.

9.2.11 Designs Based on Decompositions of K_p

Consider the linear time-invariant plant with M inputs and M outputs:

$$y(t) = G_0(D)[u](t), \tag{9.353}$$

where $y(t) = [y_1(t), \ldots, y_M(t)]^T \in R^M$, $u(t) = [u_1(t), \ldots, u_M(t)]^T \in R^M$, $G_0(D) = Z_0(D)P_0^{-1}(D)$, and $Z_0(D)$, $P_0(D)$ are $M \times M$ right coprime polynomial matrices with $P_0(D)$ being column proper (see Section 9.2.1).

The control objective is to find a feedback control $u(t)$ for the plant (9.353) with unknown $G_0(D)$ such that $y(t)$ tracks a reference output signal $y_m(t)$ asymptotically and the closed-loop plant is globally stable in the sense that all signals in the system are bounded for any bounded initial conditions, where

$$y_m(t) = W_m(D)[r](t),\ W_m(D) = \xi_m^{-1}(D), \tag{9.354}$$

with $\xi_m(D)$ being the modified interactor matrix of $G_0(D)$ (see Section 9.2.1) and $r(t)$ being a reference input signal.

The modified interactor matrix $\xi_m(D)$, which has a stable inverse, characterizes the zero structure at infinity of $G_0(D)$. The high frequency gain matrix K_p of $G_0(D)$ is defined as $K_p = \lim_{D\to\infty} \xi_m(D)G_0(D)$, being finite and nonsingular. In order to design MRAC schemes to achieve the control objective, we assume that the plant transfer matrix $G_0(D)$ is strictly proper, and of full rank and has a known interactor matrix $\xi_m(D)$, an upper bound $\bar{\nu}$ on the observability index ν of $G_0(D)$ is known, all zeros of $G_0(D)$ are stable, and

(A9.3a): All leading principal minors of the high frequency gain matrix K_p are nonzero and their signs are known.

Relaxing plant assumptions for multivariable MRAC has been an important topic of research (see [120], [124], [125], [349], [432] for relaxing the assumption of a known interactor matrix $\xi_m(D)$ and [136], [137], [141], [270], [299], [432] for relaxing the assumption of a high frequency gain matrix K_p under certain positive definiteness condition, for continuous-time multivariable MRAC designs, based on LDU, SDU, and LDS decompositions of K_p).

Next, we present three such MRAC designs (with relaxed conditions on K_p, as compared with that in Assumption (A9.4)) in a unified framework for both the continuous-time and discrete-time cases.

High Frequency Gain Matrix Decompositions

Given that all principal minors of the high frequency gain matrix K_p are nonzero, several decompositions of K_p exist and can be employed for MRAC of the plant (9.353). From Proposition 2.4, Corollary 2.1, and Corollary 2.2, the matrix K_p has the following LDU, SDU, and LDS decompositions.

LDU decomposition of K_p. Let Δ_i, $i = 1, 2, \ldots, M$, denote the leading principal minors of the high frequency gain matrix $K_p \in R^{M\times M}$ and assume that $\Delta_i \neq 0$, $i = 1, 2, \ldots, M$. Then, K_p has a unique decomposition

$$K_p = LD^*U \tag{9.355}$$

for some $M \times M$ unit (i.e., with all diagonal elements being 1) lower triangular matrix L and unit upper triangular matrix U and

$$D^* = \mathrm{diag}\{d_1^*, d_2^*, \ldots, d_M^*\} = \mathrm{diag}\left\{\Delta_1, \frac{\Delta_2}{\Delta_1}, \ldots, \frac{\Delta_M}{\Delta_{M-1}}\right\}. \tag{9.356}$$

In view of (9.356) and (A9.3a), $\mathrm{sign}[d_i^*]$ is known, $i = 1, 2, \ldots, M$, and will be used in constructing adaptive parameter update laws. Such a decomposition of K_p will be used to derive a new error model with which the parameters in L, D^*, and U can also be estimated. In addition, for the discrete-time case, similar to Assumption (A6.3), we also assume

> (A9.3aD): Some upper bounds d_i^0 of $|d_i^*|$, that is, $|d_i^*| \leq d_i^0$, $i = 1, 2, \ldots, M$, are known.

SDU decomposition of K_p. The gain matrix $K_p \in R^{M \times M}$ with all its principal minors $\Delta_i \neq 0$, $i = 1, 2, \ldots, M$, has a nonunique decomposition

$$K_p = SD_sU_s, \tag{9.357}$$

where $S \in R^{M \times M}$ is a symmetric and positive definite matrix, U is an $M \times M$ unit upper triangular matrix, and

$$\begin{aligned} D_s &= \mathrm{diag}\{s_1^*, s_2^*, \ldots, s_M^*\} \\ &= \mathrm{diag}\left\{\mathrm{sign}\left[\Delta_1\right]\gamma_1, \mathrm{sign}\left[\frac{\Delta_2}{\Delta_1}\right]\gamma_2, \ldots, \mathrm{sign}\left[\frac{\Delta_M}{\Delta_{M-1}}\right]\gamma_M\right\} \end{aligned} \tag{9.358}$$

such that $\gamma_i > 0$, $i = 1, \ldots, M$, may be arbitrary.

This SDU decomposition of K_p follows from the LDU decomposition of K_p, with $S = LD^*D_s^{-1}L^T$ and $U_s = D_s^{-1}L^{-T}D_sU$, for D_s with a set of chosen γ_i, $i = 1, 2, \ldots, M$ (that is, the choice of D_s is not unique).

LDS decomposition of K_p. The gain matrix $K_p \in R^{M \times M}$ with all its principal minors being nonzero has a nonunique decomposition

$$K_p = L_sD_sS, \tag{9.359}$$

where $S \in R^{M\times M}$ is a symmetric and positive definite matrix, L_s is an $M \times M$ unit lower triangular matrix, and

$$\begin{aligned} D_s &= \mathrm{diag}\{s_1^*, s_2^*, \ldots, s_M^*\} \\ &= \mathrm{diag}\left\{\mathrm{sign}\,[\Delta_1]\,\gamma_1, \mathrm{sign}\left[\frac{\Delta_2}{\Delta_1}\right]\gamma_2, \ldots, \mathrm{sign}\left[\frac{\Delta_M}{\Delta_{M-1}}\right]\gamma_M\right\} \end{aligned} \tag{9.360}$$

such that $\gamma_i > 0$, $i = 1, \ldots, M$, may be arbitrary.

This LDS decomposition of K_p follows from the LDU decomposition of K_p, with $L_s = LD_sU^{-T}D_s^{-1}$ and $S = U^TD_s^{-1}D^*U$, for D_s with a set of chosen γ_i, $i = 1, 2, \ldots, M$ (i.e., the choice of D_s is also not unique).

If some principal minors of K_p are zero but K_p is nonsingular, additional information is needed for such decompositions. Such information is characterized by a permutation matrix P such that PK_p has an LDU, SDU, or LDS decomposition. The knowledge of this permutation matrix can be incorporated in the design of a MRAC scheme based on a decomposition of PK_p.

Design Based on LDS Decomposition of K_p

The controller structure based on the LDS decomposition of K_p is the standard model reference controller structure used in Section 9.2.2 (see (9.102)):

$$u(t) = \Theta_1^T\omega_1(t) + \Theta_2^T\omega_2(t) + \Theta_{20}y(t) + \Theta_3 r(t), \tag{9.361}$$

where $\Theta_1 = [\Theta_{11}, \ldots, \Theta_{1\bar{\nu}-1}]^T$, $\Theta_2 = [\Theta_{21}, \ldots, \Theta_{2\bar{\nu}-1}]^T$, Θ_{20}, Θ_3, $\Theta_{ij} \in R^{M\times M}$, $i = 1, 2$, $j = 1, \ldots, \bar{\nu} - 1$, and

$$\omega_1(t) = F(D)[u](t),\ \omega_2(t) = F(D)[y](t) \tag{9.362}$$

with $F(D) = \frac{A(D)}{\Lambda(D)}$ for $A(D) = [I, DI, \ldots, D^{\bar{\nu}-2}I]^T$ and $\Lambda(D)$ being any monic stable polynomial of degree $\bar{\nu} - 1$.

With this specification of $\Lambda(D)$, $\xi_m(D)$, $P_0(D)$, $Z_0(D)$, there exist Θ_1^*, Θ_2^*, $\Theta_3^* = K_p^{-1}$ (see Remark 9.5 in Section 9.2.2) such that

$$\begin{aligned} &\Theta_1^{*T}A(D)P_0(D) + (\Theta_2^{*T}A(D) + \Lambda(D)\Theta_{20}^*)Z_0(D) \\ &= \Lambda(D)(P_0(D) - \Theta_3^*\xi_m(D)Z_0(D)). \end{aligned} \tag{9.363}$$

Hence we have the plant–model transfer matrix matching equation

$$I - \Theta_1^{*T}F(D) - \Theta_2^{*T}F(D)G_0(D) - \Theta_{20}^*G_0(D) = \Theta_3^*W_m^{-1}(D)G_0(D) \tag{9.364}$$

from which, for any $u(t)$, we have

$$u(t) - \Theta_1^{*T}\omega_1(t) - \Theta_2^{*T}\omega_2(t) - \Theta_{20}^{*}y(t) = \Theta_3^{*}W_m^{-1}(D)[y](t). \tag{9.365}$$

Using this result and the reference model expression (9.354), we obtain

$$\begin{aligned} & K_p\left(u(t) - \Theta_1^{*T}\omega_1(t) - \Theta_2^{*T}\omega_2(t) - \Theta_{20}^{*}y(t) - \Theta_3^{*}r(t)\right) \\ &= \xi_m(D)[y - y_m](t). \end{aligned} \tag{9.366}$$

Error model. With the *LDS decomposition* (9.359), we express (9.366) as

$$\begin{aligned} & D_s S\left(u(t) - \Theta_1^{*T}\omega_1(t) - \Theta_2^{*T}\omega_2(t) - \Theta_{20}^{*}y(t) - \Theta_3^{*}r(t)\right) \\ &= L_s^{-1}\xi_m(D)[y - y_m](t). \end{aligned} \tag{9.367}$$

If the controller (9.361) is implemented with $\Theta_i = \Theta_i^*$, from (9.367), we have $\xi_m(D)[y - y_m](t) = 0$, that is, $\lim_{t\to\infty}(y(t) - y_m(t)) = 0$ exponentially. Since Θ_i^*, $i = 1, 2, 20, 3$, are unknown, we use their estimates from an adaptive law. To develop such an adaptive law, we first derive an error model.

From (9.361) and (9.367), we have

$$\xi_m(D)[y - y_m](t) + \Theta_0^{*}\xi_m(D)[y - y_m](t) = D_s S\tilde{\Theta}^T(t)\omega(t), \tag{9.368}$$

where D_s is in (9.360), $S = S^T > 0$ in (9.359), $\tilde{\Theta}(t) = \Theta(t) - \Theta^*$ with $\Theta(t)$ being the estimate of $\Theta^* = [\Theta_1^{*T}, \Theta_2^{*T}, \Theta_{20}^{*}, \Theta_3^{*}]^T$, $\omega(t) = [\omega_1^T(t), \omega_2^T(t), y^T(t), r^T(t)]^T$, and, with L_s in (9.359), $\Theta_0^* = L_s^{-1} - I$ has a special form:

$$\Theta_0^* = \begin{bmatrix} 0 & 0 & 0 & \cdots & 0 \\ \theta_{21}^* & 0 & 0 & \cdots & 0 \\ \theta_{31}^* & \theta_{32}^* & 0 & \cdots & 0 \\ \vdots & \vdots & \vdots & \vdots & \vdots \\ \theta_{M-1\,1}^* & \cdots & \theta_{M-1\,M-2}^* & 0 & 0 \\ \theta_{M\,1}^* & \cdots & \theta_{M\,M-2}^* & \theta_{M\,M-1}^* & 0 \end{bmatrix}. \tag{9.369}$$

Given this structure of Θ_0^*, we define the parameter vectors

$$\begin{aligned} \theta_2^* &= \theta_{21}^* \in R, \\ \theta_3^* &= [\theta_{31}^*, \theta_{32}^*]^T \in R^2, \\ &\vdots \\ \theta_{M-1}^* &= [\theta_{M-1\,1}^*, \ldots, \theta_{M-1\,M-2}^*]^T \in R^{M-2}, \\ \theta_M^* &= [\theta_{M\,1}^*, \ldots, \theta_{M\,M-1}^*]^T \in R^{M-1} \end{aligned} \tag{9.370}$$

and let their estimates be

$$\begin{aligned}
\theta_2(t) &= \theta_{21}(t) \in R, \\
\theta_3(t) &= [\theta_{31}(t), \theta_{32}(t)]^T \in R^2, \\
&\vdots \\
\theta_{M-1}(t) &= [\theta_{M-1\,1}(t), \ldots, \theta_{M-1\,M-2}(t)]^T \in R^{M-2}, \\
\theta_M(t) &= [\theta_{M\,1}(t), \ldots, \theta_{M\,M-1}(t)]^T \in R^{M-1}.
\end{aligned} \tag{9.371}$$

Choose $f(D)$ as a stable and monic polynomial whose degree is equal to the maximum degree of the modified interactor matrix $\xi_m(D)$, introduce the filter $h(D) = \frac{1}{f(D)}$, define the filtered tracking error

$$\bar{e}(t) = \xi_m(D)h(D)[e](t) = [\bar{e}_1(t), \ldots, \bar{e}_M(t)]^T \tag{9.372}$$

with $e(t) = y(t) - y_m(t)$, and denote

$$\eta_i(t) = [\bar{e}_1(t), \ldots, \bar{e}_{i-1}(t)]^T \in R^{i-1}, \; i = 2, 3, \ldots, M. \tag{9.373}$$

Operating both sides of (9.368) by $h(D)I_M$ leads to

$$\bar{e}(t) + [0, \theta_2^{*T}\eta_2(t), \theta_3^{*T}\eta_3(t), \ldots, \theta_M^{*T}\eta_M(t)]^T = D_s S h(D)[\tilde{\Theta}^T\omega](t). \tag{9.374}$$

Based on this error equation, we introduce the estimation error

$$\epsilon(t) = \bar{e}(t) + [0, \theta_2^T(t)\eta_2(t), \theta_3^T(t)\eta_3(t), \ldots, \theta_M^T(t)\eta_M(t)]^T + \Psi(t)\xi(t), \tag{9.375}$$

where $\Psi(t)$ is the estimate of $\Psi^* = D_s S$, and

$$\xi(t) = \Theta^T(t)\zeta(t) - h(D)[\Theta^T\omega](t), \tag{9.376}$$

$$\zeta(t) = h(D)[\omega](t). \tag{9.377}$$

It then follows from (9.374)–(9.377) that

$$\begin{aligned}
\epsilon(t) = {} & [0, \tilde{\theta}_2^T(t)\eta_2(t), \tilde{\theta}_3^T(t)\eta_3(t), \ldots, \tilde{\theta}_M^T(t)\eta_M(t)]^T \\
& + D_s S\tilde{\Theta}(t)^T\zeta(t) + \tilde{\Psi}(t)\xi(t),
\end{aligned} \tag{9.378}$$

where $\tilde{\theta}_i(t) = \theta_i(t) - \theta_i^*$, $i = 2, 3, \ldots, M$, and $\tilde{\Psi}(t) = \Psi(t) - \Psi^*$.

Adaptive law. With the error model (9.378), we choose the adaptive laws

$$\left.\begin{matrix} \dot{\theta}_i(t) \\ \theta_i(t+1) - \theta_i(t) \end{matrix}\right\} = -\frac{\Gamma_{\theta i}\epsilon_i(t)\eta_i(t)}{m^2(t)}, \; i = 2, 3, \ldots, M, \tag{9.379}$$

$$\left.\begin{matrix} \dot{\Theta}^T(t) \\ \Theta^T(t+1) - \Theta^T(t) \end{matrix}\right\} = -\frac{D_s\epsilon(t)\zeta^T(t)}{m^2(t)}, \tag{9.380}$$

$$\left.\begin{matrix} \dot{\Psi}(t) \\ \Psi(t+1) - \Psi(t) \end{matrix}\right\} = -\frac{\Gamma\epsilon(t)\xi^T(t)}{m^2(t)}, \tag{9.381}$$

where $\epsilon(t) = [\epsilon_1(t), \epsilon_2(t), \ldots, \epsilon_M(t)]^T$,

$$m(t) = \sqrt{1 + \zeta^T(t)\zeta(t) + \xi^T(t)\xi(t) + \sum_{i=2}^{M} \eta_i^T(t)\eta_i(t)}, \tag{9.382}$$

and, for the *continuous-time* case, $\Gamma_{\theta i} = \Gamma_{\theta i}^T > 0$, $i = 2, 3, \ldots, M$, $\Gamma = \Gamma^T > 0$, or, for the *discrete-time* case, $0 < \Gamma_{\theta i} = \Gamma_{\theta i}^T < 2I_{i-1}$ ($\Gamma_{\theta i} \in R^{(i-1)\times(i-1)}$), $i = 2, 3, \ldots, M$, $0 < \Gamma = \Gamma^T < 2I_M$, and D_s is defined in (9.360) with $\gamma_i > 0$, $i = 1, 2, \ldots, M$, chosen to satisfy the condition $0 < D_s U^T D_s^{-1} D^* U D_s < 2I_M$.

Consider the positive definite function

$$V = \sum_{i=2}^{M} \tilde{\theta}_i^T \Gamma_{\theta i}^{-1} \tilde{\theta}_i + \mathrm{tr}[\tilde{\Psi}^T \Gamma^{-1} \tilde{\Psi}] + \mathrm{tr}[\tilde{\Theta} S \tilde{\Theta}^T]. \tag{9.383}$$

For the *continuous-time* case, we obtain the time derivative of V as

$$\begin{aligned}\dot{V} &= -\frac{2}{m^2(t)} \left(\sum_{i=2}^{M} \tilde{\theta}^T(t)\epsilon_i(t)\eta_i(t) + \mathrm{tr}[\tilde{\Psi}^T \epsilon(t)\xi^T(t)] + \mathrm{tr}[\tilde{\Theta}(t) S D_s \epsilon(t)\zeta^T(t)] \right) \\ &= -\frac{2\epsilon^T(t)\epsilon(t)}{m^2(t)}.\end{aligned} \tag{9.384}$$

In the *discrete-time* case, for

$$M_\eta(t) = \mathrm{diag}\{0, \eta_1^T(t)\Gamma_{\theta 1}\eta_1(t), \ldots, \eta_M^T(t)\Gamma_{\theta M}\eta_M(t)\}, \tag{9.385}$$

the time increment of V is

$$\begin{aligned}&V(\tilde{\theta}_i(t+1), \tilde{\Psi}(t+1), \tilde{\Theta}(t+1)) - V(\tilde{\theta}_i(t), \tilde{\Psi}(t), \tilde{\Theta}(t)) \\ &= -\sum_{i=2}^{M} \left(\frac{2\epsilon_i(t)\tilde{\theta}_i^T(t)\eta_i(t)}{m^2(t)} - \frac{\epsilon_i^2(t)\eta_i^T(t)\Gamma_{\theta i}\eta_i(t)}{m^4(t)} \right) \\ &\quad - \left(\frac{2\epsilon^T(t)\tilde{\Psi}(t)\xi(t)}{m^2(t)} - \frac{\epsilon^T(t)\Gamma\epsilon(t)\xi^T(t)\xi(t)}{m^4(t)} \right) \\ &\quad - \left(\frac{2\epsilon^T(t)D_s S\tilde{\Theta}^T(t)\zeta(t)}{m^2(t)} - \frac{\epsilon^T(t)D_s S D_s \epsilon(t)\zeta^T(t)\zeta(t)}{m^4(t)} \right) \\ &= -\epsilon^T(t) \left(2I_M - \frac{M_\eta(t) + \Gamma\xi^T(t)\xi(t) + D_s S D_s \zeta^T(t)\zeta(t)}{m^2(t)} \right) \frac{\epsilon(t)}{m^2(t)} \\ &\le -\alpha_1 \frac{\epsilon^T(t)\epsilon(t)}{m^2(t)}\end{aligned} \tag{9.386}$$

for some constant $\alpha_1 > 0$. The last inequality in (9.386) is due to the conditions that $0 < \Gamma_{\theta i} = \Gamma_{\theta i}^T < 2I_{i-1}$, $i = 2, 3, \ldots, M$, $0 < \Gamma = \Gamma^T < 2I_M$, and $0 < D_s S D_s = D_s U^T D_s^{-1} D^* U D_s < 2I_M$, and the choice of $m(t)$ in (9.382).

Design Based on LDU Decomposition of K_p

The controller structure based on an LDU decomposition of K_p is derived from (9.361) based on an LDS decomposition of K_p.

Making use of the *LDU decomposition* (9.355), we express (9.366) as

$$D^*U\left(u(t) - \Theta_1^{*T}\omega_1(t) - \Theta_2^{*T}\omega_2(t) - \Theta_{20}^*y(t) - \Theta_3^*r(t)\right) \\ = L^{-1}\xi_m(D)[y - y_m](t). \tag{9.387}$$

This equation leads to the new parametrization

$$D^*\left(u(t) - \Phi_0^*u(t) - \Phi_1^{*T}\omega_1(t) - \Phi_2^{*T}\omega_2(t) - \Phi_{20}^*y(t) - \Phi_3^*r(t)\right) \\ = L^{-1}\xi_m(D)[y - y_m](t), \tag{9.388}$$

where

$$\Phi_0^* = I - U,\ \Phi_{20}^* = U\Theta_{20}^*,\ \Phi_i = U\Theta_i,\ i = 1, 2, 3. \tag{9.389}$$

This new parametrization motivates the new controller structure

$$u(t) = \Phi_0 u(t) + \Phi_1^T\omega_1(t) + \Phi_2^T\omega_2(t) + \Phi_{20}y(t) + \Phi_3 r(t), \tag{9.390}$$

where Φ_0, Φ_1, Φ_2, Φ_{20}, and Φ_3 are the estimates of Φ_0^*, Φ_1^*, Φ_2^*, Φ_{20}^*, and Φ_3^* respectively; in particular, the parameter matrix Φ_0 has the same special triangular form as that of $\Phi_0^* = I - U$, that is,

$$\Phi_0 = \begin{bmatrix} 0 & \phi_{12} & \phi_{13} & \cdots & \phi_{1M} \\ 0 & 0 & \phi_{23} & \cdots & \phi_{2M} \\ \vdots & \vdots & \vdots & \vdots & \vdots \\ 0 & 0 & \cdots & 0 & \phi_{M-1\,M} \\ 0 & \cdots & \cdots & 0 & 0 \end{bmatrix}. \tag{9.391}$$

This special form of Φ_0 ensures that the control signal $u(t)$ is implementable from (9.390) without singularity, that is, with

$$v(t) = \Phi_1^T\omega_1(t) + \Phi_2^T\omega_2(t) + \Phi_{20}y(t) + \Phi_3 r(t) \triangleq [v_1(t), \ldots, v_M(t)]^T, \tag{9.392}$$

we can calculate $u(t) = [u_1(t), \ldots, u_M(t)]^T$ from

$$\begin{aligned} u_M(t) &= v_M(t), \\ u_{M-1}(t) &= \phi_{M-1\,M}u_M(t) + v_{M-1}(t), \\ u_{M-2}(t) &= \phi_{M-2\,M-1}u_{M-1}(t) + \phi_{M-2\,M}u_M(t) + v_{M-2}(t), \\ &\vdots \\ u_2(t) &= \phi_{23}u_3(t) + \phi_{24}u_4(t) + \cdots + \phi_{2\,M}u_M(t) + v_2(t), \\ u_1(t) &= \phi_{12}u_2(t) + \phi_{13}u_3(t) + \phi_{14}u_4(t) + \cdots + \phi_{1\,M}u_M(t) + v_1(t). \end{aligned} \tag{9.393}$$

Adaptive law. To develop an adaptive law for updating the parameters of the controller (9.390), we need to express the closed-loop system in terms of the tracking error $e(t) = y(t) - y_m(t)$ and the parameter error $\tilde{\Phi}(t) = \Phi(t) - \Phi^*$, where $\Phi(t) = [\Phi_0(t), \Phi_1^T(t), \Phi_2^T(t), \Phi_{20}(t), \Phi_3(t)]^T$ is the estimate of the unknown parameter matrix $\Phi^* = [\Phi_0^*, \Phi_1^{*T}, \Phi_2^{*T}, \Phi_{20}^*, \Phi_3^*]^T$ at time t.

From (9.388) and (9.390), we have

$$\xi_m(D)[y - y_m](t) + \Theta_0^* \xi_m(D)[y - y_m](t) = D^* \tilde{\Phi}^T(t)\bar{\omega}(t), \tag{9.394}$$

where $\Theta_0^* = L^{-1} - I$ has the same triangular structure as that in (9.369) with L in (9.355), D^* is defined in (9.356), and $\bar{\omega}(t) = [u^T(t), \omega^T(t)]^T$ with $\omega(t) = [\omega_1^T(t), \omega_2^T(t), y^T(t), r^T(t)]^T$.

Then we express (9.390) as $u(t) = \Phi^T(t)\bar{\omega}(t)$. Using (9.391), we write

$$\tilde{\Phi}^T(t)\bar{\omega}(t) = \begin{bmatrix} (\phi_1(t) - \phi_1^*)^T \chi_1(t) \\ (\phi_2(t) - \phi_2^*)^T \chi_2(t) \\ \vdots \\ (\phi_{M-1}(t) - \phi_{M-1}^*)^T \chi_{M-1}(t) \\ (\phi_M(t) - \phi_M^*)^T \chi_M(t) \end{bmatrix}, \tag{9.395}$$

where

$$\begin{aligned} \chi_1(t) &= [u_2(t), \ldots, u_M(t), \omega^T(t)]^T, \\ \chi_2(t) &= [u_3(t), \ldots, u_M(t), \omega^T(t)]^T, \\ &\vdots \\ \chi_{M-1}(t) &= [u_M(t), \omega^T(t)]^T, \\ \chi_M(t) &= \omega(t) \end{aligned} \tag{9.396}$$

and $(\phi_i(t) - \phi_i^*)^T$, $i = 1, 2 \ldots, M$, are the parts of the (row) parameter error vectors of $\tilde{\Phi}^T(t)$, corresponding to $\chi_i(t)$, that is,

$$\begin{aligned} \phi_1(t) &= [\phi_{12}, \phi_{13}, \ldots, \phi_{1M}, [\Phi_1^T(t), \Phi_2^T(t), \Phi_{20}(t), \Phi_3(t)]_1]^T, \\ \phi_2(t) &= [\phi_{23}, \phi_{24}, \ldots, \phi_{2M}, [\Phi_1^T(t), \Phi_2^T(t), \Phi_{20}(t), \Phi_3(t)]_2]^T, \\ &\vdots \\ \phi_{M-1}(t) &= [\phi_{M-1,M}, [\Phi_1^T(t), \Phi_2^T(t), \Phi_{20}(t), \Phi_3(t)]_{M-1}]^T, \\ \phi_M(t) &= [\Phi_1^T(t), \Phi_2^T(t), \Phi_{20}(t), \Phi_3(t)]_M^T, \end{aligned} \tag{9.397}$$

where $[\Phi_1^T(t), \Phi_2^T(t), \Phi_{20}(t), \Phi_3(t)]_i$ is the ith row of $[\Phi_1^T(t), \Phi_2^T(t), \Phi_{20}(t), \Phi_3(t)]$.

For Θ_0^* in the form (9.369), we also define the parameter vectors

$$\begin{aligned}
\theta_2^* &= \theta_{21}^* \in R, \\
\theta_3^* &= [\theta_{31}^*, \theta_{32}^*]^T \in R^2, \\
&\vdots \\
\theta_{M-1}^* &= [\theta_{M-11}^*, \ldots, \theta_{M-1\,M-2}^*]^T \in R^{M-2}, \\
\theta_M^* &= [\theta_{M1}^*, \ldots, \theta_{M\,M-1}^*]^T \in R^{M-1}
\end{aligned} \tag{9.398}$$

and let their estimates be

$$\begin{aligned}
\theta_2(t) &= \theta_{21}(t) \in R, \\
\theta_3(t) &= [\theta_{31}(t), \theta_{32}(t)]^T \in R^2, \\
&\vdots \\
\theta_{M-1}(t) &= [\theta_{M-11}(t), \ldots, \theta_{M-1\,M-2}(t)]^T \in R^{M-2}, \\
\theta_M(t) &= [\theta_{M1}(t), \ldots, \theta_{M\,M-1}(t)]^T \in R^{M-1}.
\end{aligned} \tag{9.399}$$

Consider the modified interactor matrix $\xi_m(D)$ given in (9.62). For $i = 1, 2, \ldots, M$, choose $f_i(D)$ (with $f_1(D) = d_1(D)$) as a stable polynomial whose degree is equal to the maximum of the degrees of the polynomials $d_j(D)$ and $h_{kl}^m(D)$, $j = 1, 2, \ldots, i$, $k = 2, \ldots, i$, $l = 1, \ldots, k-1$, and contains $d_i(D)$ as a factor, introduce the filters $h_i(D) = \frac{1}{f_i(D)}$, $i = 1, 2, \ldots, M$, and $H(D) = \text{diag}\{h_1(D), \ldots, h_M(D)\}$, and define

$$\bar{e}(t) = H(D)\xi_m(D)[y - y_m](t) = [\bar{e}_1(t), \ldots, \bar{e}_M(t)]^T \tag{9.400}$$

for $y(t) = [y_1(t), \ldots, y_M(t)]^T$ and $y_m(t) = [y_{m1}(t), \ldots, y_{mM}(t)]^T$ (by definition, it turns out that $\bar{e}_1(t) = y_1(t) - y_{m1}(t)$).[14]

For a discrete-time design, the polynomials $f_i(D)$ can be simply chosen as $f_i(D) = D^{n_i}$ for a specified degree n_i, $i = 1, 2, \ldots, M$.

Defining the auxiliary error signals

$$e_\xi(t) = \xi_m(D)[y - y_m](t) = [e_{\xi 1}(t), \ldots, e_{\xi M}(t)]^T, \tag{9.401}$$

$$e_{\theta i}(t) = [e_{\xi 1}(t), \ldots, e_{\xi i-1}(t)]^T \in R^{i-1},\ i = 2, 3, \ldots, M, \tag{9.402}$$

$$\eta_i(t) = h_i(D)[e_{\theta i}](t) \in R^{i-1},\ i = 2, 3, \ldots, M, \tag{9.403}$$

[14]A simpler (and possibly higher order) choice of the filter $H(D)$ is with $f_i(D) = f(D)$, $i = 1, 2, \ldots, M$, where $f(D)$ as a stable and monic polynomial of degree equal to the degree of the modified interactor matrix $\xi_m(D)$, as in the case of using LDS composition of K_p.

operating both sides of (9.394) by $H(D)$, and using (9.397), we have

$$\begin{aligned}
\bar{e}_1(t) &= d_1^* h_1(D)[(\phi_1 - \phi_1^*)^T \chi_1](t), \\
\bar{e}_2(t) + \theta_2^* \eta_2(t) &= d_2^* h_2(D)[(\phi_2 - \phi_2^*)^T \chi_2](t), \\
&\vdots \\
\bar{e}_{M-1}(t) + \theta_{M-1}^{*T} \eta_{M-1}(t) &= d_{M-1}^* h_{M-1}(D)[(\phi_{M-1} - \phi_{M-1}^*)^T \chi_{M-1}](t), \\
\bar{e}_M(t) + \theta_M^{*T} \eta_M(t) &= d_M^* h_M(D)[(\phi_M - \phi_M^*)^T \chi_M](t). \qquad (9.404)
\end{aligned}$$

Introduce the auxiliary signals

$$\zeta_i(t) = h_i(D)[\chi_i](t), \qquad (9.405)$$

$$\xi_i(t) = \phi_i^T(t)\zeta_i(t) - h_i(D)[\phi_i^T \chi_i](t), \qquad (9.406)$$

let $d_i(t)$ be the estimate of d_i^*, $i = 1, 2, \ldots, M$, and define the estimation errors

$$\begin{aligned}
\epsilon_1(t) &= \bar{e}_1(t) + d_1(t)\xi_1(t), \\
\epsilon_2(t) &= \bar{e}_2(t) + \theta_2(t)\eta_2(t) + d_2(t)\xi_2(t), \\
&\vdots \\
\epsilon_{M-1}(t) &= \bar{e}_{M-1}(t) + \theta_{M-1}^T(t)\eta_{M-1}(t) + d_{M-1}(t)\xi_{M-1}(t), \\
\epsilon_M(t) &= \bar{e}_M(t) + \theta_M^T(t)\eta_M(t) + d_M(t)\xi_M(t). \qquad (9.407)
\end{aligned}$$

Then, from (9.404)–(9.407), we have the error model

$$\begin{aligned}
\epsilon_1(t) &= d_1^*(\phi_1(t) - \phi_1^*)^T \zeta_1(t) + (d_1(t) - d_1^*)\xi_1(t), \\
\epsilon_2(t) &= (\theta_2(t) - \theta_2^*)\eta_2(t) + d_2^*(\phi_2(t) - \phi_2^*)^T \zeta_2(t) + (d_2(t) - d_2^*)\xi_2(t), \\
&\vdots \\
\epsilon_{M-1}(t) &= (\theta_{M-1}(t) - \theta_{M-1}^*)^T \eta_{M-1}(t) + d_{M-1}^*(\phi_{M-1}(t) - \phi_{M-1}^*)^T \zeta_{M-1}(t) \\
&\quad + (d_{M-1}(t) - d_{M-1}^*)\xi_{M-1}(t), \\
\epsilon_M(t) &= (\theta_M(t) - \theta_M^*)^T \eta_M(t) + d_M^*(\phi_M(t) - \phi_M^*)^T \zeta_M(t) \\
&\quad + (d_M(t) - d_M^*)\xi_M(t). \qquad (9.408)
\end{aligned}$$

Based on this error model, we choose the adaptive laws

$$\left.\begin{array}{c} \dot{\theta}_i(t) \\ \theta_i(t+1) - \theta_i(t) \end{array}\right\} = -\frac{\Gamma_{\theta i}\epsilon_i(t)\eta_i(t)}{m^2(t)}, \ i = 2, 3, \ldots, M, \qquad (9.409)$$

$$\left.\begin{array}{c} \dot{\phi}_i(t) \\ \phi_i(t+1) - \phi_i(t) \end{array}\right\} = -\frac{\mathrm{sign}[d_i^*]\Gamma_{\phi i}\epsilon_i(t)\zeta_i(t)}{m^2(t)}, \ i = 1, 2, \ldots, M, \qquad (9.410)$$

$$\left.\begin{array}{c} \dot{d}_i(t) \\ d_i(t+1) - d_i(t) \end{array}\right\} = -\frac{\gamma_i\epsilon_i(t)\xi_i(t)}{m^2(t)}, \ i = 1, 2, \ldots, M, \qquad (9.411)$$

where

$$m(t) = \sqrt{1 + \sum_{i=1}^{M} \zeta_i^T(t)\zeta_i(t) + \sum_{i=1}^{M} \xi_i^2(t) + \sum_{i=2}^{M} \eta_i^T(t)\eta_i(t)}, \tag{9.412}$$

and, for the *continuous-time* case, $\Gamma_{\theta i} = \Gamma_{\theta i}^T > 0$, $i = 2, 3, \ldots, M$, $\Gamma_{\phi i} = \Gamma_{\phi i}^T > 0$, $i = 1, 2, \ldots, M$, and $\gamma_i > 0$, $i = 1, 2, \ldots, M$, or for the *discrete-time* case, $0 < \Gamma_{\theta i} = \Gamma_{\theta i}^T < 2I$, $i = 2, 3, \ldots, M$, $0 < \Gamma_{\phi i} = \Gamma_{\phi i}^T < \frac{2}{d_i^0} I$, $i = 1, 2, \ldots, M$, and $0 < \gamma_i < 2$, $i = 1, 2, \ldots, M$, with $d_i^0 > 0$ as in Assumption (A9.3aD).

Introduce the parameter errors

$$\begin{aligned} \tilde{\theta}_i(t) &= \theta_i(t) - \theta_i^*, \; i = 2, 3, \ldots, M, \\ \tilde{\phi}_i(t) &= \phi_i(t) - \phi_i^*, \; i = 1, 2, \ldots, M, \\ \tilde{d}_i(t) &= d_i(t) - d_i^*, \; i = 1, 2, \ldots, M, \end{aligned} \tag{9.413}$$

and consider the positive definite function

$$V(\tilde{\theta}_i, \tilde{\phi}_i, \tilde{d}_i) = \sum_{i=2}^{M} \tilde{\theta}_i^T \Gamma_{\theta i}^{-1} \tilde{\theta}_i + \sum_{i=1}^{M} |d_i^*| \tilde{\phi}_i^T \Gamma_{\phi i}^{-1} \tilde{\phi}_i + \sum_{i=1}^{M} \gamma_i^{-1} \tilde{d}_i^2. \tag{9.414}$$

For the *continuous-time* case, the time derivative of V is

$$\begin{aligned} \dot{V} &= -\sum_{i=2}^{M} \frac{2\tilde{\theta}_i^T \eta_i(t)\epsilon_i(t)}{m^2(t)} - \sum_{i=1}^{M} \frac{2d_i^* \tilde{\phi}_i^T \zeta_i(t)\epsilon_i(t)}{m^2(t)} - \sum_{i=1}^{M} \frac{2\tilde{d}_i(t)\xi_i(t)\epsilon_i(t)}{m^2(t)} \\ &= -\sum_{i=1}^{M} \frac{2\epsilon_i^2(t)}{m^2(t)}. \end{aligned} \tag{9.415}$$

For the *discrete-time* case, the time increment of V is

$$\begin{aligned} & V(\tilde{\theta}_i(t+1), \tilde{\phi}_i(t+1), \tilde{d}_i(t+1)) - V(\tilde{\theta}_i(t), \tilde{\phi}_i(t), \tilde{d}_i(t)) \\ &= -\left(2 - \frac{|d_1^*|\zeta_1^T(t)\Gamma_{\phi 1}\zeta_1(t) + \gamma_1 \xi_1^2(t)}{m^2(t)}\right) \frac{\epsilon_1^2(t)}{m^2(t)} \\ &\quad + \sum_{i=2}^{M} \left(2 - \frac{\eta_i^T(t)\Gamma_{\theta i}\eta_i(t) + |d_i^*|\zeta_i^T(t)\Gamma_{\phi i}\zeta_i(t) + \gamma_i \xi_i^2(t)}{m^2(t)}\right) \frac{\epsilon_i^2(t)}{m^2(t)} \\ &\leq -\sum_{i=1}^{M} \frac{\alpha_i \epsilon_i^2(t)}{m^2(t)} \end{aligned} \tag{9.416}$$

for some $\alpha_i > 0$, $i = 1, 2, \ldots, M$, as from the conditions that $0 < \Gamma_{\theta i} = \Gamma_{\theta i}^T < 2I$, $i = 2, 3, \ldots, M$, $0 < \Gamma_{\phi i} = \Gamma_{\phi i}^T < \frac{2}{d_i^0} I$ with $d_i^0 \geq |d_i^*|$, $i = 1, 2, \ldots, M$, and $0 < \gamma_i < 2$, $i = 1, 2, \ldots, M$, and the choice of $m(t)$ in (9.412).[15]

[15]We can also introduce the normalization signals

$$m_1(t) = \sqrt{1 + \zeta_1^T(t)\zeta_1(t) + \xi_1^2(t)},$$

Design Based on SDU Decomposition of K_p

Using the *SDU decomposition* (9.357), similar to (9.367), from (9.366), we have

$$\begin{aligned} &SD_s\Big(u(t) - \Phi_0^* u(t) - \Phi_1^{*T}\omega_1(t) - \Phi_2^{*T}\omega_2(t) - \Phi_{20}^* y(t) - \Phi_3^* r(t)\Big) \\ &= \xi_m(D)[y - y_m](t), \end{aligned} \tag{9.417}$$

which also motivates us to use the controller structure (9.390), that is,

$$u(t) = \Phi_0 u(t) + \Phi_1^T \omega_1(t) + \Phi_2^T \omega_2(t) + \Phi_{20} y(t) + \Phi_3 r(t). \tag{9.418}$$

With $\bar{\omega}(t) = [u^T(t), \omega_1^T(t), \omega_2^T(t), y^T(t), r^T(t)]^T$, $e(t) = y(t) - y_m(t)$, and $\tilde{\Phi}(t) = \Phi(t) - \Phi^*$, where $\Phi(t) = [\Phi_0(t), \Phi_1^T(t), \Phi_2^T(t), \Phi_{20}(t), \Phi_3(t)]^T$ is the estimate of $\Phi^* = [\Phi_0^*, \Phi_1^{*T}, \Phi_2^{*T}, \Phi_{20}^*, \Phi_3^*]^T$, from (9.417) and (9.418), we have

$$\xi_m(D)[e](t) = SD_s \tilde{\Phi}^T(t)\bar{\omega}(t), \tag{9.419}$$

with $S = S^T > 0$ in (9.357) and D_s in (9.358) such that $S = LD^* D_s^{-1} L^T$.

With the special structure (9.391) of Φ_0 and Φ_0^*, we can implement the control signal $u(t)$ from (9.392) and (9.393), that is, for

$$v(t) = \Phi_1^T \omega_1(t) + \Phi_2^T \omega_2(t) + \Phi_{20} y(t) + \Phi_3 r(t) \triangleq [v_1(t), \ldots, v_M(t)]^T, \tag{9.420}$$

we can calculate $u(t) = [u_1(t), \ldots, u_M(t)]^T$ from

$$\begin{aligned} u_M(t) &= v_M(t), \\ u_{M-1}(t) &= \phi_{M-1\,M} u_M(t) + v_{M-1}(t), \\ u_{M-2}(t) &= \phi_{M-2\,M-1} u_{M-1}(t) + \phi_{M-2\,M} u_M(t) + v_{M-2}(t), \\ &\ \,\vdots \\ u_2(t) &= \phi_{23} u_3(t) + \phi_{24} u_4(t) + \cdots + \phi_{2\,M} u_M(t) + v_2(t), \\ u_1(t) &= \phi_{12} u_2(t) + \phi_{13} u_3(t) + \phi_{14} u_4(t) + \cdots + \phi_{1\,M} u_M(t) + v_1(t). \end{aligned} \tag{9.421}$$

For this consideration, similar to (9.395)–(9.397), we express

$$\tilde{\Phi}^T(t)\bar{\omega}(t) = \begin{bmatrix} (\phi_1(t) - \phi_1^*)^T \chi_1(t) \\ (\phi_2(t) - \phi_2^*)^T \chi_2(t) \\ \vdots \\ (\phi_{M-1}(t) - \phi_{M-1}^*)^T \chi_{M-1}(t) \\ (\phi_M(t) - \phi_M^*)^T \chi_M(t) \end{bmatrix}, \tag{9.422}$$

$$m_i(t) = \sqrt{1 + \eta_i^T(t)\eta_i(t) + \zeta_i^T(t)\zeta_i(t) + \xi_i^2(t)},\ i = 2, 3, \ldots, M,$$

and use $m_i(t)$ to replace $m(t)$ in (9.409)–(9.411), (9.415), and (9.416).

where

$$\begin{aligned} \chi_1(t) &= [u_2(t), \ldots, u_M(t), \omega^T(t)]^T, \\ \chi_2(t) &= [u_3(t), \ldots, u_M(t), \omega^T(t)]^T, \\ &\vdots \\ \chi_{M-1}(t) &= [u_M(t), \omega^T(t)]^T, \\ \chi_M(t) &= \omega(t) \end{aligned} \tag{9.423}$$

and $(\phi_i(t) - \phi_i^*)^T$, $i = 1, 2 \ldots, M$, are the corresponding parts of the (row) parameter error vectors from $\tilde{\Phi}^T(t)$ such that

$$\begin{aligned} \phi_1(t) &= [\phi_{12}, \phi_{13}, \ldots, \phi_{1M}, [\Phi_1^T(t), \Phi_2^T(t), \Phi_{20}(t), \Phi_3(t)]_1]^T, \\ \phi_2(t) &= [\phi_{23}, \phi_{24}, \ldots, \phi_{2M}, [\Phi_1^T(t), \Phi_2^T(t), \Phi_{20}(t), \Phi_3(t)]_2]^T, \\ &\vdots \\ \phi_{M-1}(t) &= [\phi_{M-1,M}, [\Phi_1^T(t), \Phi_2^T(t), \Phi_{20}(t), \Phi_3(t)]_{M-1}]^T, \\ \phi_M(t) &= [\Phi_1^T(t), \Phi_2^T(t), \Phi_{20}(t), \Phi_3(t)]_M^T, \end{aligned} \tag{9.424}$$

where $[\Phi_1^T(t), \Phi_2^T(t), \Phi_{20}(t), \Phi_3(t)]_i$ is the ith row of $[\Phi_1^T(t), \Phi_2^T(t), \Phi_{20}(t), \Phi_3(t)]$.

Choose $f(D)$ as a stable and monic polynomial of degree equal to the maximum degree of the modified interactor matrix $\xi_m(D)$, introduce the filter $h(D) = \frac{1}{f(D)}$, and for $e(t) = y(t) - y_m(t)$, define the filtered tracking error

$$\bar{e}(t) = \xi_m(D)h(D)[e](t) = [\bar{e}_1(t), \ldots, \bar{e}_M(t)]^T. \tag{9.425}$$

Operating both sides of (9.419) by $h(D)$ and using (9.422), we have the exactly parametrized (in terms of (9.424)) error equation

$$\bar{e}(t) = SD_s h(D)[\tilde{\phi}_1^T \chi_1(t), \tilde{\phi}_2^T \chi_2(t), \ldots, \tilde{\phi}_M^T \chi_M(t)]^T. \tag{9.426}$$

In view of this error equation, we introduce the estimation error

$$\epsilon(t) = \bar{e}(t) + \Psi(t)[\xi_1(t), \xi_2(t), \ldots, \xi_M(t)]^T, \tag{9.427}$$

where $\Psi(t)$ is the estimate of $\Psi^* = SD_s$, and

$$\xi_i(t) = \phi_i^T(t)\zeta_i(t) - h(D)[\phi_i^T \chi_i](t), \ i = 1, 2, \ldots, M, \tag{9.428}$$

$$\zeta_i(t) = h(D)[\chi_i](t), \ i = 1, 2, \ldots, M. \tag{9.429}$$

It then follows from (9.426)–(9.429) that

$$\epsilon(t) = SD_s[\tilde{\phi}_1^T \zeta_1(t), \tilde{\phi}_2^T \zeta_2(t), \ldots, \tilde{\phi}_M^T \zeta_M(t)]^T + \tilde{\Psi}(t)\xi(t), \tag{9.430}$$

where $\tilde{\Psi}(t) = \Psi(t) - \Psi^*$ and $\xi(t) = [\xi_1(t), \xi_2(t), \dots, \xi_M(t)]^T$.

Based on the error model (9.430), we choose the adaptive laws

$$\left.\begin{array}{r}\dot{\phi}_i(t) \\ \phi_i(t+1) - \phi_i(t)\end{array}\right\} = -\frac{\operatorname{sign}[d_i^*]\gamma_i \epsilon_i(t)\zeta_i(t)}{m^2(t)}, \quad i = 1, 2, \dots, M, \tag{9.431}$$

$$\left.\begin{array}{r}\dot{\Psi}(t) \\ \Psi(t+1) - \Psi(t)\end{array}\right\} = \frac{-\gamma\epsilon(t)\xi^T(t)}{m^2(t)}, \tag{9.432}$$

where

$$m(t) = \sqrt{1 + \sum_{i=1}^{M} \zeta_i^T(t)\zeta_i(t) + \sum_{i=1}^{M} \xi_i^2(t)} \tag{9.433}$$

and, for the *continuous-time* case, $\gamma_i > 0$, $i = 2, 3, \dots, M$, as in the SDU decomposition matrix D_s in (9.358) for K_p, d_i^*, $i = 2, 3, \dots, M$, are in the LDU decomposition matrix D^* in (9.355), and $\gamma > 0$ is a scalar gain, or for the *discrete-time* case, $0 < \gamma_i^2 < 2\lambda_{\min}[S^{-1}]$, $0 < \gamma < \frac{2\lambda_{\min}[S^{-1}]}{\lambda_{\max}[S^{-1}]}$, where $\lambda_{\min}[S^{-1}]$ and $\lambda_{\max}[S^{-1}]$ are the minimum and maximum eigenvalues of $S = LD^*D_s^{-1}L^T$ with L in (9.355), $D^* = \operatorname{diag}\{d_1^*, d_2^*, \dots, d_M^*\}$ in (9.356) and $D_s = \operatorname{diag}\{\operatorname{sign}[d_1^*]\gamma_1, \operatorname{sign}[d_2^*]\gamma_2, \dots, \operatorname{sign}[d_M^*]\gamma_M\}$ in (9.358).

The discrete-time design condition on γ_i and γ, more complicated than that for the continuous-time case, can be met, as

$$S^{-1} = L^{-T}D_sD^{*-1}L^{-1} = L^{-T}\operatorname{diag}\left\{\frac{\gamma_1}{|d_1^*|}, \frac{\gamma_2}{|d_2^*|}, \dots, \frac{\gamma_M}{|d_M^*|}\right\}L^{-1} \tag{9.434}$$

depends on γ_i, $i = 1, 2, \dots, M$, so that it is possible to specify γ_i, $i = 1, 2, \dots, M$, to satisfy $0 < \gamma_i^2 < 2\lambda_{\min}[S^{-1}]$ and $0 < \gamma < \frac{2\lambda_{\min}[S^{-1}]}{\lambda_{\max}[S^{-1}]}$.

Consider the positive definite function

$$V = \sum_{i=1}^{M} \tilde{\phi}_i^T \tilde{\phi}_i + \frac{1}{\gamma}\operatorname{tr}[\tilde{\Psi}^T S^{-1} \tilde{\Psi}], \tag{9.435}$$

where S is in the SDU decomposition (9.357) of K_p ($S = LD^*D_s^{-1}L^T$ for L and D^* in the LDU decomposition (9.355) of K_p). For the *continuous-time* case, we obtain the time derivative of V as

$$\begin{aligned}\dot{V} &= -\frac{2}{m^2(t)}\left(\sum_{i=1}^{M} \operatorname{sign}[d_i^*]\gamma_i\epsilon_i(t)\tilde{\phi}_i^T(t)\zeta_i(t) + \operatorname{tr}[\tilde{\Psi}^T(t)S^{-1}\epsilon(t)\xi^T(t)]\right) \\ &= -\frac{2}{m^2(t)}\left(\epsilon^T(t)D_s[\tilde{\phi}_1^T\zeta_1(t), \tilde{\phi}_2^T\zeta_2(t), \dots, \tilde{\phi}_M^T\zeta_M(t)]^T + \epsilon^T(t)S^{-1}\tilde{\Psi}(t)\xi(t)\right) \\ &= -\frac{2\epsilon^T(t)S^{-1}\epsilon(t)}{m^2(t)}.\end{aligned} \tag{9.436}$$

In the *discrete-time* case, for

$$M_\varsigma(t) = \mathrm{diag}\{\gamma_1^2\zeta_1^T(t)\zeta_1(t), \ldots, \gamma_M^2\zeta_M^T(t)\zeta_M(t)\}, \tag{9.437}$$

the time increment of V is

$$\begin{aligned}
& V(\tilde{\phi}_i(t+1), \tilde{\Psi}(t+1)) - V(\tilde{\phi}_i(t), \tilde{\Psi}(t)) \\
&= -\sum_{i=1}^{M}\left(\frac{2\,\mathrm{sign}[d_i^*]\gamma_i\epsilon_i(t)\tilde{\phi}_i^T(t)\zeta_i(t)}{m^2(t)} - \frac{\gamma_i^2\epsilon_i^2(t)\zeta_i^T(t)\zeta_i(t)}{m^4(t)}\right) \\
&\quad - \left(\frac{2\epsilon^T(t)S^{-1}\tilde{\Psi}(t)\xi(t)}{m^2(t)} - \frac{\epsilon^T(t)\gamma S^{-1}\epsilon(t)\xi^T(t)\xi(t)}{m^4(t)}\right) \\
&= -\left(\frac{2\epsilon^T(t)D_s[\tilde{\phi}_1^T\zeta_1(t), \tilde{\phi}_2^T\zeta_2(t), \ldots, \tilde{\phi}_M^T\zeta_M(t)]^T}{m^2(t)} - \frac{\epsilon^T(t)M_\varsigma(t)\epsilon(t)}{m^4(t)}\right) \\
&\quad - \left(\frac{2\epsilon^T(t)S^{-1}\tilde{\Psi}(t)\xi(t)}{m^2(t)} - \frac{\epsilon^T(t)\gamma S^{-1}\epsilon(t)\xi^T(t)\xi(t)}{m^4(t)}\right) \\
&= -\epsilon^T(t)\left(2S^{-1} - \frac{M_\varsigma(t) + \gamma S^{-1}\xi^T(t)\xi(t)}{m^2(t)}\right)\frac{\epsilon(t)}{m^2(t)} \\
&\le -\alpha_1\frac{\epsilon^T(t)\epsilon(t)}{m^2(t)}
\end{aligned} \tag{9.438}$$

for some constant $\alpha_1 > 0$ from the choice γ_i, $i = 1, 2, \ldots, M$, and λ, that is, there exists a matrix $P \in R^{M\times M}$ [298] such that $P^TP = I$ and $S^{-1} = P^TD_\lambda P$ with $D_\lambda = \mathrm{diag}\{\lambda_1, \lambda_2, \ldots, \lambda_M\}$, the eigenvalues of S^{-1}, so that

$$\begin{aligned}
& 2S^{-1} - \frac{M_\varsigma(t) + \gamma S^{-1}\xi^T(t)\xi(t)}{m^2(t)} \\
&= P^T\left(2D_\lambda - \frac{PM_\varsigma(t)P^T + \gamma D_\lambda\xi^T(t)\xi(t)}{m^2(t)}\right)P \\
&\ge P^T\left(2\lambda_{\min}[S^{-1}]I_M - \frac{\max_i\{\gamma_i^2\zeta_i^T(t)\zeta_i(t)\}I_M + \gamma\lambda_{\max}[S^{-1}]\xi^T(t)\xi(t)I_M}{m^2(t)}\right)P \\
&\ge P^T\left(2\lambda_{\min}[S^{-1}]I_M - \max\{\max_i\{\gamma_i^2\}, \gamma\lambda_{\max}[S^{-1}]\}I_M\right)P \\
&= \left(2\lambda_{\min}[S^{-1}] - \max\{\max_i\{\gamma_i^2\}, \gamma\lambda_{\max}[S^{-1}]\}\right)I_M \triangleq \alpha_1 I_M > 0.
\end{aligned} \tag{9.439}$$

Stability Properties

From the above expressions of the time derivatives or time increments of the positive definite functions V for adaptive control designs based on different decompositions of K_p, we can conclude the following stability properties of the developed adaptive laws and closed-loop control systems.

Lemma 9.10 *(i) For the adaptive law (9.379)–(9.381), $\theta_i(t) \in L^\infty$, $i = 2, 3, \ldots, M$, $\Theta(t) \in L^\infty$, and $\Psi(t) \in L^\infty$, and $\frac{\epsilon(t)}{m(t)} \in L^2 \cap L^\infty$, $\dot{\theta}_i(t) \in L^2 \cap L^\infty$ (or $\theta_i(t+1) - \theta_i(t) \in L^2 \cap L^\infty$), $i = 2, 3, \ldots, M$, $\dot{\Theta}(t) \in L^2 \cap L^\infty$ (or $\Theta(t+1) - \Theta(t) \in L^2 \cap L^\infty$), and $\dot{\Psi}(t) \in L^2 \cap L^\infty$ (or $\Psi(t+1) - \Psi(t) \in L^2 \cap L^\infty$).*

(ii) For the adaptive law (9.409)–(9.411), $\theta_i(t) \in L^\infty$, $i = 2, 3, \ldots, M$, $\phi_i(t) \in L^\infty$, $i = 1, 2, \ldots, M$, $d_i(t) \in L^\infty$, $i = 1, 2, \ldots, M$, and $\frac{\epsilon_i(t)}{m_i(t)} \in L^2 \cap L^\infty$, $\dot{\theta}_i(t) \in L^2 \cap L^\infty$ (or $\theta_i(t+1) - \theta_i(t) \in L^2 \cap L^\infty$), $i = 2, 3, \ldots, M$, $\dot{\phi}_i(t) \in L^2 \cap L^\infty$ (or $\phi_i(t+1) - \phi_i(t) \in L^2 \cap L^\infty$), $i = 1, 2, \ldots, M$, and $\dot{d}_i(t) \in L^2 \cap L^\infty$ (or $d_i(t+1) - d_i(t) \in L^2 \cap L^\infty$), $i = 1, 2, \ldots, M$.

(iii) For the adaptive law (9.431)–(9.432), $\phi_i(t) \in L^\infty$, $i = 1, 2, \ldots, M$, and $\Psi(t) \in L^\infty$, and $\frac{\epsilon(t)}{m(t)} \in L^2 \cap L^\infty$, $\dot{\phi}_i(t) \in L^2 \cap L^\infty$ (or $\phi_i(t+1) - \phi_i(t) \in L^2 \cap L^\infty$), $i = 1, 2, \ldots, M$, and $\dot{\Psi}(t) \in L^2 \cap L^\infty$ (or $\Psi(t+1) - \Psi(t) \in L^2 \cap L^\infty$).

Based on these desired properties, we can have the following results.

Theorem 9.8 *All three MRAC schemes with (9.361), (9.392), or (9.418), based on LDS, LDU, or SDU decomposition of K_p ensure closed-loop signal boundedness and asymptotic output tracking* $\lim_{t\to\infty}(y(t) - y_m(t)) = 0$.

The proof of this theorem follows from that in Section 9.2.6 for Theorem 9.2. A proof of the continuous-time results can also be found in [141].

The presented MRAC designs using parametrizations based on LDS, LDU, or SDU decomposition of K_p need the knowledge of sign$[d_i^*]$, for $K_p = LD^*U = L\,\mathrm{diag}\{d_1^*, d_2^*, \ldots, d_M^*\}\,U$ in (9.355), when all leading principal minors Δ_i, $i = 1, 2, \ldots, M$, of K_p are nonzero, where $d_i^* = \frac{\Delta_i}{\Delta_{i-1}}$, $i = 1, 2, \ldots, M$, for $\Delta_0 = 1$. As compared with Assumption (A9.3), that is, an $M \times M$ matrix S_p is known such that $K_p S_p = (K_p S_p)^T > 0$, the MRAC schemes based on LDS, LDU, or SDU decomposition of K_p use relaxed conditions on K_p.

9.3 Adaptive Backstepping Control

Backstepping is a nonlinear design method effective for control of nonlinear systems [206]. It is applicable to linear systems [205], with improved adaptive system performance. In this section, we present an adaptive backstepping control design [228] for multi-input and multi-output linear plants. Issues such as plant parametrization, state observer construction, *a priori* plant information specification, and adaptive controller design and analysis will be solved, which, due to the nonlinear nature of the backstepping control design, are different from those in Section 9.2 (also those in [93], [95], [125], [264], [282], [340], [384], [432]), where a linear controller structure is used.

9.3.1 Plant Parametrization

The control objective is to design an adaptive control scheme for a continuous-time linear multivariable time-invariant plant

$$y(t) = G(s)[u](t),\ u(t) \in R^M,\ y(t) \in R^M,\ M \geq 1, \tag{9.440}$$

where $G(s) = G_0(s)$ is an $M \times M$ strictly proper rational transfer matrix, such that the plant output $y(t)$ tracks a given bounded reference output $y_m(t)$ asymptotically and all signals in the closed-loop system are bounded,

From the left coprime polynomial matrix decomposition description of $G(s) = G_0(s) = P_l^{-1}(s)Z_l(s)$ with $P_l(s)$ row reduced (see Section 9.2.1), we can express the plant transfer matrix $G(s)$ in (9.440) as

$$G(s) = D^{-1}(s)N(s) = C_g(sI - A_g)^{-1}B_g, \tag{9.441}$$

where, for the observability index ν of $G(s)$,

$$D(s) = s^\nu I_M + A_{\nu-1}s^{\nu-1} + \cdots + A_1 s + A_0, \tag{9.442}$$

$$N(s) = B_m s^m + \cdots + B_1 s + B_0, \tag{9.443}$$

$$A_g = \begin{bmatrix} -A_{\nu-1} & I_M & 0 & \cdots & \cdots & 0 \\ -A_{\nu-2} & 0 & I_M & 0 & \cdots & 0 \\ \vdots & \vdots & \vdots & \vdots & \vdots & \vdots \\ -A_1 & 0 & \cdots & \cdots & 0 & I_M \\ -A_0 & 0 & \cdots & \cdots & 0 & 0 \end{bmatrix} \in R^{M\nu\times M\nu}, \tag{9.444}$$

$$B_g = \begin{bmatrix} 0 \\ \vdots \\ 0 \\ B_p \end{bmatrix} \in R^{M\nu\times M},\ B_p = \begin{bmatrix} B_m \\ \vdots \\ B_1 \\ B_0 \end{bmatrix} \in R^{M(m+1)\times M}, \tag{9.445}$$

$$C_g = \begin{bmatrix} I_M & 0 & \cdots & 0 & 0 \end{bmatrix} \subset R^{M\times M\nu} \tag{9.446}$$

for the $M \times M$ identity matrix I_M and some $M \times M$ parameter matrices A_i, $i = 0, 1, \ldots, \nu - 1$, and B_j, $j = 0, 1, \ldots, m$, $m \leq \nu - 1$.[16]

To verify the second equality in (9.441), defining

$$W(s) = (sI - A_g^T)^{-1}C_g^T = [W_1^T(s), W_2^T(s), \ldots, W_\nu^T(s)]^T \tag{9.447}$$

[16]The leading coefficient matrix of $D(s)$ is taken as I_M for a state-space representation of $G(s)$, convenient for a state observer design in Section 9.3.2. This follows from the left coprime decomposition of $G(s) = P_l^{-1}(s)Z_l(s)$: $P_l(s)[y](t) = Z_l(s)[u](t)$, where

$$P_l(s) = \mathrm{diag}\{s^{\nu_1}, \ldots, s^{\nu_M}\}P_\nu + \mathrm{diag}\{s^{\nu_1-1}, \ldots, s^{\nu_M-1}\}P_{\nu-1} + \cdots,$$

$$Z_l(s) = \mathrm{diag}\{s^{\nu_1-1}, \ldots, s^{\nu_M-1}\}Z_{\nu-1} + \mathrm{diag}\{s^{\nu_1-2}, \ldots, s^{\nu_M-2}\}Z_{\nu-2} + \cdots,$$

and using $sW(s) = A_g^T W(s) + C_g^T$, we obtain

$$
\begin{aligned}
sW_1(s) &= -A_{\nu-1}^T W_1(s) - A_{\nu-2}^T W_2(s) - \cdots - A_1^T W_{\nu-1}(s) - A_0^T W_\nu(s), \\
sW_i(s) &= W_{i-1}(s),\ i = 2, 3, \ldots, \nu,
\end{aligned}
\tag{9.448}
$$

from which it follows that

$$
I_N = (s^\nu I_N + A_{\nu-1}^T + A_{\nu-2}^T + \cdots + A_1^T + A_0^T) W_\nu(s) = D^T(s) W_\nu(s). \tag{9.449}
$$

Finally, using (9.445), (9.448), and (9.449), we have

$$
\begin{aligned}
& B_g^T (sI - A_g^T)^{-1} C_g^T \\
&= B_m^T W_{\nu-m}(s) + B_{m-1}^T W_{\nu-m+1}(s) + \cdots + B_1^T W_{\nu-1}(s) + B_0^T W_\nu(s) \\
&= (B_m^T s^m + B_{m-1}^T s^{m-1} + \cdots + B_1^T s + B_0^T) W_\nu(s) \\
&= N^T(s) W_\nu(s) = N^T(s)(D^T(s))^{-1}.
\end{aligned}
\tag{9.450}
$$

In view of (9.441), the plant (9.440) can be expressed as

$$
\dot{x}(t) = Ax(t) + A_p y(t) + \begin{bmatrix} 0 \\ B_p \end{bmatrix} u(t),\ y(t) = C_g x(t), \tag{9.451}
$$

where $A \in R^{M\nu \times M\nu}$ and $A_p \in R^{M\nu \times M}$ are

$$
A = \begin{bmatrix} 0 & I_M & 0 & \cdots & \cdots & 0 \\ 0 & 0 & I_M & 0 & \cdots & 0 \\ \vdots & \vdots & \vdots & \vdots & \vdots & \vdots \\ 0 & 0 & \cdots & \cdots & 0 & I_M \\ 0 & 0 & \cdots & \cdots & 0 & 0 \end{bmatrix},\ A_p = \begin{bmatrix} -A_{\nu-1} \\ -A_{\nu-2} \\ \vdots \\ -A_1 \\ -A_0 \end{bmatrix}. \tag{9.452}
$$

9.3.2 State Observer

Given that only the output $y(t)$ and input $u(t)$ in the plant model (9.451) are available for measurement, we need to construct a state observer for the multivariable plant (9.440) or (9.451). We first choose the gain matrix

$$
K = [k_1 I_M, \ldots, k_\nu I_M]^T \in R^{M\nu \times M}, \tag{9.453}
$$

---

and the definition of $D(s)$ and $N(s)$:

$$
\begin{aligned}
D(s) &= P_\nu^{-1} \mathrm{diag}\{s^{\nu-\nu_1}, \ldots, s^{\nu-\nu_M}\} P_l(s), \\
N(s) &= P_\nu^{-1} \mathrm{diag}\{s^{\nu-\nu_1}, \ldots, s^{\nu-\nu_M}\} Z_l(s).
\end{aligned}
$$

The leading row coefficient matrix P_ν of $P_l(s)$, which was defined in (9.57) as $\Gamma_r[P_l(s)] = \lim_{s\to\infty} \mathrm{diag}\{s^{-\nu_1}, \ldots, s^{-\nu_M}\} P_l(s)$, is nonsingular as $P_l(s)$ is row reduced.

where k_i, $i = 1, \ldots, \nu$, are positive scalars such that the matrix

$$A_o = A - KC_g \tag{9.454}$$

is stable (i.e., $s^\nu + k_1 s^{\nu-1} + \ldots + k_{\nu-1}s + k_\nu$ is stable), and define

$$E_i = e_i \otimes I_M, \tag{9.455}$$

where $\otimes$ is the Kronecker product[17] and e_i is the ith coordinate vector in R^ν. We then introduce the known vector signals $\xi_\nu(t)$, $\xi_i(t)$, $i = 0, 1, \ldots, \nu - 1$, $v_j(t)$, $j = 0, 1, \ldots, m$, using the filter systems

$$\dot{\xi}_\nu(t) = A_o \xi_\nu(t) + Ky(t), \tag{9.456}$$

$$\dot{\xi}_i(t) = A_o \xi_i(t) + E_{\nu-i} y(t), \tag{9.457}$$

$$\dot{v}_j(t) = A_o v_j(t) + E_{\nu-j} u(t). \tag{9.458}$$

With these signals, the parametrized state observer for the plant (9.451) is

$$\hat{x}(t) = \xi_\nu(t) - \sum_{i=0}^{\nu-1} \mathcal{A}_i \xi_i(t) + \sum_{j=0}^{m} \mathcal{B}_j v_j(t), \tag{9.459}$$

where $\mathcal{A}_i \in R^{M\nu \times M\nu}$ and $\mathcal{B}_j \in R^{M\nu \times M\nu}$ are

$$\mathcal{A}_i = \mathrm{diag}\{A_i, \ldots, A_i\}, \ \mathcal{B}_j = \mathrm{diag}\{B_j, \ldots, B_j\}. \tag{9.460}$$

This state observer has the desired convergence property.

Lemma 9.11 *The observation error $\epsilon(t) = x(t) - \hat{x}(t)$ satisfies*

$$\dot{\epsilon}(t) = A_o \epsilon(t), \ \lim_{t \to \infty} \epsilon(t) = 0 \ \textit{exponentially}. \tag{9.461}$$

Proof: From (9.451), (9.453) with scalar k_i, (9.456)–(9.459), and the facts

$$\mathcal{A}_i A_o = A_o \mathcal{A}_i, \ i = 0, 1, \ldots, \nu - 1, \ \mathcal{B}_i A_o = A_o \mathcal{B}_i, \ i = 0, 1, \ldots, m, \tag{9.462}$$

$$-\sum_{i=0}^{\nu-1} \mathcal{A}_i E_{\nu-i} y(t) = A_p y(t), \ \sum_{i=0}^{m} \mathcal{B}_i E_{\nu-i} u(t) = \begin{bmatrix} 0 \\ B_p \end{bmatrix} u(t), \tag{9.463}$$

we have the expression (9.461) where A_o is stable. ∇

[17]The Kronecker product of two matrices $A = \{a_{ij}\} \in R^{m \times n}$ and $B = \{b_{ij}\} \in R^{p \times q}$ is

$$A \otimes B = \begin{bmatrix} a_{11}B & a_{12}B & \cdots & a_{1n}B \\ a_{21}B & a_{22}B & \cdots & a_{2n}B \\ \vdots & \vdots & \vdots & \vdots \\ a_{m1}B & a_{m2}B & \cdots & a_{mn}B \end{bmatrix} \in R^{mp \times nq}.$$

Remark 9.14 The multivariable state observer (9.459) is parametrized by $\mathcal{A}_i$, $\mathcal{B}_j$, not directly by A_i, B_j. In the single-input, single-output case with $M = 1$, $A_i = a_i$ and $B_j = b_j$ are all scalars and

$$\mathcal{A}_i = \mathrm{diag}\{a_i, \ldots, a_i\} = a_i I_\nu,\ \mathcal{B}_j = \mathrm{diag}\{b_j, \ldots, b_j\} = b_j I_\nu, \tag{9.464}$$

so that the state observer (9.456)–(9.459), as in [205], is simplified as

$$\hat{x}(t) = \xi_\nu(t) - \sum_{i=0}^{\nu-1} a_i \xi_i(t) + \sum_{j=0}^{m} b_j v_j(t), \tag{9.465}$$

which is now parametrized by a_i, b_j. □

Based on the parametrized canonical form state observer (9.459), our task is to design an adaptive controller for the plant (9.451), using a multivariable version of the nonlinear backstepping design method [205].

Partitioning the state variable $x(t)$ in (9.459) as

$$x(t) = [x_1^T(t), \ldots, x_\nu^T(t)]^T,\ x_i \in R^M,\ i = 1, \ldots, \nu, \tag{9.466}$$

it follows from (9.446), (9.451), and (9.452) that $y(t) = x_1(t)$ and

$$\dot{y}(t) = \dot{x}_1(t) = x_2(t) - A_{\nu-1}x_1(t) = x_2(t) - A_{\nu-1}y(t). \tag{9.467}$$

Denoting $\epsilon(t)$ in (9.461), $\xi_i(t)$ in (9.456) and (9.457), and $v_j(t)$ in (9.458) as

$$\epsilon(t) = [\epsilon_1^T(t), \ldots, \epsilon_\nu^T(t)]^T,\ \epsilon_i(t) \in R^M,\ i = 1, \ldots, \nu, \tag{9.468}$$

$$\xi_i(t) = [\xi_{i,1}^T(t), \ldots, \xi_{i,\nu}^T(t)]^T,\ \xi_{i,k}(t) \in R^M,\ k = 1, \ldots, \nu, \tag{9.469}$$

$$v_i(t) = [v_{i,1}^T(t), \ldots, v_{i,\nu}^T(t)]^T,\ v_{i,k}(t) \in R^M,\ k = 1, \ldots, \nu, \tag{9.470}$$

and defining

$$\Theta_a = [-A_{\nu-1}, -A_{\nu-2}, \ldots, -A_1, -A_0], \tag{9.471}$$

$$\Theta_b = [B_m, B_{m-1}, \ldots, B_1, B_0], \tag{9.472}$$

$$\xi_{(2)}(t) = [\xi_{\nu-1,2}^T(t), \ldots, \xi_{0,2}^T(t)]^T, \tag{9.473}$$

$$v_{(2)}(t) = [v_{m,2}^T(t), \ldots, v_{0,2}^T(t)]^T, \tag{9.474}$$

from (9.459) and (9.461), we have

$$x_2(t) = \xi_{\nu,2}(t) + \Theta_a \xi_{(2)}(t) + \Theta_b v_{(2)}(t) + \epsilon_2(t). \tag{9.475}$$

Substituting this expression of $x_2(t)$ into (9.467) yields

$$\dot{y}(t) = \xi_{\nu,2}(t) + \Theta_a(\xi_{(2)}(t) + E_1 y(t)) + \Theta_b v_{(2)}(t) + \epsilon_2(t). \tag{9.476}$$

Remark 9.15 Equation (9.476) is the key equation in a multivariable backstepping design using the measurement of $y(t)$ only. The signal $\dot{y}(t)$, which is not available for measurement, appears at each step of the backstepping design procedure. The expression (9.476) for $\dot{y}(t)$ has the promising form because the signals $\xi_{\nu,2}(t)$, $\xi_{(2)}(t)+E_1y(t)$, and $v_{(2)}(t)$ are all available for measurement and the signal $\epsilon_2(t)$ satisfies (9.461), that is, $\lim_{t\to\infty}\epsilon_2(t)=0$, while the parameter matrices Θ_a and Θ_b contain precisely the parameters in the plant (9.451). There is no overparametrization in (9.476). Only the filter signals defined in (9.456)–(9.458) and the property (9.461) about the state observer (9.459) are used for adaptive control design and analysis. □

Notation. For the presentation in Section 9.3.3, we denote $\Theta=[\Theta_a,\Theta_b]$, $P=B_m^{-1}$ (and in Section 9.3.4 for B_m singular, $\bar{\Theta}=[\bar{\Theta}_a,\bar{\Theta}_b]$, $P=\bar{B}_m^{-1}$ for some $\bar{\Theta}_a$, $\bar{\Theta}_b$, $\bar{B}_m$) as plant parameters, their estimates as $\hat{\Theta}$, $\hat{P}$, and the parameter errors as $\tilde{\Theta}=\Theta-\hat{\Theta}$ (or $\bar{\Theta}-\hat{\Theta}$) and $\tilde{P}=P-\hat{P}$. This notation is slightly different from the rest of the book.

9.3.3 Design Procedure for B_m Nonsingular

In this section, we design an adaptive backstepping controller for the plant (9.440) under the following assumptions: $G(s)$ has full rank (that is, the plant (9.440) is output controllable); all zeros of $\det[N(s)]$ are stable; the observability index ν of $G(s)$ is known; the order m of $N(s)$ is known; the matrix B_m is nonsingular (the case when the matrix B_m is singular is addressed in the next subsection); and there is a known matrix S_m such that $S_mB_m=(S_mB_m)^T>0$. We also assume that the reference output $y_m(t)$ has bounded derivatives $y_m^{(i)}(t)$ for $i=1,2,\ldots,\rho=\nu-m$.

The following backstepping design procedure is recursive and similar to that for the single-input and single-output case in [205], but with a unique treatment of the multi-input and multi-output system.

Step 1: From (9.476), for the tracking error $z_1=y-y_m$, we obtain

$$\dot{z}_1=\xi_{\nu,2}+\Theta\bar{\omega}+B_mv_{m,2}+\epsilon_2-\dot{y}_m, \tag{9.477}$$

where

$$\Theta=[\Theta_a,\Theta_b], \tag{9.478}$$

$$\bar{\omega}=[\xi_{(2)}+E_1y,[0,v_{m-1,2}^T,\ldots v_{0,2}^T]^T]^T. \tag{9.479}$$

Defining $z_2 = v_{m,2} - \alpha_1$, with α_1 to be designed, and using (9.477), we have

$$\dot{z}_1 = B_m \alpha_1 + B_m z_2 + \xi_{\nu,2} + \Theta \bar{\omega} - \dot{y}_m + \epsilon_2. \tag{9.480}$$

For B_m nonsingular, introducing $P = B_m^{-1}$, letting $\hat{P}$ and $\hat{\Theta}$ be the estimate of P and Θ, and choosing a diagonal constant matrix C_1 with positive diagonal elements and a positive constant d_1, we write (9.480) as

$$\begin{aligned} \dot{z}_1 = &-C_1 z_1 - d_1 z_1 + B_m z_2 \\ &+ B_m \left(\alpha_1 + P(C_1 z_1 + d_1 z_1 + \xi_{\nu,2} - \dot{y}_m) + P\Theta\bar{\omega})\right) + \epsilon_2. \end{aligned} \tag{9.481}$$

Introduce the known signal

$$\phi = C_1 z_1 + d_1 z_1 + \xi_{\nu,2} - \dot{y}_m + \hat{\Theta}\bar{\omega} \tag{9.482}$$

and choose the design signal α_1 as

$$\alpha_1 = -\hat{P}\phi. \tag{9.483}$$

Then we rewrite (9.481) as

$$\dot{z}_1 = -C_1 z_1 + B_m z_2 + B_m \tilde{P}\phi + \tilde{\Theta}\bar{\omega} - d_1 z_1 + \epsilon_2, \tag{9.484}$$

where the parameter errors are

$$\tilde{\Theta} = \Theta - \hat{\Theta}, \ \tilde{P} = P - \hat{P}. \tag{9.485}$$

Choose the first Lyapunov function

$$V_1 = \frac{1}{2} z_1^T z_1 + \frac{1}{2}\,\mathrm{tr}[\tilde{\Theta}\Gamma^{-1}\tilde{\Theta}^T] + \frac{1}{2}\,\mathrm{tr}[\tilde{P}^T B_m^T S_m^T \tilde{P}] + \frac{1}{d_1}\epsilon^T P_0 \epsilon, \tag{9.486}$$

where Γ is a positive definite matrix, $P_0 = P_0^T$ is the positive definite solution of $P_0 A_0 + A_0^T P_0 = -I$ for the stable matrix A_0 in (9.461), and ${}_m B_m = (S_m B_m)^T > 0$ as assumed. With (9.461) and (9.484), the time derivative of V_1 is

$$\begin{aligned} \dot{V}_1 &= z_1^T \dot{z}_1 + \mathrm{tr}[\tilde{\Theta}\Gamma^{-1}\dot{\tilde{\Theta}}^T] + \mathrm{tr}[\tilde{P}^T B_m^T S_m^T \dot{\tilde{P}}] + \frac{2}{d_1}\epsilon^T P_0 \dot{\epsilon} \\ &= -z_1^T C_1 z_1 + z_1^T B_m z_2 + z_1^T B_m \tilde{P}\phi + z_1^T \tilde{\Theta}\bar{\omega} - d_1 z_1^T z_1 + z_1^T \epsilon_2 \\ &\quad + \mathrm{tr}[\tilde{\Theta}\Gamma^{-1}\dot{\tilde{\Theta}}^T] + \mathrm{tr}[\tilde{P}^T B_m^T S_m^T \dot{\tilde{P}}] - \frac{1}{d_1}\epsilon^T \epsilon. \end{aligned} \tag{9.487}$$

Choose the adaptive law for $\hat{P}(t)$:

$$\dot{\hat{P}} = S_m^{-T} z_1 \phi^T. \tag{9.488}$$

Then, from (9.487) and (9.488), the time derivative of V_1 becomes

$$\begin{aligned}\dot{V}_1 &= -z_1^T C_1 z_1 + z_1^T B_m z_2 + z_1^T \tilde{\Theta}\bar{\omega} - d_1 z_1^T z_1 + z_1^T \epsilon_2 + \mathrm{tr}[\tilde{\Theta}\Gamma^{-1}\dot{\tilde{\Theta}}^T] - \frac{1}{d_1}\epsilon^T\epsilon \\ &\leq -z_1^T C_1 z_1 + z_1^T B_m z_2 + z_1^T \tilde{\Theta}\bar{\omega} + \mathrm{tr}[\tilde{\Theta}\Gamma^{-1}\dot{\tilde{\Theta}}^T] - \frac{1}{d_1}\Omega(\epsilon), \end{aligned} \tag{9.489}$$

where

$$\Omega(\epsilon) = \epsilon_1^T \epsilon_1 + \frac{3}{4}\epsilon_2^T \epsilon_2 + \cdots + \epsilon_\nu^T \epsilon_\nu. \tag{9.490}$$

Step 2: Express the derivative of $z_2 = v_{m,2} - \alpha_1$ as

$$\dot{z}_2 = v_{m,3} + \beta_2 - \frac{\partial \alpha_1}{\partial y}\Theta\omega - \frac{\partial \alpha_1}{\partial y}\epsilon_2 - \delta_2, \tag{9.491}$$

where, for $\alpha_1 = [\alpha_{11}, \alpha_{12}, \ldots, \alpha_{1M}]^T$,

$$\frac{\partial \alpha_1}{\partial y} = \begin{bmatrix} \frac{\partial \alpha_{11}}{\partial y} \\ \vdots \\ \frac{\partial \alpha_{1M}}{\partial y} \end{bmatrix} = \begin{bmatrix} \frac{\partial \alpha_{11}}{\partial y_1} & \cdots & \frac{\partial \alpha_{11}}{\partial y_m} \\ \vdots & & \vdots \\ \frac{\partial \alpha_{1M}}{\partial y_1} & \cdots & \frac{\partial \alpha_{1M}}{\partial y_m} \end{bmatrix}, \tag{9.492}$$

$$\delta_2 = \left[\mathrm{tr}\left[\dot{\hat{\Theta}}\frac{\partial \alpha_{11}}{\partial \hat{\Theta}}\right], \ldots, \mathrm{tr}\left[\dot{\hat{\Theta}}\frac{\partial \alpha_{1M}}{\partial \hat{\Theta}}\right]\right]^T, \tag{9.493}$$

$$\left\{\frac{\partial \alpha_{1k}}{\partial \hat{\Theta}}\right\}_{ij} = \frac{\partial \alpha_{1k}}{\partial \hat{\Theta}_{ji}}, \; k = 1, 2, \ldots, M, \tag{9.494}$$

$$\omega = [\xi_{(2)} + E_1 y, [v_{m,2}^T, v_{m-1,2}^T, \ldots, v_{0,2}^T]^T]^T, \tag{9.495}$$

and β_2 includes all the other known terms, with $\dot{\hat{\Theta}}$ to be designed and $\left\{\frac{\partial \alpha_{1k}}{\partial \hat{\Theta}}\right\}_{ij}$ denoting the (i,j)th element of the matrix $\frac{\partial \alpha_{1k}}{\partial \hat{\Theta}}$.

Choose the second Lyapunov function as

$$V_2 = V_1 + \frac{1}{2}z_2^T z_2 + \frac{1}{d_2}\epsilon^T P_0 \epsilon. \tag{9.496}$$

In view of (9.489) and (9.491), we have the time derivative of V_2 as

$$\begin{aligned}\dot{V}_2 &\leq -z_1^T C_1 z_1 + z_2^T\left(B_m^T z_1 + z_3 + \alpha_2 + \beta_2 - \frac{\partial \alpha_1}{\partial y}\Theta\omega - \delta_2\right) \\ &\quad - z_2^T\frac{\partial \alpha_1}{\partial y}\epsilon_2 + \mathrm{tr}[\tilde{\Theta}(\omega z_1^T - \Gamma^{-1}\dot{\hat{\Theta}}^T)] - \frac{1}{d_2}\epsilon^T\epsilon - \frac{1}{d_1}\Omega(\epsilon) \\ &= -z_1^T C_1 z_1 + z_2^T\left(\hat{B}_m^T z_1 + z_3 + \alpha_2 + \beta_2 - \frac{\partial \alpha_1}{\partial y}\hat{\Theta}\omega - \delta_2\right)\end{aligned}$$

$$+ z_2^T \tilde{B}_m^T z_1 + \mathrm{tr}[\tilde{\Theta}(\omega z_1^T - \Gamma^{-1}\dot{\hat{\Theta}}^T)] - z_2^T \frac{\partial \alpha_1}{\partial y}\tilde{\Theta}\omega$$
$$- z_2^T \frac{\partial \alpha_1}{\partial y}\epsilon_2 - \frac{1}{d_2}\epsilon^T\epsilon - \frac{1}{d_1}\Omega(\epsilon). \tag{9.497}$$

Letting $\tau_1^T = \Gamma\bar{\omega}z_1^T$, we set τ_2 as

$$\tau_2^T = \tau_1^T - \Gamma\left(\omega z_2^T \frac{\partial \alpha_1}{\partial y} + [0, 0 \ldots z_2^T, 0 \ldots 0]^T z_1^T\right), \tag{9.498}$$

define the error signal $z_3 = v_{m,3} - \alpha_2$, and choose the design signal α_2,

$$\alpha_2 = -C_2 z_2 - d_2\left(\frac{\partial \alpha_1}{\partial y}\right)^T \frac{\partial \alpha_1}{\partial y} z_2 - \beta_2 + \frac{\partial \alpha_1}{\partial y}\hat{\Theta}\omega - \hat{B}_m^T z_1 + \hat{\delta}_2, \tag{9.499}$$

where

$$\hat{\delta}_2 = \left[\mathrm{tr}\left[\tau_2 \frac{\partial \alpha_{11}}{\partial \hat{\Theta}}\right], \ldots, \mathrm{tr}\left[\tau_2 \frac{\partial \alpha_{1M}}{\partial \hat{\Theta}}\right]\right]^T. \tag{9.500}$$

Substituting (9.498) and (9.499) into (9.491), we obtain

$$\dot{z}_2 = -C_2 z_2 - d_2\left(\frac{\partial \alpha_1}{\partial y}\right)^T \frac{\partial \alpha_1}{\partial y} z_2 + z_3 - B_m^T z_1$$
$$- \frac{\partial \alpha_1}{\partial y}\tilde{\Theta}\omega + \tilde{B}_m^T z_1 + \tilde{\delta}_2 - \frac{\partial \alpha_1}{\partial y}\epsilon_2, \tag{9.501}$$

where

$$\tilde{\delta}_2 = \left[\mathrm{tr}\left[(\tau_2 - \dot{\hat{\Theta}})\frac{\partial \alpha_{11}}{\partial \hat{\Theta}}\right], \ldots, \mathrm{tr}\left[(\tau_2 - \dot{\hat{\Theta}})\frac{\partial \alpha_{1M}}{\partial \hat{\Theta}}\right]\right]^T. \tag{9.502}$$

Step i: For $i = 3, \ldots, \rho - 1, \rho = \nu - m$, we express the derivative of the ith generalized error $z_i = v_{m,i} - \alpha_{i-1}$ as

$$\dot{z}_i = v_{m,i+1} + \beta_i - \frac{\partial \alpha_{i-1}}{\partial y}\Theta\omega - \frac{\partial \alpha_{i-1}}{\partial y}\epsilon_2 - \delta_i, \tag{9.503}$$

where

$$\delta_i = \left[\mathrm{tr}\left[\dot{\hat{\Theta}}\frac{\partial \alpha_{i-1\,1}}{\partial \hat{\Theta}}\right], \ldots, \mathrm{tr}\left[\dot{\hat{\Theta}}\frac{\partial \alpha_{i-1\,M}}{\partial \hat{\Theta}}\right]\right]^T, \tag{9.504}$$

$$\left\{\frac{\partial \alpha_{lk}}{\partial \hat{\Theta}}\right\}_{ij} = \frac{\partial \alpha_{lk}}{\partial \hat{\Theta}_{ji}}, \; k = 1, 2, \ldots, M, \tag{9.505}$$

and β_i represents all the known terms except $v_{m,i+1}$.

Choose the ith Lyapunov function

$$V_i = V_{i-1} + \frac{1}{2} z_i^T z_i + \frac{1}{d_i} \epsilon^T P_0 \epsilon. \tag{9.506}$$

Then the time derivative of V_{i-1} satisfies

$$\begin{aligned} \dot{V}_{i-1} \leq & -\sum_{j=1}^{i-1} z_j^T C_j z_j + z_i^T z_{i-1} + \mathrm{tr}[\tilde{\Theta}\Gamma^{-1}(\tau_{i-1}^T - \dot{\hat{\Theta}}^T)] \\ & + \sum_{j=2}^{i-1} z_j^T \tilde{\delta}_j - \sum_{j=3}^{i-1}\sum_{k=2}^{j-1} z_j^T \sigma_{k,j-1} - \sum_{j=1}^{i-1} \frac{1}{d_j} \Omega(\epsilon), \end{aligned} \tag{9.507}$$

where

$$\begin{aligned} \sigma_{k,j-1} = & \left[\mathrm{tr}\left[\left(\frac{\partial \alpha_{j-1}}{\partial y} \right)^T z_k \omega^T \Gamma \frac{\partial \alpha_{k-1,1}}{\partial \hat{\Theta}} \right], \right. \\ & \left. \ldots, \mathrm{tr}\left[\left(\frac{\partial \alpha_{j-1}}{\partial y} \right)^T z_k \omega^T \Gamma \frac{\partial \alpha_{k-1,M}}{\partial \hat{\Theta}} \right] \right]^T. \end{aligned} \tag{9.508}$$

In view of (9.503) and (9.507), we have

$$\begin{aligned} \dot{V}_i \leq & -\sum_{j=1}^{i-1} z_j^T C_j z_j + z_i^T \left(z_{i-1} + z_{i+1} + \alpha_i + \beta_i - \frac{\partial \alpha_{i-1}}{\partial y} \hat{\Theta} \omega - \delta_i \right) \\ & + \mathrm{tr}\left[\tilde{\Theta}\Gamma^{-1} \left(\tau_{i-1}^T - \Gamma \omega z_i^T \frac{\partial \alpha_{i-1}}{\partial y} - \dot{\hat{\Theta}}^T \right) \right] + \sum_{j=2}^{i-1} z_j^T \tilde{\delta}_j - \sum_{j=3}^{i-1}\sum_{k=2}^{j-1} z_j^T \sigma_{k,j-1} \\ & - z_i^T \frac{\partial \alpha_{i-1}}{\partial y} \epsilon_2 - \frac{1}{d_i} \epsilon^T \epsilon - \sum_{j=1}^{i-1} \frac{1}{d_j} \Omega(\epsilon). \end{aligned} \tag{9.509}$$

Similar to that for the SISO case in [205], we define the tuning functions

$$\tau_i^T = \tau_{i-1}^T - \Gamma \omega z_i^T \frac{\partial \alpha_{i-1}}{\partial y} \tag{9.510}$$

and choose the design signal α_i as

$$\begin{aligned} \alpha_i = & -C_i z_i - d_i \left(\frac{\partial \alpha_{i-1}}{\partial y} \right)^T \frac{\partial \alpha_{i-1}}{\partial y} z_i - z_{i-1} \\ & -\beta_i + \frac{\partial \alpha_{i-1}}{\partial y} \hat{\Theta} \omega + \hat{\delta}_i - \sum_{k=2}^{i-1} \sigma_{k,j-1}, \end{aligned} \tag{9.511}$$

where

$$\hat{\delta}_i = \left[\mathrm{tr}\left[\tau_i \frac{\partial \alpha_{i-1\,1}}{\partial \hat{\Theta}} \right], \ldots, \mathrm{tr}\left[\tau_i \frac{\partial \alpha_{i-1,M}}{\partial \hat{\Theta}} \right] \right]^T. \tag{9.512}$$

With some algebraic manipulation, it can be verified that

$$\sum_{j=3}^{i-1}\sum_{k=2}^{j-1} z_j^T \sigma_{k,j-1} = \sum_{k=2}^{i-2}\sum_{j=k+1}^{i-1} z_j^T \sigma_{k,j-1}$$
$$= \sum_{k=2}^{i-1}\sum_{j=k+1}^{i} z_j^T \sigma_{k,j-1} - \sum_{k=2}^{i-1} z_i^T \sigma_{k,i-1}. \tag{9.513}$$

Using (9.513), we see that $\dot{V}_i$ satisfies

$$\dot{V}_i \le -\sum_{j=1}^{i-1} z_j^T C_j z_j + z_i^T z_{i+1} + \mathrm{tr}[\tilde{\Theta}\Gamma^{-1}(\tau_i^T - \dot{\hat{\Theta}}^T)] + \sum_{j=2}^{i} z_j^T \tilde{\delta}_j$$
$$- \sum_{j=3}^{i}\sum_{k=2}^{j-1} z_j^T \sigma_{k,j-1} - \sum_{j=1}^{i-1} \frac{1}{d_j}\Omega(\epsilon), \tag{9.514}$$

where

$$\tilde{\delta}_j = \left[\mathrm{tr}\left[(\tau_j - \dot{\hat{\Theta}})\frac{\partial \alpha_{j-1\,1}}{\partial \hat{\Theta}}\right], \ldots, \mathrm{tr}\left[(\tau_j - \dot{\hat{\Theta}})\frac{\partial \alpha_{j-1\,M}}{\partial \hat{\Theta}}\right]\right]^T. \tag{9.515}$$

Step ρ : Finally, the derivative of $z_\rho = v_{m,\rho} - \alpha_{\rho-1}$ is expressed as

$$\dot{z}_\rho = u + \beta_\rho - \frac{\partial \alpha_{\rho-1}}{\partial y}\Theta\omega - \frac{\partial \alpha_{\rho-1}}{\partial y}\epsilon_2 - \delta_\rho, \tag{9.516}$$

where

$$\delta_\rho = \left[\mathrm{tr}\left[\dot{\hat{\Theta}}\frac{\partial \alpha_{\rho-1\,1}}{\partial \hat{\Theta}}\right], \ldots, \mathrm{tr}\left[\dot{\hat{\Theta}}\frac{\partial \alpha_{\rho-1\,M}}{\partial \hat{\Theta}}\right]\right]^T,$$
$$\left\{\frac{\partial \alpha_{lk}}{\partial \hat{\Theta}}\right\}_{ij} = \frac{\partial \alpha_{lk}}{\partial \hat{\Theta}_{ji}}, \; k = 1, 2, \ldots, M, \tag{9.517}$$

and β_ρ represents all the known terms except u.

To design the control input u and the adaptive update law $\dot{\hat{\Theta}}$ to stabilize the $(z_1, \ldots, z_\rho, \hat{P}, \hat{\Theta})$-system, we consider the Lyapunov function

$$V_\rho = V_{\rho-1} + \frac{1}{2}z_\rho^T z_\rho + \frac{1}{d_\rho}\epsilon^T P_0 \epsilon. \tag{9.518}$$

Using (9.514) and (9.516), we obtain the time derivative of V_ρ as

$$\dot{V}_\rho \le -\sum_{j=1}^{\rho-1} z_j^T C_j z_j + z_\rho^T\left(z_{\rho-1} + u + \beta_\rho - \frac{\partial \alpha_{\rho-1}}{\partial y}\hat{\Theta}\omega - \delta_\rho\right)$$
$$+ \mathrm{tr}\left[\tilde{\Theta}\Gamma^{-1}\left(\tau_{\rho-1}^T - \Gamma\omega z_\rho^T \frac{\partial \alpha_{\rho-1}}{\partial y} - \dot{\hat{\Theta}}^T\right)\right] + \sum_{j=2}^{\rho-1} z_j^T \tilde{\delta}_j$$
$$- \sum_{j=3}^{\rho-1}\sum_{k=2}^{j-1} z_j^T \sigma_{k,j-1} - z_\rho^T \frac{\alpha_{\rho-1}}{\partial y}\epsilon_2 - \frac{1}{d_\rho}\epsilon^T\epsilon - \sum_{j=1}^{\rho-1}\frac{1}{d_j}\Omega(\epsilon). \tag{9.519}$$

To eliminate $\tilde{\Theta}$ from (9.519), we choose the update law for $\hat{\Theta}$ as

$$\dot{\hat{\Theta}}^T = \tau_\rho^T = \tau_{\rho-1}^T - \Gamma \omega z_\rho^T \frac{\partial \alpha_{\rho-1}}{\partial y}. \tag{9.520}$$

Similar to that in the ith step design, we have

$$\begin{aligned}\dot{V}_\rho \leq & -\sum_{j=1}^{\rho-1} z_j^T C_j z_j + z_\rho^T \left(z_{\rho-1} + u + \beta_\rho - \frac{\alpha_{\rho-1}}{\partial y}\hat{\Theta}\omega - \hat{\delta}_\rho + \sum_{k=2}^{\rho-1} \sigma_{k,\rho-1} \right) \\ & -z_\rho^T \frac{\partial \alpha_{\rho-1}}{\partial y}\epsilon_2 - \frac{1}{d_\rho}\epsilon^T \epsilon - \sum_{j=1}^{\rho-1} \frac{1}{d_\rho}\Omega(\epsilon),\end{aligned} \tag{9.521}$$

where

$$\hat{\delta}_\rho = \left[\operatorname{tr}\left[\tau_\rho \frac{\partial \alpha_{\rho-1\,1}}{\partial \hat{\Theta}} \right], \ldots, \operatorname{tr}\left[\tau_\rho \frac{\partial \alpha_{\rho-1\,M}}{\partial \hat{\Theta}} \right] \right]^T. \tag{9.522}$$

In view of (9.521), we choose the control input u as

$$\begin{aligned} u = & -C_\rho z_\rho - d_\rho \left(\frac{\partial \alpha_{\rho-1}}{\partial y} \right)^T \frac{\partial \alpha_{\rho-1}}{\partial y} z_\rho - z_{\rho-1} \\ & - \beta_\rho + \frac{\partial \alpha_{\rho-1}}{\partial y}\hat{\Theta}\omega + \delta_\rho - \sum_{k=2}^{\rho-1} \sigma_{k,\rho-1}. \end{aligned} \tag{9.523}$$

With this choice of u, we have

$$\begin{aligned}\dot{V}_\rho \leq & -\sum_{j=1}^{\rho} z_j^T C_j z_j - \sum_{j=1}^{\rho} d_j \left(\frac{\partial \alpha_{j-1}}{\partial y} z_j + \frac{1}{2d_j}\epsilon_2 \right)^T \left(\frac{\partial \alpha_{j-1}}{\partial y} z_j + \frac{1}{2d_j}\epsilon_2 \right) \\ & -\sum_{j=1}^{\rho} \frac{1}{d_j}\Omega(\epsilon),\end{aligned} \tag{9.524}$$

where

$$\tilde{\delta}_\rho = \left[\operatorname{tr}\left[(\tau_\rho - \dot{\hat{\Theta}}) \frac{\partial \alpha_{\rho-1\,1}}{\partial \hat{\Theta}} \right], \ldots, \operatorname{tr}\left[(\tau_\rho - \dot{\hat{\Theta}}) \frac{\partial \alpha_{\rho-1\,M}}{\partial \hat{\Theta}} \right] \right]^T. \tag{9.525}$$

The complete $(z_1, \ldots, z_\rho, \hat{P}, \hat{\Theta})$-system is

$$\begin{aligned}\dot{z}_1 &= -C_1 z_1 - d_1 z_1 + B_m z_2 + \tilde{\Theta}\bar{\omega} + B_m \tilde{P}\phi + \epsilon_2, \\ \dot{z}_2 &= -C_2 z_2 - d_2 \left(\frac{\partial \alpha_1}{\partial y} \right)^T \left(\frac{\partial \alpha_1}{\partial y} \right) z_2 + z_3 - B_m^T z_1 - \left(\frac{\partial \alpha_1}{\partial y} \right) \tilde{\Theta}\omega + \tilde{B}_m^T z_1 \\ &\quad + \tilde{\delta}_2 - \left(\frac{\partial \alpha_1}{\partial y} \right) \epsilon_2, \quad \ldots, \\ \dot{z}_j &= -C_j z_j - d_j \left(\frac{\partial \alpha_{j-1}}{\partial y} \right)^T \left(\frac{\partial \alpha_{j-1}}{\partial y} \right) z_j + z_{j+1} - z_{j-1} - \left(\frac{\partial \alpha_{j-1}}{\partial y} \right) \tilde{\Theta}\omega\end{aligned}$$

$$+ \tilde{\delta}_j - \sum_{k=2}^{j-1} \sigma_{k,j-1} - \frac{\partial \alpha_{j-1}}{\partial y} \epsilon_2, \quad \ldots,$$

$$\dot{z}_\rho = -C_\rho z_\rho - d_\rho \left(\frac{\partial \alpha_\rho}{\partial y} \right) z_\rho - z_{\rho-1} - \frac{\partial \alpha_\rho}{\partial y} \tilde{\Theta} \omega - \sum_{k=2}^{\rho-1} \sigma_{k,\rho-1} - \frac{\partial \alpha_{\rho-1}}{\partial y} \epsilon_2,$$

$$\dot{\hat{P}} = S_m^{-T} z_1 \phi^T,$$

$$\dot{\hat{\Theta}}^T = \Gamma \left(\bar{\omega} z_1^T - \sum_{j=2}^{\rho} \omega z_2^T \frac{\partial \alpha_{j-1}}{\partial y} + [0, 0, \ldots, 0, z_2^T, \ldots, 0]^T z_1^T \right). \tag{9.526}$$

From (9.524), we conclude that $z_1, \ldots, z_\rho, \hat{P}$, $\hat{\Theta}$ and ϵ all are bounded, and so is the plant output $y(t)$. With the boundedness of $y(t)$, from (9.456) and (9.457), ξ_i, $i = 0, 1, \ldots, \nu - 1$, are bounded. Using the plant expression (9.440)–(9.441) and filter (9.458), we find that

$$v_j(t) = (sI - A_o)^{-1} E_{\nu-j} N^{-1}(s) D(s)[y](t). \tag{9.527}$$

Since $y(t)$ is bounded and $(sI - A_o)^{-1} E_{\nu-j} N^{-1}(s) D(s)$ is stable and proper, for $j \le m \le \nu - 1$, $v_{0,1}, \ldots, v_{0,m+1}, v_{1,1}, \ldots, v_{1,m}, \ldots, v_{m,\rho-1}$ are all bounded from (9.527). Then we have the boundedness of ϕ in (9.482) and that of $\alpha_i, i = 1, \ldots, \rho-1$. Thus, $v_{m,\rho}$ and $\sigma_{k,j-1}$, $k = 2, \ldots, j-1$, $j = 3, \ldots, i-1$, $i = 2, \ldots, \rho-1$, are bounded, and so are the control u in (9.523), $\hat{x}$ in (9.459), and $x(t)$ in (9.451) (see Lemma 9.11), that is, all closed-loop signals are bounded.

From (9.524) and the boundedness of V_ρ, it follows that

$$\int_0^t \sum_{j=1}^{\rho} z_j^T C_j z_j \, d\tau \le - \int_0^t \dot{V}_\rho \, d\tau = V_\rho(0) - V_\rho(t) \le c_0, \; \forall t \ge 0, \tag{9.528}$$

for some constant $c_0 > 0$, which implies that $z_i(t) \in L^2$, $i = 1, 2, \ldots, \rho$. In particular, the tracking error $z_1 = y - y_m$ belongs to L^2. This is a key feature of the backstepping design which ensures that the error signals $z_i(t)$, $i = 1, 2, \ldots, \rho$, all have finite energy and provides a useful performance measure for the adaptive control system. To show that the tracking error $z_1 = y - y_m$ converges to zero, we use the closed-loop boundedness and the expression of $\dot{z}_1$ in (9.527) to conclude that $\dot{z}_1$ is bounded. This fact, together with the L^2 property of $z_1(t)$, implies that $\lim_{t \to \infty} z_1(t)$ exists and is zero.

9.3.4 Design Based on SDU Decomposition of B_m

The design condition used above, that there is a known matrix S_m such that $S_m B_m = (S_m B_m)^T > 0$, can be relaxed if all leading principal minors of B_m

are nonzero. Similar to that in Section 9.2.11, the SDU decomposition of B_m (see (9.357)) is employed to reparametrize the error system (9.477).

To proceed, assume that all the leading principal minors Δ_i, $i = 1, 2, \ldots, M$, of $B_m \in R^{M\times M}$ are nonzero and let B_m have the SDU decomposition (9.357): $B_m = SD_sU_s$, where $D_s = \mathrm{diag}\{\mathrm{sign}[\Delta_1]\gamma_1, \mathrm{sign}[\frac{\Delta_2}{\Delta_1}]\gamma_2, \ldots, \mathrm{sign}[\frac{\Delta_M}{\Delta_{M-1}}]\gamma_M\}$, $S = LD^*D_s^{-1}L^T$, and $U_s = D_s^{-1}L^{-T}D_sU$ for L and U from the LDU decomposition (3.555) of B_m, with $\gamma_i > 0$, $i = 1, 2, \ldots, M$, being arbitrary.

Step 1. With $B_m = SD_sU_s$ and $z_2 = v_{m,2} - \alpha_1$, we rewrite (9.477) as

$$\begin{aligned}\dot{z}_1 &= \xi_{\nu,2} + \Theta\bar{\omega} + SD_sU_sv_{m,2} + \epsilon_2 - \dot{y}_m \\ &= SD_sz_2 + SD_s(\alpha_1 + (U_s - I)v_{m,2}) + \xi_{\nu,2} + \Theta\bar{\omega} - \dot{y}_m + \epsilon_2 \qquad (9.529)\\ &= SD_sz_2 + SD_s\left(\alpha_1 + [\theta_1^T\eta_1, \theta_2^T\eta_2, \ldots, \theta_{M-1}^T\eta_{M-1}, 0]^T\right) \\ &\quad + \xi_{\nu,2} + \Theta\bar{\omega} - \dot{y}_m + \epsilon_2, \qquad (9.530)\end{aligned}$$

where, since U_s is unit upper triangular, the matrix $U_s - I$ has the form

$$U_s - I = \begin{bmatrix} 0 & \theta_{12} & \theta_{13} & \cdots & \theta_{1M} \\ 0 & 0 & \theta_{23} & \cdots & \theta_{2M} \\ \vdots & \vdots & \vdots & \vdots & \vdots \\ 0 & 0 & \cdots & 0 & \theta_{M-1\,M} \\ 0 & \cdots & \cdots & 0 & 0 \end{bmatrix} \qquad (9.531)$$

and the parameter vectors θ_i and regressors $\eta_i(t)$, $i = 1, 2, \ldots, M-1$, are

$$\begin{aligned}\theta_1 &= [\theta_{12}, \theta_{13}, \ldots, \theta_{1M}]^T \in R^{M-1}, \\ \theta_2 &= [\theta_{23}, \theta_{24}, \ldots, \theta_{2M}]^T \in R^{M-2}, \\ &\vdots \\ \theta_{M-2} &= [\theta_{M-2\,M-1}, \theta_{M-2\,M}]^T \in R^2, \\ \theta_{M-1} &= \theta_{M-1\,M} \in R, \qquad (9.532)\\ \eta_i(t) &= \left[0_{(M-i)\times i}, I_{M-i}\right] v_{m,2}(t),\ i = 1, 2, \ldots, M-1 \qquad (9.533)\end{aligned}$$

(i.e., $\eta_i(t)$ is the vector containing the last $M - i$ elements of $v_{m,2}(t)$).

Introduce the new parameter matrices $P = (SD_s)^{-1}$ and $\Psi = P\Theta$ and denote their estimates as $\hat{P}$ and $\hat{\Theta}$. Let the estimates of θ_i be $\hat{\theta}_i$, $i = 1, 2, \ldots, M-1$, and choose the design signal α_1 as

$$\alpha_1 = -\hat{P}\phi - [\hat{\theta}_1^T\eta_1, \hat{\theta}_2^T\eta_2, \ldots, \hat{\theta}_{M-1}^T\eta_{M-1}, 0]^T - \hat{\Phi}\bar{\omega}, \qquad (9.534)$$

where $c_1 >$ and $d_1 >$ are design parameters and

$$\phi = c_1 z_1 + d_1 z_1 + \xi_{\nu,2} - \dot{y}_m + \hat{\Theta}\bar{\omega}. \tag{9.535}$$

Then, similar to (9.484), we express (9.530) as

$$\begin{aligned}\dot{z}_1 = &-c_1 z_1 - d_1 z_1 + SD_s z_2 + SD_s \tilde{P}\phi \\ &+ SD_s[\tilde{\theta}_1^T \eta_1, \tilde{\theta}_2^T \eta_2, \ldots, \tilde{\theta}_{M-1}^T \eta_{M-1}, 0]^T + SD_s \tilde{\Phi}\bar{\omega} + \epsilon_2,\end{aligned} \tag{9.536}$$

where $\tilde{\Psi} = \Psi - \hat{\Psi}$, $\tilde{P} = P - \hat{P}$, $\tilde{\theta}_i = \theta_i - \hat{\theta}_i$, $i = 1, 2, \ldots, M-1$.

In view of this error equation, we choose the adaptive laws

$$\dot{\hat{P}} = D_s z_1 \phi^T, \tag{9.537}$$

$$\dot{\hat{\Psi}} = D_s z_1 \bar{\omega}^T, \tag{9.538}$$

$$\dot{\hat{\theta}}_i = \text{sign}[d_i^*]\gamma_i z_{1i}\eta_i, \; i = 1, 2, \ldots, M-1, \tag{9.539}$$

where $\text{sign}[d_i^*]\gamma_i$, $i = 1, 2, \ldots, M-1$, are the diagonal elements of D_s, with $\gamma_i > 0$, and z_{1i}, $i = 1, 2, \ldots, M-1$, are the ith components of z_1.

Consider the first Lyapunov function

$$V_1 = \frac{1}{2} z_1^T S^{-1} z_1 + \frac{1}{2}\,\text{tr}[\tilde{\Psi}\tilde{\Psi}^T] + \frac{1}{2}\,\text{tr}[\tilde{P}^T \tilde{P}] + \frac{1}{d_1}\epsilon^T P_1 \epsilon + \frac{1}{2}\sum_{i=1}^{M-1} \tilde{\theta}_i^T \tilde{\theta}_i, \tag{9.540}$$

where $P_1 = P_1^T > 0$ such that $P_1 A_0 + A_0^T P_1 = -\text{diag}\{S^{-1}, S^{-1}, \ldots, S^{-1}\}$ for the stable matrix A_0 in (9.461). It can be verified that

$$\dot{V}_1 = -c_1 z_1^T S^{-1} z_1 + z_1^T D_s z_2 - \frac{1}{d_1}\Omega(\epsilon), \tag{9.541}$$

where

$$\Omega(\epsilon) = \epsilon_1^T S^{-1}\epsilon_1 + \frac{3}{4}\epsilon_2^T S^{-1}\epsilon_2 + \cdots + \epsilon_\nu^T S^{-1}\epsilon_\nu. \tag{9.542}$$

Step 2. With $z_3 = v_{m,3} - \alpha_2$, from (9.491), we have

$$\dot{z}_2 = \alpha_2 + z_3 + \bar{\beta}_2 - \frac{\partial \alpha_1}{\partial y}\Theta\omega - \frac{\partial \alpha_1}{\partial y}\epsilon_2 \tag{9.543}$$

for some known $\bar{\beta}_2$ including $-\delta_2$ which, dependent on $\dot{\Theta}$ and Θ in (9.941), now depends on $\dot{\Psi}$ and Ψ. For the second Lyapunov function

$$V_2 = V_1 + \frac{1}{2} z_2^T z_2 + \frac{1}{2}\,\text{tr}[\tilde{\Theta}\Gamma^{-1}\tilde{\Theta}^T] + \frac{1}{d_2}\epsilon^T P_0 \epsilon, \tag{9.544}$$

where $\Gamma = \Gamma > 0$, $P_0 = P_0^T$ such that $P_0 A_0 + A_0^T P_0 = -I$ for the stable matrix A_0 in (9.461), and $\tilde{\Theta} = \Theta - \hat{\Theta}$ with $\hat{\Theta}$ being the estimate of Θ, we derive

$$\dot{V}_2 \leq -c_1 z_1^T S^{-1} z_1 + z_2^T \left(D_1 z_1 + z_3 + \alpha_2 + \bar{\beta}_2 - \frac{\partial \alpha_1}{\partial y} \hat{\Theta} \omega \right) + \mathrm{tr}\left[\tilde{\Theta} \left(-\omega z_2^T \frac{\partial \alpha_1}{\partial y} - \Gamma^{-1} \dot{\hat{\Theta}}^T \right) \right] - z_2^T \frac{\partial \alpha_1}{\partial y} \epsilon_2 - \frac{1}{d_2} \epsilon^T \epsilon - \frac{1}{d_1} \Omega(\epsilon). \quad (9.545)$$

With $C_2 > 0$ being a diagonal matrix, we choose the design signal

$$\alpha_2 = -C_2 z_2 - d_2 \left(\frac{\partial \alpha_1}{\partial y} \right)^T \frac{\partial \alpha_1}{\partial y} z_2 - \beta_2 + \frac{\partial \alpha_1}{\partial y} \hat{\Theta} \bar{\omega} - D_s z_1 \quad (9.546)$$

and define the tuning function τ_2 as

$$\tau_2^T = -\Gamma \omega z_2^T \frac{\partial \alpha_1}{\partial y}. \quad (9.547)$$

Then, the error equation (9.543) becomes

$$\dot{z}_2 = -C_2 z_2 - d_2 \left(\frac{\partial \alpha_1}{\partial y} \right)^T \frac{\partial \alpha_1}{\partial y} z_2 + z_3 - \frac{\partial \alpha_1}{\partial y} \tilde{\Theta} \omega - \frac{\partial \alpha_1}{\partial y} \epsilon_2. \quad (9.548)$$

From here on, the steps $3, 4, \ldots, \rho = \nu - m$ of the design procedure presented in Section 9.3.3 can be used to complete the adaptive control scheme (note that $C_1 = c_1 S^{-1}$ for (9.509) and $\tau_1 = 0$ for (9.510)) [59].

9.3.5 Design Procedure for B_m Singular

When B_m is singular, the adaptive control problem is essentially more difficult. In this case, additional information about the plant transfer matrix $G(s)$ in (9.440) is needed. A form of such information is the right interactor matrix $\xi^r(s)$ [419] of $G(s)$, which is the dual concept of the interactor matrix $\xi(s)$ [443], as defined in (2.31) (see Proposition 2.2), of $G(s)$, applied to $G^T(s)$ (see Section 9.2.7), characterizing the infinity zero structure of the plant (9.440) from the right. Similar to that in Proposition 2.3 or Lemma 9.1 for the modified left interactor (MLI) matrix $\xi_m(s)$, from the knowledge of the right interactor matrix $\xi^r(s)$ of $G(s)$, one can construct a modified right interactor (MRI) matrix $\xi_m^r(s)$ [384] (see Section 9.2.7) which has the form

$$\xi_m^r(s) = \begin{bmatrix} d_1^r(s) & h_{12}^r(s) & \cdots & \cdots & h_{1M}^r(s) \\ 0 & d_2^r(s) & h_{23}^r(s) & \cdots & h_{2M}^r(s) \\ \vdots & \vdots & \vdots & \vdots & \vdots \\ 0 & \cdots & 0 & d_{M-1}^r(s) & h_{M-1\,M}^r(s) \\ 0 & \cdots & \cdots & 0 & d_M^r(s) \end{bmatrix}, \quad (9.549)$$

where $h_{ij}^r(s)$, $i = 1, \ldots, M-1$, $j = 2, 3, \ldots, M$, are polynomials, and $d_i^r(s)$, $i = 1, \cdots, M$, are any monic stable polynomials of certain degrees $r_i > 0$ such that $\lim_{D \to \infty} G_0(s)\xi_m^r(s) = K_p^r$ is nonsingular and finite.

To develop a backstepping design for B_m singular, we assume: $G(s)$ has full rank; all zeros of $\det[N(s)]$ are stable; the observability index ν of $G(s)$ is known; the modified right interactor matrix $\xi_m^r(s)$ of the plant (9.440) is known; and there is a known matrix $\bar{S}$ such that $\bar{S}K_p^r = (\bar{S}K_p^r)^T > 0$.

Precompensator. With the knowledge of the MRI matrix $\xi_m^r(s)$ of $G(s)$, we construct the precompensator for the plant input:

$$u(t) = \frac{\xi_m^r(s)}{d(s)}[\bar{u}](t), \tag{9.550}$$

where $d(s)$ is a monic stable polynomial of degree n_a, the maximum degree of the elements of $\xi_m^r(s)$, and $\bar{u}(t) \in R^M$ is a new control signal to be designed. With this compensation scheme, the resulting plant transfer matrix $G(s)\frac{\xi_m^r(s)}{d(s)}$, from $\bar{u}(t)$ to $y(t)$, has a simple infinity zero structure: $d(s)I_M$.

Parametrization. After precompensation, we have the new plant

$$y(t) = \bar{G}(s)[\bar{u}](t), \tag{9.551}$$

where the transfer matrix $\bar{G}(s)$ is

$$\bar{G}(s) = G(s)\frac{\xi_m^r(s)}{d(s)} = \bar{D}^{-1}(s)\bar{N}(s) = \bar{C}_g(sI - \bar{A}_g)^{-1}\bar{B}_g, \tag{9.552}$$

similar to that in (9.441) but with new components:

$$\bar{D}(s) = d(s)D(s), \ \bar{N}(s) = N(s)\xi_m(s), \tag{9.553}$$

$$\bar{A}_g = \begin{bmatrix} -\bar{A}_{\bar{\nu}-1} & I_M & 0 & \cdots & \cdots & 0 \\ -\bar{A}_{\bar{\nu}-2} & 0 & I_M & 0 & \cdots & 0 \\ \vdots & \vdots & \vdots & \vdots & \vdots & \vdots \\ -\bar{A}_1 & 0 & \cdots & \cdots & 0 & I_r \\ -\bar{A}_0 & 0 & \cdots & \cdots & 0 & 0 \end{bmatrix} \in R^{M\bar{\nu} \times M\bar{\nu}}, \tag{9.554}$$

$$\bar{B}_g = \begin{bmatrix} 0 \\ \vdots \\ 0 \\ \bar{B}_p \end{bmatrix} \in R^{M\bar{\nu} \times M}, \ \bar{B}_p = \begin{bmatrix} \bar{B}_{\bar{m}} \\ \vdots \\ \bar{B}_1 \\ \bar{B}_0 \end{bmatrix} \in R^{M(\bar{m}+1) \times M}, \tag{9.555}$$

$$\bar{C}_g = \begin{bmatrix} I_M & 0 & \cdots & 0 & 0 \end{bmatrix} \in R^{M \times M\bar{\nu}}, \tag{9.556}$$

with $\bar{A}_i$, $i = 0, 1, \ldots, \bar{\nu} - 1$, and $\bar{B}_j$, $j = 0, 1, \ldots, \bar{m}$, being $M \times M$ parameter matrices. Since $d(s)$ is a monic polynomial, and so is $D(s)$ a monic polynomial matrix (i.e., with the leading coefficient matrix being an identity matrix), we have $\bar{\nu} \triangleq \partial[\bar{D}(s)] = \partial[d(s)D(s)] = \partial[D(s)] + \partial[d(s)] = \nu + n_a$. Since $\lim_{s\to\infty} G(s)\xi_m^r(s) = K_p^r(s)$, $\bar{m} = \partial[N(s)\xi_m^r(s)] = \partial[D(s)] = \nu$.

For the precompensated plant (9.551), we can apply the adaptive backstepping design procedure to construct an adaptive controller. To show some design ideas, next we give an illustrative example.

Design example. Consider the plant (9.441) with

$$D(s) = s^2 I + A_1 s + A_0, \; N(s) = B_1 s + B_0, \tag{9.557}$$

with B_1 *singular* and *nonzero*. Assume that the plant transfer matrix $G(s) = D^{-1}(s)N(s)$ has a MRI matrix of the form

$$\xi_m^r(s) = R_2 s^2 + R_1 s + R_0, \tag{9.558}$$

so we can choose

$$d(s) = s^2 + d_1 s + d_0 I \tag{9.559}$$

for the precompensator design (9.550). For $\xi_m^r(s)$ being a MRI matrix of $G(s)$, we have $B_1 R_2 = 0$, $N(s)\xi_m^r(s) = \bar{B}_2 s^2 + \bar{B}_1 s + \bar{B}_0$, $\bar{B}_2 = K_p^r = B_1 R_1 + B_0 R_2$, and $\bar{m} = \nu = 2$. For the compensated plant (9.551) with $\bar{\nu} = 4$, we define

$$\begin{aligned} \bar{\Theta}_a &= [-\bar{A}_3, -\bar{A}_2, -\bar{A}_1, -\bar{A}_0] \\ &= [-(A_1 + d_1 I), -(A_0 + d_1 A_1 + d_0 I), -(d_1 A_0 + d_0 A_1), -d_0 A_0], \end{aligned} \tag{9.560}$$

$$\begin{aligned} \bar{\Theta}_b &= [\bar{B}_2, \bar{B}_1, \bar{B}_0] \\ &= [B_1 R_1 + B_0 R_2, B_1 R_0 + B_0 R_1, B_0 R_0]. \end{aligned} \tag{9.561}$$

Similar to (9.476) for the case of a nonsingular B_m, we have

$$\dot{y} = \bar{\xi}_{4,2} + \bar{\Theta}_a \bar{\xi}_{(2)} + \bar{\Theta}_b \bar{v}_{(2)} + \epsilon_2 - \bar{A}_3 y. \tag{9.562}$$

Since the parameters in $\bar{\Theta}_a$ are the combinations of the parameters in $\Theta_a = [-A_1, -A_0]$ and the known constants d_1, d_0, we can regroup these parameters to estimate Θ_a instead of the overparametrized $\bar{\Theta}_a$. Introducing

$$\hat{\xi}_{(2)} = [\bar{\xi}_{3,2}^T + d_1 \bar{\xi}_{2,2}^T + d_0 \bar{\xi}_{1,2}^T, \bar{\xi}_{2,2}^T + d_1 \bar{\xi}_{1,2}^T + d_0 \bar{\xi}_{0,2}^T]^T, \tag{9.563}$$

from (9.560), (9.562), and (9.563), we obtain the desired expression for $\dot{y}$:

$$\begin{aligned} \dot{y} &= \bar{\xi}_{4,2}(t) + \Theta_a(\hat{\xi}_{(2)} + E_1 y(t)) + \bar{\Theta}_b \bar{v}_{(2)}(t) \\ &\quad - d_1 \bar{\xi}_{3,2}^T(t) - d_0 \bar{\xi}_{2,2}^T(t) - d_0 y(t) + \epsilon_2(t). \end{aligned} \tag{9.564}$$

With this new parametrization, the backstepping design has two steps.

Step 1: From (9.564), defining the measured signal $f(t) = d_1\bar{\xi}_{3,2}^T(t) + d_0\bar{\xi}_{2,2}^T(t) + d_0 y(t)$, we find the derivative of the tracking error $z_1 = y - y_m$ as

$$\dot{z}_1 = \bar{\xi}_{4,2} + \bar{\Theta}\bar{\omega} + \bar{B}_2\bar{v}_{3,2} + \epsilon_2 - \dot{y}_r - f, \tag{9.565}$$

where

$$\bar{\Theta} = [\Theta_a, \bar{\Theta}_b], \tag{9.566}$$

$$\bar{\omega} = [\bar{\xi}_{(2)} + E_1 y, [0, \bar{v}_{1,2}^T, \bar{v}_{0,2}^T]^T]^T. \tag{9.567}$$

Defining $z_2 = \bar{v}_{2,2} - \alpha_1$, with α_1 to be designed, and using (9.565), we have

$$\dot{z}_1 = \bar{B}_2\alpha_1 + \bar{B}_2 z_2 + \bar{\xi}_{4,2} + \bar{\Theta}\bar{\omega} - \dot{y}_r + \epsilon_2 - f. \tag{9.568}$$

Letting $\hat{P}$ and $\hat{\Theta}$ be the estimates of $P = \bar{B}_2^{-1}$ and $\bar{\Theta}$ and choosing a diagonal matrix $C_1 > 0$ and scalar $d_1 > 0$, we rewrite (9.568) as

$$\begin{aligned}\dot{z}_1 = &-C_1 z_1 - d_1 z_1 + \bar{B}_2 z_2 \\ &+ \bar{B}_2\left(\alpha_1 + P(C_1 z_1 + d_1 z_1 + \bar{\xi}_{4,2} - \dot{y}_r - f) + P\bar{\Theta}\bar{\omega}\right) + \epsilon_2\end{aligned} \tag{9.569}$$

(note that $\bar{B}_2 = K_p^r$ is nonsingular). To proceed, we define

$$\bar{\phi} = C_1 z_1 + d_1 z_1 + \bar{\xi}_{4,2} - \dot{y}_r + \hat{\Theta}\omega - f \tag{9.570}$$

and choose the design signal

$$\alpha_1 = -\hat{P}\bar{\phi} \tag{9.571}$$

and the adaptive update law for $\hat{P}$ as

$$\dot{\hat{P}} = \bar{S}^{-T} z_1 \bar{\phi}^T. \tag{9.572}$$

Then for V_1 in (9.486) with $B_2^T S^T$ replaced by $\bar{B}_2^T \bar{S}^T$, we have

$$\dot{V}_1 \leq -z_1^T C_1 z_1 + z_1^T \bar{B}_2 z_2 + z_1^T \tilde{\Theta}\bar{\omega} + \operatorname{tr}\left[\tilde{\Theta}\Gamma^{-1}\dot{\tilde{\Theta}}^T\right] - \frac{1}{d_1}\Omega(\epsilon), \tag{9.573}$$

where $\tilde{\Theta} = \bar{\Theta} - \hat{\Theta}$ and

$$\Omega(\epsilon) = \epsilon_1^T\epsilon_1 + \frac{3}{4}\epsilon_2^T\epsilon_2 + \epsilon_3^T\epsilon_3 + \epsilon_4^T\epsilon_4. \tag{9.574}$$

Step 2: Express the derivative of $z_2 = \bar{v}_{2,2} - \alpha_1$ as

$$\dot{z}_2 = \bar{u} + \beta_2 - \frac{\partial\alpha_1}{\partial y}\Theta\omega - \frac{\partial\alpha_1}{\partial y}\epsilon_2 - \delta_2, \tag{9.575}$$

where

$$\frac{\partial \alpha_1}{\partial y} = \begin{bmatrix} \frac{\partial \alpha_{11}}{\partial y} \\ \vdots \\ \frac{\partial \alpha_{1M}}{\partial y} \end{bmatrix} = \begin{bmatrix} \frac{\partial \alpha_{11}}{\partial y_1} & \cdots & \frac{\partial \alpha_{11}}{\partial y_m} \\ \vdots & & \vdots \\ \frac{\partial \alpha_{1M}}{\partial y_1} & \cdots & \frac{\partial \alpha_{1M}}{\partial y_m} \end{bmatrix}, \tag{9.576}$$

$$\delta_2 = \left[\text{tr}\left[\dot{\hat{\Theta}} \frac{\partial \alpha_{11}}{\partial \hat{\Theta}}\right], \ldots, \text{tr}\left[\dot{\hat{\Theta}} \frac{\partial \alpha_{1M}}{\partial \hat{\Theta}}\right]\right]^T, \tag{9.577}$$

$$\left\{ \frac{\partial \alpha_{1k}}{\partial \hat{\Theta}} \right\}_{ij} = \frac{\partial \alpha_{1k}}{\partial \hat{\Theta}_{ji}}, \; k = 1, 2, \ldots, M, \tag{9.578}$$

$$\omega = [\bar{\xi}_{(2)} + E_1 y, [\bar{v}_{2,2}^T, \bar{v}_{1,2}^T, \bar{v}_{0,2}^T]^T]^T, \tag{9.579}$$

and β_2 represents all other known terms.

For V_2 in (9.496), in view of (9.573) and (9.575), we have

$$\begin{aligned} \dot{V}_2 = & -z_1^T C_1 z_1 + z_2^T \left(\hat{B}_2^T z_1 + u + \beta_2 - \frac{\partial \alpha_1}{\partial y} \hat{\Theta} \omega - \delta_2 \right) + z_2^T \tilde{B}_2^T z_1 \\ & + \text{tr}\left[\tilde{\Theta}(\omega z_1^T - \Gamma^{-1} \dot{\hat{\Theta}}^T) \right] - z_2^T \frac{\partial \alpha_1}{\partial y} \tilde{\Theta} \omega - z_2^T \frac{\partial \alpha_1}{\partial y} \epsilon_2 - \frac{\epsilon^T \epsilon}{d_2} - \frac{\Omega(\epsilon)}{d_1}. \end{aligned} \tag{9.580}$$

With the choice of the adaptive update law for $\hat{\Theta}$,

$$\dot{\hat{\Theta}}^T = \Gamma \left(\bar{\omega} z_1^T - \omega z_2^T \frac{\partial \alpha_1}{\partial y} + [0, \ldots z_2^T, \ldots 0]^T z_1^T \right) \tag{9.581}$$

and the precompensator input $\bar{u}$,

$$\bar{u} = -C_2 z_2 - d_2 \left(\frac{\partial \alpha_1}{\partial y} \right)^T \frac{\partial \alpha_1}{\partial y} z_2 - \beta_2 + \frac{\partial \alpha_1}{\partial y} \hat{\Theta} \omega - \hat{B}_2^T z_1 + \hat{\delta}_2, \tag{9.582}$$

where $\hat{\delta}_2 = \delta_2$ for $\bar{m} = 2$, we have

$$\dot{V}_2 \le -z_1^T C_1 z_1 - z_2^T C_2 z_2 - d_2 \left\| \frac{\partial \alpha_1}{\partial y} z_2 + \frac{1}{2d_2} \epsilon_2 \right\|_2^2 - \left(\frac{1}{d_1} + \frac{1}{d_2} \right) \Omega(\epsilon). \tag{9.583}$$

From this inequality it follows that $z_1, z_2, \hat{P}$, and $\hat{\Theta}$ are bounded, and so is the plant output $y(t)$. Similar to the case when B_m is nonsingular, we can show the boundedness of $\bar{v}_{0,1}$, $\bar{v}_{0,2}$, $\bar{v}_{0,3}$, $\bar{v}_{1,1}$, $\bar{v}_{1,2}$, $\bar{v}_{2,1}$, as well as the boundedness of all other closed-loop signals. Using (9.583) and signal boundedness, we have that $z_1 = y - y_m \in L^2$, $z_2 \in L^2$, and from (9.569), it follows that $\dot{z}_1$ is also bounded, so that $\lim_{t\to\infty} z_1(t) = 0$ (see Lemma 2.14).

9.4 Adaptive Control of Robotic Systems

In this section, we design and analyze robust adaptive control schemes for robot manipulator systems with parameter uncertainties, parameter time variations, and unmodeled dynamics, demonstrating that adaptive control techniques developed so far are readily applicable to this class of multivariable systems whose dynamics are nonlinear in nature, even in the presence of parameter variations and unmodeled dynamics whose size may not be small.

9.4.1 Robotic System Modeling

We first present the mathematical models of robot manipulators with time-varying parameters or unmodeled dynamics and their parametrized forms and then use a 2-link planar manipulator to illustrate these models.

Euler-Lagrange Equations

To derive the dynamic equations of a robot manipulator whose parameters may explicitly depend on time t, we use the Euler-Lagrange equations [1], [217], [302], [357] of a robot manipulator system, which are expressed as

$$\frac{d}{dt}\frac{\partial L}{\partial \dot{q}} - \frac{\partial L}{\partial q} = u, \tag{9.584}$$

where $q = [q_1, \ldots, q_n]^T$ is a set of position variables of n ($n > 0$) joints of the robot manipulator, $u = [u_1, \ldots, u_n]^T$ is the applied joint torque vector, and L is the Lagrangian defined as $L = K - P$, the difference between the kinetic energy K and the potential energy P of the form

$$K = \frac{1}{2}\dot{q}^T D(q,t)\dot{q}, \; P = P(q,t) \tag{9.585}$$

with $D(q,t) \in R^{n\times n}$ being the symmetric and positive definite manipulator inertia matrix. For (9.584) and (9.585) to represent the manipulator dynamics with time-varying parameters, the mass and the moment of inertia of each link of the manipulator should not explicitly depend on $\dot{q}$.

Letting d_{ij} be the (i,j)th element of $D(q,t)$ and $\phi(q,t) = \frac{\partial P(q,t)}{\partial q}$ and substituting (9.585) into (9.584), we obtain the manipulator dynamic equation

$$D(q,t)\ddot{q} + \frac{\partial D(q,t)}{\partial t}\dot{q} + C(q,\dot{q},t)\dot{q} + \phi(q,t) = u, \tag{9.586}$$

where the kjth element of $C(q,\dot{q},t) \in R^{n\times n}$ is

$$c_{kj} = \sum_{i=1}^{n} \frac{1}{2}\left(\frac{\partial d_{kj}}{\partial q_i} + \frac{\partial d_{ki}}{\partial q_j} - \frac{\partial d_{ij}}{\partial q_k}\right)\dot{q}_i. \tag{9.587}$$

A key feature of the manipulator model (9.586) is that the inertia matrix $D(q,t)$ and the potential energy $P(q,t)$ are explicitly time dependent, which takes into account the effect of changes in environment of the robot system or changes of the manipulator dynamics with time.

Moreover, an important property of the manipulator model (9.586) is

$$x^T\left(M(q,\dot{q},t) - 2C(q,\dot{q},t)\right)x = 0,\ \forall x \in R^n, \tag{9.588}$$

where $M(q,\dot{q},t) = \frac{dD(q,t)}{dt} - \frac{\partial D(q,t)}{\partial t}$, whose (i,j)th element is $(\frac{\partial d_{ij}}{\partial q})^T\dot{q}$. When $D(q,t) = D(q)$ does not explicitly depend on t, that is, $\frac{\partial D(q,t)}{\partial t} = 0$, (9.588) becomes the well-known equation $x^T(\frac{dD(q)}{dt} - 2C(q,\dot{q}))x = 0$ [357].

A manipulator with unmodeled dynamics may be modeled as

$$\begin{aligned} D(q)\ddot{q} + C(q,\dot{q})\dot{q} + \phi(q) = u &+ H_1[g_1(\dot{q})](t) + H_2[g_2(q)](t) \\ &+ H_3[g_3(u)](t), \end{aligned} \tag{9.589}$$

where H_i, $i = 1,2,3$, with appropriate dimensions, are linear operators with rational transfer matrices, representing the unmodeled part of the robot dynamics, and $g_1(\dot{q})$, $g_2(q)$, $g_3(u)$ are certain vector functions of $\dot{q}$, q, u. For simplicity, $D(q)$, $C(q,\dot{q})$, and $\phi(q)$ are assumed not explicitly time dependent in the unmodeled dynamics problem. The model (9.589) is generalized from some practical robot systems [302], [330], [356].

System Parametrization

We use the transformation technique developed in [350] to parametrize the manipulator model (9.586) or (9.589). Let Λ_0 be any $n \times n$ constant matrix whose eigenvalues have positive real parts and define

$$v = \dot{q}_d - \Lambda_0(q - q_d),\ s = \dot{q} - v,\ e = q - q_d \tag{9.590}$$

Clearly, it follows from (9.590) that

$$s = \dot{e} + \Lambda_0 e,\ \dot{s} = \ddot{q} - \dot{v} \tag{9.591}$$

and s, v, $\dot{v}$ depend only on q, q_d, $\dot{q}$, $\dot{q}_d$, $\ddot{q}_d$ and not on the joint acceleration vector $\ddot{q}(t)$. Using (9.591), we express the manipulator model (9.586) as

$$\begin{aligned} &D(q,t)\dot{s} + C(q,\dot{q},t)s \\ &= u - D(q,t)\dot{v} - C(q,\dot{q},t)v - \phi(q,t) - \frac{\partial D(q,t)}{\partial t}\dot{q} \\ &\triangleq u - Y(q,q_d,\dot{q},\dot{q}_d,\ddot{q}_d,t)\theta^*(t) - \frac{\partial D(q,t)}{\partial t}\dot{q}, \end{aligned} \tag{9.592}$$

where $Y(q, q_d, \dot{q}, \dot{q}_d, \ddot{q}_d, t)$ is an $n \times n_\theta$ matrix of known functions for some $n_\theta > 0$ and $\theta^*(t) \in R^{n_\theta}$ contains parameters that may be time varying. In (9.592), the regressor $Y(q, q_d, \dot{q}, \dot{q}_d, \ddot{q}_d, t)$ is bounded for bounded q, q_d, $\dot{q}$, $\dot{q}_d$, $\ddot{q}_d$.

Similarly, for the manipulator model (9.589), we obtain

$$D(q)\dot{s} + C(q, \dot{q})s = u - Y(q, q_d, \dot{q}, \dot{q}_d, \ddot{q}_d)\theta^* + H_1[g_1(\dot{q})](t) + H_2[g_2(q)](t) + H_3[g_3(u)](t), \quad (9.593)$$

where, for this unmodeled dynamics problem, $\theta^* \in R^{n_\theta}$ is a constant vector, $Y(q, q_d, \dot{q}, \dot{q}_d, \ddot{q}_d)$ is not explicitly time dependent.

Notation

To present adaptive control designs and analysis, we use $\|\cdot\|$ to denote a Euclidean (l^2) vector norm or its induced matrix norm and $\|x(\cdot)\|_\infty$ (or $\|x(\cdot)\|_1$, $\|x(\cdot)\|_2$) to denote the L^∞ (or L^1, L^2) signal norm of $x(t)$ (see Section 2.3.2). We also use the operator norms (see Section 2.8), summarized as follows.

Considering a linear time-invariant operator H represented by an $n \times n$ stable proper rational matrix $H(s)$ whose impulse response matrix (i.e., its inverse Laplace transform) is $h(t) = \{h_{ij}(t)\}$ and a vector signal $x(t) = [x_1(t), \ldots, x_n(t)]^T$, we have the signal and operator norms

$$\|x(\cdot)\|_\infty = \max_{1 \le i \le n} \sup_{t \ge 0} |x_i(t)|, \ \|H\|_\infty = \max_{1 \le i \le n} \int_0^\infty \sum_{j=1}^n |h_{ij}(\tau)|\, d\tau,$$

$$\|x(\cdot)\|_2 = \sqrt{\int_0^\infty \sum_{i=1}^n |x_i(t)|^2\, dt}, \ \|H\|_2 = \sqrt{\sup_{\omega \ge 0} \max_{1 \le i \le n} \lambda_i[H^T(-j\omega)H(j\omega)]},$$

$$\|x(\cdot)\|_1 = \int_0^\infty \sum_{i=1}^n |x_i(t)|\, dt, \ \|H\|_1 = \max_{1 \le j \le n} \int_0^\infty \sum_{i=1}^n |h_{ij}(\tau)|\, d\tau. \quad (9.594)$$

Note that the L^2 induced operator norm $\|H\|_2 = \|H(s)\|_\infty$, the H^∞ norm of the transfer matrix $H(s)$, for a stable $H(s)$.

Let $y(t) = H(s)[x](t)$, that is, $y(t) = \int_0^t h(t-\tau)x(\tau)\, d\tau$. These norms satisfy $\|y(\cdot)\|_\infty \le \|\bar{H}\|_\infty \|x(\cdot)\|_\infty$, $\|y(\cdot)\|_1 \le \|H\|_1 \|x(\cdot)\|_1$, and $\|y(\cdot)\|_2 \le \|H\|_2 \|x(\cdot)\|_2$ [78]. Each inequality can be made an equality by some choice of $x(t)$. The case when $H(s)$ is not square can be similarly formulated.

For a signal $x(t)$, let x_t be the truncated version of x: $x_t(\tau) = x(\tau)$ for $0 \le \tau \le t$ and $x_t(\tau) = 0$ for $\tau > t$. If follows that

$$\|x_t(\cdot)\|_\infty = \max_{1 \le i \le n} \sup_{0 \le \tau \le t} |x_i(\tau)|. \quad (9.595)$$

Control Objective

The control objective is, for a given reference position vector $q_d(t)$ with $q_d(t)$, $\dot{q}_d(t)$, $\ddot{q}_d(t)$ bounded, to generate the applied torque u for the manipulator (9.586) or (9.589) with unknown parameters so that all signals in the closed-loop robot system are bounded and the joint position vector q tracks q_d as close as possible.

For the manipulator model (9.592), we assume:

(A9m.1): $\|\theta^*(t)\| \leq \rho_0$, $\|\dot{\theta}^*(t)\| \leq \rho$ for some constants $\rho_0 > 0$, $\rho > 0$, and all $t \geq 0$.

(A9m.2): $\|\frac{\partial D(q,t)}{\partial t}\| \leq \gamma f(q)$ for some constant $\gamma > 0$ and known function $f(q)$ which is bounded for any bounded q.

For the manipulator model (9.593), we assume

(A9m.3): For any $t \geq 0$, $\|(g_1(\dot{q}(\cdot)))_t\|_\infty \leq f_1(\dot{q}(t))$, $\|(g_2(q(\cdot)))_t\|_\infty \leq f_2(q(t))$, for some known functions $f_1(\dot{q})$, $f_2(q)$, which are bounded for bounded $\dot{q}$ and q; and $\|H_1\|_\infty \leq \gamma_1$, $\|H_2\|_\infty \leq \gamma_2$, for some constants $\gamma_1 > 0$, $\gamma_2 > 0$.

(A9m.4): $\|(g_3(u(\cdot)))_t\|_\infty \leq \|u_t(\cdot)\|_\infty$ and $\|H_3^i\|_\infty \leq \mu_i$, where H_3^i is the ith row of H_3, $i = 1, \ldots, n$.

Assumption (A9m.1) requires only the boundedness of the manipulator parameters and their derivatives, not the smallness of the time variations of the parameters. Smallness of the parameter variations is usually an assumption for the design of adaptive control schemes for time-varying plants, which is not needed here because of the special structure of the robot manipulator dynamics. Assumption (A9m.2) requires that $\frac{\partial D(q,t)}{\partial t}$ satisfies certain relative boundedness condition. Assumption (A9m.3) requires that the L^∞ gains of H_1, H_2 are finite and $g_1(\dot{q})$, $g_2(q)$ satisfy certain relative boundedness conditions. Assumption (A9m.4) is similar to (A9m.3) but $\mu_i \geq 0$, $i = 1, \ldots, n$, are to be specified for the robust stability of an adaptive robot control system. We note that the bounds γ, γ_1, γ_2 are not needed for an adaptive controller design.

9.4.2 Illustrative Example

We now present a 2-link planar manipulator [357] (see Figure 9.1) as an illustrative example for the robot system modeling and parametrization.

The manipulator configuration may be described as: there are two revolute joints with joint angles q_1, q_2, two links with masses M_1, M_2, lengths l_1, l_2,

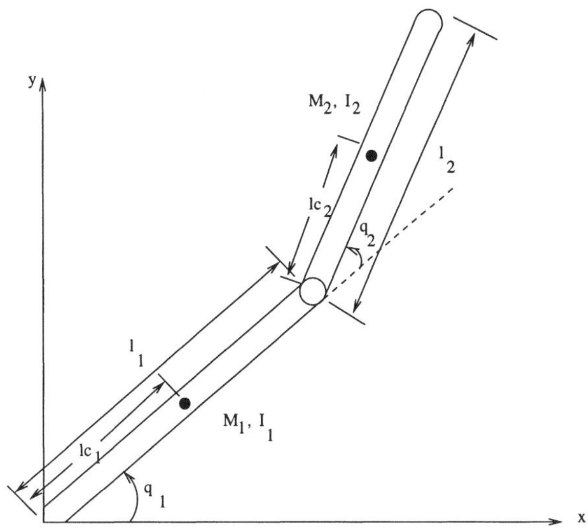

Figure 9.1: The 2-link planar manipulator.

distances l_{c1}, l_{c2} from the joints to the mass centers, and rotational inertias I_1, I_2. The inertia matrix $D(q,t)$ has four elements,

$$
\begin{aligned}
d_{11} &= M_1 l_{c1}^2 + M_2(l_1^2 + l_{c2}^2 + 2l_1 l_{c2}\cos q_2) + I_1 + I_2, \\
d_{12} = d_{21} &= M_2(l_{c2}^2 + l_1 l_{c2}\cos q_2) + I_2, \\
d_{22} &= M_2 l_{c2}^2 + I_2,
\end{aligned}
\tag{9.596}
$$

and the potential energy is

$$
P(q,t) = (M_1 l_{c1} + M_2 l_1) g \sin q_1 + M_2 l_{c2} g \sin(q_1 + q_2), \tag{9.597}
$$

where g is the gravitational acceleration constant. The matrix $C(q, \dot{q}, t)$ in (9.586) has four elements,

$$
c_{11} = h\dot{q}_2,\ c_{12} = (\dot{q}_1 + \dot{q}_2)h,\ c_{21} = -\dot{q}_1 h,\ c_{22} = 0, \tag{9.598}
$$

where $h = -M_2 l_1 l_{c2} \sin q_2$ [357].

Without parameter variations and unmodeled dynamics, we have

$$
D(q)\dot{s} + C(q,\dot{q})s = u - Y(q, q_d, \dot{q}, \dot{q}_d, \ddot{q}_d)\theta^* \tag{9.599}
$$

as the model of the planar robot manipulator, where, for $v = [v_1, v_2]^T$,

$$
\theta^* = [\theta_1^*, \ldots, \theta_9^*]^T
$$

$$\triangleq [M_1 l_{c1}^2, M_2 l_1^2, M_2 l_{c2}^2, M_2 l_1 l_{c2}, I_1, I_2, M_1 l_{c1} g, M_2 l_1 g, M_2 l_{c2} g]^T, \tag{9.600}$$

$$Y(q, q_d, \dot{q}, \dot{q}_d, \ddot{q}_d) \triangleq Y_0(q, \dot{q}, v, \dot{v})$$
$$= \begin{bmatrix} \dot{v}_1 & \dot{v}_1 & \dot{v}_1 + \dot{v}_2 & \cos q_2(2\dot{v}_1 + \dot{v}_2) - \sin q_2(\dot{q}_2 v_1 + (\dot{q}_1 + \dot{q}_2)v_2) \\ 0 & 0 & \dot{v}_1 + \dot{v}_2 & \cos q_2 \dot{v}_1 + \sin q_2 \dot{q}_1 v_1 \end{bmatrix.$$
$$\begin{matrix} \dot{v}_1 & \dot{v}_1 + \dot{v}_2 & \cos q_1 & \cos q_1 & \cos(q_1 + q_2) \\ 0 & \dot{v}_1 + \dot{v}_2 & 0 & 0 & \cos(q_1 + q_2) \end{matrix}\Bigg]. \tag{9.601}$$

When $\theta^* = \theta^*(t)$ is time varying, the manipulator model is (9.592) with the same $Y(q, q_d, \dot{q}, \dot{q}_d, \ddot{q}_d)$ in (9.601) and

$$\frac{\partial D(q,t)}{\partial t}$$
$$= \begin{bmatrix} 2\dot{\theta}_4^*(t)\cos q_2 + \sum_{i=1, i\neq 4}^{6} \dot{\theta}_i^*(t) & \dot{\theta}_3^*(t) + \dot{\theta}_4^*(t)\cos q_2 + \dot{\theta}_6^*(t) \\ \dot{\theta}_3^*(t) + \dot{\theta}_4^*(t)\cos q_2 + \dot{\theta}_6^*(t) & \dot{\theta}_3^*(t) + \dot{\theta}_6^*(t) \end{bmatrix}, \tag{9.602}$$

which represents parameter variations. Assuming that $|\dot{\theta}_i(t)| \leq \rho_i$, we obtain the bound ρ in (A9m.1) as $\rho = \sqrt{\sum_{i=1}^{9} \rho_i^2}$ and the bounds γ, $f(q)$ in (A9m.2) as $\gamma = \sqrt{\rho_1^2 + \rho_2^2 + 4\rho_3^2 + 6\rho_4^2 + \rho_5^2 + 4\rho_6^2}$, $f(q) = 1$.

9.4.3 Design for Parameter Variations

In this section we present an adaptive control scheme for robot manipulators modeled by (9.592) and analyze its stability and tracking properties.

When the inertia $D(q,t) = D(q)$ and the potential energy $P(q,t) = P(q)$ are not explicitly time dependent, that is, $\frac{\partial D(q,t)}{\partial t} = 0$, $\theta^*(t) = \theta^*$ is constant in (9.592), the adaptive control scheme [350]

$$u(t) = Y(q, q_d, \dot{q}, \dot{q}_d, \ddot{q}_d)\theta(t) - K_D s(t),\ 0 < K_D = K_D^T \in R^{n \times n}, \tag{9.603}$$

$$\dot{\theta}(t) = -\Gamma^{-1} Y^T(q, q_d, \dot{q}, \dot{q}_d, \ddot{q}_d) s(t),\ 0 < \Gamma = \Gamma^T \in R^{n_\theta \times n_\theta}, \tag{9.604}$$

guarantees that the closed-loop system is globally stable and convergent in the sense that $q(t)$, $\dot{q}(t)$, $\theta(t)$ are bounded and $\lim_{t\to\infty} e(t) = 0$. This can be proved by considering the positive definite function

$$V(s, \tilde{\theta}) = \frac{1}{2}(s^T D s + \tilde{\theta}^T \Gamma \tilde{\theta}),\ \tilde{\theta}(t) = \theta(t) - \theta^*,\ D = D(q(t)), \tag{9.605}$$

whose time derivative along (9.592) and (9.604) is $\dot{V} = -s^T(t) K_D s(t) \leq 0$.

When $D(q,t)$, $P(q,t)$ are both explicitly time dependent, we have obtained the manipulator model as (9.592) in which $\theta^*(t)$ is time varying and the term $\frac{\partial D(q,t)}{\partial t}\dot{q}$ appears. If the parameters in $D(q,t)$, $P(q,t)$ were known, then $\theta^*(t)$, $\frac{\partial D(q,t)}{\partial t}\dot{q}$ could be calculated so that the control law

$$u(t) = Y(q, q_d, \dot{q}, \dot{q}_d, \ddot{q}_d, t)\theta^*(t) + \frac{\partial D(q,t)}{\partial t}\dot{q} - K_D s \tag{9.606}$$

could be implemented, which guarantees global stability and asymptotic tracking. For unknown $D(q,t)$, $P(q,t)$, next we develop an adaptive control scheme which is robust with respect to the time variation of $\theta^*(t)$ and $\frac{\partial D(q,t)}{\partial t}\dot{q}$.

With $Y(q, q_d, \dot{q}, \dot{q}_d, \ddot{q}_d, t)$, s, K_D, Γ defined before, we propose the following feedback controller structure for the manipulator (9.592):

$$u(t) = Y(q, q_d, \dot{q}, \dot{q}_d, \ddot{q}_d, t)\theta(t) - m(t)\psi(t) - K_D s(t), \tag{9.607}$$

$$m(t) = k_0\|\dot{q}(t) + v(t)\|f(q), \; \psi(t) = m(t)s(t) \tag{9.608}$$

and the following update law for $\theta(t)$:

$$\dot{\theta}(t) = -\Gamma^{-1}\left(Y^T(q, q_d, \dot{q}, \dot{q}_d, \ddot{q}_d, t)s + \sigma(t)\theta(t)\right), \tag{9.609}$$

where $\sigma(t)$ is the switching signal [151], [154] (see Section 5.5.2) using *a priori* knowledge of the upper bound M on $\sup_{t\geq 0}\|\theta^*(t)\|$ and a $\sigma_0 > 0$:

$$\sigma(t) = \begin{cases} 0 & \text{if } \|\theta(t)\| < M, \\ \sigma_0(\frac{\|\theta(t)\|}{M} - 1) & \text{if } M \leq \|\theta(t)\| < 2M, \\ \sigma_0 & \text{if } \|\theta(t)\| \geq 2M. \end{cases} \tag{9.610}$$

This adaptive controller has the desired stability and tracking properties.

Theorem 9.9 *All signals in the closed-loop system are bounded, and the tracking error $e(t) = q(t) - q_d(t)$ satisfies*

$$\int_{t_1}^{t_2} \|e(t)\|^2\,dt \leq \alpha_0\left(\frac{\gamma^2}{k_0^2} + \rho\right)(t_2 - t_1) + \beta_0 \tag{9.611}$$

for some constants $\alpha_0 > 0$, $\beta_0 > 0$ and any $t_2 > t_1 \geq 0$. Moreover, $e(t) \in L^2$ and $\lim_{t\to\infty} e(t) = 0$ in the absence of parameter time variations, that is, when $\dot{\theta}^*(t) = 0$ and $\frac{\partial D(q,t)}{\partial t} = 0$.

Proof: Considering the positive definite function

$$V(s, \tilde{\theta}) = \frac{1}{2}(s^T D s + \tilde{\theta}^T\Gamma\tilde{\theta}), \; \tilde{\theta}(t) = \theta(t) - \theta^*(t), \; D = D(q(t), t) \tag{9.612}$$

from (9.588), (9.591), (9.592), (9.607)–(9.609), (A9m.1), (A9m.2), we have

$$\begin{aligned}
\dot{V} &= -s^T(t)K_D s(t) - m^2(t)s^T(t)s(t) \\
&\quad -\frac{1}{2}s^T(t)\frac{\partial D(q,t)}{\partial t}(\dot{q}(t)+v(t)) - \sigma(t)\tilde{\theta}^T(t)\theta(t) - \tilde{\theta}^T(t)\Gamma\dot{\theta}^*(t) \\
&\leq -s^T(t)K_D s(t) - \left(m(t)\|s(t)\| - \frac{\gamma}{4k_0}\right)^2 + \frac{\gamma^2}{16k_0^2} \\
&\quad -\sigma(t)\tilde{\theta}^T(t)\theta(t) - \tilde{\theta}^T(t)\Gamma\dot{\theta}^*(t) \\
&\leq -s^T(t)K_D s(t) - \left(m(t)\|s(t)\| - \frac{\gamma}{4k_0}\right)^2 + \frac{\gamma^2}{16k_0^2} - (\sigma(t)-\sigma_0)\tilde{\theta}^T(t)\theta(t) \\
&\quad -\sigma_0\left(\|\theta(t)\| - \left(\frac{\rho_0}{2} + \frac{\rho\|\Gamma\|}{2\sigma_0}\right)\right)^2 + \frac{\sigma_0}{4}\left(\rho_0 + \frac{\rho\|\Gamma\|}{\sigma_0}\right)^2 + \rho_0\rho\|\Gamma\|. \qquad (9.613)
\end{aligned}$$

Since γ, ρ, ρ_0 are constants and $\sigma(t)$ defined in (9.610) satisfies

$$|(\sigma(t)-\sigma_0)\tilde{\theta}^T(t)\theta(t)| \leq 12\sigma_0 M^2, \qquad (9.614)$$

it follows from the second inequality of (9.613) that $\dot{V} \leq 0$ for $\theta(t)$, $s(t)$ outside a certain bounded set. Therefore $s(t)$ and $\theta(t)$ are bounded, which, in view of (9.591), (9.607), implies that $q(t)$, $\dot{q}(t)$, $u(t)$ are also bounded.

Since $\sigma(t)\tilde{\theta}^T(t)\theta(t) \geq 0$, using the first inequality of (9.613), we obtain

$$\dot{V} \leq -s^T(t)K_D s(t) + \frac{\gamma^2}{16k_0^2} + k_1\rho \qquad (9.615)$$

for some $k_1 > 0$. Since $V(\epsilon(t), \tilde{\Theta}(t), \tilde{\Psi}(t))$ is bounded, from (9.615), we have

$$\int_{t_1}^{t_2} \|s(t)\|^2\,dt \leq \alpha_1\left(\frac{\gamma^2}{k_0^2} + \rho\right)(t_2 - t_1) + \beta_1 \qquad (9.616)$$

for some constants $\alpha_1 > 0$, $\beta_1 > 0$, and any $t_2 > t_1 \geq 0$.

To show that (9.616) implies (9.611), let us consider the relation $s(t) = \dot{e}(t) + \Lambda_0 e(t)$, where $-\Lambda_0$ is a stable matrix (see (9.591)), and denote H as the linear operator or the impulse-response matrix from $s(t)$ to $e(t)$ as the case may be, that is, $e(t) = H[s](t) = \int_0^t H(t-\tau)s(\tau)\,d\tau$. It follows that

$$\begin{aligned}
&\int_{t_1}^{t_2} \|e(t)\|^2\,dt = \int_{t_1}^{t_2} \|\int_0^t H(t-\tau)s(\tau)d\tau\|^2\,dt \\
&\leq 2\int_{t_1}^{t_2} \|\int_{t_1}^{t} H(t-\tau)s(\tau)\,d\tau\|^2 dt + 2\int_{t_1}^{t_2} \|\int_0^{t_1} H(t-\tau)s(\tau)d\tau\|^2\,dt \\
&\leq 2\|H\|_2^2 \int_{t_1}^{t_2} \|s(t)\|^2\,dt + 2\int_{t_1}^{t_2} \|H(t-t_1)e(t_1)\|^2\,dt. \qquad (9.617)
\end{aligned}$$

Since the operator H is stable and $e(t)$ is bounded, both $\|H\|_2$ and $\int_{t_1}^{t_2} \|H(t-t_1)e(t_1)\|^2 \, dt$ are finite for any $t_2 > t_1 \geq 0$ (see the proof of Lemma 2.8 in Section 2.5.3). Hence from (9.616) and (9.617) we prove (9.611).

When $\dot{\theta}^*(t) = 0$, $\frac{\partial D(q,t)}{\partial t} = 0$, that is, $\rho = \gamma = 0$, it follows from (9.611) that $e(t) \in L^2$. Since $\dot{e}(t) = s(t) - \Lambda_0 e(t)$ is bounded, $\lim_{t\to\infty} e(t) = 0$. ∇

To implement the controller (9.607), we need the knowledge of $f(q)$ to generate the bounding signal $m(t)$ in (9.608) but not the knowledge of the bounds γ, ρ. For a chosen $f(q)$, different choices of k_0 in generating $m(t)$ may have different effects on the tracking performance, while increasing k_0 may reduce the effect of γ in the mean error (9.611).

It is interesting to note that for the signal boundedness and the mean tracking error (9.611), the size of parameter variations, characterized by γ and ρ, is not required to be small. This is an important feature in adaptive control of robotic systems. With $\dot{q}$, q available for measurement, the manipulator model (9.586) is equivalent to the first-order model (9.592) for which an adaptive controller admits "large" parameter variations.

9.4.4 Design for Unmodeled Dynamics

We now consider the manipulator (9.593) with unmodeled dynamics. If the terms $H_1[g_1(\dot{q})](t)$, $H_2[g_2(q)](t)$, $H_3[g_3(u)](t)$ were available for measurement and θ^* were known, the control law

$$\begin{aligned} u(t) = {} & Y(q, q_d, \dot{q}, \dot{q}_d, \ddot{q}_d)\theta^* \\ & - H_1[g_1(\dot{q})](t) - H_2[g_2(q)](t) - H_3[g_3(u)](t) - K_D s(t) \end{aligned} \quad (9.618)$$

could be implemented so that $\frac{d}{dt}(s^T(t)D(q)s(t)) = -2s^T(t)K_D s(t)$, showing the boundedness of $s(t)$ and exponentially fast tracking. However, to ensure the boundedness of $u(t)$, one needs $\|H_3\|_\infty = \max_{i=1,\ldots,n} \mu_i < 1$.

To solve the adaptive control problem in which $H_1[g_1(\dot{q})](t)$, $H_2[g_2(q)](t)$, $H_3[g_3(u)](t)$ and θ^* are unknown, with $Y(q, q_d, \dot{q}, \dot{q}_d, \ddot{q}_d)$, K_D, $s(t)$, $\sigma(t)$, Γ defined before, we use the control scheme for the manipulator (9.593):

$$\begin{aligned} u(t) = {} & Y(q, q_d, \dot{q}, \dot{q}_d, \ddot{q}_d)\theta(t) \\ & - m_1(t)\phi_1(t) - m_2(t)\phi_2(t) - m_3(t)\phi_3(t) - K_D s(t), \end{aligned} \quad (9.619)$$

where the update law for $\theta(t)$ is

$$\dot{\theta}(t) = -\Gamma^{-1}(Y^T(q, q_d, \dot{q}, \dot{q}_d, \ddot{q}_d)s(t) + \sigma(t)\theta(t)) \quad (9.620)$$

with $\sigma(t)$ defined in (9.610), and, for $s(t) = [s_1(t), \ldots, s_n(t)]^T$, the design signals are

$$m_1(t) = k_{10}\sqrt{n} f_1(\dot{q}(t)),\ m_2(t) = k_{20}\sqrt{n} f_2(q(t)),\ m_3(t) = \|u_t(\cdot)\|_\infty, \tag{9.621}$$

$$\phi_1(t) = m_1(t)s(t),\ \phi_2(t) = m_2(t)s(t),\ \phi_3(t) = [\phi_{31}(t), \ldots, \phi_{3n}(t)]^T, \tag{9.622}$$

$$\phi_{3i}(t) = \begin{cases} \lambda_i \operatorname{sign}[s_i(t)m_3(t)] & \text{if } |s_i(t)m_3(t)| \geq \delta_i, \\ \frac{\lambda_i s_i(t) m_3(t)}{\delta_i} & \text{if } |s_i(t)m_3(t)| < \delta_i, \end{cases} \tag{9.623}$$

$$k_{j0} > 0,\ j = 1, 2,\ \delta_i > 0,\ 0 < \lambda_i < 1,\ i = 1, \ldots, n. \tag{9.624}$$

This adaptive control scheme has the stability and tracking properties:

Theorem 9.10 *All closed-loop signals are bounded for any $\mu_i \in [0, \lambda_i]$, $i = 1, \ldots, n$, and the tracking error $e(t)$ satisfies*

$$\int_{t_1}^{t_2} \|e(t)\|^2\, dt \leq \alpha_0 \left(\frac{\gamma_1^2}{k_{10}^2} + \frac{\gamma_2^2}{k_{20}^2} + \delta \right)(t_2 - t_1) + \beta_0,\ \delta = \max_{i=1,\ldots,n} \delta_i \tag{9.625}$$

for some constants $\alpha_0 > 0$, $\beta_0 > 0$. Moreover, $e(t) \in L^2$ and $\lim_{t\to\infty} e(t) = 0$ in the absence of the unmodeled dynamics, that is, when $H_1 = H_2 = H_3 = 0$.

Proof: For the positive definite function V in (9.605), from (9.593), (9.619), (9.620), and Assumptions (A9m.3) and (A9m.4), we have

$$\begin{aligned} \dot{V} &= -s^T(t)K_D s(t) - (m_1^2(t) + m_2^2(t))s^T(t)s(t) - m_3(t)s^T(t)\phi_3(t) \\ &\quad - \sigma(t)\tilde{\theta}^T(t)\theta(t) + s^T(t)H_1[g_1(\dot{q})](t) \\ &\quad + s^T(t)H_2[g_2(q)](t) + s^T(t)H_3[g_3(u)](t) \\ &\leq -s^T(t)K_D s(t) - \left(m_1(t)\|s(t)\| - \frac{\gamma_1}{2k_{10}}\right)^2 + \frac{\gamma_1^2}{4k_{10}^2} - \left(m_2(t)\|s(t)\| - \frac{\gamma_2}{2k_{20}}\right)^2 \\ &\quad + \frac{\gamma_2^2}{4k_{20}^2} - (\sigma(t) - \sigma_0)\tilde{\theta}^T(t)\theta(t) + \frac{1}{4}\sigma_0\|\theta^*\|^2 - \sigma_0\left(\|\theta(t)\| - \frac{\|\theta^*\|}{2}\right)^2 \\ &\quad - \sum_{i=1}^n m_3(t)(\lambda_i - \mu_i)|s_i(t)| + \sum_{i=1}^n m_3(t)(|s_i(t)|\lambda_i - s_i(t)\phi_{3i}(t)). \end{aligned} \tag{9.626}$$

From (9.623), we see that

$$|m_3(t)(|s_i(t)|\lambda_i - s_i(t)\phi_{3i}(t))| \leq 2\lambda_i\delta_i,\ i = 1, \ldots, n. \tag{9.627}$$

It follows from (9.626), (9.627) that, for $0 \leq \mu_i \leq \lambda_i$, $i = 1, \ldots, n$, the time derivative $\dot{V} \leq 0$ whenever $\theta(t)$ and $s(t)$ are outside a certain bounded set, that is, $s(t)$, $\theta(t)$ are bounded, and so are $q(t)$, $\dot{q}(t)$. Then, from (9.619)–(9.623) and Assumptions (A9m.3) and (A9m.4), we have

$$\|u_t(\cdot)\|_\infty \leq k_2 + \max_i \lambda_i \|u_t(\cdot)\|_\infty \tag{9.628}$$

for some constant $k_2 > 0$. From (9.624) and (9.627), we have that $u(t) \in L^\infty$.

Using (9.626), (9.627) and the fact that $\sigma(t)\tilde{\theta}^T(t)\theta(t) \geq 0$, we obtain

$$\dot{V} \leq -s^T(t)K_D s(t) + \frac{\gamma_1^2}{4k_{10}^2} + \frac{\gamma_2^2}{4k_{20}^2} + 2\sum_{i=1}^{n} \lambda_i \delta_i. \tag{9.629}$$

For $\delta = \max_{i=1,\ldots,n} \delta_i$, (9.629) implies that

$$\int_{t_1}^{t_2} \|s(t)\|^2\, dt \leq \alpha_1 \left(\frac{\gamma_1^2}{k_{10}^2} + \frac{\gamma_2^2}{k_{20}^2} + \delta \right)(t_2 - t_1) + \beta_1 \tag{9.630}$$

for some constants $\alpha_1 > 0$, $\beta_1 > 0$ and any $t_2 > t_1 \geq 0$, which implies (9.625).

When $H_1 = 0$, $H_2 = 0$, $H_3 = 0$, the expression (9.626) for $\dot{V}$ becomes

$$\begin{aligned}\dot{V} = &-s^T(t)K_D s(t) - (m_1^2(t) + m_2^2(t))s^T(t)s(t)\\ &- m_3(t)s^T(t)\phi_3(t) - \sigma(t)\tilde{\theta}^T(t)\theta(t).\end{aligned} \tag{9.631}$$

Since $s^T(t)\phi_3(t) \geq 0$, (9.631) shows that $s(t) \in L^2$. Hence, from (9.591), it follows that $e(t) \in L^2$ and $\dot{e}(t)$ is bounded, so that $\lim_{t\to\infty} e(t) = 0$. ∇

We have thus proved the signal boundedness of the closed-loop system in the presence of $H_1[g_1(q)](t)$, $H_2[g_2(q)](t)$, $H_3[g_3(u)](t)$. The gains of the linear operators H_1, H_2 are assumed to be finite but not small. The gain of H_3 is required small to ensure the boundedness of $u(t)$.

Remark 9.16 The modifying term $\sigma(t)\theta(t)$ in (9.620) can be replaced by a fixed-σ modification [147] (see Section 3.7.2): $\sigma_0\theta(t)$, $\sigma_0 > 0$. The closed-loop signal boundedness still holds, and, however, the tracking performance description (9.625) is changed as

$$\int_{t_1}^{t_2} \|e(t)\|^2\, dt \leq \alpha_0 \left(\frac{\gamma_1^2}{k_{10}^2} + \frac{\gamma_2^2}{k_{20}^2} + \delta + \sigma_0 \right)(t_2 - t_1) + \beta_0. \tag{9.632}$$

Due to the presence of σ_0 in (9.632), this scheme cannot guarantee asymptotic tracking: $\lim_{t\to\infty} e(t) = 0$, in the absence of the unmodeled dynamics, though the scheme does not need the knowledge of the upper bound on $\|\theta^*\|$. □

Remark 9.17 The use of the bounding signals $m_i(t)$ defined in (9.621) is the key to ensure signal boundedness in the presence of the unmodeled dynamics satisfying (A9m.3), (A9m.4). To generate these signals, the knowledge of the stability margin of the unmodeled dynamics is not needed.

Alternate bounding signals may be used under other assumptions for the unmodeled dynamics. For example, if $\|\bar{H}_3^i\|_\infty \leq \mu_i$, where $\bar{H}_3^i$ is the ith row of $H_3(s)(s+a_0)$ (this s is the Laplace variable), $i = 1, \ldots, n$, for some known constant $a_0 > 0$, and $\|(\frac{1}{s+a_0}[g_3(u)])_t(\cdot)\|_\infty \leq \|(\frac{1}{s+a_0}[u])_t(\cdot)\|_\infty$, then we can choose $m_3(t) = \|(\frac{1}{s+a_0}[u])_t(\cdot)\|_\infty$ and set $0 < \lambda_i \leq a_0$. Another choice is the bounding signal $m_3(t) = \frac{\sqrt{n}}{s+\delta_0}[\|u\|](t)$ when $\|g_3(u)\| \leq \|u\|$ and $\|\hat{H}_3^i\|_1 \leq \mu_i$, where $\hat{H}_3^i$ is the ith row of $H_3(s-\delta_0)s$ with the Laplace variable s for some known constant $\delta_0 > 0$. For this $m_3(t)$, the condition on λ_i is $\|(\lambda_1, \ldots, \lambda_m)^T\|\sqrt{n} < \delta_0$ and $\lambda_i > 0$ for $i = 1, \ldots, n$. We note that a similar L^1 norm condition can be derived for H_1 and H_2 for the design of a robust adaptive controller.

When θ^* is known and $H_1[g_1(\dot{q})](t)$, $H_2[g_2(q)](t)$, $H_3[g_3(u)](t)$ are present, the controller (9.619) with $\theta(t) = \theta^*$ becomes a robust nonadaptive controller which results in a tracking error $e(t)$ exponentially converging to a residual set whose size is of the order of $\frac{\gamma_1^2}{k_{10}^2} + \frac{\gamma_2^2}{k_{20}^2} + \delta$, which implies (9.625). □

Alternative Design

Next, we present the robust adaptive control design assuming that the upper bounds on the gains of the unmodeled dynamics H_1, H_2 are known:

> (A9m.3a): $g_1(\dot{q})$, $g_2(q)$ satisfy the same conditions in (A9m.3), and $\|H_j^i\|_\infty \leq \gamma_{ji}$ for some known constants $\gamma_{ji} > 0$, where H_j^i is the ith row of H_j, $j = 1, 2$, $i = 1, \ldots, n$.

We choose $\phi_j(t) = [\phi_{j1}(t), \ldots, \phi_{jn}(t)]^T$, $j = 1, 2, 3$, in (9.619), as

$$\phi_{ji}(t) = \begin{cases} \lambda_{ji}\,\mathrm{sign}[s_i(t)m_j(t)] & \text{if } |s_i(t)m_j(t)| \geq \delta_{ji}, \\ \frac{\lambda_{ji}s_i(t)m_j(t)}{\delta_{ji}} & \text{if } |s_i(t)m_j(t)| < \delta_{ji}, \end{cases} \tag{9.633}$$

$$\delta_{ji} > 0,\ \lambda_{1i} \geq \frac{\gamma_{1i}}{k_{10}},\ \lambda_{2i} \geq \frac{\gamma_{2i}}{k_{10}},\ 0 < \lambda_{3i} < 1. \tag{9.634}$$

This scheme ensures that all signals in the closed-loop system are bounded for $0 \leq \mu_i \leq \lambda_{3i}$, $i = 1, \ldots, n$, and the tracking error $e(t)$ satisfies

$$\int_{t_1}^{t_2} \|e(t)\|^2\, dt \leq \alpha_0 \delta(t_2 - t_1) + \beta_0,\ \delta = \max_{j=1,2,3,i=1,\ldots,n} \delta_{ji} \tag{9.635}$$

for some constants $\alpha_0 > 0$, $\beta_0 > 0$, $\forall t_2 > t_1 \geq 0$. Moreover, $e(t) \in L^2$ and $\lim_{t\to\infty} e(t) = 0$ in the absence of the unmodeled dynamics H_1, H_2, and H_3.

Remark 9.18 We see from (9.635) that the mean tracking error explicitly depends only on the design parameter δ, not on the bounds γ_{ji} defined in (A9m.3a). A smaller δ may result in a smaller mean tracking error (9.635). Hence, with the knowledge of the unmodeled dynamics bounds, the control signals in (9.633) may improve tracking performance. □

Another result is the adaptive controller with a so-called variable structure [424]: letting $\delta_{ji} \to 0$ in (9.633), $j = 1, 2, 3$, $i = 1, \ldots, n$, we obtain

$$\phi_{ji}(t) = \begin{cases} \lambda_{ji} \, \text{sign}[s_i(t) m_j(t)] & \text{if } s_i(t) m_j(t) \neq 0, \\ 0 & \text{if } s_i(t) m_j(t) = 0. \end{cases} \tag{9.636}$$

Using variable structure systems theory [424], it can be shown that the system solution exists [102] and that for $0 \leq \mu_i \leq \lambda_{3i}$, $i = 1, \ldots, n$, all closed-loop signals are bounded and the sliding surface $e(t)q(t) - q_d(t) = 0$ is reached asymptotically (with possible chattering of the control signal $u(t)$). For a variable structure controller, $\sigma(t) = 0$ can be used in the update law (9.620).

Finally, we note that the proposed designs can be combined to handle both the parameter variations and unmodeled dynamics.

9.5 Discussion

Literature Review

Since the 1970s, adaptive control of multi-input, multi-output (MIMO) systems has made significant progress [17], [70], [73], [80], [81], [84], [92], [93], [94], [95], [120], [122], [124], [125], [137], [138], [141], [185], [178], [182], [191], [212], [222], [228], [264], [266], [270], [282], [295], [296], [297], [299], [340], [347], [349], [377], [379], [380], [384], [411], [419], [432], [441]. Most of the developed multivariable adaptive control schemes are based on a model reference control approach. Different model reference adaptive control schemes have been developed for multivariable linear time-invariant systems under different assumptions and are reported in the literature either for the continuous-time case or for the discrete-time case. In this chapter, we have presented a unified framework of multivariable model reference adaptive control schemes for both continuous-time and discrete-time systems, clarified the class of unmodeled dynamics admissible to robust multivariable model reference adaptive control based on different controller parametrizations, derived adaptive laws for updating the controller parameters, and analyzed stability, robustness, and tracking performance of multivariable adaptive control systems. There are other parametric designs, including backstepping (see Section 9.3), decentralized adaptive control [66], [112], [142], [144], command generator tracker approach [178], and pole placement control [81], [94], [95] (see next subsection).

Pole Placement Designs

We first present a direct adaptive pole placement control design for multivariable linear time-invariant plants and then discuss some key issues with multivariable indirect adaptive pole placement control designs.

Direct adaptive pole placement design. A direct adaptive pole placement control scheme using output feedback for a multi-input, multi-output linear time-invariant plant was developed in [94], [95], as the multivariable version of that in [92] (see Section 7.4). For the plant $y(t) = G(D)[u](t)$, where $y(t) \in R^M$ and $u(t) \in R^M$ and $G(D)$ is nonsingular with the observability index ν and has the form $G(D) = Z(D)P^{-1}(D)$ for some $M \times M$ right coprime polynomial matrices $Z(D)$ and $P(D)$ with $P(D)$ being column proper and of degree equal to the controllability index of $G(D)$ (see Section 9.2.1), the nominal pole placement controller structure [94], [95] is

$$u(t) = \Theta_1^{*T}\omega_1(t) + \Theta_2^{*T}\omega_2(t) + \Theta_{20}^* y(t) + \Theta_3^* r(t), \tag{9.637}$$

where

$$\omega_1(t) = \frac{A(D)}{\Lambda(D)}[u_d](t),\ \omega_2(t) = \frac{A(D)}{\Lambda(D)}[y](t), \tag{9.638}$$

$$A(D) = [I, DI, \ldots, D^{\nu-2}I]^T, \tag{9.639}$$

$\Lambda(D)$ is a monic and stable polynomial of degree $\nu - 1$, and $\Theta_1^* \in R^{M(\nu-1)\times M}$, $\Theta_2^* \in R^{M(\nu-1)\times M}$, $\Theta_{20}^* \in R^{M\times M}$, and $\Theta_3^* \in R^{M\times M}$ satisfy

$$\begin{aligned}&\Theta_1^{*T}A(D)P(D) + (\Theta_2^{*T}A(D) + \Theta_{20}^*\Lambda(D))A(D)Z(D)\\ &= \Lambda(D)(P(D) - \Theta_3^* P_d(D))\end{aligned} \tag{9.640}$$

for a given $M \times M$ stable polynomial matrix $P_d(D)$ and a chosen parameter matrix Θ_3^* such that $\Gamma_c[P(D)] = \Theta_3^*\Gamma_c[P_d(D)]$ and $\mu_j = \partial_{cj}[P(D)] = \max_{1\le i\le M}\{\partial[p_{ij}(D)]\} = \partial_{cj}[\Theta_3^* P_d(D)]$, $j = 1, 2, \ldots, M$,[18] where $\Gamma_c[P(D)]$ and $\Gamma_c[P_d(D)]$ are the leading column matrices of $P(D)$ and $P_d(D)$ (see (9.55)),

[18] Recall from Section 9.2.2 that for a model reference design we had the choice of $P_d(D) = \xi_m(D)Z(D)$ and $\Theta_3^* = K_p^{-1}$, where $Z(D)$ is stable and $\xi_m(D)$ is a modified (stable) left interactor matrix of $G(D)$ such that $\lim_{D\to\infty}\xi_m(D)G(D) = K_p$ is finite and nonsingular (see Proposition 2.3 and Section 9.2.2).

which can be assumed to be unit upper triangular [442]. With this design, the closed-loop pole placement equation (9.640) has a solution under the right coprimeness of $Z(D)$ and $P(D)$, and the closed-loop system becomes

$$y(t) = Z(D)P_d^{-1}(D)[r](t). \qquad (9.641)$$

For $G(D)$ unknown, we use the adaptive version of (9.637):

$$u(t) = \Theta_1^T(t)\omega_1(t) + \Theta_2^T(t)\omega_2(t) + \Theta_{20}(t)y(t) + \Theta_3(t)r(t), \qquad (9.642)$$

where Θ_i, $i = 1, 2, 20, 3$, are the estimates of Θ_i^*, $i = 1, 2, 20, 3$. An adaptive scheme was proposed in [95] for updating the parameter estimates Θ_i, $i = 1, 2, 20, 3$, which is the multivariable version of that in [92] (see Section 7.4), based on a diagonal choice of $P_d(D) = \text{diag}\{p_1(D), \ldots, p_M(D)\}U(D)$ for some stable polynomials $p_i(D)$ and unimodular polynomial matrix $U(D)$ (i.e., $U^{-1}(D)$ is also a polynomial matrix). Stability of such a direct adaptive pole placement control design is addressed in [62], [95]. An adaptive scheme can also be developed for the choice of $P_d(D)$ in (9.640). With $Z(D)$ and $P(D)$ being right coprime, there exist polynomial matrices $B^*(D)$ and $A^*(D)$ of degree $\nu - 1$ [13], [282] such that $B^*(D)Z(D) + C^*(D)P(D) = I_M$ (in [95], this equation is modified with $U(D)$ replacing I_M). Using this equation, an error model suitable for parameter adaptation can be derived based on a modified parametrization developed in Section 10.4 (see (10.103)–(10.114)).

Issues with indirect adaptive pole placement designs. Indirect adaptive control schemes are designed using the estimates of the parameters in the plant transfer matrix $G(D)$ in the form $G(D) = P_l^{-1}(D)Z_l(D)$ convenient for parameter estimation using input and output measurements $u(t)$ and $y(t) = G(D)[u](t)$, based on the model $P_l(D)[y](t) = Z_l(D)[u](t)$. Estimates of $P_l(D)$ and $Z_l(D)$ are obtainable using proper parametrizations of $P_l(D)$ and $Z_l(D)$ and standard adaptive laws [94]. For adaptive control, the adaptive version of the closed-loop pole placement equation (9.640), that is,

$$\begin{aligned}&\Theta_1^T A(D)\hat{P}(D) + (\Theta_2^T A(D) + \Theta_{20}\Lambda(D))A(D)\hat{Z}(D)\\ &= \Lambda(D)(\hat{P}(D) - \Theta_3 P_d(D)),\end{aligned} \qquad (9.643)$$

needs to be solved on-line for Θ_i, $i = 1, 2, 20, 3$, for the on-line estimates $\hat{P}(D)$ and $\hat{Z}(D)$ of $P(D)$ and $Z(D)$ such that $G(D) = Z(D)P^{-1}(D)$. The adaptive version (also see the last subsection of this section: Indirect Multivariable MRAC) of the method used in Lemma 9.3 and Remark 9.5 for model reference control, implicitly in terms of $P(D)$ and $Z(D)$ (explicitly in terms of $P_l(D)$ and

$Z_l(D)$), may be employed for solving (9.643) under the coprimeness condition on $\hat{P}(D)$ and $\hat{Z}(D)$ or $\hat{P}_l(D)$ and $\hat{Z}_l(D)$ (the estimates of $P_l(D)$ and $Z_l(D)$). An alternative controller structure and closed-loop pole placement equation for indirect adaptive pole placement control are also proposed in [94].

Another type of indirect adaptive pole placement control design is the multivariable version of the internal mode controller structure for output tracking in the SISO case (see Section 7.2 and Section 7.3.2). In this case, two key issues are the derivation of the internal model principle for multivariable pole placement control (i.e., a pole placement design for an M-input M-output plant, which ensures that $\lim_{t\to\infty}(y(t)-y_m(t))=0$ exponentially, for a reference output $y_m(t)\in R^M$ such that $Q_m(s)[y_m](t) =$ for some $M\times M$ polynomial matrix $Q_m(s)$) and the on-line solution of a closed-loop pole placement equation similar to (7.137) and (7.302) (i.e., an on-line (adaptive) estimation scheme for the unknown plant parameters plus an algorithm for a multivariable Diophantine equation with a guaranteed solution).

MRAC Based on Input Error Method

An input error method was used in [73] and [340] for deriving an adaptive law for updating the parameters of a model reference adaptive controller. To illustrate this method, we consider the controller structure (9.102):

$$u(t) = \Theta_1^T\omega_1(t) + \Theta_2^T\omega_2(t) + \Theta_{20}y(t) + \Theta_3 r(t), \tag{9.644}$$

where $\Theta_1 = [\Theta_{11},\ldots,\Theta_{1\nu-1}]^T$, $\Theta_2 = [\Theta_{21},\ldots,\Theta_{2\nu-1}]^T$, with $\Theta_{ij}\in R^{M\times M}$, $i=1,2$, $j=1,\ldots,\nu-1$, $\Theta_{20}\in R^{M\times M}$, and $\Theta_3\in R^{M\times M}$, and

$$\omega_1(t) = F(D)[u](t),\ \omega_2(t) = F(D)[y](t), \tag{9.645}$$

$$F(D) = \frac{A(D)}{\Lambda(D)},\ A(D) = [I_M, DI_M,\ldots,D^{\nu-2}I_M]^T, \tag{9.646}$$

with a monic stable polynomial $\Lambda(D)$ of degree $\nu-1$ for the plant $y(t) = G_0(D)[u](t)$. With the reference model system $y_m(t) = W_m(D)[r](t)$ for $W_m(D)=\xi_m^{-1}(D)$, the associated matching equation is (9.106), that is,

$$I - \Theta_1^{*T}F(D) - (\Theta_2^{*T}F(D)+\Theta_{20}^*)G_0(D) = \Theta_3^*W_m^{-1}(D)G_0(D). \tag{9.647}$$

Operating both sides of (9.647) on $u(t)$, we have

$$u(t) - \Theta_1^{*T}\omega_1(t) - \Theta_2^{*T}\omega_2(t) - \Theta_{20}^*y(t) = \Theta_3^*W_m^{-1}(D)[y](t). \tag{9.648}$$

Introducing the filter $h(D) = \frac{1}{f(D)}$ for a monic stable polynomial $f(D)$ of degree equal to that of $\xi_m(D)$ and the parameter matrix Θ^* and signal $\zeta(t)$,

$$\Theta^* = [\Theta_1^{*T}, \Theta_2^{*T}, \Theta_{20}^*, \Theta_3^*]^T, \tag{9.649}$$

$$\zeta(t) = [h(D)[\omega_1^T](t), h(D)[\omega_2^T](t), \\ h(D)[y^T](t), h(D)[(\xi_m(D)[y])^T](t)]^T, \tag{9.650}$$

we express (9.648) as

$$\bar{u}(t) \triangleq h(D)[u](t) = \Theta^{*T}\zeta(t). \tag{9.651}$$

Based on this linear model, we define the estimation error

$$\epsilon(t) = \Theta^T(t)\zeta(t) - \bar{u}(t), \tag{9.652}$$

where $\Theta(t)$ is the estimate of Θ^*, and for the continuous-time case, choose the adaptive law for $\Theta(t)$ as

$$\dot{\Theta}^T(t) = -\frac{\Gamma\epsilon(t)\zeta^T(t)}{1 + \kappa\zeta^T(t)\zeta(t)}, \tag{9.653}$$

where $\kappa > 0$ and $\Gamma = \Gamma^T > 0$. For the positive definite function $V(\tilde{\Theta}) = \text{tr}[\tilde{\Theta}\Gamma^{-1}\tilde{\Theta}^T]$, where $\tilde{\Theta}(t) = \Theta(t) - \Theta^*$, using $\epsilon(t) = \tilde{\Theta}^T(t)\zeta(t)$, we have

$$\dot{V} = -\frac{2\epsilon^T(t)\epsilon(t)}{1 + \kappa\zeta^T(t)\zeta(t)}, \tag{9.654}$$

which implies that $\Theta(t) \in L^\infty$, $\dot{\Theta}(t) \in L^\infty$, and $\frac{\epsilon(t)}{m(t)} \in L^\infty$ and $\frac{\epsilon(t)}{m(t)} \in L^2$ and $\dot{\Theta}(t) \in L^2$, where $m(t) = \sqrt{1 + \kappa\zeta^T(t)\zeta(t)}$.

For this input error adaptive scheme[19] to ensure closed-loop signal boundedness, the parameter matrix $\Theta_3(t)$, which is the estimate of $K_3^* = K_p^{-1}$ (for

[19]It is called an *input error* method as $\epsilon(t)$ defined in (9.652) is related to the input error $e_i(t) = r_p(t) - r(t)$ for $r_p(t) \triangleq \xi_m(D)[y](t)$, which satisfies

$$r_p(t) - r(t) = \xi_m(D)[y - y_m](t) = \Theta_3^{*-1}\tilde{\Theta}^T(t)\omega(t),$$

where $\omega(t) = [\omega_1^T(t), \omega_2^T(t), y^T(t), r^T(t)]^T$ and $\tilde{\Theta}^T(t)\omega(t) = u(t) - u^*(t)$ is the control error for the actual control $u(t) = \Theta^T(t)\omega(t)$ as from (9.644) and the ideal control $u^*(t) \triangleq \Theta^{*T}\omega(t)$ which achieves plant–model matching: $\xi_m(D)[y - y_m](t) = 0$. It can be verified that

$$e_i(t) = \Theta_3^{-1}(t)\tilde{\Theta}^T(t)f(D)[\zeta](t).$$

the plant high frequency gain matrix K_p), needs to be nonsingular for all $t \geq 0$, indicated by the error equation (similar to (9.129) without normalization and with $\mu = 0$ and $\Psi = I$, also see (5.88) for the SISO version)

$$\begin{aligned} \bar{e}(t) &= h(D)\xi_m(D)[y - y_m](t) = h(D)[\Theta_3^{*-1}\tilde{\Theta}^T\omega](t) \\ &= h(D)[\Theta_3^{-1}\tilde{\Theta}^T f(D)[\zeta]](t) = \Theta_3^{-1}(t)\epsilon(t) - \xi(t), \end{aligned} \tag{9.655}$$

where $\epsilon(t) = \tilde{\Theta}^T(t)\zeta](t)$ and

$$\xi(t) = \Theta_3^{-1}(t)\tilde{\Theta}^T(t)\zeta](t) - h(D)[\Theta_3^{-1}\tilde{\Theta}^T f(D)[\zeta]](t). \tag{9.656}$$

The adaptive law (9.653) does not automatically guarantee the boundedness of Θ_3^{-1}, and certain modification is needed for ensuring such a boundedness property needed for closed-loop signal boundedness.

For the case when $M = 1$, $\Theta(t) = \theta_3(t)$ is a scalar, and with the sign of $\theta_3^* = \frac{1}{k_p}$ and a lower bound $\theta_3^0 \in (0, |\theta_3^*|]$, a parameter projection can be implemented to ensure $\theta_3(t) \geq \theta_3^0$, $t \geq 0$ (see Section 3.8). When $M > 1$, the task of ensuring boundedness of Θ_3^{-1}, in addition to the above L^∞ and L^2 properties, is much more difficult. In [73], a switching algorithm is used to modify the estimate $\Theta_3(t)$ so that $\det[\Theta_3(t)] \neq 0$, $t \geq 0$.

Similar designs can be derived for the discrete-time case.

Designs Based on Prediction Models

Another type of adaptive control algorithms are developed in [124], [125] (for the discrete-time case) based on the parametrization (as from (9.468))

$$\begin{aligned} &\Theta_3^{*-1}u(t) - \Theta_3^{*-1}\Theta_1^{*T}\omega_1(t) - \Theta_3^{*-1}\Theta_2^{*T}\omega_2(t) - \Theta_3^{*-1}\Theta_{20}^{*}y(t) \\ &= W_m^{-1}(D)[y](t), \end{aligned} \tag{9.657}$$

which is a prediction model for $W_m^{-1}(D)[y](t) = \xi_m(D)[y](t)$ [125]. Introducing the filter $h(D) = \frac{1}{f(D)}$ for a monic stable polynomial $f(D)$ of degree equal to that of $\xi_m(D)$ and the new parameter matrix Φ^* and signal $\bar{\zeta}(t)$,

$$\Phi^* = [-\Theta_3^{*-1}\Theta_1^{*T}, -\Theta_3^{*-1}\Theta_2^{*T}, -\Theta_3^{*-1}\Theta_{20}^{*}, \Theta_3^{*-1}]^T, \tag{9.658}$$

$$\bar{\zeta}(t) = [h(D)[\omega_1^T](t), h(D)[\omega_2^T](t), h(D)[y^T](t), h(D)[u^T](t)]^T, \tag{9.659}$$

we express (9.657) as

$$\bar{y}(t) \triangleq h(D)\xi_m(D)[y](t) = \Phi^{*T}\bar{\zeta}(t). \tag{9.660}$$

Based on this linear model, we define the estimation error

$$\epsilon(t) = \Phi^T(t)\bar{\zeta}(t) - \bar{y}(t), \tag{9.661}$$

where $\Phi(t)$ is the estimate of Φ^*. Then adaptive laws for $\Phi(t)$ can be derived (the discrete-time algorithms were given in [124], [125]; see (9.664)) to ensure desired L^2 and L^∞ properties for a normalized $\epsilon(t)$, $\dot{\Phi}(t)$, and $\Phi(t)$.

The control law [125] in this case is given as

$$\Phi^T(t)\bar{\omega}(t) = \xi_m(D)[y_m](t) \tag{9.662}$$

for a given reference output $y_m(t)$, where $\bar{\omega}(t) = [\omega_1^T(t), \omega_2^T(t), y^T(t), u^T(t)]^T$. The explicit expression for $u(t)$ is

$$u(t) = \Phi_3^{-1}\left(\xi_m(D)[y_m](t) - \Phi_1^T(t)\omega_1(t) - \Phi_2^T(t)\omega_2(t) - \Phi_{20}(t)y(t)\right). \tag{9.663}$$

To ensure closed-loop signal boundedness, the parameter matrix $\Phi_3(t)$, which is the estimate of $\Phi_3^* = K_p$, needs to be nonsingular for all $t \geq 0$, for $u(t)$ to be nonsingular. For this purpose, the following adaptive law was used in [124], [125] for a discrete-time design:

$$\Phi(t) = \Phi(t-1) + \frac{a(t)\bar{\omega}(t-d)\left(\bar{y}^T(t) - \bar{\omega}^T(t-d)\Phi(t-1)\right)}{c + \bar{\omega}^T(t-d)\bar{\omega}(t-d)}, \tag{9.664}$$

where $\bar{y}(t) = \xi_m(D)[y](t-d)$ with $f(D) = f(z) = z^d$ (for $\xi_m(D)$ to have degree d), $c > 0$, and $a(t) \in (0, 2)$ is chosen such that

$$\Phi_3(t) = \Phi_3(t-1) + \frac{a(t)u(t-d)\left(\bar{y}^T(t) - \bar{\omega}^T(t-d)\Phi(t-1)\right)}{c + \bar{\omega}^T(t-d)\bar{\omega}(t-d)} \tag{9.665}$$

is nonsingular for $\Phi_3(t-1)$ nonsingular, given that $\Phi_3(0)$ is chosen to be nonsingular. This is possible because for $\Phi_3(t-1)$ nonsingular there are only M values of $a(t)$ to make $\Phi_3(t) \in R^{M \times M}$ singular.

Switching Adaptive Control

What is *a priori* plant knowledge needed for adaptive control of a multivariable plant? It depends on the complexity of a controller structure and an associated adaptive law for updating the controller parameters, as seen from the comparison of a MRAC scheme without or with using an LDU, SDU, or LDS decomposition of the plant high frequency gain matrix K_p.

Employing a family of estimators, a hysteresis switching control strategy [271] was used in [432] for adaptive control of linear multivariable plants. Running a total of $M\bar{n} \times M! \times 2^M$ multivariable estimators (each of them based on a possible LR ($R = DU$) decomposition of a high frequency gain related matrix) in parallel (where $\bar{n}$ is an upper bound of the plant order n) for all possible plant parametrizations and using a switching algorithm to select an ultimate estimator over a finite time interval based on a prediction error measure, a multivariable model reference adaptive control scheme was designed in [432] for M-input, M-output linear time-invariant minimum phase plants under the assumptions that a lower bound $d_{\min} > 0$ is known for $|d_i^*|$, $i = 1, 2, \ldots, M$, where d_i^*, $i = 1, 2, \ldots, M$, are the diagonal elements of the D matrix in the LDU decomposition of PK_p for some (unknown) permutation matrix P such that all principal minors of PK_p are nonzero, and the plant parameters are in a bounded and known convex region.

Indirect Multivariable MRAC

As the multivariable version of that in Sections 7.1 and 7.3.1 for the SISO case, an indirect multivariable MRAC scheme can be developed for an $M \times M$ plant $y(t) = P_l^{-1}(D)Z_l(D)[u](t)$, where

$$\begin{aligned} Z_l(D) &= \mathrm{diag}\{D^{\nu_1 - 1}, \ldots, D^{\nu_M - 1}\}Z_{\nu-1} + \mathrm{diag}\{D^{\nu_1 - 2}, \ldots, D^{\nu_M - 2}\}Z_{\nu-2} + \cdots, \\ P_l(D) &= \mathrm{diag}\{D^{\nu_1}, \ldots, D^{\nu_M}\}P_\nu + \mathrm{diag}\{D^{\nu_1 - 1}, \ldots, D^{\nu_M - 1}\}P_{\nu-1} + \cdots, \end{aligned} \quad (9.666)$$

with P_ν being a unit lower triangular matrix [94]. Filtering both sides of $P_l(D)[y](t) = Z_l(D)[u](t)$ by $\frac{1}{\Lambda_e(D)}$ for a monic and stable polynomial of degree ν, with some algebraic manipulation, we can obtain

$$\bar{y}(t) \triangleq \mathrm{diag}\{D^{\nu_1}, \ldots, D^{\nu_M}\}\frac{1}{\Lambda_e(D)}[y](t) = [\theta_1^{*T}\zeta_1(t), \ldots, \theta_M^{*T}\zeta_M(t)]^T \quad (9.667)$$

for some parameter vectors θ_i^* and vector signals $\zeta_i(t)$, $i = 1, 2, \ldots, M$, such that (9.667) constitutes an exact parametrization of $P_l(D)$ and $Z_l(D)$.

To design an adaptive scheme for updating the estimates $\theta_i(t)$ of θ_i^*, $i = 1, 2, \ldots, M$, we introduce the estimation errors

$$\epsilon_i(t) = \theta_i^T(t)\zeta_i(t) - \bar{y}_i(t),\ i = 1, 2, \ldots, M, \quad (9.668)$$

for $\bar{y}(t) = [\bar{y}_1(t), \ldots, \bar{y}_M(t)]^T$ and have the error equations

$$\epsilon_i(t) = \tilde{\theta}_i^T(t)\zeta_i(t),\ \tilde{\theta}_i(t) = \theta_i(t) - \theta_i^*,\ i = 1, 2, \ldots, M, \quad (9.669)$$

based on which adaptive laws can be derived for $\theta_i(t)$, $i = 1, 2, \ldots, M$.

To solve an on-line (adaptive) version of (9.82) or (9.105), we need to assume that the estimated plant transfer matrix $\hat{P}_l^{-1}\hat{Z}_l(D)$ at each time t has the same interactor matrix $\xi_m(D)$ with a possibly different estimated high frequency gain matrix $\hat{K}_p$ and that $\hat{K}_p(t)$ is nonsingular for all t.

Problems

9.1 Verify (9.13) and show that $\sqrt{\text{trace}[A^T M A]}$ for a matrix $M = M^T > 0$ is also a norm of the matrix A, based on Definition 2.1.

9.2 Derive (9.17) and verify (9.18), (9.19), and (9.22).

9.3 Derive (9.26), (9.32), and (9.36).

9.4 Study the $L^{1+\alpha}$ robustness (see Section 5.4) of the adaptive state feedback control scheme developed in Section 9.1.1.

9.5 Develop robust adaptive control schemes (see Section 5.5) using state feedback for state tracking (see Section 9.1.1).

9.6 Study the system identification scheme of Section 9.1.3 by an adaptive system design and a simulation of an $n = 2$ case.

9.7 Compare the controller structure (9.79) with (9.102) in terms of admitting unmodeled dynamics for robust adaptive control, analogous to that for comparing the controller structure (5.5) with (5.333) (see (5.335)–(5.337)).

9.8 Verify (9.111). (For $G(D) = G_0(D) + \mu\,(G_0(D)\Delta_m(D) + \Delta_a(D))$, derive

$$
\begin{aligned}
& G(D)\left(I - \Theta_1^{*T}F(D) - \Theta_2^{*T}F(D)G(D)\right)^{-1}\Theta_3^* \\
& - G_0(D)\left(I - \Theta_1^{*T}F(D) - \Theta_2^{*T}F(D)G_0(D)\right)^{-1}\Theta_3^* \\
= & \left(I + G_0(D)\left(I - \Theta_1^{*T}F(D) - \Theta_2^{*T}F(D)G_0(D)\right)^{-1}\Theta_2^{*T}F(D)\right) \\
& \times \mu(G_0(D)\Delta_m(D) + \Delta_a(D)) \\
& \times \left(I - \Theta_1^{*T}F(D) - \Theta_2^{*T}F(D)\left(G_0(D) + \mu\left(G_0(D)\Delta_m(D) + \Delta_a(D)\right)\right)\right)^{-1}\Theta_3^*,
\end{aligned}
$$

and use (9.86): $I - \Theta_1^{*T}F(D) - \Theta_2^{*T}F(D)G_0(D) = \Theta_3^* W_m^{-1}(D)G_0(D)$).

9.9 Derive the result of Theorem 9.1 for the controller structure (9.102).

9.10 Derive an error equation, similar to that in (9.126), for the controller structure (9.102) and design adaptive laws for MRAC based on (9.102).

9.11 Verify (9.139).

9.12 Verify (9.142) for the alternative estimation error model (9.145C).

9.13 Derive $\bar{\Delta}_1$ and D_1 in (9.141) and verify their properties.

9.14 Use the developed multi-input, multi-output version of Lemma 2.6 (see the footnote for (9.142)) to verify (9.142).

9.15 Develop a discrete-time multi-input, multi-output version of Lemma 2.6, and use it to verify the discrete-time version of (9.142).

9.16 Verify (9.151C)–(9.154C).

9.17 Verify (9.151D)–(9.154D).

9.18 Verify (9.160) for $d_m = 2, 3$, respectively, for both the CT and DT cases.

9.19 Verify (9.183D). *Hint*: Use the equality

$$D\left[\theta^T \frac{1}{f(D)}[\omega]\right](t) = \theta^T(t+1)\frac{1}{f(D)}[\omega](t+1) = D[\theta^T](t)\frac{D}{f(D)}[\omega](t)$$

to express

$$\begin{aligned}
& f(D)\left[\theta^T \frac{1}{f(D)}[\omega]\right](t) \\
&= \left(D^{d_m} + a_{d_m-1}D^{d_m-1} + \cdots + a_1 D + a_0\right)\left[\theta^T \frac{1}{f(D)}[\omega]\right](t) \\
&= \left(D^{d_m-1} + a_{d_m-1}D^{d_m-2} + \cdots + a_1\right) D\left[\theta^T \frac{1}{f(D)}[\omega]\right](t) + \theta^T(t)\frac{a_0}{f(D)}[\omega](t) \\
&= \left(D^{d_m-1} + a_{d_m-1}D^{d_m-2} + \cdots + a_1\right)\left[D[\theta^T]\frac{D}{f(D)}[\omega]\right](t) + \theta^T(t)\frac{a_0}{f(D)}[\omega](t) \\
&= \left(D^{d_m-1} + a_{d_m-1}D^{d_m-2} + \cdots + a_1\right)\left[(D-1)[\theta^T]\frac{D}{f(D)}[\omega]\right](t) \\
&\quad + \left(D^{d_m-1} + a_{d_m-1}D^{d_m-2} + \cdots + a_1\right)\left[\theta^T \frac{D}{f(D)}[\omega]\right](t) + \theta^T(t)\frac{a_0}{f(D)}[\omega](t),
\end{aligned}$$

where $\left(D^{d_m-1} + a_{d_m-1}D^{d_m-2} + \cdots + a_1\right)[\theta^T \frac{D}{f(D)}[\omega]](t)$ can be "swapped" in a way similar to that $\left(D^{d_m} + a_{d_m-1}D^{d_m-1} + \cdots + a_1 D + a_0\right)\left[\theta^T \frac{1}{f(D)}[\omega]\right](t)$.

9.20 Prove closed-loop signal boundedness from (9.184C) in continuous time.

9.21 Prove closed-loop signal boundedness from (9.184D) in discrete time.

9.22 Develop MRAC schemes using modified right interactor matrices.

9.23 Prove that $\xi_m(s) = \xi_1 s + \xi_0$ is strictly positive real (and so is $\xi_m^{-1}(s)$) if and only if $\xi_1 = \xi_1^T \geq 0$ and $\xi_0 + \xi_0^T > 0$, by direct verification of Definitions 9.1 and 9.2 (with a complex vector $z = x + jy$ for real x and y).

9.24 Show that if $\xi_m(s) = \xi_1 s + \xi_0$ is a modified interactor matrix of a strictly proper transfer matrix $G_0(s)$, then ξ_1 is nonsingular.

9.25 Verify (9.205) and (9.209).

9.26 Verify (9.225) and (9.226).

9.27 Verify (9.232) and (9.233).

9.28 Verify the last statement in Remark 9.12 for the fixed control law (9.245) and compare the derived g_0^* with that for Theorem 9.5 (to meet (9.234)).

9.29 Verify (9.259) and (9.261).

9.30 Develop a version of the robust adaptive control scheme (9.215)–(9.221) for the plant (9.254) with feedback unmodeled dynamics.

9.31 Develop a robust adaptive control scheme, similar to that described in (9.215)–(9.221) and (9.258), for the plant

$$y(t) = G_0(s)(I + \mu\Delta_m(s))[u](t) + \mu\Delta_f(s)[y](t) + d(t),$$

with multiplicative and feedback unmodeled dynamics $\mu\Delta_m(s)$ and $\mu\Delta_f(s)$ as well as output disturbance $d(t)$.

9.32 Verify (9.285) and (9.288).

9.33 Verify (9.300) and (9.307).

9.34 Consider the error model $\epsilon(t) = K_p\tilde{\Theta}^T(t)\zeta(t)$ first.

(i) Analyze the adaptive law $\dot{\tilde{\Theta}}^T(t) = -\Gamma\epsilon(t)\zeta^T(t)$ using the positive definite function $V = \mathrm{tr}[\tilde{\Theta}(t)\tilde{\Theta}^T(t)]$, $t \in [0, \infty)$, with $\Gamma K_p + K_p^T\Gamma^T > 0$.

(ii) Analyze the adaptive law $\dot{\tilde{\Theta}}^T(t) = -\epsilon(t)\zeta^T(t)$ using the positive definite function $V = \mathrm{tr}[\tilde{\Theta}(t)\Gamma\tilde{\Theta}^T(t)]$ for $\Gamma = \Gamma^T > 0$, $t \in [0, \infty)$, and the condition $\Gamma K_p + K_p^T\Gamma^T > 0$ [282].

(iii) Analyze the adaptive law $\tilde{\Theta}^T(t+1) = \tilde{\Theta}^T(t) - \Gamma\epsilon(t)\zeta^T(t)$ using the positive definite function $V = \mathrm{tr}[\tilde{\Theta}(t)\tilde{\Theta}^T(t)]$, $t \in \{0, 1, \ldots,\}$, and the condition that $\Gamma K_p + K_p^T\Gamma^T - K_p^T\Gamma^T\Gamma K_p > 0$ [120], [124].

(iv) Analyze the adaptive law $\tilde{\Theta}^T(t+1) = \tilde{\Theta}^T(t) - \epsilon(t)\zeta^T(t)$ using the positive definite function $V = \mathrm{tr}[\tilde{\Theta}(t)\Gamma\tilde{\Theta}^T(t)]$, $\Gamma = \Gamma^T > 0$, $t \in \{0, 1, \ldots, \}$, and a similar condition to that in part (iii).

(v) Repeat the above four cases for the corresponding adaptive laws with normalization, for example, $\dot{\tilde{\Theta}}^T(t) = -\frac{\Gamma\epsilon(t)\zeta^T(t)}{1+\zeta^T(t)\zeta(t)}$.

(vi) Try to work on the case with $\epsilon(t) = K_p\tilde{\Theta}^T(t)\zeta(t) + \tilde{\Psi}(t)\xi(t)$.

9.35 Develop a version of the adaptive control scheme with (9.201) based on the LDU decomposition of K_p, with L known, for the plant (9.197) (see the design in Section 9.1.2).

9.36 Redesign the adaptive control schemes in Section 9.2.11, incorporating a known permutation matrix P such that PK_p has an LDU, SDU, or LDS decomposition (see the comment after (9.360)).

9.37 Verify (9.378), (9.384), and (9.386).

9.38 Verify (9.404), (9.408), (9.415), and (9.416).

9.39 Verify (9.430), (9.436), and (9.438).

9.40 Prove Lemma 9.10.

9.41 Apply the backstepping design in Section 9.3.3 to the plant (9.440):

$$y(t) = G(s)[u](t), \; G(s) = \begin{bmatrix} \frac{1}{s-1} & \frac{1}{s} \\ \frac{1}{s+3} & \frac{2}{s+1} \end{bmatrix}.$$

Hint: Use $G(s) = P_l^{-1}(s)Z_l(s)$ for

$$P_l(s) = \begin{bmatrix} s(s-1) & \\ 0 & (s+3)(s+1) \end{bmatrix}, \; Z_l(s) = \begin{bmatrix} s & s-1 \\ s+1 & 2(s+3) \end{bmatrix}.$$

9.42 Complete the design and analysis of the adaptive backstepping control scheme using SDU decomposition of B_m, as proposed in Section 9.3.4.

9.43 Apply the backstepping design in Section 9.3.5 to the plant (9.440):

$$y(t) = G(s)[u](t), \; G(s) = \begin{bmatrix} \frac{1}{s-1} & \frac{1}{s} \\ \frac{1}{s+3} & \frac{1}{s+1} \end{bmatrix}.$$

Hint: Verify and use the fact that $G(s)$ has a modified right interactor matrix $\xi_m^r(s)$ and an associated high frequency gain matrix K_p^r:

$$\xi_m^r(s) = \begin{bmatrix} s+1 & -s(s+1) \\ 0 & (s+1)(s+1) \end{bmatrix}, \; K_p^r = \begin{bmatrix} 1 & 0 \\ 1 & 3 \end{bmatrix}.$$

9.44 Derive the backstepping design for the case when $M = 1$ [205].

9.45 For the two-link planar manipulator example in Section 9.4.2, verify the parametrized plant model (9.599) with (9.600)–(9.602).

9.46 For the two-link planar manipulator example in Section 9.4.2, derive an adaptive control scheme with a reduced-dimension parameter vector, assuming that the lengths l_1 and l_2 of the two links are known.

9.47 Consider the manipulator model

$$D(q)\ddot{q} + C(q, \dot{q}, t)\dot{q} + \phi(q, t) = K_u u,$$

where K_u is an unknown constant $n \times n$ matrix for $u \in R^n$ and $q \in R^n$. Derive an adaptive control scheme for (i) K_u being diagonal and positive definite, (ii) K_u being diagonal and nonsingular, or (iii) K_u being nondiagonal and nonsingular (the cases (ii) and (iii) may just be for theory test).

9.48 Verify the closed-loop system (9.641) with a pole placement controller.

9.49 Develop an adaptive scheme for updating the parameters of the controller structure (9.642). (*Hint*: Use $B^*(D)Z(D)+C^*(D)P(D) = I_M$ for some polynomial matrices $B^*(D)$ and $A^*(D)$ of degree $\nu - 1$, as compared with that in Section 7.4 for the SISO case.)

9.50 Develop a parameter projection algorithm for (9.653) to ensure that $|\Theta_3(t)| \geq \theta_3^0 > 0$, $t \geq 0$ and some θ_3^0, for the case when $M = 1$.

9.51 Develop the discrete-time version of (9.653) and derive a parameter projection to ensure $|\Theta_3(t)| \geq \theta_3^0 > 0$ for the case when $M = 1$.

9.52 Develop an adaptive algorithm for updating the parameter estimates $\theta_i(t)$, $i = 1, 2, \ldots, M$, based on the error model (9.669) and derive a solution scheme for solving an adaptive version of (9.82) or (9.105) for implementing the adaptive controller structure (9.79) or (9.102).

Advanced Topics

State feedback design and its parametrization. To generalize the results of parametrizing a state feedback control law using plant input output measurements for the SISO case (see Section 4.4) to the MIMO case, we consider the continuous-time linear time-invariant plant

$$\dot{x}(t) = Ax(t) + Bu(t), \; y(t) = Cx(t), \tag{9.670}$$

where $A \in R^{n\times n}$, $B \in R^{n\times M}$, $C \in R^{M\times n}$, $u(t) \in R^M$, and $y(t) \in R^M$.

The multivariable version of the adaptive control problem solved in Section 4.2 is to use the state feedback controller structure

$$u(t) = K_1^T(t)x(t) + K_2(t)r(t), \tag{9.671}$$

where $K_1(t) \in R^{n\times M}$ and $K_2(t) \in R^{M\times M}$ are updated by a stable adaptive law to be derived, to ensure that all closed-loop signals are bounded and the plant output $y(t)$ tracks a reference output $y_m(t)$ asymptotically, where

$$y_m(t) = W_m(s)[r](t), \; W_m(s) = \xi_m^{-1}(s) \tag{9.672}$$

with $\xi_m(s)$ being the modified left interactor matrix of $G(s) = C(sI - A)^{-1}B$ and $r(t)$ being a bounded reference input.

As in the SISO case, the key step is to establish the existence of constant matrices $K_1^* \in R^{n\times M}$ and $K_2^* \in R^{M\times M}$ such that the plant–model matching

$$C(sI - A - BK_1^{*T})^{-1}BK_2^* = \xi_m^{-1}(s) \tag{9.673}$$

is achieved. From this matching condition, the error equation

$$y(t) - y_m(t) = \xi_m^{-1}(s)[K_2^{*-1}(K_1 - K_1^*)^T x + (K_2 - K_2^*)r(t)](t), \tag{9.674}$$

similar to that in (4.48), can be derived. Based on this error equation, adaptive laws for updating the parameter estimates $K_1(t)$ and $K_2(t)$ can be derived, similar to that in Section 9.2, to solve the adaptive control problem.

Analogous to that in Section 4.4, to develop adaptive control schemes using output feedback, we need to parametrize the state feedback term $K_1^{*T}x(t)$ as

$$K_1^{*T}x(t) = \Theta_1^{*T}\omega_1(t) + \Theta_2^{*T}\omega_2(t), \tag{9.675}$$

where $\Theta_1^* = [\Theta_{11}^*, \ldots, \Theta_{1\nu}^*]^T \in R^{\nu M\times M}$, $\Theta_2^* = [\Theta_{21}^*, \ldots, \Theta_{2\nu}^*]^T \in R^{\nu M\times M}$, $\Theta_{ij} \in R^{M\times M}$, $i = 1, 2$, $j = 1, \ldots, \nu$, for the observability index ν of $G(s)$, and

$$\omega_1(t) = F(s)[u](t), \; \omega_2(t) = F(s)[y](t), \tag{9.676}$$

$$F(s) = \frac{A(s)}{\Lambda(s)},\ A(s) = [I_M, sI_M, \ldots, s^{\nu-1}I_M]^T, \tag{9.677}$$

for a monic stable polynomial $\Lambda(s)$ of degree ν, based on a "full"-order observer (using input $u(t)$ and output $y(t)$) for $K_1^{*T}x(t)$;[21] or as

$$K_1^{*T}x(t) = \Theta_1^{*T}\omega_1(t) + \Theta_2^{*T}\omega_2(t) + \Theta_{20}^* y(t), \tag{9.678}$$

for $\Theta_{20}^* \in R^{M\times M}$ and Θ_1^*, $\omega_1(t)$, Θ_2^*, and $\omega_2(t)$ defined above but with ν replaced by $\nu - 1$, based on a reduced-order observer for $K_1^{*T}x(t)$.

Solutions to the above matching and parametrization problems may be derivable using multivariable linear systems theory [13], [171], [241], [242], [442], [444], in view of model reference control theory (see Section 9.2).

Backstepping design. Is there a backstepping design without using a precompensator (see (9.550)) for the case when B_m is singular in the plant (9.440)? As compared with the model reference control designs of Section 9.2, which make use of the plant left interactor matrix $\xi_m(s)$ for generating the reference output $y_m(t)$, the present backstepping design does not use an interactor matrix for $y_m(t)$ and could be modified to do so in order to avoid increasing control system order by using an input precompensator. Note that plants with B_m nonsingular have a uniform interactor matrix $s^{\nu-m}I$.

Another issue with a multivariable backstepping design is the repeated estimation of Θ as an individual parameter matrix as well as a part of $\Psi = (SD_s)^{-1}\Theta$ (see (9.536)). How to get rid of this overparametrization?

Adaptive control with unknown interactor matrices. As we have seen, an interactor matrix of a controlled multi-input, multi-output plant plays an important role in a stable adaptive control design. On the other hand, interactor matrices heavily depend on their associated transfer matrices. Reducing the knowledge of an interactor matrix for multivariable adaptive control has been an important research topic [63], [80], [85], [93], [120], [347], [349], [367], [419], [432], which needs further investigation. From the design and analysis of MRAC schemes, one may see that any polynomial matrix $\xi(D)$ such that $\lim_{D\to\infty}\xi(D)G(D) = K_l$ or $\lim_{D\to\infty}G(D)\xi(D) = K_r$ is finite and nonsingular and that $\xi^{-1}(D)$ is stable could play the role of a modified interactor matrix for MRAC or adaptive backstepping control of a multivariable plant.

[21] A full-order observer for $x(t)$ has an order equal to n, while a reduced-order observer for $x(t)$ has an order equal to $n - M$ for $y(t) \in R^M$. The difference here is we need an observer for $K_1^{*T}x(t) = [k_{11}^{*T}x(t), k_{12}^{*T}x(t), \ldots, k_{1M}^{*T}x(t)]^T$. It was shown by Luenberger in [241] that an observer of order $\nu - 1$ (and with arbitrary poles if (A, C) is observable) can be constructed for $k^{*T}x(t)$ with $k^* \in R^n$ (see also [13], [171], [242], [444]).

Use of Nussbaum gains for multivariable adaptive control. Reducing the knowledge of the plant high frequency gain matrix K_p is also important for adaptive control of multivariable plants. Based on LDU, SDU, and LDS decompositions of K_p (see Section 9.2.11), we may employ Nussbaum gains to replace the knowledge of the signs of the diagonal elements of D in the adaptive laws (9.380), (9.410), and (9.431) to develop MRAC schemes in the continuous-time case (note that the use of Nussbaum gains for discrete-time adaptive control has not been addressed in the literature).

This idea can be used for Lyapunov designs (see Section 9.1.2, Remark 9.6, and Section 9.2.8), where the issue with an unknown L matrix is still open.

More applications of LDU decompositions of K_p. The LDU, LDS, or SDU decomposition of Θ_3^* in (9.648) or that of $\Theta_3^{*-1} = K_p$ in (9.657) may be employed to construct a modified parametrization of (9.651) or (9.660), based on which a modified adaptive parameter update law may be developed to ensure the nonsingularity of the estimate of Θ_3^* or Θ_3^{*-1}, in addition to other desired adaptive law properties, for closed-loop signal boundedness.

Multivariable adaptive pole placement control. As discussed in the second subsection of Section 9.5, there are some key open issues for multivariable adaptive pole placement control, including stability of direct and indirect adaptive control schemes and design and analysis of indirect adaptive control schemes based on the internal model principle for output tracking.

Adaptive control of multivariable nonlinear systems. Despite some preliminary results in the literature (e.g., [157], [191], [206], [251], [340], [341]), adaptive control of an M-input, M-output nonlinear system

$$\dot{x} = f(x) + \sum_{i=1}^{M} g_i(x)u_i, \; x \in R^n,$$
$$y = [y_1, \ldots, y_M]^T = [h_1(x), \ldots, h_M(x)]^T,$$

with $f(x), g_1(x), \ldots, g_M(x) \in R^n$, and $h_1(x), \ldots, h_M(x) \in R$ being smooth (differentiable) functions, is an open, challenging, and important area of research. Solutions are yet to be developed for different classes of systems.

Other topics. Similar to that in the SISO case, adaptive control of MIMO time-varying systems, adaptive control of MIMO systems with actuator failures, and adaptive control of MIMO systems with feedback unmodeled dynamics also need to be further investigated.

Chapter 10

Adaptive Control of Systems with Nonlinearities

This chapter presents a general adaptive inverse approach for control of plants with actuator imperfections caused by nonsmooth nonlinearities such as dead-zone, backlash, hysteresis, and other piecewise-linear characteristics. An adaptive inverse is employed for cancelling the effect of an actuator nonlinearity with unknown parameters, and a linear or nonlinear feedback control law is used for controlling the dynamics of a linear or smooth nonlinear part following the actuator nonlinearity. State feedback and output feedback control designs are presented, which all lead to linearly parametrized error models suitable for developing adaptive laws to update the inverse parameters. This approach suggests that control systems with commonly used linear or nonlinear feedback controllers such as those with model reference, PID, pole placement, or other dynamic compensation designs can be combined with an adaptive inverse to handle uncertain nonsmooth actuator nonlinearities.

In Section 10.1 we present a general parametrized actuator nonlinearity model illustrated by a dead-zone characteristic and a backlash characteristic and introduce a parametrized inverse for cancelling the actuator nonlinearity, illustrated by a dead-zone inverse and a backlash inverse. In Section 10.2 we design state feedback adaptive inverse control schemes, while in Section 10.3 we develop a general output feedback adaptive inverse control scheme and present three output feedback designs: model reference, pole placement, and PID (as illustrated by an adaptive backlash compensation example with simulation results showing improved system performance by an adaptive backlash inverse). We also develop feedback adaptive inverse control schemes in Section 10.4 for multivariable linear plants with actuator nonlinearities using adaptive parameter update laws based on a coupled estimation error model or a Lyapunov design. In Section 10.5 we address the adaptive actuator nonlinearity compensation control problems for cases when both the plant dynamics and actuator nonlinearities are unknown. In Section 10.6 we present several solutions to adaptive compensation of unknown actuator nonlinearities in plants with nonlinear dynamics using adaptive nonlinearity inverses and nonlinear feedback control designs such as feedback linearization and backstepping.

10.1 Actuator Nonlinearity Compensation

Nonsmooth nonlinear characteristics such as dead-zones, backlash, and hysteresis are common in actuators, such as mechanical connections, hydraulic servovalves, piezoelectric translators, and electric servomotors, and muscular systems [82], [130], [195], [210], [211], [219], [255], [309], [326], [328], [344], [421]. They are usually poorly known and may vary with time. They often severely limit system performance, giving rise to undesirable inaccuracy or oscillations or even leading to instability. The development of adaptive compensation control schemes for systems with unknown actuator nonlinearities has been a task of major practical interest as well as theoretical significance.

10.1.1 Actuator Nonlinearities

Consider a plant with a nonlinearity $N(\cdot)$ at the input of a linear part $G(D)$:

$$y(t) = G(D)[u](t), \; u(t) = N(v(t)), \tag{10.1}$$

where $N(\cdot)$ represents an actuator uncertainty such as a dead-zone, backlash, hysteresis, or piecewise-linear characteristic, $v(t)$ is the applied control, $u(t)$ is neither accessible for control nor available for measurement,[1] and $G(D)$ is a rational transfer function in continuous time (when $D = s$ denotes the Laplace transform variable or the time differentiation operator $s[x](t) = \dot{x}(t)$) or in discrete time (when $D = z$ denotes the z-transform variable or the time advance operator $z[x](t) = x(t+1)$) for a unified presentation.

The control objective is to design an adaptive compensator to cancel the effect of the uncertain actuator nonlinearity $N(\cdot)$ so that a commonly used control scheme for the linear part $G(D)$ can be used to ensure desired system performance. To achieve such an objective, there are two key tasks: one is the

[1]If $u(t)$ is accessible for control, then the actuator nonlinearity can be bypassed; if $u(t)$ is available for measurement, then with information of $v(t)$, the actuator nonlinearity $N(\cdot)$ can be accurately identified so that an exact inversion of $N(\cdot)$ can be applied to completely cancel the effects of $N(\cdot)$. In both cases, $N(\cdot)$ can be made to "disappear."

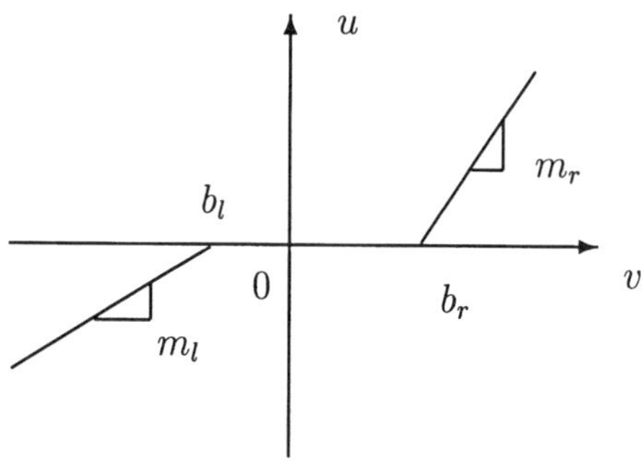

Figure 10.1: Dead-zone characteristic.

clarification of the class of actuator nonlinearities for which such compensators can be developed and the other is the design of adaptive laws which can effectively update the compensator parameters. In this subsection, we fulfill the first task by presenting a parametrized nonlinearity model suitable for adaptive compensation schemes to be developed in the next sections.

Nonlinearity model. Dead-zone, backlash, hysteresis, and piecewise-linear characteristics are typical examples of the actuator nonlinearity $N(\cdot)$. These nonlinear characteristics have break-points so that they are nondifferentiable (nonsmooth) but can be parametrized [326], [395], [401]. The parametrized models of these nonlinearities can be unified as

$$u(t) = N(v(t)) = N(\theta^*; v(t)) = -\theta^{*T}\omega^*(t) + a_s^*(t), \tag{10.2}$$

where $\theta^* \in R^{n_\theta}$ ($n_\theta \geq 1$) is an unknown parameter vector and $\omega^*(t) \in R^{n_\theta}$ and $a_s^*(t) \in R$, whose components are determined by the signal motion in the nonlinear characteristic $N(\cdot)$ and therefore are also unknown [395].

Dead-zone characteristic. To illustrate the nonlinearity model (10.2), consider a dead-zone characteristic $DZ(\cdot)$ described as

$$u(t) = N(v(t)) = DZ(v(t)) = \begin{cases} m_r(v(t) - b_r) & \text{if } v(t) \geq b_r \\ 0 & \text{if } b_l < v(t) < b_r \\ m_l(v(t) - b_l) & \text{if } v(t) \leq b_l, \end{cases} \tag{10.3}$$

where $m_r > 0$, $m_l > 0$, $b_r > 0$, and $b_l < 0$ (see Figure 10.1).

Introducing the indicator function $\chi[X]$ of the event X as

$$\chi[X] = \begin{cases} 1 & \text{if } X \text{ is true,} \\ 0 & \text{otherwise,} \end{cases} \tag{10.4}$$

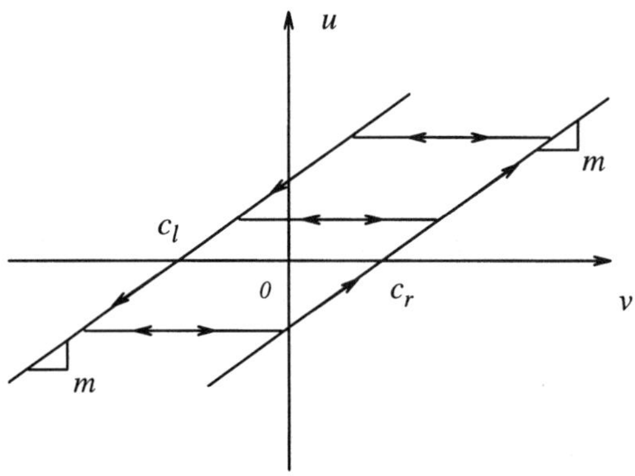

Figure 10.2: Backlash characteristic.

we define the dead-zone indicator functions

$$\chi_r(t) = \chi[u(t) > 0], \tag{10.5}$$

$$\chi_l(t) = \chi[u(t) < 0]. \tag{10.6}$$

Then, introducing the dead-zone parameter vector and its regressor,

$$\theta^* = [m_r, m_r b_r, m_l, m_l b_l]^T, \tag{10.7}$$

$$\omega^*(t) = [-\chi_r(t)v(t), \chi_r(t), -\chi_l(t)v(t), \chi_l(t)]^T, \tag{10.8}$$

we obtain (10.2) with $a_s^*(t) = 0$ for the dead-zone characteristic (10.3).

Backlash characteristic. A backlash characteristic $u(t) = B(v(t)) = B(m, c_r, c_l; v(t))$, which is a dynamic nonlinearity, is described by two parallel straight lines connected with horizontal line segments (see Figure 10.2). The upward side is active when both $v(t)$ and $u(t)$ increase:

$$u(t) = m(v(t) - c_r), \ \dot{v}(t) > 0, \ \dot{u}(t) > 0. \tag{10.9}$$

The downward side is active when both $v(t)$ and $u(t)$ decrease:

$$u(t) = m(v(t) - c_l), \ \dot{v}(t) < 0, \ \dot{u}(t) < 0, \tag{10.10}$$

where $m > 0$, $c_l < c_r$ are constant parameters. The motion on any inner segment is characterized by $\dot{u}(t) = 0$ even if $v(t)$ increases or decreases.

A compact description of the backlash characteristic $B(\cdot)$ is

$$\dot{u}(t) = \begin{cases} m\dot{v}(t) & \text{if } \dot{v}(t) > 0 \text{ and } u(t) = m(v(t) - c_r) \text{ or} \\ & \text{if } \dot{v}(t) < 0 \text{ and } u(t) = m(v(t) - c_l), \\ 0 & \text{otherwise.} \end{cases} \tag{10.11}$$

A backlash characteristic in *discrete time* can be described as

$$u(t) = B(v(t)) = \begin{cases} m(v(t) - c_l) & \text{if } v(t) \leq v_l, \\ m(v(t) - c_r) & \text{if } v(t) \geq v_r, \\ u(t-1) & \text{if } v_l < v(t) < v_r, \end{cases} \tag{10.12}$$

where the v-axis projections v_l and v_r of the intersections of the two parallel lines of slope m with the horizontal inner segment containing $u(t-1)$ are

$$v_l = \frac{u(t-1)}{m} + c_l, \; v_r = \frac{u(t-1)}{m} + c_r. \tag{10.13}$$

For a parametrized nonlinearity $N(\cdot)$ in (10.2), we will develop an adaptive inverse as a compensator for cancelling $N(\cdot)$ with unknown parameters.

10.1.2 Parametrized Nonlinearity Inverses

The essence of the adaptive inverse approach is to employ an inverse

$$v(t) = \widehat{NI}(u_d(t)) \tag{10.14}$$

to cancel the effect of the unknown nonlinearity $N(\cdot)$, where the inverse characteristic $\widehat{NI}(\cdot)$ is parametrized by an estimate θ of θ^* and $u_d(t)$ is a desired control signal from a feedback law. The key requirement for such an inverse is that its parameters can be updated from an adaptive law and should stay in a prespecified region needed for implementation of an inverse. In our designs, such an adaptive law is developed based on a linearly parametrized error model and the parameter boundaries are ensured by parameter projection.

Inverse model. A desirable inverse (10.14) should be parametrizable as

$$u_d(t) = -\theta^T(t)\omega(t) + a_s(t) \tag{10.15}$$

for some known signals $\omega(t) \in R^{n_\theta}$ and $a_s(t) \in R$ whose components are determined by the signal motion in the nonlinearity inverse $\widehat{NI}(\cdot)$ such that $v(t)$, $\omega(t)$, and $a_s(t)$ are bounded if $u_d(t)$ is bounded.

The error due to an uncertain actuator nonlinearity $N(\cdot)$ is

$$d_N(t) = \theta^{*T}(\omega(t) - \omega^*(t)) + a_s^*(t) - a_s(t), \tag{10.16}$$

which should satisfy the conditions that $d_N(t)$ is bounded, $t \geq 0$, and $d_N(t) = 0$, $t \geq t_0$, if $\theta(t) = \theta^*$, $t \geq t_0$, and $\widehat{NI}(\cdot)$ is correctly initialized: $d_N(t_0) = 0$ (such initialization is needed for inversion of a dynamic nonlinearity $N(\cdot)$ such as a backlash or hysteresis characteristic).

As shown in [326], [395], [401], the inverses for a dead-zone, backlash, hysteresis, and piecewise-linearity have such desired properties.

Dead-zone inverse. Let the estimates of the dead-zone parameters $m_r b_r$, m_r, $m_l b_l$, m_l be $\widehat{m_r b_r}$, $\widehat{m_r}$, $\widehat{m_l b_l}$, $\widehat{m_l}$, respectively. Then the inverse for the dead-zone characteristic (10.3) is described by

$$v(t) = \widehat{NI}(u_d(t)) = \widehat{DI}(u_d(t)) = \begin{cases} \frac{u_d(t) + \widehat{m_r b_r}}{\widehat{m_r}} & \text{if } u_d(t) > 0, \\ 0 & \text{if } u_d(t) = 0, \\ \frac{u_d(t) + \widehat{m_l b_l}}{\widehat{m_l}} & \text{if } u_d(t) < 0. \end{cases} \tag{10.17}$$

For the dead-zone inverse (10.17), to arrive at the desired form (10.15), we introduce the inverse indicator functions

$$\widehat{\chi_r}(t) = \chi[v(t) > 0], \tag{10.18}$$

$$\widehat{\chi_l}(t) = \chi[v(t) < 0] \tag{10.19}$$

and the inverse parameter vector and regressor

$$\theta = [\widehat{m_r}, \widehat{m_r b_r}, \widehat{m_l}, \widehat{m_l b_l}]^T, \tag{10.20}$$

$$\omega(t) = [-\widehat{\chi_r}(t)v(t), \widehat{\chi_r}(t), -\widehat{\chi_l}(t)v(t), \widehat{\chi_l}(t)]^T. \tag{10.21}$$

Then, the dead-zone inverse (10.17) is

$$\begin{aligned} u_d(t) &= \widehat{m_r}\widehat{\chi_r}(t)v(t) - \widehat{m_r b_r}\widehat{\chi_r}(t) + \widehat{m_l}\widehat{\chi_l}(t)v(t) - \widehat{m_l b_l}\widehat{\chi_l}(t) \\ &= -\theta^T\omega(t), \end{aligned} \tag{10.22}$$

that is, $a_s(t) = 0$ in (10.15). It follows from (10.7), (10.8), (10.16), and (10.21) that the error term $d_N(t)$ in (10.16) is

$$\begin{aligned} d_N(t) &= \theta^{*T}\omega(t)\chi[u(t) = 0] \\ &= -m_r\chi[0 < v(t) < b_r](v(t) - b_r) \\ &\quad - m_l\chi[b_l < v(t) < 0](v(t) - b_l), \end{aligned} \tag{10.23}$$

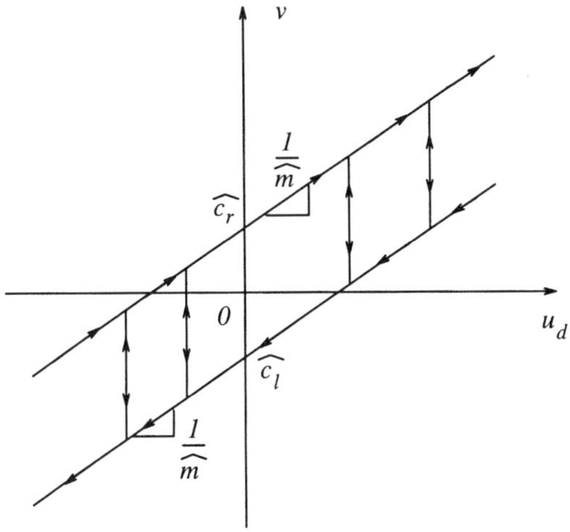

Figure 10.3: Backlash inverse.

which has the desired properties that $d_N(t)$ is bounded for all $t \geq 0$ and $d_N(t) = 0$ whenever $\theta = \theta^*$. Furthermore, $d_N(t) = 0$ whenever $v(t) \geq b_r$ or $v(t) \leq b_l$, that is, when $u(t)$ and $v(t)$ are outside the dead-zone, which is the case when $\widehat{b_r} \triangleq \frac{\widehat{m_r b_r}}{\widehat{m_r}} \geq b_r$ and $\widehat{b_l} \triangleq \frac{\widehat{m_l b_l}}{\widehat{m_l}} \leq b_l$ [395].

Backlash inverse. Let the estimates of the backlash parameters m, c_r, c_l be $\widehat{m}$, $\widehat{c}_l$, $\widehat{c}_r$. A backlash inverse $v(t) = \widehat{BI}(u_d(t))$ is described by

$$\dot{v}(t) = \begin{cases} \frac{1}{\widehat{m}}\dot{u}_d(t) & \text{if } \dot{u}_d(t) > 0 \text{ and } v(t) = \frac{u_d(t)}{\widehat{m}} + \widehat{c}_r \text{ or} \\ & \text{if } \dot{u}_d(t) < 0 \text{ and } v(t) = \frac{u_d(t)}{\widehat{m}} + \widehat{c}_l, \\ 0 & \text{if } \dot{u}_d(t) = 0, \\ \widehat{g}(t,t) & \text{if } \dot{u}_d(t) > 0 \text{ and } v(t) = \frac{u_d(t)}{\widehat{m}} + \widehat{c}_l, \\ -\widehat{g}(t,t) & \text{if } \dot{u}_d(t) < 0 \text{ and } v(t) = \frac{u_d(t)}{\widehat{m}} + \widehat{c}_r, \end{cases} \tag{10.24}$$

where, with the Dirac δ-function $\delta(t)$ (see (2.15)),

$$\widehat{g}(\tau, t) = \delta(\tau - t)(\widehat{c}_r - \widehat{c}_l). \tag{10.25}$$

As shown in Figure 10.3, the signal motion of $v(t)$ and $u_d(t)$ of the backlash inverse (3.14) is on two straight lines and vertical jumps between the lines, where the downward side line is

$$v(t) = \frac{u_d(t)}{\widehat{m}} + \widehat{c}_l, \ \dot{u}_d(t) < 0, \tag{10.26}$$

and the upward side line is

$$v(t) = \frac{u_d(t)}{\widehat{m}} + \widehat{c}_r, \ \dot{u}_d(t) > 0. \tag{10.27}$$

Vertical jumps of $v(t)$ occur whenever $\dot{u}_d(t)$ changes its sign: from the downward line to the upward line if from $\dot{u}_d(t) < 0$ to $\dot{u}_d(t) > 0$ or from the upward line to the downward line if from $\dot{u}_d(t) > 0$ to $\dot{u}_d(t) < 0$.

The *discrete-time* backlash inverse $v(t) = \widehat{BI}(u_d(t))$ is

$$v(t) = \begin{cases} v(t-1) & \text{if } u_d(t) = u_d(t-1), \\ \frac{u_d(t)}{\widehat{m}} + \widehat{c}_l & \text{if } u_d(t) < u_d(t-1), \\ \frac{u_d(t)}{\widehat{m}} + \widehat{c}_r & \text{if } u_d(t) > u_d(t-1). \end{cases} \tag{10.28}$$

The backlash inverse (10.24) or (10.28) has the desired property [395]

$$u_d(t_0) = B(BI(u_d(t_0))) \Rightarrow B(BI(u_d(t))) = u_d(t), \ \forall t \geq t_0, \tag{10.29}$$

where $BI(\cdot) = \widehat{BI}(\cdot)|_{\widehat{m}=m, \widehat{c}_l=c_l, \widehat{c}_r=c_r}$, the exact backlash inverse.

To parametrize the above backlash inverse, using the indicator function χ defined in (10.4), we introduce the backlash inverse indicator functions

$$\widehat{\chi_r}(t) = \chi[v(t) = \tfrac{u_d(t)}{\widehat{m}} + \widehat{c}_r], \tag{10.30}$$

$$\widehat{\chi_l}(t) = \chi[v(t) = \tfrac{u_d(t)}{\widehat{m}} + \widehat{c}_l] \tag{10.31}$$

and the backlash indicator functions

$$\chi_r(t) = \chi[\dot{u}(t) > 0], \ \chi_l(t) = \chi[\dot{u}(t) < 0], \ \chi_s(t) = \chi[\dot{u}(t) = 0]. \tag{10.32}$$

With $\widehat{mc_r} = \widehat{m}\widehat{c}_r$, $\widehat{mc_l} = \widehat{m}\widehat{c}_l$, we also introduce

$$\theta = [\widehat{mc_r}, \widehat{m}, \widehat{mc_l}]^T, \tag{10.33}$$

$$\theta^* = [mc_r, m, mc_l]^T, \tag{10.34}$$

$$\omega(t) = [\widehat{\chi_r}(t), -v(t), \widehat{\chi_l}(t)]^T. \tag{10.35}$$

It can be verified that

$$u_d(t) = -\theta^T \omega(t), \tag{10.36}$$

$$u(t) - u_d(t) = (\theta - \theta^*)^T \omega(t) + d_N(t), \tag{10.37}$$

where, with u_s being the value of $u(t)$ when $\dot{u}(t) = 0$,

$$\begin{aligned} d_N(t) &= (\chi_r(t) - \widehat{\chi_r}(t))(m(v(t) - c_r)) \\ &\quad + (\chi_l(t) - \widehat{\chi_l}(t))(m(v(t) - c_l)) + \chi_s(t)u_s. \end{aligned} \tag{10.38}$$

Control error. It is important to see that the inverse (10.15), when applied to the nonlinearity (10.2), results in the control error

$$u(t) - u_d(t) = (\theta - \theta^*)^T \omega(t) + d_N(t), \tag{10.39}$$

which is suitable for developing an adaptive inverse compensator. For the adaptive designs to be presented in the next sections, we assume that the inverse block (10.14) has the form (10.15) and $d_N(t)$ in (10.16) has the stated properties. We should note that the signals $a_s^*(t)$ in (10.2) and $a_s(t)$ in (10.15) are nonzero if the nonlinearity $N(\cdot)$ and its inverse $NI(\cdot)$ have inner loops as in the hysteresis case [391], [395].

10.2 State Feedback Inverse Control

Consider the plant (10.1), where $G(D)$ has a controllable state-space form

$$\begin{aligned} D[x](t) &= Ax(t) + Bu(t), \\ y(t) &= Cx(t) \end{aligned} \tag{10.40}$$

for some known constant matrices $A \in R^{n\times n}$, $B \in R^{n\times 1}$, and $C \in R^{1\times n}$, $n > 0$, that is, $G(D) = C(DI - A)^{-1}B$, and $r(t)$ is a bounded reference input.

To control the plant (10.40) with an actuator nonlinearity $u(t) = N(v(t))$ parametrizable as that in (10.2), we propose the control scheme shown in Figure 10.4, consisting of the nonlinearity inverse $v(t) = \widehat{NI}(u_d(t))$ parametrizable as in (10.15) and the state feedback control law

$$u_d(t) = K^* x(t) + r(t), \tag{10.41}$$

where $K^* \in R^{1\times n}$ is a constant-gain vector such that the eigenvalues of $A + BK^*$ are equal to some desired closed-loop system poles. The choice of such a K^* can be made from a pole placement [335] or linear quadratic optimal control design [216]. A pole placement design can be used to match the closed-loop transfer function $C(DI - A - BK^*)^{-1}B$ to a chosen reference model system if the zeros of $G(D)$ have good stability (see Sections 4.2 and 4.5).

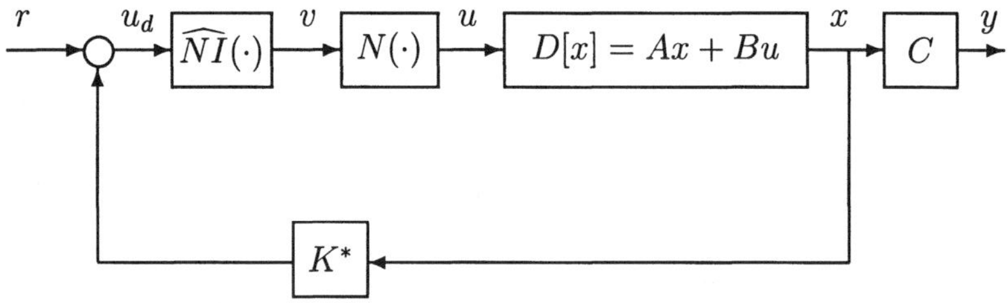

Figure 10.4: State feedback inverse control.

Error model. With the controller (10.41), the closed-loop system is

$$y(t) = W_m(D)[r](t) + W_m(D)[(\theta - \theta^*)^T \omega + d_N](t) + \epsilon_0(t), \tag{10.42}$$

where $\epsilon_0(t)$ is an exponentially decaying term due to initial conditions and

$$W_m(D) = C(DI - A - BK^*)^{-1}B. \tag{10.43}$$

The ideal system output is that of the closed-loop system without the actuator nonlinearity $N(\cdot)$ and its inverse $\widehat{NI}(\cdot)$, that is, with $u(t) = u_d(t)$:

$$y(t) = W_m(D)[r](t) + \epsilon_0(t). \tag{10.44}$$

In view of this output, we introduce the reference model system

$$y_m(t) = W_m(D)[r](t). \tag{10.45}$$

Ignoring the effect of $\epsilon_0(t)$, introducing $d(t) = W_m(D)[d_N](t)$, which is bounded because $d_N(t)$ in (10.16) is, and using (10.42) and (10.45), we have

$$\begin{aligned} y(t) - y_m(t) &= W_m(D)[(\theta - \theta^*)^T \omega](t) + d(t) \\ &= W_m(D)[\theta^T \omega](t) - \theta^{*T} W_m(D)[\omega](t) + d(t). \end{aligned} \tag{10.46}$$

In view of (10.46), we define the estimation error

$$\epsilon(t) = y(t) - y_m(t) + \xi(t), \tag{10.47}$$

where

$$\xi(t) = \theta^T(t)\zeta(t) - W_m(D)[\theta^T \omega](t), \tag{10.48}$$

$$\zeta(t) = W_m(D)[\omega](t). \tag{10.49}$$

Using (10.47)–(10.49), we arrive at the error equation

$$\epsilon(t) = (\theta(t) - \theta^*)^T \zeta(t) + d(t), \tag{10.50}$$

which is linear in the parameter error $\theta(t) - \theta^*$, with an additional bounded error $d(t)$ due to the uncertainty of the nonlinearity $N(\cdot)$ in the plant (10.1).

Gradient projection adaptive law. The adaptive update law for $\theta(t) = [\theta_1(t), \ldots, \theta_{n_\theta}(t)]^T$ can be developed using a gradient algorithm to minimize a cost function $J(\theta)$ (see Sections 3.3 and 3.6.2) subject to certain constraints on the parameter estimates.

The related optimization problem is to find an iterative algorithm to generate a sequence of θ which will asymptotically

$$\text{minimize } J(\theta), \text{ subject to } \theta_i \in [\theta_i^a, \theta_i^b],\ i = 1, 2, \ldots, n_\theta, \tag{10.51}$$

for some known parameter bounds $\theta_i^a < \theta_i^b$ such that $\theta_i^* \in [\theta_i^a, \theta_i^b]$, $i = 1, 2, \ldots, n_\theta$, for $\theta^* = [\theta_i^*, \ldots, \theta_{n\theta}^*]^T$, and that for any set of $\theta_i \in [\theta_i^a, \theta_i^b]$, $i = 1, 2, \ldots, p$, the characteristic $N(\theta; \cdot)$ renders an actuator uncertainty model in the same class to which the true model $N(\theta^*; \cdot)$ belongs, for example, for $N(\theta^*; \cdot)$ being a dead-zone of the form (10.3), $N(\theta; \cdot)$ should also be a dead-zone of the form (10.3) with possibly different parameters. The constraint that $\theta_i \in [\theta_i^a, \theta_i^b]$, $i = 1, 2, \ldots, n_\theta$, is to ensure that the nonlinearity inverse $\widehat{NI}(\cdot)$ is implementable as an inverse for the same type of nonlinearity $N(\cdot)$.

In the dead-zone case, the parameter boundaries are $m_{r1} \leq m_r \leq m_{r2}$, $m_{l1} \leq m_l \leq m_{l2}$, $0 \leq b_r \leq b_{r0}$, $-b_{l0} \leq b_l \leq 0$, for some known positive constants m_{r1}, m_{l1}, m_{r2}, m_{l2}, b_{r0}, b_{l0}. The above constraint condition, in terms of $\theta^* = [m_r, m_r b_r, m_l, m_l b_l]^T$ as in (10.7), is $\theta_i^* \in [\theta_i^a, \theta_i^b]$, $i = 1, 2, 3, 4$, for $\theta_1^a = m_{r1}$, $\theta_2^a = 0$, $\theta_3^a = m_{l1}$, $\theta_4^a = -m_{l2} b_{l0}$, $\theta_1^b = m_{r2}$, $\theta_2^b = m_{r2} b_{r0}$, $\theta_3^b = m_{l2}$, $\theta_4^b = 0$. Then for any $\theta_i \in [\theta_i^a, \theta_i^b]$, $i = 1, 2, 3, 4$, $DZ(\theta; \cdot)$ is a dead-zone characteristic described by (10.3).

For the formulated optimization problem, the gradient projection method [151], [325], [395] suggests the adaptive update law for $\theta(t)$ as

$$\left.\begin{array}{l} \dot{\theta}(t) \\ \theta(t+1) - \theta(t) \end{array}\right\} = -\Gamma \frac{\partial J}{\partial \theta}(t) + f(t), \tag{10.52}$$

where $f(t)$ is a parameter projection term and

$$\Gamma = \text{diag}\{\gamma_1, \ldots, \gamma_{n_\theta}\} \tag{10.53}$$

with $\gamma_i > 0$ for a continuous-time design and $0 < \gamma_i < 2$ for a discrete-time design. To ensure $\theta_i \in [\theta_i^a, \theta_i^b]$, $i = 1, 2, \ldots, n_\theta$, we introduce

$$g(t) = [g_1(t), \ldots, g_{n_\theta}(t)]^T = -\Gamma \frac{\partial J}{\partial \theta}(t), \tag{10.54}$$

initialize the adaptive law (10.52) with

$$\theta_i(0) \in [\theta_i^a, \theta_i^b], \ i = 1, 2, \ldots, n_\theta, \tag{10.55}$$

and set the ith component of the modification term $f(t)$ as

$$f_i(t) = \begin{cases} 0 & \text{if } \theta_i(t) \in (\theta_i^a, \theta_i^b) \text{ or} \\ & \text{if } \theta_i(t) = \theta_i^a, \ g_i(t) \geq 0 \text{ or} \\ & \text{if } \theta_i(t) = \theta_i^b, \ g_i(t) \leq 0, \\ -g_i(t) & \text{otherwise} \end{cases} \tag{10.56}$$

for a continuous-time design or

$$f_i(t) = \begin{cases} 0 & \text{if } \theta_i(t) + g_i(t) \in [\theta_i^a, \theta_i^b], \\ \theta_i^b - \theta_i(t) - g_i(t) & \text{if } \theta_i(t) + g_i(t) > \theta_i^b, \\ \theta_i^a - \theta_i(t) - g_i(t) & \text{if } \theta_i(t) + g_i(t) < \theta_i^a \end{cases} \tag{10.57}$$

for a discrete-time design for $i = 1, 2, \ldots, n_\theta$.

For the commonly used cost function

$$J(\theta) = \frac{\epsilon^2}{2m^2}, \tag{10.58}$$

where $\epsilon = (\theta - \theta^*)^T\zeta + d$ (see (10.50)), and

$$m(t) = \sqrt{1 + \zeta^T(t)\zeta(t)} \tag{10.59}$$

such that $\frac{\epsilon^2}{2m^2}$ is bounded if $\theta(t)$ is bounded, it follows that

$$\frac{\partial J}{\partial \theta}(t) = \frac{\epsilon(t)\zeta(t)}{m^2(t)}. \tag{10.60}$$

Therefore, a desired adaptive update law for $\theta(t)$ is

$$\left.\begin{array}{l} \dot{\theta}(t) \\ \theta(t+1) - \theta(t) \end{array}\right\} = -\frac{\Gamma\epsilon(t)\zeta(t)}{m^2(t)} + f(t), \tag{10.61}$$

where $\theta(t)$ is initialized by (10.55) and projected with $f(t)$ in (10.56) or (10.57) for $g(t) = -\frac{\Gamma\epsilon(t)\zeta(t)}{m^2(t)}$ in (10.54).

Using the analysis technique of Section 3.8, it can be verified that this adaptive law ensures that $\theta_i(t) \in [\theta_i^a, \theta_i^b]$, $i = 1, 2, \ldots, n_\theta$, and that $\frac{\epsilon(t)}{m(t)}$ and $\dot{\theta}(t)$ or $\theta(t+1) - \theta(t)$ are bounded by $\frac{d(t)}{m(t)}$ in the mean square sense: $\int_{t_1}^{t_2} \frac{\epsilon^2(t)}{m^2(t)}\, dt \leq a_1 + b_1 \int_{t_1}^{t_2} \frac{d^2(t)}{m^2(t)}\, dt$, $\int_{t_1}^{t_2} \|\dot{\theta}(t)\|_2^2\, dt \leq a_2 + b_2 \int_{t_1}^{t_2} \frac{d^2(t)}{m^2(t)}\, dt$, in continuous time, or

$\sum_{t=t_1}^{t_2} \frac{\epsilon^2(t)}{m^2(t)} \leq a_1 + b_1 \sum_{t=t_1}^{t_2} \frac{d^2(t)}{m^2(t)}$, $\sum_{t=t_1}^{t_2} \|\theta(t+1) - \theta(t)\|_2^2 \leq a_2 + b_2 \sum_{t=t_1}^{t_2} \frac{d^2(t)}{m^2(t)}$, in discrete time, for some constants $a_i, b_i > 0$, $i = 1, 2$, and all $t_2 > t_1 \geq 0$.

Continuous-time Lyapunov design. In continuous time, with $\tilde{\theta}(t) = \theta(t) - \theta^*$, the closed-loop system (10.40)–(10.41) is

$$\begin{aligned} \dot{x}(t) &= (A + BK^*)x(t) + Br(t) + B(\tilde{\theta}(t)\omega(t) + d_N(t)), \\ y(t) &= Cx(t). \end{aligned} \tag{10.62}$$

The ideal system performance is that with $u(t) = u_d(t)$, which means $\tilde{\theta}(t)\omega(t) + d_N(t) = 0$. This motivates us to introduce the reference system

$$\begin{aligned} \dot{x}_m(t) &= (A + BK^*)x_m(t) + Br(t), \\ y_m(t) &= Cx_m(t). \end{aligned} \tag{10.63}$$

Defining the state error vector $\tilde{x}(t) = x(t) - x_m(t)$ and the output error $e(t) = y(t) - y_m(t)$, from (10.62) and (10.63), we obtain

$$\begin{aligned} \dot{\tilde{x}}(t) &= (A + BK^*)\tilde{x}(t) + B(\tilde{\theta}(t)\omega(t) + d_N(t)), \\ e(t) &= C\tilde{x}(t). \end{aligned} \tag{10.64}$$

Consider the positive definite function

$$V(\tilde{x}, \tilde{\theta}) = \frac{1}{2}(\tilde{x}^T P \tilde{x} + \tilde{\theta}^T \Gamma^{-1} \tilde{\theta}), \tag{10.65}$$

where $P \in R^{n \times n}$ with $P = P^T > 0$ satisfying the Lyapunov equation

$$P(A + BK^*) + (A + BK^*)^T P = -2Q \tag{10.66}$$

for some $n \times n$ matrix $Q = Q^T > 0$ and $\Gamma = \text{diag}\{\gamma_1, \ldots, \gamma_{n_\theta}\}$ with $\gamma_i > 0$, $i = 1, 2, \ldots, n_\theta$. With (10.64) and (10.66), the time derivative of $V(\tilde{x}, \tilde{\theta})$ is

$$\begin{aligned} \dot{V} &= \frac{d}{dt}V(\tilde{x}, \tilde{\theta}) = \tilde{x}^T(t)P\dot{\tilde{x}}(t) + \tilde{\theta}^T(t)\Gamma^{-1}\dot{\tilde{\theta}}(t) \\ &= -\tilde{x}^T(t)Q\tilde{x}(t) + \tilde{x}^T(t)PB(\tilde{\theta}^T(t)\omega(t) + d_N(t)) + \tilde{\theta}^T(t)\Gamma^{-1}\dot{\tilde{\theta}}(t). \end{aligned} \tag{10.67}$$

If $d_N(t)$ were absent, then the choice of $\dot{\tilde{\theta}}(t) = \dot{\theta}(t) = -\Gamma\omega(t)\tilde{x}^T(t)PB$ would make $\dot{V} = -\tilde{x}^T(t)Q\tilde{x}(t) \leq 0$, as desired. In the presence of $d_N(t)$ and with the need of parameter projection, we choose the adaptive law

$$\dot{\theta}(t) = -\Gamma\omega(t)\tilde{x}^T(t)PB + f(t), \tag{10.68}$$

where $\theta(t)$ is initialized by (10.55) and projected with $f(t)$ in (10.56) for $g(t) = -\Gamma\omega(t)\tilde{x}^T(t)PB = [g_1(t), \ldots, g_{n_\theta}]^T$ to replace that in (10.55).

With the adaptive law (10.68), we have

$$\dot{V} = -\tilde{x}^T(t)Q\tilde{x}(t) + 2\tilde{x}^T(t)PBd_N(t) + 2\tilde{\theta}^T(t)\Gamma^{-1}f(t). \tag{10.69}$$

From the parameter projection algorithm (10.56), it follows that

$$\theta_i(t) \in [\theta_i^a, \theta_i^b],\ i = 1, 2, \ldots, n_\theta, \tag{10.70}$$

$$(\theta_i(t) - \theta_i^*)f_i(t) \leq 0,\ i = 1, 2, \ldots, n_\theta, \tag{10.71}$$

that is, $\tilde{\theta}^T(t)\Gamma^{-1}f(t) \leq 0$. Since $Q = Q^T > 0$ and $d_N(t)$ is bounded, we have, from (10.69) and (10.71), the boundedness of $\tilde{x}(t)$, from (10.63), that of $x_m(t)$ and, in turn, from $\tilde{x}(t) = x(t) - x_m(t)$, that of $x(t)$, from (10.41), that of $u_d(t)$, and from (10.15), that of $\omega(t)$, $a(t)$ and $v(t)$. Thus all closed-loop signals are bounded. Finally, from (10.68), (10.69), and (10.71), $\tilde{x}(t)$, $e(t) = y(t) - y_m(t)$, and $\dot{\theta}(t)$ are all bounded by $d_N(t)$ in a mean square sense (e.g., $\int_{t_1}^{t_2} e^2(t)\,dt \leq \gamma_0 + k_0 \int_{t_1}^{t_2} d_N^2(t)\,dt$ for some constant γ_0, $k_0 > 0$).

10.3 Output Feedback Inverse Control

When only the output $y(t)$ of the plant (10.1) is available for measurement, we need to use an output feedback control law to generate $u_d(t)$ for the inverse (10.14). As shown in Figure 10.5, such a controller structure is

$$u_d(t) = C_1(D)[w](t),\ w(t) = r(t) - C_2(D)[y](t), \tag{10.72}$$

where $r(t)$ is a reference input and $C_1(D)$ and $C_1(D)$ are linear blocks which are designed as if the nonlinearity $N(\cdot)$ were absent, that is, when $u(t) = u_d(t)$. The block $\widehat{NI}(\cdot)$ is the inverse of $N(\cdot)$, whose parameters are to be updated by an adaptive law designed based on a closed-loop error equation.

General Design Guidelines

Reference system. The desired reference output $y_m(t)$ to be tracked by the plant output $y(t)$ now is defined as the output of the closed-loop system with the controller (10.41) applied to the plant (10.1) without the actuator nonlinearity $N(\cdot)$, that is, when $u(t) = u_d(t)$.

With $u(t) = u_d(t)$, the closed-loop output is

$$y(t) = y_m(t) \triangleq \frac{G(D)C_1(D)}{1 + G(D)C_1(D)C_2(D)}[r](t). \tag{10.73}$$

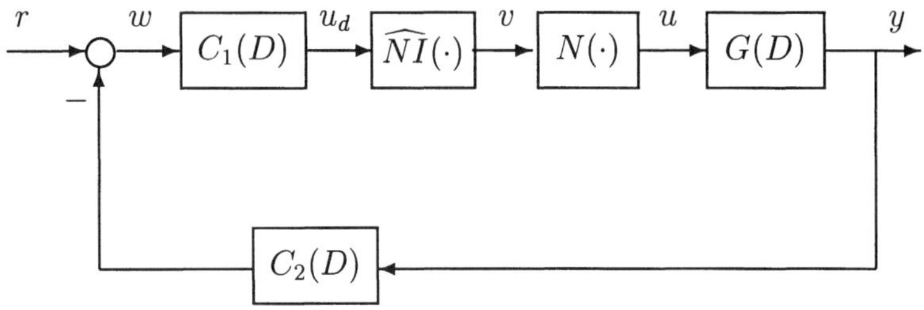

Figure 10.5: Output feedback inverse control.

Hence $C_1(D)$ and $C_2(D)$ should be chosen such that the transfer function

$$W_m(D) = \frac{G(D)C_1(D)}{1 + G(D)C_1(D)C_2(D)} \tag{10.74}$$

has all its poles at some desired stable locations and that the closed-loop system is internally stable when $u(t) = u_d(t)$.

Error model. To develop an adaptive law for updating the parameters of the inverse $\widehat{NI}(\cdot)$, we express the closed-loop system as

$$y(t) = y_m(t) + W(D)[(\theta - \theta^*)^T\omega](t) + d(t), \tag{10.75}$$

where

$$W(D) = \frac{G(D)}{1 + G(D)C_1(D)C_2(D)}, \tag{10.76}$$

$$d(t) = W(D)[d_N](t). \tag{10.77}$$

To use (10.75) to derive a desirable error model, $W(D)$ in (10.76) is required to have all its poles located in a desired stable region. Then the error signal $d(t)$ in (10.77) is bounded because $d_N(t)$ in (10.16) is bounded.

Similar to (10.46)–(10.48), in view of (10.75), we define the estimation error

$$\epsilon(t) = y(t) - y_m(t) + \xi(t), \tag{10.78}$$

where

$$\xi(t) = \theta^T(t)\zeta(t) - W(D)[\theta^T\omega](t), \tag{10.79}$$

$$\zeta(t) = W(D)[\omega](t). \tag{10.80}$$

Using (10.75) and (10.78)–(10.80), we arrive at the error equation

$$\epsilon(t) = (\theta - \theta^*)^T\zeta(t) + d(t), \tag{10.81}$$

which has the same form as that in (10.50) for a state feedback design.

Adaptive law. Based on the linearly parametrized error equation (10.81) and the gradient projection optimization method described in Section 10.2, similar to that in (10.61), we choose the adaptive update law for $\theta(t)$:

$$\left.\begin{array}{l}\dot{\theta}(t)\\ \theta(t+1)-\theta(t)\end{array}\right\} = -\frac{\Gamma\epsilon(t)\zeta(t)}{m^2(t)} + f(t), \tag{10.82}$$

which is initialized by (10.55) and projected with $f(t)$ in (10.56) or (10.57) for $g(t) = -\frac{\Gamma\epsilon(t)\zeta(t)}{m^2(t)}$, where Γ is given in (10.53). This adaptive law also ensures that $\theta_i(t) \in [\theta_i^a, \theta_i^b]$, $i = 1, 2, \ldots, n_\theta$, and that the normalized error $\frac{\epsilon(t)}{m(t)}$ and parameter variation $\dot{\theta}(t)$ or $\theta(t+1) - \theta(t)$ are bounded by $\frac{d(t)}{m(t)}$ in a mean sense.

Summary. An output feedback adaptive inverse control design for the plant (10.1) with an actuator nonlinearity $N(\cdot)$ requires:

(i) $N(\cdot)$ should be parametrized by $\theta^* = [\theta_1^*, \ldots, \theta_{n_\theta}^*]^T$ as in (10.2);

(ii) the nonlinearity inverse $\widehat{NI}(\cdot)$ in (10.14), as a compensator to cancel $N(\cdot)$, should be parametrizable as (10.15) with the stated properties;

(iii) the true nonlinearity parameters $\theta_i^* \in [\theta_i^a, \theta_i^b]$ for some known constants $\theta_i^a < \theta_i^b$, $i = 1, 2, \ldots, n_\theta$, such that for any set of $\theta_i \in [\theta_i^a, \theta_i^b]$, $i = 1, 2, \ldots, n_\theta$, the characteristic $N(\theta; \cdot)$ renders a nonlinearity model of the same type as that of the true nonlinearity model $N(\theta^*; \cdot)$; and

(iv) the compensators $C_1(D)$ and $C_2(D)$ in (10.72) should ensure that both $W_m(D)$ in (10.74) and $W(D)$ in (10.76) have good stability properties.

As a comparison, the conditions (i)–(iii) are also needed for a state feedback adaptive inverse control design in Section 10.2, while the state feedback control law (10.41) can always place the eigenvalues of $A + BK^*$ at some desired locations when the plant (10.40) is controllable.

Output Feedback Designs

Many existing control design methods for the plant (10.1) without the actuator nonlinearity $N(\cdot)$ may be used to construct the compensators $C_1(D)$ and $C_2(D)$ in the controller (10.72) to be incorporated in the above adaptive inverse control scheme for a plant with an actuator nonlinearity. Those designs include model reference, PID, and pole placement, which will be given in this section. To parametrize the linear output feedback controller (10.72), we express $G(D) = k_p \frac{Z(D)}{P(D)}$, where k_p is a scalar gain, and $Z(D)$ and $P(D)$

are monic polynomials of degrees n and m. Using the notation D, we present continuous-time and discrete-time designs in a unified framework.

Model reference design [395]. A model reference control design for the adaptive inverse compensation control scheme (10.72) is

$$u_d(t) = \phi_1^{*T}\omega_1(t) + \phi_2^{*T}\omega_2(t) + \phi_{20}^* y(t) + \phi_3^* r(t), \tag{10.83}$$

where $\omega_1(t)$, $\omega_2(t)$ are the signals from two identical filters:

$$\omega_1(t) = \frac{a(D)}{\Lambda(D)}[u_d](t), \ \omega_2(t) = \frac{a(D)}{\Lambda(D)}[y](t) \tag{10.84}$$

with $a(D) = [1, D, \ldots, D^{n-2}]^T$ and $\Lambda(D)$ being a stable polynomial of degree $n-1$, and $\phi_1^* \in R^{n-1}$, $\phi_2^* \in R^{n-1}$, $\phi_{20}^* \in R$, $\phi_3^* \in R$ satisfy $\phi_3^* = k_p^{-1}$ and

$$\begin{aligned} &\phi_1^{*T} a(D)P(D) + (\phi_2^{*T} a(D) + \phi_{20}^*\Lambda(D))k_p Z(D) \\ &= \Lambda(D)(P(D) - Z(D)P_m(D)), \end{aligned} \tag{10.85}$$

where $P_m(D)$ is a stable polynomial of degree $n-m$.

In the form of the controller (10.72), we identify

$$C_1(D) = \frac{\phi_3^*}{1 - \phi_1^{*T}\frac{a(D)}{\Lambda(D)}}, \ C_2(D) = k_p\left(\phi_2^{*T}\frac{a(D)}{\Lambda(D)} + \phi_{20}^*\right). \tag{10.86}$$

With this choice of $C_1(D)$ and $C_2(D)$, we have

$$W_m(D) = \frac{1}{P_m(D)}, \ W(D) = \frac{k_p}{P_m(D)}\left(1 - \phi_1^{*T}\frac{a(D)}{\Lambda(D)}\right). \tag{10.87}$$

It shows that both $W_m(D)$ and $W(D)$ have good stability properties from that of $P_m(D)$ and $\Lambda(D)$ independent of $G(D)$. For internal stability, the zero polynomial $Z(D)$ should be stable because model reference control cancels the dynamics of $Z(D)$ to lead to the closed-loop system $y(t) = W_m(D)[r](t)$ in the absence of $N(\cdot)$ and its inverse $\widehat{NI}(\cdot)$, that is, when $u(t) = u_d(t)$.

Pole placement design [400]. A pole placement control design is

$$u_d(t) = \phi_1^{*T}\omega_1(t) + \phi_2^{*T}\omega_2(t) + \phi_{20}^* y(t) + \phi_3^* r(t), \tag{10.88}$$

where $\omega_1(t)$ and $\omega_2(t)$ are the same as that defined in (10.84) and $\phi_1^* \in R^{n-1}$, $\phi_2^* \in R^{n-1}$, $\phi_{20}^* \in R$, $\phi_3^* \in R$ satisfy $\phi_3^* = k_p^{-1}$ and

$$\begin{aligned} &\phi_1^{*T} a(D)P(D) + (\phi_2^{*T} a(D) + \phi_{20}^*\Lambda(D))k_p Z(D) \\ &= \Lambda(D)(P(D) - P_d(D)), \end{aligned} \tag{10.89}$$

where $P_d(D)$ is a stable polynomial of degree n which contains the desired closed-loop poles. Then $C_1(D)$ and $C_2(D)$ are also expressed in (10.86) with

$$W_m(D) = \frac{Z(D)}{P_d(D)},\ W(D) = \frac{k_p Z(D)}{P_d(D)} \left(1 - \phi_1^{*T} \frac{a(D)}{\Lambda(D)}\right). \tag{10.90}$$

It also shows that both $W_m(D)$ and $W(D)$ have good stability properties from that of $P_d(D)$ and $\Lambda(D)$ independent of $G(D)$. However, the reference output $y_m(t) = W_m(D)[r](t)$ now depends on the plant zero polynomial $Z(D)$, which is allowed to be unstable. To solve (10.89) for an arbitrary but stable $P_d(D)$, it is required that $Z(D)$ and $P(D)$ are coprime.

PID design. A typical PID design for the controller structure (10.72) has $C_1(D) = C(D)$ and $C_2(D) = 1$, so that

$$u_d(t) = C(D)[r - y](t), \tag{10.91}$$

where the PID controller $C(D)$ has the form

$$C(D) = C(s) = \alpha + \frac{\beta}{s} + \gamma\, s \tag{10.92}$$

for a continuous-time design and

$$C(D) = C(z) = \alpha + \frac{\beta}{z-1} + \frac{\gamma\,(z-1)}{z} \tag{10.93}$$

for a discrete-time design, where α, β, and γ are design parameters.

Unlike either a model reference or pole placement scheme, which can be systematically designed for a general nth-order plant $G(D)$, a PID controller only applies to certain classes of plants. In other words, given a linear plant $G(D)$, one needs to make sure that a choice of $C(D)$ is able to ensure both $W_m(D)$ in (10.74) and $W(D)$ in (10.76) with good stability properties.

Illustrative example. Many simulation results for model reference and pole placement designs were presented in [395], [400], [401], which indicate that control systems with an adaptive inverse have significantly improved tracking performance in addition to closed-loop signal boundedness. As an additional illustrative example, we now present a detailed PI design [376].

An example of systems with actuator backlash is a liquid tank where the backlash is in the valve control mechanism. The liquid level $y(t)$ is

$$y(t) = G(s)[u](t),\ u(t) = B(v(t)), \tag{10.94}$$

where $u(t)$ is the flow, $G(s) = \frac{k_p}{s}$ with a constant k ($k_p = 1$, for simplicity), $B(\cdot)$ represents the actuator backlash, and $v(t)$ is the control input.

The backlash characteristic $u(t) = B(v(t))$ (see Figure 10.2) is described by two parallel straight lines connected with inner horizontal line segments of length $2c > 0$. The upward line is active when both $v(t)$ and $u(t)$ increase:

$$u(t) = v(t) - c, \ \dot{v}(t) > 0, \ \dot{u}(t) > 0, \tag{10.95}$$

and the downward line is active when both $v(t)$ and $u(t)$ decrease:

$$u(t) = v(t) + c, \ \dot{v}(t) < 0, \ \dot{u}(t) < 0. \tag{10.96}$$

The motion on any inner segment is characterized by $\dot{u}(t) = 0$.

As shown in [395], in the presence of backlash, a proportional controller $v = -\alpha e$ cannot reduce the error $e(t) = y(t) - r$ to zero, and a PI controller $v(t) = -(e(t) + \int_0^t e(\tau)\, d\tau)$ may lead to a limit cycle (e, v) trajectory.

To cancel the effects of backlash, we use an adaptive backlash inverse

$$v(t) = \widehat{BI}(u_d(t)), \tag{10.97}$$

shown in Figure 10.3 and described by two straight lines and vertical jumps between the lines. With $\widehat{c}(t)$ being the estimate of c, the downward side is

$$v(t) = u_d(t) - \widehat{c}(t), \ \dot{u}_d(t) < 0, \tag{10.98}$$

and the upward side is

$$v(t) = u_d(t) + \widehat{c}(t), \ \dot{u}_d(t) > 0. \tag{10.99}$$

Vertical jumps of $v(t)$ occur whenever $\dot{u}_d(t)$ changes its sign.

Introducing the backlash parameter and its estimate

$$\theta^* = c, \ \theta(t) = \widehat{c}(t) \tag{10.100}$$

and defining the backlash inverse regressor

$$\omega(t) = 2\widehat{\chi}(t) - 1, \ \widehat{\chi}(t) = \begin{cases} 1 & \text{if } v(t) = u_d(t) + \widehat{c}(t), \\ 0 & \text{otherwise}, \end{cases} \tag{10.101}$$

we have the control error expression (10.39), where $d_N(t) \in [-2c, 2c]$ and $d_N(t) = 0$, $t \geq t_0$, if $\theta = \theta^*$ and the backlash inverse is correctly initialized: $u(t_0) = u_d(t_0)$, that is, the backlash inverse $\widehat{BI}(\cdot)$ has the desired property

$$u_d(t_0) = B(\widehat{BI}(\theta^*; u_d(t_0))) \Rightarrow B(\widehat{BI}(\theta^*; u_d(t))) = u_d(t), \ \forall t \geq t_0. \tag{10.102}$$

Combined with the backlash inverse (10.97) is the PI controller

$$u_d(t) = \left(\alpha + \frac{\beta}{s}\right)[r - y](t), \ \alpha > 0, \ \beta > 0, \tag{10.103}$$

which leads to the closed-loop error system

$$y(t) - \frac{\alpha s + \beta}{s^2 + \alpha s + \beta}[r](t) = \frac{s}{s^2 + \alpha s + \beta}[(\theta - \theta^*)\omega + d_N](t). \tag{10.104}$$

In view of (10.73), (10.75), and (10.104), we have

$$W_m(s) = \frac{\alpha s + \beta}{s^2 + \alpha s + \beta}, \ W(s) = \frac{s}{s^2 + \alpha s + \beta}, \tag{10.105}$$

which have desired stability properties for some α and β. Then, we can update $\theta(t)$ from the continuous-time adaptive law (10.82) with $f(t)$ chosen to ensure $\widehat{c}(t) \in [c_1, c_2]$, where $0 < c_1 < c < c_2$ with c_1 and c_2 being known constants.

For the choice of $\alpha = \beta = 1$, $r = 10$, and $y(0) = 5$, which lead to large tracking error $e(t)$ without a backlash inverse, what happens in the closed-loop system with an adaptive backlash inverse $\widehat{BI}(\cdot)$ is shown in Figures 10.6(a)–(f) for $\widehat{c}(0) = 1.5$ (similar to other choices of $\widehat{c}(0)$). These results clearly show that an adaptive backlash inverse is able to adaptively cancel the effects of the unknown backlash so that the system tracking performance is significantly improved: the tracking error $e(t) = y(t) - r$ reduces to very small values, and the limit cycle, which appeared with a PI controller alone, is eliminated.

10.4 Designs for Multivariable Systems

Many control systems are multivariable. Coupling dynamics in a multivariable system often make it more difficult to control than a single-variable system. In this section, state feedback and output feedback adaptive inverse control schemes are presented for compensation of unknown nonsmooth nonlinearities at the input of a known multivariable linear plant, based on a multivariable parameter estimation algorithm for coupled error models.

Consider a multivariable plant with actuator nonlinearities, described by

$$y(t) = G(D)[u](t), \ \ u(t) = N(v(t)), \tag{10.106}$$

where $G(D)$ is an $M \times M$ strictly proper rational transfer matrix, and $u(t) = [u_1(t), \ldots, u_M(t)]^T$ and $v(t) = [v_1(t), \ldots, v_M(t)]^T$ are the output and input of the multivariable actuator nonlinearity $N(\cdot) = [N_1(\cdot), \ldots, N_M(\cdot)]^T$ such that

$$u_i(t) = N_i(v_i(t)), \ i = 1, 2, \ldots, M. \tag{10.107}$$

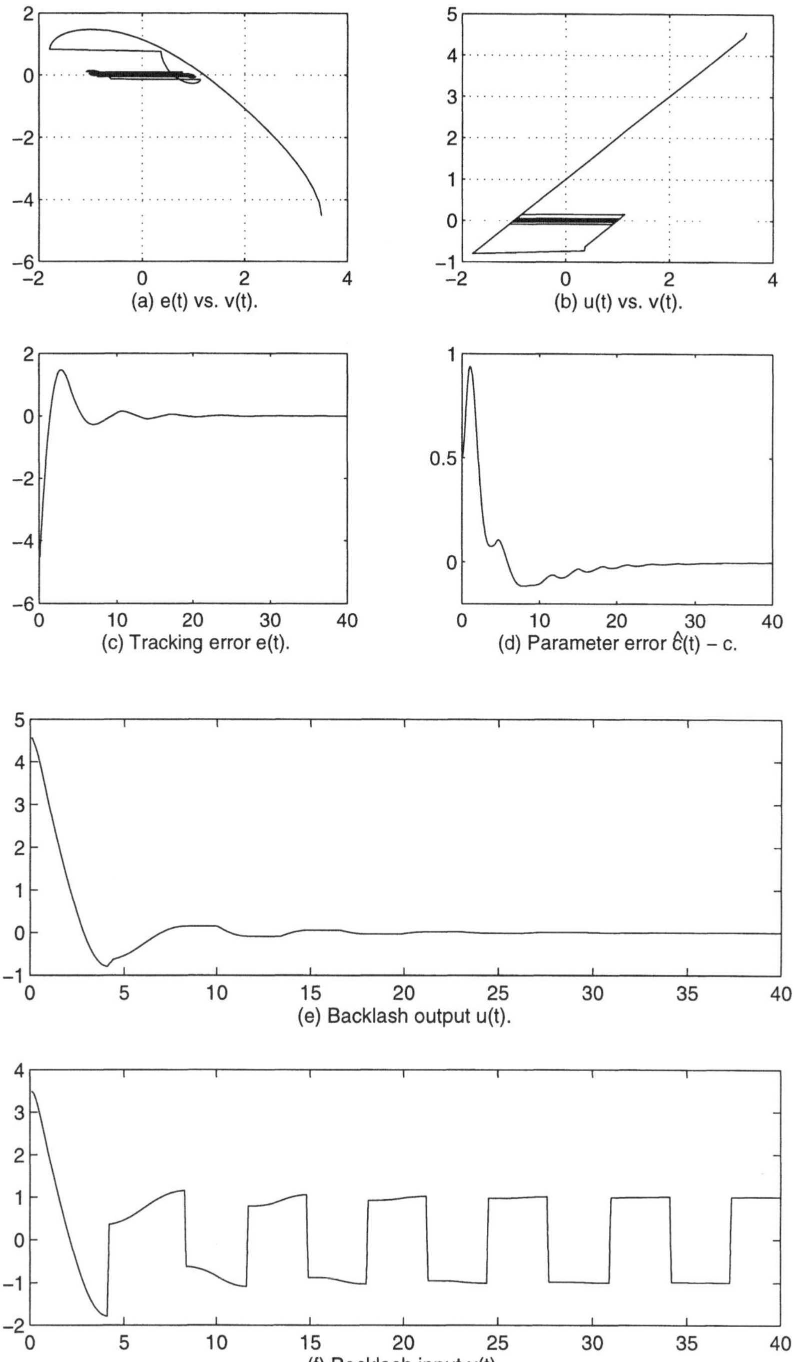

(a) e(t) vs. v(t).

(b) u(t) vs. v(t).

(c) Tracking error e(t).

(d) Parameter error $\hat{c}(t) - c$.

(e) Backlash output u(t).

(f) Backlash input v(t).

Figure 10.6: Adaptive backlash inverse control system responses.

The control problem is to design adaptive inverses for cancelling the nonlinearities $N_i(\cdot)$, to be combined with commonly used multivariable control schemes for the linear part $G(D)$ for desired tracking performance. In this problem, $v(t)$ is the control and $u(t)$ is not accessible for either control or measurement.

Nonlinearity model and its inverse. Similar to that in Section 10.1 for $M = 1$, we assume that each $N_i(\cdot)$, which may be a dead-zone, backlash, hysteresis, or other piecewise-linear characteristic, can be parametrized as

$$u_i(t) = N_i(v_i(t)) = N_i(\theta_i^*; v_i(t)) = -\theta_i^{*T}\omega_i^*(t) + a_i^*(t) \tag{10.108}$$

for some unknown parameter vectors $\theta_i^* \in R^{n_i}$, $n_i \geq 1$, $i = 1, \ldots, M$, and some unknown regressor vector signals $\omega_i^*(t) \in R^{n_i}$ and scalar signals $a_i^*(t)$. To cancel such nonlinearities, we use a multivariable nonlinearity inverse

$$v(t) = \widehat{NI}(u_d(t)), \tag{10.109}$$

where $u_d(t) = [u_{d1}(t), \ldots, u_{dm}(t)]^T$ is a design vector signal to be generated for $\widehat{NI}(\cdot) = [\widehat{NI}_1(\cdot), \ldots, \widehat{NI}_m(\cdot)]^T$, that is, the inverse (10.109) is equivalent to

$$v_i(t) = \widehat{NI}_i(u_{di}(t)),\ i = 1, \ldots, M. \tag{10.110}$$

We also assume that each $\widehat{NI}_i(\cdot)$ can be parametrized as

$$u_{di}(t) = -\theta_i^T(t)\omega_i(t) + a_i(t),\ i = 1, \ldots, M, \tag{10.111}$$

where $\theta_i \in R^{n_i}$ is an estimate of θ_i^* and $\omega_i(t) \in R^{n_i}$ and $a_i(t)$ are some known signals, as in the case of an inverse for a dead-zone, backlash, or hysteresis.

The uncertainties in $N_i(\cdot)$ cause a control error at each input channel:

$$u_i(t) - u_{di}(t) = \tilde{\theta}_i^T(t)\omega_i(t) + d_i(t), \tag{10.112}$$

where $\tilde{\theta}_i(t) = \theta_i(t) - \theta_i^*$, $i = 1, 2, \ldots, M$, and the unparametrized error is

$$d_i(t) = \theta_i^{*T}(\omega_i(t) - \omega_i^*(t)) + a_i^*(t) - a_i(t), \tag{10.113}$$

which should satisfy the same condition as that for $d_N(t)$ in (10.16).

In the vector form, the control error (10.112) is

$$u(t) - u_d(t) = \tilde{\Theta}^T(t)\omega(t) + d_N(t), \tag{10.114}$$

where

$$\tilde{\Theta}^T(t) = \begin{bmatrix} \tilde{\theta}_1^T(t) & 0 & 0 & \cdots & 0 \\ 0 & \tilde{\theta}_2^T(t) & 0 & \cdots & 0 \\ \vdots & \vdots & \vdots & \vdots & \vdots \\ 0 & 0 & \cdots & 0 & \tilde{\theta}_M^T(t) \end{bmatrix}, \tag{10.115}$$

$$\omega(t) = [\omega_1^T(t), \ldots, \omega_M^T(t)]^T, \tag{10.116}$$

$$d_N(t) = [d_1(t), \ldots, d_M(t)]^T. \tag{10.117}$$

State feedback designs. For (10.106) in the state variable form

$$D[x](t) = Ax(t) + Bu(t),$$
$$y(t) = Cx(t), \tag{10.118}$$

where $A \in R^{n\times n}$, $B \in R^{n\times M}$, and $C \in R^{M\times n}$, with (A, B) being controllable. To generate $u_d(t)$ for the inverse (10.109), we use the state feedback law

$$u_d(t) = K^*x(t) + r(t), \tag{10.119}$$

where $K^* \in R^{M\times n}$ is such that the eigenvalues of $A + BK^*$ are placed at some desired locations, and $r(t)$ is a bounded reference input signal.

Using (10.114), (10.118), and (10.119), we obtain

$$D[x](t) = (A + BK^*)x(t) + Br(t) + B\tilde{\Theta}^T(t)\omega(t) + Bd_N(t),$$
$$y(t) = Cx(t). \tag{10.120}$$

The ideal system performance is that of (10.120) with $\tilde{\Theta}^T(t)\omega(t) + d_N(t) = 0$. Hence, we can define the reference output $y_m(t)$ for $y(t)$ as

$$y_m(t) = W_m(D)[r](t), \tag{10.121}$$

where

$$W_m(D) = C(DI - A - BK^*)^{-1}B. \tag{10.122}$$

Then, with $d(t) = W_m(D)[d_N](t)$, the output tracking error is

$$e(t) = [e_1(t), \ldots, e_M(t)]^T = y(t) - y_m(t) = W_m(D)[\tilde{\Theta}^T\omega](t) + d(t). \tag{10.123}$$

Since $W_m(D)$ is an $M \times M$ transfer matrix, the error equation (10.123) has coupled dynamics, for which the parameter estimation algorithm of [397] can be applied. To proceed, we define the estimation errors

$$\epsilon_i(t) = e_i(t) + \xi_{i1}(t) + \cdots + \xi_{iM}(t), \tag{10.124}$$

where

$$\xi_{ij}(t) = \theta_j^T(t)\zeta_{ij}(t) - w_{ij}(D)[\theta_j^T\omega_j](t), \tag{10.125}$$

$$\zeta_{ij}(t) = w_{ij}(D)[\omega_j](t) \tag{10.126}$$

with $w_{ij}(D)$ being the (i,j)th element of $W_m(D)$ for $i,j=1,2,\ldots,M$.

Substituting (10.123), (10.125), and (10.126) into (10.124) gives

$$\epsilon_i(t) = \sum_{j=1}^{M} (\theta_j(t) - \theta_j^*)^T \zeta_{ij}(t) + d_i(t), \tag{10.127}$$

where d_i is the ith component of $d(t)$. Introducing

$$\theta(t) = [\theta_1^T(t), \ldots, \theta_M^T(t)]^T, \tag{10.128}$$

$$\theta^* = [\theta_1^{*T}, \ldots, \theta_M^{*T}]^T, \tag{10.129}$$

$$\zeta_i(t) = [\zeta_{i1}^T(t), \ldots, \zeta_{iM}^T(t)]^T, \tag{10.130}$$

we express the error model (10.127) as

$$\epsilon_i(t) = (\theta(t) - \theta^*)^T \zeta_i(t) + d_i(t),\ i = 1, 2, \ldots, M. \tag{10.131}$$

This is a set of linear error models with bounded "disturbances" $d_i(t)$. One important feature of these equations is that dynamic coupling in the multivariable error equation (10.123) leads to a set of M estimation errors which all contain the overall parameter vectors $\theta(t)$ and θ^*.

Introducing the cost function

$$J(\theta) = \frac{\epsilon_1^2 + \cdots + \epsilon_M^2}{2m^2}, \tag{10.132}$$

$$m^2(t) = 1 + \sum_{i=1}^{M} \zeta_i^T(t)\zeta_i(t) \tag{10.133}$$

and applying the gradient projection optimization technique of Section 10.2, we have the adaptive law for $\theta(t)$:

$$\left.\begin{array}{l} \dot{\theta}(t) \\ \theta(t+1) - \theta(t) \end{array}\right\} = -\frac{\Gamma}{m^2(t)} \sum_{i=1}^{m} \epsilon_i(t)\zeta_i(t) + f(t), \tag{10.134}$$

where $f(t)$ is for parameter projection and Γ is an adaptation gain matrix. Both $f(t)$ and Γ can be chosen similar to that in (10.56) or (10.57) and (10.53).

To develop the multivariable counterpart of the continuous-time adaptive scheme (10.68), with the same notation as that in (10.62)–(10.66) except that $\tilde{\theta}^T(t)\omega(t)$ is replaced by $\tilde{\Theta}^T(t)\omega(t)$, we have the time derivative of $V(\tilde{x}, \tilde{\theta})$ in (10.65), where $\tilde{\theta}(t) = \theta(t) - \theta^*$ in (10.128) and (10.129), as

$$\dot{V} = -\tilde{x}^T(t)Q\tilde{x}(t) + \tilde{x}^T(t)PB(\tilde{\Theta}^T(t)\omega(t) + d_N(t)) + \tilde{\theta}^T(t)\Gamma^{-1}\dot{\tilde{\theta}}(t). \tag{10.135}$$

Introducing the known signals

$$\bar{x}(t) = [\bar{x}_1(t), \ldots, \bar{x}_M(t)]^T = (\tilde{x}^T P B)^T, \tag{10.136}$$

$$\bar{\omega}(t) = [\bar{x}_1(t)\omega_1^T(t), \ldots, \bar{x}_M(t)\omega_M^T(t)]^T, \tag{10.137}$$

we express (10.135) as

$$\dot{V} = -\tilde{x}^T(t)Q\tilde{x}(t) + \tilde{\theta}^T(t)\bar{\omega}(t) + \tilde{x}^T(t)PBd_N(t) + \tilde{\theta}^T(t)\Gamma^{-1}\dot{\tilde{\theta}}(t), \tag{10.138}$$

which suggests the following adaptive law for $\theta(t)$:

$$\dot{\theta}(t) = -\Gamma\bar{\omega}(t) + f(t), \tag{10.139}$$

where $f(t)$ is for parameter projection and Γ is diagonal and positive.

Output feedback designs. To develop a multivariable output feedback adaptive inverse control scheme, we assume that $G(D)$ is nonsingular with the observability index ν and has the form $G(D) = Z(D)P^{-1}(D)$ for some $M \times M$ known polynomial matrices $Z(D)$ and $P(D)$.

The controller structure for $u_d(t)$ is

$$u_d(t) = \Phi_1^{*T}\omega_1(t) + \Phi_2^{*T}\omega_2(t) + \Phi_{20}^* y(t) + \Phi_3^* r(t), \tag{10.140}$$

where

$$\omega_1(t) = \frac{A(D)}{\Lambda(D)}[u_d](t), \ \omega_2(t) = \frac{A(D)}{\Lambda(D)}[y](t), \tag{10.141}$$

$$A(D) = [I, DI, \ldots, D^{\nu-2}I]^T, \tag{10.142}$$

$\Lambda(D)$ is a monic and stable polynomial of degree $\nu - 1$, and $\Phi_1^* \in R^{M(\nu-1)\times M}$, $\Phi_2^* \in R^{M(\nu-1)\times M}$, $\Phi_{20}^* \in R^{M\times M}$, and $\Phi_3^* \in R^{M\times M}$ satisfy

$$\begin{aligned} &\Phi_1^{*T}A(D)P(D) + (\Phi_2^{*T}A(D) + \Phi_{20}^*\Lambda(D))A(D)N(D) \\ &= \Lambda(D)(P(D) - \Phi_3^* P_d(D)) \end{aligned} \tag{10.143}$$

for a given $M \times M$ stable polynomial matrix $P_d(D)$.

A model reference design needs a stable $Z(D)$. The choice to meet (10.143) is $P_d(D) = \xi_m(D)Z(D)$ and $\Phi_3^* = K_p^{-1}$, where $\xi_m(D)$ is a modified left interactor matrix (which is a stable polynomial matrix) of $G(D)$ such that

$$\lim_{D\to\infty} \xi_m(D)G(D) = K_p \tag{10.144}$$

is finite and nonsingular (see Proposition 2.3 and Section 9.2.2). For a pole placement design which does not need a stable $Z(D)$, $\Phi_3^* P_d(D)$ is chosen such that $P_d^{-1}(D)$ is stable, $\Gamma_c[P(D)] = \Phi_3^*\Gamma_c[P_d(D)]$, and $\mu_j = \partial_{cj}[P(D)] = \max_{1\le i\le M}\{\partial[p_{ij}(D)]\} = \partial_{cj}[\Theta_3^* P_d(D)]$, $j = 1, 2, \ldots, M$, where $\Gamma_c[P(D)]$ and $\Gamma_c[P_d(D)]$ are the leading column matrices of $P(D)$ and $P_d(D)$ (see (9.55) in Section 9.2.1), which can be assumed to be unit upper triangular [442]. With this choice, the closed-loop pole placement equation (10.143) has a solution under the right coprimeness of $Z(D)$ and $P(D)$ (see Section 2.2).

With the controller structure (10.140) and the inverse (10.109), using the control error (10.114), we can derive the tracking error equation

$$e(t) = y(t) - y_m(t) = W(D)[\tilde{\Theta}^T\omega](t) + d(t), \tag{10.145}$$

where

$$y_m(t) = Z(D)P_d^{-1}(D)[r](t), \tag{10.146}$$

$$W(D) = Z(D)P_d^{-1}(D)\Phi_3^{*-1}\left(I - \Phi_1^{*T}\frac{A(D)}{\Lambda(D)}\right), \tag{10.147}$$

and $d(t) = W(D)[d_N](t)$ is bounded because $d_N(t)$ is and $W(D)$ is stable. Since the error equation (10.145) has the same form as that in (10.109), similar to (10.124)–(10.134), we can also develop an adaptive update law for $\theta_i(t)$, $i = 1, \ldots, M$, to implement the adaptive inverse (10.109).

10.5 Designs for Unknown Linear Dynamics

When both the nonlinearity $N(\cdot)$ and the linear part $G(D)$ in the plant (10.1) or (10.106) are unknown, the parameters of $G(D)$ and $N(\cdot)$ appear bilinearly and the model reference controller structure (10.83) or (10.140) results in a nonlinear parametrization not suitable for parameter estimation. To achieve a linear parametrization, we need a new controller structure.

10.5.1 Designs for SISO Plants

For the single-input, single-output (SISO) plant (10.1), as in [395], such a new controller structure is obtained by using $u_d(t) = -\theta^{*T}\omega(t)$, an inverse with true parameters (see (10.15)), to reparametrize the term $\phi_1^{*T}\omega_1(t)$:

$$\phi_1^{*T}\omega_1(t) = \phi_1^{*T}\frac{a(D)}{\Lambda(D)}[u_d](t) = \phi_4^{*T}\omega_4(t), \tag{10.148}$$

where $\phi_4^* = -\phi_1^* \otimes \theta^* \triangleq -[\phi_{11}^*\theta^{*T}, \phi_{11}^*\theta^{*T}, \ldots, \phi_{1n-1}^*\theta^{*T}]^T \in R^{(n-1)n_\theta}$ with $\otimes$ being the Kronecker product and the corresponding regressor vector is

$$\omega_4(t) = \frac{A(D)}{\Lambda(D)}[\omega](t). \tag{10.149}$$

As before, $\Lambda(D)$ is a stable polynomial of degree $n-1$, while $A(D)$ analogous to the vector $a(D)$ now is a matrix:

$$A(D) = [I_{n_\theta}, DI_{n_\theta}, \ldots, D^{n-2}I_{n_\theta}]^T, \tag{10.150}$$

where I_{n_θ} is the $n_\theta \times n_\theta$ identity matrix for $\theta^* \in R^{n_\theta}$. With $\omega_2(t) = \frac{a(D)}{\Lambda(D)}[y](t)$ in (10.84) and ϕ_2, ϕ_{20}, ϕ_3, and ϕ_4 being the estimates of $\phi_2^* \in R^{n-1}$, $\phi_{20}^* \in R$, $\phi_3^* \in R$, and $\phi_4^* \in R^{n_\theta(n-1)}$, the reparametrized controller structure is

$$u_d(t) = \phi_2^T\omega_2(t) + \phi_{20}y(t) + \phi_3 r(t) + \phi_4^T\omega_4(t). \tag{10.151}$$

With the new controller (10.151), we obtain the tracking error equation

$$e(t) = k_p W_m(D)[(\phi - \phi^*)^T\psi](t) + d(t), \tag{10.152}$$

where

$$\psi(t) = [\omega^T(t), \omega_2^T(t), y(t), r(t), \omega_4^T(t)]^T, \tag{10.153}$$

$$\phi(t) = [\theta^T(t), \phi_2^T(t), \phi_{20}(t), \phi_3(t), \phi_4^T(t)]^T, \tag{10.154}$$

$$\phi^* = [\theta^{*T}, \phi_2^{*T}, \phi_{20}^*, \phi_3^*, \phi_4^{*T}]^T, \tag{10.155}$$

and $d(t)$ is as in (10.77). The error equation (10.152) also suggests the following update law for $\phi(t)$ and $\rho(t)$ (the estimate of $\rho^* = k_p$):

$$\left.\begin{array}{l}\dot{\phi}(t)\\ \phi(t+1)-\phi(t)\end{array}\right\} = -\frac{\text{sign}[k_p]\Gamma\zeta(t)\epsilon(t)}{1+\zeta^T(t)\zeta(t)+\xi^2(t)} + f_\phi(t), \tag{10.156}$$

$$\left.\begin{array}{l}\dot{\rho}(t)\\ \rho(t+1)-\rho(t)\end{array}\right\} = -\frac{\gamma_\rho\xi(t)\epsilon(t)}{1+\zeta^T(t)\zeta(t)+\xi^2(t)} + f_\rho(t), \tag{10.157}$$

where

$$\epsilon(t) = e(t) + \rho(t)\xi(t), \tag{10.158}$$

$$\xi(t) = \phi^T(t)\zeta(t) - W_m(D)[\phi^T\psi](t), \tag{10.159}$$

$$\zeta(t) = W_m(D)[\psi](t), \tag{10.160}$$

$$\Gamma = \text{diag}\{\gamma_1, \ldots, \gamma_{n_\phi}\},\ \gamma_\rho \in R, \tag{10.161}$$

with $\gamma_i > 0$ and $\gamma_\rho > 0$ for a continuous-time design and $0 < \gamma_i < 2/k_p^0$ ($|k_p| \le k_p^0$) and $0 < \gamma_\rho < 2$ for a discrete-time design and n_ϕ being the dimension of the parameter estimate $\phi(t)$. The terms $f_\phi(t)$ and $f_\rho(t)$ are to be chosen to achieve robustness with respect to $d(t)$. Parameter projection similar to (10.56) or (10.57) can be used as a choice for $f_\phi(t)$ and $f_\rho(t)$. Then, the adaptive laws (10.156) and (10.157) along with the respective parameter projections $f_\phi(t)$ and $f_\rho(t)$ guarantee that $\phi(t)$, $\rho(t)$, $\frac{\epsilon(t)}{m(t)}$, $\dot{\phi}(t)$, $\dot{\rho}(t) \in L^\infty$ and that the normalized error $\frac{\epsilon(t)}{m(t)}$ and parameter variations $\dot{\phi}(t)$, $\dot{\rho}(t)$ or $\phi(t+1)-\phi(t)$, $\rho(t+1)-\rho(t)$ are bounded by $\frac{d(t)}{m(t)}$ in a mean sense.

10.5.2 Designs for MIMO Plants

For the multi-input, multi-output (MIMO) plant (10.106), the adaptive actuator nonlinearity compensation control problem is still open. It is mainly the high frequency gain matrix K_p defined in (10.144) that causes the difficulty in solving this problem. In this subsection, we formulate the control problems for both the case when K_p is known and when K_p is unknown and present some solution schemes for such problems.

Open-Loop Design

Consider the plant (10.106), that is, $y(t) = G(D)[u](t)$, with $G(D) = G_0(D)K_p$, where $G_0(D)$ is known and with good stability, and the reference model (10.121), $y_m(t) = W_m(D)[r](t)$, with $W_m(D) = G_0(D)$. Then we have

$$e(t) = y(t) - y_m(t) = G_0(D)K_p[\tilde{\Theta}^T\omega](t) + d(t), \tag{10.162}$$

where $d(t) = G_0(D)K_p[d_N](t)$ is bounded.

For $K_p = \{k^*_{ij}\} \in R^{M\times M}$ unknown and nonsingular, there is a nonzero element in each column of K_p. Denoting these nonzero elements as $k^*_{m_i i}, i = 1,2,\ldots,M$, for some $m_i \in \{1,2,\ldots,M\}$, we identify the following set of conditions [229] on the plant high frequency gain matrix K_p:

(C1): $\frac{k^*_{ji}}{k^*_{m_i i}}$ and $\mathrm{sign}[k^*_{m_i i}]$ are known, $i,j = 1,2,\ldots,M, j \neq m_i$, for some $k^*_{m_i i} \neq 0, m_i \in \{1,2,\ldots,M\}, i = 1,2,\ldots,M$.

To develop adaptive update laws for θ_i, $i = 1,2,\ldots,M$, we let k_{ij} be the estimates of k^*_{ij}, $i,j = 1,2,\ldots,M$, and define the estimation errors

$$\begin{aligned}\epsilon_j(t) = e_j(t) &+ \sum_{l=1}^{M} k_{l1}(\theta_1^T g_{jl}(D)[\omega_1](t) - g_{jl}(D)[\theta_1^T\omega_1](t)) + \cdots \\ &+ \sum_{l=1}^{M} k_{lM}(\theta_M^T g_{ij}(D)[\omega_M](t) - g_{jl}(D)[\theta_M^T\omega_M](t)),\end{aligned} \tag{10.163}$$

where $e_j(t)$ is the ith element of the tracking error $e(t)$. It can be verified that, with (C1) satisfied, the adaptive laws

$$\dot{\theta}_i(t) = -\text{sign}[k_{m_i i}^*] \sum_{l=1}^{M} \sum_{j=1}^{M} \frac{k_{li}^*}{k_{m_i i}^*} \epsilon_j(t) g_{jl}(D)[\omega_i](t), \tag{10.164}$$

$$\dot{k}_{li}(t) = -\sum_{j=1}^{M} \epsilon_j(t) \left(\theta_i^T g_{jl}(D)[\omega_i](t) - g_{jl}(D)[\theta_l^T \omega_i](t) \right) \tag{10.165}$$

for $l, i = 1, 2, \ldots, M$, ensure the desired properties (i) $\theta_i(t) \in L^\infty, k_{li}(t) \in L^\infty, i, j = 1, 2, \ldots, M$, and (ii) $\epsilon_j(t) \in L^2, \dot{\theta}_i(t) \in L^2, \dot{k}_{li}(t) \in L^2, i, j, l = 1, \ldots, M$, for $d(t) = 0$, and can be made robust with respect to $d(t) \neq 0$ but bounded by a parameter projection similar to that in (10.56).

The discrete-time counterparts of these results can be also derived.

State Feedback Designs

Consider the M-input plant

$$D[x](t) = Ax(t) + Bu(t), \ \ u(t) = N(v(t)), \tag{10.166}$$

where $A \in R^{n\times n}$ and $B \in R^{n\times M}$ are unknown parameter matrices and $N(\cdot)$ is the unknown actuator nonlinearity block as before.

Define a reference system

$$D[x_m](t) = A_m x_m(t) + B_m r(t), \tag{10.167}$$

where $A_m \in R^{n\times n}$ is a stable matrix, $B_m \in R^{n\times M}$, and $r(t) \in R^M$ is bounded.

Choose the adaptive controller

$$u_d(t) = K_1(t)x(t) + K_2(t)r(t), \tag{10.168}$$

where K_1 and K_2 are estimates of some $K_1^* \in R^{M\times n}$ and $K_2^* \in R^{M\times M}$.

A continuous-time Lyapunov design is based on the matching condition that there exist $K_1^* \in R^{M\times n}$ and $K_2^* \in R^{M\times M}$ such that

$$A_m = A + BK_1^*, \ B_m = BK_2^*. \tag{10.169}$$

Then we have the state tracking error equation for $\tilde{x}(t) = x(t) - x_m(t)$:

$$\begin{aligned} \dot{\tilde{x}}(t) = A_m\tilde{x}(t) + B_m(K_2^*)^{-1}\left(\tilde{K}_1 x(t) + \tilde{K}_2 r(t)\right) \\ + B_m(K_2^*)^{-1}\tilde{\Theta}_N^T(t)\omega_N(t) + Bd(t), \end{aligned} \tag{10.170}$$

where $\tilde{K}_1 = K_1 - K_1^*$ and $\tilde{K}_2 = K_2 - K_2^*$. Similar to (10.135)–(10.139), a stable adaptive scheme can be developed if the gain matrix K_2^* is known.

What if the gain matrix K_2^* is unknown?

When there is an output signal $y(t) = Cx(t) \in R^M$, we may define the reference output $y_m(t) = Cx(t)$. If there exist K_1^* and K_2^* such that

$$C(sI - A - BK_1^*)^{-1}BK_2^* = C(sI - A_m)^{-1}B_m \triangleq W_m(D), \tag{10.171}$$

then we can derive the output tracking error equation

$$\begin{aligned} e(t) = {} & W_m(D)(K_2^*)^{-1}\left(\tilde{K}_1 x(t) + \tilde{K}_2 r(t)\right) \\ & + W_m(D)(K_2^*)^{-1}\tilde{\Theta}_N^T(t)\omega_N(t) + W_m(D)(K_2^*)^{-1}[d](t). \end{aligned} \tag{10.172}$$

A stable gradient adaptive scheme can be developed if K_2^* is known.

Again, what if the gain matrix K_2^* is unknown?

Output Feedback Designs

When the parameter matrix Φ_1^* in the controller structure (10.140) is unknown, the term $\Phi_1^{*T}\omega_1(t)$ needs to be reparametrized, similar to the SISO case (see Section 10.5.1). Using the inverse expression (10.111) with true parameters, $u_d(t) = -[\theta_1^{*T}\omega_1, \ldots, \theta_M^{*T}\omega_M]^T = -\Theta^{*T}\omega(t)$, we express

$$\Phi_1^{*T}\omega_1(t) = -\Phi_1^{*T}\frac{A(D)}{\Lambda(D)}[\Theta^{*T}\omega](t) = \Phi_4^{*T}\psi_4(t), \tag{10.173}$$

where, for $\Phi_1^{*T} = [\Phi_{11}^{*T}, \Phi_{12}^{*T}, \ldots, \Phi_{1\nu-1}^{*T}]$ and $\Phi_{1i}^* \in R^{M\times M}$,

$$\Phi_4^{*T} = -[\Phi_{11}^{*T}\Theta^{*T}, \Phi_{12}^{*T}\Theta^{*T}, \ldots, \Phi_{1\nu-1}^{*T}\Theta^{*T}] \in R^{Mm\times n_\theta(\nu-1)}, \tag{10.174}$$

$$\psi_4(t) = \frac{A_\theta(D)}{\Lambda(D)}[\omega_N](t),\ A_\theta = [I_{n_\theta}, DI_{n_\theta}, D^2I_{n_\theta}, \ldots, D^{\nu-2}I_{n_\theta}]^T, \tag{10.175}$$

with I_{n_θ} being the $n_\theta \times n_\theta$ identity matrix for $n_\theta = \sum_{i=1}^M n_i$, with $\theta_i^* \in R^{n_i}$.

Then, the adaptive version of (10.140) is

$$u_d(t) = \Phi_4^T\omega_4(t) + \Phi_2^T\omega_2(t) + \Phi_{20}y(t) + \Phi^3 r(t), \tag{10.176}$$

where Φ_4, Φ_2, Φ_{20}, and Φ_3 are the estimates of Φ_4^*, Φ_2^*, Φ_{20}^*, and Φ_3^*, respectively. With this controller, for the reference model system (10.146), $y_m(t) = W_m(D)[r](t)$, with $W_m(D) = \xi_m^{-1}(D)$ and $\xi_m(D)$ being the modified left interactor matrix of $G(D)$, we can derive the tracking error equation

$$\begin{aligned} \xi_m(D)[y - y_m](t) = {} & K_p\left(\tilde{\Phi}_4^T\psi_4(t) + \tilde{\Phi}_2^T\psi_2(t) + \tilde{\Phi}_{20}^T y(t) + \tilde{\Theta}_3 r(t)\right) \\ & + K_p\tilde{\Theta}^T(t)\omega(t) + d_a(t), \end{aligned} \tag{10.177}$$

where $\tilde{\Phi}_4^T = \Phi_4 - \Phi_4*$, $d_a(t) = \Phi_3^{*-1}\left(I - \Phi_1^{*T}F(D)\right)[d](t)$, and $\Phi_3^* = K_p^{-1}$. Based on this error equation, using the design method of Section 9.2, a stable gradient adaptive scheme can be developed if K_p^* is known.

The same quation is: What if the gain matrix K_p^* is unknown?

Remark 10.1 For the case of unknown K_p, under the condition (C1), stable adaptive actuator nonlinearity compensation designs can be developed, that is, both the controller parameters and inverse parameters can be adaptively updated to ensure closed-loop signal boundedness. Reducing the *a priori* knowledge of K_p (i.e., relaxing the condition (C1)) is an important issue in adaptive actuator nonlinearity compensation control, as the presence of actuator nonlinearities needs to be handled in many applications and introduces more technical challenges as well, compared with that in Section 9.2. □

10.6 Designs for Nonlinear Dynamics

The adaptive inverse approach developed in Sections 10.1–10.5 can also be applied to systems with nonsmooth nonlinearities at the inputs of smooth nonlinear dynamics. Employing adaptive inverse compensation, state feedback or output feedback adaptive control schemes may be designed for some classes of nonlinear plants with actuator nonlinearities:

$$\begin{aligned} \dot{x}(t) &= f(x(t)) + g(x(t))u(t), \ u(t) = N(v(t)), \\ y(t) &= h(x(t)), \end{aligned} \tag{10.178}$$

where $f(x) \in R^n$, $g(x) \in R^n$, and $h(x) \in R$ are smooth functions of $x \in R^n$, $N(\cdot)$ represents an unknown nonsmooth actuator uncertainty as a dead-zone, backlash, hysteresis or piecewise-linear characteristic, and $v(t) \in R$ is the control input, while $u(t)$ is not accessible for either control or measurement.

Recall from (10.2) that the actuator nonlinearity $N(\cdot)$ is parametrized as

$$u(t) = N(v(t)) = N(\theta^*; v(t)) = -\theta^{*T}\omega^*(t) + a_s^*(t), \tag{10.179}$$

where $\theta^* \in R^{n_\theta}$ $(n_\theta \geq 1)$ is an unknown parameter vector and $\omega^*(t) \in R^{n_\theta}$ and bounded $a_s^*(t) \in R$ whose components are determined by the signal motion in the nonlinear characteristic $N(\cdot)$ and are also unknown [395].

The key to controlling such plants is to use an adaptive inverse

$$v(t) = \widehat{NI}(u_d(t)) : \quad u_d(t) = -\theta^T(t)\omega(t) + a_s(t) \tag{10.180}$$

to cancel the unknown nonlinearity $N(\cdot)$ so that feedback control schemes designed for (10.178) without $N(\cdot)$, with the help of $\widehat{NI}(\cdot)$, can be applied to (10.178) with $N(\cdot)$ to achieve desired system performance. In (10.180) (see (10.15)), the signals $\omega(t) \in R^{n_\theta}$ and $a_s(t) \in R$ are known, whose components are determined by the signal motion in the nonlinearity inverse $\widehat{NI}(\cdot)$ such that $v(t)$, $\omega(t)$, and $a_s(t)$ are bounded if $u_d(t)$ is, and $\theta(t)$ is an estimate of the unknown parameter vector θ^* in the actuator nonlinearity $N(\cdot)$.

10.6.1 Design for Feedback Linearizable Systems

In this subsection, an adaptive inverse controller is developed for feedback linearizable nonlinear systems [157], [251] with nonsmooth actuator nonlinearities. The use of an actuator nonlinearity inverse and a feedback linearizing controller leads to an error equation suitable for deriving an adaptive update law for the inverse parameters. An illustrative example is given. Such adaptive control schemes are also developed for multivariable nonlinear systems with actuator nonlinearities. For nonlinear systems that do not possess a relative degree, dynamic extension is employed to realize adaptive inverse compensation designs for actuator nonlinearities. These adaptive designs ensure closed-loop signal boundedness in the presence of uncertain actuator nonlinearities.

Preliminaries

For two functions $h(x) \in R$ and $f(x) = [f_1(x), f_2(x), \ldots, f_n(x)]^T \in R^n$, $x \in R^n$, their *Lie derivative* [157] is defined as

$$L_f h = \frac{\partial h}{\partial x} f = \frac{\partial h}{\partial x_1} f_1 + \cdots + \frac{\partial h}{\partial x_n} f_n. \tag{10.181}$$

For $g(x) \in R^n$, and recursively with $L_f^0 h = h$, it follows that

$$L_g L_f h(x) = \frac{\partial L_f h}{\partial x} g, \tag{10.182}$$

$$L_f^{i+1} h = L_f(L_f^i h) = \frac{\partial L_f^i h}{\partial x} f, i = 1, 2, \ldots. \tag{10.183}$$

A single-input, single-output (SISO) nonlinear system

$$\begin{aligned} \dot{x} &= f(x) + g(x)u, \ x \in R^n, \ u \in R, \\ z &= h(x), \ z \in R, \end{aligned} \tag{10.184}$$

has *relative degree* ρ [157] at a point x_0 if (i) $L_g L_f^k h(x) = 0$ for $\forall x \in N(x_0)$ (a neighborhood of x_0), $\forall k = 0, 1, 2, \ldots, \rho - 2$, and (ii) $L_g L_f^{\rho-1} h(x_0) \neq 0$.

Assume that system (10.184) has relative degree n at x and define

$$y = T(x) = \begin{bmatrix} T_1(x) \\ T_2(x) \\ T_3(x) \\ \vdots \\ T_n(x) \end{bmatrix} = \begin{bmatrix} h(x) \\ L_f h(x) \\ L_f^2 h(x) \\ \vdots \\ L_f^{n-1} h(x) \end{bmatrix}, \tag{10.185}$$

which is a diffeomorphism [157], [426].[2] It follows that

$$\begin{aligned} \dot{T}_{k-1} &= T_k, k = 2, 3, \ldots, n, \\ \dot{T}_n &= L_f^n h(x) + L_g L_f^{n-1} h(x) u. \end{aligned} \tag{10.186}$$

With the new state $y = T(x)$, the feedback control law

$$u = \frac{1}{L_g L_f^{n-1} h(x)} (w - L_f^n h(x)) \tag{10.187}$$

linearizes system (10.184), that is, leads to the system

$$\dot{y}_i = y_{i+1}, \; i = 1, 2, \ldots, n-1, \; \dot{y}_n = w, \tag{10.188}$$

which can be put in a matrix form $\dot{y} = Ay + bw$, where $w \in R$ is a new input. Here we note that $T(x)$ should be a diffeomorphism in the region which is reachable by $x(t)$ driven by $w(t)$ in the control law (10.187).

If system (10.184) has relative degree $\rho < n$ at x, system (10.184) can be transformed into the system (instead of (10.188))

$$\begin{aligned} &\dot{y}_1 = y_2, \;\; \dot{y}_2 = y_3, \ldots, \;\; \dot{y}_{\rho-1} = y_\rho, \\ &\dot{y}_\rho = b(\xi, \xi_z) + a(\xi, \xi_z) u, \;\; \dot{\xi}_z = q(\xi, \xi_z), \end{aligned} \tag{10.189}$$

where $\xi = [y_1, y_2, \ldots, y_\rho]^T$ and $\xi_z = [y_{\rho+1}, \ldots, y_n]^T$ (here ξ is defined from (10.185) with n replaced by ρ, while ξ_z is not but always exists to make $y = T(x)$ a diffeomorphism [157]). The zero dynamics system is $\dot{\xi}_z = q(\xi, \xi_z(t))$ [157], [206]. A feedback linearizing controller $u = \frac{1}{a(\xi,\xi_z)}(w - b(\xi, \xi_z))$ may ensure that ξ is bounded. However, even if the system zero dynamics $\dot{\xi}_z = q(0, \xi_z(t))$ are stable, a bounded ξ may not ensure that ξ_z is also bounded unless

[2] A continuously differentiable vector function $T(x) \in R^n$ for $x \in R^n$ is a diffeomorphism of U onto V for some open subsets U and V of R^n if (i) $T(U) = V$, (ii) T is one-to-one, and (iii) the inverse function T^{-1} from V to U is also continuously differentiable. If $\bar{y} = T(\bar{x})$ and $\frac{\partial T}{\partial x}|_{x=\bar{x}}$ is nonsingular, then there exist open sets U containing $\bar{x}$ and V containing $\bar{y}$ such that T is a diffeomorphism of U onto V [426].

$\dot{\xi}_z = q(\xi, \xi_z)$ is bounded-input–bounded-state stable [206] with ξ as the input and ξ_z as the state. For a tracking problem, $y_1, \ldots, y_\rho$ do not converge to zero. The zero dynamics subsystem is driven by the "inputs" y_i, $i = 1, 2, \ldots, \rho$, with its own states as y_j, $j = \rho + 1, \rho + 2, \ldots, n$. For bounded inputs y_i, $i = 1, 2, \ldots, \rho$, and with arbitrary initial conditions $y_j(0)$, for $y_j(t)$ to be bounded, the condition is that the zero dynamic subsystem $\dot{y}_j(t) = q_j(y(t))$, $j = \rho + 1, \rho + 2, \ldots, n$, is bounded-input–bounded-state stable.

There are similar concepts [157] for an M-input, M-output system

$$\dot{x} = f(x) + \sum_{i=1}^{M} g_i(x) u_i,$$
$$z_i = h_i(x),\ i = 1, 2, \ldots, M. \tag{10.190}$$

Such a system has a (vector) relative degree $\{r_1, \ldots, r_M\}$ at a point x_0 if (i) $L_{g_j} L_f^k h_i(x) = 0$, $\forall 1 \le j \le M, k < r_i - 1, 1 \le i \le M$, and $\forall x$ in a neighborhood of x_0, and (ii) the $M \times M$ matrix

$$\alpha(x) = \begin{bmatrix} L_{g_1} L_f^{r_1-1} h_1(x) & \cdots & L_{g_M} L_f^{r_1-1} h_1(x), \\ L_{g_1} L_f^{r_2-1} h_2(x) & \cdots & L_{g_M} L_f^{r_2-1} h_2(x), \\ \vdots & \vdots & \vdots \\ L_{g_1} L_f^{r_M-1} h_M(x) & \cdots & L_{g_M} L_f^{r_M-1} h_M(x) \end{bmatrix} \tag{10.191}$$

is nonsingular at $x = x_0$. For each output $z_i = h_i(x), i = 1, \ldots, M$, define

$$\begin{aligned} \phi_1^i(x) &= h_i(x), \\ \phi_2^i(x) &= L_f h_i(x), \\ &\vdots \\ \phi_{r_i}^i(x) &= L_f^{r_i-1} h_i(x), \end{aligned} \tag{10.192}$$

where r_i is such that $L_{g_j} L_f^{r_i} h_i(x) \neq 0$, $1 \le j \le M$, and introduce the new state variable vector

$$y = T(x) = [\phi_1^1(x), \ldots, \phi_{r_1}^1(x), \ldots, \phi_1^M(x), \ldots, \phi_{r_M}^M(x)]^T. \tag{10.193}$$

If $r_1 + \cdots + r_M = n$ and $\alpha(x)$ is nonsingular, then $T(x) \in R^n$ is a diffeomorphism and system (10.190) can be exactly feedback linearized by the control $u = \alpha(x)^{-1}(w - \beta(x))$, where

$$\beta(x) = \begin{bmatrix} L_f^{r_1} h_1(x) \\ L_f^{r_2} h_2(x) \\ \vdots \\ L_f^{r_M} h_M(x) \end{bmatrix}. \tag{10.194}$$

If $r_1 + \cdots + r_M = \rho < n$ and $\alpha(x)$ is nonsingular, then there exists a diffeomorphism $T(x)$ but system (10.190) can only be partially linearized and there exist zero dynamics which need to be stable for closed-loop stability.

Adaptive Inverse Control

From (10.179) and (10.180), we have

$$\dot{x} = f(x) + g(x)\left(u_d + \tilde{\theta}^T\omega + d_N\right), \tag{10.195}$$

where $d_N(t)$ is given in (10.16) and is bounded. The control task now is to design a feedback law for u_d and an adaptive update law for θ to ensure desired closed-loop system properties.

Design with exact feedback linearization. In this case, the assumption is that system (10.184) has relative degree n. Introduce $y = T(x) \in R^n$, where $T(x)$ is a diffeomorphism defined in (10.185). Then it follows that

$$\dot{y} = \frac{\partial T}{\partial x}\left(f(x) + g(x)u_d + g(x)\left(\tilde{\theta}^T\omega + d_N(t)\right)\right). \tag{10.196}$$

Choose the feedback linearizing control input u_d as

$$u_d = \frac{1}{L_g T_n}(w - L_f T_n) + u_N = \alpha^{-1}(x)\left(w - \beta(x)\right) + u_N, \tag{10.197}$$

where u_N is a design signal to ensure robustness with respect to the bounded error term d_N, $T_n(x)$ is the nth row of $T(x)$, $\alpha(x) = L_g T_n$, and $\beta(x) = L_f T_n$. This control law has the properties

$$\frac{\partial T}{\partial x}\left(f(x) - g(x)\alpha^{-1}(x)\beta(x)\right) = Ay, \tag{10.198}$$

$$\frac{\partial T}{\partial x}g(x)\alpha^{-1}(x) = b, \tag{10.199}$$

where A and b are defined from (10.188):

$$A = \begin{bmatrix} 0 & 1 & 0 & \cdots & 0 \\ 0 & 0 & 1 & & 0 \\ \vdots & \vdots & & \ddots & \\ 0 & 0 & 0 & & 1 \\ 0 & 0 & 0 & \cdots & 0 \end{bmatrix}, \quad b = \begin{bmatrix} 0 \\ 0 \\ \vdots \\ 0 \\ 1 \end{bmatrix}. \tag{10.200}$$

With the controller (10.197), we can rewrite (10.196) as

$$\dot{y} = Ay + bw + \frac{\partial T}{\partial x}g(x)\left(\tilde{\theta}^T\omega + d_N(t) + u_N\right). \tag{10.201}$$

Then we choose the linear feedback control law

$$w = -Fy + r, \tag{10.202}$$

where $F \in R^{1\times n}$ is such that $A - bF$ has desired stable eigenvalues and r is a reference input, and introduce the reference system as

$$\dot{y}_m = (A - bF)y_m + b\,r. \tag{10.203}$$

For the tracking error $\tilde{y} = y - y_m$, it follows from (10.201)–(10.203) that

$$\dot{\tilde{y}} = \dot{y} - \dot{y}_m = (A - bF)\tilde{y} + \frac{\partial T}{\partial x} g(x)(\tilde{\theta}^T \omega + d_N + u_N). \tag{10.204}$$

Choose the adaptive law

$$\dot{\theta}(t) = -\alpha(x)\Gamma\omega(t)\tilde{y}^T(t)Pb + f_\theta(t), \tag{10.205}$$

where $f_\theta(t)$ is a parameter projection for $\theta(t)$ (see Section 10.2), for which Γ is chosen diagonal with positive diagonal elements, to ensure that $\theta(t)$ is bounded and $\tilde{\theta}^T(t)\Gamma^{-1}f_\theta(t) \le 0$. Then, we have

Consider the positive definite function

$$V(\tilde{y}, \tilde{\theta}) = \frac{1}{2}\left(\tilde{y}^T P \tilde{y} + \tilde{\theta}^T \Gamma^{-1} \tilde{\theta}\right), \tag{10.206}$$

where $P \in R^{n\times n}$ with $P = P^T > 0$ satisfies the Lyapunov equation $P(A - bF) + (A - bF)^T P = -2Q$ for a chosen $n \times n$ matrix $Q = Q^T > 0$, and $\Gamma = \operatorname{diag}\{\gamma_1, \ldots, \gamma_{n_\theta}\}$ with $\gamma_i > 0$, $i = 1, 2, \ldots, n_\theta$. With (10.204) and (10.205), the time derivative of $V(\tilde{y}, \tilde{\theta})$ is

$$\dot{V} = -\tilde{y}^T Q \tilde{y} + \alpha(x)\tilde{y}^T Pb(d_N(t) + u_N) + \tilde{\theta}^T \Gamma^{-1} f_\theta. \tag{10.207}$$

Case (i): $d = 0$. In this case (as in the piecewise linearity case [402]), we choose $u_N = 0$ in the control law (10.197) and obtain $\dot{V} = -\tilde{y}^T(t)Q\tilde{y}(t) \le 0$. This means that V is bounded and $\tilde{y} = y - y_m \in L^2$, i.e., $\tilde{y}$ and $\tilde{\theta}$ are bounded, and so are y and θ. Since $y = T(x)$ is a diffeomorphism, $x(t)$ is also bounded, and so are u_d and $\dot{\tilde{y}}$, which implies that $\lim_{t\to\infty}(y(t) - y_m(t)) = 0$.

Case (ii): $d \neq 0$. We choose the design signal

$$u_N = -\gamma\alpha(x)\tilde{y}^T Pb, \ \gamma > 0. \tag{10.208}$$

Then, from (10.207), we have

$$\dot{V} = -\tilde{y}^T Q\tilde{y} + \tilde{y}^T Pb\alpha(x)d_N + \tilde{\theta}^T(t)\Gamma^{-1}f_\theta(t) - \gamma\left(\alpha(x)\tilde{y}^T Pb\right)^2$$
$$\leq -\tilde{y}^T(t)Q\tilde{y}(t) + \tilde{\theta}^T(t)\Gamma^{-1}f_\theta(t) + \frac{1}{4\gamma}d_N^2(t). \tag{10.209}$$

Since $d_N(t)$ is bounded, we conclude that $\tilde{y}$ is bounded, and so is $y(t)$. Then $x(t)$ is bounded because $T(x)$ is a diffeomorphism, and u_d is also bounded.

In summary, we have the following results.

Theorem 10.1 *For the nonlinear system*

$$\dot{x} = f(x) + g(x)u, \; u = N(v), \; z = h(x), \tag{10.210}$$

whose relative degree is n, if the actuator nonlinearity is parametrized as $u(t) = -\theta^{*T}\omega^*(t) + a_s^*(t)$, then the feedback linearization control law (10.197) plus (10.202) combined with the adaptive inverse (10.180) with the adaptive law (10.205) ensures that all closed-loop signals are bounded. Furthermore, if the uncertainty $d = 0$, the tracking error $y - y_m$ goes to zero asymptotically.

Design for systems with zero dynamics. When the system relative degree is $\rho < n$, the system can only be partially feedback linearized. Let $y = T(x) \in R^n$ be a diffeomorphism. In this case, there are two parts of $y = [\xi^T, \xi_z^T]^T = T(x) = [T_c^T(x), T_z^T(x)]^T$: $\xi = T_c(x) = [y_1, \ldots, y_\rho]^T$, which is the feedback linearizable part, and $\xi_z = T_z(x) = [y_{\rho+1}, \ldots, y_n]^T$, which is the zero dynamics part. Similar to (10.196), we can obtain

$$\dot{\xi} = \frac{\partial T_c}{\partial x}(f(x) + g(x)u_d + g(x)(\tilde{\theta}^T\omega + d_N(t)))$$
$$= A_\rho\xi + b_\rho w + \frac{\partial T_c}{\partial x}g(x)(\tilde{\theta}^T\omega + d_N(t)) \tag{10.211}$$

by the control law $u_d = \alpha^{-1}(x)(w - \beta(x)) + u_N$, which has the properties

$$\frac{\partial T_c}{\partial x}(f(x) - g(x)\alpha^{-1}(x)\beta(x)) = A_\rho\xi, \; \frac{\partial T_c}{\partial x}g(x)\alpha^{-1}(x) = b_\rho. \tag{10.212}$$

The zero dynamics system is $\dot{\xi}_z = q(\xi, \xi_z)$. The reference model is similar to that in (10.203) but with $y_m \in R^\rho$. Defining the new tracking error $\tilde{y} = \xi - y_m$, the adaptive law is of the same form as that in (10.205), which ensures that ξ and θ are bounded. If the zero dynamics system is bounded-input–bounded-state stable, the system state y is bounded, and so is x.

Multivariable Adaptive Compensation

Consider the M-input, M-output nonlinear plant

$$\dot{x}(t) = f(x) + g(x)u, u = N(v),$$
$$z = h(x), \tag{10.213}$$

with state variables $x = [x_1, \ldots, x_n]^T$, input $v = [v_1, \ldots, v_M]^T$, output $z = [z_1, \ldots, z_M]^T = [h_1(x), \ldots, h_M(x)]^T$, and actuator nonlinearities $u_i = N_i(v_i)$, $i = 1, 2, \ldots, M$, for $u = [u_1, \ldots, u_M]^T$ and $N(\cdot) = \text{diag}\{N_1(\cdot), \ldots, N_M(\cdot)\}$, such as dead-zone, backlash, or hysteresis. The models of the actuator nonlinearity $N(\cdot)$ and its inverse $\widehat{NI}(\cdot)$ are described in (10.107)–(10.117).

Design for systems with relative degrees. We first assume that the plant (10.213) has a relative degree, that is $\alpha(x)$ in (10.191) is nonsingular. With the control error given in (10.114), the plant (10.213) becomes

$$\dot{x} = f(x) + g(x)(u_d + \tilde{\Theta}^T\omega + d_N(t)). \tag{10.214}$$

When the plant (10.213) has a vector relative degree $\{r_1, \ldots, r_M\}$ and $r_1 + \cdots + r_M = \rho \leq n$, we can use the feedback linearizing control law

$$u_d = \alpha^{-1}(x)(w - \beta(x)) + u_N, \tag{10.215}$$

where $\beta(x)$ is in (10.194), $u_N \in R^M$ is to compensate the bounded error term $d \in R^M$, and $r \in R^M$ is a signal to be designed. In this case, there exists a transformation $\xi = T(x) \in R^\rho$ such that

$$\dot{\xi} = A\xi + Bw + \frac{\partial T(x)}{\partial x}g(x)(\tilde{\Theta}^T\omega + d_N + u_N), \tag{10.216}$$

where $A \in R^{\rho\times\rho}$ and $B \in R^{M\times\rho}$ are in the Brunowsky canonical form [157]

$$A = \text{block diag}\{A_1, \ldots, A_M\},\ B = \text{block diag}\{B_1, \ldots, B_M\}, \tag{10.217}$$

where $A_i \in R^{r_i\times r_i}$, $B_i \in R^{r_i}$ in the form of A, B in (10.200) with different dimensions, as the result of the feedback linearization:

$$\frac{\partial T(x)}{\partial x}f(x) - \frac{\partial T(x)}{\partial x}g(x)\alpha^{-1}(x)\beta(x) = A\xi, \tag{10.218}$$
$$\frac{\partial T(x)}{\partial x}g(x)\alpha^{-1}(x) = B. \tag{10.219}$$

Choose the design signal $w = -F\xi + r$, where $F \in R^{M\times n}$ is such that $A - BF$ is stable and r is a reference input, and define the reference model system

$$\dot{y}_m = (A - BF)y_m + Br. \tag{10.220}$$

For the state error vector $\tilde{y} = \xi - y_m \in R^\rho$, from (10.216) and (10.220), we obtain the error equation

$$\dot{\tilde{y}} = (A - BF)\tilde{y} + \frac{\partial T}{\partial x} g(x)(\tilde{\Theta}^T \omega + d_N + u_N). \tag{10.221}$$

Define the positive definite function

$$V(\tilde{y}, \tilde{\theta}_1, \ldots, \tilde{\theta}_M) = \frac{1}{2}\left(\tilde{y}^T P \tilde{y} + \sum_{i=1}^{M} \tilde{\theta}_i^T \Gamma_i^{-1} \tilde{\theta}_i\right), \tag{10.222}$$

where $P \in R^{\rho\times\rho}$ with $P = P^T > 0$ satisfies $P(A-BF)+(A-BF)^T P = -2Q$ for a chosen $\rho \times \rho$ matrix $Q = Q^T > 0$ and $\Gamma_i = \mathrm{diag}\{\gamma_{i1}, \ldots, \gamma_{in_{\theta_i}}\}$ with $\gamma_{ij} > 0, j = 1, 2, \ldots, n_{\theta_i}$, $i = 1, 2, \ldots, M$.

Choose the adaptive laws

$$\dot{\theta}_i(t) = -\Gamma_i e_i(t)\omega_i(t) + f_{\theta_i}(t),\ i = 1, 2, \ldots M, \tag{10.223}$$

where $f_{\theta_i}(t)$ is a parameter projection which ensures that θ_i is bounded and $\tilde{\theta}_i^T \Gamma_i f_{\theta_i} \leq 0$ and $e_i, i = 1, 2, \ldots, M$, is the ith component of $\tilde{y}^T PB\alpha(x)$.

Set the design signals u_N as

$$u_{Ni} = -\gamma_i e_i,\ \gamma_i > 0, i = 1, 2, \ldots, M, \tag{10.224}$$

with u_{Ni} being the ith component of u_N.

From (10.221)–(10.223), the time derivative of $V(\tilde{y}, \tilde{\theta}_1, \ldots, \tilde{\theta}_M)$ is

$$\begin{aligned}\dot{V} &= -\tilde{y}^T Q \tilde{y} + \sum_{i=1}^{M} e_i d_i + \sum_{i=1}^{M} \tilde{\theta}_i^T \Gamma_i^{-1} f_{\theta_i} - \sum_{i=1}^{M} \gamma_i e_i^2 \\ &\leq -\tilde{y}^T Q \tilde{y} + \sum_{i=1}^{M} \tilde{\theta}_i^T \Gamma_i^{-1} f_{\theta_i} + \sum_{i=1}^{M} \frac{1}{4\gamma_i} d_i^2.\end{aligned} \tag{10.225}$$

Since d_i is bounded, $i = 1, 2, \ldots, M$, $\tilde{y}$ is bounded, and so is $\xi(t)$. If $\rho < n$, system zero dynamics exist. If the zero dynamic subsystem is bounded-input–bounded-state stable (BIBS), x is also bounded.

In summary, we have the following result.

Theorem 10.2 *For the plant (10.213), with actuator nonlinearities (10.108) and with a relative degree and a BIBS stable zero dynamic system, the adaptive compensation control scheme, consisting of the feedback control law (10.215) with $w = -F\xi + r$ and adaptive inverse (10.111) updated from the adaptive law (10.223), ensures that all closed-loop signals are bounded.*

Design with dynamic extension. We now develop an adaptive actuator nonlinearity compensation scheme for multivariable nonlinear systems without relative degree. Such a scheme uses a dynamic extension technique [157], combined with adaptive inversion and feedback linearization.

A multivariable nonlinear system (10.213), in general, may have no relative degree. In this case, a dynamic state feedback or dynamic extension [157],

$$\dot{\kappa} = \gamma(x, \kappa) + \delta(x, \kappa)\tilde{v}, \quad u = a(x, \kappa) + b(x, \kappa)\tilde{v}, \tag{10.226}$$

where $\tilde{v}$ is a new input and $\kappa \in R^{n_\kappa}$ is the internal state vector of this dynamic compensator driven by $\tilde{v}$ and x, can be employed to make the extended system

$$\begin{aligned} &\dot{x} = f(x) + g(x)a(\eta) + g(x)b(\eta)\tilde{v}, \quad \dot{\kappa} = \gamma(\eta) + \delta(\eta)\tilde{v}, \\ &z = h(x) \end{aligned} \tag{10.227}$$

with $\eta = [x^T, \kappa^T]^T, \in R^{n+n_\kappa}$, to have a relative degree $\rho \leq n + n_\kappa$.

Consider system (10.213) with actuator nonlinearities $u_i = N_i(v_i)$, which does not have relative degree, and assume that there exists a dynamic extension (10.226) such that the extended system (10.227) has relative degree. Then for the extended system (10.227), functions $\alpha(\eta)$ and $\beta(\eta)$, similar to those in (10.191) and (10.194), exist such that $\alpha(\eta)$ is nonsingular. For actuator nonlinearity compensation when actuator nonlinearities $u_i = N_i(v_i)$ are present and u is not accessible for control, we use the inverse (10.111), u_d as the dynamic extension input, and $\tilde{v}$ as feedback linearizing control,

$$u_d = a(\eta) + b(\eta)\tilde{v}, \ \tilde{v} = \alpha^{-1}(\eta)(w - \beta(\eta)) + u_N. \tag{10.228}$$

Then, we can find a transformation $\xi = T(\eta) \in R^\rho$ such that

$$\dot{\xi} = A\xi + Bw + \frac{\partial T(\eta)}{\partial \eta}\begin{bmatrix} g(x)(\tilde{\Theta}^T\omega + d_N + u_N) \\ 0 \end{bmatrix}. \tag{10.229}$$

Since $\eta = [x^T, \kappa^T]^T$, we have

$$\frac{\partial T(\eta)}{\partial \eta}\begin{bmatrix} g(x)(\tilde{\Theta}^T\omega + d_N + u_N) \\ 0 \end{bmatrix} = \frac{\partial T(\eta)}{\partial x} g(x)(\tilde{\Theta}^T\omega + d_N + u_N). \tag{10.230}$$

With a reference model similar to (10.220), the tracking error equation is

$$\dot{\tilde{y}} = (A - BF)\tilde{y} + \frac{\partial T(\eta)}{\partial x} g(x)(\tilde{\Theta}^T\omega + d_N + u_N). \tag{10.231}$$

From here on, the adaptive law design and signal boundedness analysis are similar to (10.221)–(10.225), leading to results similar to Theorem 10.2.

In summary, we developed an adaptive compensation control scheme for feedback linearizable nonlinear systems with unknown nonsmooth actuator nonlinearities. A combination of an adaptive inverse for the actuator nonlinearity with a feedback linearizing control law leads to a linearly parametrized error model suitable for adaptively updating the parameters of the inverse. The adaptive scheme is applicable to multivariable nonlinear systems with relative degrees and can be used, with dynamic extension, for multivariable systems without relative degrees. The case when the smooth nonlinear dynamics are also unknown needs further investigation.

10.6.2 Design for Parametric-Strict-Feedback Systems

In this subsection, an adaptive inverse control design is presented for nonlinear dynamics in the *parametric-strict-feedback* form [206] with an unknown actuator nonlinear characteristic such as a dead-zone, backlash, or hysteresis. The adaptive controller employs an adaptive inverse to compensate the actuator nonlinearity and an adaptive feedback law designed using the backstepping method [206] for adaptive nonlinear control. The adaptive control scheme ensures closed-loop signal boundedness in the presence of uncertainties in both dynamics parameters and actuator nonlinearity parameters.

Problem Statement

Consider the *parametric-strict-feedback* plant with an actuator nonlinearity:

$$\begin{aligned} \dot{x}_i &= x_{i+1} + \theta_p^{*T} \varphi_i(x_1, \ldots, x_i), \; i = 1, \ldots, n-1, \\ \dot{x}_n &= \varphi_0(x) + \theta_p^{*T} \varphi_n(x) + \beta_0(x) u, \\ u &= N(v), \; v \in R, \; u \in R, \end{aligned} \tag{10.232}$$

where $\theta_p^* \in R^p$ is the vector of unknown constant parameters, φ_0, β_0, and $\varphi_i, 1 \leq i \leq n$, are known smooth nonlinear functions in R^n, $\beta_0(x) \neq 0, \forall x \in R^n$, $x_i, 1 \leq i \leq n$, are the state variables, and $N(\cdot)$ is the actuator nonlinearity described by (10.2) illustrated by a dead-zone characteristic (10.3). As in [176], [204], [206], $x_1(t)$ is considered as the plant output: $y(t) = x_1(t)$.

The control objective is to design a feedback control $v(t)$ for the plant with unknown parameters in both θ_p^* and θ^*, to ensure that all closed-loop signals are bounded and the output $y(t)$ tracks a reference output signal $y_m(t)$.

To achieve the control objective, the following assumptions are made:

(A10.1): Full state measurements of $x_i(t), 1 \leq i \leq n$ are available.

(A10.2): The nonlinearity parameters $\theta^* = [\theta_1^*, \ldots, \theta_{n_\theta}^*]^T$ have some known bounds θ_i^a and θ_i^b: $\theta_i^* \in [\theta_i^a, \theta_i^b]$, $i = 1, 2, \ldots, n_\theta$.

(A10.3): $0 < \beta_a < |\beta_0(x)| < \beta_b$ for some positive constants β_a, β_b.

(A10.4): The reference output signal $y_m(t)$ and its first n derivatives are known and bounded.

(A10.5): An upper bound M_p on $\|\theta_p^*\|$ is known, where $\|\cdot\|$ is the Euclidean vector norm.

Assumption (A10.1) is needed for the backstepping design [204]; Assumption (A10.2) is used to project the estimates of the components of θ^* for implementing an adaptive inverse (see Section 10.2); Assumption (A10.3) is needed for a nonsingular control; Assumption (A10.4) is needed for tracking; and Assumption (A10.5) is for robust adaptation of the estimate of θ_p^*.

Adaptive Control Design

The adaptive control scheme for the plant (10.232) consists of an adaptive inverse $\widehat{NI}(\cdot)$ to cancel the actuator nonlinearity $N(\cdot)$ and an adaptive backstepping nonlinear feedback law for stabilization and tracking.

Adaptive Inverse. The adaptive inverse is described by (10.180): $v(t) = \widehat{NI}(u_d(t))$, that is, $u_d(t) = -\theta^T(t)\omega(t) + a_s(t)$, where $\omega(t)$ and $a_s(t)$ are known signals ($a_s(t) = 0$ for a dead-zone or backlash inverse [395]), and $\theta(t)$ is the estimate of θ^*, to be updated from an adaptive law. The control error is

$$u(t) - u_d(t) = (\theta(t) - \theta^*)^T\omega(t) + d_N(t), \tag{10.233}$$

where $d_N(t)$ given in (10.16) is bounded.

Backstepping feedback control. The backstepping design technique has been extensively used for nonlinear control. Briefly, such a technique, for the dynamics part of the plant (10.232), is described by a design procedure of n steps. At each i step, an "error" coordinate z_i is defined and a stabilizing function α_i is designed to stabilize an ith-order subsystem with respect to a Lyapunov function V_i. Finally, at step n, a feedback control law for u_d, an adaptive update law for the estimate $\theta_p(t)$ of θ_p^*, and an adaptive update law for the estimate $\theta(t)$ are designed to stabilize the closed-loop system. The backstepping design procedure [204], [206] for our problem is given next.

Step 1: Defining $z_1 = x_1 - y_m = y - y_m$ and $z_2 = x_2 - \alpha_1$, we have

$$\dot{z}_1 = \dot{x}_1 - \dot{y}_m = z_2 + \alpha_1 + \theta_p^{*T}\varphi_1 - \dot{y}_m. \tag{10.234}$$

At this step, (10.234) is viewed as a first-order system to be stabilized by α_1 with respect to the first positive definite function

$$V_1 = \frac{1}{2}z_1^2 + \frac{1}{2}(\theta_p - \theta_p^*)^T\Gamma^{-1}(\theta_p - \theta_p^*), \tag{10.235}$$

where $\Gamma = \Gamma^T > 0$. The time derivative of V_1 is

$$\dot{V}_1 = z_1(z_2 + \alpha_1 + \theta_p^T\varphi_1 - \dot{y}_m) + (\theta_p - \theta_p^*)^T\Gamma^{-1}(\dot{\theta}_p - \Gamma z_1\varphi_1). \tag{10.236}$$

Choosing the first stabilizing function as

$$\alpha_1 = -c_1 z_1 - \theta_p^T\varphi_1 + \dot{y}_m \tag{10.237}$$

and defining the first tuning function as

$$\tau_1 = \Gamma z_1 \varphi_1, \tag{10.238}$$

we obtain

$$\dot{z}_1 = z_2 - c_1 z_1 - (\theta_p - \theta_p^*)^T\varphi_1, \tag{10.239}$$

$$\dot{V}_1 = -c_1 z_1^2 + z_1 z_2 + (\theta_p - \theta_p^*)^T\Gamma^{-1}(\dot{\theta}_p - \tau_1). \tag{10.240}$$

The term $z_1 z_2$ will be cancelled at the next step, and $\dot{\theta}_p$ will be given later on.

Step 2: Introducing $z_3 = x_3 - \alpha_2$, we have

$$\begin{aligned} \dot{z}_2 = \dot{x}_2 - \dot{\alpha}_1 = x_3 + \theta_p^{*T}\varphi_2 - \frac{\partial \alpha_1}{\partial x_1}\left(x_2 + \theta_p^{*T}\varphi_1\right) \\ -\frac{\partial \alpha_1}{\partial \theta_p}\dot{\theta}_p - \frac{\partial \alpha_1}{\partial y_m}\dot{y}_m - \frac{\partial \alpha_1}{\partial \dot{y}_m}\ddot{y}_m. \end{aligned} \tag{10.241}$$

Now, (10.239) and (10.241) can be viewed as a second-order system to be stabilized by α_1 given in (10.237) and α_2 with respect to the second positive definite function $V_2 = V_1 + \frac{1}{2}z_2^2$. The time derivative of V_2 is

$$\begin{aligned} \dot{V}_2 = -c_1 z_1^2 + z_2 \Bigg(z_1 + z_3 + \alpha_2 - \frac{\partial \alpha_1}{\partial x_1}x_2 - \frac{\partial \alpha_1}{\partial \theta_p}\dot{\theta}_p \\ - \frac{\partial \alpha_1}{\partial y_m}\dot{y}_m - \frac{\partial \alpha_1}{\partial \dot{y}_m}\ddot{y}_m + \theta_p^T\left(\varphi_2 - \frac{\partial \alpha_1}{\partial x_1}\varphi_1\right)\Bigg) \\ + (\theta_p - \theta_p^*)^T\Gamma^{-1}\left(\dot{\theta}_p - \tau_1 - \Gamma z_2\left(\varphi_2 - \frac{\partial \alpha_1}{\partial x_1}\varphi_1\right)\right). \end{aligned} \tag{10.242}$$

Introduce the second tuning function τ_2 as

$$\tau_2 = \tau_1 + \Gamma z_2 \left(\varphi_2 - \frac{\partial \alpha_1}{\partial x_1}\varphi_1\right) \tag{10.243}$$

and choose the second stabilizing function α_2 as

$$\begin{aligned}\alpha_2 = & -z_1 - c_2 z_2 + \frac{\partial \alpha_1}{\partial x_1}x_2 + \frac{\partial \alpha_1}{\partial \theta_p}\left(\tau_2 - \Gamma f_m(\|\theta_p\|)\theta_p\right) \\ & + \frac{\partial \alpha_1}{\partial y_m}\dot{y}_m + \frac{\partial \alpha_1}{\partial \dot{y}_m}\ddot{y}_m - \theta_p^T\left(\varphi_2 - \frac{\partial \alpha_1}{\partial x_1}\varphi_1\right),\end{aligned} \tag{10.244}$$

where

$$f_m(\theta_p) = \begin{cases} 0 & \text{if } \|\theta_p\| \le M_p, \\ f_0(1 - e^{-\alpha_0(\|\theta_p\| - M_p)^{n_m}}) & \text{if } \|\theta_p\| > M_p, \end{cases} \tag{10.245}$$

$$f_0 > 0, \alpha_0 > 0,\ n_m \ge n-1,\ M_p > \|\theta_p^*\|. \tag{10.246}$$

This modification term $f_m(\cdot)$ is introduced to ensure robust adaptation of θ_p with respect to the error term $d_N(t)$ in (10.233). Then we rewrite $\dot{V}_2$ as

$$\begin{aligned}\dot{V}_2 = & -c_1 z_1^2 - c_2 z_2^2 + z_2 z_3 + -z_2 \frac{\partial \alpha_1}{\partial \theta_p}\left(\dot{\theta}_p - \tau_2 + \Gamma f_m(\theta_p)\theta_p\right) \\ & + (\theta_p - \theta_p^*)^T \Gamma^{-1}(\dot{\theta}_p - \tau_2)\end{aligned} \tag{10.247}$$

and $\dot{z}_2$ as

$$\begin{aligned}\dot{z}_2 = & -z_1 - c_2 z_2 + z_3 - (\theta_p - \theta_p^*)^T\left(\varphi_2 - \frac{\partial \alpha_1}{\partial x_1}\varphi_1\right) \\ & - \frac{\partial \alpha_1}{\partial \theta_p}\left(\dot{\theta}_p - \tau_2 + \Gamma f_m(\theta_p)\theta_p\right).\end{aligned} \tag{10.248}$$

Step i ($i = 3, \ldots, n-1$): For $z_{i+1} = x_{i+1} - \alpha_i$ and $V_i = V_{i-1} + \frac{1}{2}z_i^2$, we have

$$\begin{aligned}\dot{z}_i = \dot{x}_i - \dot{\alpha}_{i-1} = & \ x_{i+1} + \theta_p^{*T}\varphi_i - \sum_{k=1}^{i-1}\frac{\partial \alpha_{i-1}}{\partial x_k}(x_{k+1} + \theta_p^{*T}\varphi_k) \\ & - \frac{\partial \alpha_{i-1}}{\partial \theta_p}\dot{\theta}_p - \sum_{k=1}^{i}\frac{\partial \alpha_{i-1}}{\partial y_m^{(k-1)}}y_m^{(k)}.\end{aligned} \tag{10.249}$$

Now, the ith-order system $(z_1, \ldots, z_i)$ is to be stabilized by α_i with respect to $V_i = V_{i-1} + \frac{1}{2}z_i^2$. To proceed, we derive

$$\dot{V}_i = -\sum_{k=1}^{i-1} c_k z_k^2 - \left(\sum_{k=1}^{i-2} z_{k+1}\frac{\partial \alpha_k}{\partial \theta_p}\right)\left(\dot{\theta}_p - \tau_{i-1} + \Gamma f_m(\theta_p)\theta_p\right)$$

$$+ z_i \left(z_{i-1} + z_{i+1} + \alpha_i - \sum_{k=1}^{i-1} \frac{\partial \alpha_{i-1}}{\partial x_k} x_{k+1} - \frac{\partial \alpha_{i-1}}{\partial \theta_p} \dot{\theta}_p \right.$$
$$\left. - \sum_{k=1}^{i} \frac{\partial \alpha_{i-1}}{\partial - y_m^{(k-1)}} y_m^{(k)} + \theta_p^T \left(\varphi_i - \sum_{k=1}^{i-1} \frac{\partial \alpha_{i-1}}{\partial x_k} \varphi_k \right) \right)$$
$$+ (\theta_p - \theta_p^*)^T \Gamma^{-1} \left(\dot{\theta}_p - \Gamma \sum_{l=1}^{i} z_l \left(\varphi_l - \sum_{k=1}^{l-1} \frac{\partial \alpha_{l-1}}{\partial x_k} \varphi_k \right) \right). \quad (10.250)$$

Introducing the ith tuning function τ_i as

$$\tau_i = \tau_{i-1} + \Gamma z_i \left(\varphi_i - \sum_{k=1}^{i-1} \frac{\partial \alpha_{i-1}}{\partial x_k} \varphi_k \right) = \Gamma \sum_{l=1}^{i} z_l \left(\varphi_l - \sum_{k=1}^{l-1} \frac{\partial \alpha_{l-1}}{\partial x_k} \varphi_k \right) \quad (10.251)$$

and choosing the ith stabilizing function α_i as

$$\alpha_i = -z_{i-1} - c_i z_i + \sum_{k=1}^{i-1} \frac{\partial \alpha_{i-1}}{\partial x_k} x_{k+1} + \frac{\partial \alpha_{i-1}}{\partial \theta_p} (\tau_i - \Gamma f_m(\theta_p)\theta_p) + \sum_{k=1}^{i} \frac{\partial \alpha_{i-1}}{\partial y_m^{(k-1)}} y_m^{(k)}$$
$$- \left(\theta_p^T - \sum_{k=1}^{i-2} z_{k+1} \frac{\partial \alpha_k}{\partial \theta_p} \Gamma \right) \left(\varphi_i - \sum_{k=1}^{i-1} \frac{\partial \alpha_{i-1}}{\partial x_k} \varphi_k \right), \quad (10.252)$$

we rewrite $\dot{V}_i$ as

$$\dot{V}_i = - \sum_{k=1}^{i} c_k z_k^2 + z_i z_{i+1} - \sum_{k=1}^{i-1} z_{k+1} \frac{\partial \alpha_k}{\partial \theta_p} \left(\dot{\theta}_p - \tau_i + \Gamma f_m(\theta_p)\theta_p \right)$$
$$+ (\theta_p - \theta_p^*)^T \Gamma^{-1} (\dot{\theta}_p - \tau_i) \quad (10.253)$$

and $\dot{z}_i$ as

$$\dot{z}_i = -z_{i-1} - c_i z_i + z_{i+1} - \frac{\partial \alpha_{i-1}}{\partial \theta_p} \left(\dot{\theta}_p - \tau_i + \Gamma f_m(\theta_p)\theta_p \right)$$
$$- (\theta_p - \theta_p^*)^T \left(\varphi_i - \sum_{k=1}^{i-1} \frac{\partial \alpha_{i-1}}{\partial x_k} \varphi_k \right)$$
$$+ \left(\sum_{k=1}^{i-2} z_{k+1} \frac{\partial \alpha_k}{\partial \theta_p} \Gamma \right) \left(\varphi_i - \sum_{k=1}^{i-1} \frac{\partial \alpha_{i-1}}{\partial x_k} \varphi_k \right). \quad (10.254)$$

Step n: Introducing $V_n = V_{n-1} + \frac{1}{2} z_n^2$, we have

$$\dot{V}_n = - \sum_{k=1}^{n-1} c_k z_k^2 + \left(\sum_{k=1}^{n-2} z_{k+1} \frac{\partial \alpha_k}{\partial \theta_p} \right) \left(\tau_{n-1} - \dot{\theta}_p + \Gamma f_m(\theta_p)\theta_p \right)$$
$$+ z_n \left(z_{n-1} + \beta_0 u + \varphi_0 - \sum_{k=1}^{n-1} \frac{\partial \alpha_{n-1}}{\partial x_k} x_{k+1} - \frac{\partial \alpha_{n-1}}{\partial \theta_p} \dot{\theta}_p \right.$$

$$
\begin{aligned}
&- \sum_{k=1}^{n} \frac{\partial \alpha_{n-1}}{\partial y_m^{(k-1)}} y_m^{(k)} + \theta_p^T \left(\varphi_n - \sum_{k=1}^{n-1} \frac{\partial \alpha_{n-1}}{\partial x_k} \varphi_k \right) \Bigg) \\
&+ (\theta_p - \theta_p^*)^T \Gamma^{-1} \left(\dot{\theta}_p - \Gamma \sum_{l=1}^{n} z_l \left(\varphi_l - \sum_{k=1}^{l-1} \frac{\partial \alpha_{l-1}}{\partial x_k} \varphi_k \right) \right). \quad (10.255)
\end{aligned}
$$

At this step, if the actuator nonlinearity $N(\cdot)$ were absent and $u(t)$ were the control, then we could design the adaptive update law for θ_p as

$$
\begin{aligned}
\dot{\theta}_p &= \tau_n - \Gamma f_m(\theta_p)\theta_p \\
&= \tau_{n-1} + \Gamma z_n \left(\varphi_n - \sum_{k=1}^{n-1} \frac{\partial \alpha_{n-1}}{\partial x_k} \varphi_k \right) - \Gamma f_m(\theta_p)\theta_p \\
&= \Gamma \sum_{l=1}^{n} z_l \left(\varphi_l - \sum_{k=1}^{l-1} \frac{\partial \alpha_{l-1}}{\partial x_k} \varphi_k \right) - \Gamma f_m(\theta_p)\theta_p \quad (10.256)
\end{aligned}
$$

(a choice is $\alpha_0 = 0$ in (10.246) so that $f_m(\theta_p) = 0$), and the control u as

$$
\begin{aligned}
u = \frac{1}{\beta_0} \Bigg(&-z_{n-1} - c_n z_n - \varphi_0 + \sum_{k=1}^{n-1} \frac{\partial \alpha_{n-1}}{\partial x_k} x_{k+1} + \frac{\partial \alpha_{n-1}}{\partial \theta_p} \dot{\theta}_p + \sum_{k=1}^{n} \frac{\partial \alpha_{n-1}}{\partial y_m^{(k-1)}} y_m^{(k)} \\
&- \left(\theta_p^T - \sum_{k=1}^{n-2} z_{k+1} \frac{\partial \alpha_k}{\partial \theta_p} \Gamma \right) \left(\varphi_n - \sum_{k=1}^{n-1} \frac{\partial \alpha_{n-1}}{\partial x_k} \varphi_k \right) \Bigg). \quad (10.257)
\end{aligned}
$$

For the resulting closed-loop system, we would have

$$
\dot{V}_n = - \sum_{k=1}^{n} c_k z_k^2 - (\theta_p - \theta_p^*)^T f_m(\theta_p)\theta_p, \quad (10.258)
$$

from which, the closed-loop signal boundedness and the asymptotic tracking $\lim_{t\to\infty}(y(t) - y_m(t)) = 0$ could be established [206].

However, in our control problem, the actuator nonlinearity $N(\cdot)$ is present and the control is $v(t) = \widehat{DI}(u_d(t))$, generated from the adaptive inverse whose parameter θ is also to be updated. Thus we need to design $u_d(t)$ and choose an adaptive law for θ to stabilize the closed-loop system with respect to

$$
V = V_n + \frac{1}{2}(\theta - \theta^*)^T \Gamma_N^{-1} (\theta - \theta^*), \ \Gamma_N = \Gamma_N^T > 0. \quad (10.259)
$$

As the feedback control law, we choose

$$
\begin{aligned}
u_d = \frac{1}{\beta_0} \Bigg(&-z_{n-1} - c_n z_n - \varphi_0 + \sum_{k=1}^{n-1} \frac{\partial \alpha_{n-1}}{\partial x_k} x_{k+1} + \frac{\partial \alpha_{n-1}}{\partial \theta_p} \dot{\theta}_p + \sum_{k=1}^{n} \frac{\partial \alpha_{n-1}}{\partial y_m^{(k-1)}} y_m^{(k)} \\
&- \left(\theta_p^T - \sum_{k=1}^{n-2} z_{k+1} \frac{\partial \alpha_k}{\partial \theta_p} \Gamma \right) \left(\varphi_n - \sum_{k=1}^{n-1} \frac{\partial \alpha_{n-1}}{\partial x_k} \varphi_k \right) \\
&- \Gamma f_m(\theta_p)\theta_p \sum_{k=1}^{n-2} z_{k+1} \frac{\partial \alpha_k}{\partial \theta_p} \Bigg). \quad (10.260)
\end{aligned}
$$

In view of (10.233), (10.255), and (10.260), we have

$$\dot{V} = -\sum_{k=1}^{n} c_k z_k^2 - f_m(\theta_p)(\theta_p - \theta_p^*)^T\theta_p + (\theta - \theta^*)^T\Gamma_N^{-1}(\dot{\theta} + \Gamma_N z_n \beta_0 \omega) + z_n \beta_0 d_N. \tag{10.261}$$

Then, the adaptive law for $\theta = [\theta_1, \theta_2, \ldots, \theta_{n_\theta}]^T$ is chosen as

$$\dot{\theta} = -\Gamma_N z_n \beta_0 \omega + f_N \tag{10.262}$$

with $\theta_i(0) \in [\theta_i^a, \theta_i^b]$ (see Assumption (A10.2)), where

$$\Gamma_N = \operatorname{diag}\{\gamma_1, \gamma_2, \ldots, \gamma_{n_\theta}\},\ \gamma_i > 0, \tag{10.263}$$

and the ith component of $f_N(t)$, $i = 1, 2, 3, 4$, is

$$f_i(t) = \begin{cases} 0 & \text{if } \theta_i(t) \in (\theta_i^a, \theta_i^b) \text{ or} \\ & \text{if } \theta_i(t) = \theta_i^a,\ g_i(t) \geq 0 \text{ or} \\ & \text{if } \theta_i(t) = \theta_i^b,\ g_i(t) \leq 0, \\ -g_i(t) & \text{otherwise} \end{cases} \tag{10.264}$$

with $g_i(t)$ being the ith component of

$$g_N(t) = -\Gamma_N z_n \beta_0 \omega. \tag{10.265}$$

The parameter projection term $f_N(t)$ is for robustness of the parameter update law with respect to the bounded $d_N(t)$ as well as for the estimate θ to satisfy $\theta_i \in [\theta_i^a, \theta_i^b]$, $i = 1, 2, \ldots, n_\theta$, to implement the inverse (10.180).

With the above adaptive laws (10.256), (10.262), and the feedback control law (10.260), the overall closed-loop system is described by

$$\begin{aligned}
\dot{z}_1 &= z_2 - c_1 z_1 - (\theta_p - \theta_p^*)^T\omega_1, \\
\dot{z}_2 &= -z_1 - c_2 z_2 + z_3 - (\theta_p - \theta_p^*)^T\omega_2 - \sum_{k=3}^{n} \frac{\partial \alpha_1}{\partial \theta_p}\Gamma z_k \omega_k, \\
&\vdots \\
\dot{z}_i &= -z_{i-1} - c_i z_i + z_{i+1} - (\theta_p - \theta_p^*)^T\omega_i \\
&\quad - \sum_{k=i+1}^{n} \frac{\partial \alpha_{i-1}}{\partial \theta_p}\Gamma z_k \omega_k + \sum_{k=1}^{i-2} z_{k+1}\frac{\partial \alpha_k}{\partial \theta_p}\Gamma \omega_i, \\
&\vdots \\
\dot{z}_n &= -z_{n-1} - c_n z_n - (\theta_p - \theta_p^*)^T\omega_n + \sum_{k=1}^{n-2} z_{k+1}\frac{\partial \alpha_k}{\partial \theta_p}\Gamma \omega_n, \\
\dot{\theta}_p &= \Gamma \sum_{l=1}^{n} z_l \omega_l - \Gamma f_m(\theta_p)\theta_p, \\
\dot{\theta} &= -\Gamma_N z_n \beta_0 \omega + f_N,
\end{aligned} \tag{10.266}$$

where

$$\omega_i(x_1, \ldots, x_i, \theta_p) = \varphi_i - \sum_{k=1}^{i-1} \frac{\partial \alpha_{i-1}}{\partial x_k} \varphi_k. \tag{10.267}$$

This adaptive control system has the following properties.

Theorem 10.3 *The adaptive inverse control scheme, consisting of the feedback law (10.260) updated from (10.256) and the inverse (10.180) updated from (10.262), ensures that all closed-loop signals are bounded.*

Proof: From (10.261) and (10.262), we have

$$\begin{aligned}\dot{V} = &- \sum_{k=1}^{n-1} c_k z_k^2 - f_m(\theta_p)(\theta_p - \theta_p^*)^T \theta_p - c_n z_n^2 \\ &+ z_n \beta_0 d_N + (\theta - \theta^*)^T \Gamma_N^{-1} f_N. \end{aligned} \tag{10.268}$$

The modification (10.245) and the parameter projection (10.264) ensure that

$$-f_m(\theta_p)(\theta_p - \theta_p^*)^T \theta_p \leq 0, \ (\theta - \theta^*)^T \Gamma_N^{-1} f_N \leq 0. \tag{10.269}$$

Since d_N and β_0 are bounded and $f_m(\theta_p)(\theta_p - \theta_p^*)^T \theta_p$ and/or $-(\theta - \theta^*)^T \Gamma_N^{-1} f_N$ grows unboundedly if θ_p and/or θ grows unboundedly, it follows from (10.259), (10.268), and Assumption (A10.3) that $\theta_p, \theta \in L^\infty$, and $z_k \in L^\infty$, $k = 1, \ldots, n$.

Since $z_1 = x_1 - y_m = y - y_m$ is bounded, we have $x_1 = y \in L^\infty$ and hence $\varphi_1(x_1) \in L^\infty$. It follows from (10.237) that $\alpha_1 \in L^\infty$. By $z_2 = x_2 - \alpha_1$, we have $x_2 \in L^\infty$ and hence $\varphi_2(x_1, x_2) \in L^\infty$. It follows from (10.244) that $\alpha_2 \in L^\infty$. Continuing in the same way, we prove that $\alpha_i \in L^\infty, i = 1, \ldots, n-1, x_i \in L^\infty, i = 1, \ldots, n$. Then from (10.260), we have $u_d \in L^\infty$.

Since the projection algorithm (10.260) ensures $\theta_i \in [\theta_i^a, \theta_i^b]$, $i = 1, 2, \ldots, n_\theta$, we have $v(t) \in L^\infty$ so that $u(t) \in L^\infty$. Therefore, we conclude that all signals in the closed loop system (10.266) are bounded. ∇

Remark 10.2 For the case when $d_N(t) = 0$ (e.g., the piecewise linear case [402] or the actuator nonlinearity-free case without an inverse: $u(t) = u_d(t)$ [206]), from (10.268), we have that $z_i(t) \in L^2$, $i = 1, 2, \ldots, n$, and, with $z_1(t) = y(t) - y_m(t) \in L^2$ and $\dot{z}_1(t) \in L^\infty$, that $\lim_{t \to \infty}(y(t) - y_m(t)) = 0$. $\square$

10.6.3 Design for Output-Feedback Systems

Adaptive output feedback compensation schemes can also be developed for nonlinear systems with actuator nonlinearities, whose state variables are not available for measurements. In this subsection, we present such a scheme for nonlinear systems in the *output-feedback canonical* form [206]. The adaptive inverse compensation design employs an adaptive inverse to compensate the actuator nonlinearity and uses a backstepping design for stabilization and output tracking control. The main tasks for such a design are to develop an augmented parametrization of the state observer which involves two sets of parameters—one from the actuator nonlinearity and the other from a smooth nonlinear dynamic part—to implement an output feedback control law, and to design a new robust adaptive control law using nonlinear damping to make the adaptive backstepping design admissible to a parameter projection needed to implement an adaptive actuator nonlinearity inverse.

Problem Statement

Consider the nonlinear plant in the output-feedback canonical form [206] with an actuator nonlinear characteristic $N(\cdot)$:

$$\begin{aligned}
\dot{x}_1 &= x_2 + \varphi_{01}(y) + \sum_{i=1}^{q} a_i \varphi_{i1}(y), \\
\dot{x}_2 &= x_3 + \varphi_{02}(y) + \sum_{i=1}^{q} a_i \varphi_{i2}(y), \\
&\vdots \\
\dot{x}_{\rho-1} &= x_\rho + \varphi_{0\rho-1}(y) + \sum_{i=1}^{q} a_i \varphi_{i\rho-1}(y), \\
\dot{x}_\rho &= x_{\rho+1} + \varphi_{0\rho}(y) + \sum_{i=1}^{q} a_i \varphi_{i\rho}(y) + b_m \sigma(y) u, \\
&\vdots \\
\dot{x}_n &= \varphi_{0n}(y) + \sum_{i=1}^{q} a_i \varphi_{in}(y) + b_0 \sigma(y) u, \\
y &= x_1, \ u = N(v),
\end{aligned} \tag{10.270}$$

where $a_1, \ldots, a_q, b_0, \ldots, b_m$ are constant parameters, $m = n - \rho$, $\varphi_{ij}, i = 0, \ldots, q, j = 1, \ldots, n$, and σ are known smooth nonlinear functions, $x_i, i = 1, \ldots, n$, are the state variables, $v(t)$ is the control, and $y(t)$ is the output. The actuator nonlinearity $N(\cdot)$ is described by (10.2) or (10.179) as a dead-zone, backlash, hysteresis, or piecewise-linear characteristic.

The control objective is to design an output feedback control $v(t)$ for the nonlinear plant (10.270) with actuator nonlinearity $N(\cdot)$ to ensure that all closed-loop signals are bounded and the plant output $y(t) = x_1(t)$ tracks a given reference signal $y_m(t)$ under the following assumptions:

(A10.1a): $y(t) = x_1(t)$ is available for measurement.

(A10.2): The nonlinearity parameters $\theta^* = [\theta_1^*, \ldots, \theta_{n\theta}^*]^T$ have some known bounds θ_i^a and θ_i^b: $\theta_i^* \in [\theta_i^a, \theta_i^b]$, $i = 1, 2, \ldots, n_\theta$.

(A10.3a): $\sigma(y)$ is bounded and $\sigma(y) \neq 0$, $\forall y$.

(A10.4a): The reference output signal $y_m(t)$ and its first ρ derivatives are known and bounded.

(A10.5a): Upper bounds on the Euclidean vector norms of $p = b_m^{-1}$, $a = [a_1, \ldots, a_q]^T$ and $\phi_i^* = b_i\theta^*$, $i = 0, 1, 2, \ldots, m$, are known.

(A10.6): The smooth nonlinearity parameters $a_1, \ldots, a_q, b_0, \ldots, b_m$ are unknown, but the sign of b_m is known and the polynomial $B(s) = b_m s^m + \cdots + b_1 s + b_0$ is stable.

State Observer

Since $x_2, \ldots, x_n$ are not measurable, we need an observer for the state variables $x_1, \ldots, x_n$ with a new parametrization to take into account the presence of the unknown actuator nonlinearity $N(\cdot)$, based on which an adaptive control scheme employing an adaptive dead-zone inverse and an adaptive backstepping output feedback control design can be developed.

We first write the plant (10.270) in the compact form

$$\begin{aligned} \dot{x} &= Ax + \varphi_0(y) + \sum_{i=1}^{q} a_i \varphi_i(y) + b\sigma(y)u, \\ y &= cx, \; x = [x_1, \ldots, x_n]^T, \\ u &= N(v), \end{aligned} \tag{10.271}$$

where

$$A = \begin{bmatrix} 0 & 1 & 0 & \cdots & 0 \\ 0 & 0 & 1 & & 0 \\ \vdots & \vdots & & \ddots & \\ 0 & 0 & 0 & & 1 \\ 0 & 0 & 0 & \cdots & 0 \end{bmatrix}, \; c = [1, 0, \ldots, 0], \; b = \begin{bmatrix} 0 \\ \vdots \\ 0 \\ b_m \\ \vdots \\ b_0 \end{bmatrix}, \tag{10.272}$$

$$\varphi_i(y) = [\varphi_{i1}(y), \ldots, \varphi_{in}(y)]^T, \quad i = 0, \ldots, q. \tag{10.273}$$

To construct an observer for the plant (10.271), we choose $k = [k_1, \ldots, k_n]^T$ such that all eigenvalues of $A_0 = A - kc$ are at some desired stable locations.

As in [206], an estimate of $x(t)$ can be defined as

$$\hat{x}(t) = \xi_0 + \sum_{i=1}^{q} a_i \xi_i + \sum_{i=0}^{m} b_i v_i, \tag{10.274}$$

where

$$\dot{\xi}_0 = A_0 \xi_0 + ky + \varphi_0(y) + \kappa, \tag{10.275}$$

$$\dot{\xi}_i = A_0 \xi_i + \varphi_i(y), \ i = 1, \ldots, q, \tag{10.276}$$

$$\dot{v}_i = A_0 v_i + e_{n-i} \sigma(y) u, \ i = 0, 1, \ldots, m, \tag{10.277}$$

κ is a design signal [177], and e_i denotes the ith coordinate vector in R^n. With (10.274)–(10.277), the state estimation error $\epsilon(t) = x(t) - \hat{x}(t)$ satisfies

$$\dot{\epsilon} = A_0 \epsilon - \kappa. \tag{10.278}$$

In our control problem, the signal $u(t)$ is not measured so that the signals $v_0, \ldots, v_m$ are not available for control design and need to be reparametrized.

Recall from (10.2) or (10.179) that the actuator nonlinearity $N(\cdot)$ is

$$u(t) = N(v(t)) = N(\theta^*; v(t)) = -\theta^{*T} \omega^*(t) + a_s^*(t), \tag{10.279}$$

from (10.15) or (10.180) that the nonlinearity inverse $\widehat{NI}(\cdot)$ is

$$v(t) = \widehat{NI}(u_d(t)) : \quad u_d(t) = -\theta^T(t) \omega(t) + a_s(t), \tag{10.280}$$

and from (10.16) that the control error is

$$u(t) - u_d(t) = (\theta - \theta^*)^T \omega(t) + d_N(t), \tag{10.281}$$

where the bounded error term $d_N(t)$ is defined in (10.16).

Substituting (10.46) into (10.281) and using (10.16), we obtain

$$u(t) = -\theta^{*T} \omega(t) + a_s(t) + d_N(t), \tag{10.282}$$

where $a_s(t)$ is known. With $\Lambda(s) = \det(sI - A_0)$, from (10.277), we express

$$\begin{aligned} v_i(t) &= [v_{i1}(t), v_{i2}(t), \ldots, v_{in}(t)]^T \\ &= [p_{i1}(s), p_{i2}(s), \ldots, p_{in}(s)]^T \frac{1}{\Lambda(s)} [\sigma(y) u](t), \ i = 0, \ldots, m, \end{aligned} \tag{10.283}$$

for some known polynomials $p_{ij}(s)$, $i = 0, \ldots, m$, $j = 1, \ldots, n$. Using (10.282) with $a_s(t) = 0$ [3] and (10.283), we obtain

$$v_{ij}(t) = -\theta^{*T} \omega_{ij}(t) + d_{ij}(t), \tag{10.284}$$

[3] Since $a_s(t)$ in (10.282) is known and is zero for a dead-zone, backlash, or piecewise-linear characteristic case, for simplicity of presentation, we assume $a_s(t) = 0$.

where

$$\omega_{ij}(t) = \frac{p_{ij}(s) I_{n_\theta}}{\Lambda(s)}[\sigma(y)\omega](t), \tag{10.285}$$

$$d_{ij}(t) = \frac{p_{ij}(s)}{\Lambda(s)}[\sigma(y) d_N](t) \tag{10.286}$$

with I_{n_θ} being the $n_\theta \times n_\theta$ identity matrix for $\theta \in R^{n_\theta}$.[4] In particular, denoting the second component of ξ_i as $\xi_{i2}, i = 0, \dots, q$, we have

$$\begin{aligned} \hat{x}_2 &= \xi_{02} + \sum_{i=1}^{q} a_i \xi_{i2} + \sum_{i=0}^{m} b_i v_{i2} \\ &= \xi_{02} + \sum_{i=1}^{q} a_i \xi_{i2} - \sum_{i=0}^{m} b_i \theta^{*T} \omega_{i2} + \sum_{i=0}^{m} b_i d_{i2}(t), \end{aligned} \tag{10.287}$$

$$\omega_{m2}(t) = \frac{(s^{m+1} + k_1 s^m) I_4}{s^n + k_1 s^{n-1} + \cdots + k_{n-1} s + k_n}[\sigma(y)\omega](t). \tag{10.288}$$

The signal $\hat{x}_2$, the second component of $\hat{x}$ in (10.274), plays a key role in a backstepping design, similar to the adaptive nonlinear control case [206]. The expressions (10.284) and (10.285) suggest that in the presence of the actuator nonlinearity $N(\cdot)$, a natural parametrization of $\hat{x}_2$ is to estimate $b_0\theta^*, b_1\theta^*, \dots, b_m\theta^*$. It is important to note that this augmented parametrization, similar to that in (10.148)–(10.151), is not an overparametrization.

To implement the nonlinearity inverse (10.15), we also need the estimate of θ^*. The expressions (10.285) and (10.286) are crucial for our adaptive dead-zone inverse controller design: the reparametrization of $\hat{x}_2$ by $b_0\theta^*$, $b_1\theta^*$, ..., $b_m\theta^*$ leads to a linearly parametrized closed-loop system despite the nonlinear appearance of the parameters in the smooth dynamics and the nonsmooth actuator nonlinearity $N(\cdot)$ and $\theta^T \omega_{m2}^{(\rho-1)}$ renders a place to the signal u_d at the last step of backstepping, as shown next.

Remark 10.3 In the absence of the actuator nonlinearity and without the use of a nonlinearity inverse, it follows that $u(t) = u_d(t)$. The signals $v_{ij}(t)$ used for an adaptive backstepping control design are simply from

$$v_{ij}(t) = \frac{p_{ij}(s)}{\Lambda(s)}[\sigma(y)u](t), \tag{10.289}$$

and only the first equation of (10.285) is used for design [177], [206]. □

[4]Recall that the notation $y_2(t) = W(s)[y_1](t)$ is used to denote that y_2 is the output of the linear system with input y_1 and transfer function $W(s)$.

Adaptive Design

We now apply the backstepping technique [206], with the new parametrization (10.283) and (10.284) of the state observer filters (10.275)–(10.277), to derive an adaptive control design for the plant (10.270):

(i) The design procedure has ρ steps, and at each step, an error z_i is defined and a stabilizing function α_i is designed to stabilize an ith-order subsystem.

(ii) The adaptive law for θ is given at the first step.

(iii) The feedback control law for u_d, the design signal κ, the adaptive law for $\hat{a}$, the estimate of a, and the adaptive laws for $\phi_0, \ldots, \phi_m$, the estimates of $b_0\theta^*, \ldots, b_m\theta^*$, are given at the last step.

There are two key techniques to be used: one is a nonlinear damping method [206] to incorporate a parameter projection algorithm for θ in the adaptive backstepping design, and the other is a smooth robustness modification term f_m (see Section 10.6.2) suitable for backstepping. In view of Remark 10.3, the design can be easily simplified for the case when $u(t) = u_d(t)$.

Step 1: Using (10.270), (10.287), and (10.284), we express

$$\dot{y} = \dot{x}_1 = x_2 + \varphi_{01} + \sum_{i=1}^{q} a_i \varphi_{i1} = \xi_{02} + \sum_{i=1}^{q} a_i \xi_{i2} + \sum_{i=0}^{m} b_i v_{i2} + \varphi_{01} + \sum_{i=1}^{q} a_i \varphi_{i1} + \epsilon_2$$

$$= \xi_{02} + \sum_{i=1}^{q} a_i \xi_{i2} - \sum_{i=0}^{m} b_i \theta^{*T} \omega_{i2} + \sum_{i=0}^{m} b_i d_{i2} + \varphi_{01} + \sum_{i=1}^{q} a_i \varphi_{i1} + \epsilon_2. \tag{10.290}$$

Using this expression, for the tracking error $z_1 = y - y_m = x_1 - y_m$, we obtain

$$\dot{z}_1 = \xi_{02} + \sum_{i=1}^{q} a_i \xi_{i2} - \sum_{i=0}^{m-1} b_i \theta^{*T} \omega_{i2} + d$$
$$- b_m \theta^{*T} \omega_{m2} + \varphi_{01} + \sum_{i=1}^{q} a_i \varphi_{i1} - \dot{y}_m + \epsilon_2, \tag{10.291}$$

where

$$d(t) = \sum_{i=0}^{m} b_i d_{i2}(t). \tag{10.292}$$

Note that, in view of Assumption (A10.3a), (10.286), and (10.292), $d(t)$ is bounded, which is favorable for our robust adaptive control design.

Letting θ, $\hat{p}$, $\hat{a}$, $\phi_i, i = 0, \ldots, m-1$, be the estimates of θ^*, $p = b_m^{-1}$, $a = [a_1, \ldots, a_q]^T$, $\phi_i^* = b_i \theta^*$, respectively, and α_1 be a design signal to be determined, and defining $z_2 = -\theta^T \omega_{m2} - \alpha_1$ and

$$\psi = c_1 z_1 + r_1 z_1 + \xi_{02} + \varphi_{01} - \dot{y}_m + \hat{a}^T(\xi_{(2)} + \varphi_{(1)})$$
$$- \sum_{i=0}^{m-1} \phi_i^T \omega_{i2}, \; c_1 > 0, \; r_1 > 0, \tag{10.293}$$

where $\xi_{(2)} = [\xi_{12}, \ldots, \xi_{q2}]^T, \varphi_{(1)} = [\varphi_{11}, \ldots, \varphi_{q1}]^T$, we rewrite (10.291) as

$$\dot{z}_1 = -c_1 z_1 - r_1 z_1 + b_m z_2 + b_m(\alpha_1 + \hat{p}\psi) + b_m(p - \hat{p})\psi + (a - \hat{a})^T(\xi_{(2)} + \varphi_{(1)}) \\ - \sum_{i=0}^{m-1}(\phi_i^* - \phi_i)^T \omega_{i2} - b_m(\theta^* - \theta)^T \omega_{m2} + \epsilon_2 + d, \tag{10.294}$$

which can be viewed as a first-order system to be stabilized by α_1 with respect to the positive definite function

$$V_1 = \frac{1}{2}\left(z_1^2 + \frac{|b_m|}{\gamma_p}(p - \hat{p})^2 + \frac{1}{\gamma_a}(a - \hat{a})^T(a - \hat{a}) + |b_m|(\theta^* - \theta)^T \Gamma_N^{-1}(\theta^* - \theta) \right. \\ \left. + \sum_{i=0}^{m-1}(\phi_i^* - \phi_i)^T \Gamma_i^{-1}(\phi_i^* - \phi_i) + \frac{1}{d_1}\epsilon^T P_0 \epsilon\right). \tag{10.295}$$

Here we introduce the constants $\gamma_a > 0$, $d_1 > 0$ and the $n_\theta \times n_\theta$ constant matrices $\Gamma_i = \Gamma_i^T > 0, i = 0, \ldots, m-1$, and $P_0 = P_0^T > 0$, which satisfies the equation $P_0 A_0 + A_0^T P_0 = -2I$.

To begin our adaptive design, we choose:

- the design signal α_1,

$$\alpha_1 = -\hat{p}\psi; \tag{10.296}$$

- the adaptive update law for $\hat{p}$,

$$\dot{\hat{p}} = \gamma_p \,\mathrm{sign}[b_m]\psi z_1 - \gamma_p f_m(\hat{p})\hat{p}, \ \gamma_p > 0, \tag{10.297}$$

  where $f_m(\cdot)$ is the modification function

$$f_m(\hat{\phi}) = \begin{cases} 0 & \text{if } \|\hat{\phi}(t)\| \le M_\phi, \\ f_0(1 - e^{-\alpha_0(\|\hat{\phi}\| - M_\phi)^{n_m}}) & \text{if } \|\hat{\phi}(t)\| > M_\phi, \end{cases} \tag{10.298}$$

  $f_0 > 0, \alpha_0 > 0$, $n_m = p - 1$, for $\hat{\phi}(t)$ being the estimate of a parameter vector ϕ, $M_\phi > \|\phi\|$, and $\|\cdot\|$ being the Euclidean vector norm;

- the adaptive law for $\theta = [\theta_1, \ldots, \theta_{n_\theta}]^T \in R^{n_\theta}$,

$$\dot{\theta} = -\mathrm{sign}[b_m]\Gamma_N z_1 \omega_{m2} + f_N \tag{10.299}$$

  with $\theta_i(0) \in [\theta_i^a, \theta_i^b]$, where

$$\Gamma_N = \mathrm{diag}\{\gamma_1, \gamma_2, \ldots, \gamma_{n_\theta}\}, \ \gamma_i > 0, \ i = 1, \ldots, n_\theta, \tag{10.300}$$

  and the ith component of $f_N(t)$, $i = 1, 2, 3, 4$, is

$$f_i(t) = \chi_i g_i \tag{10.301}$$

with $g_i(t)$ being the ith component of

$$g_N(t) = -\mathrm{sign}[b_m]\Gamma_N z_1 \omega_{m2}, \tag{10.302}$$

$$\chi_i = \begin{cases} 0 & \text{if } \theta_i(t) \in (\theta_i^a, \theta_i^b) \text{ or} \\ & \text{if } \theta_i(t) = \theta_i^a \text{ and } g_i(t) \geq 0 \text{ or} \\ & \text{if } \theta_i(t) = \theta_i^b \text{ and } g_i(t) \leq 0, \\ -1 & \text{otherwise} \end{cases} \tag{10.303}$$

and θ_i^a, θ_i^b the parameter bounds of θ_i^*: $\theta_i^* \in [\theta_i^a, \theta_i^b]$, $i = 1, 2, \ldots, n_\theta$, for $\theta^* = [\theta_1^*, \ldots, \theta_{n_\theta}^*]^T$ (see Assumption (A10.2)).

The projection term $f_N(t)$ is to ensure that the components of θ also satisfy the condition (A10.2), which is needed for implementing the adaptive inverse (10.280). We should note that the modification term $f_m(\hat{\phi})$ in (10.298) has a $(\rho - 2)$th-order partial derivative with respect to $\hat{\phi}$, which is crucial for the adaptive backstepping design with modified adaptive laws.

From (10.295)–(10.303), we obtain the time derivative of V_1 as

$$\begin{aligned} \dot{V}_1 = & -c_1 z_1^2 - r_1 z_1^2 + b_m z_1 z_2 + |b_m|(p - \hat{p}) f_m(\hat{p})\hat{p} - |b_m|(\theta^* - \theta)^T \Gamma_N^{-1} f_N \\ & + (a - \hat{a})^T \frac{1}{\gamma_a}\left(\gamma_a z_1(\xi_{(2)} + \varphi_{(1)}) - \dot{\hat{a}}\right) - \sum_{i=0}^{m-1} (\phi_i^* - \phi_i)^T \Gamma_i^{-1}(\Gamma_i z_1 \omega_{i2} + \dot{\phi}_i) \\ & - \frac{1}{d_1}\epsilon^T \epsilon + \frac{1}{d_1}\epsilon^T P_0(d_1 P_0^{-1} e_2 z_1 - \kappa) + z_1 d, \end{aligned} \tag{10.304}$$

where $e_2 = [0, 1, 0, \ldots, 0]^T$. The term "$b_m z_1 z_2$" will be cancelled at the next step, and $\dot{\hat{a}}$, $\dot{\phi}_i$, and κ are to be given later on.

Step 2: Using (10.275), (10.290), and (10.296), we write

$$\begin{aligned} \dot{z}_2 = & -\frac{d}{dt}\left(\theta^T \omega_{m2}\right) - \dot{\alpha}_1 \\ = & -\dot{\theta}^T \omega_{m2} - \theta^T \dot{\omega}_{m2} - \frac{\partial \alpha_1}{\partial y}\Bigg(\xi_{02} + \sum_{i=1}^{q} a_i \xi_{i2} - \sum_{i=0}^{m} b_i \theta^{*T} \omega_{i2} + d + \varphi_{01} \\ & + \sum_{i=1}^{q} a_i \varphi_{i1} + \epsilon_2\Bigg) - \frac{\partial \alpha_1}{\partial \hat{p}}\dot{\hat{p}} - \frac{\partial \alpha_1}{\partial y_m}\dot{y}_m - \frac{\partial \alpha_1}{\partial \dot{y}_m}\ddot{y}_m - \frac{\partial \alpha_1}{\partial \xi_0}(A_0 \xi_0 + ky \\ & + \varphi_0(y) + \kappa) - \sum_{i=1}^{q} \frac{\partial \alpha_1}{\partial \xi_i}\dot{\xi}_i - \frac{\partial \alpha_1}{\partial \hat{a}}\dot{\hat{a}} - \sum_{i=0}^{m-1} \frac{\partial \alpha_1}{\partial \phi_i}\dot{\phi}_i - \sum_{i=0}^{m-1} \frac{\partial \alpha_1}{\partial \omega_{i2}}\dot{\omega}_{i2}. \end{aligned} \tag{10.305}$$

Letting β_2 contain all known terms in (10.305) except $-\theta^T \dot{\omega}_{m2}$ and $-f_N^T \omega_{m2} = -\sum_{i=1}^{n_\theta} \chi_i g_i \omega_{m2i}$ (note that $\dot{\theta} = g_N + f_N$) and defining $z_3 = -\theta^T \dot{\omega}_{m2} - \alpha_2$, we express $\dot{z}_2$ as

$$\begin{aligned}
\dot{z}_2 &= -\theta^T \dot{\omega}_{m2} - f_N^T \omega_{m2} - \frac{\partial \alpha_1}{\partial y}\left(\sum_{i=1}^{q} a_i \xi_{i2} - \sum_{i=0}^{m} b_i \theta^{*T} \omega_{i2} + d + \sum_{i=1}^{q} a_i \varphi_{i1} + \epsilon_2\right) \\
&\quad + \beta_2 - \frac{\partial \alpha_1}{\partial \xi_0}\kappa - \frac{\partial \alpha_1}{\partial \hat{a}}\dot{\hat{a}} - \sum_{i=0}^{m-1} \frac{\partial \alpha_1}{\partial \phi_i}\dot{\phi}_i \\
&= z_3 + \alpha_2 - \sum_{i=1}^{n_\theta} \chi_i g_i \omega_{m2i} + \beta_2 - \frac{\partial \alpha_1}{\partial y} a^T(\xi_{(2)} + \varphi_{(1)}) + \sum_{i=0}^{m} \frac{\partial \alpha_1}{\partial y}\phi_i^{*T}\omega_{i2} \\
&\quad - \frac{\partial \alpha_1}{\partial \hat{a}}\dot{\hat{a}} - \sum_{i=0}^{m-1} \frac{\partial \alpha_1}{\partial \phi_i}\dot{\phi}_i - \frac{\partial \alpha_1}{\partial y} d - \frac{\partial \alpha_1}{\partial y}\epsilon_2 - \frac{\partial \alpha_1}{\partial \xi_0}\kappa. \qquad (10.306)
\end{aligned}$$

To continue our design, introducing $\phi_m^* = b_m \theta^*$ and letting ϕ_m, $\hat{b}_m$ be the estimates of ϕ_m^*, b_m, respectively, we choose:

- the design signal α_2,

$$\begin{aligned}
\alpha_2 &= -c_2 z_2 - r_2 z_2 \left(\frac{\partial \alpha_1}{\partial y}\right)^2 - z_2 \sum_{i=1}^{n_\theta} s_{2i} g_i^2 \omega_{m2i}^2 - \beta_2 - \hat{b}_m z_1 \\
&\quad + \frac{\partial \alpha_1}{\partial y}\hat{a}^T(\xi_{(2)} + \varphi_{(1)}) - \sum_{i=0}^{m} \frac{\partial \alpha_1}{\partial y}\phi_i^T \omega_{i2} \\
&\quad + \frac{\partial \alpha_1}{\partial \hat{a}}\gamma_a \left(\left(z_1 - \frac{\partial \alpha_1}{\partial y} z_2\right)(\xi_{(2)} + \varphi_{(1)}) - f_m(\hat{a})\hat{a}\right) \\
&\quad + \sum_{i=0}^{m-1} \frac{\partial \alpha_1}{\partial \phi_i}\Gamma_i \left(-\left(z_1 - \frac{\partial \alpha_1}{\partial y} z_2\right)\omega_{i2} - f_m(\phi_i)\phi_i\right) \\
&\quad + \frac{\partial \alpha_1}{\partial \xi_0} d_1 P_0^{-1} e_2 \left(z_1 - \frac{\partial \alpha_1}{\partial y} z_2\right) \qquad (10.307)
\end{aligned}$$

  for some constant $s_{2i} > 0$, $1 = 1, 2, \ldots, n_\theta$;

- the adaptive update law for $\hat{b}_m$,

$$\dot{\hat{b}}_m = -\gamma_m z_1 z_2 - \gamma_m f_m(\hat{b}_m)\hat{b}_m, \ \gamma_m > 0, \qquad (10.308)$$

  where $f_m(\cdot)$ is defined in (10.298).

It is important to note that the term $-z_2 \sum_{i=1}^{n_\theta} s_{2i} g_i^2 \omega_{m2i}^2$ in α_2 of (10.307) is a "damping" factor for "robustness" of the adaptive backstepping design with respect to the "disturbance" $-\sum_{i=1}^{n_\theta} \chi_i g_i \omega_{m2i}$ in $\dot{z}_2$ of (10.306). Such a disturbance is due to the parameter projection $f_N(t)$ in (10.299). Although it is a known term, for an adaptive backstepping design, $-\sum_{i=1}^{n_\theta} \chi_i g_i \omega_{m2i}$ cannot be included in β_2, that is, in α_2, because it is nonsmooth (nondifferentiable).

To proceed, we then express

$$\begin{aligned}
\dot{z}_2 = & -c_2 z_2 - r_2 z_2 \left(\frac{\partial \alpha_1}{\partial y}\right)^2 - z_2 \sum_{i=1}^{n_\theta} s_{2i} g_i^2 \omega_{m2i}^2 - \sum_{i=1}^{n_\theta} \chi_i g_i \omega_{m2i} - \hat{b}_m z_1 + z_3 \\
& - \frac{\partial \alpha_1}{\partial y}(a - \hat{a})^T (\xi_{(2)} + \varphi_{(1)}) + \sum_{i=0}^{m} \frac{\partial \alpha_1}{\partial y}(\phi_i^* - \phi_i)^T \omega_{i2} \\
& + \frac{\partial \alpha_1}{\partial \hat{a}} \left(\gamma_a \left(\left(z_1 - \frac{\partial \alpha_1}{\partial y} z_2\right)(\xi_{(2)} + \varphi_{(1)}) - f_m(\hat{a})\hat{a}\right) - \dot{\hat{a}}\right) \\
& + \sum_{i=0}^{m-1} \frac{\partial \alpha_1}{\partial \phi_i} \left(\Gamma_i \left(-\left(z_1 - \frac{\partial \alpha_1}{\partial y} z_2\right) \omega_{i2} - f_m(\phi_i)\phi_i\right) - \dot{\phi}_i\right) \\
& - \frac{\partial \alpha_1}{\partial y} d - \frac{\partial \alpha_1}{\partial y} \epsilon_2 + \frac{\partial \alpha_1}{\partial \xi_0} \left(d_1 P_0^{-1} e_2 \left(z_1 - \frac{\partial \alpha_1}{\partial y} z_2\right) - \kappa\right).
\end{aligned} \tag{10.309}$$

Defining the positive definite function

$$V_2 = V_1 + \frac{1}{2}\left(z_2^2 + \frac{1}{\gamma_m}(b_m - \hat{b}_m)^2 + (\phi_m^* - \phi_m)^T \Gamma_m^{-1} (\phi_m^* - \phi_m)\right), \tag{10.310}$$

where $\Gamma_m = \Gamma_m^T > 0$, we obtain the time derivative of V_2 as

$$\begin{aligned}
\dot{V}_2 = & -c_1 z_1^2 - r_1 z_1^2 - c_2 z_2^2 - r_2 z_2^2 \left(\frac{\partial \alpha_1}{\partial y}\right)^2 - z_2^2 \sum_{i=1}^{n_\theta} s_{2i} g_i^2 \omega_{m2i}^2 - z_2 \sum_{i=1}^{n_\theta} \chi_i g_i \omega_{m2i} \\
& + z_2 z_3 + |b_m|(p - \hat{p}) f_m(\hat{p})\hat{p} - |b_m|(\theta^* - \theta)^T \Gamma_N^{-1} f_N \\
& + (b_m - \hat{b}_m) f_m(\hat{b}_m)\hat{b}_m + (a - \hat{a})^T \frac{1}{\gamma_a}\left(\gamma_a \left(z_1 - \frac{\partial \alpha_1}{\partial y} z_2\right)(\xi_{(2)} + \varphi_{(1)}) - \dot{\hat{a}}\right) \\
& - \sum_{i=0}^{m-1} (\phi_i^* - \phi_i)^T \Gamma_i^{-1} \left(\Gamma_i \left(z_1 - \frac{\partial \alpha_1}{\partial y} z_2\right) \omega_{i2} + \dot{\phi}_i\right) \\
& + (\phi_m^* - \phi_m)^T \Gamma_m^{-1} \left(\Gamma_m \frac{\partial \alpha_1}{\partial y} z_2 \omega_{m2} - \dot{\phi}_m\right) \\
& + z_2 \frac{\partial \alpha_1}{\partial \hat{a}} \left(\gamma_a \left(\left(z_1 - \frac{\partial \alpha_1}{\partial y} z_2\right)(\xi_{(2)} + \varphi_{(1)}) - f_m(\hat{a})\hat{a}\right) - \dot{\hat{a}}\right) \\
& + z_2 \sum_{i=0}^{m-1} \frac{\partial \alpha_1}{\partial \phi_i} \left(\Gamma_i \left(-\left(z_1 - \frac{\partial \alpha_1}{\partial y} z_2\right) \omega_{i2} - f_m(\phi_i)\phi_i\right) - \dot{\phi}_i\right) \\
& - \frac{1}{d_1}\epsilon^T \epsilon + \frac{1}{d_1}\epsilon^T P_0 \left(d_1 P_0^{-1} e_2 \left(z_1 - \frac{\partial \alpha_1}{\partial y} z_2\right) - \kappa\right) \\
& + d\left(z_1 - \frac{\partial \alpha_1}{\partial y} z_2\right) + \frac{\partial \alpha_1}{\partial \xi_0} z_2 \left(d_1 P_0^{-1} e_2 \left(z_1 - \frac{\partial \alpha_1}{\partial y} z_2\right) - \kappa\right).
\end{aligned} \tag{10.311}$$

Step $j = 3, \ldots, \rho - 1$: The time derivative of $z_j = -\theta^T \omega_{m2}^{(j-2)} - \alpha_{j-1}$ is

$$\dot{z}_j = -\frac{d}{dt}\left(\theta^T \omega_{m2}^{(j-2)}\right) - \dot{\alpha}_{j-1} = -\dot{\theta}^T \omega_{m2}^{(j-2)} - \theta^T \omega_{m2}^{(j-1)} - \dot{\alpha}_{j-1}. \tag{10.312}$$

Introducing $z_{j+1} = -\theta^T \omega_{m2}^{(j-1)} - \alpha_j$, we have

$$\begin{aligned}
\dot{z}_j &= z_{j+1} + \alpha_j - f_N^T \omega_{m2}^{(j-2)} + \beta_j \\
&\quad - \frac{\partial \alpha_{j-1}}{\partial y}\left(\sum_{i=1}^{q} a_i \xi_{i2} - \sum_{i=0}^{m} b_i \theta^{*T} \omega_{i2} + d + \sum_{i=1}^{q} a_i \varphi_{i1} + \epsilon_2\right) \\
&\quad - \frac{\partial \alpha_{j-1}}{\partial \xi_0}\kappa - \frac{\partial \alpha_{j-1}}{\partial \hat{a}}\dot{\hat{a}} - \sum_{i=0}^{m} \frac{\partial \alpha_{j-1}}{\partial \phi_i}\dot{\phi}_i \\
&= z_{j+1} + \alpha_j - \sum_{i=1}^{n_\theta} \chi_i g_i \omega_{m2i}^{(j-2)} + \beta_j - \frac{\partial \alpha_{j-1}}{\partial y} a^T(\xi_{(2)} + \varphi_{(1)}) + \sum_{i=0}^{m} \frac{\partial \alpha_{j-1}}{\partial y}\phi_i^{*T}\omega_{i2} \\
&\quad - \frac{\partial \alpha_{j-1}}{\partial \hat{a}}\dot{\hat{a}} - \sum_{i=0}^{m} \frac{\partial \alpha_{j-1}}{\partial \phi_i}\dot{\phi}_i - \frac{\partial \alpha_{j-1}}{\partial y} d - \frac{\partial \alpha_{j-1}}{\partial y}\epsilon_2 - \frac{\partial \alpha_{j-1}}{\partial \xi_0}\kappa, \qquad (10.313)
\end{aligned}$$

where β_j contains all known terms in (10.313) except the term $-f_N^T \omega_{m2}^{(j-2)} = -\sum_{i=1}^{n_\theta} \chi_i g_i \omega_{m2i}^{(j-2)}$.

The design signal α_j is chosen as

$$\begin{aligned}
\alpha_j &= -c_j z_j - r_j z_j\left(\frac{\partial \alpha_{j-1}}{\partial y}\right)^2 - z_j \sum_{i=1}^{n_\theta} s_{ji} g_i^2 (\omega_{m2i}^{(j-2)})^2 - \beta_j - z_{j-1} \\
&\quad + \frac{\partial \alpha_{j-1}}{\partial y}\hat{a}^T(\xi_{(2)} + \varphi_{(1)}) - \sum_{i=0}^{m} \frac{\partial \alpha_{j-1}}{\partial y}\phi_i^T \omega_{i2} \\
&\quad + \frac{\partial \alpha_{j-1}}{\partial \hat{a}}\gamma_a\left(\left(z_1 - \sum_{i=1}^{j-1}\frac{\partial \alpha_i}{\partial y} z_{i+1}\right)(\xi_{(2)} + \varphi_{(1)}) - f_m(\hat{a})\hat{a}\right) \\
&\quad + \sum_{i=0}^{m-1} \frac{\partial \alpha_{j-1}}{\partial \phi_i}\Gamma_i\left(-\left(z_1 - \sum_{k=1}^{j-1}\frac{\partial \alpha_k}{\partial y} z_{k+1}\right)\omega_{i2} - f_m(\phi_i)\phi_i\right) \\
&\quad + \frac{\partial \alpha_{j-1}}{\partial \phi_m}\Gamma_m\left(\sum_{i=1}^{j-1}\frac{\partial \alpha_i}{\partial y} z_{i+1}\omega_{m2} - f_m(\phi_m)\phi_m\right) \\
&\quad + \frac{\partial \alpha_{j-1}}{\partial \xi_0} d_1 P_0^{-1} e_2\left(z_1 - \sum_{i=1}^{j-1}\frac{\partial \alpha_i}{\partial y} z_{i+1}\right) - \sum_{k=2}^{j-1} z_k \frac{\partial \alpha_{k-1}}{\partial \xi_0} d_1 P_0^{-1} e_2 \frac{\partial \alpha_{j-1}}{\partial y} \\
&\quad - \sum_{k=2}^{j-1} z_k \frac{\partial \alpha_{k-1}}{\partial \hat{a}}\gamma_a \frac{\partial \alpha_{j-1}}{\partial y}(\xi_{(2)} + \varphi_{(1)}) \\
&\quad + \sum_{k=2}^{j-1} z_k \sum_{i=0}^{m} \frac{\partial \alpha_{k-1}}{\partial \phi_i}\Gamma_i \frac{\partial \alpha_{j-1}}{\partial y}\omega_{i2}. \qquad (10.314)
\end{aligned}$$

Then we express $\dot{z}_j$ as

$$\dot{z}_j = -c_j z_j - r_j z_j\left(\frac{\partial \alpha_{j-1}}{\partial y}\right)^2 - z_j \sum_{i=1}^{n_\theta} s_{ji} g_i^2 (\omega_{m2i}^{(j-2)})^2 - \sum_{i=1}^{n_\theta} \chi_i g_i \omega_{m2i}^{(j-2)}$$

$$
\begin{aligned}
&- z_{j-1} + z_{j+1} - \frac{\partial \alpha_{j-1}}{\partial y}(a - \hat{a})^T(\xi_{(2)} + \varphi_{(1)}) + \sum_{i=0}^{m} \frac{\partial \alpha_{j-1}}{\partial y}(\phi_i^* - \phi_i)^T \omega_{i2} \\
&+ \frac{\partial \alpha_{j-1}}{\partial \hat{a}} \left(\gamma_a \left(\left(z_1 - \sum_{i=1}^{j-1} \frac{\partial \alpha_i}{\partial y} z_{i+1} \right) (\xi_{(2)} + \varphi_{(1)}) - f_m(\hat{a})\hat{a} \right) - \dot{\hat{a}} \right) \\
&+ \sum_{i=0}^{m-1} \frac{\partial \alpha_{j-1}}{\partial \phi_i} \left(\Gamma_i \left(- \left(z_1 - \sum_{k=1}^{j-1} \frac{\partial \alpha_k}{\partial y} z_{k+1} \right) \omega_{i2} - f_m(\phi_i)\phi_i \right) - \dot{\phi}_i \right) \\
&+ \frac{\partial \alpha_{j-1}}{\partial \phi_m} \left(\Gamma_m \left(\sum_{i=1}^{j-1} \frac{\partial \alpha_i}{\partial y} z_{i+1}\omega_{m2} - f_m(\phi_m)\phi_m \right) - \dot{\phi}_m \right) \\
&- \frac{\partial \alpha_{j-1}}{\partial y} d - \frac{\partial \alpha_{j-1}}{\partial y} \epsilon_2 + \frac{\partial \alpha_{j-1}}{\partial \xi_0} \left(d_1 P_0^{-1} e_2 \left(z_1 - \sum_{i=1}^{j-1} \frac{\partial \alpha_i}{\partial y} z_{i+1} \right) - \kappa \right) \\
&- \sum_{k=2}^{j-1} z_k \left(\frac{\partial \alpha_{k-1}}{\partial \xi_0} d_1 P_0^{-1} e_2 + \frac{\partial \alpha_{k-1}}{\partial \hat{a}} \gamma_a (\xi_{(2)} + \varphi_{(1)}) \right. \\
&\left. - \sum_{i=0}^{m} \frac{\partial \alpha_{k-1}}{\partial \phi_i} \Gamma_i \omega_{i2} \right) \frac{\partial \alpha_{j-1}}{\partial y}. \qquad (10.315)
\end{aligned}
$$

For the positive definite function $V_j = V_{j-1} + \frac{1}{2}z_j^2$, we obtain

$$
\begin{aligned}
\dot{V}_j = &- \sum_{i=1}^{j} c_i z_i^2 - r_1 z_1^2 - \sum_{i=2}^{j} r_i z_i^2 (\frac{\partial \alpha_{i-1}}{\partial y})^2 + z_j z_{j+1} \\
&- \sum_{k=2}^{j} z_k^2 \sum_{i=1}^{n_\theta} s_{ki} g_i^2 (\omega_{m2i}^{(k-2)})^2 - \sum_{k=2}^{j} z_k \sum_{i=1}^{n_\theta} \chi_i g_i \omega_{m2i}^{(k-2)} \\
&+ |b_m|(p - \hat{p}) f_m(\hat{p})\hat{p} - |b_m|(\theta^* - \theta)^T \Gamma_N^{-1} f_N + (b_m - \hat{b}_m) f_m(\hat{b}_m)\hat{b}_m \\
&+ (a - \hat{a})^T \frac{1}{\gamma_a} \left(\gamma_a \left(z_1 - \sum_{i=1}^{j-1} \frac{\partial \alpha_i}{\partial y} z_{i+1} \right) (\xi_{(2)} + \varphi_{(1)}) - \dot{\hat{a}} \right) \\
&- \sum_{i=0}^{m-1} (\phi_i^* - \phi_i)^T \Gamma_i^{-1} \left(\Gamma_i \left(z_1 - \sum_{k=1}^{j-1} \frac{\partial \alpha_k}{\partial y} z_{k+1} \right) \omega_{i2} + \dot{\phi}_i \right) \\
&+ (\phi_m^* - \phi_m)^T \Gamma_m^{-1} \left(\Gamma_m \sum_{i=1}^{j-1} \frac{\partial \alpha_i}{\partial y} z_{i+1} \omega_{m2} - \dot{\phi}_m \right) \\
&+ \sum_{l=2}^{j} z_l \frac{\partial \alpha_{l-1}}{\partial \hat{a}} \left(\gamma_a \left(\left(z_1 - \sum_{i=1}^{l-1} \frac{\partial \alpha_i}{\partial y} z_{i+1} \right) (\xi_{(2)} + \varphi_{(1)}) - f_m(\hat{a})\hat{a} \right) - \dot{\hat{a}} \right) \\
&+ \sum_{l=2}^{j} z_l \sum_{i=0}^{m-1} \frac{\partial \alpha_{l-1}}{\partial \phi_i} \left(\Gamma_i \left(- \left(z_1 - \sum_{k=1}^{l-1} \frac{\partial \alpha_k}{\partial y} z_{k+1} \right) \omega_{i2} - f_m(\phi_i)\phi_i \right) - \dot{\phi}_i \right) \\
&+ \sum_{l=2}^{j} z_l \frac{\partial \alpha_{l-1}}{\partial \phi_m} \left(\Gamma_m \left(\sum_{k=1}^{l-1} \frac{\partial \alpha_k}{\partial y} z_{k+1} \omega_{m2} - f_m(\phi_m)\phi_m \right) - \dot{\phi}_m \right)
\end{aligned}
$$

$$
\begin{aligned}
&- \frac{1}{d_1}\epsilon^T\epsilon + \frac{1}{d_1}\epsilon^T P_0 \left(d_1 P_0^{-1} e_2 \left(z_1 - \sum_{i=1}^{j-1} \frac{\partial \alpha_i}{\partial y} z_{i+1} \right) - \kappa \right) \\
&+ d \left(z_1 - \sum_{i=1}^{j-1} \frac{\partial \alpha_i}{\partial y} z_{i+1} \right) + \sum_{l=2}^{j} \frac{\partial \alpha_{l-1}}{\partial \xi_0} z_l \left(d_1 P_0^{-1} e_2 \left(z_1 - \sum_{i=1}^{l-1} \frac{\partial \alpha_i}{\partial y} z_{i+1} \right) - \kappa \right) \\
&- \sum_{l=3}^{j} z_l \sum_{k=2}^{l-1} z_k \left(\frac{\partial \alpha_{k-1}}{\partial \xi_0} d_1 P_0^{-1} e_2 + \frac{\partial \alpha_{k-1}}{\partial \hat{a}} \gamma_a (\xi_{(2)} + \varphi_{(1)}) \right. \\
&\left. - \sum_{i=0}^{m} \frac{\partial \alpha_{k-1}}{\partial \phi_i} \Gamma_i \omega_{i2} \right) \frac{\partial \alpha_{l-1}}{\partial y}.
\end{aligned} \tag{10.316}
$$

Step ρ: Using (10.280) with $a_s(t) = 0$ and (10.288), we have

$$
\begin{aligned}
\theta^T \omega_{m2}^{(\rho-1)} &= \theta^T \frac{(s^n + k_1 s^{n-1}) I_{n_\theta}}{s^n + k_1 s^{n-1} + \cdots + k_{n-1}s + k_n} [\sigma(y)\omega](t) \\
&= -\sigma(y) u_d + \omega_0,
\end{aligned} \tag{10.317}
$$

where $\omega_0(t)$ is known:

$$
\omega_0(t) = -\frac{(k_2 s^{n-2} + \cdots + k_{n-1}s + k_n) I_{n_\theta}}{s^n + k_1 s^{n-1} + \cdots + k_{n-1}s + k_n} [\sigma(y)\omega](t). \tag{10.318}
$$

With this expression, the derivative of $z_\rho = -\theta^T \omega_{m2}^{(\rho-2)} - \alpha_{\rho-1}$ is

$$
\begin{aligned}
\dot{z}_\rho &= \sigma(y) u_d + \beta_\rho - \frac{\partial \alpha_{\rho-1}}{\partial y} \left(\sum_{i=1}^{q} a_i \xi_{i2} - \sum_{i=0}^{m} b_i \theta^{*T} \omega_{i2} + d + \sum_{i=1}^{q} a_i \varphi_{i1} + \epsilon_2 \right) \\
&\quad - \frac{\partial \alpha_{\rho-1}}{\partial \xi_0} \kappa - \frac{\partial \alpha_{\rho-1}}{\partial \hat{a}} \dot{\hat{a}} - \sum_{i=0}^{m} \frac{\partial \alpha_{\rho-1}}{\partial \phi_i} \dot{\phi}_i \\
&= \sigma(y) u_d + \beta_\rho - \frac{\partial \alpha_{\rho-1}}{\partial y} a(\xi_{(2)} + \varphi_{(1)}) + \sum_{i=0}^{m} \frac{\partial \alpha_{\rho-1}}{\partial y} \phi_i^{*T} \omega_{i2} \\
&\quad - \frac{\partial \alpha_{\rho-1}}{\partial \xi_0} \kappa - \frac{\partial \alpha_{\rho-1}}{\partial \hat{a}} \dot{\hat{a}} - \sum_{i=0}^{m} \frac{\partial \alpha_{\rho-1}}{\partial \phi_i} \dot{\phi}_i - \frac{\partial \alpha_{\rho-1}}{\partial y} d - \frac{\partial \alpha_{\rho-1}}{\partial y} \epsilon_2,
\end{aligned} \tag{10.319}
$$

where β_ρ contains all known terms in (2.999).

Defining the positive definite function

$$
V_\rho = V_{\rho-1} + \frac{1}{2} z_\rho^2 \tag{10.320}
$$

and using (10.316) with $j = \rho - 1$ and (10.319), we have

$$
\dot{V}_\rho = -\sum_{i=1}^{\rho-1} c_i z_i^2 - r_1 z_1^2 - \sum_{i=2}^{\rho-1} r_i z_i^2 (\frac{\partial \alpha_{i-1}}{\partial y})^2
$$

$$
\begin{aligned}
&- \sum_{k=2}^{\rho-1} z_k^2 \sum_{i=1}^{n_\theta} s_{ki} g_i^2 (\omega_{m2i}^{(k-2)})^2 - \sum_{k=2}^{\rho-1} z_k \sum_{i=1}^{n_\theta} \chi_i g_i \omega_{m2i}^{(k-2)} \\
&+ |b_m|(p-\hat{p}) f_m(\hat{p})\hat{p} - |b_m|(\theta^* - \theta)^T \Gamma_N^{-1} f_N + (b_m - \hat{b}_m) f_m(\hat{b}_m)\hat{b}_m \\
&+ (a-\hat{a})^T \frac{1}{\gamma_a} \left(\gamma_a \left(z_1 - \sum_{i=1}^{\rho-1} \frac{\partial \alpha_i}{\partial y} z_{i+1} \right) (\xi_{(2)} + \varphi_{(1)}) - \dot{\hat{a}} \right) \\
&- \sum_{i=0}^{m-1} (\phi_i^* - \phi_i)^T \Gamma_i^{-1} \left(\Gamma_i \left(z_1 - \sum_{k=1}^{\rho-1} \frac{\partial \alpha_k}{\partial y} z_{k+1} \right) \omega_{i2} + \dot{\phi}_i \right) \\
&+ (\phi_m^* - \phi_m)^T \Gamma_m^{-1} \left(\Gamma_m \sum_{i=1}^{\rho-1} \frac{\partial \alpha_i}{\partial y} z_{i+1} \omega_{m2} - \dot{\phi}_m \right) \\
&+ \sum_{l=2}^{\rho-1} z_l \frac{\partial \alpha_{l-1}}{\partial \hat{a}} \left(\gamma_a \left(\left(z_1 - \sum_{i=1}^{l-1} \frac{\partial \alpha_i}{\partial y} z_{i+1} \right) (\xi_{(2)} + \varphi_{(1)}) - f_m(\hat{a})\hat{a} \right) - \dot{\hat{a}} \right) \\
&+ \sum_{l=2}^{\rho-1} z_l \sum_{i=0}^{m-1} \frac{\partial \alpha_{l-1}}{\partial \phi_i} \left(\Gamma_i \left(- \left(z_1 - \sum_{k=1}^{l-1} \frac{\partial \alpha_k}{\partial y} z_{k+1} \right) \omega_{i2} - f_m(\phi_i)\phi_i \right) - \dot{\phi}_i \right) \\
&+ \sum_{l=2}^{\rho-1} z_l \frac{\partial \alpha_{l-1}}{\partial \phi_m} \left(\Gamma_m \left(\sum_{k=1}^{l-1} \frac{\partial \alpha_k}{\partial y} z_{k+1} \omega_{m2} - f_m(\phi_m)\phi_m \right) - \dot{\phi}_m \right) \\
&- \frac{1}{d_1} \epsilon^T \epsilon + \frac{1}{d_1} \epsilon^T P_0 \left(d_1 P_0^{-1} e_2 \left(z_1 - \sum_{i=1}^{\rho-2} \frac{\partial \alpha_i}{\partial y} z_{i+1} \right) - \kappa \right) \\
&+ d \left(z_1 - \sum_{i=1}^{\rho-2} \frac{\partial \alpha_i}{\partial y} z_{i+1} \right) + \sum_{l=2}^{\rho-1} \frac{\partial \alpha_{l-1}}{\partial \xi_0} z_l \left(d_1 P_0^{-1} e_2 \left(z_1 - \sum_{i=1}^{l-1} \frac{\partial \alpha_i}{\partial y} z_{i+1} \right) - \kappa \right) \\
&+ z_\rho \left(\sigma(y) u_d + z_{\rho-1} + \beta_\rho - \frac{\partial \alpha_{\rho-1}}{\partial y} \hat{a}^T (\xi_{(2)} + \varphi_{(1)}) + \sum_{i=0}^{m} \frac{\partial \alpha_{\rho-1}}{\partial y} \phi_i^T \omega_{i2} \right. \\
&\left. - \frac{\partial \alpha_{\rho-1}}{\partial \hat{a}} \dot{\hat{a}} - \sum_{i=0}^{m} \frac{\partial \alpha_{\rho-1}}{\partial \phi_i} \dot{\phi}_i - \frac{\partial \alpha_{\rho-1}}{\partial \xi_0} \kappa - \frac{\partial \alpha_{\rho-1}}{\partial y} d - \frac{\partial \alpha_{\rho-1}}{\partial y} \epsilon_2 \right) \\
&- \sum_{l=3}^{\rho-1} z_l \sum_{k=2}^{l-1} z_k \left(\frac{\partial \alpha_{k-1}}{\partial \xi_0} d_1 P_0^{-1} e_2 + \frac{\partial \alpha_{k-1}}{\partial \hat{a}} \gamma_a (\xi_{(2)} + \varphi_{(1)}) \right. \\
&\left. - \sum_{i=0}^{m} \frac{\partial \alpha_{k-1}}{\partial \phi_i} \Gamma_i \omega_{i2} \right) \frac{\partial \alpha_{l-1}}{\partial y}. \qquad (10.321)
\end{aligned}
$$

In view of (10.321), we choose:

- the adaptive update law for $\hat{a}$,

$$
\dot{\hat{a}} = \gamma_a \left(z_1 - \sum_{i=1}^{\rho-1} \frac{\partial \alpha_i}{\partial y} z_{i+1} \right) (\xi_{(2)} + \varphi_{(1)}) - \gamma_a f_m(\hat{a})\hat{a}; \qquad (10.322)
$$

- the adaptive update law for ϕ_i,

$$\dot{\phi}_i = -\Gamma_i \left(z_1 - \sum_{k=1}^{\rho-1} \frac{\partial \alpha_k}{\partial y} z_{k+1} \right) \omega_{i2} - \Gamma_i f_m(\phi_i)\phi_i, \quad i = 0, \ldots, m-1; \tag{10.323}$$

- the adaptive update law for ϕ_m,

$$\dot{\phi}_m = \Gamma_m \sum_{i=1}^{\rho-1} \frac{\partial \alpha_i}{\partial y} z_{i+1} \omega_{m2} - \Gamma_m f_m(\phi_m)\phi_m. \tag{10.324}$$

With this choice, we have

$$\begin{aligned}
\dot{V}_\rho =& -\sum_{i=1}^{\rho-1} c_i z_i^2 - r_1 z_1^2 - \sum_{i=2}^{\rho-1} r_i z_i^2 (\frac{\partial \alpha_{i-1}}{\partial y})^2 \\
& - \sum_{k=2}^{\rho-1} z_k^2 \sum_{i=1}^{n_\theta} s_{ki} g_i^2 (\omega_{m2i}^{(k-2)})^2 - \sum_{k=2}^{\rho-1} z_k \sum_{i=1}^{n_\theta} \chi_i g_i \omega_{m2i}^{(k-2)} \\
& + |b_m|(p - \hat{p}) f_m(\hat{p})\hat{p} - |b_m|(\theta^* - \theta)^T \Gamma_N^{-1} f_N + (b_m - \hat{b}_m) f_m(\hat{b}_m)\hat{b}_m \\
& + (a - \hat{a})^T f_m(\hat{a})\hat{a} + \sum_{i=0}^{m} (\phi_i^* - \phi_i)^T f_m(\phi_i)\phi_i - \frac{1}{d_1}\epsilon^T \epsilon \\
& + \frac{1}{d_1}\epsilon^T P_0 \left(d_1 P_0^{-1} e_2 \left(z_1 - \sum_{i=1}^{\rho-1} \frac{\partial \alpha_i}{\partial y} z_{i+1} \right) - \kappa \right) + d \left(z_1 - \sum_{i=1}^{\rho-1} \frac{\partial \alpha_i}{\partial y} z_{i+1} \right) \\
& + \sum_{l=2}^{\rho-1} \frac{\partial \alpha_{l-1}}{\partial \xi_0} z_l \left(d_1 P_0^{-1} e_2 \left(z_1 - \sum_{i=1}^{\rho-1} \frac{\partial \alpha_i}{\partial y} z_{i+1} \right) - \kappa \right) \\
& + z_\rho \left(\sigma(y) u_d + z_{\rho-1} + \beta_\rho - \frac{\partial \alpha_{\rho-1}}{\partial y} \hat{a}^T (\xi_{(2)} + \varphi_{(1)}) + \sum_{i=0}^{m} \frac{\partial \alpha_{\rho-1}}{\partial y} \phi_i^T \omega_{i2} \right. \\
& - \frac{\partial \alpha_{\rho-1}}{\partial \hat{a}} \dot{\hat{a}} - \sum_{i=0}^{m} \frac{\partial \alpha_{\rho-1}}{\partial \phi_i} \dot{\phi}_i - \frac{\partial \alpha_{\rho-1}}{\partial \xi_0} \kappa + \sum_{k=2}^{\rho-1} z_k \left(\frac{\partial \alpha_{k-1}}{\partial \xi_0} d_1 P_0^{-1} e_2 \right. \\
& \left. \left. + \frac{\partial \alpha_{k-1}}{\partial \hat{a}} \gamma_a (\xi_{(2)} + \varphi_{(1)}) - \sum_{i=0}^{m} \frac{\partial \alpha_{k-1}}{\partial \phi_i} \Gamma_i \omega_{i2} \right) \frac{\partial \alpha_{\rho-1}}{\partial y} \right).
\end{aligned} \tag{10.325}$$

This expression motivates us to choose:

- the design signal κ,

$$\kappa = d_1 P_0^{-1} e_2 \left(z_1 - \sum_{i=1}^{\rho-1} \frac{\partial \alpha_i}{\partial y} z_{i+1} \right); \tag{10.326}$$

- the control signal u_d,

$$u_d = \frac{1}{\sigma(y)}\Bigg(-z_{\rho-1} - \beta_\rho - c_\rho z_\rho - r_\rho z_\rho(\frac{\partial\alpha_{\rho-1}}{\partial y})^2 + \frac{\partial\alpha_{\rho-1}}{\partial y}\hat{a}^T(\xi_{(2)} + \varphi_{(1)})$$
$$- \sum_{i=0}^{m} \frac{\partial\alpha_{\rho-1}}{\partial y}\phi_i^T \omega_{i2} + \frac{\partial\alpha_{\rho-1}}{\partial\hat{a}}\dot{\hat{a}} + \sum_{i=0}^{m}\frac{\partial\alpha_{\rho-1}}{\partial\phi_i}\dot{\phi}_i + \frac{\partial\alpha_{\rho-1}}{\partial\xi_0}\kappa$$
$$- \sum_{k=2}^{\rho-1} z_k \frac{\partial\alpha_{\rho-1}}{\partial y}\left(\frac{\partial\alpha_{k-1}}{\partial\xi_0} d_1 P_0^{-1} e_2 + \frac{\partial\alpha_{k-1}}{\partial\hat{a}}\gamma_a(\xi_{(2)} + \varphi_{(1)})\right.$$
$$\left.- \sum_{i=0}^{m} \frac{\partial\alpha_{k-1}}{\partial\phi_i}\Gamma_i\omega_{i2}\right)\Bigg). \tag{10.327}$$

With this choice, $\dot{V}_\rho$ becomes

$$\dot{V}_\rho = -\sum_{i=1}^{\rho} c_i z_i^2 + |b_m|(p-\hat{p})f_m(\hat{p})\hat{p} - |b_m|(\theta^* - \theta)^T\Gamma_N^{-1} f_N$$
$$+ (b_m - \hat{b}_m)f_m(\hat{b}_m)\hat{b}_m + (a-\hat{a})^T f_m(\hat{a})\hat{a} + \sum_{i=0}^{m}(\phi_i^* - \phi_i)^T f_m(\phi_i)\phi_i - \frac{\epsilon^T\epsilon}{d_1}$$
$$- r_1\left(z_1 - \frac{d}{2r_1}\right)^2 + \frac{d^2}{4r_1} - \sum_{i=2}^{\rho} r_i\left(\frac{\partial\alpha_{i-1}}{\partial y}z_i + \frac{d}{2r_i}\right)^2 + \sum_{i=2}^{\rho}\frac{d^2}{4r_i}$$
$$- \sum_{k=2}^{\rho-1}\sum_{i=1}^{n_\theta} s_{ki}\left(g_i\omega_{m2i}^{(k-2)} z_k + \frac{\chi_i}{2s_{ki}}\right)^2 + \sum_{k=2}^{\rho-1}\sum_{i=1}^{n_\theta}\frac{\chi_i^2}{4s_{ki}}. \tag{10.328}$$

Remark 10.4 In view of (10.303), the last term in (10.328), $\sum_{k=2}^{\rho-1}\sum_{i=1}^{n_\theta}\frac{\chi_i^2}{4s_{ki}}$, becomes zero if all components of θ fall in the known range (θ_i^a, θ_i^b). It is nonzero if any component of θ is leaving (θ_i^a, θ_i^b). This perturbation term can be avoided by a different form of nonlinear damping [206]. To see this, noting that g_N has z_1 as a factor in (10.302), we denote $g_N = \bar{g}_N z_1$ and replace the damping term $-z_2\sum_{i=1}^4 s_{2i}g_i^2\omega_{m2i}^2$ by $-z_2\sum_{i=1}^{n_\theta} s_{2i}\bar{g}_i^2\omega_{m2i}^2$ for α_2 in (10.307). Then for $\dot{V}_2$ in (10.311), we have $-z_2^2\sum_{i=1}^4 s_{2i}\bar{g}_i^2\omega_{m2i}^2 - z_2\sum_{i=1}^4 \chi_i g_i\omega_{m2i} = -\sum_{i=1}^4 s_{2i}(\bar{g}_i\omega_{m2i}z_2 + \frac{\chi_i}{2s_{2i}}z_1)^2 + \sum_{i=1}^4\frac{z_1^2}{4s_{2i}}$, where $\frac{z_1^2}{4s_{2i}}$ can be accommodated by $-c_1z_1^2$ with sufficiently large s_{2i}. Similarly, the last term of $\dot{V}_\rho$ can also be avoided [140]. Note that all terms associated with d still remain. □

To show the stability of this adaptive design, we observe the following facts:

- It follows from (10.298), (10.307), and (10.314) that all these partial derivatives $\frac{\partial\alpha_j}{\partial\hat{a}}$, $\frac{\partial\alpha_j}{\partial\phi_0}$, ..., $\frac{\partial\alpha_j}{\partial\phi_m}$ are well defined.

- It follows from (10.298) that

$$f_m(\hat{\phi} - \phi)^T \hat{\phi} \geq 0 \tag{10.329}$$

and $f_m(\hat{\phi} - \phi)^T \hat{\phi}$ grows unbounded if $\hat{\phi}$ grows unbounded.

- The adaptive law (10.299)–(10.303) ensures that

$$(\theta^* - \theta)^T \Gamma_N^{-1} f_N \geq 0, \tag{10.330}$$

$$\theta_i \in [\theta_i^a, \theta_i^b],\ i = 1, 2, \ldots, n_\theta, \tag{10.331}$$

which is crucial for implementing the adaptive inverse (10.280).

Based on these observations and the fact that $d(t)$ in (10.291) and (10.16) is bounded, and so is $\sum_{k=2}^{\rho-1} \sum_{i=1}^{n_\theta} \frac{\chi_i^2}{4s_{ki}}$ (recall that $\chi_i = -1$ or $\chi_i = 0$ in (10.303)), from (10.328), we conclude that V_ρ is bounded. Thus we have:

Lemma 10.1 *The adaptive backstepping design ensures that* $z_1, \ldots, z_\rho$, $\hat{p}$, $\hat{b}_m$, $\hat{a}$, θ, $\phi_0, \ldots, \phi_m$ *are all bounded. Furthermore,* $y = z_1 + y_m$ *is bounded.*

Analysis for the closed-loop system stability is given for the case of an actuator dead-zone $DZ(\cdot)$ (10.3) with equal slopes $m_r = m_l = m_s$, for which an adaptive dead-zone inverse (10.17) with equal slope estimates $\widehat{m_r} = \widehat{m_l} = \widehat{m_s}$ is used. In this case, we have the parametrization

$$\theta^* = [m_s, m_s b_r, m_s b_l]^T,\ \theta = [\widehat{m_s}, \widehat{m_s b_r}, \widehat{m_s b_l}]^T, \tag{10.332}$$

$$\omega = [-v(t), \widehat{\chi_r}(t), 1 - \widehat{\chi_r}(t)]^T. \tag{10.333}$$

Theorem 10.4 *The adaptive backstepping control design with the adaptive feedback law (10.327) and the adaptive dead-zone inverse (10.17), applied to the plant (10.270) with an equal slope dead-zone (10.3), that is,* $m_r = m_l$, *ensures that all closed-loop signals are bounded.*

Proof: With $m_r = m_l$ and (10.282), (10.283), (10.332), (10.333), we have

$$u(t) = -\theta^{*T} \omega(t) + d_N(t). \tag{10.334}$$

From Lemma 10.1, we have that $z_1, \ldots, z_\rho, \hat{p}, \hat{b}_m, \hat{a}, \theta, \phi_0, \ldots, \phi_m$ are bounded. From (10.275), (10.276), and (10.326), we have that $\xi_0, \ldots, \xi_q$ are bounded.

Rewriting the plant (10.270) as

$$s^n[y] - \sum_{i=1}^{n} s^{n-i} \left[\varphi_{0i}(y) + \sum_{j=1}^{q} a_j \varphi_{ji}(y) \right] = \sum_{i=0}^{m} b_i s^i [\sigma(y) u] \tag{10.335}$$

and using (10.283), we have

$$v_{k2} = \frac{p_{k2}(s)}{\Lambda(s)}[\sigma(y)u]$$
$$= \frac{s^n p_{k2}(s)}{B(s)\Lambda(s)}[y] - \sum_{i=1}^{n} \frac{s^{n-i} p_{k2}(s)}{B(s)\Lambda(s)} \left[\varphi_{0i}(y) + \sum_{j=1}^{q} a_j \varphi_{ji}(y)\right], \quad (10.336)$$

$k = 0, \ldots, m-1$, where $p_{k2}(s)$ are polynomials of order $k+1$, respectively.

Since both $\Lambda(s) = s^n + k_1 s^{n-1} + \cdots + k_n$ and $B(s) = b_m s^m + \cdots + b_1 s + b_0$ are stable, $\frac{s^i p_{k2}(s)}{B(s)\Lambda(s)}, i = 0, \ldots, n, k = 0, \ldots, m-1$, are all stable and proper transfer functions. From (10.336), we have that $v_{k2}, k = 0, \ldots, m-1$, are all bounded because y is. From (10.334) and (10.284), we have

$$v_{k2}(t) = -\theta^{*T} \omega_{k2}(t) + d_{k2}(t), \quad k = 0, \ldots, m-1, \quad (10.337)$$

where

$$\omega_{k2}(t) = \frac{p_{k2}(s) I_3}{\Lambda(s)}[\sigma(y)\omega](t), \quad (10.338)$$

$$d_{k2}(t) = \frac{p_{k2}(s)}{\Lambda(s)}[\sigma(y) d_N](t). \quad (10.339)$$

From (10.337) and $d_{k2}(t) \in L^\infty$, we have $\theta^{*T}\omega_{k2}(t) \in L^\infty$. Expressing

$$\omega_{k2}(t) = \left[\frac{p_{k2}(s)}{\Lambda(s)}[-\sigma(y)v](t), \frac{p_{k2}(s)}{\Lambda(s)}[\sigma(y)\widehat{\chi_r}](t), \frac{p_{k2}(s)}{\Lambda(s)}[\sigma(y)(1-\widehat{\chi_r})](t)\right]^T,$$

$$\theta^{*T}\omega_{k2}(t) = m_s \frac{p_{k2}(s)}{\Lambda(s)}[-\sigma(y)v](t) + m_s b_r \frac{p_{k2}(s)}{\Lambda(s)}[\sigma(y)\widehat{\chi_r}](t)$$
$$+ m_s b_l \frac{p_{k2}(s)}{\Lambda(s)}[\sigma(y)(1-\widehat{\chi_r})](t) \quad (10.340)$$

and using $\theta^{*T}\omega_{k2} \in L^\infty, \widehat{\chi_r} \in L^\infty$, we see that $\omega_{k2} \in L^\infty, k = 0, \ldots, m-1$.

From (10.296), $\alpha_1 = -\hat{p}\psi$ and (10.293),

$$\psi = c_1 z_1 + r_1 z_1 + \xi_{02} + \varphi_{01} - \dot{y}_m + \hat{a}^T(\xi_{(2)} + \varphi_{(1)}) - \sum_{i=0}^{m-1} \phi_i^T \omega_{i2}, \quad (10.341)$$

we have $\alpha_1 \in L^\infty$. Since $z_2 = -\theta^T \omega_{m2} - \alpha_1$, we have $\theta^T \omega_{m2} \in L^\infty$. Similar to the proof of $\omega_{k2} \in L^\infty$ from $\theta^{*T}\omega_{k2} \in L^\infty, k = 0, \ldots, m-1$, we can obtain the boundedness of ω_{m2} from the boundedness of $\theta^T \omega_{m2}$.

Since $\omega_{02}, \omega_{12}, \ldots, \omega_{m2}$ are bounded, we have, from (10.307) and (10.314), that $\alpha_2, \ldots, \alpha_{\rho-1}$ are bounded, and so is κ. Since $\sigma(y) \neq 0$, from (10.327) we have that u_d is bounded, and so are $v = \widehat{DI}(u_d)$ and $u = DZ(v)$. It follows from (10.3) and (10.285) that $\omega, \omega_{ij} \in L^\infty, i = 0, \ldots, m, j = 1, \ldots, n$. From (10.277), we have that $v_0, \ldots, v_m$ are bounded. Then $\hat{x}(t)$ is bounded from (10.274) and, finally, $x(t) = \hat{x}(t) + \epsilon(t)$ is bounded from (10.274)–(10.278). ∇

Design Example

We now illustrate the proposed adaptive backstepping design procedure for a third-order nonlinear plant [206] with an actuator dead-zone:

$$\dot{x}_1 = x_2 + ay^2,\ \dot{x}_2 = x_3,\ \dot{x}_3 = u,\ u = DZ(v),\ y = x_1, \tag{10.342}$$

where the parameter a is unknown and the dead-zone parameters m_r, b_r, m_l, b_l are unknown but with known bounds.

First we choose the dead-zone inverse $v(t) = \widehat{DI}(u_d(t))$ and the filters

$$\dot{\xi}_0 = A_0\xi_0 + ky + \kappa, \tag{10.343}$$

$$\dot{\xi}_1 = A_0\xi_1 + \varphi_1(y), \tag{10.344}$$

$$\omega_{02} = \frac{s+k_1}{s^3 + k_1 s^2 + k_2 s + k_3} I_4[\omega], \tag{10.345}$$

where

$$\varphi_1(y) = [y^2, 0, 0]^T,\ k = [k_1, k_2, k_3]^T = [1, 3, 1]^T, \tag{10.346}$$

$$A_0 = \begin{bmatrix} -k_1 & 1 & 0 \\ -k_2 & 0 & 1 \\ -k_3 & 0 & 0 \end{bmatrix} = \begin{bmatrix} -1 & 1 & 0 \\ -3 & 0 & 1 \\ -1 & 0 & 0 \end{bmatrix}. \tag{10.347}$$

Then we apply our adaptive control design to the plant (10.342) as follows.

Step 1: Derivation of the first subsystem:

$$z_1 = y - y_m, \tag{10.348}$$

$$\alpha_1 = \dot{y}_m - \xi_{02} - \hat{a}(\xi_{12} + \varphi_{11}) - c_1 z_1 - r_1 z_1, \tag{10.349}$$

$$\dot{\theta} = -\Gamma_N z_1 \omega_{02} + f_N. \tag{10.350}$$

Step 2: Derivation of the second subsystem:

$$z_2 = -\theta^T\omega_{02} - \alpha_1, \tag{10.351}$$

$$\begin{aligned}
\alpha_2 = {} & -z_1 - c_2 z_2 - r_2 z_2 (\frac{\partial\alpha_1}{\partial y})^2 - z_2 \sum_{i=1}^{4} s_{2i} g_i^2 \omega_{02i}^2 + \Gamma_N z_1 \omega_{02}^T \omega_{02} \\
& + \frac{\partial\alpha_1}{\partial y}\xi_{02} + \frac{\partial\alpha_1}{\partial y_m}\dot{y}_m + \frac{\partial\alpha_1}{\partial \dot{y}_m}\ddot{y}_m + \frac{\partial\alpha_1}{\partial\xi_0}(A_0\xi_0 + ky) + \frac{\partial\alpha_1}{\partial\xi_1}(A_0\xi_1 + \varphi_1(y)) \\
& + \frac{\partial\alpha_1}{\partial y}\hat{a}^T(\xi_{12} + \varphi_{11}) - \frac{\partial\alpha_1}{\partial y}\phi_0^T\omega_{02} \\
& + \frac{\partial\alpha_1}{\partial\hat{a}}\gamma_a\left(\left(z_1 - \frac{\partial\alpha_1}{\partial y}z_2\right)(\xi_{12} + \varphi_{11}) - f_m(\hat{a})\hat{a}\right) \\
& + \frac{\partial\alpha_1}{\partial\xi_0}d_1 P_0^{-1} e_2\left(z_1 - \frac{\partial\alpha_1}{\partial y}z_2\right).
\end{aligned} \tag{10.352}$$

Step 3: Derivation of the third subsystem:

$$z_3 = -\theta^T \dot{\omega}_{02} - \alpha_2, \tag{10.353}$$

$$\dot{\hat{a}} = \gamma_a \left(z_1 - \frac{\partial \alpha_1}{\partial y} z_2 - \frac{\partial \alpha_1}{\partial y} z_3 \right) (\xi_{12} + \varphi_{11}) - \gamma_a f_m(\hat{a})\hat{a}, \tag{10.354}$$

$$\dot{\phi}_0 = \Gamma_0 \left(\frac{\partial \alpha_1}{\partial y} z_2 + \frac{\partial \alpha_2}{\partial y} z_3 \right) \omega_{02} - \Gamma_0 f_m(\phi_0)\phi_0, \tag{10.355}$$

$$\kappa = d_1 P_0^{-1} e_2 \left(z_1 - \frac{\partial \alpha_1}{\partial y} z_2 - \frac{\partial \alpha_2}{\partial y} z_3 \right), \tag{10.356}$$

$$\begin{aligned} u_d = & -z_2 - c_3 z_3 - r_3 (\frac{\partial \alpha_2}{\partial y})^2 z_3 + \frac{\partial \alpha_2}{\partial \xi_0}\kappa + \frac{\partial \alpha_2}{\partial \hat{a}}\dot{\hat{a}} + \frac{\partial \alpha_2}{\partial y}\hat{a}(\xi_{12} + \varphi_{11}) \\ & - \frac{\partial \alpha_2}{\partial y}\phi_0^T \omega_{02} + \frac{\partial \alpha_2}{\partial \phi_0}\dot{\phi}_0 + \dot{\theta}^T \dot{\omega}_{02} + \theta^T \omega_0 + \frac{\partial \alpha_2}{\partial y}\xi_{02} + \frac{\partial \alpha_2}{\partial y_m}\dot{y}_m + \frac{\partial \alpha_2}{\partial \dot{y}_m}\ddot{y}_m \\ & + \frac{\partial \alpha_2}{\partial \ddot{y}_m}\frac{d^3 y_m}{dt^3} + \frac{\partial \alpha_2}{\partial \xi_0}(A_0 \xi_0 + ky) \\ & + \frac{\partial \alpha_2}{\partial \xi_1}(A_0 \xi_1 + \varphi_1(y)) + \frac{\partial \alpha_2}{\partial \omega_{02}}\dot{\omega}_{02} \\ & - z_2 \frac{\partial \alpha_2}{\partial y}\left(\frac{\partial \alpha_2}{\partial \xi_0} d_1 P_0^{-1} e_2 + \frac{\partial \alpha_2}{\partial \hat{a}}\gamma_a(\xi_{12} + \varphi_{11}) \right), \end{aligned} \tag{10.357}$$

where

$$\omega_0 = \frac{-3s - 1}{s^3 + 3s^2 + 3s + 1} \; I_4[\omega], \;\; P_0 = \begin{bmatrix} 7 & -1 & -3 \\ -1 & 3 & -1 \\ -3 & -1 & 5 \end{bmatrix}. \tag{10.358}$$

With this choice of the control law and adaptive laws, we have

$$\begin{aligned} \dot{z}_1 = & -c_1 z_1 - r_1 z_1 + z_2 + (a - \hat{a})(\xi_{12} + \varphi_{11}) - (\theta^* - \theta)^T \omega_{02} + \epsilon_2 + d_{02}, \\ \dot{z}_2 = & \; z_3 - z_1 - c_2 z_2 - r_2 z_2 (\frac{\partial \alpha_1}{\partial y})^2 - z_2 \sum_{i=1}^{4} s_{2i} g_i^2 \omega_{02i}^2 - \sum_{i=1}^{4} \chi_i g_i \omega_{02i} \\ & - \frac{\partial \alpha_1}{\partial y}(a - \hat{a})(\xi_{12} + \varphi_{11}) + \frac{\partial \alpha_1}{\partial y}(\phi_0^* - \phi_0)^T \omega_{02} \\ & + \frac{\partial \alpha_1}{\partial \hat{a}}\gamma_a \frac{\partial \alpha_2}{\partial y} z_3(\xi_{12} + \varphi_{11}) + \frac{\partial \alpha_1}{\partial \xi_0} d_1 P_0^{-1} e_2 \frac{\partial \alpha_2}{\partial y} z_3 - \frac{\partial \alpha_1}{\partial y} d_{02} - \frac{\partial \alpha_1}{\partial y}\epsilon_2, \\ \dot{z}_3 = & -z_2 - c_3 z_3 - r_3 z_3 (\frac{\partial \alpha_2}{\partial y})^2 - \frac{\partial \alpha_2}{\partial y}(a - \hat{a})(\xi_{12} + \varphi_{11}) + \frac{\partial \alpha_2}{\partial y}(\phi_0^* - \phi_0)^T \omega_{02} \\ & - z_2 \frac{\partial \alpha_2}{\partial y}\left(\frac{\partial \alpha_2}{\partial \xi_0} d_1 P_0^{-1} e_2 + \frac{\partial \alpha_2}{\partial \hat{a}}\gamma_a(\xi_{12} + \varphi_{11}) \right) - \frac{\partial \alpha_2}{\partial y} d_{02} - \frac{\partial \alpha_2}{\partial y}\epsilon_2. \end{aligned} \tag{10.359}$$

Then the time derivative of the positive definite function

$$V = \frac{1}{2}\left(z_1^2 + z_2^2 + z_3^2 + \frac{1}{\gamma_a}(a - \hat{a})^2 + (\theta^* - \theta)^T \Gamma_N^{-1}(\theta^* - \theta) \right.$$

$$+ (\phi_0^* - \phi_0)^T \Gamma_0^{-1} (\phi_0^* - \phi_0) + \frac{1}{d_1} \epsilon^T P_0 \epsilon \Big) \tag{10.360}$$

can be derived as

$$\dot{V} = -\sum_{i=1}^{3} c_i z_i^2 - (\theta^* - \theta)^T \Gamma_N^{-1} f_N + (\phi_0^* - \phi_0)^T f_m(\phi_0)\phi_0 + (a - \hat{a}) f_m(\hat{a})\hat{a}$$
$$- r_1 \left(z_1 - \frac{d_{02}}{2r_1} \right)^2 + \frac{d_{02}^2}{4r_1} - r_2 \left(\frac{\partial \alpha_1}{\partial y} z_2 + \frac{d_{02}}{2r_2} \right)^2 + \frac{d_{02}^2}{4r_2} - r_3 \left(\frac{\partial \alpha_2}{\partial y} z_3 + \frac{d_{02}}{2r_3} \right)^2$$
$$+ \frac{d_{02}^2}{4r_3} - \sum_{i=1}^{4} s_{2i} \left(g_i \omega_{02i} z_2 + \frac{\chi_i}{2s_{2i}} \right)^2 + \sum_{i=1}^{4} \frac{\chi_i^2}{4s_{2i}} - \frac{1}{d_1} \epsilon^T \epsilon. \tag{10.361}$$

From here on, the results of Lemma 10.1 and Theorem 10.4 can be verified.

10.7 Discussion

Summary

We have presented a general adaptive inverse approach for control of plants with unknown nonsmooth actuator nonlinearities such as dead-zone, backlash, hysteresis, and other piecewise-linear characteristics. This approach combines an adaptive inverse with a feedback control law. The adaptive inverse is to cancel the effect of the actuator nonlinearity, while the feedback control law can be designed as if the actuator nonlinearity were absent. For parametrizable actuator nonlinearities which have parametrizable inverses, state or output feedback adaptive inverse controllers were developed which led to linearly parametrized error models suitable for the developments of gradient projection or Lyapunov adaptive laws for updating the inverse parameters.

The adaptive inverse approach is an algorithm-based compensation approach for cancelling unknown actuator nonlinearities caused by component imperfections. This approach can be incorporated with existing control designs such as model reference, PID, pole placement, linear quadratic, backstepping, and other dynamic compensation techniques. An adaptive inverse can be added into a control system loop without the need to change a feedback control design for known linear or nonlinear dynamics following the actuator nonlinearity. It also is applicable to feedback linearizable, parametric-strict-feedback, and output-feedback nonlinear plants with actuator nonlinearities.

Improvements of system tracking performance by an adaptive inverse have been shown by simulation results. Some signal boundedness properties of adaptive laws and closed-loop systems have been established. However, despite the existence of a true inverse which completely cancels the actuator nonlinearity, an analytical proof of a tracking error convergent to zero with an adaptive inverse is still not available for a general adaptive inverse control design. Moreover, adaptive inverse control designs for more general multivariable or nonlinear dynamics are still open issues under investigation.

Literature Review

To compensate for actuator imperfections caused by nonlinear characteristics to improve control system performance, an adaptive inverse compensation approach was proposed and developed in [326], [328], [395]. More recent results of [229], [396], [397], [398], [400], [412], [401], [413] have been summarized in this chapter. The essence of this approach is to use adaptive algorithms to modify control signals to compensate for the effects of possible actuator imperfections. The compensation signal is generated from an adaptive inverse of the actuator nonlinearity. Such an adaptive inverse is so constructed that, when implemented with the true parameters of the actuator nonlinearity, it completely cancels the actuator nonlinearity and, when implemented with adaptive estimates of these true parameters, it asymptotically cancels the actuator nonlinearity. This approach has been developed in several directions with applications [30], [40], [61], [113], [181], [184], [211], [218], [219], [220], [245], [260], [327], [343], [345], [353], [354], [355], [358], [409], [429], [430], [431].

Problems

10.1 Verify the properties of the error term $d_N(t)$ in (10.23).

10.2 For the backlash model (10.11), derive (10.2).

10.3 Verify (10.35)–(10.37).

10.4 Derive (10.42).

10.5 Verify the properties of the adaptive law (10.61).

10.6 Design and simulate a second-order plant with an actuator dead-zone for each of the three adaptive designs in Section 10.2.

10.7 Verify (10.152) and derive (10.156) and (10.157).

10.8 Derive the adaptive law (10.164)–(10.165) and verify its properties.

10.9 Develop a stable gradient adaptive control scheme based on (10.145).

10.10 Develop a stable gradient adaptive control scheme based on (10.172).

10.11 Verify (10.177) and use it to develop a stable gradient adaptive control scheme (also use the design techniques of Section 9.2).

10.12 Consider a single-link manipulator with joint flexibility and damping:

$$I\ddot{q}_1 + MgL\sin(q_1) + k(q_1 - q_2) + b(\dot{q}_1 - \dot{q}_2) = 0,$$
$$J\ddot{q}_2 - k(q_1 - q_2) - b(\dot{q}_1 - \dot{q}_2) = u,$$

where q_1, q_2 are the angular positions of the link and motor with inertia I and J, k is the elasticity constant of the joint spring, M and L represent the mass and the position of the center of gravity of the link, and u is the torque applied by the motor [217] (u is proportional to motor current). With the state variables $x_1 = q_1, x_2 = \dot{q}_1, x_3 = q_2, x_4 = \dot{q}_2$, the plant has the form (10.184). Study adaptive compensation of an actuator dead zone (due to, e.g., imperfections in a power amplifier [210]) with unknown parameters [243].

(i) For the plant with flexibility only, that is, $k > 0$, $b = 0$, as in [357], develop and simulate an adaptive actuator dead-zone compensation scheme.

(ii) Repeat part (i) for both flexibility and damping, that is, $k > 0, b > 0$.

10.13 Study adaptive dead-zone compensation for the second-order plant

$$\begin{aligned}\dot{x}_1 &= x_2 + 1.5x_1^2,\\ \dot{x}_2 &= u,\\ u &= DZ(v),\end{aligned}$$

for an actuator dead-zone $DZ(\cdot)$ with parameters $m_r = 1.2, b_r = 0.5, m_l = 2, b_l = -0.8$ by designing and simulating an adaptive state feedback design (see Section 10.6.2). Develop an output feedback adaptive dead-zone compensation controller for this plant (see Section 10.6.3).

10.14 The plant considered in Section 10.6.3 can be assumed to have $b_m = 1$ because the actuator nonlinearity has its own gain. In this case, we do not need the estimates $\hat{p}$ and $\hat{b}_m$. However, even if b_m is known, the parameter estimate ϕ_m is still needed when $\rho \geq 4$. Verify this statement.

10.15 For the plant considered in Section 10.6.3, if the states $x_i, i = 1, \ldots, n$, are available for measurement, due to the fact that the output-feedback canonical systems belong to the class of parametric-strict-feedback systems, one can directly apply the result of Section 10.6.2 to derive an adaptive design which ensures the closed-loop signal boundedness.

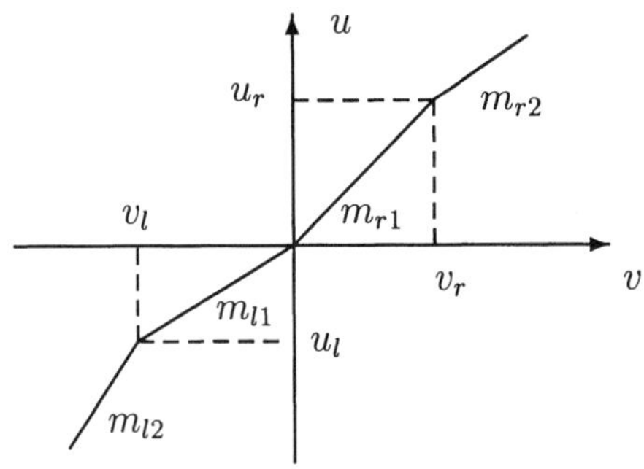

Figure 10.7: Piecewise-linear characteristic.

10.16 The boundedness assumption (A10.3a) for $\sigma(y)$ in Section 10.6.3 is for a simple presentation of the adaptive control design. If $\sigma(y)$ is only bounded for y bounded, which is yet to be established, then we can use a bounding signal and a nonlinear damping term in the control law to enhance robustness with respect to a bounded disturbance multiplied by $\sigma(y)$ due to the presence of the actuator nonlinearity. Develop such an adaptive control scheme.

10.17 With the help of Remark 10.3, develop an adaptive control scheme for the plant (10.170) without actuator nonlinearity $N(\cdot)$ [206].

10.18 Consider the piecewise-linear characteristic shown in Figure 10.5 as an example of the actuator nonlinearity $N(\cdot)$, which is described as

$$u(t) = N(v(t)) = \begin{cases} m_{r1}v(t) & \text{if } 0 \leq v(t) \leq v_r, \\ m_{r2}(v(t) - v_r) + m_{r1}v_r & \text{if } v(t) > v_r, \\ m_{l1}v(t) & \text{if } v_l \leq v(t) < 0, \\ m_{l2}(v(t) - v_l) + m_{l1}v_l & \text{if } v(t) < v_l, \end{cases}$$

where $m_{r1}, m_{r2}, m_{l1}, m_{l2}$, v_r, v_l (and u_r, u_l, as indicated in Figure 10.7) are constant parameters. Develop a parametrization scheme (10.2) for this nonlinearity, as that in (10.4)–(10.8) for a dead-zone, for the case when v_r and v_l are known or the case when u_r and u_l are known and propose and parametrize an inverse for this actuator nonlinearity.

Also design and simulate the developed adaptive inverse control schemes for different plants with this type of actuator nonlinearity.

Advanced Topics

Actuator compensation for multivariable systems. The problem is open and appears to be challenging when the gain matrix K_p or interactor matrix $\xi_m(D)$ is unknown (see Section 10.5.2).

Actuator compensation for feedback linearizable systems. The problem is to develop a stable adaptive actuator nonlinearity compensation control scheme for the case when the nonlinear dynamics, in addition to the actuator nonlinearities, are also unknown. Adaptive estimation of the unknown nonlinear dynamics parametrized by a linear model [341] should be employed, in combination with adaptive estimation of the uncertain actuator nonlinearity parameters, to construct an adaptive controller for an unknown nonlinear feedback linearizable plant with actuator nonlinearities. The multivariable version of this problem is also important and challenging.

Actuator compensation for nonminimum phase systems. When the zeros (or zero dynamics) of the controlled plant are unstable, model reference adaptive control designs cannot be used for adaptive actuator nonlinearity compensation. Other designs such as pole placement and LQ have to be investigated for possible solutions to the problem.

Control of sandwich systems. A new and more challenging problem is to control sandwich nonlinear systems. To introduce this problem, we consider the three-block sandwich system with a nonlinear block $N(\cdot)$ between two dynamic blocks $G_i(s)$ and $G_o(s)$:

$$y(t) = G_o(s)[u](t),\ u(t) = N(x(t)),\ x(t) = G_i(s)[v](t),$$

where s is the differentiation operator $s[x](t) = \dot{x}(t)$ or the Laplace transform operator as the case may be, $v(t)$ is the accessible control input, $y(t)$ is the measured output, $G_o(s)$ and $G_i(s)$ are rational transfer functions with their denominator degrees being n_o and n_i, respectively, and $N(\cdot)$ is the non-smooth nonlinearity such as a dead-zone, backlash, or hysteresis characteristic.

For this sandwich system, the control objective is to cancel the effect of the sandwiched nonlinearity as quickly as possible so that the system output $y(t)$ tracks a desired trajectory satisfactorily. This is possible if the sandwiched nonlinearity is effectively compensated. However, the dynamics of $G_i(s)$ limit the effectiveness of a dead-zone inverse on the sandwiched dead-zone.

Discrete-time, nonlinear, and multivariable versions of this problem can also be formulated (together with the problem of actuator failures as well).

Bibliography

[1] Abdallah, A., D. Dawson, P. Dorato and M. Jamshidi, "Survey of robust control for rigid robots," *IEEE Control Systems Magazine*, vol. 11, no. 2, pp. 24–30, 1991.

[2] Allin, A. and G. F. Inbar, "FNS control schemes for the upper limb," *IEEE Transactions on Biomedical Engineering*, vol. 33, no. 9, pp. 818–827, 1986.

[3] Ambrosino, G., G. Celantano and F. Garofalo, "Variable structure model reference adaptive control systems," *International Journal of Control*, vol. 39, no. 6, pp. 1339–1349, 1984.

[4] Anderson, B. D. O., "Exponential stability of linear equations arising in adaptive identification," *IEEE Transactions on Automatic Control*, vol. AC-22, no. 1, pp. 83–88, 1977.

[5] Anderson, B. D. O., "Adaptive systems, lack of persistency of excitation and bursting phenomena," *Automatica*, vol. 21, no. 3, pp. 247–258, 1985.

[6] Anderson, B. O. D., R. R. Bitmead, C. R. Johanson, P. V. Kokotović, R. L. Kosut, I. M. Y. Mareels, L. Praly and B. D. Riedle, *Stability of Adaptive Systems: Passivity and Averaging Analysis*, MIT Press, Cambridge, MA, 1986.

[7] Anderson, B. D. O. and R. M. Johnstone, "Adaptive systems and time-varying plants," *International Journal of Control*, vol. 37. no. 2, pp. 367–377, 1983.

[8] Anderson, B. D. O. and R. M. Johnstone, "Global adaptive pole positioning," *IEEE Transactions on Automatic Control*, vol. AC-30, no. 1, pp. 11–22, 1985.

[9] Anderson, B. D. O. and J. B. Moore, *Optimal Control: Linear Quadratic Methods*, Prentice-Hall, Englewood Cliffs, NJ, 1990.

[10] Anderson, B. O. D. and S. Vongpanitlerd, *Network Analysis and Synthesis: A Modern Systems Theory Approach*, Prentice-Hall, Englewood Cliffs, NJ, 1973.

[11] Annaswamy, A. M. and K. S. Narendra, "Adaptive control of simple time-varying systems," *Proceedings of the 28th IEEE Conference on Decision and Control*, Tampa, FL, 1989.

[12] Annaswamy, A. M., F. P. Skantze and A. P. Loh, "Adaptive control of continuous-time systems with convex/concave parametrization," *Automatica*, vol. 14, no. 1, pp. 33–49, 1998.

[13] Antsaklis, P. J. and A. N. Michel, *Linear Systems*, McGraw-Hill, New York, 1997.

[14] Arcak, M. and P. V. Kokotović, "Robust nonlinear control of systems with input unmodeled dynamics," *Systems and Control Letters*, vol. 41, no. 2, pp. 115–122, October 2000.

[15] Arcak, M. and P. V. Kokotović, "Redesign of backstepping for robustness against unmodeled dynamics," *International Journal of Robust Nonlinear Control*, vol. 11, no. 7, pp. 633–643, 2001.

[16] Armstrong-Helouvry, B., P. Dupont and C. Canudas de Wit, "A survey of models, analysis tools and compensation methods for the control of machines with friction," *Automatica*, vol. 30, no. 7, pp. 1083–1138, 1994.

[17] Asano, T., K. Yoshikawa and S. Suzuki, "The design of a precompensator for multivariable adaptive control—A network-theoretic approach," *IEEE Transactions on Automatic Control*, vol. AC-35, no. 6, pp. 706–710, 1990.

[18] Åström, K. J., "Direct methods for nonminimum phase systems," *Proceedings of the 19th IEEE Conference on Decision and Control*, Albuquerque, NM, 1980.

[19] Åström, K. J., "Analysis of rohrs counterexample to adaptive control," *Proceedings of the 22nd IEEE Conference on Decision and Control*, pp. 982–987, San Antonio, TX, 1983.

[20] Åström, K. J. and T. Hagglund, *PID Controllers: Theory, Design and Tuning*, 2nd ed., Instrument Society of America, Research Triangle Park, NC, 1995.

[21] Åström, K. J., T. Hagglund, C. C. Hang and W. K. Ho, "Automatic tuning and adaptation for PID controllers: A survey," *Proceedings of*

the IFAC International Symposium on Adaptive Systems in Control and Signal Processing, pp. 121–126, Grenoble, France, July 1992.

[22] Åström, K. J. and B. Wittenmark, *Computer-Controlled Systems: Theory and Design*, Prentice-Hall, Englewood Cliffs, NJ, 1990.

[23] Åström, K. J. and B. Wittenmark, *Adaptive Control*, 2nd ed., Addison-Wesley, Reading, MA, 1995.

[24] Athans, M. and P. L. Falb, *Optimal Control*, McGraw-Hill, New York, 1966.

[25] Bai, E.-W. and S. Sastry, "Persistency of excitation, sufficient richness and parameter convergence in discrete-time adaptive control," *Systems and Control Letters*, vol. 6, pp. 153–163, 1985.

[26] Bai, E.-W. and S. Sastry, "Discrete-time adaptive control utilizing prior information," *IEEE Transactions on Automatic Control*, vol. AC-31, no. 8, pp. 779–782, 1986.

[27] Bai, E.-W. and S. Sastry, "Global stability proofs for continuous-time indirect adaptive control schemes," *IEEE Transactions on Automatic Control*, vol. AC-32, no. 4, pp. 537–543, 1987.

[28] Bartolini, G. and A. Ferrara, "On the parameter convergence properties of a combined VS/adaptive control scheme during sliding motion," *IEEE Transactions on Automatic Control*, vol. AC-44, no. 1, pp. 70–76, 1999.

[29] Bartolini, G. and T. Zolezzi, "The V.S.S. approach to the model reference control of nonminimum phase linear plants," *IEEE Transactions on Automatic Control*, vol. AC-33, no. 9, pp. 859–864, 1988.

[30] Belbas, S. A. and I. D. Mayergoyz, "Optimal control of dynamic systems with hysteresis," *International Journal of Control*, vol. 73, no. 1, pp. 22–28, January 2000.

[31] Bellman, R. E., *Dynamic Programming*, Princeton University Press, Princeton, NJ, 1957.

[32] Bernotas, L. A., P. E. Crago and H. J. Chizeck, "Adaptive control of electrically stimulated muscle," *IEEE Transactions on Biomedical Engineering*, vol. BE-34, no. 2, pp. 140–147, 1987.

[33] Bitmead, R. R., M. Gevers and V. Wertz, *Adaptive Optimal Control*, Prentice-Hall, Englewood Cliffs, NJ, 1990.

[34] Bohm, M., M. A. Demetriou, S. Reich and I. G. Rosen, "Model reference adaptive control of distributed parameter systems," *SIAM Journal of Control and Optimization*, vol. 36, no. 1, pp. 33–83, 1998.

[35] Borisson, U., "Self-tuning regulators for a class of multivariable systems," *Automatica*, vol. 15, no. 2, pp. 209–217, 1979.

[36] Boyd, S. and S. Sastry, "On parameter convergence in adaptive control," *Systems and Control Letters*, vol. 3, no. 6, pp. 311–319, 1983.

[37] Boyd, S. and S. S. Sastry, "Necessary and sufficient conditions for parameter convergence in adaptive control," *Automatica*, vol. 22, no. 6, pp. 629–639, 1986.

[38] Bryson, A. E. and Y-C. Ho, *Applied Optimal Control: Optimization, Estimation, and Control*, Hemisphere Publishing, New York, 1975.

[39] Campion, G. and G. Bastin, "Indirect adaptive state feedback control of linearly parametrized nonlinear systems," *International Journal of Adaptive Control and Signal Processing*, vol. 4, pp. 345–358, 1990.

[40] Campos, J. and F. L. Lewis, "Deadzone compensation in discrete time using adaptive fuzzy logic," *IEEE Transactions on Fuzzy Systems*, vol. 7, no. 6, pp. 697–707, December 1999.

[41] Canudas de Wit, C. A., *Adaptive Control for Partially Known Systems: Theory and Applications*, Elsevier Science Publishers, Amsterdam, 1988.

[42] Chalam, V. V., *Adaptive Control Systems: Techniques and Applications*, Marcel Dekker, New York, 1987.

[43] Chen, B. M., *Robust and H_∞ Control*, Springer-Verlag, London, 2000.

[44] Chen, C. T., *Linear System Theory and Design*, Holt, Rinehart and Winston, New York, 1984.

[45] Chen, D. G. and Paden, B., "Nonlinear adaptive torque-ripple cancellation for step motors," *Proceedings of the 29th IEEE Conference on Decision and Control*, pp. 3319–3324, Honolulu, HI, 1990.

[46] Chen, F. C. and H. K. Khalil, "Adaptive control of a class of nonlinear discrete-time systems using neural networks," *IEEE Transactions on Automatic Control*, vol. AC-40, no. 5, pp. 791–801, 1995.

[47] Chen, H. F., X. B. Cao and H. T. Fang, "Stability of adaptively stabilized stochastic systems," *IEEE Transactions on Automatic Control*, vol. AC-46, no. 11, pp. 1732–1736, 2001.

[48] Chen, H. F. and L. Guo, "Stochastic adaptive pole-zero assignment with convergence analysis," *Systems and Control Letters*, vol. 7, pp. 159–164, 1986.

[49] Chen, H. F. and L. Guo, "A robust stochastic adaptive controller," stochastic systems," *IEEE Transactions on Automatic Control*, vol. AC-33, no. 11, pp. 1035–1043, 1988.

[50] Chen, H. F. and L. Guo, *Identification and Stochastic Adaptive Control*, Birkhauser, Boston, 1991.

[51] Chen, W. and L. Guo, "Prediction-based discrete-time adaptive nonlinear stochastic control," *IEEE Transactions on Automatic Control*, vol. AC-44, no. 9, pp. 1725–1729, 1999.

[52] Chen, Y. H., "Robust control system design: Nonadaptive versus adaptive," *International Journal of Control*, vol. 51, no. 6, pp. 1457–1477, 1990.

[53] Chen, Y. H., "Adaptive robust control of uncertain systems with measurement noise," *Automatica*, vol. 28, no. 4, pp. 715–728, July 1992.

[54] Chen, Z. and P. Cook, "Robustness of model reference adaptive control systems with unmodeled dynamics," *International Journal of Control*, vol. 39, no. 1, pp. 201–214, 1984.

[55] Cheng, J.-W. J. and Y. M. Wang, "Adaptive control for systems with input constraints: A survey," in *Adaptive Control of Nonsmooth Dynamic Systems*, G. Tao and F. L. Lewis, eds., pp. 311–332, Springer-Verlag, London, 2001.

[56] Chua, L. O. and S. C. Bass, "A generalized hysteresis model," *IEEE Transactions on Circuit Theory*, vol. CT-19, no. 1, pp. 36–48, 1972.

[57] Clary, J. P. and G. F. Franklin, "Self-tuning control with a priori plant knowledge," *Proceedings of the 23rd IEEE Conference on Decision and Control*, pp. 369–374, Las Vegas, NV, 1984.

[58] Costa, R. R. and L. Hsu, "Robustness of VS-MRAC with respect to unmodeled dynamics and external disturbances," *International Journal of Adaptive Control and Signal Processing*, vol. 6, no. 1, pp. 19–33, 1992.

[59] Costa, R. R., L. Hsu, A. K. Imai and G. Tao, "Adaptive backstepping design for MIMO plants using factorization," *Proceedings of the 2002 American Control Conference*, pp. 4601–4606, Anchorage, AK, 2002.

[60] Cristi, R., "Internal persistency of excitation in indirect adaptive control," *IEEE Transactions on Automatic Control*, vol. AC-32, no. 12, pp. 1101–1103, 1987.

[61] Cruz-Hernande, J. M. and V. Hayward, "Phase control approach to hysteresis reduction," *IEEE Transactions on Control Systems Technology*, vol. 9 no. 1, pp. 17–26, 2001.

[62] Das, M, "A proof of global stability of a direct adaptive pole placement scheme for multivariable systems," *Proceedings of the 1985 American Control Conference*, pp. 71–76, Boston, 1985.

[63] Das, M., "Multivariable adaptive model matching using less a priori information," *Journal of Dynamic Systems, Measurement, and Control*, vol. 108, pp. 151–153, 1986.

[64] Dasgupta, S. and A. S. Bhagwat, "Conditions for designing strictly positive real functions for adaptive output error identification," *IEEE Transactions on Circuits and Systems*, vol. CAS-34, no. 7, pp. 731–736, 1987.

[65] Datta, A., M.-T. Ho and S. P. Bhattacharyya, *Structure and Synthesis of PID Controllers*, Springer-Verlag, London, 2000.

[66] Datta, A. and P. A. Ioannou, "Decentralized adaptive control," in *Control and Dynamics: Advances in Theory and Applications*, vol. 53, C. T. Leondes, ed., pp. 125–171, Academic Press, New York, 1992.

[67] Datta, A. and P. A. Ioannou, "Performance improvement versus robust stability in model reference adaptive control," *IEEE Transactions on Automatic Control*, vol. AC-39, no. 12, pp. 2370–2388, 1994.

[68] Dawson, D. M., Z. Qu and F. L. Lewis, "Hybrid adaptive robust-control for a robot manipulator," *International Journal of Adaptive Control and Signal Processing*, vol. 6, no. 6, pp. 537–545, November 1992.

[69] DeCarlo, R. A., S. H. Zak and G. P. Matthews, "Variable structure control of nonlinear multivariable systems: A tutorial," *Proceedings of IEEE*, vol. 76, pp. 212–232, March 1988.

[70] De La Sen, M. and M. B. Paz, "Discrete multivariable adaptive control," *International Journal of Control*, vol. 42, no. 5, pp. 1071–1097, 1985.

[71] De Larminat, Ph., "On the stabilizability condition in indirect adaptive control," *Automatica*, vol. 20, no. 7, pp. 793–795, 1984.

[72] De Mathelin, M. and M. Bodson, "Frequency domain conditions for parameter convergence in multivariable recursive identification," *Automatica*, vol. 26, no. 4, pp. 757–767, 1990.

[73] De Mathelin, M. and M. Bodson, "Multivariable model reference adaptive control with unknown high frequency gain," *Proceedings of the 32nd IEEE Conference on Decision and Control*, pp. 3078–3083, San Antonio, TX, 1993.

[74] Demetriou, M. A. and I. G. Rosen, "Model reference adaptive control of abstract hyperbolic distributed parameter system," *Proceedings of the 32nd IEEE Conference on Decision and Control*, pp. 2424–2429, San Antonio, TX, 1993.

[75] Demetriou, M. A. and I. G. Rosen, "On the persistence of excitation in the adaptive estimation of distributed parameter systems," *IEEE Transactions on Automatic Control*, vol. AC-39, no. 5, pp. 1117–1123, 1994.

[76] Demetriou, M. A. and I. G. Rosen, "Adaptive parameter estimation for degenerate parabolic systems," *Journal of Mathematical Analysis and Applications*, vol. 189, no. 3, pp. 815–847, 1995.

[77] Desoer, C. A., "Slowly varying discrete systems," *Electronics Letters*, vol. 6, no. 11, pp. 339–340, May 1970.

[78] Desoer, C. A. and M. Vidyasagar, *Feedback Systems: Input-Output Properties*, Academic Press, New York, 1975.

[79] Deif, A. S., *Advanced Matrix Theory for Scientists and Engineers*, John Wiley & Sons, New York, 1982.

[80] Dion, J. M., L. Dugard and J. Carrillo, "Interactor and multivariable adaptive model matching," *IEEE Transactions on Automatic Control*, vol. AC-33, no. 4, pp. 399–401, 1988.

[81] Djaferis, T., M. Das and H. Elliott, "Reduced order adaptive pole placement for multivariable systems," *IEEE Transactions on Automatic Control*, vol. AC-29, no. 7, pp. 637–638, 1984.

[82] Dorf, R. C. and R. H. Bishop, *Modern Control Systems*, 8th ed., Addison-Wesley, Reading, MA, 1998.

[83] Duarte, M. A. and K. S. Narendra, "A new approach to model reference adaptive control," *International Journal of Adaptive Control and Signal Processing*, vol. 3, no. 1, pp. 53–73, 1989.

[84] Dugard, L. and J. M. Dion, "Direct adaptive control for linear multivariable systems," *International Journal of Control*, vol. 42, no. 6, pp. 1251–1281, 1985.

[85] Dugard, L., G. C. Goodwin and C. E. de Souza, "Prior knowledge in model reference adaptive control of multivariable systems," *IEEE Transactions on Automatic Control*, vol. AC-29, no. 8, pp. 761–764, 1984.

[86] Duncan, T. E., L. Guo and B. Pasik-Duncan, "Adaptive continuous time linear quadratic Gaussian control," *IEEE Transactions on Automatic Control*, vol. AC-44, no. 9, pp. 1653–1662, 1999.

[87] Duncan, T. E., B. Pasik-Duncan and Q. Zhang, "Adaptive control of stochastic manufacturing systems with hidden Markovian demands and small noise," *IEEE Transactions on Automatic Control*, vol. AC-44, no. 2, pp. 427–430, 1999.

[88] Duncan, T. E. and B. Pasik-Duncan, "Adaptive control of continuous-time linear stochastic systems," *Mathematics of Control, Signals, and Systems*, vol. 3, pp. 45–60, 1990.

[89] Egardt, B., *Stability of Adaptive Controllers*, Springer-Verlag, Berlin, 1979.

[90] Egardt, B., "Stability analysis of adaptive control systems with disturbances," *Proceedings of the 1980 Joint American Control Conference*, San Francisco, CA, 1980.

[91] Elliott, H., "Hybrid adaptive control of continuous time systems," *IEEE Transactions on Automatic Control*, vol. AC-27, no. 2, pp. 419–426, 1982.

[92] Elliott, H., R. Cristi and M. Das, "Global stability of adaptive pole placement algorithms," *IEEE Transactions on Automatic Control*, vol. AC-30, no. 4, pp. 348–356, 1985.

[93] Elliott, H. and W. A. Wolovich, "A parameter adaptive control structure for linear multivariable systems," *IEEE Transactions on Automatic Control*, vol. AC-27, no. 5, pp. 340–352, 1982.

[94] Elliott, H. and W. A. Wolovich, "Parametrization issues in multivariable adaptive control," *Automatica*, vol. AC-20, no. 5, pp. 533–545, 1984.

[95] Elliott, H., W. A. Wolovich and M. Das, "Arbitrary adaptive pole placement for linear multivariable systems," *IEEE Transactions on Automatic Control*, vol. AC-29, no. 3, pp. 221–229, 1984.

[96] Ezal, K., P. V. Kokotović, A. R. Teel and T. Basar, "Disturbance attenuating output-feedback control of nonlinear systems with local optimality," *Automatica*, vol. 37, no. 6, pp. 805–817, 2001.

[97] Feng, G., "Analysis of a new algorithm for continuous time robust adaptive control," *IEEE Transactions on Automatic Control*, vol. AC-44, no. 9, pp 1764–1768, 1999.

[98] Feng, G., "An approach to adaptive control of fuzzy dynamic systems," *IEEE Transactions on Fuzzy Systems*, vol. 10, no. 2, pp. 268–275, 2002.

[99] Feng, G. and R. Lozano, eds., *Adaptive Control Systems*, Butterworth-Heinemann, Oxford, 1999.

[100] Feng, G. and M. Palaniswami, "A stable adaptive implementation of the internal model principle," *IEEE Transactions on Automatic Control*, vol. AC-37, no. 8, pp. 1220–1225, 1992.

[101] Feuer, A. and A. S. Morse, "Adaptive control of single-input, single-output linear systems," *IEEE Transactions on Automatic Control*, vol. AC-23, no. 4, pp. 557–569, 1978.

[102] Filippov, A. F., "Differential equations with discontinuous right hand sides," *American Mathematical Society Translations*, vol. 42, pp. 199–231, 1964.

[103] Fortmann, T. E. and K. L. Hitz, *An Introduction to Linear Control Systems*, Marcel Dekker, New York, 1977.

[104] Franklin, G. F., J. D. Powell and A. Emani-Naeini, *Feedback Control of Dynamic Systems*, 3rd ed., Addison-Wesley, Reading, MA, 1994.

[105] Franklin, G. F., J. D. Powell and M. Workman, *Digital Control of Dynamic Systems*, 3rd ed., Addison-Wesley, Reading, MA, 1998.

[106] Francis, B. A. and P. P. Khargonekar, eds., *Robust Control Theory*, Springer-Verlag, New York, 1995.

[107] Freeman, R. A. and P. V. Kokotovi'c, *Robust Nonlinear Control Design: State-Space and Lyapunov Techniques*, Birkhauser, Boston, 1996.

[108] Fu, L.-C. and S. Sastry, "Slow drift instability in model reference adaptive systems—an averaging analysis," *International Journal of Control*, vol. 45, no. 2, pp. 503–527, 1987.

[109] Fu, L.-C., M. Bodson and S. Sastry, "New stability theorems for averaging and their applications to the convergence analysis of adaptive identification and control schemes," *Proceedings of the 24th IEEE Conference on Decision and Control*, pp. 473–477, Ft. Lauderdale, FL, 1985.

[110] Fu, M. and B. R. Barmish, "Adaptive stabilization of linear systems via switching control," *IEEE Transactions on Automatic Control*, vol. AC-31, no. 12, pp. 1097–1103, 1986.

[111] Garrido, R. and R. Lozano, "Singularity-free multivariable model reference adaptive control," *IEEE Transactions on Automatic Control*, vol. AC-39, no. 9, pp. 1856–1861, 1994.

[112] Gavel, D. T. and D. D. Siljak, "Decentralized adaptive control: Structural conditions for stability," *IEEE Transactions on Automatic Control*, vol. AC-34, no. 4, pp. 413–425, 1989.

[113] Ge, P. and M. Jouaneh, "Modeling of the backlash hysteresis nonlinearity," *Journal of Mechanical Design*, vol. 120 no. 3, pp. 408–413, September 1998.

[114] Ge, S. S., T. H. Lee and C. J. Harris, *Adaptive Neural Network Control of Robot Manipulators*, World Scientific, River Edge, NJ, 1998.

[115] Ge, S. S., C. C. Hang, T. H. Lee and T. Zhang, *Stable Adaptive Neural Network Control*, Kluwer Academic, Boston, 2001.

[116] Gevers, M., G. C. Goodwin and V. Wertz, "Continuous time stochastic adaptive control," *SIAM Journal on Control and Optimization*, vol. 29, no. 2, pp. 264–282, March 1991.

[117] Giri, F., J. M. Dion, M. M'Saad and L. Dugard, "A globally convergent pole placement indirect adaptive controller," *Proceedings of the 26th IEEE Conference on Decision and Control*, pp. 1–6, Los Angeles, 1987.

[118] Giri, F., F. Ahmed-Zaid and P. A. Ioannou, "Stable indirect adaptive control of continuous-time systems with no a priori knowledge about the parameters," *Proceedings of the IFAC International Symposium on Adaptive Systems in Control and Signal Processing*, Grenoble, France, July 1992.

[119] Giri, F., M. M'Saad, L. Dugard and J. M. Dion, "Pole placement direct adaptive control for time-varying ill-modeled plants," *IEEE Transactions on Automatic Control*, vol. AC-35, no. 6, pp. 723–726, 1990.

[120] Goodwin, G. C. and R. S. Long, "Generalization of results on multivariable adaptive control," *IEEE Transactions on Automatic Control*, vol. AC-25, no. 6, pp. 1241–1245, 1980.

[121] Goodwin, G. C., R. Lozano, D. Q. Mayne and R. H. Middleton, "Rapprochement between continuous and discrete model reference adaptive control," *Automatica*, vol. 22, no. 2, pp. 199–207, 1986.

[122] Goodwin, G. C., B. C. McInnis and J. C. Wang, "Model reference adaptive control for systems having non-square transfer functions," *Proceedings of the 21st IEEE Conference on Decision and Control*, pp. 744–749, Orlando, FL, 1982.

[123] Goodwin, G. C. and D. Q. Mayne, "A parameter estimation perspective of continuous time model reference adaptive control," *Automatica*, vol. 23, no. 1, pp. 57–70, 1987.

[124] Goodwin, G. C., P. J. Ramadge and P. E. Caines, "Discrete-time multivariable adaptive control," *IEEE Transactions on Automatic Control*, vol. AC-25, no. 6, pp. 449–456, 1980.

[125] Goodwin, G. C. and K. S. Sin, *Adaptive Filtering Prediction and Control*, Prentice-Hall, Englewood Cliffs, NJ, 1984.

[126] Golub, G. H. and C. F. van Loan, *Matrix Computations*, Johns Hopkins University Press, Baltimore, MD, 1983.

[127] Guo, L., "Self-convergence of weighted least squares with applications to stochastic adaptive control," *IEEE Transactions on Automatic Control*, vol. AC-41, no. 1, pp. 79–89, 1996.

[128] Guo, L., "On critical stability of discrete-time adaptive nonlinear control," *IEEE Transactions on Automatic Control*, vol. AC-42, no. 11, pp. 1488–1499, 1997.

[129] Hang, C. C., T. H. Lee and W. K. Ho, *Adaptive Control*, Instrument Society of America, Research Triangle Park, NC, 1993.

[130] Hatwell, M. S., B. J. Oderkerk, C. A. Sacher and G. F. Inbar, "The development of a model reference adaptive controller to control the knee joint of paraplegics," *IEEE Transactions on Automatic Control*, vol. AC-36, no. 6, pp. 683–691, 1991.

[131] Hsu, L., "Variable structure model reference adaptive control using only input and output measurements: The general case," *IEEE Transactions on Automatic Control*, vol. 35, no. 11, pp. 1238–1243, 1990.

[132] Hsu, L., A. D. Araujo and R. R. Costa, "Analysis and design of I/O based variable structure adaptive control," *IEEE Transactions on Automatic Control*, vol. 39, no. 1, pp. 4–21, 1994.

[133] Hsu, L., A. D. Araujo and F. Lizarralde, "New results on input/output variable structure model reference adaptive control: Design and stability analysis," *IEEE Transactions on Automatic Control*, vol. 42, no. 3, pp. 386–393, 1997.

[134] Hsu, L. and R. R. Costa, "Bursting phenomena in continuous-time adaptive systems with a σ-modification," *IEEE Transactions on Automatic Control*, vol. AC-32, no. 1, pp. 84–86, 1987.

[135] Hsu, L. and R. R. Costa, "Variable structure model reference adaptive control using only input and output measurement: Part I," *International Journal of Control*, vol. 49, no. 2, pp. 399–416, 1989.

[136] Hsu, L. and R. R. Costa, "IMO direct adaptive control with reduced prior knowledge of the high frequency gain," *Proceedings of the 38th IEEE Conference on Decision and Control*, pp. 3303–3308, Phoenix, AZ, 1999.

[137] Hsu, L., R. R. Costa, A. K. Imai and P. Kokotović, "Lyapunov based adaptive control of MIMO systems," *Proceedings of 2001 American Control Conference*, pp. 4808–4813, Arlington, VA, 2001.

[138] Ilchmann, A., D. H. Owens and D. Pratzel-Wolters, "High gain robust adaptive controllers for multivariable systems," *Systems and Control Letters*, vol. 8, pp. 397–404, 1987.

[139] Ichikawa, K., "Adaptive control of delay systems," *International Journal of Control*, vol. 43, no. 6, pp. 1653–1659, 1986.

[140] Ikhouane, F. and M. Krstić, "Adaptive backstepping with parameter projection: Robustness and asymptotic performance," *Automatica*, vol. 34, no. 4, pp. 429–435, 1998.

[141] Imai, A. K., R. R. Costa, L. Hsu, G. Tao and P. V. Kokotović, "Multivariable MRAC using high frequency gain matrix factorization," *Proceedings of the 40th IEEE Conferences on Decision and Control*, pp. 1193–1198, Orlando, FL, 2001.

[142] Ioannou, P. A., "Design of decentralized adaptive systems," in *Advances in Large Scale Systems—Theory and Applications*, vol. 1, J. B. Cruz, ed., pp. 117–141, JAI Press, Greenwich, CT, 1984.

[143] Ioannou, P. A., "Adaptive stabilization of not necessarily minimum phase plants," *Systems and Control Letters*, vol. 7, pp. 281–287, 1986.

[144] Ioannou, P. A., "Decentralized adaptive control of interconnected systems," *IEEE Transactions on Automatic Control*, vol. AC-31, no. 3, pp. 291–298, 1986.

[145] Ioannou, P. A., "Robust adaptive controller with zero residual tracking errors," *IEEE Transactions on Automatic Control*, vol. AC-31, no. 8, pp. 773–776, 1986.

[146] Ioannou, P. A. and and A. Datta, "Robust adaptive controller: A unified approach," *Proceedings of the IEEE*, vol. 79, pp. 1736–1768, 1991.

[147] Ioannou, P. A. and P. V. Kokotović, *Adaptive Systems with Reduced Models*, Springer-Verlag, Berlin, 1983.

[148] Ioannou, P. A. and P. V. Kokotović, "Instability analysis and improvement of robustness of adaptive control," *Automatica*, vol. 20, no. 5, pp. 583–594, 1984.

[149] Ioannou, P. A. and P. V. Kokotović, "Robust redesign of adaptive control," *IEEE Transactions on Automatic Control*, vol. AC-29, no. 3, pp. 202–211, 1984.

[150] Ioannou, P. A. and and J. Sun, "Theory and design of robust direct and indirect adaptive control schemes," *International Journal of Control*, vol. 47, no. 3, pp. 775–813, 1988.

[151] Ioannou, P. A. and J. Sun, *Robust Adaptive Control*, Prentice-Hall, Englewood Cliffs, NJ, 1996.

[152] Ioannou, P. A. and G. Tao, "Frequency domain conditions for strictly positive functions," *IEEE Transactions on Automatic Control*, vol. AC-32, no. 1, pp. 53–54, 1987.

[153] Ioannou, P. A. and G. Tao, "Dominant richness and improvement of performance of robust adaptive control," *Automatica*, vol. 25, no. 2, pp. 287–291, 1989.

[154] Ioannou, P. A. and K. Tsakalis, "A robust direct adaptive controller," *IEEE Transactions on Automatic Control*, vol. AC-31, no. 11, pp. 1033–1043, 1986.

[155] Ioannou, P. A. and K. Tsakalis, "Robust discrete time adaptive control," in *Adaptive and Learning Systems: Theory and Applications*, K. S. Narendra, ed., Plenum Press, New York, 1986.

[156] Ioannou, P. A. and K. Tsakalis, "The class of unmodeled dynamics in robust adaptive control," *Proceedings of the 1988 American Control Conference*, Atlanta, GA, June 1988.

[157] Isidori, A., *Nonlinear Control Systems*, 3rd ed., Springer-Verlag, Berlin, 1995.

[158] Isidori, A., S. M. Joshi and A. G. Kelkar, "Asymptotic stability of interconnected passive non-linear systems," *International Journal of Robust Nonlinear Control*, vol. 9, no. 5, pp. 261–273, 1999.

[159] Isidori, A. and T. J. Tarn, "Robust regulation for nonlinear-systems with gain-bounded uncertainties," *IEEE Transactions on Automatic Control*, vol. AC-40, no. 10, pp. 1744–1754, 1995.

[160] Janecki, D., "Persistency of excitation for continuous-time systems—time-domain approach," *Systems and Control Letters*, vol. 8, pp. 333-344, 1987.

[161] Janecki, D., "Stability analysis of Elliott's direct adaptive pole placement," *Systems and Control Letters*, vol. 11, pp. 19–26, 1988.

[162] Jerome, N. F. and W. H. Ray, "Model-predictive control of linear-multivariable systems having time delays and right-half-plane zeros," *Chemical Engineering Science*, vol. 47, no. 4, pp. 763–785, March 1992.

[163] Jiang, Z. P. "A combined backstepping and small-gain approach to adaptive output feedback control," *Automatica*, vol. 35, no. 6, pp. 1131–1139, 1999.

[164] Jiang, Z. P. "Decentralized and adaptive nonlinear tracking of large-scale systems via output feedback," *IEEE Transactions on Automatic Control*, vol. 45, no. 11, pp. 2122–2128, 2000.

[165] Johansson, R., "Parametric models of linear multivariable systems for adaptive control," *IEEE Transactions on Automatic Control*, vol. AC-32, no. 3, pp. 303–313, 1987.

[166] Johnstone, R. M. and B. D. O. Anderson, "Exponential convergence of recursive least squares with exponential forgetting factor—adaptive control," *Systems and Control Letters*, vol. 2, no. 2., pp. 69–76, 1982.

[167] Joshi, S. M., *Control of Large Flexible Space Structures*, Springer Verlag, New York, 1989.

[168] Joshi, S. M. and S. Gupta, "On a class of marginally stable positive real systems," *IEEE Transactions on Automatic Control*, vol. AC-41, no. 1, pp. 152–155, 1996.

[169] Joshi, S. M. and A. G. Kelkar, "Inner loop control of supersonic aircraft in the presence of aeroelastic modes," *IEEE Transactions on Control Systems Technology*, vol. 6, no. 6, pp. 730–739, 1998.

[170] Joshi, S. M. and A. G. Kelkar, "Design of norm-bounded and sector-bounded LQG controllers for uncertain systems," *Journal of Optimization Theory and Applications*, vol. 111, no. 2, pp. 269–282, May 2002.

[171] Kailath, T., *Linear Systems*, Prentice-Hall, Englewood Cliffs, NJ, 1980.

[172] Kalman, R. E., "Contributions to the theory of optimal control," *Bol. Soc. Mat. Mexicana*, vol. 5, pp. 102–119, 1960.

[173] Kalman, R. E. and R. S. Bucy, "New results in linear filtering and prediction theory," *J. Basic Eng., Transactions ASME*, Ser. D, vol. 83, pp. 95–108, 1961.

[174] Kanellakopoulos, I., "Passive adaptive control of nonlinear systems," *International Journal of Adaptive Control and Signal Processing*, vol. 7, no. 5, pp. 339–352, 1993.

[175] Kanellakopoulos, I., P. V. Kokotović and R. Marino, "An extended direct scheme for robust adaptive nonlinear control," *Automatica*, vol. 27, no. 2, pp. 247–255, 1991.

[176] Kanellakopoulos, I., P. V. Kokotović and A. S. Morse, "Systematic design of adaptive controllers for feedback linearizable systems," *IEEE Transactions on Automatic Control*, vol. AC-36, no. 11, pp. 1241–1253, 1991.

[177] Kanellakopoulos, I., M. Krstić and P. V. Kokotović, "κ-adaptive control of output-feedback nonlinear systems," *Proceedings of the 32nd IEEE Conference on Decision and Control*, pp. 1061–1066, San Antonio, TX, 1993.

[178] Kaufman, H., I. Barkana and K. Sobel, *Direct Adaptive Control Algorithms: Theory and Applications*, Springer, New York, 1998.

[179] Khalil, H. K., *Nonlinear Systems*, 2nd ed., Prentice-Hall, Upper Saddle River, NJ, 1996.

[180] Khalil, H. K., "Adaptive output feedback control of nonlinear systems represented by input-output models," *IEEE Transactions on Automatic Control*, vol. AC-41, no. 2, pp. 177–188, 1996.

[181] Kim, J.-H., J.-H. Park, S.-W. Lee and E. K. P. Chong, "A two-layered fuzzy logic controller for systems with dead-zones," *IEEE Transactions on Industrial Electronics*, vol. 41, no. 2, pp. 155-162, April 1994.

[182] Kinnaert, M., R. Hanus and J. L. Henrotte, "A new decoupling precompensator for indirect adaptive control of multivariable linear systems," *IEEE Transactions on Automatic Control*, vol. AC-32, no. 5, pp. 455–459, 1987.

[183] Kirk, D. E., *Optimal Control Theory: An Introduction*, Prentice-Hall, Englewood Cliffs, NJ, 1970.

[184] Knohl, T. and H. Unbehauen, "Adaptive position control of electrohydraulic servo systems using ANN," *Mechatronics*, vol. 10, no. 1-2, pp. 127–143, 2000.

[185] Koivo, H. N., "A multivariable self-tuning controller," *Automatica*, vol. 16, no. 3, pp. 354–366, 1980.

[186] Kokotović, P. V., ed., *Foundations of Adaptive Control*, Springer-Verlag, Berlin, 1991.

[187] Kokotović, P. V. and M. Arcak, "Constructive nonlinear control: A historical perspective," *Automatica*, vol. 37, no. 5, pp. 637–662, 2001.

[188] Kokotović, P. V., H. K. Khalil and J. O'Reilly, *Singular Perturbations in Systems and Control: Analysis and Design*, Academic Press, New York, 1986.

[189] Kokotović, P. V., B. Riedle and L. Praly, "On a stability criterion for continuous slow adaptation," *Systems and Control Letters*, vol. 6, pp. 7–14, 1985.

[190] Kokotović, P. V. and H. J. Sussmann, "A positive real condition for global stabilization of nonlinear systems," *Systems and Control Letters*, vol. 13, pp. 125–133, 1989.

[191] Kosmatopoulos, E. B. and P. A. Ioannou, "Robust switching adaptive control of multi-input nonlinear systems," *IEEE Transactions on Automatic Control*, vol. AC-47, no. 4, pp. 610–624, 2002.

[192] Kosut, R. L., B. D. O. Anderson and I. M. Y. Mareels, "Stability theory for adaptive systems: Method of averaging and persistency of excitation," *IEEE Transactions on Automatic Control*, vol. AC-32, no. 1, 26–34, 1987.

[193] Kosut, R. L. and B. Friedlander, "Robust adaptive control: Conditions for global stability," *IEEE Transactions on Automatic Control*, vol. AC-30, no. 7, pp. 610–624, 1985.

[194] Kosut, R. L. and C. R. Johnson, Jr., "An input-output view of robustness in adaptive control,"' *Automatica*, vol. 20, no. 5, pp. 569–581, 1984.

[195] Krasnoselskii, M. A. and A. V. Pokrovskii, *Systems with Hysteresis*, Springer-Verlag, Berlin, 1983.

[196] Krause, J., M. Athans, S. Sastry and L. Valavani, "Robustness studies in adaptive control," *Proceedings of the 22nd IEEE Conference on Decision and Control*, pp. 977–981, San Antonio, TX, 1983.

[197] Kreisselmeier, G., "An approach to stable indirect adaptive control," *Automatica*, vol. 21, no. 4, pp. 425–431, 1985.

[198] Kreisselmeier, G., "A robust indirect adaptive controller approach," *International Journal of Control*, vol. 43, no. 1, pp. 161–175, 1986.

[199] Kreisselmeier, G., "Adaptive control of a class of slow time-varying plants," *Systems and Control Letters*, vol. 8, no. 2, pp. 97–103, 1986.

[200] Kreisselmeier, G. and B. D. O. Anderson, "Robust model reference adaptive control," *IEEE Transactions on Automatic Control*, vol. AC-31, no. 2, pp. 127–133, 1986.

[201] Kreisselmeier, G. and K. S. Narendra, "Stable model reference adaptive control in the presence of bounded disturbances," *IEEE Transactions on Automatic Control*, vol. AC-27, no. 6, pp. 1169–1175, 1982.

[202] Kreisselmeier, G. and G. Rietze-Augst, "Richness and excitation on an interval—with application to continuous-time adaptive control," *IEEE Transactions on Automatic Control*, vol. AC-35, no. 2, pp. 165–171, 1990.

[203] Kreisselmeier, G. and M. C. Smith, "Stable adaptive regulation of arbitrary nth-order plants," *IEEE Transactions on Automatic Control*, vol. AC-31, no. 4, pp. 299–305, 1988.

[204] Krstić, M., I. Kanellakopoulos and P. V. Kokotović, "Adaptive nonlinear control without overparametrization," *Systems and Control Letters*, vol. 19, pp. 177–185, 1992.

[205] Krstić, M., I. Kanellakopoulos and P. V. Kokotović, "Nonlinear design of adaptive controllers for linear systems," *IEEE Transactions on Automatic Control*, vol. AC-39, no. 4, pp. 738–752, 1994.

[206] Krstić, M., I. Kanellakopoulos and P. V. Kokotović, *Nonlinear and Adaptive Control Design*, John Wiley & Sons, New York, 1995.

[207] Krstić, M. and P. V. Kokotović, "Adaptive nonlinear design with controller-identifier separation and swapping," *IEEE Transactions on Automatic Control*, vol. AC-40, no. 3, pp. 426–440, 1995.

[208] Kumar, P. R. and P. Varaiya, *Stochastic Systems: Estimation, Identification, and Adaptive Control*, Prentice-Hall, Englewood Cliffs, NJ, 1986.

[209] Kuo, B. C., *Digital Control Systems*, 2nd ed., Saunders College Publishing, Ft. Worth, PA, 1992.

[210] Kuo, B. C., *Automatic Control Systems*, 7th ed., Prentice-Hall, Englewood Cliffs, NJ, 1995.

[211] Kurdila, A. J. and G. Webb, "Compensation for distributed hysteresis operators in active structural systems," *Journal of Guidance Control and Dynamics*, vol. 20, no. 6, pp. 1133–1140, November-December 1997.

[212] Landau, Y. D., *Adaptive Control: The Model Reference Approaches*, Marcel Dekker, New York, 1979.

[213] Landau, Y. D., R. Lozano and M. M'Saad, *Adaptive Control*, Springer, London, 1998.

[214] Lee, T.-H. and K. S. Narendra, "Robust adaptive control of discrete-time systems using persistent excitation," *Automatica*, vol. 24, no. 6, pp. 781–788, 1988.

[215] Lefschetz, S., *Stability of Nonlinear Control Systems*, Academic Press, New York, 1963.

[216] Lewis, F. L., *Applied Optimal Control and Estimation*, Prentice-Hall, Englewood Cliffs, NJ, 1992.

[217] Lewis, F. L., C. T. Abdallah and D. M. Dawson, *Control of Robot Manipulators*, Macmillan, New York, 1993.

[218] Lewis, F. L., J. Campos and R. Selmic, *Neuro-Fuzzy Control of Industrial Systems with Actuator Nonlinearities*, SIAM: Frontiers in Applied Mathematics, Philadelphia, PA, 2002.

[219] Lewis, F. L., K. Liu, R. R. Selmic and L. X. Wang, "Adaptive fuzzy logic compensation of actuator dead-zones," *Journal of Robotic Systems*, vol. 14, no. 6, pp. 501–511, 1997.

[220] Lewis, F. L., W. K. Tim, L. Z. Wang and Z. X. Li, "Deadzone compensation in motion control systems using adaptive fuzzy logic control," *IEEE Transactions on Control Systems Technology*, vol. 7, no. 6, pp. 731–742, November 1999.

[221] Limanond, S. and K. S. Tsakalis, "Model reference adaptive and non-adaptive control of linear time-varying plants," *IEEE Transactions on Automatic Control*, vol. AC-45, no. 7, pp. 1290–1300, 2000.

[222] Limanond, S. and K. S. Tsakalis, "Adaptive and non-adaptive 'pole placement' control of multivariable linear time-varying plants," *International Journal of Control*, vol. 74, no. 5, pp. 507–523, 2001.

[223] Lin, W. and C. Qian, "Adaptive regulation of high-order lower-triangular systems: Adding a power integrator technique," *Systems and Control Letters*, vol. 39, pp. 353–364, 2000.

[224] Lin, W. and C. Qian, "Adaptive control of nonlinearly parameterized systems: A nonsmooth feedback framework," *IEEE Transactions on Automatic Control*, vol. AC-47, no. 5, pp. 757–774, 2002.

[225] Lin, W. and C. Qian, "Adaptive control of nonlinearly parameterized systems: The smooth feedback case," *IEEE Transactions on Automatic Control*, vol. AC-47, no. 8, pp. 1249–1266, 2002.

[226] Lin, Z. L., *Low Gain Feedback*, Lecture Notes in Control and Information Sciences, vol. 240, Springer-Verlag, London, 1998.

[227] Lin, Z. L. and G. Tao, "Adaptive control of a nonminimum phase linear system," *IEEE Transactions on Automatic Control*, vol. AC-45, no. 4, pp. 824–829, 2000.

[228] Ling, Y. and G. Tao, "Adaptive backstepping control design for linear multivariable plants," *International Journal of Control*, vol. 68, no. 6, pp. 1289–1304, 1997.

[229] Ling, Y. and G. Tao, "Multivariable adaptive actuator compensation with partially known high frequency gain matrix," *Proc. of the 37th IEEE Conf. on Decision and Control*, pp. 3990–3995, Tampa, FL, 1998.

[230] Ljung, L., *System Identification: Theory for the User*, 2nd ed., Prentice-Hall, Upper Saddle River, NJ, 1999.

[231] Ljung, L. and T. Soderstrom, *Theory and Practice of Recursive Identification*, MIT Press, Cambridge, MA, 1983.

[232] Loh, A. P., A. M. Annaswamy and F. P. Skantze, "Adaptation in the presence of a general nonlinear parametrization: An error model approach," *IEEE Transactions on Automatic Control*, vol. AC-44, no. 9, pp. 1634–1652, 1999.

[233] Lozano-Leal, R., "Robust adaptive regulation with persistent excitation," *IEEE Transactions on Automatic Control*, vol. AC-34, no. 12, pp. 1260–1267, 1989.

[234] Lozano-Leal, R., J. Collado and S. Mondie, "Model reference robust adaptive control without a priori knowledge of the high frequency gain," *IEEE Transactions on Automatic Control*, vol. AC-35, no. 1, pp. 71–78, 1990.

[235] Lozano-Leal, R. and G. C. Goodwin, "A globally convergent adaptive pole placement algorithm without a persistency of excitation requirement," *IEEE Transactions on Automatic Control*, vol. AC-30, no. 8, pp. 795–797, 1985.

[236] Lozano-Leal, R. and S. M. Joshi, "Strictly positive real transfer-functions revisited," *IEEE Transactions on Automatic Control*, vol. AC-35, no. 11, pp. 1243–1245, 1990.

[237] Lozano-Leal, R. and X. H. Zhao, "Adaptive pole placement without excitation probing signals," *IEEE Transactions on Automatic Control*, vol. AC-39, no. 1, pp. 47–58, 1994.

[238] Lu, C. Y., J. A. Stankovic, G. Tao and S. H. Son, "Design and evaluation of a feedback control EDF scheduling algorithm," *Proceedings of the 20th IEEE Real-Time Systems Symposium*, pp. 56–67, Phoenix, AZ, 1999.

[239] Lu, Y., T. Abdelzaher, C. Y. Lu and G. Tao, "An adaptive control framework for QoS guarantees and its application to differentiated caching services," *Proceedings of the 2002 International Workshop on Quality of Service*, Miami Beach, FL, May 2002.

[240] Lu, Y., G. Tao and T. Abdelzaher, "LDU parametrized discrete-time multivariable MRAC and application to a web cache system," *Proceedings of the 41st IEEE Conference on Decision and Control*, pp. 4492–4497, Las Vegas, NV, 2002.

[241] Luenberger, D. G., "Observers for multivariable systems," *IEEE Transactions on Automatic Control*, vol. AC-11, no. 2, pp. 190–199, 1966.

[242] Luenberger, D. G., "An introduction to observers," *IEEE Transactions on Automatic Control*, vol. AC-16, no. 6, pp. 596–603, 1971.

[243] Ma, X. L. and G. Tao, "Adaptive actuator compensation control with feedback linearization," *IEEE Transactions on Automatic Control*, vol. AC-45, no. 9, pp. 1705–1711, 2000.

[244] Maciejowski, J. M., *Multivariable Feedback Design*, Addison-Wesley, Reading, MA, 1989.

[245] Majima, S., K. Kodama and T. Hasegawa, "Modeling of shape memory alloy actuator and tracking control system with the model," *IEEE Transactions on Control Systems Technology*, vol. 9 no. 1, pp. 54–59, 2001.

[246] Manitius, A. and A. Olbrot, "Finite spectrum assignment problem for systems with delays," *IEEE Transactions on Automatic Control*, vol. AC-24, no. 4, pp. 541–553, 1979.

[247] Mareels, I. M. Y. and R. R. Bitmead, "Nonlinear dynamics in adaptive control: Chaotic and periodic stabilization," *Automatica*, vol. 22, no. 6, pp. 641–655, 1986.

[248] Mareels, I. M. Y., R. R. Bitmead, M. Gevers, C. R. Johnson, Jr., R. L. Kosut and M. A. Poubelle, "How exciting can a signal really be?" *Systems and Control Letters*, vol. 8, pp. 197–304, 1987.

[249] Mareels, I. M. Y. and J. W. Polderman, *Adaptive Systems: An Introduction*, Birkhauser, Boston, 1996.

[250] Marino, R. and P. Tomei, "Global adaptive output feedback control of nonlinear systems—Part I: linear parametrization, " *IEEE Transactions on Automatic Control*, vol. AC-38, no. 1, pp. 17–32, 1993.

[251] Marino, R. and P. Tomei, *Nonlinear Control Design: Geometric, Adaptive and Robust*, Prentice-Hall, Hemel Hempstead, UK, 1995.

[252] Martin-Sanches, J. M., "A globally stable APCS in the presence of bounded noises and disturbances," *IEEE Transactions on Automatic Control*, vol. AC-29, no. 5, pp. 461–464, 1984.

[253] Martin-Sanches, J. M., "Adaptive control for time-varying systems," *Proceedings of the 1985 American Control Conference*, p. 1260, 1985.

[254] Mayergoyz, I. D., *Mathematical Models of Hysteresis*, Springer-Verlag, Berlin, 1991.

[255] Merritt, H. E., *Hydraulic Control Systems*, John Wiley & Sons, New York, 1967.

[256] Middleton, R. H. and G. C. Goodwin, "Adaptive control of time-varying linear systems," *IEEE Transactions on Automatic Control*, vol. AC-33, no. 2, pp. 150–155, 1988.

[257] Middleton, R. H., G. C. Goodwin, D. J. Hill and D. Q. Mayne, "Design issues in adaptive control," *IEEE Transactions on Automatic Control*, vol. AC-33, no. 1, pp. 50–58, 1988.

[258] Miller, D. E. and E. J. Davison, "An adaptive controller which provides an arbitrary good transient and steady-state response," *IEEE Transactions on Automatic Control*, vol. AC-36, no. 1, pp. 68–81, 1991.

[259] Miller, R. K. and A. N. Michel, *Ordinary Differential Equations*, Academic Press, New York, 1982.

[260] Mittal, S. and C. H. Menq, "Hysteresis compensation in electromagnetic actuators through Preisach model inversion," *IEEE-ASME Transactions on Mechatronics*, vol. 5, no. 4, pp. 394–409, December 2000.

[261] Miyasato, Y., "A model reference adaptive controller for systems with uncertain relative degree r, $r+1$ or $r+2$ and unknown signs of high frequency gains," *Automatica*, vol. 36, no. 6, pp. 889–896, 2000.

[262] Mo, L. and M. M. Bayoumi, "Adaptive pole-placement control of MIMO systems," *IEEE Transactions on Automatic Control*, vol. AC-38, no. 6, pp. 967–970, 1993.

[263] Monopoli, R. V., "Model reference adaptive control with an augmented error signal," *IEEE Transactions on Automatic Control*, vol. AC-19, no. 5, pp. 474–484, 1974.

[264] Monopoli, R. V. and C. C. Hsing, "Parameter adaptive control of multivariable systems," *International Journal of Control*, vol. 22, no. 3, pp. 313–327, 1975.

[265] Morse, A. S., "Global stability of parameter adaptive control systems," *IEEE Transactions on Automatic Control*, vol. AC-25, no. 6, pp. 433–439, 1980.

[266] Morse, A. S., "Parametrizations for multivariable adaptive control," *Proceedings of the 20th IEEE Conference on Decision and Control*, pp. 970–972, San Diego, CA, 1981.

[267] Morse, A. S., "A $4(n+1)$-dimensional model reference adaptive stabilizer for any relative degree one or two, minimum phase system of dimension n or less," *Automatica*, vol. 23, no. 1, pp. 123–125, 1987.

[268] Morse, A. S., "Towards a unified theory of parameter adaptive control: Tunability," *IEEE Transactions on Automatic Control*, vol. AC-35, no. 9, pp. 1002–1012, 1990.

[269] Morse, A. S., "Towards a unified theory of parameter adaptive control—Part II: certainty equivalence and input tuning," *IEEE Transactions on Automatic Control*, vol. AC-37, no. 1, pp. 15–29, 1992.

[270] Morse, A. S., "A gain matrix decomposition and some of its applications," *Systems and Control Letters*, vol. 21, no. 1, pp. 1–10, 1993.

[271] Morse, A. S., D. Q. Mayne and G. C. Goodwin, "Applications of hysteresis switching in parameter adaptive control," *IEEE Transactions on Automatic Control*, vol. AC-37, no. 9, pp. 1343–1354, 1992.

[272] Morse, A. S. and F. M. Pait, "MIMO design models and internal regulators for cyclicly-switched parameter-adaptive control systems," *Proceedings of the 1993 American Control Conference*, pp. 1349–1353, San Francisco, CA, 1993.

[273] Mudgett, D. R and A. S. Morse, "Adaptive stabilization of linear systems with unknown high-frequency gains," *IEEE Transactions on Automatic Control*, vol. AC-30, no. 6, pp. 549–554, 1985.

[274] Muth, E. J., *Transform Methods: With Applications to Engineering and Operations Research*, Prentice-Hall, Englewood Cliffs, NJ, 1977.

[275] Mosca, E., *Optimal, Predictive, and Adaptive Control*, Prentice-Hall, Englewood Cliffs, NJ, 1995.

[276] Naik, S. M. and P. R. Kumar, "Robust indirect adaptive-control of time-varying plants with unmodeled dynamics and disturbances," *SIAM Journal of Control and Optimization*, vol. 32, no. 6, pp. 1696–1725, 1994.

[277] Naik, S. M., P. R. Kumar and B. E. Ydstie, "Robust continuous-time adaptive control by parameter projection," in *Foundations of Adaptive Control*, P. V. Kokotović, ed., pp. 153–199, Springer-Verlag, Berlin, 1991.

[278] Nam, K. and A. Arapostathis, "A model reference adaptive control scheme for pure-feedback nonlinear systems," *IEEE Transactions on Automatic Control*, vol. AC-33, no. 9, pp. 803–811, 1988.

[279] Narendra, K. S and A. M. Annaswamy, "Robust adaptive control in the presence of bounded disturbances," *IEEE Transactions on Automatic Control*, vol. AC-31, no. 4, pp. 306–315, 1986.

[280] Narendra, K. S and A. M. Annaswamy, "Robust adaptive control," in *Adaptive and Learning Systems: Theory and Applications*, K. S. Narendra, ed., Plenum Press, New York, 1986.

[281] Narendra, K. S. and A. M. Annaswamy, "A new adaptive law for robust adaptation without persistent excitation," *IEEE Transactions on Automatic Control*, vol. AC-32, no. 2, pp. 134–145, 1987.

[282] Narendra, K. S. and A. M. Annaswamy, *Stable Adaptive Systems*, Prentice-Hall, Englewood Cliffs, NJ, 1989.

[283] Narendra, K. S. and J. Balakrishnan, "Adaptive control using multiple models," *IEEE Transactions on Automatic Control*, vol. AC-42, no. 2, pp. 171–187, 1997.

[284] Narendra, K. S. and Y-H. Lin, "Stable discrete adaptive control," *IEEE Transactions on Automatic Control*, vol. AC-25, no. 6, pp. 456–461, 1980.

[285] Narendra, K. S., Y. H. Lin and L. S. Valavani, "Stable adaptive controller design—Part II: proof of stability," *IEEE Transactions on Automatic Control*, vol. AC-25, no. 3, pp. 440–448, 1980.

[286] Narendra, K. S. and S. Mukhopadhyay, "Adaptive control of nonlinear multivariable systems," *Neural Networks*, vol. 7, no. 5, pp. 737–752, 1994.

[287] Narendra, K. S. and S. Mukhopadhyay, "Adaptive control using neural networks and approximate models," *IEEE Transactions on Neural Networks*, vol. 8, no. 3, pp. 475–485, 1997.

[288] Narendra, K. S. and J. H. Taylor, *Frequency Domain Criteria for Absolute Stability*, Academic Press, New York, 1973.

[289] Narendra, K. S. and L. S. Valavani, "Stable adaptive controller design—direct control," *IEEE Transactions on Automatic Control*, vol. AC-23, no. 4, pp. 570–583, 1978.

[290] Nassiri-Toussi, K. and W. Ren, "Indirect adaptive pole-placement control of MIMO stochastic systems: Self-tuning results," *IEEE Transactions on Automatic Control*, vol. AC-42, no. 1, pp. 38–52, 1997.

[291] Nazaruddin, Y. Y. and H. Unbehauen, "Design of multivariable state-space adaptive controllers, and its application to a turbo-generator pilot plant," *Control Engineering Practice*, vol. 4, no. 5, pp. 699–704, 1996.

[292] Netushil, A., *Theory of Automatic Control*, Mir Publishers, Moscow, 1973.

[293] Nussbaum, R. D., "Some remarks on a conjecture in parameter adaptive control," *Systems and Control Letters*, vol. 3, pp. 243–246, 1983.

[294] Ogata, K., *Modern Control Engineering*, Prentice-Hall, Englewood Cliffs, NJ, 1997.

[295] Ohkawa, F., "A model reference adaptive control system for a class of discrete-time linear time-varying systems with time delay," *International Journal of Control*, vol. 42, no. 5, pp. 1227–1238, 1985.

[296] Ohkawa, F. and Y. Yonezawa, "A model reference adaptive control system for discrete multivariable systems with time delays," *International Journal of Control*, vol. 36, no. 6, pp. 925–934, 1982.

[297] Ordonez, R. and K. M. Passino, "Stable multiple-input multiple-output adaptive fuzzy/neural control," *IEEE Transactions on Fuzzy Systems*, vol. 7, no. 3, pp. 345–353, 1999.

[298] Ortega, J. M., *Matrix Theory*, Plenum Press, New York, 1987.

[299] Ortega, R., L. Hsu and A. Astolfi, "Adaptive control of multivariable systems with reduced prior knowledge," *Proceedings of the 40th IEEE Conference on Decision and Control*, pp. 4198–4203, Orlando, FL, 2001.

[300] Ortega, R. and R. Lozano, "Globally stable adaptive controller for systems with delay," *International Journal of Control*, vol. 47, no. 1, pp. 17–23, 1988.

[301] Ortega, R., L. Praly and Y. Tang, "Direct adaptive tuning of robust controllers with guaranteed stability properties," *Systems and Control Letters*, vol. 8, pp. 321–326, 1987.

[302] Ortega, R. and M. Spong, "Adaptive motion control of rigid robots: A tutorial," *Automatica*, vol. 25, no. 6, pp. 877–888, 1989.

[303] Ortega, R. and Y. Tang, "Robustness of adaptive controllers: A survey," *Automatica*, vol. 25, no. 5, pp. 651–677, 1989.

[304] Pan, Z. and T. Basar, "Adaptive controller design for tracking and disturbance attenuation in parametric strict-feedback nonlinear systems," *IEEE Transactions on Automatic Control*, vol. AC-43, no. 8, pp. 1066–1083, 1998.

[305] Pappadakis, I. N. M. and S. C. A. Thomopoulous, "Modified output error method with parameter mismatch compensation for improved performance model reference adaptive control," *Proceedings of the 31st IEEE Conference on Decision and Control*, pp. 3249–3254, Tucson, AZ, 1992.

[306] Parks, P. C., "Lyapunov redesign of model reference adaptive control systems," *IEEE Transactions on Automatic Control*, vol. AC-11, no. 3, pp. 262–267, 1966.

[307] Peterson, B. B. and K. S. Narendra, "Bounded error adaptive control," *IEEE Transactions on Automatic Control*, vol. AC-27, no. 6, pp. 1161–1168, 1982.

[308] Phillips, C. L. and H. T. Nagle, *Digital Control System Analysis and Design*, 3rd ed., Prentice-Hall, Englewood Cliffs, NJ, 1995.

[309] Physik Instrumente, *Products for Micropositioning*, Catalogue 108–12, ed. E, 1990.

[310] Polderman, J. W., "A state space approach to the problem of adaptive pole placement assignment," *Mathematics of Control, Signals, and Systems*, vol. 2, pp. 71–94, 1989.

[311] Polycarpou, M., "Stable adaptive neural control scheme for nonlinear systems," *IEEE Transactions on Automatic Control*, vol. AC-41, no. 3, pp. 447–451, 1996.

[312] Polycarpou, M. and P. A. Ioannou, "On the existence and uniqueness of solutions in adaptive control systems," *IEEE Transactions on Automatic Control*, vol. AC-38, no. 3, pp. 474–479, 1993.

[313] Polycarpou, M. and P. Ioannou, "A robust adaptive nonlinear control design," *Automatica*, vol. 32, no. 3, pp. 423–427, March 1996.

[314] Polycarpou, M. and J. Conway, "Indirect adaptive nonlinear control of drug delivery systems," *IEEE Transactions on Automatic Control*, vol. AC-43, no. 6, pp. 849–856, 1998.

[315] Pomet, J.-B. and L. Praly, "Adaptive nonlinear control: An estimation-based algorithm," in *New Trends in Nonlinear Control Theory*, J. Descusse, M. Fliess, A. Isidori and D. Leborgne, eds., Springer-Verlag, Berlin, 1989.

[316] Pomet, J.-B. and L. Praly, "Adaptive nonlinear regulation: Estimation from the Lyapunov equation, " *IEEE Transactions on Automatic Control*, vol. AC-37, no. 6, pp. 729–740, 1992.

[317] Pontryagin, L. S., V. G. Boltyanskii, R. V. Gamkrelidze and E. F. Mishchenko, *The Mathematical Theory of Optimal Processes*, Wiley-Interscience, New York, 1962.

[318] Popov, V. M., *Hyperstability of Control Systems*, Springer-Verlag, Berlin, 1973.

[319] Praly, L., "Towards a globally stable direct adaptive control scheme for not necessarily minimum phase systems," *IEEE Transactions on Automatic Control*, vol. AC-29, no. 10, pp. 946–949, 1984.

[320] Praly, L., "Robust model reference adaptive controllers—Part I: stability analysis," *Proceedings of the 23rd IEEE Conference on Decision and Control*, pp. 1009–1014, Las Vegas, NV, 1984.

[321] Praly, L., "Global stability of a direct adaptive control scheme robust with respect to a graph topology," in *Adaptive and Learning Systems: Theory and Applications*, K. S. Narendra, ed., Plenum Press, New York, 1986.

[322] Praly, L., G. Bastin, J.-B. Pomet and Z. P. Jiang, "Adaptive stabilization of nonlinear systems," in *Foundations of Adaptive Control*, P. V. Kokotović, ed., Springer-Verlag, Berlin, 1991.

[323] Qu, Z., *Robust Control of Nonlinear Uncertain Systems*, John Wiley & Sons, New York, 1998.

[324] Rao, M. R. M., *Ordinary Differential Equations: Theory and Applications*, Edward Arnold, London, 1980.

[325] Rao, S. S., *Optimization: Theory and Applications*, Wiley Eastern, New Delhi, 1984.

[326] Recker, D., *Adaptive Control of Systems Containing Piecewise Linear Nonlinearities*, Ph.D. Thesis, University of Illinois, Urbana, 1993.

[327] Recker, D. A., "Indirect adaptive non-linear control of discrete time systems containing a dead-zone," *International Journal of Adaptive Control and Signal Processing*, vol. 11, no. 1, pp. 33–48, February 1997.

[328] Recker, D., P. V. Kokotović, D. Rhode and J. Winkelman, "Adaptive nonlinear control of systems containing a dead-zone," *Proceedings of the 30th IEEE Conference on Decision and Control*, pp. 2111–2115, Brighton, England, 1991.

[329] Recker, D. and P. V. Kokotović, "Indirect adaptive nonlinear control of discrete-time systems containing a dead-zone," *Proceedings of the 32nd IEEE Conference on Decision and Control*, pp. 2647–2653, San Antonio, TX, 1993.

[330] Reed, J. S. and P. A. Ioannou, "Instability analysis and robust adaptive control of robotic manipulators," *IEEE Transactions on Robotics and Automation*, vol. 5., no. 3, pp. 381–386, 1989.

[331] Riedle, B. D., B. Cyr and P. V. Kokotović, "Disturbance instabilities in an adaptive system," *IEEE Transactions on Automatic Control*, vol. AC-29, no. 9, pp. 822–824, 1984.

[332] Riedle, B. D. and P. V. Kokotović, "A stability-instability boundary for disturbance-free slow adaptation and unmodeled dynamics," *IEEE Transactions on Automatic Control*, vol. AC-30, no. 10, pp. 1027–1030, 1985.

[333] Rudin, W., *Principles of Mathematical Analysis*, McGraw-Hill, New York, 1976.

[334] Rohrs, C. E., L. Valavani, M. Athans and G. Stein, "Robustness of adaptive control algorithms in the presence of unmodeled dynamics," *Proceedings of the 21st IEEE Conference on Decision and Control*, pp. 3–11, Orlando, FL, 1982.

[335] Rugh, W. J., *Linear System Theory*, 2nd ed., Prentice-Hall, Englewood Cliffs, NJ, 1996.

[336] Saberi, A., B. M. Chen and P. Sannuti, *Loop Transfer Recovery: Analysis and Design*, Springer-Verlag, London, 1993.

[337] Samson, C., "Stability analysis of adaptively controlled system subject to bounded disturbances," *Automatica*, vol. 19, no. 1, pp. 81–86, 1983.

[338] Sacher, C. A. and G. F. Inbar, "Tracking of the muscle recruitment characteristic during adaptive control of the electrically stimulated knee," *Proceedings of the 1990 International Conference of the IEEE Engineering in Medicine and Biology Society*, vol. 12, no. 5, pp. 2315–2317, 1990.

[339] Sastry, S. S., "Model reference adaptive control: Stability, parameter convergence and robustness," *I. M. A. J. of Mathematical Control and Information*, vol. 1, pp. 27–66, 1984.

[340] Sastry, S. and M. Bodson, *Adaptive Control: Stability, Convergence, and Robustness*, Prentice-Hall, Englewood Cliffs, NJ, 1989.

[341] Sastry, S. S. and A. Isidori, "Adaptive control of linearizable systems," *IEEE Transactions on Automatic Control*, vol. AC-34, no. 11, pp. 1123–1131, 1989.

[342] Sayed, A. H. and P. V. Kokotović, "A note on a Lyapunov argument for stochastic gradient methods in the presence of noise," *Proceedings of the 1995 American Control Conference*, pp. 587-588, Seattle, WA.

[343] Seidl, D. R., S. L. Lam, J. A. Putman and R. D. Lorenz, "Neural network compensation of gear backlash hysteresis in position-controlled mechanisms," *IEEE Transactions on Industry Applications*, vol. 31, no. 6, pp. 1475–1483, 1995.

[344] Selmic, R. R. and F. L. Lewis, "Dead-zone compensation in motion control systems using neural networks," *IEEE Transactions on Automatic Control*, vol. AC-45, no. 4, pp. 602–613, 2000.

[345] Selmic, R. R. and F. L. Lewis, "Neural net backlash compensation with Hebbian tuning using dynamic inversion," *Automatica*, vol. 37, no. 8, pp. 1269–1277, August 2001.

[346] Sepulchre, R., M. Jankovic and P. V. Kokotović, *Constructive Nonlinear Control*, Springer Verlag, London, 1997.

[347] Shah, S. L., C. Mohtadi and D. W. Clarke, "Multivariable adaptive control without a prior knowledge of the delay matrix," *Systems and Control Letters*, vol. 9, pp. 295–306, 1987.

[348] Shimkin, N. and A. Feuer, "Persistency of excitation in continuous-time systems," *Systems and Control Letters*, vol. 9, pp. 225–233, 1987.

[349] Singh, R. P. and K. S. Narendra, "Prior information in the design of multivariable adaptive controllers," *IEEE Transactions on Automatic Control*, vol. AC-29, no. 12, pp. 1108–1111, 1984.

[350] Slotine, J. and W. Li, "On the adaptive control of robot manipulators," *International Journal of Robotics Research*, vol. 6, no. 3, pp. 49–59, 1987.

[351] Slotine, J.-J. E. and W. Li, *Applied Nonlinear Control*, Prentice-Hall, Englewood Cliffs, NJ, 1991.

[352] Slotine, J. J. and S. S. Sastry, "Tracking control of non-linear systems using sliding surface, with application to robot manipulators," *International Journal of Control*, vol. 38, no. 2, pp. 465–492, 1983.

[353] Smith, R. C., "A nonlinear optimal control method for magnetostrictive actuators," *Journal of Intelligent Material Systems and Structures*, vol. 9, no. 6, pp. 468–486, June 1998.

[354] Smith, R. C., "Inverse compensation for hysteresis in magnetostrictive transducers," *Mathematical and Computer Modelling*, vol. 33, no. 1-3, pp. 285–298, January-February 2001.

[355] Smith, R. C. and Z. Ounaies, "A domain wall model for hysteresis in piezoelectric materials," *Journal of Intelligent Material Systems and Structures*, vol. 11, no. 1, pp. 62–79, January 2000.

[356] Spong, M. W., "Adaptive control of flexible joint manipulators," *Systems and Control Letters*, vol. 13, pp. 15–21, 1989.

[357] Spong, M. W. and M. Vidyasagar, *Robot Dynamics and Control*, John Wiley & Sons, New York, 1989.

[358] Su, C. Y., Y. Stepanenko, J. Svoboda and T. P. Leung, "Robust adaptive control of a class of nonlinear systems with unknown backlash-like hysteresis," *IEEE Transactions on Automatic Control*, vol. AC-45, no. 12, pp. 2427–2433, 2000.

[359] Suarez, D. A. and R. Lozano, "Adaptive control of nonminimum phase systems subject to bounded disturbances," *IEEE Transactions on Automatic Control*, vol. AC-41, no. 12, pp. 1830–1836, 1996.

[360] Sun, J., "A modified model reference adaptive control scheme for improved transient performance," *IEEE Transactions on Automatic Control*, vol. AC-38, no. 8, pp. 1255–1259, 1993.

[361] Sun, J., "A hybrid adaptive control scheme using sampled data and intersampling compensation," *Automatica*, vol. 33, no. 2, pp. 203–207, 1997.

[362] Sun, J. and P. A. Ioannou, "Robust adaptive LQ control schemes," *IEEE Transactions on Automatic Control*, vol. AC-37, no. 1, pp. 100–106, 1992.

[363] Sun, J., A. W. Olbrot and M. P. Polis, "Robust stabilization and robust performance using model-reference control and modeling error compensation," *IEEE Transactions on Automatic Control*, vol. AC-39, no. 3, pp. 630–635, 1994.

[364] Tan, K. K., Q.-G. Wang and C. C. Hang (with T. Hagglund), *Advances in PID Control*, Springer-Verlag, London, 1999.

[365] Tang, Y., "Adaptive frequency response identification using the Lagrange filter," *Automatica*, vol. 29, no. 2, pp. 451–455, 1993.

[366] Tao, G., *Model Reference Adaptive Control: Stability, Robustness and Performance Improvement*, Ph.D. Thesis, University of Southern California, August 1989.

[367] Tao, G., "Model reference adaptive control of multivariable plants with unknown interactor matrix," *Proceedings of the 29th IEEE Conference on Decision and Control*, pp. 2780–2785, Honolulu, HI, 1990.

[368] Tao, G., "Robust adaptive control with reduced knowledge of unmodeled dynamics," *Proceedings of the 29th IEEE Conference on Decision and Control*, pp. 3214–3219, Honolulu, HI, 1990.

[369] Tao, G., "New normalizing signals for robust adaptive control," *Proceedings of the 1991 American Control Conference*, pp. 138–143, Boston.

[370] Tao, G., "Unified robustness analysis of model reference adaptive control schemes with new normalizing signals," Technical Report, May 1991.

[371] Tao, G., "Model reference adaptive control of multivariable plants with delays," *International Journal of Control*, vol. 55, no. 2, pp. 393–414, 1992.

[372] Tao, G., "On robust adaptive control of robot manipulators," *Automatica*, vol. 28, no. 4, pp. 803–807, 1992.

[373] Tao, G., "Adaptive control systems with $L^{1+\alpha}$ tracking," *International Journal of Control*, vol. 64, no. 5, pp. 859–870, 1996.

[374] Tao, G., "Adaptive control of partially known systems," *IEEE Transactions on Automatic Control*, vol. AC-40, no. 10, pp. 1813–1818, 1995.

[375] Tao, G., "Adaptive control of systems with nonsmooth input and output nonlinearities," *Proceedings of the 34th IEEE Conference on Decision and Control*, New Orleans, LA, 1995.

[376] Tao, G., "Adaptive inverse for actuator compensation," in *Adaptive Control Systems: Advanced Techniques for Applications*, G. Feng and R. Lozano, eds., Butterworth-Heinemann, London, 1999.

[377] Tao, G. and P. A. Ioannou, "Robust model reference adaptive control for multivariable plants," *International Journal of Adaptive Control and Signal Processing*, vol. 2, no. 3, pp. 217–248, 1988.

[378] Tao, G. and P. A. Ioannou, "Strictly positive real matrices and the Lefschetz-Kalman-Yakubovich lemma," *IEEE Transactions on Automatic Control*, vol. AC-33, no. 12, pp. 1183–1185, 1988.

[379] Tao, G. and P. A. Ioannou, "Robust stability and performance improvement of multivariable adaptive control systems," *International Journal of Control*, vol. 50, no. 5, pp. 1825–1855, 1989.

[380] Tao, G. and P. A. Ioannou, "A model reference adaptive controller for multivariable plants with zero residual tracking error," *Proceedings of the 28th IEEE Conference on Decision and Control*, pp. 1597–1600, Tampa, FL, 1989.

[381] Tao, G. and P. A. Ioannou, "Persistency of excitation and overparametrization in model reference adaptive control," *IEEE Transactions on Automatic Control*, vol. AC-35, no. 2, pp. 254–256, 1990.

[382] Tao, G. and P. A. Ioannou, "Robust adaptive control: A modified scheme," *International Journal of Control*, vol. 54, no. 1, pp. 241–256, 1991.

[383] Tao, G. and P. A. Ioannou, "Robust adaptive control of plants with unknown order and high frequency gain," *International Journal of Control*, vol. 53, no. 3, pp. 559–578, 1991.

[384] Tao, G. and P. A. Ioannou, "Stability and robustness of model reference adaptive control schemes," in *Control and Dynamics: Advances in Theory and Applications*, vol. 53, C. T. Leondes, ed., pp. 99–123, Academic Press, New York, 1992.

[385] Tao, G. and P. A. Ioannou, "Model reference adaptive control of plants with unknown relative degree," *IEEE Transactions on Automatic Control*, vol. AC-38, no. 6, pp. 976–982, 1993.

[386] Tao, G. and P. V. Kokotović, "Adaptive control of systems with backlash," *Automatica*, vol. 29, no. 2, pp. 323–335, 1993.

[387] Tao, G. and P. V. Kokotović, "Adaptive control of plants with unknown dead-zones," *IEEE Transactions on Automatic Control*, vol. AC-39, no. 1, pp. 59–68, 1994.

[388] Tao, G. and P. V. Kokotović, "Adaptive control of systems with unknown output hystereses," *Proceedings of the 1994 American Control Conference*, pp. 870–874, Baltimore.

[389] Tao, G. and P. V. Kokotović, "Discrete-time adaptive control of systems with unknown nonsmooth input nonlinearities," *Proceedings of the 33rd IEEE Conference on Decision and Control*, pp. 1171–1176, Lake Buena Vista, FL, 1994.

[390] Tao, G. and P. V. Kokotović, "Discrete-time adaptive control of systems with unknown dead-zones," *International Journal of Control*, vol. 61, no. 1, pp. 1–17, 1995.

[391] Tao, G. and P. V. Kokotović, "Adaptive control of plants with unknown hystereses," *IEEE Transactions on Automatic Control*, vol. AC-40, no. 2, pp. 200–212, 1995.

[392] Tao, G. and P. V. Kokotović, "Adaptive control of plants with unknown output backlash," *IEEE Transactions on Automatic Control*, vol. AC-40, no. 2, pp. 326–330, 1995.

[393] Tao, G. and P. V. Kokotović, "Adaptive control of plants with unknown output dead-zones," *Automatica*, vol. 31, no. 2, pp. 287–291, 1995.

[394] Tao, G. and P. V. Kokotović, "Continuous-time adaptive control of systems with unknown backlash," *IEEE Transactions on Automatic Control*, vol. AC-40, no. 6, pp. 1083–1087, 1995.

[395] Tao, G. and P. V. Kokotović, *Adaptive Control of Systems with Actuator and Sensor Nonlinearities*, John Wiley & Sons, New York, 1996.

[396] Tao, G. and Y. Ling, "Multivariable adaptive actuator nonlinearity compensation," *Proceedings of the 36th IEEE Conference on Decision and Control*, pp. 2858–2863, San Diego, 1997.

[397] Tao, G. and Y. Ling, "Parameter estimation for coupled multivariable error models," *International Journal of Adaptive Control and Signal Processing*, vol. 13, no. 2, pp. 145–159, 1999.

[398] Tao, G., Y. Ling and A. Taware, "Parametrizations for adaptive control of multivariable systems with actuator nonlinearities," *Proceedings of the 2000 American Control Conference*, pp. 2659–2663, Chicago, 2000.

[399] Tao, G., X. D. Tang and S. M. Joshi, "Adaptive output rejection of unmatched input disturbances," *Systems and Control Letters*, vol 47, no. 1, pp. 25-35, September 2002.

[400] Tao, G. and M. Tian, "Design of adaptive dead-zone inverse for nonminimum phase plants," *Proceedings of the 1995 American Control Conference*, pp. 2059–2063, Seattle, WA.

[401] Tao, G. and M. Tian, "Discrete-time adaptive control of systems with multi-segment piecewise-linear nonlinearities," *IEEE Transactions on Automatic Control*, vol. AC-43, no. 5, pp. 719–723, 1998.

[402] Taylor, D., P. V. Kokotović, R. Marino and I. Kanellakopoulos, "Adaptive regulation of nonlinear systems with unmodeled dynamics," *IEEE Transactions on Automatic Control*, vol. AC-34, no. 3, pp. 405–421, 1989.

[403] Tarn, T. J., A. K. Bejczy and S. Ganguly, "Feedback-control of robot arm with discrete nonlinear model," *Journal of Robotic Systems*, vol. 9, no. 4, pp. 529–559, June 1992.

[404] Tarn, T. J., N. Xi and A. K. Bejczy, "Path-based approach to integrated planning and control for robotic systems," *Automatica*, vol. 32, no. 12, pp. 1675–1687, 1996.

[405] Teel, A. R., "Error-based adaptive nonlinear control and regions of feasibility," *International Journal of Adaptive Control and Signal Processing*, vol. 6, no. 4, pp. 319–327, July 1992.

[406] Teel, A. R., "A nonlinear small gain theorem for the analysis of control systems with saturation," *IEEE Transactions on Automatic Control*, vol. AC-41, no. 9, pp. 1256–1270, 1996.

[407] Teel, A. R., R. Kadiyala, P. V. Kokotović and S. Sastry, "Indirect techniques for adaptive input output linearization of nonlinear-systems," *International Journal of Control*, vol. 53, no. 1, pp. 193–222, 1991.

[408] Teolis, C., et al., Internal Technical Report, Techno-Sciences, Lanham, MD, 1997.

[409] Teolis, C., H. G. Kwatny and M. Mattice, "Symbolic computing tools for nonsmooth dynamics and control," in *Variable Structure Systems, Sliding Mode and Nonlinear Control, Lecture Notes in Control and Information Sciences*, Series no. 247, K. D. Young and U. Ozguner, eds., pp. 237–262, Springer-Verlag, Berlin, 1999.

[410] Thaler, G. J. and M. P. Pastel, *Analysis and Design of Nonlinear Feedback Control Systems*, McGraw-Hill, New York, 1962.

[411] Thathachar, M. A. L. and F. Gajendran, "A global convergence criterion for discrete-time multivariable adaptive systems," *International Journal of Control*, vol. 34, no. 1, pp. 127–142, 1981.

[412] Tian, M. and G. Tao, "Adaptive control of a class of nonlinear systems with unknown dead-zones," *Proceedings of the 13th World Congress of IFAC*, vol. E, pp. 209–213, San Francisco, July 1996.

[413] Tian, M. and G. Tao, "Adaptive dead-zone compensation for output-feedback canonical systems," *International Journal of Control*, vol. 67, no. 5, pp. 791–812, 1997.

[414] Tsakalis, K. S., "Robustness of model reference adaptive controllers: An input-output approach," *IEEE Transactions on Automatic Control*, vol. AC-37, no. 5, pp. 556–565, 1992.

[415] Tsakalis, K. S. and P. A. Ioannou, "Adaptive control of time-varying plants," *Automatica*, vol. 23, no. 4, pp. 459–468, July 1987.

[416] Tsakalis, K. S. and P. A. Ioannou, "Adaptive control of linear time-varying plants: A new model reference controller structure," *IEEE Transactions on Automatic Control*, vol. AC-34, no. 10, pp. 1038–1047, 1989.

[417] Tsakalis, K. S. and P. A. Ioannou, "A new indirect adaptive control scheme for time-varying plants," *IEEE Transactions on Automatic Control*, vol. AC-35, no. 6, pp. 697–705, 1990.

[418] Tsakalis, K. and P. A. Ioannou, *Linear Time Varying Systems: Control and Application*, Prentice-Hall, Englewood Cliffs, NJ, 1993.

[419] Tsiligiannis, C. A. and S. A. Svoronos, "Multivariable self-tuning control via right interactor matrix," *IEEE Transactions on Automatic Control*, vol. AC-31, no. 4, pp. 987–989, 1986.

[420] Truxal, J. G., *Automatic Feedback Control System Synthesis*, McGraw-Hill, New York, 1955.

[421] Truxal, J. G., *Control Engineers' Handbook*, McGraw-Hill, New York, 1958.

[422] Tsypkin, Y. Z., *Adaptation and Learning in Automatic Systems*, Academic Press, New York, 1971.

[423] Utkin, V. I.,"Variable structure systems with sliding modes," *IEEE Transactions on Automatic Control*, vol. AC-22, no. 2, pp. 212–222, 1977.

[424] Utkin, V. I., *Sliding Modes in Control Optimization*, Springer-Verlag, Berlin, 1991.

[425] Vaughan, N. D. and J. B. Gamble, "The modeling and simulation of a proportional solenoid valve," presented at the ASME Winter Annual Meeting, Dallas, TX, November 1990.

[426] Vidyasagar, M., *Nonlinear Systems Analysis*, 2nd ed., Prentice-Hall, Englewood Cliffs, NJ, 1993.

[427] Visintin, A., "Mathematical models of hysteresis," in *Topics in Nonsmooth Mechanics*, J. J. Moreau, P. D. Panagiotopoulos and G. Strang, eds., pp. 295–326, Birkhauser Verlag, Berlin, 1988.

[428] Wang, H., G. P. Liu, C. J. Harris and M. Brown, *Advanced Adaptive Control*, Pergamon Press, New York, 1995.

[429] Webb, G., A. Kurdila and D. Lagoudas, "Adaptive hysteresis model for model reference control with actuator hysteresis," *Journal of Guidance Control and Dynamics*, vol. 23 no. 3, pp. 459–465, May-June 2000.

[430] Webb, G., D. Lagoudas and A. J. Kurdila, "Hysteresis modeling of SMA actuators for control applications," *Journal of Intelligent Material Systems and Structures*, vol. 9, no. 6, pp. 432–448, June 1998.

[431] Webb, G., D. Lagoudas and A. Kurdila, "Adaptive hysteresis compensation for SMA actuators with stress-induced variations in hysteresis," *Journal of Intelligent Material Systems and Structures*, vol. 10, no. 11, pp. 845–854, November 1999.

[432] Weller, S. R. and G. C. Goodwin, "Hysteresis switching adaptive control of linear multivariable systems," *IEEE Transactions on Automatic Control*, vol. AC-39, no. 7, pp. 1360–1375, 1994.

[433] Wen, C., "A robust adaptive controller with minimal modification for discrete time-varying systems," *IEEE Transactions on Automatic Control*, vol. AC-39, no. 5, pp. 987–991, 1994.

[434] Wen, C., "An indirect robust continuous-time adaptive controller with minimal modifications," *Automatica*, vol. 31, no. 2, pp. 293–296, 1995.

[435] Wen, C., "Robust adaptive tracking using internal model principle," *International Journal of Control*, vol. 64, no. 1, pp. 127–140, 1996.

[436] Wen, C., "Robust adaptive control of proper systems," *IEEE Transactions on Automatic Control*, vol. AC-43, no. 11, pp. 1579–1584, 1998.

[437] Wen, J. T. and M. Balas, "Robust adaptive control in Hilbert spaces," *Journal of Mathematical Analysis and Applications*, vol. 143, pp. 1–26, 1989.

[438] Wen, J. T., "Time domain and frequency-domain conditions for strict positive realness," *IEEE Transactions on Automatic Control*, vol. AC-33, no. 10, pp. 988–992, 1988.

[439] Wen, J. T., K. Kreutzdelgado and D. S. Bayard, "Lyapunov function-based control laws for revolute robot arms: Tracking control, robustness, and adaptive-control," *IEEE Transactions on Automatic Control*, vol. AC-37, no. 2, pp. 231–237, 1992.

[440] Willems, J. C. and C. I. Byrnes, "Global adaptive stabilization in the absence of information on the high frequency gain," *Proceedings of the 6th International Conference on Analysis and Optimization of Systems*, Springer Lecture Notes in Control and Information Sciences, vol. 62, pp. 49–57, June 1984.

[441] Willner, Y., M. Heymann and M. Bodson, "Parameter convergence and uniqueness in parametrizations for multivariable adaptive control," *SIAM Journal on Control and Optimization*, vol. 30, pp. 800–815, 1992.

[442] Wolovich, W. A., *Linear Multivariable Systems*, Springer-Verlag, New York, 1974.

[443] Wolovich, W. A and P. L. Falb, "Invariants and canonical forms under dynamic compensation," *SIAM Journal of Control and Optimization*, vol. 14, no. 6, pp. 996–1008, 1976.

[444] Wonham, W. M. and A. S. Morse, "Feedback invariants of linear multivariable systems," *Automatica*, vol. 8, no. 1, pp. 93–100, 1972.

[445] Wu, Y., P. E. Allaire and G. Tao, "An adaptive controller for a left ventricle assist device," *Proceedings of the 2nd Joint EMBS-BMES Conference*, Houston, TX, October 2002.

[446] Xie, L. L. and L. Guo, "Fundamental limitations of discrete-time adaptive nonlinear control," *IEEE Transactions on Automatic Control*, vol. AC-44, no. 9, pp. 1777–1782, 1999.

[447] Yao, B. and M. Tomizuka, "Adaptive robust control of SISO nonlinear systems in a semi-strict-feedback form," *Automatica*, vol. 33, no. 5, pp. 893–900, 1997.

[448] Yao, B. and M. Tomizuka, "Adaptive robust control of MIMO nonlinear systems in semi-strict-feedback forms," *Automatica*, vol. 37, no. 9, pp. 1305–1321, 2001.

[449] Ydstie, B. E., "Transient performance and robustness of direct adaptive control," *IEEE Transactions on Automatic Control*, vol. AC-37, no. 8, pp. 1091–1105, 1992.

[450] Yeung, K. S. and Y. P. Chen, "A New controller design for manipulators using the theory of variable structure systems," *IEEE Transactions on Automatic Control*, vol. AC-33, no. 2, pp. 200–206, 1988.

[451] Young, K. K. D, "Design of variable structure model following control systems," *IEEE Transactions on Automatic Control*, vol. AC-23, no. 6, pp. 1079–1085, 1978.

[452] Young, K. K. D., P. V. Kokotović and V. I. Utkin, "A singular perturbation analysis of high-gain feedback systems," *IEEE Transactions on Automatic Control*, vol. AC-22, no. 6, pp. 931–938, 1977.

[453] Zhang, C. S. and R. J. Evans, "Amplitude constrained adaptive control," *International Journal of Control*, vol. 46, no. 1, pp. 53–64, 1987.

[454] Zhang, C. S. and R. J. Evans, "Rate constrained direct adaptive control," *International Journal of Control*, vol. 48, no. 6, pp. 2179–2187, 1988.

[455] Zhang, J. F., "Robust stabilization of LTI systems via indirect adaptive controllers," *IEEE Transactions on Automatic Control*, vol. AC-40, no. 5, pp. 900–906, 1995.

[456] Zhang, J. F. and P. E. Caines, "Adaptive control via a simple switching algorithm," *SIAM Journal of Control and Optimization*, vol. 34, no. 1, pp. 365–388, 1996.

[457] Zhang, J. F. and L. Y. Wang, "Performance lower bounds in stochastic robust and adaptive control," *IEEE Transactions on Automatic Control*, vol. AC-46, no. 7, pp. 1137–1141, 2001.

[458] Zhang, Y. and P. A. Ioannou. "A new class of nonlinear robust adaptive controller," *International Journal of Control*, vol. 65, no. 5, pp. 745–769, 1996.

[459] Zhou, K., J. C. Doyle and K. Glover, *Robust and Optimal Control*, Prentice-Hall, Upper Saddle River, NJ, 1996.

[460] Zinober, A. S. I., O. M. E. El-Ghezawi and S. A. Billings, "Multivariable variable-structure adaptive model-following control systems," *Proceedings of IEE*, Part D, vol. 129, pp. 6–12, 1982.

Index